KARL MARAMOROSCH

is Program Director of Insect Physiology and Virology at the Boyce Thompson Institute for Plant Research in Yonkers, New York.

Viruses, Vectors, and Vegetation

# Viruses, Vectors, and Vegetation

EDITED BY

Karl Maramorosch

*Boyce Thompson Institute for Plant Research*
*Yonkers, New York*

INTERSCIENCE PUBLISHERS
A DIVISION OF JOHN WILEY & SONS
New York · London · Sydney · Toronto

Library of Congress Catalog Card Number 68-24680
SBN 471 567868

PRINTED IN THE UNITED STATES OF AMERICA

# Preface

Plant viruses are among the most important, most complex, and most extensively distributed plant disease agents in the world. The intriguing mechanisms by which these viruses propagate and survive in plant and vector reservoirs and the economically important diseases they cause have attracted numerous workers to this vast and expanding field.

Knowledge of the interactions between viruses, their carriers, and plants is steadily gaining in importance. Application of accomplishments in this field has already transcended the boundaries of the agricultural sciences and entered advanced frontiers of medicine, biochemistry, biophysics, and ultrastructure research. Since there seemed to be an obvious need for a publication that would bring together these studies and findings, the multiauthored book *Viruses, Vectors, and Vegetation* was initiated.

The decision to publish this book was made in 1966, following a United States–Japan Conference on Interactions between Arthropods and Plant-Pathogenic Viruses. Under the sponsorship of the United States–Japan Cooperative Science Program, nine Americans and ten Japanese met for four days in October 1965 in Tokyo, Japan, to discuss the latest information concerning viruses and vectors. The Tokyo conference was initiated by Paul Oman, who was then Assistant Director of Research of the Entomology Division, U.S. Department of Agriculture and is presently Professor of Entomology at Oregon State University, Corvallis, Oregon, and by Karl Maramorosch, Program Director of Insect Physiology and Virology at the Boyce Thompson Institute for Plant Research in Yonkers, New York. Professor Hidefumi Asuyama of Tokyo University acted as Japanese coordinator, and Karl Maramorosch as U.S. coordinator. The major objectives of the conference were the direct exchange of information on the interactions between viruses and arthropods and the evaluation of techniques of study that would result in a better understanding of the work in progress in the two countries. At the Tokyo conference there was a certain degree of partiality toward leafhopper-borne viruses, while aphid-borne viruses, which are far more numerous, were discussed in only 7 of the 23 submitted papers. This was due, in part, to the fact that leafhopper-borne viruses were the main interest of the organizers of the conference. It was also due undoubtedly to the fact that these viruses present a number of intriguing aspects, such as transovarian passage, multiplication, and deleterious effects on the vector. Japanese workers were the first to discover congenital infection of insect vectors by plant-pathogenic viruses, and, in fact, were the first to find that viruses are transmitted by insect vectors. They also pioneered in the study of viruses that multiply in plants as well as in insects. Professor Fukushi and co-workers published the first electron micrographs of one of these viruses inside cells of insect hosts, as well as in cells of infected plants.

Since the rules of the United States–Japan Cooperative Science Program defined not only the total number, but also the nationality of the contributors, limiting the participation to nine U.S. and ten Japanese scientists, it was not possible to cover in the Tokyo conference the subject of vectors adequately and to discuss all aspects of virus–host interactions. This limitation was easily overcome when the decision was made to publish a comprehensive book on plants, viruses, and vectors. All but three of the original participants at the conference were able to submit chapters for the book, and more than a dozen additional chapters were prepared by outstanding experts who did not participate in the conference. In addition to the topics discussed in Tokyo, which were brought up to date and broadened considerably, additional chapters now cover virus purification and bioassay methods, specificity in aphid-borne viruses, aphid vector morphology, nematode, fungus, and mite vectors, nonpersistent leafhopper-borne viruses, fluorescent antibody studies, virus diseases of maize, and disease control through vector control. Contributors for *Viruses, Vectors, and Vegetation* were invited from all over the world, from universities, research institutes, and experiment stations in the United States, Canada, The United Kingdom, Australia, Brazil, and The Philippines, where significant work is being carried out. Each of the authors and co-authors is well known in his field and each has prepared a thoughtful treatment of his subject. Some of the chapters deal with as yet unpublished experimental data, others with recent literature, and still others are more historical in nature. All contribute some unpublished information and personal interpretations and conclusions.

It was the Editor's intent to provide readers interested in one particular aspect with a self-contained chapter describing that subject. As a result of this decision, it was sometimes impossible to avoid the overlap of information in some chapters. The Editor felt that completeness of description warranted this occasional duplication. Consequently, each of the 29 chapters contains a detailed account of experiments performed, interpretation of the results, and a report of the most recent advances in the field. The chapters bring into sharp focus current direction of research. The topics covered range from methodology to application, and explore every aspect of virus transmission to plants. On the whole, an attempt has been made to collate the large body of information and the vast array of data which continue to accumulate in the field of virus transmission studies. Most of the representative work in this field has been included, as far as possible, although certain groups of vectors have been omitted. This was done because in some groups, particularly in beetles, thrips, and mealybugs, less progress has been made in the past decade, and adequate reviews of earlier work have been published.

It may be of interest to note that the last published textbook in this field was by W. Carter: *Insects in Relation to Plant Disease* (Interscience, 1962, 705 p.). A symposium volume *Biological Transmission of Disease Agents* (Academic Press, 1962, K. Maramorosch, Ed., 192 p.) and a symposium published in 1963 (Annals of the New York Academy of Sciences, 105:681–764, by the same editor) are now out of date. The latter was of limited scope, since it included only four articles.

It is hoped that the primary purpose of the present book will be achieved, namely to provide a stimulating forum for discussion of new ideas and observations in plant–virus–vector research.

The Editor wishes to pay special tribute to the National Science Foundation in Washington, D.C. and to the Japan Society for the Promotion of Science in Tokyo for sponsoring the Tokyo Conference, and to all of the authors for their excellent contributions.

KARL MARAMOROSCH

*Boyce Thompson Institute for Plant Research*
Yonkers, New York
January 16, 1969

# Contributors

MYRON K. BRAKKE, *Plant Pathology Department, University of Nebraska, Lincoln, Nebraska*

LEONARD BROADBENT, *Bristol College of Science & Technology, University of Bath, Bath, England*

C. H. CADMAN, *Scottish Horticultural Research Institute, Invergowrie, Dundee, Scotland*

A. S. COSTA, *Secao de Virologia, Instituto Agronomico, Campinas, Brasil*

TRAVIS R. EVERETT, *The Ford Foundation, Entomology Program, New Delhi, India*

A. R. FORBES, *Canada Department of Agriculture, Research Station, Vancouver, British Colombia*

J. H. FREITAG, *Department of Entomology and Parasitology, University of California, Berkeley, California*

TEIKICHI FUKUSHI, *Department of Botany, Faculty of Agriculture, Hokkaido University, Sapporo, Japan*

ROBERT R. GRANADOS, *Boyce Thompson Institute for Plant Research, Yonkers, New York*

TAMOTSU ISHIHARA, *College of Agriculture, Ehime University, Matsuyama, Japan*

D. D. JENSEN, *Division of Entomology, University of California, Berkeley, California*

H. A. LAMEY, *Crops Research Division, U. S. Department of Agriculture, Baton Rouge, Louisiana*

K. D. LING, *International Rice Research Institute, Los Banos, Laguna, Philippines*

H. R. MACCARTHY, *Canada Department of Agriculture, Research Station, Vancouver, British Columbia*

KARL MARAMOROSCH, *Boyce Thompson Institute for Plant Research, Yonkers, New York*

JUN MITSUHASHI, *Division of Entomology, National Institute of Agricultural Sciences, Nishigahara, Kita-ku, Tokyo, Japan*

SOCHO NASU, *Division of Entomology, National Institute of Agricultural Sciences, Tokyo, Japan*

YASUMICHI, NISHI, *Kyushu Agricultural Experiment Station, Chikugo, Fukuoka, Japan*

PAUL OMAN, *Department of Entomology, Oregon State University, Corvallis, Oregon*

THOMAS P. PIRONE, *Department of Plant Pathology, University of Kentucky, Lexington, Kentucky*

W. F. ROCHOW, *Department of Plant Pathology, Cornell University, Ithaca, New York*

YASUO SAITO, *Institute for Plant Virus Research, Chiba, Japan*

EISHIRO SHIKATA, *Department of Botany, Faculty of Agriculture, Hokkaido University, Sapporo, Japan*

R. C. SINHA, *Department of Agriculture, Plant Research Institute, Experimental Farm, Ottawa, Ontario, Canada*

J. T. SLYKHUIS, *Canada Department of Agriculture, Cell Biology Research Institute, Research Branch, Ottawa, Canada*

FLOYD F. SMITH, *Agricultural Research Service, U. S. Department of Agriculture, Beltsville, Maryland*

NAOJI SUZUKI, *Institute for Plant Virus Research, Chiba, Japan*

K. G. SWENSON, *Entomology Department, Oregon State University, Corvallis, Oregon*

EDWARD S. SYLVESTER, *Department of Entomology, University of California, Berkeley, California*

C. E. TAYLOR, *Scottish Horticultural Research Institute, Invergowrie, Dundee, Scotland*

DAVID S. TEAKLE, *Department of Microbiology, University of Queensland, Medical School, Herston, Brisbane, Queensland, Australia*

RAYMOND E. WEBB, *Agricultural Research Service, U. S. Department of Agriculture, Beltsville, Maryland*

ROBERT F. WHITCOMB, *Plant Virus Vector Pioneering Research Laboratory, Entomology Research Division, Agricultural Research Service, U. S. Department of Agriculture, Beltsville, Maryland*

# Contents

# Criteria of Specificity in Virus–Vector Relationships

PAUL OMAN

*Department of Entomology*
*Oregon State University*
*Corvallis, Oregon*

Specificity, the condition of being peculiar to a particular organism or group of organisms, is a widespread and familiar biological phenomenon. As the concept applies to the relationships between plant pathogenic viruses and their arthropod vectors, specificity implies that a given virus may be transmitted only by a single vector species, or by a group of phylogenetically related species.

The concept of specificity in virus–vector relationships—both group specificity and species specificity—is of long standing. Storey (1931) pointed out that the mosaic viruses are largely transmitted by aphids, and the yellows viruses by leafhoppers. Evidence derived from the early work with viruses of rice dwarf, western curly-top of beets, aster yellows, Argentine curly-top of beets, and peach yellows, each of which appeared to be transmitted by a single species of leafhopper, naturally encouraged the idea of specificity in virus–vector relationships at the species level. It was presumably evidence such as this that led Leach (1940, p. 295), to comment that "There are numerous well-demonstrated cases of specificity between insect vectors and viruses they transmit. The most striking examples are found in those virus diseases transmitted by leafhoppers, . . ." But he went on to remark, "One should remember, however, that the evidence for specificity of transmission is generally of a negative character."

Intensive studies by numerous investigators, particularly during the past two decades, have made it clear that most arthropod-borne plant pathogenic viruses can be transmitted, at least experimentally, by more than one species. Although this evidence has tended to erode the concept of vector–virus specificity, there remain a few instances of seemingly absolute or near absolute specificity. Moreover, the recently accumulated evidence seems to strengthen the idea of group specificity, and thus reinforce the general concept of specificity in a vector–virus relationship as a biological phenomenon. Our current understanding of these relationships contains nothing that seriously conflicts with the broad generalizations made by Black (1959) in his illuminating discussion of the biological cycles of plant viruses in insect vectors. He pointed out (*1*) that the vectors of a single plant virus are almost always restricted to one of the major arthropod taxa such as aphids, leafhoppers, white flies, mealybugs, thrips, mites, or beetles; and (*2*) that a virus is almost always transmitted by only one of the principal types of transmission recognized.

It seems obvious that there must be a vast difference in the biological processes involved in the transmission of nonpersistent viruses by aphids, for example, and transmission of the circulatory-propagative viruses by leafhoppers. In the former the required conditions may be no more than the absence of an inhibitor and/or an inactivator in the saliva of the vector. In contrast, the transmission of a circulatory-propagative virus requires that all the vector's various tissues between the gut and salivary glands be permeable to the virus, and that a suitable environment for multiplication exist in at least one of the tissues. Just how complicated a process the passage of virus through these tissues may be is uncertain, but Bawden (1964) concludes that "permeability [of tissues] cannot be a chance event, but must depend on the virus acting specifically on the tissues to render them permeable." Presumably the specific requirements for propagation are comparable or even more rigorous. Considering these circumstances it would be expected that the factors influencing specificity, and therefore the criteria for determining specificity, would be quite different in these two major groups of vectors.

The excellent review of problems of specificity in anthropod vectors of plant viruses by Day and Bennetts (1954), and contributions to this subject by several other authors (Black, 1954; Sylvester, 1954; Black, 1959; Maramorosch, 1963; Bawden, 1964) make a further detailed discussion of this general topic unnecessary. It is generally recognized that many factors, individually or in combination, may influence specificity, or what appears to be specificity, and that much of the evidence is of a negative nature. Among these factors are the behavior of potential vectors, ability of virus to multiply in the arthropod host, presence or absence of inhibitors in vector tissues or secretions, permeability of vector tissues through which viruses must pass in circulative transmission, and vector immunity to virus infection. One of the great needs today, as it was when Day and Bennetts reviewed the situation prior to 1954, is for advances in the fundamental aspects of insect physiology and biochemistry as they relate to plant pathogenic viruses.

The concept of specificity in vector–virus relationships does not necessarily include the idea of evolutionary association of the two biological entities. However, it appears that most authors who have addressed themselves to this problem have implied such an association, at least at the group specificity level. According to Black (1959, p. 182) Bawden considered the evidence of virus multiplication in both plant and arthropod host to favor the symptomless insect vectors as the original evolutionary source. Black's hypothesis (1959, p. 183), that viruses with reproductive cycles in both plants and animals evolved by retrograde evolution from organisms that at one time were saprophytic or commensal in both plant and insect, clearly embraces the idea of evolutionary association at or near the species level with respect to vectors. The suggestion (Bawden, 1964, p. 314) that viruses could have arisen as normal components of organisms that acquired transmissibility and pathogenicity does not exclude the idea that evolutionary adjustment to a vector would naturally follow. Black's work (1944) with two strains of potato yellow dwarf specifically transmitted by related leafhoppers, and similar evidence with respect to other groups of viruses and their vectors, strongly supports the idea that specificity and evolutionary association are related phenomena.

The idea that vector relationships might be better indicators of virus relationships than disease symptomatology in the plant host, and useful in virus classification, has apparently intrigued numerous workers. Kunkel (1935) felt that the biological relationships of plant viruses with their vectors, when better known, might furnish a good basis for differentiation and classification. Four of the ten "families" of viruses treated by Holmes (1939) were characterized by their insect host relationships. Black's (1944) speculations on the possible existence of parallelisms between the relationships of viruses and the relationships of their vectors embodies this concept. However, Black (1956) later concluded that vector relationships are not dependable clues to relationships of the viruses themselves, except in a very general way, i.e., group relationship. Bawden (1964, p. 316), although emphasizing the need to base ideas of relationships among viruses on serological evidence, apparently (p. 324) does not reject the idea that groups of what he calls "collective species" might have evolved from a common stock. Recently summarized evidence (Simpson and Hauser, 1966) on the comparative morphology of certain viruses exhibiting quite different biological affinities suggests that there are some cases in which viruses have "jumped" to unrelated hosts to which they subsequently became well adapted. Thus, although vector associations seem unreliable as a basis for arriving at a meaningful classification of viruses, it still seems clear that there are valid parallelisms within or among groups of collective species of viruses and their primary vectors. But in order to evaluate the significance of evidence of this sort, it is necessary first to determine, if we can, if intrinsic relationships or accidental associations are involved. The conclusions drawn should obviously be greatly influenced by determinations of this sort.

Let us assume that some arthropod-transmitted plant viruses evolved in association with their plant hosts, that some evolved in association with their arthropod hosts, and that some, which have reproductive cycles in both plants and arthropods, originated in association with both types of hosts (Black, 1959, p. 183). If this is the case, and there is validity to the idea that organisms so associated tend to adjust their relationship toward a peaceful coexistence, then we would expect the degree of adjustment to be a factor of time and the rapidity with which genetic drift and selective pressure occurred in one or both of the associated organisms. Through man's activities and naturally occurring environmental changes, potential hosts (both plant and arthropod) of viruses are constantly being brought into contact with new strains or species of viruses. It naturally follows that we would expect to find all degrees of adaptation represented in vector–virus association. This is exactly the situation that exists. The nature of arthropod involvement in the transmission of plant pathogenic viruses apparently includes almost every conceivable type or combination of interrelationship (Black, 1959; Kennedy *et al.*, 1962; Maramorosch, 1963). The effects of the viruses on their arthropod hosts range from harmful to beneficial (Maramorosch and Jensen, 1963). Although there appears to be an increasing tendency among workers in this field to regard virus infection of an arthropod host as usually involving deleterious effects on the vector, rather than the virus being nonpathogenic, as was long believed, there still seems no sound basis for rejecting the idea that peaceful coexistence between vector and virus is not the general rule.

To better understand the nature of vector–virus specificity, we need to distinguish among these many different types or degrees of relationships. Although it is probably safe to assume that evolutionary association results in some degree of vector–virus specificity, and therefore in efficiency in perpetuation of the virus, the reverse assumption, that efficiency in transmission implies evolutionary association, is not warranted beyond the vector group level. There is convincing evidence that with some viruses that are transmitted with a high degree of efficiency, the association between vector and virus can only be of recent origin.

Whether or not these cases are included in our concept of vector–virus specificity is perhaps immaterial; the important point is that if we are properly to understand the nature and significance of vector–virus relationships, we must be able to discriminate between those relationships that derive from an evolutionary association and those that are fortuitous and relatively recent. It is to this problem that this attempt to develop or define criteria for measuring specificity is directed.

Our very uncertain understanding of vector–virus relationships is based, at best, upon only a small fraction of the potentially useful information. Much of the information available is of a negative nature, and therefore of indeterminate value. As mentioned earlier, there is great need for additional knowledge of the biochemistry of the vector–virus relationship. But in spite of the paucity of reliable data, there are types of useful information that have been generally overlooked but which may help in attempts to understand vector–virus relationships. However much we need to know about the physiology and biochemistry of viruses in their arthropod hosts, we should not permit preoccupation with those problems to result in neglect of other evidence which appears to be particularly useful in guiding future studies along potentially profitable lines. The following discussion deals with examples of such possibilities.

One of the most intensively studied of the vector–virus relationships is that of the beet leafhopper and the North American curly-top virus (WCTV). It was long considered that WCTV was transmitted by but a single vector species, *Circulifer tenellus,* and the relationship of these two organisms has often been mentioned as being one of the most specific. But in view of the many anomalies in the WCTV–*Circulifer tenellus* relationship it is rather surprising that the idea of a high degree of specificity in the relationship has not long since been challenged. A close scrutiny of various aspects of the relationship makes it evident that only to a limited extent does this case fit what should be expected of a vector–virus association resulting from a long period of joint evolutionary development.

Curly-top of beets was first recorded from North America in 1889. Extended inquiry among European plant pathologists in the years following discovery of curly-top failed to reveal the presence of any curly-top-like disease of beets in the Old World prior to the report by Bennett and Tanrisever (1957, 1958) of a strain of curly-top in Turkey. Although *Circulifer tenellus,* for many years the only known vector of curly-top, was long thought to be native to North America, Oman (1936, 1948, 1949) called attention to evidence that indicates the species to be native to the Mediterranean basin of the

Old World, and presumably adventive to the New World. The first positive record of the occurrence of *C. tenellus* in North America was in 1895 (Gillette and Baker, 1895) although in all likelihood it was present considerably prior to that time. It now has a wide distribution, being known from western North America from Mexico northward to Canada; from Florida and the West Indies; from South Africa; from the regions in North Africa and Eurasia that border the Mediterranean; and, recently, from the Hawaiian Islands. The same or a very closely related species occurs in the northwestern part of the Indian subcontinent. A very closely related species, *Circulifer opacipennis,* the established vector of curly-top in Turkey (Bennett and Tanrisever, 1957), is known from North Africa and Eurasia. The question of whether *C. tenellus* and *C. opacipennis* are distinct species is moot, but they are so considered at present.

Although curly-top has been considered to be a disease essentially limited to western North America, recently there have been several reports of curly-top in the eastern United States (Heggested and Moore, 1959; Schneider, 1959; Troutman and Fenne, 1959; Webb, 1963), well outside the area in which either recorded vector is known to occur. However, curly-top has not been reported from Florida or the West Indies, or from North or South Africa or Hawaii, even though populations of *C. tenellus* are established in those places. Attempts by Severin and other workers, (summarized by Day and Bennetts, 1954, pp. 25–26) to transmit WCTV with other North American leafhopper species have been unsuccessful. But as noted above, Bennett and Tanrisever (1957, 1958) found that an Old World species, *Circulifer opacipennis,* transmitted a strain of curly-top virus in Turkey. Interestingly enough, Black (1953) predicted the findings of Bennett and Tanrisever, for he commented that "other species in the genus (*Circulifer*) that live around the eastern Mediterranean might prove to be vectors if tested."

The relatives of *C. tenellus* are indigenous to the Old World, but the viruses presumed to be related to WCTV are apparently all indigenous to the New World. These are the South American leafhopper-transmitted viruses—the Argentine curly-top virus of beets (ACTV), transmitted by *Agalliana ensigera* (Bennett *et al.*, 1946); the Brazilian curly-top virus of beets (BCTVB), transmitted by *Agallia albidula;* and the Brazilian curly-top virus that attacks tomatoes (BCTVS), transmitted by *Agalliana ensigera* and *Agalliana sticticollis* (Bennett and Costa, 1949; Costa, 1952). In addition, Adsuar (1955) reported a curly-top disease of tomatoes from Puerto Rico, and Zaumeyer and Smith (1964) encountered a curly-top-like disease in beans in El Salvador. Simons (1962) discussed in considerable detail the possible relationships of WCTV, ACTV, BCTVB, and BCTVS with pseudo-curly-top virus (PCTV) (Simons and Coe, 1958) that is transmitted to solanaceous plants in Florida by a membracid, *Micrutalis malleifera.* Simons concluded, "If the vector of PCTV in Florida were a leafhopper, the virus would almost certainly be classified as another strain of curly-top." It is difficult to disagree with his analysis.

How do we rationalize this situation in which the causative organism of a disease is transmitted in North America only by a species that is presumed to be adventive from the Old World, while at the same time the disease apparently has but limited occurrence in the region of origin of its supposed sole vectors? Either misidentifications of viruses are involved, or there are gross inconsistencies in the occurrence records of curly-top virus and its vectors. If we assume the identifications to be correct, and that WCTV accompanied *C. tenellus* to the New World, we need to explain why the virus has not accompanied this same species to some other part of the world—to South Africa, to the West Indies, or to Hawaii—or why curly-top is not more prevalent in the Old World. If, on the other hand, we assume WCTV to be native to North America, how was it perpetuated prior to the arrival of its vector, *C. tenellus,* presumably at some post-Columbian time?

I have suggested (Maramorosch and Oman, 1966), purely as a working hypothesis, that WCTV is native to North America, that it evolved and was perpetuated here in association with some native leafhopper, probably an agallian leafhopper, and that it found a suitable vector in the immigrant species, *Circulifer tenellus*. Subsequent disturbance of the vector–virus ecosystem in the western United States, with resultant enormous increase in plant hosts of both the virus and its new vector, caused the disease to become of economic importance.

Clarification of the questions involved in this one problem will do much to improve our understanding of the origin of arthropod-borne plant-pathogenic viruses, and the nature of vector–virus relationships. Two lines of investigation would serve to test the above outlined hypothesis. First, explore anew and more thoroughly the relationship between *C. tenellus* and WCTV. There appears to be no valid reason to believe that the relationships between these two entities is of a fundamental nature, but the question of multiplication of WCTV in *C. tenellus* needs to be reexamined. Find out whether or not populations of *C. tenellus* in various regions outside of western North America —in Florida, Puerto Rico, Jamaica, Hawaii, or South Africa—are infected with WCTV. If they are not, determine if they can transmit it, and with the approximate efficiency of populations from the western United States. The method used by Freitag *et al.* (1955), in their classic experiments with specimens of *C. tenellus* from Morocco, would suffice.

Second, and perhaps this line of inquiry is the more important, conduct a systematic search among the agallian leafhoppers, with special reference to species of *Agalliopsis* of the *novella* complex that occur in the central and eastern parts of the United States and in the Carribean to ascertain whether any of them can transmit one or more of the strains of WCTV. Curly-top diseased plants that occur naturally on occasion in the eastern United States are suggested as a source of virus for testing.

More exact knowledge of the Old World distribution of the curly-top virus would provide collateral information that would be helpful in clarifying anomalies in the curly-top picture. Gibson (1966) has encountered what appears to be curly-top in Iran. Should it develop that curly-top, in the collective species sense, has a wide distribution in remote areas of Asia Minor, near the center of origin of *Circulifer*, the hypothesis of a New World origin for the curly-top viruses would scarcely be tenable.

Admittedly much of the beet leafhopper–curly-top association has become obscured by the passage of time and may never be clarified because of the rather uncertain chronology of pertinent events. A much more detailed record can be assembled with regard to another virus, X disease, and its vectors, one of which is known to have been translocated from the Old World to the New.

In the western United States X disease of stone fruits has been transmitted by *Colladonus geminatus, Colladonus montanus, Fieberiella florii, Keonolla confluens,* and *Scaphytopius acutus* (Wolfe *et al.*, 1950; Anthon and Wolfe, 1951; Wolfe, 1955). The record of transmission by *Keonolla confluens* appears questionable and needs to be reinvestigated. Gilmer (1954) reported transmission in New York with *Colladonus clitellarius;* Gilmer and McEwen (1958) reported transmission with *Fieberiella florii, Scaphytopius acutus,* and *Gyponana striata* [=*Gyponana lamina?*]. Gilmer *et al.* (1966) reported transmission in New York with the additional vectors *Norvellina seminuda* and *Paraphlepsius irroratus.*

Of the species recorded as vectors of X disease, all are native to North America except *Fieberiella florii.* X disease has not been recorded from the Old World, whence came *F. florii.* In the western United States, *C. geminatus* is considered to be the primary vector, although Wolfe (1955) pointed out that the role of the nymphal stages of leafhoppers in the spread of X disease was not determined, and that although *C. geminatus* will reproduce and thrive on diseased peach and cherry when confined to the trees, he had not observed the natural occurrence of *C. geminatus* nymphs on those hosts in the field. On the other hand, Phillips (1951) and Wolfe (1955) both indicated that nymphs of *F. florii* are commonly found associated with these stone fruits, and Wolfe stated that transmission of X disease with nymphs of *F. florii* was less difficult than with the nymphs of *C. geminatus*. Gilmer *et al.* (1966) consider *S. acutus* to be the most important of the several vectors in New York.

X disease was first reported on peach in Connecticut in 1933, although evidence from growers indicated that it had been present several years earlier. It was identified in Massachusetts and New York in 1938; in Ontario, Canada, Michigan, and Pennsylvania in 1941, and in Ohio in 1944. It has since been reported in Illinois, Indiana, Maine, New Hampshire, Rhode Island, and Vermont. (Records from numerous sources are summarized by Stoddard *et al.*, 1951.)

Western X disease was recognized in Idaho in 1936. What was subsequently determined to be the same disease was observed in Washington in 1935 and 1936 and in Utah in 1937. It was apparently present in Oregon in 1939 and in Colorado in 1941. It also occurs in British Columbia and is probably the same as diseases of stone fruits recorded from California. (Records from numerous sources are summarized by Reeves *et al.*, 1951.)

*Fieberiella florii* is known to have been established in Connecticut as early as 1918 and evidently was rather widely distributed in the eastern United States prior to 1930. Records of its early occurrence, compiled from the U.S. National Insect Collection and from literature, are as follows: Connecticut, 1918 and 1920; Massachusetts, 1921; District of Columbia, 1925; Rhode Island, 1926; New York, 1934; Utah, 1934; North Carolina, 1940; Tennessee, 1941; and Washington and California, 1944. No doubt a critical search of other insect collections will reveal other, and possibly earlier, records for some regions of North America.

X disease virus appears to be primarily associated with stone fruits, although Kunkel (1944) demonstrated that it could be experimentally transmitted to herbaceous plants by dodder. Of the proved leafhopper vectors of X disease virus, only *F. florii* appears to be consistently associated with stone fruits in essential phases of its life cycle. Thus, on circumstantial evidence, *F. florii* would seem to be the most logical choice to show a specific relationship with the virus. Gilmer *et al.*

(1966) report it to be an efficient vector. Whitcomb *et al.* (1966) demonstrated the multiplication of X disease virus in *Colladonus montanus,* but it has not been demonstrated that the virus multiplies in other of its leafhopper vectors, or that it passes through the egg stage of any vector.

These circumstances provide material for interesting speculation relative to the association between X disease and its vectors. Four possible interpretations suggest themselves, namely:

*1.* X-disease virus originated in North America and several species of leafhoppers, both native and adventive, proved capable of transmitting it.

*2.* X-disease virus was present in North America prior to the introduction of *Fieberiella florii,* but unrecognized. When associated with the virus, *F. florii* was able to transmit it.

*3.* X-disease virus was brought to the New World by *F. florii,* and native species were found to be capable of transmitting it.

4. X-disease virus was brought to the New World independently of *F. florii,* and both native and adventive species proved capable of transmitting it.

Each of these possible interpretations leaves questions unanswered, but available information points to the desirability of directing additional attention to the relationship between the virus and *F. florii.* If X-disease virus is found to pass through the eggs of infected *F. florii* females to the progeny, a strong circumstantial case will have been made for alternative *3,* above. But this would still leave unexplained the apparent absence of X disease from the Old World, the original home of *F. florii.*

It is not often that we are able to trace backwards in time the tenuous chain of events leading to outbreaks of virus diseases that attract the attention of investigators. One such series of events about which there seems to be a fair degree of reliable information is the introduction of corn stunt virus (CSV) into southeastern United States and its subsequent spread and increase to levels of economic importance.

The virus of corn stunt (CSV), as known in Mexico and adjacent United States, is transmitted by two species of deltocephaline leafhoppers, *Dalbulus maidis* and *D. elimatus.* The vector–virus relationships have been considered to be specific. When corn stunt appeared in southeastern United States in 1962, it was therefore natural that a search would be made for the known vectors, particularly for *D. maidis,* which seemed a more likely candidate to survive in a warm, humid area than did *D. elimatus.* However, it quickly became apparent that the incidence of corn stunt could not be accounted for by the very limited occurrence of *D. maidis* in the area. The existence of an alternate vector was therefore suspected.

Among the prime suspects as a vector of CSV was *Graminella nigrifrons,* an ubiquitous species native to the southeastern states but not previously incriminated, or even tested, as a vector of CSV. Subsequent studies (Granados *et al.,* 1966) have established that *G. nigrifrons* is capable of transmitting CSV, with which it has apparently only recently become associated. *Graminella nigrifrons* and *D. maidis* are not closely related, although they are assigned to the same subfamily of leafhoppers. This appears to be a clear case of "accidental" association between a virus and a leafhopper that proved to be a capable vector, and one that deserves further study of the vector–virus relationships. Further studies of the virus–vector interactions of *G. nigrifrons* and CSV in comparison with the interactions between CSV and *D. maidis,* as suggested by Granados *et al.,* should provide the type of information needed to improve our understanding of the nature of specificity.

Members of the family Delphacidae are often referred to by workers in the field of plant virus transmission as leafhoppers. This is understandable, for even though a taxonomically quite different group is involved, the delphacid plant hoppers exhibit a remarkable similarity to the leafhoppers in their relationships with plant-pathogenic viruses. Although there are relatively few known cases of delphacid transmission of viruses, the available evidence indicates a comparatively high degree of vector specificity. The vector relationships of maize rough dwarf virus (MRDV) (Harpaz, 1966), for example, are comparable to relationships that often occur among the leafhoppers. The primary vector of MRDV is *Delphacodes striatella,* although Harpaz *et al.* (1965) have shown *Delphacodes pellucida* to be capable of transmitting the virus in Europe, and Harpaz (1966) has demonstrated transmission with *Sogatella vibix* in Israel. *Delphacodes striatella* also transmits several other viruses (Harpaz, 1964), although the number is questionable because of uncertainty as to the exact identity of some of the virus diseases studied.

Among the delphacids there are a few instances of passage of the virus through the egg to progeny of infected females. Of those that show transovarial passage of virus, in three there is some evidence of cytological changes or of deleterious action of the virus on the vector. Sukhov (1940), reported inclusion bodies in *Delphacodes striatella* infected with "Zakuklivanie" virus; Watson and Sinha (1959) report the death of some embryos of *Delphacodes pellucida* infected with virus of European wheat striate mosaic; and Showers and Everett (1967, in press) working with hoja blanca virus, found that *Sogata orizicola* progeny resulting from matings of nontransmitting (NT) males with transmitting (TT) females had a shorter average life span than progeny resulting from the reciprocal cross. Transovarial passage was found to occur in a small percentage of the NT male × TT female crosses, but not in the reciprocal cross. The work of Kisimoto and Watson (1965) showing the deleterious effect of inbreeding on embryonic development of two different species of *Delphacodes* emphasizes the need to make critical evaluations of cases suspected of showing adverse effects of a virus on its vector.

The aphids, in marked contrast to the leafhoppers and delphacids, exhibit little if any tendency toward vector specificity except at the group level. Kennedy *et al.* (1962), following their penetrating analysis of the problems of vector specificity among aphids, concluded that "circulative" viruses transmitted by aphids are more likely to be vector specific than are the "stylet-borne" viruses. Yet even this very general observation is based to a considerable extent upon negative evidence, for the ten circulative viruses listed by those authors as being transmitted by a single vector, nonvectors were not listed for five, indicating a lack of tests with other potential vectors. Of the additional 20 circulative viruses listed, there are four for which no nonvectors are shown. There is, of course, as those authors point out, the tendency for investigators not to report negative evidence even when such is available. However, whatever bias of that sort may be represented in the above data would seem to be compensated for by the paucity of investigations designed to determine if these circulative viruses can be transmitted by other species.

Although there is this lack of evidence of vector specificity among the aphid transmitted viruses, there is at the same time an abundance of evidence of group specificity with these same viruses. It seems clear that a very large group of viruses has evolved in association with aphids, but not in association with any particular species or group of aphids. The evidence indicates the normal relationship between aphids and the viruses they transmit to be one that involves extreme genetic plasticity, with a tendency for genetic drift to occur in either the vector or the virus, or both.

The logical explanation of this situation would seem to lie in the intricate biology of the aphids, which includes an elaborate system of polymorphism and usually a succession of plant hosts, including obligatory or facultative alternation between taxonomically quite different plant species. As Kennedy and Stroyan (1959) point out, among the aphids, "Field observations and the results of insectary experiment suggest that the raw material of evolutionary change is produced every season . . . ." It seems reasonable to suppose that viruses would need to have great genetic plasticity to accommodate to the many changes of host and host physiology that would result from association with aphids as vectors. Aphid clones would seem to provide the uniform conditions under which natural selection of a virus would operate mainly as a stabilizing factor. When conditions change, the pressure would operate to select any variant that is better fitted to survive under new conditions, as suggested by Bawden (1964).

Other than group specificity, whatever vector specificity exists between aphids and viruses appears to be in the nature of adaptation between "strains" of viruses and species or clones of aphids. These adaptations are not necessarily in conflict with the generalization outlined above, for Rochow (1963) in his discussion of variation within and among aphid vectors of plant viruses concludes that, "Results of the various studies on vector specificity of barley yellow dwarf virus are perhaps best explained by assuming that the virus is extremely variable and that the existing strains differ in the efficiency with which each aphid species can transmit them."

But with the aphid-borne viruses, the safest course would seem to be to extend to this field of speculation the admonition of Bodenheimer and Swirski (1957) to students of aphids "never to generalize."

At this very preliminary stage of development of knowledge about vector–virus relationships it is extremely unlikely that any very reliable criteria for distinguishing between intrinsic and fortuitous vector–virus associations can be defined. Still, in our efforts to interpret the significance of experimental results it seems worthwhile to scrutinize critically certain aspects of vector–virus relationships. Analyses of concordance and discordance in the evidence before us, as it relates to the total vector–virus relationship, are a legitimate and necessary activity. The following areas are among those that may warrant exploration.

## I. CORRELATIONS OR DISCONTINUITIES IN DISTRIBUTION OF VECTOR AND VIRUS

We can find many cases in which the distribution of a virus exceeds the known distribution of its presumed vector; for others, the presumed "primary" vectors of a virus may be different in different regions. Although we cannot consider the negative evidence involved in such cases as indicative of lack of specificity, a critical examination of the distribution of both virus and vector in time and space, with due consideration being given to the presumed endemicity of each, may supply important corroborative evidence. Ecological preferences of the arthropod host, and its "normal" plant host associations, should be an integral part of this analysis. If the "normal" plant host of the vector is not also a host for the virus, the fundamental nature of the vector–virus relationship becomes suspect.

As mentioned earlier, the occasional occurrence of curly-top in the eastern United States represents a case in which distribution of the virus exceeds the known distribution of its vector(s). The apparent absence of *Circulifer tenellus* from this region is, of course, negative evidence. It is legitimate to question how reliable such evidence may be. Speculations leading to the conclusion that some other vector must be involved are as follows.

The incidence of curly-top infection in field populations of beet leafhoppers in the western United States, whence migrants to the eastern United States are presumed to come, may be expected to be in the order of 10%. Thus, for every infectious leafhopper that migrated to the eastern seaboard and transmitted curly-top, there would be nine others that were not infectious—unless, of course, infected leafhoppers have a greater tendency to migrate, a point on which we have no information. Further, it seems reasonable to assume that for every observed and reported case of curly-top, there must be many others that go undetected or unidentified. Thus, to account for the incidence of curly-top that presumably occurs on occasion in the eastern states, there would need to be relatively large numbers of leafhoppers. If this were the case, it is likely that some specimens of *C. tenellus* would long since have been found and identified. Such is not the case.

There are other considerations that argue against the idea that *C. tenellus* periodically invades the eastern United States. Chief among these is the fact that unusually dry summers such as have occurred several times during the period in which cases of curly-top have been reported provide ecological conditions that should permit survival and reproduction of migrant beet leafhoppers. Had a second generation of *C. tenellus* occurred, the chances of detection of the leafhopper would have been enhanced, as would the chances of the secondary spread of any curly-top infections. There is no convincing evidence that *C. tenellus* has ever penetrated further eastward than the central states, west of the Appalachian Mountains, although other migrant species of leafhoppers, such as *Dalbulus maidis,* have been detected well outside their normal range.

### *A. Efficiency of Transmission of Virus by Vector*

Because numerous factors other than the intrinsic vector–virus relationships influence the efficiency with which a vector may transmit a virus under natural conditions, evidence of this sort needs to be experimentally evaluated. Species that may be highly efficient so far as the acquisition, propagation, and transmission of virus are concerned could be very inefficient vectors from the standpoint of disease epidemics.

The converse is also true; species that are relatively inefficient vectors of a virus may be primarily responsible for its spread in nature because of their behavior. This appears to be the situation with WCTV and the beet leafhopper. The apparent lack of multiplication of WCTV in *Circulifer tenellus,* and the relatively low incidence of natural infection of leafhopper populations in nature suggests, inefficiency of transmission, yet the propensity of massive populations of *C. tenellus* to make long-range dispersal movements during which feeding occurs on many kinds of plants, makes the beet leafhopper responsible for widespread outbreaks of curly-top disease of cultivated crops in western North America.

Could it be that differences in the behavior of populations of the vector accounts for the apparent absence of curly-top from most parts of the Old World where *C. tenellus* or *C. opacipennis* are known to occur? Frazier (1953), in his report of explorations for the beet leafhopper in the Mediterranean region, commented as follows with respect to species of the *C. tenellus* group:

> Populations were not found in great abundance and the distribution of some of the forms appeared to be sharply limited by undetermined factors. Often a population would be found only in a small sampling area but not in surrounding or adjacent areas that appeared equally, or more, favorable. In several instances distribution appeared limited to a narrow zone bordering lake or sea shores. Specimens were usually collected in very moderate or low numbers.

and further,

> In general the ecological characteristics of the genus *Circulifer,* and especially the *C. tenellus* group, as observed in the Mediterranean region, were very similar to those of the beet leafhopper in western North America so far as habitat and host plants are concerned. However, the Mediterranean populations seemed to differ in three important respects: (*1*) In no area of the Mediterranean was sugar beet or wild beet a good host plant of *Circulifer* although good populations were occasionally found on weeds growing in and around sugar beet plantings. (*2*) Russian thistle, although not found abundantly, was never observed to be more than a relatively mediocre host of any species of *Circulifer*. Only adults were found on this host. (*3*) Most *C. tenellus* group populations appeared to have a restricted distribution locally."

### *B. Persistence of Virus in Arthropod Host*

That some viruses survive in quite unrelated nonvector arthropods indicates that relatively little significance should be attached to this phenomenon. Curly-top virus survived for up to 24 hr after being injected into the blood of a caterpillar (Smith, 1941) and for as long as 14 days in nonvector species of leafhoppers (Bennett and Wallace, 1938; Smith, 1941).

## II. DEGREE OF BIOCHEMICAL OR PHYSIOLOGICAL ADAPTATION BETWEEN VECTOR AND VIRUS

Compatibility, or the tendency to adjust toward a "peaceful coexistence" relationship, is widely accepted as a measure of specificity in many biological relationships; the opposite condition, antagonism between the biological entities involved, is usually considered to indicate the absence of association during evolutionary development. With viruses and their arthropod vectors we have such anomalous circumstances as apparently compatible relationships between virus and adult and nymphal stages (*e.g., Delphacodes striatella* and European wheat streak mosaic virus) but not between virus and egg stage of the vector. Considering that the physiological processes of embryonic development are different from those involved in postembryonic development, this situation is not necessarily inconsistent with the general idea that a relatively advanced state of mutual adaptation has been attained. While antagonistic interactions between virus and embryos would not be expected to assist in perpetuating the vector–virus relationship, other circumstances of the relationship might well more than offset this factor.

### *A. Propagation and Circulation of Virus in Arthropod Host*

Initial doubts that so-called "plant" viruses ever multiplied in their arthropod hosts have been dispelled by the work of Fukushi (1935, 1940), Black (1950), and Maramorosch (1952a). For most leafhopper transmitted viruses, however, this essential bit of information is not yet available. It may be difficult to obtain because negative evidence is often inconclusive. For example, the beet leafhopper and curly-top virus have been critically investigated (Freitag, 1936; Bennett and Wallace, 1938). The evidence indicates that the virus does not increase, but being of a negative nature it is not considered entirely conclusive. Bennett (1962) reported that the curly-top virus content of the beet leafhopper was influenced by the virus concentration in the plants upon which it fed, and concluded that there was probably no measurable increase of virus in the vector.

The situation is further complicated by the fact that a virus can clearly persist for some time, and may even multiply in an arthropod host that is unable to transmit it. Maramorosch (1952b) showed that aster yellows virus was acquired by *Dalbulus maidis* fed on infected plants, and although not detectable two days after acquisition, became detectable 17 days later. Bawden (1964, pp. 122–123) intepreted this as evidence of multiplication of the virus in a nonvector arthropod host. Considered together then, the evidence of circulation and transmission without propagation and propagation without transmission, indicate that circulation and propagation are independent phenomena. Neither alone should be considered indicative of a high degree of specificity, but among the sucking arthropods that transmit viruses a close intrinsic association of vector and virus can scarcely be assumed without both these conditions having been met.

In most cases it would probably be legitimate to accept adaptations of the sort necessary for propagation and circulation as having been derived from evolutionary association, but this should not be an unquestioned assumption. Considering the vast number of species of potential vectors that could be brought into association with viruses, it would be remarkable if some of these chance associations did not provide the kind of biochemical or physiological environment necessary to permit virus propagation. It follows then that other criteria should also be considered in judging the significance of evidence of propagation and circulation of viruses in their arthropod hosts.

### *B. Transovarial Passage of Virus*

In certain vector–virus relationships (*e.g.*, *Nephotettix cincticeps* and rice dwarf virus; *Austroagallia torrida* and rugose leaf curl virus; *Agalliopsis novella* and clover club leaf virus), passage of the virus from infective females through the eggs to their progeny appears to be a relatively comon occurrence, if not the general rule. Among the leafhoppers, so far as I am aware, transovarial passage of virus is always correlated with other evidence that supports the idea of true specificity in the relationship, or at least is not in conflict with that interpretation. It is probably the most important single bit of evidence by which we can measure degrees of specificity.

It will be quite obvious that it is impossible to apply the criteria outlined in the foregoing discussion to many of the vector–virus situations with any degree of exactness. In part, at least, this is because our knowledge of arthropod-transmitted plant-pathogenic viruses is largely limited to those viruses that cause diseases in economically important plants, and these are the cases in which endemicity and host associations of the biological elements are most uncertain. In effect, we have permitted the insects to choose our problems for us, and by so doing have attacked what are probably the more complex of the problems from the standpoint of a better understanding of the probable origin of some viruses. Rarely have there been deliberate attempts to find viruses in their insect vectors as Black (1944) did with the wound tumor and clover club leaf viruses. Little attention has been given to the several translocated vectors that have been shown to transmit viruses in regions where they are not native. Examples of such translocated vectors of plant-pathogenic viruses, in addition to those discussed earlier, are *Euscelidius variegatus,* a European species adventive to North America, where it has been shown capable of transmitting California aster yellows (Severin, 1947), and *Scaphoideus littoralis,* a North American species adventive to Europe where it transmits "Flavescence dorée" of grape (Schvester *et al.,* 1961). Neither of these species has been critically studied in its native home for ability to transmit viruses.

Among scientists interested in problems of arthropod-borne plant pathogenic viruses there has been considerable discussion of the need for "isolation" facilities where plants, arthropods and viruses could be brought together for experimental purposes (Miller, 1966). We should remind ourselves that translocation of vectors has sometimes given us opportunities to investigate certain problems that would be undertaken at isolation laboratories, if they existed. By being opportunistic and focusing attention on situations of this sort, when they occur, and by conducting systematic searches among selected groups of insects for evidence of infection with viruses that can be transmitted to plants, I believe our understanding of the complex vector–virus relationship can be much accelerated. Some of these studies will not necesasrily require elaborate or expensive facilities or equipment, or a high degree of sophistication in the investigative technics. And as Black (1944) has pointed out, the search among potential vectors for viruses has the obvious advantage of providing a means of virus transmission simultaneously with the discovery of the virus. By judicious selection of the potential vector groups to be explored, with particular attention to the plant host association of candidate vectors, the method may also provide accurate information as to the probable plant source of the virus in nature.

# BIBLIOGRAPHY


Adsuar, J. 1955. A disease of tomato in Puerto Rico resembling the Brazilian curly-top of tomatoes. J. Agr. Univ. Puerto Rico 39:113–114.

Anthon, E. W., and Wolfe, H. R. 1951. Additional insect vectors of western X-disease. Plant Disease Reporter 35(8):345–6.

Bawden, F. C. 1964. Plant Viruses and Virus Diseases. 4th ed., Ronald Press, N. Y. 361 p.

Bennett, C. W. 1962. Curly-top virus content of the beet leafhopper influenced by virus concentration in diseased plants. Phytopathology 52(6):538–541.

Bennett, C. W., Carnsner, E., Coons, G. H., and Brandes, E. W. 1946. The Argentine curly-top of sugar beet. J. Agr. Res. 72:19–47.

Bennett, C. W., and Costa, A. S. 1949. The Brazilian curly-top of tomato and tobacco resmbling North American and Argentine curly-top of sugar beet. J. Agr. Res. 78:675–693.

Bennett, C. W., and Tanrisever, A. 1957. Sugar beet curly-top disease in Turkey. Plant Disease Reporter 41(9):721–725.

——— 1958. Curly-top disease in Turkey and its relationship to curly-top in North America. J. Amer. Soc. Sugar Beet Technol. 10(3):189–211.

Bennett, C. W., and Wallace, H. E. 1938. Relation of the curly-top virus to the vector, *Eutettix tenellus*. J. Agr. Res. 56:31–51.

Black, L. M. 1944. Some viruses transmitted by agallian leafhoppers. Proc. Amer. Phil. Soc. 88(2):132–144.

——— 1950. A plant virus that multiplies in its insect vector. Nature 166:852–853.

——— 1953. Transmission of plant viruses by cicadellids. Advances Virus Res. 1:69–89.

——— 1954. Parasitological Reviews. Arthropod transmission of plant viruses. Exp. Parasitol. 3(1):72–104.

——— 1956. Transmission of viruses by leafhoppers. 10th Int. Congr. Entomol. Proc. 3:201–204.

——— 1959. Biological cycles of plant viruses in insect vectors. In The Viruses, Vol. 2 (3 vols.). Burnet, F. M., and Stanley, W. M., Eds., Academic Press, New York, pp. 157–185.

Bodenheimer, F. S., and Swirski, E. 1957. Aphidoidea of the Middle East. Weizmann Scientific Press, Jerusalem, Israel. 378 pp.

Costa, A. S. 1952. Further studies on tomato curly-top in Brazil. Phytopathology 42:396–403.

Day, M. F., and Bennets, M. J. 1954. A review of problems of specificity in arthropod vectors of plant and animal viruses. Commonwealth Sci. Res. Organ. Canberra, Australia. 172 p.

Frazier, N. W. 1951. A survey of the Mediterranean region for the beet leafhopper. J. Econ. Entomol. 46(4):551–554.

Freitag, J. H. 1936. Negative evidence on multiplication of curly-top virus in the beet leafhopper, *Eutettix tenellus*. Hilgardia 10:305–342.

Freitag, J. H., Frazier, N. W., and Huffaker, C. B. 1955. Crossbreeding beet leafhoppers from California and French Morocco. J. Econ. Entomol. 48(3): 341–342.

Fukushi, T. 1935. Multiplication of virus in its insect vector. Proc. Imp. Acad. (Tokyo) 11:301–303.

——— 1940. Further studies on the dwarf disease of rice plant. J. Faculty Agr. Hokkaido Univ. 45:83–154.

Gibson, K. E. 1966. Personal communication.

Gillette, C. P., and Baker, C. F. 1895. A preliminary list of the Hemiptera of Colorado. Colorado Agr. Exp. Sta. Bull. 31:1–137.

Gilmer, R. M. 1954. Insect transmission of X-disease virus in New York. Plant Disease Reporter 38(9):628–629.

Gilmer, R. M., and McEwen, F. L. 1958. Insect transmission of X-disease virus. Phytopathology 48(5):262.

Gilmer, R. M., Palmiter, D. H., Schaeffers, G. A., and McEwen, F. L. 1966. Leafhopper transmission of X-disease virus of stone fruits in New York. New York State Agr. Exp. Sta., Geneva. 22 pp.

Granados, R. R., Maramorosch, K., Everett, T., and Pirone, T. P. 1966. Transmission of corn stunt virus by a new leafhopper vector, *Graminella nigrifrons* Forbes. Contributions Boyce Thompson Inst. 23(7):275–280.

Harpaz, I. 1964. Host plant–vector and Host plant–virus relationships in the rough dwarf disease of maize. Maydica 9:16–20.

——— 1966. Further studies on the vector relations of the maize rough dwarf virus (MRDV). Maydica 11:18–26.

Harpaz, I., Vidano, C., Lovisolo, O., and Conti, M. 1965. Indagini Comparative su *Javesella pellucida* Fabricius e *Laodelphax striatellus* Fallen quali vettori del virus del nanismo ruvido del mais (maize rough dwarf virus). Atti Accad. Sci. Torino 99:885–901.

Heggestad, H. E., and Moore, E. L. 1959. Occurrence of curly-top in tobacco in Maryland and North Carolina in 1958. Plant Disease Reporter 43(7):682–684.

Holmes, F. O. 1939. Handbook of Phytopathogenic Viruses. Burgess Publishing Co., Minneapolis. 221 p.

Kennedy, J. S., Day, M. F., and Eastop, V. F. 1962. A conspectus of aphids as vectors of plant viruses. Commonwealth Inst. Entomol., London. 114 p.

Kennedy, J. S., and Stroyan, H. L. G. 1959. Biology of aphids. Ann. Rev. Entomol. 4:139–160.

Kisimoto, R., and Watson, M. A. 1965. Abnormal development of embryos induced by inbreeding in *Delphacodes pellucida* Fabricius and *Delphacodes dubia* Kirschbaum (Araeopidae, Homoptera), vectors of European wheat striate mosaic virus. J. Invertebrate Pathol. 7(3):297–305.

Kunkel, L. O. 1935. Possibilities in plant virus classification. Bot. Rev. 1:1–17.

Kunkel, L. O. 1944. Transmission of virus from X-diseased peach trees to herbaceous plants. Phytopathology 34:1006.

Leach, J. G. 1940. Insect transmission of plant diseases. McGraw Hill, New York. 615 p.

Maramorosch, K. 1952a. Direct evidence for multiplication of aster-yellows virus in its insect vector. Phytopathology 42:59–64.

——— 1952b. Studies on the nature of the specific transmission of aster-yellows and corn stunt viruses. Phytopathology 42(12):663–668.

——— 1963. Arthropod transmission of plant viruses. Ann. Rev. Entomol. 8:369–414.

Maramorosch, K., and Jensen, D. D. 1963. Harmful and beneficial effects of plant viruses in insects. Ann. Rev. Microbiol. 17:495–530.

Maramorosch, K., and Oman, P. 1966. U.S.-Japan Joint Conference on Arthropod-Borne Plant Viruses. BioScience 16(9):608–610.

Miller, P. R. 1966. International usefulness of an isolation laboratory for plant pathogens, especially viruses and their vectors. Plant Disease Reporter 50(11):803–805.

Oman, P. W. 1936. Distributional and synonymical notes on the beet leafhopper, *Eutettix tenellus* Baker. Proc. Entomol. Soc. Washington 38 (7): 164–165.

——— 1948. Notes on the beet leafhopper, *Circulifer tenellus* Baker and its relatives (Homoptera: Cicadellidae). J. Kansas Entomol. Soc. 21(1):10–14.

——— 1949. The nearctic leafhoppers (Homoptera: Cicadellidae). A generic classification and check list. Washington Entomol. Soc. Mem. 3:1–253.

Phillips, J. H. H. 1951. An annotated list of Hemiptera inhabiting sour cherry orchards in the Niagara Peninsula, Ontario. Can. Entomol. 83:194–205.

Reeves, E. L., Blodgett, E. C., Lott, T. B., Milbrath, J. A., Richards, B. L., and Zeller, S. M. 1951. Western X-Disease *In* Virus diseases and other disorders with viruslike symptoms of stone fruits in North America. Agricultural Handbook 10. 276 pp. U.S. Dept. Agr. p. 43–52.

Rochow, W. F. 1963. Variation within and among aphid vectors of plant viruses. Ann. N.Y. Acad. Sci. 105:713–729.

Schneider, C. L. 1959. Occurrence of curly-top of sugar beets in Maryland in 1958. Plant Disease Reporter 43(7):681.

Schvester, A., Carle, P., and Moutous, G. 1961. Sur la transmission de la flavescence dorée des vignes par une cicadelle. Compt. Rend. Acad. Agr. France, p. 1021–1024.

Severin, H. H. P. 1947. Newly discovered leafhopper vectors of California aster-yellows virus. Hilgardia 17(16):511–523.

Showers, W. B., and Everett, T. R. 1967. Transovarial acquisition of hoja blanca virus by the rice delphacid. J. Econ. Entomal. 60(3):757–760.

Simons, J. N. 1962. The pseudo-curly top disease in south Florida. J. Econ. Entomol. 55(3):358–363.

Simons, J. N., and Coe, D. M. 1958. Transmission of the pseudo-curly-top virus in Florida by a treehopper. Virology 6:43–48.

Simpson, R. W., and Hauser, R. E. 1966. Structural components of vesicular stomatitis virus. Virology 29(4):654–667.

Smith, K. M. 1941. Some notes on the relationship of plant viruses with vector and nonvector insects. Parasitology 33(1):110–116.

Stoddard, E. M., Hildebrand, E. M., Palmiter, D. H., and Parker, K. G. 1951. X-Disease. *In* Virus Diseases and Other Disorders with Virus-like Symptoms of Stone Fruits in North America. Agricultural Handbook 10. U.S. Dept. Agr. 276 p.

Storey, H. H. 1931. The bearing of insect vectors on the differentiation and classification of plant viruses. 2nd Congr. Int. Pathol. Compt. Remd. Communications p. 471–479.

Sukhov, K. S. 1940. X-bodies in salivary glands of *Delphax striatella* Fallen, the carrier of zakuklivanie. Compt. Rend Acad. Sci. URSS 27:377–379.

Sylvester, E. S. 1954. Aphid transmission of nonpersistent viruses with special reference to the *Brassica nigra* virus. Hilgardia 23:53–98.

Troutman, J. L., and Fenne, S. B. 1959. The occurrence of curly-top in Virginia. Plant Disease Reporter 43(2):155–156.

Watson, M. A., and Sinha, R. C. 1959. Studies on the transmission of European wheat striata mosaic virus by *Delphacodes pellucida* Fabricius. Virology 8:139–163.

Webb, R. E. 1963. Tomato curly-top in Maryland. Plant Disease Reporter 47(1):53.

Whitcomb, R. F., Jensen, D. D., and Richardson, J. 1966. The infection of leafhoppers by western X-disease virus. II. Fluctuation of virus concentration in the hemolymph after I injection. Virology 28(3):454–458.

Wolfe, H. R. 1955. Relation of leafhopper nymphs to the western X-disease virus. J. Econ. Entomol. 48:588–590.

Wolfe, H. R., Anthon, E. W., and Jones, L. S. 1950. Transmission of western X-disease of peaches by the leafhopper *Colladonus germinatus* (Van D.). Phytopathology 40(10):971.

Zaumeyer, W. J., and Smith, F. F. 1964. Unpublished reports.

# Fungi as Vectors and Hosts of Viruses

David S. Teakle

*Department of Microbiology, University of Queensland, Medical School, Herston, Queensland, Australia*

## I. INTRODUCTION

The study of fungi as virus vectors is a rapidly expanding area of research. Before 1965, only one fungus vector, *Olpidium brassicae,* was known. By 1967, three further vectors, *Synchytrium endobioticum, Polymyxa graminis,* and *Spongospora subterranea,* had been reported.

The possibility that fungi might transmit viruses was first suggested by McKinney (1930), who stated, "It is possible that subterranean vectors such as nematodes, soil-borne insects, other animal forms, or even fungi, may carry the (wheat mosaic) virus." McKinney, Webb, and Dungan (1925) had inoculated wheat plants with cultures of *Helminthosporium sativum* isolated from diseased plants and had failed to produce mosaic leaf symptoms. Linford and McKinney (1954) noted a high correlation between the presence of *Polymyxa graminis* and mosaic in wheat, but concluded that there was no evidence that the fungus was acting as a virus vector or a reservoir host. More recent work, however, indicates that *P. graminis* is a vector and the reservoir host of wheat mosaic virus (Estes and Brakke, 1966). Fry (1958) and Grogan, Zink, Hewitt, and Kimble (1958) reported the constant association of *Olpidium brassicae* with lettuce big vein disease. However, they considered that *O. brassicae* did not act as a virus vector but caused the disease directly, possibly by producing a toxin which was translocated from the roots to the leaves. Again, more recent work indicates that *O. brassicae* is a vector of lettuce big vein virus (Campbell, Grogan, and Purcifull, 1961; Tomlinson, Smith, and Garrett, 1962).

Hidaka (1960) reported a correlation between the incidence of *Olpidium brassicae* and tobacco stunt virus in tobacco seedlings growing in soil subjected to chemical and physical treatments. Teakle (1960) found that the same fungus was usually present in roots of plants

TABLE 1
FAILURE OF CERTAIN FUNGI TO TRANSMIT VIRUSES

| Virus | Particle dimensions | Nonvector | Inoculum | References |
|---|---|---|---|---|
| Arabis mosaic | 30 mμ diam | *Endogone* sp. | Healthy and virus-infected plants grown together in inoculated soils | Jha, 1961 |
| Bean common mosaic | 750 × 15 mμ | *Uromyces phaseoli typica* | Uredospores from virus infected plants | Nelson, 1932 |
| Bean southern mosaic | 30 mμ diam | *Olpidium brassicae* | Mixture of zoospores and virus | Teakle and Gold, 1963 |
| Beet yellows | 1250 × 10 mμ | *Peronospora schachtii* | Spores from virus-infected plants | Hansen, 1947 |
| | | *Ramularia beticola* | Spores from virus-infected plants | Hansen, 1947 |
| | | *Uromyces betae* | Spores from virus-infected plants | Hansen, 1947 |
| Citrus psorosis A | ? | *Phytophthora citrophthora* | Cultures from virus-infected plants | Koltz and DeWolfe, 1961 |
| | | *Phytophthora parasitica* | Cultures from virus-infected plants | Koltz and DeWolfe, 1961 |
| Pea enation mosaic | 30 mμ diam | *Ascochyta pisi* | Cultures from virus-infected plants | Johnson and Jones, 1943 |
| | | *Cladosporium pisicola* | Cultures from virus-infected plants | Johnson and Jones, 1943 |

| | | | | |
|---|---|---|---|---|
| | | *Fusarium* sp. | Cultures from virus-infected plants | Johnson and Jones, 1943 |
| | | *Pythium* sp. | Cultures from virus-infected plants | Johnson and Jones, 1943 |
| | | *Rhizoctonia* sp. | Cultures from virus-infected plants | Johnson and Jones, 1943 |
| | | *Erysiphe polygoni* | Conidia from virus-infected plants | Johnson and Jones, 1943 |
| Pea mosaic | $750 \times 15$ mμ | *Ascochyta pisi* | Cultures from virus-infected plants | Johnson and Jones, 1943 |
| | | *Cladosporium pisicola* | Cultures from virus-infected plants | Johnson and Jones, 1943 |
| | | *Fusarium* sp. | Cultures from virus-infected plants | Johnson and Jones, 1943 |
| | | *Pythium* sp. | Cultures from virus-infected plants | Johnson and Jones, 1943 |
| | | *Rhizoctonia* sp. | Cultures from virus-infected plants | Johnson and Jones, 1943 |
| | | *Erysiphe polygoni* | Conidia from virus-infected plants | Johnson and Jones, 1943 |
| Potato X | $515 \times 13$ mμ | *Rhizoctonia solani* | Cultures from virus-infected plants | Bawden and Kassanis, 1947 |
| | | *Thielaviopsis basicola* | Cultures from virus-infected plants | Bawden and Kassanis, 1947 |
| Potato Y | $730 \times 15$ mμ | *Synchytrium endobioticum* | Zoospores from virus-infected plants; mixture of zoospores and virus | Nienhaus and Stille, 1965 |

*(continued)*

TABLE 1 (*continued*)

| Virus | Particle dimensions | Nonvector | Inoculum | References |
|---|---|---|---|---|
| Tobacco mosaic | 300×18 mμ | *Olpidium brassicae* | Mixture of zoospores and virus | Teakle and Gold, 1963 |
| Tobacco necrosis | 30 mμ diam | *Aphanomyces raphani* | Mixture of zoospores and virus | Teakle, 1963 |
| | | *Plasmodiophora brassicae* | Mixture of resting spores and virus | Teakle, 1963 |
| | | *Olpidium brassicae* (*Brassica* strains) | Mixture of zoospores and virus | Mowat, 1961, 1963<br>Teakle, 1962 |
| | | ***Rhizoctonia solani*** | Cultures from virus-infected plants | Bawden and Kassanis, 1947 |
| | | *Thielaviopsis basicola* | Endoconidia, chlamydospores, mycelia cultured from virus-infected plants, with or without TNV added | Szirmai, 1939, 1962<br>Bawden and Kassanis, 1947<br>Hecht, 1962 |
| Tobacco rattle | 180×25 mμ | *Olpidium brassicae* | Mixture of zoospores and virus | Teakle and Gold, 1963 |
| Tobacco ringspot | 30 mμ diam | *Olpidium brassicae* | Mixture of zoospores and virus | Teakle and Gold, 1963 |
| Tomato bushy stunt | 30 mμ diam | *Olpidium brassicae* | Mixture of zoospores and virus | Teakle and Gold, 1963 |
| Tomato ringspot | 30 mμ diam | *Olpidium brassicae* | Mixture of zoospores and virus | Teakle and Gold, 1963 |
| Wheat mosaic | 160×25 mμ | *Helminthosporium sativum* | Cultures from virus-infected plants | McKinney *et al.*, 1925 |

naturally infected with tobacco necrosis virus (TNV). Tobacco necrosis virus suspensions regularly caused infection in seedling roots only when zoospores of *O. brassicae* were present. Both authors suggested that a vector-like relationship was involved. Both these claims have been independently confirmed (Hiruki, 1965; Kassanis and Macfarlane, 1964).

The tardiness in discovering that certain fungi are virus vectors was not due to the absence of transmission tests. At least 18 fungi have been tested unsuccessfully as vectors by a variety of inoculation methods (Table 1). Factors which slowed the discovery include:

*1.* Four of the seven fungus-transmitted viruses are either difficult or impossible to transmit mechanically.

*2.* The viruses and fungus vectors are soil borne. The soil contains an almost unlimited array of microorganisms available to test as vectors.

*3.* The known fungus vectors are difficult, if not impossible, to culture on artificial media.

*4.* Fungi are known to cause disease by themselves, sometimes at a distance, by the production of toxins.

That fungi are hosts of viruses has been recognized only since Hollings (1962) demonstrated the presence of three viruses in the cultivated mushroom, *Agaricus bisporus,* suffering from die-back. Other studies have associated virus-like agents with diseased *Helminthosporium victoriae* (Lindberg, 1959) and yeasts (Lindegren, Bang, and Hirano, 1962), but in these cases further evidence is desirable in order to confirm the virus etiology.

With the exception of mushroom viruses, research on fungi as hosts of viruses has proved to be disappointingly difficult.

## II. THE VECTORS AND VIRUSES

### *A. Economic Importance*

The economic losses caused by the four fungus vectors include (*1*) losses caused by the fungi themselves, and (*2*) losses caused by the viruses transmitted by the fungi. An attempt to estimate these losses is made in Table 2. Such an estimate must take into account the geographic distribution, the incidence and severity of the disease, and the efficacy and cost of control measures. Unfortunately, only limited quantitative information is available.

The four fungus vectors all have a wide geographic distribution (Table 2). This is particularly the case with *Olpidium brassicae,* which

TABLE 2

ECONOMIC IMPORTANCE AND GEOGRAPHIC DISTRIBUTION OF FUNGUS-TRANSMITTED VIRUSES AND THEIR VECTORS

| Virus or vector | Diseases caused | Economic importance | Geographic distribution | Control |
|---|---|---|---|---|
| *Olpidium brassicae* | Slight stunting and yellowing in heavily infected plants | Slight | All continents | Rarely attempted |
| Lettuce big vein | Lettuce big vein | Moderate | Widespread in Australasia, Europe, and North America | Soil treatments to eliminate *Olpidium* and tolerant lettuce varieties; rarely attempted |
| Tobacco necrosis | French bean stipple streak; potato superficial tuber lesions; tulip necrotic disease | Slight–moderate | Japan, Australasia, Europe, and North America | Rarely attempted since disease occurrence usually sporadic |
| Tobacco stunt | Tobacco stunt | Considerable | Japan | Resistant tobacco varieties and sterilization of tobacco seed-beds |
| *Synchytrium endobioticum* | Potato wart | Considerable | Africa, Asia, Europe, and North and South America [a] | Resistant varieties and quarantine |

| | | | | |
|---|---|---|---|---|
| Potato X | Mild mosaic | Moderate | All continents | Resistant varieties and selection of virus-free plants |
| *Polymyxa graminis* | Slight stunting | Slight | Denmark, Italy, Japan, and North America | Rarely attempted |
| Wheat mosaic | Wheat mosaic | Considerable | Italy, Japan and North America [b] | Resistant varieties |
| *Spongospora subterranea* | Potato powdery scab | Considerable | All continents [c] | Resistant varieties |
| Potato mop top | Potato mop top | Considerable | Europe | Resistant varieties |

[a] See Commonwealth Mycological Institute Distribution Map of Plant Diseases No. 1.
[b] See Commonwealth Mycological Institute Distribution May of Plant Diseases No. 84.
[c] See Commonwealth Mycological Institute Distribution Map of Plant Diseases No. 34.

has been reported from all continents except Antarctica (Teakle, 1963). Records of its occurrence are most numerous from Europe (15 out of 30 countries) where cool conditions suitable for its development occur and where there has been a long mycological tradition. The paucity of records from Africa (1 record), Asia (2 records), and South America (1 record) probably does not reflect the true situation. One reason for the wide geographical distribution of *O. brassicae* is the wide range of its host, which includes more than 55 genera of phanerogams. Although different strains of *O. brassicae* usually have distinct host preferences, single strains are usually able to multiply in plants belonging to at least several different families (Sahtiyanci, 1962), It must be remembered, however, that although vector strains of *O. brassicae* are widespread and common, some of the records probably refer to nonvector strains.

*Spongospora subterranea* and *Synchytrium endobioticum* also are widely distributed. They are readily disseminated in infected potato tubers used for planting. Both fungi apparently infect only potato and a few other solanaceous hosts.

*Polymyxa graminis* has been reported only from three areas, but these are widely separated. This may merely reflect the inconspicuous nature of the fungus and the absence of attempts to detect it. The known host range is confined to certain grasses and red clover (Gerdemann, 1955). If the red clover is confirmed as a host, however, a much wider non-grass host range could be indicated.

The geographic distribution of the viruses (Table 2) varies from narrow (e.g., tobacco stunt virus) to wide (e.g., potato virus X). Potato virus X is the only fungus-transmitted virus known to persist under natural conditions in the absence of its vector, and which is more widespread than its vector.

*Olpidium brassicae* and *Polymyxa graminis* are of little importance by themselves and cause no easily recognizable disease. At most they probably cause a small amount of superficial root necrosis, and slight chlorosis and stunting in the tops. However, the four viruses they transmit are of moderate to considerable importance.

Lettuce big vein virus delays maturity and decreases the size of head lettuce in Arizona (Marlatt and McKittrick, 1962). In subtropical Queensland, Australia, the disease occurs only during the cooler months and, as in Arizona, does not normally prevent heading. In other countries, however, a proportion of the crop may be rendered unmarketable.

Tobacco necrosis virus may cause losses in tulips of 20–50% of the crop, or even, on occasion, the entire crop (Kassanis, 1949; Mowat,

1966). The virus commonly infects the roots, but under certain undefined conditions, may reach the tops and cause severe leaf and flower necrosis.

Tobacco necrosis virus also causes a necrotic disease of the pods, stems, and foliage of French bean (*Phaseolus vulgaris*). This disease, "stipple streak," commonly occurs in peaty soils in Holland (Bawden and van der Want, 1949). Root rot may be severe on susceptible varieties, whereas resistant varieties remain healthy (Hubbeling, 1957).

In potato, TNV causes superficial lesions on the tubers (Noordam, 1957) which affect their suitability for market.

In some localities the occurrence of diseases caused by TNV is sporadic and cannot be predicted with any certainty (Mowat, 1966). In other localities, however, the diseases occur regularly. Noordam (1960) reported that potatoes planted in fields where beans and tulips always became infected remained healthy, whereas beans and tulips remained free of TNV in fields where potatoes became diseased. Whether the different hosts were favorable to different strains of the virus or of the vector was not established.

Before 1950, tobacco stunt caused heavy losses to tobacco growers in Japan. In the Hatano district, there was an estimated 10% loss in financial return to the growers. Even heavier losses in subsequent years were prevented by the use of such control measures as the selection of disease-free seedlings for transplantation to the field and sterilization of the seedbeds (Hidaka *et al.*, 1956). Plants which reach the six-leaf stage before exposure to soil infection do not usually exhibit symptoms.

The wheat mosaic virus transmitted by *Polymyxa* formerly caused heavy losses in wheat in eastern United States during years favorable to the disease. The use of resistant varieties has resulted in reduced damage, but as recently as 1957 the disease was estimated to have caused $4 million loss to the wheat crop in the state of Kansas alone (Sill and King, 1958).

In contrast to *Olpidium brassicae* and *Polymyxa graminis*, *Synchytrium endobioticum* and *Spongospora subterranea* cause considerable losses on their own account, especially by disfiguring potato tubers. In many countries, quarantine controls are exercized against their introduction and spread. Infection occurs most readily under cold, wet conditions, and may be inhibited under less favorable environmental conditions (Walker, 1952).

Potato mop top virus, transmitted by *Spongospora subterranea,* occasionally causes considerable losses in potato through the unsightly necrotic effects on the tubers (Calvert and Harrison, 1966). Since the virus is tuber transmitted, it is probably important in both "seed" potatoes and potatoes intended for human consumption. There is insufficient information to determine if the virus occurs outside western Europe, but this is likely. In the past it has been confused with tobacco rattle virus (Calvert and Harrison, 1966).

The potato virus X transmitted experimentally by *Synchytrium endobioticum* may cause a 30% loss in the yield of potatoes (Bald and Norris, 1940). However, there is no evidence that the fungus plays any part in the spread or persistence of the virus in the field. Transmission by the rubbing together of leaves or other mechanical methods is usually sufficient to explain the spread, and tuber transmission from season to season is the rule.

### *B. Properties of the Vectors*

The four fungi known to transmit viruses belong to two orders, namely, the Chytridiales (*Olpidium brassicae* and *Synchytrium endobioticum*) and the Plasmodiophorales (*Polymyxa graminis* and *Spongospora subterranea*). In the past (*e.g.*, see Butler and Jones, 1949), these orders have been placed together in the class Phycomycetes, subclass Archimycetes. However, more recently (*e.g.*, see Alexopoulos, 1962), these orders have been placed in different classes, the Chytridiomycetes and the Plasmodiophoromycetes. Whether or not the fungi in the two orders are closely related, they have a number of important characters in common. Sexual reproduction in both groups typically follows the conjugation of zoospores (isogametes), while asexual reproduction occurs by means of zoosporangia. In neither group is there a true mycelium. In both, the fungal thallus has a relatively naked developmental stage in the host plant. On the other hand, fungi in the two orders are sharply differentiated by the flagellation of the zoospores; species in the Chytridiales have one long, posterior, whiplash flagellum, whereas species in the Plasmodiophorales have two whiplash flagella of unequal length.

The life cycles of the four fungus vectors are known imperfectly, and some aspects are controversial. Possible schemes are illustrated diagrammatically in Figures 1–4.

Techniques for handling the four fungi have been, or are being, developed. This is an important prerequisite to a study of the virus–vector relations. Techniques of handling *O. brassicae* and the viruses it transmits have been reviewed by Teakle (1967). The techniques involved with the three other vectors are dealt with either in recent papers on their transmission of viruses or in earlier papers concerning their life histories, *e.g.*, with *Polymyxa graminis* (Ledingham, 1939), with

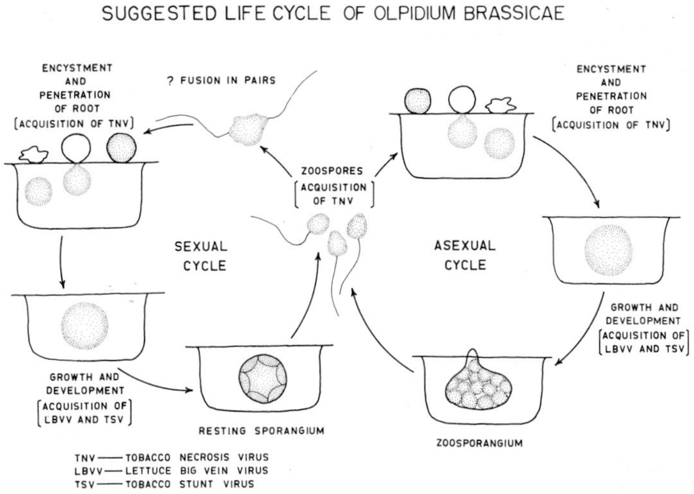


*Fig.* 1

*Spongospora subterranea* (Kole, 1954), and with *Synchytrium endobioticum* (Curtis, 1921).

Since the viruses presumably all are transmitted following penetration of the host plant by zoospores, the ability to prepare highly infective zoospore suspensions is essential. With *O. brassicae,* Teakle (1962) and Teakle and Gold (1964) found that TNV infection of roots was induced only when actively swimming zoospores were able to

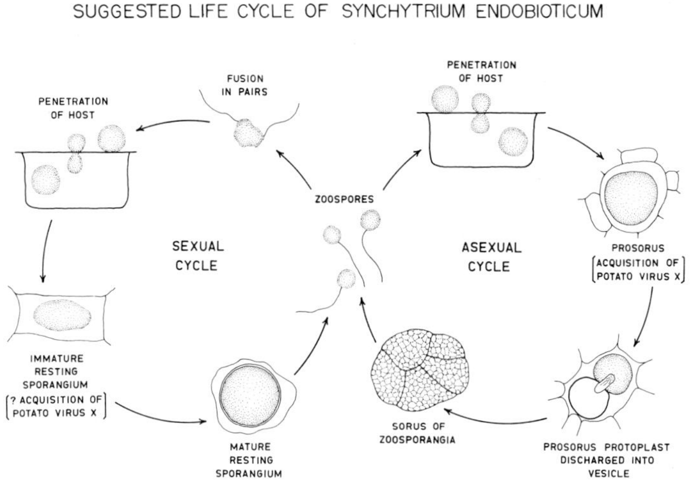


*Fig.* 2.

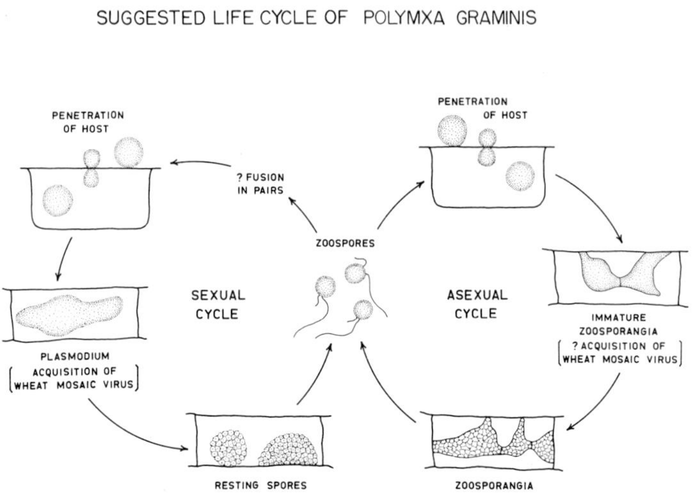


*Fig.* 3.

encyst on roots in the presence of TNV. Mild treatments, such as aging for 0.5 hr or more, heating to 35°C for 10 min, or the addition of $4\times10^{-5}$ M $CuSo_4\cdot5H_2O$ permanently inactivated the delicate zoospores, and prevented the TNV infection of plants subsequently placed in the suspensions. Although numerical data on root infection by *O. brassicae* were not given, numerous microscopic examinations confirmed a correlation between zoospore motility and infectivity. Similarly, when

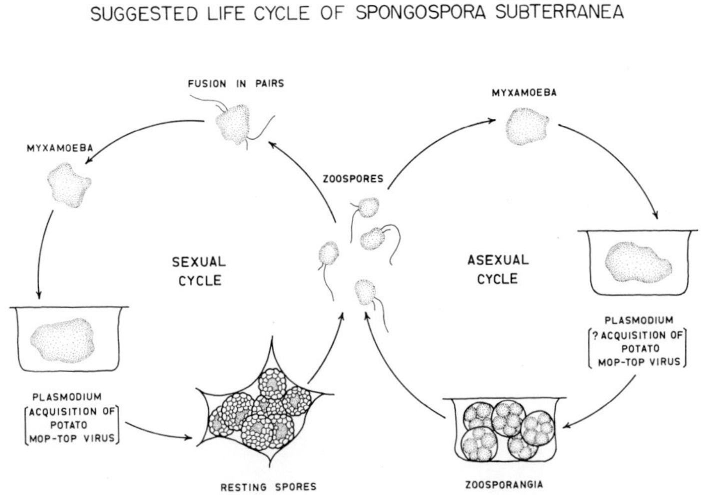


*Fig.* 4.

Hiruki (1965) added certain dilute ($10^{-3}$ M) poisons to zoospore suspensions, those which destroyed motility before the plants were added destroyed the ability of the zoospores to infect or to transmit virus.

Campbell and Grogan (1963, 1964) have challenged the concept that *O. brassicae* has a delicate zoospore. In comparing the properties of the *O. brassicae* zoospore and the factor involved in TNV transmission (not in determining "the thermal death point" of *O. brassicae* zoospores), Teakle (1962) found that zoospores heated to 35°C for 10 min lost motility permanently and were "inactivated" (not "destroyed") when retained in aqueous suspension. Campbell and Grogan found that when zoospores were heated to 55°C for 10 min, they were immobilized, but still infectious when applied to plants growing in soil. Assuming that Campbell and Grogan's results were not due to the inadvertent inclusion of resting sporangia with the zoospores, it appears that inactivated zoospores may be reactivated when added to plants growing in soil, whereas they remain inactivated in aqueous suspension. The reason for this difference is unknown. Teakle (unpublished data) has found that zoospore suspensions immobilized by mild heat treatments often have considerably reduced infectivity when inoculated in soil. There appears to be good reason, therefore for retaining as high a degree of motility as possible under all inoculation conditions if a heavy root infection by both fungus and virus is desired. Grogan and Campbell's (1966) suggestion that the mild treatments used by Teakle (1962) affect the attachment of TNV to the zoospore plasma membrane rather than the ability of the zoospore to infect the plant is unsupported by any evidence.

The effect of mild treatments on the infectivity of zoospores of the vectors other than *O. brassicae* is less certain. However, Brakke and Estes (1967) reported that certain wheat root washings presumably containing zoospores of *Polymyxa graminis,* heated to 35°C for 10 min, could not transmit wheat mosaic virus, whereas those heated to 30°C could. The most likely interpretation is that the zoospores of *P. graminis,* like those of *O. brassicae,* were inactivated by heating to 35°C for 10 min and were unable to infect the test plants under the conditions of the experiment.

### *C. Properties of the Viruses*

The viruses transmitted by fungi exhibit interesting differences in properties. Some are rod shaped (potato virus X, potato mop top virus, and wheat mosaic virus), whereas others are spherical (tobacco necrosis virus, tobacco satellite virus, and probably tobacco stunt virus). The morphology of lettuce big vein virus is doubtful. Potato virus X and most strains of tobacco necrosis virus (TNV) are stable and are readily sap transmissible. Tobacco stunt virus and other strains of TNV are unstable and difficult to transmit. Lettuce big vein virus has not yet been sap transmitted. Unstable strains of TNV occur with little or no protective protein coat (Kassanis and Welkie, 1963).

## III. VIRUS–VECTOR RELATIONSHIPS

Figures 1–4 contain information on the stages during which the viruses are acquired. The evidence is reasonably strong in the case of *Olpidium*-transmitted viruses, but less so in the case of the other viruses.

### *A. Olpidium brassicae and Lettuce Big Vein Virus (LBVV)*

Lettuce big vein was recognized as being distinct from the lettuce brown blight disease in California about 1926–1929 when Jagger developed varieties of lettuce resistant to brown blight but not to big vein (Jagger and Chandler, 1934; Jagger, 1940). The symptoms of the disease included a pronounced yellow vein banding, leaf puckering, and stunting of the plant.

The big vein disease was reported by Jagger and Chandler (1934) to resemble the brown blight disease in many ways, *e.g.*, in field distribution, in being soil borne, and in being eliminated by partial sterilization of the soil with either steam or formaldehyde. A soil borne virus was suggested as a possible cause of each disease. Jagger and Chandler (1934) did not report investigations on the association of *O. brassicae* with big vein, although previously Jagger (1928) had reported its association with brown blight. Selection of resistant varieties led to the practical elimination of brown blight, but not lettuce big vein.

Subsequent workers associated the big vein disease with low temperatures (Thompson and Doolittle, 1942), high soil moisture (Pryor, 1944), a root aphid (Thompson, Doolittle, and Smith, 1944), and tobacco necrosis virus in the roots (Fry, 1952; Yarwood, 1954) without convincing evidence of a causal relationship. It remained for Fry (1958) and Grogan, Zink, Hewitt, and Kimble (1958) to establish its relationship with *O. brassicae.* The epidermal cells of the roots of big vein lettuce plants were found to contain large numbers of resting sporangia and zoosporangia. The fungus was absent, however, from plants in areas free of the big vein disease. Further, treatments which eliminated *O. brassicae* from roots, soil, or zoospore suspensions eliminated the big vein disease. These workers considered the possibility that a virus was involved, but decided that the evidence failed to substantiate this hypothesis. They concluded that *O. brassicae* was the cause of big vein, possibly through the production of a toxic substance which was translocated to the leaves.

In the following years, the association of *O. brassicae* with lettuce big vein disease was repeatedly confirmed by workers who either accepted the fungus toxin hypothesis or withheld judgement. However, following the evidence of Hidaka (1960) and Teakle (1960) of a correlation between infection of roots by *O. brassicae* and the tobacco stunt and tobacco necrosis viruses, the possible virus etiology of lettuce big vein disease was reinvestigated. Campbell, Grogan, and Purcifull (1961) and Tomlinson, Smith, and Garrett (1962) independently reported graft transmitting the disease in lettuce in the absence of any visible pathogen, and postulated that *O. brassicae* was acting as a vector of a lettuce big-vein virus. Tomlinson *et al.* (1962) reported that they had successfully graft transmitted big vein in 1957 and 1958. Their grafting work apparently lapsed when *O. brassicae* was suggested as the sole cause of the disease, and was resumed only in 1961. Thus, the discovery that *O. brassicae* was associated with lettuce big vein disease paradoxically delayed the production of evidence that a virus was involved. On the other hand, establishment of the association of *O. brassicae* with lettuce big vein led to the discovery of the vector role of this fungus with tobacco necrosis virus and possibly with tobacco stunt virus.

Studies on the relationship of lettuce big vein virus and *O. brassicae* indicate that the virus survives inside the resting sporangium. Campbell (1962) reported that *O. brassicae* in a lettuce root macerate which passed a 400-mesh screen could survive and transmit lettuce big vein virus despite acidification to *p*H 0.7 for 2 hr with hydrochloric acid or incubation at *p*H 11.3–12.3 for 30 min with trisodium phosphate.

The multiplication of the lettuce big vein virus in *O. brassicae* was considered unlikely by Campbell (1962). The LBVV was lost during four consecutive transfers of *O. brassicae* in sugar beet, a probable nonhost of the virus. Under the same conditions it was perpetuated in lettuce, a known host of the virus.

Acquisition of LBVV by *O. brassicae* apparently occurs inside the host plant. When virus-free *O. brassicae* was allowed to multiply in lettuce plants inoculated with LBVV by grafting, the *O. brassicae* became viruliferous (Tomlinson and Garrett, 1962, 1964).

The work on the relationship of *O. brassicae* and LBVV indicates a superficially simple interaction. The virus is merely acquired by the fungus protoplast in the virus-infected root. It is carried within the zoospores into fresh roots and there released. Whether virus acquisition and release is by pinocytosis, phagocytosis, or some other method has not been established.

A large gap in our knowledge concerns the nature of LBVV. Although the virus is sufficiently stable to survive for more than 8 years in dried, infested soil (Pryor, 1946), presumably within the resting sporangia of *O. brassicae,* the virus has not yet been purified and characterized.

The known host range of LBVV is restricted to certain species in the family Compositae. Although zoospores of *Olpidium* obtained from a wide range of other species within the Compositae and from other families have transmitted LBVV, this could indicate the persistence of the virus in the fungus rather than its multiplication in the plant (Campbell, 1965).

### *B. Olpidium brassicae and Tobacco Stunt Virus (TSV)*

The tobacco stunt disease was first found in tobacco seedbeds in Japan in 1931, and by 1956 had been reported throughout the country (Hidaka *et al.*, 1956). The symptoms commonly occur at the 6–7-leaf stage and include vein clearing progressing to vein necrosis, necrotic leaf spotting, stem necrosis at the ground level, and stunting. The disease can be controlled by soil sterilization.

The causal virus was first thought to be sap transmissible and 25–35 $m\mu$ in diameter (Hidaka, 1951). Later, however, Hidaka *et al.* (1956) reported that the disease could be artificially transmitted only by grafting and that the virus particles were about 18 $m\mu$ in diameter. Recently, Hiruki (1964) showed that the virus could be sap transmitted in the presence of certain chelating agents. Further work is required to demonstrate convincingly the nature and morphology of the virus particles.

Hidaka *et al.* (1956) presented evidence which indicated that the tobacco stunt virus had a fungus vector. Treating infected soil for 40 days with oxygen or carbon dioxide had little influence on its infectivity. Since tobacco mosaic virus in soil is inactivated by aeration (Johnson, 1926; Johnson, and Hoggan, 1937), these results were interpreted as indicating the carriage of tobacco stunt virus within the body of a resistant vector rather than simple adsorption to soil particles. Since treatment of the soil with certain fungicides reduced infectivity by 90% or more, whereas insecticides had no appreciable effect, it was suggested that the vector was a fungus.

The association of *O. brassicae* with tobacco stunt disease was reported by Hidaka (1960, 1961). A close correlation was observed between the incidence of tobacco stunt and the prevalence of *O. brassicae* in the roots as affected by soil temperature, treatment of the soil by fungicides or heat, and fractionation of the soil by centrifugation and sedimentation.

In later work, tobacco stunt disease was induced by inoculating seedlings with suspensions containing either zoospores or resting spores (Hidaka and Tagawa, 1962). The disease was not induced by inoculating with zoospore suspensions treated by Seitz filtration, dilution by more than $10^3$, or heat at 45°C for 20 min.

Hiruki (1965) confirmed and extended the work, indicating the constant association of *O. brassicae* with tobacco stunt. The inability of the delicate zoospore to withstand exposure to freezing and certain dilute ($10^{-3}$ M) poisons was correlated with its inability to transmit TSV.

In work on the vector relationship, Hiruki (1965) showed that the virus was lost when the fungus was subjected to serial transfer in cowpea. The virus was regained by the fungus after culture for 10 days in the roots of tobacco plants mechanically inoculated in the leaves with TSV. No virus was transmitted by zoospores exposed to root leachates of TSV-infected plants.

The available evidence indicates an intimate relationship of *O. brassicae* and TSV. Probably, like LBVV, TSV is carried internally in the fungus protoplast. It is acquired or transmitted only during the direct contact of the fungus and plant cell protoplast.

### *C. Olpidium brassicae and Tobacco Necrosis Virus (TNV)*

TNV was described by Smith and Bald (1935) as a virus which caused the necrosis of the lower leaves and stem of glasshouse-grown tobacco seedlings. Subsequently, the virus was found commonly to infect the roots of a wide range of plants growing in infested soil. Smith (1937) presented evidence that in his glasshouse the virus was waterborne. Since infection commonly did not occur in a water culture, however, he concluded that the wounding of root hairs as the root makes its way through the soil was necessary for infection.

Teakle (1960) investigated *O. brassicae* as a possible vector of TNV for two reasons. *Olpidium brassicae* and TNV had each been shown to be associated with lettuce big vein disease (Fry, 1952, 1958; Yarwood, 1954; Grogan, Zink, Hewitt, and Kimble, 1958); and there was evidence that a biological factor was involved in TNV transmission, since soils naturally infested with TNV were sometimes highly infective, whereas steamed soils supplemented with TNV from leaves were consistently only slightly infective (Yarwood, 1960). *Olpidium brassicae* was commonly detected in the roots of plants that were naturally infected with TNV. Furthermore, combined suspensions of *O. brassicae* zoospores from the roots and TNV from the leaves induced heavy TNV infection when added to seedlings in soil or petri dishes. Also, heavy TNV infection was induced by zoospore suspensions taken from lettuce roots previously inoculated with both TNV and *O. brassicae* zoospores. Subsequent work (Teakle, 1962) produced evidence that *O. brassicae* entered the roots at the same time and place as the TNV. It was claimed that the fungus was acting as a virus vector and not merely predisposing roots to virus infection.

The ability of *O. brassicae* to function as a vector of TNV was not accepted without question. Mowat (1961), using a cabbage strain of *O. brassicae,* failed to confirm the work of Teakle (1960). Later it was recognized, however, that certain strains of *O. brassicae* from species of *Brassica* differ from certain strains of *O. brassicae* from lettuce in being nonvectors of TNV (Teakle, 1962; Mowat, 1963; Teakle and Hiruki, 1964; Kassanis and Macfarlane, 1965).

Campbell and Grogan (1963) regarded the relationship of *O. brassicae* to TNV as uncertain, because no data showed that the virus was retained during the repeated serial transfer of *O. brassicae.* However, virus retention during the serial transfer of *O. brassicae* may or may not occur both with TNV (Campbell and Grogan, 1963; Teakle, 1963) and with LBVV (Campbell, 1962). Since these viruses apparently have different virus–vector relationships, this criterion obviously is inadequate for determining virus–vector relationships. The criterion of a virus vector would appear to be simply the ability to successfully transmit a virus.

Work with TNV antisera has been useful in elucidating the virus–vector relationship of *O. brassicae* and TNV. Teakle and Gold (1963) reported that virus transmission was achieved by zoospore suspensions which were exposed to TNV and then treated with dilute TNV antiserum at a concentration which inactivated TNV alone. This protective action of the *O. brassicae* zoospores on the TNV was taken to indicate virus acquisition by the zoospores. Although Teakle and Gold (1963) speculated that the virus might be taken into the zoospore protoplast, this was thought to be unlikely by Kassanis and Macfarlane (1964), since use of concentrated antisera completely eliminated TNV transmission. They suggested that TNV was merely adsorbed on the zoospore surface where it was less vulnerable to the antiserum. Teakle and Gold (unpublished data) found that zoospores immobilized or nearly immobilized by exposure to 10% $CO_2$ in air or a temperature of approximately 1°C apparently can still acquire TNV and protect it from inactivation by dilute antiserum. These results also are interpreted as a possible indication of the surface acquisition of TNV.

The acquisition of TNV by *O. brassicae* inside the plant root has not been clearly demonstrated. Although Teakle and Gold (1963) reported that zoospores from TNV-infected plants released into dilute antiserum still transmitted virus, although at a reduced rate, Kassanis and Macfarlane (1964) and Campbell and Fry (1966) eliminated transmission by this means. Furthermore, Campbell and Fry (1966) showed that a hydrochloric acid treatment of resting sporangium preparations from virus-infected plants eliminated TNV transmission but not LBVV transmission. Also, when roots were air dried for several weeks, TNV infectivity was lost whereas *O. brassicae* infectivity and LBVV infectivity were not (Mowat, 1964; Campbell and Fry, 1966). Thus, the weight of evidence indicates that TNV is borne by *O. brassicae* in a manner different from LBVV. Presumably it is acquired outside the root, following the independent release of zoospores and TNV. Why TNV is not borne internally by *O. brassicae* is unknown. It is possible, however, that the inhibitory effect of TNV infection of the root on the development of *O. brassicae* (Teakle, 1963, 1964; Fry and Campbell, 1966) is involved.

### *D. Olpidium brassicae and Tobacco Satellite Virus*

This virus is morphologically and antigenically unrelated to TNV, but is able to multiply only in TNV-infected leaves (Kassanis, 1962). It is detected by its ability to reduce the size and number of local lesions produced by TNV, as well as by its physical and serological properties. Kassanis suggested the 17 mμ diameter particle of the satellite virus is deficient in the information required for self-replication and requires the assistance of the larger 30 mμ diameter TNV particle.

The relationship of satellite virus and *Olpidium brassicae* has not been reported. Teakle (1962) observed that a culture of TNV containing satellite virus gave a mixture of large and small lesions before and after transmission by *O. brassicae*. Kassanis and Macfarlane (unpublished data cited in Kassanis, 1964) reported that satellite virus was transmitted successfully only in association with TNV. Presumably the tobacco satellite virus is transmitted by *O. brassicae* in a way similar to that of TNV.

### *E. Synchytrium endobioticum and Potato Virus X (PVX)*

PVX probably occurs universally where potatoes are grown commercially. The virus is readily transmitted by leaf contact between healthy and diseased plants. Spread may also occur between plants whose only contact is below the ground (Roberts, 1950). Occasional spread by insects, such as grasshoppers, may occur (Walters, 1952; Norris, 1954).

Although there was no evidence to implicate *Synchytrium endobioticum* as a vector, Nienhaus and Stille (1965) tested the ability of this fungus to transmit PVX. The virus was transmitted by zoospores released from PVX-infected tubers. Potato virus X was not transmitted when zoospore suspensions from virus-free tubers were mixed with partially purified PVX.

Presumably PVX entered the immature thallus of the fungus during its development in the virus-infected plant, and later was discharged from the plant inside the zoospores. Apparently the zoospores, once discharged from a PVX-free plant, did not acquire PVX as a surface or internal contaminant, or, if acquired, the virus was not released when the zoospores entered the plant.

No evidence was published on the origin of the zoospores, but presumably they were discharged from asexual zoosporangia. Whether PVX can be transmitted through the sexual resting sporangia, as is lettuce big vein virus, was not tested.

### *F. Polymyxa graminis and Wheat Mosaic Virus*

Wheat mosaic virus occurs in eastern United States, Japan, and Italy. It causes mosaic, stunting, rosetting, and yield loss in susceptible varieties. The virus is rather difficult to transmit by sap inoculation. When soil transmission occurs, the virus may stay localized in the roots or move to the tops. High soil moisture and low temperatures favor soil transmission.

Although McKinney (1930) suggested that "even fungi" might transmit the soil-borne wheat mosaic virus, it was not until 1954 that Linford and McKinney (1954) associated a specific fungus, *P. graminis,* with infected plants. However, Linford and McKinney's report lost the impact it deserved by their conclusion that ". . . there is no evidence that *Polymyxa graminis* can function as a virus vector or as a reservoir host. Association recorded here between this fungus and the mosaic disease is imperfect." Linford's untimely death prevented the completion and publication of further work done from 1954 to 1958. Certain findings made by Linford, however, have been summarized by Brakke, Estes, and Schuster (1965). It was found that transmission was achieved by placing a thoroughly washed, naturally infected wheat root on one side of a petri dish containing distilled water and a young wheat seedling on the other side. After 2 days at 15°C, the wheat seedling was planted and kept at 15°C for several months. Mosaic leaf symptoms were frequently observed.

There was an indication of a positive correlation between virus transmission and the presence of *P. graminis* in the naturally infected source root.

Recent work by Estes and Brakke (1966) has confirmed the correlation indicated by Linford and McKinney (1954). Plants that developed virus infection after soil inoculation were frequently (60%) infected with *P. graminis,* whereas similarly exposed plants which did not become infected with virus were seldom (2%) infected with *P. graminis*. Furthermore, wheat mosaic virus was transmitted from wheat roots only if they contained resting spores of *P. graminis*. Thus, there is circumstantial evidence that *P. graminis* is acting in a vector-like capacity.

The nature of the presumed virus–vector relationship is uncertain, but speculations can be made. Since soil retains virus infectivity despite drying and treating with toluene (McKinney, Paden, and Koehler, 1957), the virus may survive within the resting spores of the fungus. Resting spores in root debris may be a major source of virus in soil (Brakke and Estes, 1967). Also, washings from living roots infected by both *P. graminis* and the virus are infective. Whether the zoospores transmitting the virus from living roots originate from zoosporangia or resting spores or both has not been determined (Brakke and Estes, 1967).

Recently Canova (1964, 1965) reported that *O. brassicae* was associated with transmission of wheat mosaic virus in Italy. This claim has been withdrawn, however, and *P. graminis* has been reported to be the vector (Canova, 1966). Whether *O. brassicae* is an additional vector has yet to be established.

### *G. Spongospora subterranea and Potato Mop Top virus (PMTV)*

Potato mop top virus has recently been described as a rod-shaped virus of potatoes causing four main types of symptoms (Calvert and Harrison, 1966): (*1*) shortened internodes and crowded leaves leading to a dwarfed and bunched growth habit; (*2*) irregular yellow blotches or ring and line patterns, especially on the lower leaves; (*3*) chlorotic chevrons (inverted V's) on the upper leaves; and (*4*) necrotic rings on the tubers. The virus was transmitted to only 40–70% of plants derived from seed tubers showing symptoms, and therefore in the absence of fresh infections would tend to be self-eliminating. However, when healthy potato or tobacco plants were grown in naturally infested soil, at least half became infected with PMTV.

The vector of PMTV is reported to be *Spongospora subterranea,* with supporting evidence yet to be published (Calvert and Harrison,

1966). Harrison (private communication, 1966) has evidence that the virus is carried through the resting spores of *S. subterranea*. Whether the virus is transmitted by zoospores released from the zoosporangia has not been reported.

## IV. VIRUSES INFECTING FUNGI

Three diseases of fungi are caused by viruses or virus-like agents: die-back of mushroom (Gandy, 1960; Hollings, 1962), degeneration of *Helminthosporium victoriae* (Lindberg, 1959), and lysis and degeneration of yeasts (Lindegren and Bang, 1961). These diseases occur in a basidiomycete (mushroom) and several ascomycetes (*Helminthosporium victoriae* and yeasts), *i.e.*, fungal classes different from those of the virus vectors.

In addition, other examples of "mutation" and "sectoring" in fungi may prove to be worthy of investigation by virologists. One such case is the "vegetative death" of *Aspergillus glaucus,* which Jinks (1959) showed to be transmitted following hyphal anastomosis between affected and unaffected thalli. Virus etiology would appear as likely as the "cytoplasmic inheritance" suggested by Jinks.

### A. *Mushroom Viruses*

Productivity of the cultivated edible mushroom, *Agaricus bisporus,* is drastically reduced by a number of disorders, including watery stipe, La France, and mummy. Although unfavorable growing conditions can produce such disorders in healthy mushrooms, viruses also can be responsible.

Gandy (1960) showed that the disease (watery stipe or die-back) was transmitted when a healthy mycelium in compost was inoculated with diseased mycelium. The severity of attack varied with the time of inoculation, being the greatest at spawning and least at fruiting. As the disease progressed, the mycelium deteriorated and turned from white to brown. The culture of such affected mycelium on agar was difficult and the resulting colonies had a growth rate only one-seventh of that of a healthy mycelium.

A complex of three viruses was associated with die-back disease by Hollings (1962). These viruses had spherical particles of 25 m$\mu$ and 29 m$\mu$ diameter and bacilliform particles of $19 \times 50$ m$\mu$. The spherical parties were shown to be antigenic. The 29 m$\mu$ diameter particles showed an ultraviolet absorption spectrum similar to that of tomato bushy stunt and tobacco mosaic viruses. Purified preparations were infective when injected into the base of young sporophores or when applied to the cut surfaces of expanding sporophores. They were not infective when discs of healthy mushroom mycelium on agar were dipped in them (Hollings *et al.*, 1963). Virus purification was facilitated by the use of ultrasonic waves to disrupt the mushroom mycelium; using this technique, the three viruses could be detected in suspensions containing as little as 1.0 mg fresh weight of mycelium (Hollings *et al.*, 1965).

Observations have yielded some preliminary information on the relative importance of the viruses in reduced yields of mushrooms. The virus with 29 mμ diameter particles seems more damaging than that with 25 mμ diameter particles. The virus with bacilliform particles has been found only in combination with one or both of the other viruses, and increases damage over that caused by the spherical viruses alone.

Recently, attention has been drawn to the similarity in symptoms and transmission between the die-back disease, occurring in England, and the La France disease, occurring in the U.S.A. (Schisler, Sinden, and Sigel, 1967). The La France disease was first reported from Pennsylvania, U.S.A., in 1950 (Linden and Hauser, 1950). Later, under the name of X-disease, it was found to be widespread in the U.S.A., occurring from coast to coast (Lambert, 1961; Kneebone, Lockard, and Hager, 1962).

Laboratory tests, both in England and the U.S.A., showed that spores from diseased sporophores germinated and grew on agar to produce diseased mycelia (Schisler *et al.*, 1967). If diseased spores germinated on agar and the germ tubes anastomosed with the tips of hyphae of healthy mycelia, the mycelia became diseased. Further, tests under commercial conditions showed that spraying diseased spores onto the compost during spawning induced the disease. It was speculated that diseased spores carried by air currents were a major sources of natural infection both within and between mushroom farms.

The English and American mushroom diseases are so similar that it seems likely that the same viruses are present in both countries. As yet, however, the viruses responsible for the American disease have not yet been described.

Since diseased spores transmitted the disease between mycelia of the same variety or of different varieties of mushroom, Schisler *et al.*, (1967) regarded the spores as "vectors". This is an unusual vector relationship in that the hosts and vector all belong to the one species. An analogous situation exists in the pollen transmission of viruses, where the term "vector" for infected pollen has not come into general use.

### *B. Suspected Yeast Virus ("Zymophage")*

Possible viruses of yeasts, including *Saccharomyces* sp., have been reported by Lindegren and associates (Lindegren and Bang, 1961; Lindegren, Bang, and Hirano, 1962; Hirano, Lindegren, and Bang, 1962). Plaques similar to those produced by bacterial viruses occurred sporadically on agar plates thickly seeded with certain strains of yeast. Disintegration (lysis) and the internal disorganization of yeast cells were involved. Other signs of disease included multiple budding instead of single budding and frequent transformation from ovoid to mycelial growth. The filtrates of affected cultures contained spherical particles 80–270 mμ in diameter, and rod-shaped particles 200–900 mμ long and 30–110 mμ wide. Spherical particles approximately 200 mμ in diameter were observed in sections of diseased cells. In further work (Bang, 1963; Lindegren, 1966), the reduced size of colonies of *Saccharomyces* on agar was found to be correlated with infectivity both to *Saccharomyces* and *Rhodotorula gracilis*. A healthy culture of *Saccharomyces* which produced uniformly large colonies was converted to a culture producing a mixture of large and small colonies after exposure to a culture filtrate obtained from a mixed culture of a lytic and the healthy culture. Furthermore, the infectivity survived heating to 100°C for 30 min and could be passed through a millipore filter containing pores of 0.45 μ in diameter.

Whether the observed particles consist of, or contain, nucleoprotein has not been determined. The various particles have not yet been obtained in a highly purified form for the determination of their physical and chemical properties. Until more precise data on the properties of the postulated virus are obtained, doubt will remain on its viral nature.

### *C. Suspected Virus of Helminthosporium victoriae*

A stunting disease of certain newly isolated cultures of *Helminthosporium victoriae* from oats was studied by Lindberg (1959). Stunted cultures always remained stunted when transferred, whereas healthy, fast-growing cultures remained healthy unless allowed to contact a diseased mycelium. In this case, the healthy culture stopped growing within a few hours, and the aerial mycelium lysed or collapsed. Furthermore, healthy fungus was infected whether dipped into either a viable suspension of a macerated diseased culture, or a nonviable clarified extract of sap from a diseased culture. The disease agent was separated from the fungus by fractionation with ammonium sulfate and centrifugation, using techniques that would separate agents with the size and properties of viruses. Tests to determine if a toxin or bacterial pathogen were present were consistently negative.

In later work, Lindberg and Pirone (1963) found that the disease agent was antigenic. At present, attempts are being made to detect virus-like particles in sections of diseased hyphae of *Helminthosporium victoriae* by electron microscopy (Werner and Lindberg, 1966).

## V. DISCUSSION

That certain fungi in the Chytridiales and Plasmodiophorales are able to transmit viruses could be related to the fact that all these fungi are in a naked or nearly naked condition at certain stages of their life cycles (Figs. 1–4). These are the zoospore stage, active outside the host plant, and the plasmodial stage, active inside the host plant. Only tobacco necrosis virus has been shown to be acquired by the zoospore stage of its vector, *Olpidium brassicae*. In this case the carriage may be superficial. All the other viruses apparently are acquired only inside the host plant, possibly by the viruses entering the fungus protoplast.

When the viruses enter the fungus protoplast has not been determined, but presumably it is early in the development of the fungus in the host plant. At this time, resistance to virus penetration would be least because of the absence of fungal walls. With *Olpidium brassicae,* Sahtiyanci (1962) observed that both sexual and asexual thalli were naked for one day after zoospore penetration. Thereafter, a thin membrane was visible. This developed into a zoosporangial wall in two days or a thick, three-layered, resting-sporangial wall in four days. Lettuce big vein and tobacco stunt viruses apparently are acquired readily by both sexual and asexual stages of the fungus.

The situation with *Polymyxa graminis* is less clear. Ledingham (1939) reported that even at a very early stage a delicate, but definite membrane surrounded the developing zoosporangium, whereas the plasmodium of the developing resting spores was naked. Wheat mosaic virus is transmitted by zoospores from resting spores, but whether by zoospores from zoosporangia has not been established (Brakke and Estes, 1967).

Although the possession of one or more relatively naked stages may facilitate virus acquisition and transmission, this is not the only factor involved, as is indicated by the phenomena of virus and vector specificity. For instance, when suspensions of viruses and *Olpidium brassicae* zoospores are mixed, transmission will occur only in the case of the tobacco necrosis virus and the tobacco satellite virus.

Also, when *Synchytrium endobioticum* multiplies in virus-infected potatoes, potato virus X is transmitted while potato virus Y is not (Nienhaus and Stille, 1965). Furthermore, only certain strains of *O. brassicae* are vectors of tobacco necrosis virus.

The basis for vector specificity in *O. brassicae* has not been convincingly demonstrated. Teakle and Hiruki (1964) reported that a lettuce strain of *O. brassicae* could transmit TNV to both favorable and unfavorable hosts of the *O. brassicae*, whereas a *Brassica* strain of *O. brassicae* could transmit TNV to neither. They presented evidence suggesting that more virus was acquired by the lettuce strain than by the *Brassica* strain. Mowat (1966) showed that zoospores of a transmitting and a nontransmitting isolate both carried net negative surface charges differing significantly in magnitude. It is possible that such surface charges affect the acquisition of TNV by the zoospore.

On the other hand, Kassanis and Macfarlane (1965) have presented evidence indicating that the host plant–fungus interaction affects the ability of the virus to infect the plant. Although this suggestion may have some merit in the case of *O. brassicae* and TNV, it probably does not apply in the case of *Synchytrium endobioticum* and PVX. In this case, transmission occurs when zoospores are taken from a virus-infected plant, but not when zoospores are mixed with a virus suspension. Since the host plant–fungus interaction would be the same in both instances, a difference in the virus–fungus interaction obviously is involved.

One of the difficulties in interpreting the available evidence on virus–vector relationships is that few comparative tests have been done. For instance, since TNV transmission is prevented by treatment of the motile zoospores of *Olpidium brassicae* with concentrated TNV antiserum, it is postulated that the zoospores carry TNV externally where it is exposed to the antiserum (Kassanis and Macfarlane, 1964). This interpretation would be strengthened if concentrated antiserum to a virus believed to be internally borne (*i.e.*, tobacco stunt virus or lettuce big vein virus) did not prevent transmission of the virus by *O. brassicae* zoospores. Unfortunately, such antisera are not yet available.

Campbell and Fry (1966) have made a useful start in comparative work by contrasting the ability of *O. brassicae* resting sporangia to retain lettuce big vein virus infectivity following acid treatment with their inability to retain TNV infectivity. The interpretation of this as a difference in virus–vector relationships would be strengthened, however, if the relative abilities of the two viruses to withstand acid treatment were known. This evidence, like most other evidence dealing with virus–vector relationships, is indirect. More direct and convincing evidence will be provided when the viruses can be detected on or in the fungi by electron microscopy.

The work by Nienhaus and Stille (1965), in which potato virus X was transmitted by *Synchytrium endobioticum,* is of interest because this was a laboratory-glasshouse discovery. There appears to be no evidence that transmission occurs in the field. This discovery raises the question of whether other viruses, such as white clover mosaic virus and tobacco mosaic virus, might be transmitted by fungi as well as by mechanical means.

Of those virus diseases and virus-like diseases occurring in the field for which no vector has been reported, fungi could be associated with the transmission of oat mosaic and sugarcane chlorotic streak. Although Ledingham (1939) reported that *Polymyxa graminis* did not infect oats, and Linford and McKinney (1954) did not find this fungus in the roots of two oat mosaic virus-infected plants, Bruehl and Damsteegt (1961) reported both *Polymyxa graminis* and *Olpidium brassicae* in the roots of oat mosaic virus-infected oats. Oat mosaic virus apparently has particles which are flexuous rods $660 \times 14$ mμ (Gold, McKinney, and Scott, 1957). These dimensions make it unlike any of the fungus-transmitted viruses which have yet been reported. Antoine (1965) reported that a chytrid was frequently found in roots of sugarcane infected with chlorotic streak virus. Its relationship with the disease remains to be established.

In conclusion, considerable scope exists for the study of fungi, both as hosts and as vectors of viruses. This will involve the integration of detailed knowledge concerning fungi and viruses, and, in the case of vectors, plants. Hence, the field offers a challenge to teams or individuals with wide biological interests.

## BIBLIOGRAPHY

Alexopoulos, C. J. 1962. Introductory mycology, 2nd ed. Wiley, New York. 613 p.

Antoine, R. 1965. Cane diseases. Rept. Mauritania Sugar Ind. Res. Inst. 1964: 51–68; Rev. Appl. Mycol. 44:3139.

Bald, J. G., and Norris, D. O. 1940. The effect of the latent virus (virus X) on the yield of potatoes. J. Council Sci. Ind. Res. 13:252–254.

Bang, Y. N. 1963. Some aspects of the viral syndrome in Saccharomyces. Ph.D. thesis, Southern Illinois University, Carbondale.

Bawden, F. C., and Kassanis, B. 1947. *Primula obconica,* a carrier of tobacco necrosis viruses. Annu. Appl. Biol. 34:127–135.

Bawden, F. C., and van der Want, J. P. H. 1949. Bean stipple-streak caused by a tobacco necrosis virus. Tijdschr. Plantenzietken 55:142–150.

Brakke, M. K., and Estes, A. P. 1967. Some properties of the vector of soil borne wheat mosaic virus in root washings and soil debris. Phytopathology 57 (in press).

Brakke, M. K., Estes, A. P., and Schuster, M. L. 1965. Transmission of soil borne wheat mosaic virus. Phytopathology 45:79–86.

Bruehl, G. W., and Damsteegt, V. D. 1961. Soil borne mosaic of fall-seeded oats in western Washington. Plant Disease Reporter 45:884–888.

Butler, E. J., and Jones, S. G. 1949. Plant pathology. MacMillan, London. 979 pp.

Calvert, E. L., and Harrison, B. D. 1966. Potato mop top, a soil borne virus. Plant Pathol. 15:134–139.

Campbell, R. N. 1962. Relationship between the lettuce big vein virus and its vector, *Olpidium brassicae.* Nature 195:675–677.

——— 1965. Weeds as reservoir hosts of the lettuce big vein virus. *Can. J. Bot.* 43:1141–1149.

Campbell, R. N., and Fry, P. R. 1966. The nature of the associations between *Olpidium brassicae* and lettuce big vein and tobacco necrosis viruses. Virology 29:222–233.

Campbell, R. N., and Grogan, R. G. 1963. Big vein virus of lettuce and its transmission by *Olpidium brassicae.* Phytopathology 53:252–259.

——— 1964. Acquisition and transmission of lettuce big vein virus by *Olpidium brassicae.* Phytopathology 54:681–690.

Campbell, R. N., Grogan, R. G., and Purcifull, D. E. 1961. Graft transmission of big vein of lettuce. Virology 15:82–85.

Canova, A. 1964. Researches on virus diseases of Gramineae. I. Wheat mosaic transmissible through the soil. Phytopathol. Mediterranea 3:86–94; Rev. Appl. Mycol. 45:1746.

——— 1965. Researches on virus diseases of Gramineae. II. Transmission of wheat mosaic virus. Phytopathol. Mediterranea 4:122–124; Reg. Appl. Mycol. 45:2448.

——— 1966. Researches on virus diseases of Gramineae. III. Polymyxa graminis vector of wheat mosaic virus. Phytopathol. Mediterranea 5:53–58; Rev. Appl. Mycol. 46:662.

Curtis, K. M. 1921. The life history and cytology of *Synchytrium endobioticum* (Schilb.) Pers., the cause of the wart disease in potato. Phil. Trans. Roy. Soc. (London) B210:409–478.

Estes, A. P., and Brakke, M. K. 1966. Correlation of *Polymyxa graminis* with transmission of soil borne wheat mosaic virus. Virology 28:772–774.

Fry, P. R. 1952. Note on occurrence of a tobacco-necrosis virus in roots of lettuce showing big vein. New Zealand J. Sci. Tech. Sect. A, 34:224–225.

——— 1958. The relationship of *Olpidium brassicae* (Wor.) Dang. to the big vein disease of lettuce. New Zealand J. Agr. Res. 1:301–304.

Fry, P. R., and Campbell, R. N. 1966. Transmission of a tobacco necrosis virus by *Olipidium brassicae.* Virology 30:517–527.

Gandy, D. G. 1960. A transmissible disease of cultivated mushrooms ("watery stripe"). Annu. Appl. Biol. 48:427–430.

Gerdemann, J. W. 1955. Occurrence of *Polymyxa graminis* in red clover roots. Plant Disease Reporter 39:859.

Gold, A. H., McKinney, H. H., and Scott, H. A. 1957. A comparative study of virus particles infecting some grasses. Proc. Int. Cong. Crop Protection, 4th, Hamburg, 1957, 1:351–354.

Grogan, R. G., and Campbell, R. N. 1966. Fungi as vectors and hosts of viruses. Annu. Rev. Phytopathol. 4:29–52.

Grogan, R. G., Zink, F. W., Hewitt, W. B., and Kimble, K. A. 1958. The association of *Olpidium* with the big vein disease of lettuce. Phytopathology 48: 292–297.

Hansen, H. P. 1947. Meeting Inst. Int. Récherche Betteravières, Brussels. (Cited by Bawden, F. C. 1956. Plant viruses and virus diseases. Chronica Botanica Co., Waltham, Mass. 335 p.).

Hecht, E. I. 1962. Nonspecific acquired resistance to pathogens resulting from localized infections by *Thielaviopsis basicola* or viruses in tobacco leaves. M.S. thesis, Cornell University, Ithaca, New York.

Hidaka, Z. 1951. On a new virus disease "Tobacco stunt." [in Japanese]. Annu. Phytopathol. Soc. Japan 15:40–41. (Abstr.)

——— 1960. The behaviour of tobacco stunt virus in soils, particularly supposing *Olpidium brassicae* (Wor.) Dang. as the vector. Proc. Symp. Soil-Borne Viruses, Dundee, Scotland.

——— 1961. Studies on the tobacco stunt disease. Vector of tobacco stunt disease [in Japanese]. Annu. Rep. Hatano Tobacco Exp. Sta. 1961, p. 110.

Hidaka, Z., Nakano, K., Uozumi, T., Shimizu, T., and Hiruki, C. 1956. Studies on the tobacco stunt disease. Bull. Hatano Tobacco Exp. Sta. 40:1–80.

Hidaka, Z., and Tagawa, A. 1962. The relationship between the occurrence of tobacco stunt disease and *Olpidium brassicae* [In Japanese]. Annu. Phytopathol. Soc. Japan 27:77–78. (Abstr.)

Hirano, T., Lindegren, C. C., and Bang, Y. N. 1962. Electron microscopy of virus-infected yeast cells. J. Bacteriol. 83:1363–1364.

Hiruki, C. 1964. Mechanical transmission of tobacco stunt virus. Virology 23:288–290.

——— 1965. Transmission of tobacco stunt virus by *Olpidium brassicae*. Virology 25:541–549.

Hollings, M. 1962. Viruses associated with a die-back disease of cultivated mushroom. Nature 196:962–965.

Hollings, M., Gandy, D. G., and Last, F. T. 1963. A virus disease of a fungus: die-back of cultivated mushroom. Endeavour 22:112–117.

Hollings, M., Stone, O. M., and Last, F. T. 1965. Detection and identification of viruses in mushroom sporophores and mycelium disrupted with ultrasound. Rept. Glasshouse Crops. Res. Inst. 1964:151–154.

Hubbeling, N. 1957. New aspects of breeding for disease resistance in beans. Euphytica 6:111–141.

Jagger, I. C. 1928. The brown blight disease of lettuce. Phytopathology 18:949–950. (Abstr.)

——— 1940. Brown blight of lettuce. Phytopathology 30:53–64.

Jagger, I. C., and Chandler, N. 1934. Big vein, a disease of lettuce. Phytopathology 24:1253–1256.

Jha, A. 1961. Arabis mosaic virus in strawberry. J. Hort. Sci. 36:219–227.

Jinks, J. L. 1959. Lethal suppressive cytoplasms in aged clones of *Aspergillus glaucus*. J. Gen. Microbiol. 21:397–409.

Johnson, F., and Jones, L. K. 1943. A report on a study of virus transmission by fungi and nodule bacteria of peas. Plant Disease Reporter 27:656–658.

Johnson, J. 1926. The attenuation of plant viruses and the inactivating influence of oxygen. Science 64:210.

Johnson, J., and Hoggan, I. A. 1937. The inactivation of the ordinary tobacco-mosaic virus by microorganisms. Phytopathology 27:1014–1027.

Kassanis, B. 1949. A necrotic disease of forced tulips caused by tobacco necrosis viruses. Annu. Appl. Biol. 36:14–17.

——— 1962. Properties and behaviour of a virus depending for its multiplication on another. J. Gen. Microbiol. 27:477–488.

——— 1964. Properties of tobacco necrosis virus and its association with satellite virus. Annu. Inst. Phytopathol. Benaki NS 6:7–26.

Kassanis, B., and Macfarlane, I. 1964. Transmission of tobacco necrosis virus by zoospores of *Olpidium brassicae*. J. Gen. Microbiol. 36:79–93.

——— 1965. Interaction of virus strain, fungus isolate, and host species in the transmission of tobacco necrosis virus. Virology 26:603–612.

Kassanis, B., and Welkie, G. W. 1963. The nature and behaviour of unstable variants of tobacco necrosis virus. Virology 21:540–550.

Klotz, L. J., and DeWolfe, T. A. 1961. Can *Phytopathora* spp. transmit psorosis? Proc. Conf. Int. Organ. Citrus Virol., 2nd, p. 56.

Kneebone, L. R., Lockard, J. D., and Hager, R. A. 1962. Infectivity studies with X-disease. Mushroom Sci. 5:461–467.

Kole, A. P. 1954. A contribution to the knowledge of *Spongospora subterranea* (Wallr.) Lagerh., the cause of powdery scab of potatoes. Tijdschr. Plantenzietken 60: 1–65.

Lambert, E. B. 1961. Progress report on X-disease. Mushroom News 9(4):4–7.

Ledingham, G. A. 1939. Studies on *Polymyxa graminis,* n. gen. n. sp., a plasmodiophoraceous root parasite of wheat. Can. J. Res. 17:38–51.

Lindberg, G. D. 1959. A transmissible disease of *Helminthosporium victoriae.* Phytopathology 49:29–32.

Lindberg, G. D., and Pirone, T. P. 1963. Serological differentiation of normal and diseased *Helminthosporium victoriae*. Phytopathology 53:881. (Abstr.)

Lindegren, C. C. 1966. Private communication.

Lindegren, C. C., and Bang, Y. N. 1961. The zymophage. Antonie van Leeuwenhoek 27:1–18.

Lindegren, C. C., Bang, Y. N., and Hirano, T. 1962. Progress report on the zymophage. N.Y. Acad. Sci. Trans. Ser. II, 24:540–566.

Linford, M. B., and McKinney, H. H. 1954. Occurrence of *Polymyxa graminis* in roots of small grains in the United States. Plant Disease Reporter 38:711–713.

Marlatt, R. B., and McKittrick, R. T. 1962. Effect of big vein on the irrigated head lettuce crop. Plant Disease Reporter 46:428–429.

McKinney, H. H. 1930. A mosaic of wheat transmissible to all cereal species in the tribe Hordeae. J. Agr. Res. 40:547–556.

McKinney, H. H., Paden, W. R., and Koehler, B. 1957. Studies on chemical control and overseasoning of, and natural inoculation with, the soil borne viruses of wheat and oats. Plant Disease Reporter 41:256–266.

McKinney, H. H., Webb, R. W., and Dungan, G. H. 1925. Wheat rosette and its control. Illinois Agr. Exp. Sta. Bull. 264:275–296.

Mowat, W. P. 1961. Other soil borne viruses. Annu. Rept. Scot. Hort. Res. Inst., 8th, p. 57.

——— 1963. Chytrid-borne viruses. Annu. Rept. Scot. Hort. Res. Inst., 10th, p. 71–72.

——— 1964. Chytrid-borne viruses. Annu. Rept. Scot. Hort. Res. Inst., 11th, p. 66–67.

——— 1966. Tobacco necrosis virus and *Olpidium brassicae*. Annu. Rept. Scot. Hort. Res. Inst., 12th, p. 47–48.

Nelson, R. 1932. Investigations on the mosaic disease of bean (*Phaseolus vulgaris* L.). Michigan Agr. Exp. Sta. Tech. Bull. 118:1–17.

Nienhaus, F., and Stille, B. 1965. Übertragung des Kartoffel-X-Virus durch Zoosporen von *Synchytrium endobioticum*. Phytopathol. Z. 54:335–337.

Noordam, D. 1957. Tobacco necrosis virus associated with a superficial affection of potato tubers [in Dutch]. Tijdscher. Plantenziekten 63:237-241.

——— 1960. Tobacco necrosis viruses in the Netherlands. Proc. Symp. Soil-Borne Viruses, Dundee, Scotland.

Norris, D. O. 1954. A note on the transmission of potato virus X in the field. J. Australian Inst. Agr. Sci. 20: 56–57.

Pryor, D. E. 1944. The big vein disease of lettuce in relation to soil moisture. J. Agr. Res. 68:1–9.

——— 1946. Exploratory experiments with the big vein disease of lettuce. Phytopathology 36:264–272.

Roberts, F. M. 1950. The infection of plants by viruses through roots. Annu. Appl. Biol. 37:385–396.

Sahtiyanci, S. 1962. Studien über einige wurzelparasitäre Olpidiaceen. Arch. Mikrobiol. 41:187–228.

Schisler, L. C., Sinden, J. W., and Sigel, E. M. 1967. Etiology, symptomatology, and epidemiology of a virus disease of cultivated mushrooms. Phytopathology 57:519–526.

Sill, W. H., and King, C. L. 1958. The 1957 soil borne wheat mosaic epiphytotic in Kansas. Plant Disease Reporter 42:513–516.

Sinden, J. W., and Hauser, E. 1950. Report on two new mushroom diseases. Mushroom Sci. 1:96–100.

Smith, K. M. 1937. Further studies on a virus found in the roots of certain normal-looking plants. Parasitology 29:86–95.

Smith, K. M., and Bald, J. G. 1935. A description of a necrotic virus disease affecting tobacco and other plants. Parasitology 27:231–245.

Szirmai, J. 1939. Untersuchungen und Beobachtungen an "Necrotic Virus" in Zusammenhang mit dem Pilz *Thielavia basicola* Zopf. Phytophthol. Z. 12:219–227.

——— 1962. Private communication.

Teakle, D. S. 1960. Association of *Olpidium brassicae* and tobacco necrosis virus. Nature 188:431–432.

——— 1962. Transmission of tobacco necrosis virus by a fungus, *Olpidium brassicae*. Virology 18:224–231.

——— 1963. Transmission of tobacco necrosis virus by *Olpidium brassicae*. Ph.D. thesis, University of California, Berkeley. 122 p.

——— 1964. Transmission of tobacco necrosis virus by *Olpidium brassicae*. Diss. Abstr. 24:4904.

——— 1967. Fungus transmission of plant viruses. *In* K. Maramorosch and H. Koprowski [eds.] Methods in virology. Academic Press, New York, p. 369–391.

Teakle, D. S., and Gold, A. H. 1963. Further studies of *Olpidium* as a vector of tobacco necrosis virus. Virology 19:310–315.

——— 1964. Prolonging the motility and virus-transmitting ability of *Olpidium* zoospores with chemicals. Phytopathology 54:29–32.

Teakle, D. S., and Hiruki, C. 1964. Vector specificity in *Olpidium*. Virology 24: 539–544.

Thompson, R. C., and Doolittle, S. P. 1942. Influence of temperature on the expression of big vein symptoms in lettuce. Phytopathology 32:542–544.

Thompson, R. C., Doolittle, S. P., and Smith, F. F. 1944. Investigations on the transmission of big vein of lettuce. Phytopathology 34:900–904.

Tomlinson, J. A., and Garrett, R. G. 1962. Role of *Olpidium* in the transmission of big vein disease of lettuce. Nature 194:249–250.

——— 1964. Studies on the lettuce big vein virus and its vector *Olpidium brassicae* (Wor.) Dang. Ann. Appl. Biol. 54:45–61.

Tomlinson, J. A., Smith, B. R., and Garrett, R. G. 1962. Graft transmission of lettuce big vein. Nature 193:599–600.

Walker, J. C. 1952. Diseases of vegetable crops. McGraw-Hill, New York. 529 p.

Walters, H. J. 1952. Some relationships of three plant viruses to the differential grasshopper, *Melanoplus differentialis* (Thos.). Phytopathology 42:355–362.

Werner, H. J., and Lindberg, G. D. 1966. Electron microscope observations of *Helminthosporium victoriae*. J. Gen. Microbiol. 45:123–125.

Yarwood, C. E. 1954. Tobacco-necrosis virus on lettuce. Plant Disease Reporter 38:263.

—— 1960. Soil inhibits infection with tobacco necrosis virus. Plant Disease Reporter 44:639–642.

# Nematode Vectors

C. E. TAYLOR AND C. H. CADMAN

*Scottish Horticultural Research Institute, Invergowrie, Dundee, Scotland*

## I. INTRODUCTION

That this chapter can be written at all is the best testimony to the pace of research on nematodes as virus vectors, for their role as such was discovered only a decade ago (Hewitt, Raski, and Goheen, 1958). The idea of soil being a source of infection dates back to the work of Beijerinck (1898) with tobacco mosaic and Behrens (1899) with tobacco "mauche" disease. The possibility that nematodes might be concerned in transmission was not overlooked, but suspicion fell on the wrong sorts of nematodes and experiments failed to implicate any as vectors. Percival (1895), for example, claimed that the hop cyst nematode, *Heterodera humuli,* was responsible for "nettlehead" disease in hops (*Humulus lupulus*); Duffield (1925) subsequently exonerated *H. humuli* but Bock (1966) has recently shown that the disease is associated with infection by *Prunus* necrotic ringspot virus and the nematode-borne *Arabis* mosaic virus. Johnson (1945) showed that treatment of wheat mosaic-infected soil with calcium cyanide, carbon disulfide, chloropicrin, or methyl bromide prevented transmission of the virus and surmised from this that nematodes might be involved as vectors, but further experiments (McKinney, Webb, and Dungan, 1957) failed to associate nematodes with transmission. Attempts to transmit lettuce big vein were made with several nematode species, but with negative results (Allen, 1948). Schnidler (1958) failed to transmit tobacco mosaic and cucumber mosaic viruses with *Meloidogyne* sp. and carnation mottle virus with *Helicotylenchus nannus* or *Paratylenchus* sp.

It was, however, the convincing demonstration by Hewitt and his colleagues in 1958 that *X. index* transmitted fanleaf virus in grapevine which stimulated the search for nematode vectors for other soil-borne viruses known in Britain and continental Europe. Discoveries quickly followed: *X. diversicaudatum* was associated with transmission of

*Arabis* mosaic virus (Harrison and Cadman, 1959; Jha and Posnette, 1959) and *Longidorus elongatus* with tomato black ring and raspberry ringspot viruses (Harrison, Mowat, and Taylor, 1961; Taylor, 1962a); species of *Trichodorus* were shown to transmit tobacco rattle virus (Harrison, 1961b; Sol and Seinhorst, 1961) and the similar pea early browning virus (Gibbs and Harrison, 1964a; van Hoof, 1962, 1964). Eight serologically distinct viruses are now known to have nematodes as vectors; antigenically distinct strains of some of these viruses have different nematode species as vectors. Table 1 gives a conspectus of this information.

Outbreaks of disease caused by nematode-borne viruses are characterized in three ways that help to distinguish them from those caused by viruses that have aerial vectors:

*1.* Diseased plants appear in patches in the crop.

*2.* Infection spreads slowly from the periphery of the patches to the rest of the crop.

*3.* Infection persists in the soil for long periods, even during a fallow or when crops that resist infection are grown.

Nematode-borne viruses are readily transmitted by mechanical inoculation of sap to a wide range of herbaceous test plants and this has been instrumental in the rapid advancement of our knowledge about the physical and biological properties of the viruses. Their wide host ranges and geographical distribution have, in fact, led to an embarrassing proliferation of names for diseases and viruses which the application of serological techniques has now done much to eliminate and rationalize.

## II. THE NEMATODE VECTORS

Thus far, Dorylaimid nematodes in the genera *Xiphinema, Longidorus,* and *Trichodorus* are the only ones known to transmit plant viruses. These genera are fairly cosmopolitan and have been reported from most parts of the world where nematode surveys have been undertaken. Where the host ranges have been investigated, the vector species seem to be polyphagous, although they multiply at widely different rates on different hosts. Species concerned in virus transmission are probably obligate parasites and are migratory ectoparasites feeding mainly at the root tips and to a lesser degree along the sides of the roots.

Morphologically the genera *Xiphinema* and *Longidorus* appear to be very closely related and are commonly placed together in the subfamily Longidorinae. The species of both genera are relatively large nematodes and are characterized by the presence of a long, hollow stylet or onchiostyle with which they penetrate plant roots and feed on the cell contents (Fig. 1). The stylet is of the order of 100 μ long in the adult, but its full length may not be exerted in feeding (Davis and Jenkins, 1960). However, there is evidence that *X. diversicaudatum* can reach the vascular tissues of the fine feeder roots of *Petunia hybrida* (Pitcher and Posnette, 1963), and from measurements of a series of feeding tracts in the root tips of grapevine seedlings Radewald (1962) estimated the depth of penetration of the stylets of *X. index* to be between 30 and 90 μ.

Feeding by *Xiphinema* species usually results in considerable distortion and necrosis of the root tips. Christie (1952) found *X. americanum* associated with extensive necrosis and destruction of feeder roots of laurel oak (*Quercus laurifolia*) in Florida, and Perry (1958) demonstrated that the species is pathogenic in strawberry (*Fragaria ananassa*) and associated with reduction of the root system and extreme black root rot. In Wisconsin, *X. americanum* causes stunting in nursery stocks of ornamental spruce (*Picea pungens, P. glauca densata*) (Griffin and Epstein, 1964) and severe decline of periwinkle (*Vinca minor*) (Epstein and Barker, 1966). In the eastern United States, Schindler (1957) and Schindler and Braun (1957) reported that *X. diversicaudatum* caused severe galling on the roots of fig (*Ficus carica*), roses (*Rosa* sp.), strawberry, and groundnut (*Arachis hypogaea*) and in Great Britain, Harrison and Winslow (1961) found galls on the roots of celery (*Apium dulce*), raspberry (*Rubus idaeus*), strawberry, and chickweed (*Stellaria media*) grown in infested soil. Raski and Radewald (1958) described the symptoms on grapevines (*Vitis vinifera* var. Thompson Seedless) associated with parasitism by *X. index* as a pronounced malformation of feeder rootlets and deformation of the growing points of many laterals, together with bending, swelling, and extensive cortical necrosis of the roots. Galling by *Longidorus elongatus* appears to be less severe than that caused by *Xiphinema* but in sandy soils Seinhorst (1966) and Sharma (1965) found the growth of strawberry to be affected by moderate infestations, and in Oregon, *L. elongatus* was associated with severe necrosis or "dieback" of the roots of peppermint (*Mentha piperita*) (Horner and Jensen, 1954; Jensen, 1961). In Great Britain *L. attenuatus,* with *Trichodorus* sp., has been associated with root damage to sugar beets in soils where "docking disorder" occurs (Whitehead, 1965).

The genus *Trichodorus* is taxonomically and morphologically distinct from *Longidorus* and *Xiphinema* but resembles them in parasitic associations, particularly in causing root galling (Fig. 2). *Trichodorus*

TABLE 1
NEMATODE VECTORS OF PLANT VIRUSES

| Virus | | Vector | Reference |
|---|---|---|---|
| | | *A. NEPO viruses* | |
| *Arabis* mosaic | Type strain | *Xiphinema diversicaudatum* | Harrison and Cadman, 1959 |
| | | | Jha and Posnette, 1959 |
| *Arabis* mosaic | Grapevine fanleaf strain | *X. index* | Hewitt *et al.*, 1958 |
| *Arabis* mosaic | Type strain and rhubarb mosaic isolate | *X. coxi* | Fritzsche, 1964 |
| Cherry leaf roll | Type strain | *X. coxi* | Fritzsche, 1964 |
| | | *X. diversicaudatum* | Fritzsche and Kegler, 1964 |
| Raspberry ringspot | Type strain | *Longidorus elongatus* | Taylor, 1962a |
| Raspberry ringspot | English strain | *L. macrosoma* | Harrison, 1962 |
| Strawberry latent ringspot | Type strain | *X. diversicaudatum* | Lister, 1964 |
| Tobacco ringspot | Arkansas isolate | *X. americanum* | Fulton, 1962 |
| Tomato black ring | English strain | *L. attenuatus* | Harrison *et al.*, 1961 |
| Tomato black ring | Scottish strain | *L. elongatus* | Harrison *et al.*, 1961 |
| Tomato ringspot | Peach yellow-bud isolate | *X. americanum* | Breece and Hart, 1959 |
| Tomato ringspot | Grape yellow-vein isolate | *X. americanum* | Teliz *et al.*, 1966 |

*B. NETU viruses*

| | | | |
|---|---|---|---|
| Pea early browning | Dutch isolate | *Trichodorus pachydermus*<br>*T. teres* | van Hoof, 1962 |
| Pea early browning | English isolate | *T. anemones* | Harrison, 1967 |
| | | *T. pachydermus* | Gibbs and Harrison, 1964 |
| | | *T. viruliferus* | Gibbs and Harrison, 1963 |
| Tobacco rattle | Dutch isolate | *T. pachydermus* | Sol and Seinhorst, 1961 |
| Tobacco rattle | Dutch (*Gladiolus* notchleaf) isolate | *T. similis* | Cremer and Kooistra, 1964 |
| Tobacco rattle | English isolate | *T. primitivus* | Harrison, 1961b |
| Tobacco rattle | German isolate | *T. primitivus* | Sänger, 1961 |
| Tobacco rattle | Scottish isolate | *T. primitivus* | Mowat and Taylor, 1962 |
| Tobacco rattle | U.S.A.: California isolate | *T. allius*<br>*T. christiei*<br>*T. porosus* | Ayala and Allen, 1966 |
| Tobacco rattle | U.S.A.: Oregon isolate | *T. allius* | Jensen and Allen, 1964 |
| Tobacco rattle | U.S.A.: Wisconsin isolate | *T. christiei* | Walkinshaw *et al.*, 1961 |

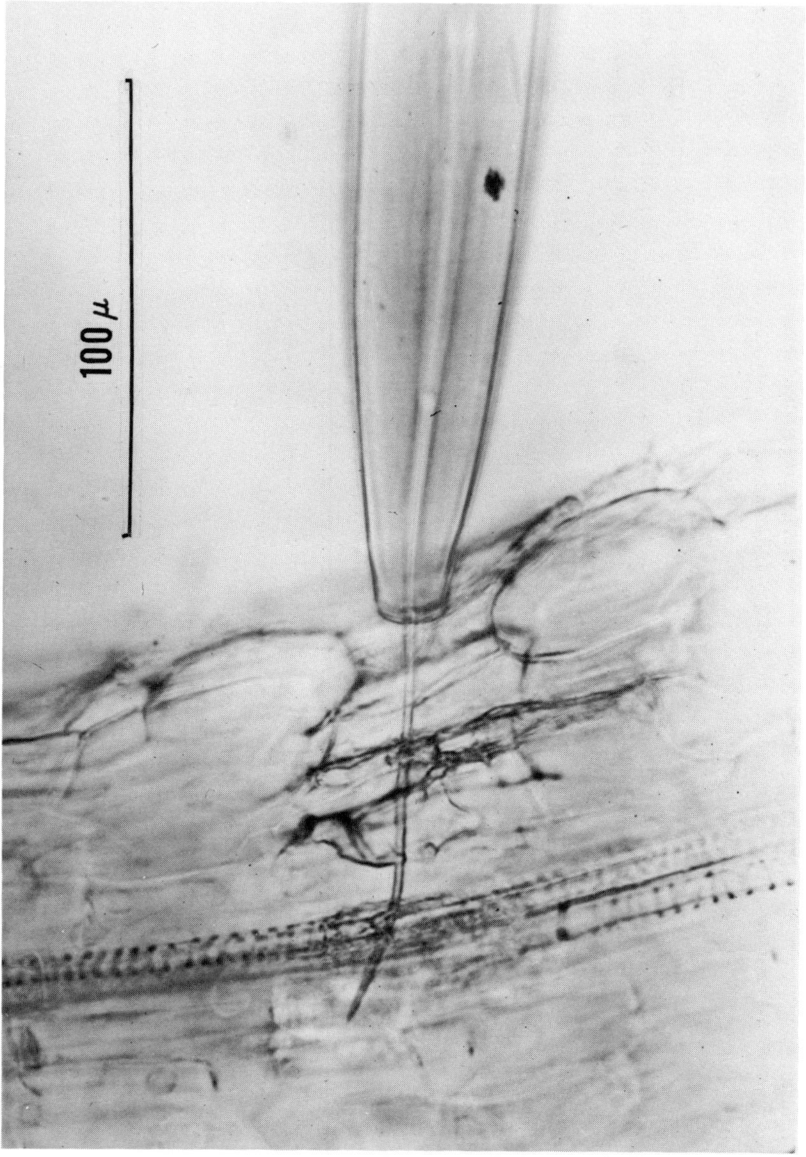


*Fig. 1.* The protracted spear of *Xiphinema diversicaudatum* inserted in the root of *Petunia* sp. (Published by permission of E. J. Brill, Ltd., Leiden. ref. Pitcher and Posnette, Nematologica, 1963.)

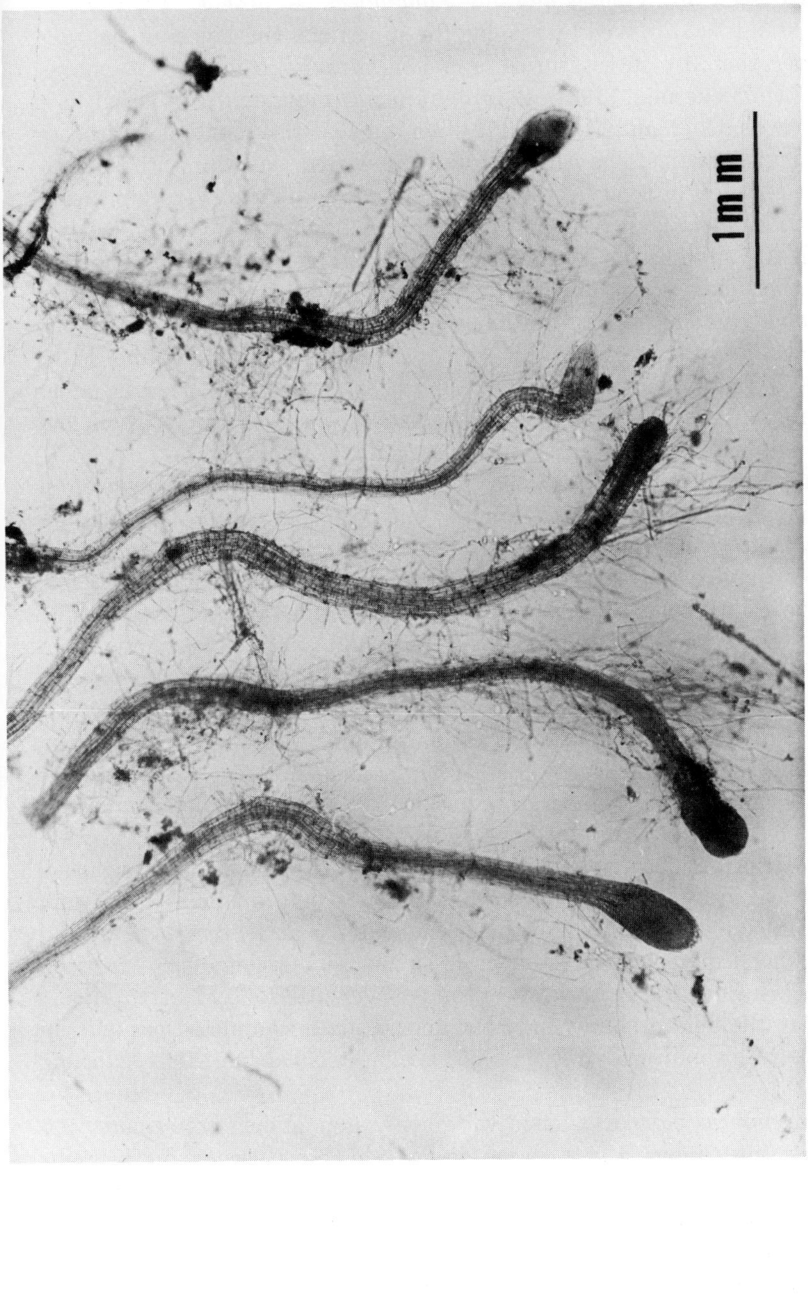


*Fig. 2.* Galls caused by the feeding of *Longidorus elongatus* on the roots of *Stellaria media*.

species are relatively short, thick nematodes with the stylet in the form of a non-axial, dorsal tooth from which the functional lumen of the stylet seen in *Xiphinema* and *Longidorus* is absent. *Trochodorus christiei* is said to feed by rasping the superficial cell layers of the root, rather than by direct thrust and penetration (Rohde and Jenkins, 1957a; Zuckerman, 1961, 1962) but using individually isolated corn root cells, Chen and Mai (1965) observed that the nematode placed its tip region against the cell wall and punctured the wall with smooth, direct thrusts of its stylet. "Stubby root" is the typical symptom caused by *T. christiei* and results from mild cellular hypertrophy associated with the suppression of mitosis in the root meristem (Flegg, 1965); whether this symptom is a typical reaction to feeding by other *Trichodorus* species is questionable, however. Usually parasitism by *Trichodorus* is not associated with necrosis, but Pitcher and Flegg (1965) found that feeding by *T. viruliferus* produced necrotic areas on apple roots, halted growth, and caused the root tip to swell, but did not induce the proliferation of secondary roots.

*Xiphinema americanum* and *X. diversicaudatum* have a wide host range, mostly among woody plants, but both species infest herbaceous crops and weeds and transmit viruses to them. *Xiphinema index* on the other hand, apparently multiplies successfully only on a restricted range of woody plants that includes grapevine (*Vitis vinifera*) and other *Vitis* species, fig, and rose (Radewald, 1962; Raski and Lider, 1959); slight to moderate population increases occur on *Parthenocissus* sp., peach (*Prunus persica*), Upland cotton (*Gossypium hirsutum*), *Pistacia* sp., and *Ampelopsis* (Radewald, 1962; Radewald and Raski, 1962b; Weiner and Raski, 1961). On a good host such as grapevine and at an optimum temperature of 30°C, *X. index* completes its development from oviposition to the adult stage in 22–27 days (Radewald and Raski, 1962a) and population increases of 2000-fold have been recorded in the glasshouse after 6–12 months on grapevine (Raski and Hewitt, 1960). Available evidence indicates that *X. americanum* and *X. diversicaudatum* multiply more slowly (Fritzsche, 1966; Harrison and Winslow, 1961; Lownsbery and Maggenti, 1963; Norton, 1963) and in the field probably only one generation is completed per year in temperate conditions (Flegg, 1966; Griffin and Darling, 1964; Norton, 1963).

*Longidorus elongatus* has a wide host range among crop plants and weeds: in Oregon, U.S.A., large populations have been found in crops of peppermint (Horner and Jensen, 1954), and in eastern Scotland, population increases have been recorded on rye grass (*Lolium perenne*), chickweed, strawberry, blackcurrant (*Ribes nigrum*), and sugar beet (*Beta vulgaris*) (Taylor, 1967). In the laboratory the life cycle takes about 16 weeks at 20°C (Yassin, 1966a) but in the field there is probably only one generation per year with a three- to fourfold increase of population per year on favored hosts (Murant and Taylor, 1965; Taylor and Murant, 1968). Little is known about the ecology of the other *Longidorus* vector species, *L. attenuatus* and *L. macrosoma,* which have only recently been described (Hooper, 1961), but both are widely distributed in Europe (Dalmasso and Caubel, 1966; Debrot, 1964; Harrison *et al.*, 1961; Sturhan, 1966; Weischer, 1966; Whitehead, 1965) and they probably have a wide host range and similar habits to *L. elongatus.*

Allen (1957) reviewed the genus *Trichodorus* and described ten new species; several other new species were later described by Hooper (1962, 1963) and Seinhorst (1963), but little is known about their biology except that some transmit tobacco rattle and pea early browning viruses. Christie and Perry (1951) provided the first evidence of parasitism in *Trichodorus* when they found *T. christiei* injuring the roots of beet, celery, and sweet corn (*Zea mays*) in Florida, U.S.A. This species is polyphagous and crops known to be injured by it include maize, cabbage (*Brassica oleracea*), cauliflower (*B. oleracea*), fig, various grasses, blueberry (*Vaccinium corymbosum*), onion (*Allium cepa*), peach, tomato (*Lycopersicon esculentum*), and many others (Alhassan and Hollis, 1966; Christie, 1959; Hoff and Mai, 1962; Rohde and Jenkins, 1957a; Zuckerman, 1961). Rohde and Jenkins (1957a, 1957b) found that the species completed its life cycle in 16–17 days at 30°C and in 21–22 days at 22°C and that populations could increase tenfold or more in 60 days. *Trichodorus pachydermus* also has a wide host range and has been found in association with ornamental bulbous crops, pea (*Pisum sativum*), potato (*Solanum tuberosum*), sugar beet, and tobacco (*Nicotiana tabacum*) (Cremer and Kooistra, 1964; Sol and Seinhorst, 1961; Taconis and Kuiper, 1963; Whitehead, 1965). *Trichodorus teres* and *T. viruliferus* have been found in association with peas (Gibbs and Harrison, 1964a; van Hoof, 1962) and the latter species has been observed feeding on the roots of apple trees (Pitcher and Flegg, 1965). *Trichodorus allius* is a vector of tobacco rattle virus in Oregon, U.S.A. and, as its name implies, feeds on onions, with tobacco and tomato as other probable hosts (Jensen and Allen, 1964).

## III. THE VIRUSES

All the nematode-transmitted viruses infect a wide range of woody and herbaceous hosts and some are seed and pollen borne (Cadman, 1963). They are sap transmissible to herbaceous test plants and relatively stable *in vitro* (Table 2). The viruses divide into two groups on the basis of particle shape. Harrison (1960) called those with polyhedral particles "soil-borne ringspot viruses" but as the ability to cause ringspot symptoms is not an invariable feature of, or unique to, this group of viruses, Cadman (1963) suggested avoiding this ambiguity by naming them "NEPO viruses," i.e., NEmatode-transmitted viruses with POlyhedral particles. The second group includes the viruses with rod-shaped or tubular particles for which Harrison (1964a) derived the name NETU, i.e., NEmatode-transmitted viruses with TUbular particles.

### A. NEPO Viruses

This group includes *Arabis* mosaic, strawberry latent ringspot, raspberry ringspot, tomato black ring, cherry leaf-roll, tobacco ringspot, and tomato ringspot viruses. These viruses share no antigenic determinants but they form a group with close natural affinities. All possess polyhedral particles of about 30 m$\mu$ in diameter, have similar chemicophysical properties (Table 2), and cause closely similar symptoms in herbaceous test plants. They are transmitted by species of *Xiphinema* or *Longidorus.*

*Arabis* mosaic virus was first isolated from a single plant of *Arabis* growing in a glasshouse at Cambridge, England (Smith and Markham, 1944). Subsequently, the viruses associated with raspberry yellow-dwarf and strawberry yellow-crinkle diseases were identified with the type strain of *Arabis* mosaic virus (Cadman, 1960) and were shown to be transmitted by *X. diversicaudatum* (Harrison and Cadman, 1959; Jha and Posnette, 1959). In Germany, *X. diversicaudatum* (=*para-elongatum*) and *X. coxi* were found associated with rhubarb infected with *Arabis* mosaic virus (Fritzsche and Schmidt, 1963;) and both species transmitted the virus experimentally (Fritzsche, 1964). The type strain of *Arabis* mosaic virus has thus far been found only in Europe, but a distantly related strain, or serotype, of the virus which infects grapevines probably occurs wherever this crop is cultivated. In the United States the two principal diseases caused by this strain of *Arabis* mosaic are called grapevine fanleaf and grapevine yellow mosaic; in Europe the same diseases are named arricciamento, roncet, urticado, or court-noué, and Reisigkrankheit, panachure, or clorose infecciosa, respectively. Hewitt *et al.* (1958) were the first to show that *X. index* transmits grapevine fanleaf virus (Fig. 3) and subsequently the nematode has been found to be abundant in Europe in association with infected grapevines (Amici *et al.,* 1964; Cadman, Dias, and Harrison, 1960; Lamberti and Raski, 1964; Martelli and Raski, 1963;

TABLE 2

PHYSICAL PROPERTIES OF NEMATODE-TRANSMITTED (NEPO) VIRUSES

| Virus | Thermal endpoint, °C | Dilution endpoint (reciprocal), $M^{-1}$ | Longevity in sap at 18–20°C, days | Reference |
|---|---|---|---|---|
| *Arabis* mosaic (type strain) | 52–55 | $10^3$–$10^4$ | 5–10 | Cropley, 1961a |
| *Arabis* mosaic (grapevine fanleaf strain) | 58–65 | $10^3$–$10^4$ | 10–28 | Cadman *et al.*, 1960 |
| Cherry leaf roll | 52–55 | $10^3$–$10^4$ | 5–10 | Cropley, 1961a |
| Raspberry ringspot (type strain) | 65–70 | $10^3$–$10^4$ | 15–21 | Harrison, 1958 |
| Raspberry ringspot (English strain) | 70–74 | $10^4$–$10^5$ | 28–35 | Debrot, 1964 |
| Strawberry latent ringspot | 54–56 | $10^3$–$10^5$ | 50 | Lister, 1964 |
| Tobacco ringspot | 55–65 | $10^3$–$10^4$ | 10–14 | Lister *et al.*, 1963 |
| Tomato ringspot | 58–60 | $10^3$–$10^4$ | 3–6 | Cadman and Lister, 1961 |
| Tomato ringspot (grape yellow-vein) | 60–62 | $10^3$–$10^4$ | 6–8 | Gooding and Hewitt, 1962 |
| Tomato black ring (Scottish strain) | 52–57 | $10^4$–$10^5$ | 50 | Harrison, 1957 |

*Fig. 3.* Healthy (left) and fanleaf virus-infected (right) Mission grapevine.

Vuittenez, 1961; Vuittenez and Legin, 1964). In contrast, tobacco ringspot and tomato ringspot viruses seem restricted to the American continent although their vector, *X. americanum,** is cosmopolitan in distribution. Breece and Hart (1959) were the first to associate *X. americanum* with the transmission of the virus causing peach yellow-bud mosaic, subsequently identified with tomato ringspot virus (Cadman and Lister, 1961). Tomato ringspot virus has since been identified in many herbaceous (McLean, 1962) and woody hosts (Gooding, 1963; Milbrath and Reynolds, 1961; Stace-Smith, 1962). Recently, Teliz, Grogan, and Lownsbery (1966) have shown that strains of tomato ringspot virus which cause yellow-bud mosaic in peach and yellow-vein in grapevine are not identical serologically with one another and that both are transmitted by *X. americanum.*

Tobacco ringspot virus causes diseases in a great variety of crop plants, but despite its wide host range and the cosmopolitan distribution of its vector, *X. americanum,* the virus has only occasionally been recorded outside North America; infection of *Gladiolus,* grown in Scotland but derived from imported Canadian material (Cadman, 1963),

* *Xiphinema americanum* is now regarded as a complex of at least seven species (Lima, 1968) and differences between populations within the United States have been described (Tarjan, 1968).

and of *Anemone coronaria* in Somerset, England (Hollings, 1965) are examples.

Two other viruses are known to be transmitted by *Xiphinema*. The first is cherry leaf roll, a virus which occurs in English cherry trees (Cropley, 1961a) and which Fritzsche and Kegler (1964) implicate both *X. diversicaudatum* (=*X. paraelongatum*) and *X. coxi* as vectors. The second, strawberry latent ringspot virus, infects raspberry, cherry, blackcurrant, rose, and several weed hosts, as well as strawberry (Lister, 1964; C. E. Taylor and P. R. Thomas, unpublished; B. D. Harrison, unpublished) and was discovered because it failed to react with antiserum to *Arabis* mosaic virus. Lister (1964) transmitted the virus to cucumber and *Petunia* with *X. diversicaudatum*.

Strains of raspberry ringspot virus cause different diseases in different countries. In Scotland the virus causes the economically important "raspberry leaf-curl" disease (Cadman, 1956). An antigenically similar strain is associated with spoonleaf disease of redcurrants in the Netherlands (Harrison, 1961a; Maat, 1965; Maat, van der Meer, and Pfaeltzer, 1962). Each of these strains is transmitted by *L. elongatus* (Taylor, 1962a; van der Meer, 1965) but they differ serologically from strains of the virus isolated from blackberry and raspberry in England which are transmitted by *L. macrosoma* (Harrison, 1964b), and from those associated with cherry diseases of the rasp-leaf Pfeffinger type (Cadman, 1960). Strains of tomato black ring virus cause beet ringspot disease in Scotland, celery yellow-vein and lettuce ringspot diseases in England, and potato bouquet and potato pseudoaucuba in Germany. As with raspberry ringspot virus, variants that possess few antigenic groups in common have proved to have different *Longidorus* species as vectors; for example, the lettuce ringspot and beet ringspot strains are serologically distinct and are transmitted by *L. attenuatus* and *L. elongatus*, respectively (Harrison *et al.*, 1961).

NEPO viruses are usually readily transmissible by mechanical inoculation of sap from woody or tannin-containing hosts to herbaceous plants, provided the pH level of the inoculum is alkaline. Under these conditions acids and some kinds of tannins are prevented from combining with and precipitating the viruses (Cadman, 1959). For this purpose alkaloids, such as nicotine or caffeine, have proved useful additives because they precipitate tannins from solution as well as increase the pH level (Cadman, 1959; Diener and Weaver, 1959). Extracts so prepared sometimes contain sufficient virus for serological tests or for the preparation of specific antisera. Roots are sometimes better sources of infective preparations than leaves because their virus content may be less affected by inhibitors and by interactions between day length,

*Fig. 4.* Strawberry var. Redgauntlet infected with raspberry ringspot virus (foreground) compared with healthy plants (background).

light intensity, and temperature which affect the concentration of both viruses and substances that inhibit infectivity. Grapevine fanleaf virus, for example, is readily transmissible from infected grapevines in spring and autumn but only with difficulty in summer (Dias, 1963), and both raspberry ringspot and cherry leaf-roll viruses are easier to transmit from cherry leaves in spring than in late summer (Cropley, 1961a, 1961b).

Diseases caused by NEPO viruses typically show the "shock" and "recovery" phases so characteristic of tobacco ringspot disease in tobacco; *i.e.*, the first parts of a plant to become invaded systemically show severe symptoms, whereas those formed subsequent to infection contain the virus but appear to be healthy. Grapevines newly infected with fanleaf virus display ringspot patterns on their leaves (Vuittenez, 1961) but the leaves subsequently produced show only the typical symptoms of fanleaf (Hewitt, 1950). Chlorotic rings precede the development of rasp-leaf symptoms in sweet cherry (Blumer, 1954, 1955) and the typical symptoms of leaf curl in raspberry (Cadman and Harris, 1952). In strawberry, on the other hand, ringspot symptoms develop annually on the new foliage of infected plants, while leaves produced in summer are normal in appearance. Lister (1962) suggests that this may result from failure of the virus to infect plants systemically; a proportion of crown cuttings made from infected plants are often free from virus.

NEPO viruses are transmitted through the seeds and pollen of infected crop and weed plants (Lister and Murant, 1967). The proportion of infected seeds set by diseased plants seems to depend on the virus involved, the host plant, and the age at which plants first become infected. Several viruses are transmitted highly efficiently through the seed of soybean (*Glycine soja*) if the plants are inoculated when young, but experiments with tobacco ringspot virus have shown that few seeds become infected if the plants are inoculated at flowering time (Crowley, 1959; Athow and Bancroft, 1959). Germination seems unaffected and seedlings from infected seeds rarely or never appear diseased. Although transmission through the seed seems typical of the whole group of NEPO viruses, the relevance of this to the ecology of the viruses varies. Lister and Murant (1961, 1967) consider that infected weed seeds are important reservoirs of tomato black ring and raspberry ringspot viruses in soils in Scotland, but found no evidence that this was so with *Arabis* mosaic virus in soils in England: only a single infection of *Arabis* mosaic virus was detected in a total of 444 weed seedlings in soils from different outbreaks, compared with over 10% infection of seedlings in samples of soils from outbreaks of tomato black ring virus in Scotland. Similarly, soil from an outbreak of tobacco ringspot virus in Indiana, U.S.A., yielded not a single infection among 202 weed seedlings tested. It may be significant that under natural conditions the viruses transmitted by *Longidorus* species are more consistently seed borne than those transmitted by *Xiphinema* species (Lister and Murant, 1967; Murant and Lister, 1967); this is referred to again below.

### *B. NETU Viruses*

Tobacco rattle virus occurs in many parts of Europe, and in North and South America and has an extraordinarily wide natural and experimental host range. The disease in tobacco was first described and named "mauche" by Behrens in 1899, but it is nowadays better known as the cause of economically important diseases in bulbous crops in the Netherlands and internal necrosis (spraing or corky ringspot) of potato tubers in Britain, continental Europe, and the United States. Preparations of the virus contain rigid tubular particles of two main lengths, about 75 mμ and 185 mμ, respectively, only the longer of which are infective (Harrison and Nixon, 1959; Harrison and Woods, 1966). On their own, the shorter particles are not infective, but recent work (Lister, 1966; Frost, Harrison, and Woods, 1967) has demonstrated that they are essential to the production of stable virus. Both experimentally and in the field, plants may develop severe and persistent symptoms, but contain no detectable virus antigen and few, or no, characteristic virus particles. Sänger and Brandenburg (1961) and Cadman (1962) showed that such infections were associated with unstable material which they concluded was virus nucleic acid unassociated with virus protein. The frequency of occurrence and the unusual properties of such unstable forms of tobacco rattle virus account for past difficulties in identifying the causes of stem mottle and spraing diseases in potato.

Pea early browning virus occurs naturally in pea and lucerne crops in the Netherlands and in Britain, but has not been reported from the United States (Bos and van der Want, 1962; Gibbs and Harrison, 1964a, 1964b). The virus also has long and short particles, respectively about 210 m$\mu$ and 105 m$\mu$, and, like tobacco rattle virus, it does not readily infect plants systemically. In pea, for example, the symptoms may be limited to part of the stem or to one or two of the leaflets. In Europe, several species of *Trichodorus* have been implicated as vectors of pea early browning and tobacco rattle viruses. In the United States, tobacco rattle virus has been transmitted by *T. christiei, T. porosus,* and *T. allius* (Ayala and Allen, 1966; Jensen and Allen, 1964; Walkinshaw, Griffen, and Larson, 1961).

## IV. VIRUS AND VECTORS

### A. *Identifying the Vector*

Usually the first indication of the existence of nematode-borne virus infection in a crop is the development of a patch or patches of diseased plants. In perennial crops these usually appear 2–3 years after planting—rarely in the first year of growth. The patches may vary in extent from a few square yards to several acres, depending on the previous cropping of the soil, but the subsequent spread of infection from the periphery of the patches is slow and rarely exceeds a few feet each year. The problem of establishing an association with a vector organism has been approached in different ways, but the several procedures and techniques that have been evolved are now routinely and widely used in the course of investigating outbreaks of soil-borne virus.

1. SOIL TREATMENTS. Virus infection in the soil may be abolished by air drying (Cadman and Harrison, 1960), by grinding to create friction between soil particles (Sol, van Heuven, and Seinhorst, 1960), or by treatment with chemicals that do not inactivate the virus *in vitro*

(Cadman and Harrison, 1960; Harrison and Cadman, 1959). The loss of infectivity is indicative of an association with a biological agent but none of these treatments can define the nature of the agent involved. Erroneous conclusions may indeed be drawn from the results of chemical treatments if these are rigidly regarded as having fungicidal, insecticidal, or nematicidal action and their wide biocidal spectrum is ignored. For example, infectivity of tomato black-ring-infected soil was abolished after treatment with quintozene (pentachloronitrobenzene) (Cadman and Harrison, 1960), a chemical thought to possess only fungicidal properties but one which also kills *L. elongatus,* the nematode vector of the virus (Taylor and Murant, 1966a).

2. TESTING FIELD SOIL. Initially the soil-borne nature of several viruses was demonstrated by comparing the behavior of healthy plants grown in pots of soil taken from around the roots of diseased plants with comparable plants grown in soil taken from around healthy plants, or alternatively in infective soil that had been autoclaved or chemically treated. Wagnon and Breece (1955) showed that peach seedlings became infected within one year when grown in soil taken from around the roots of trees infected with peach yellow-bud mosaic virus but remained uninfected when grown in soil from adjacent healthy trees. The transmission of fanleaf virus to grapevines (Hewitt *et al.,* 1958), tomato black ring virus (Scottish strain) to peaches (Harrison, 1957), and *Arabis* mosaic virus to strawberries (Jha and Posnette, 1959) were similarly demonstrated. The technique suffers, however, from the delay—sometimes months, or even years—between the exposure to infection and the development of disease symptoms, but this impediment can be overcome by the use of "bait plants" (Cadman and Harrison, 1960). The term "bait plant"—acquired from mycological terminology—connotes seedlings, preferably of herbaceous hosts of the virus, which, when exposed to infection by planting in soils, can be used to assay the presence of infectivity. Usually bait plants are exposed for 3–6 weeks and are then assayed for virus by inoculating sap from roots and/or tops to suitable test plants. Thence the viruses can, if necessary, be identified by serological means. A great variety of bait and test plants has been used for different virus/nematode combinations and information about these is in the literature.

3. SOIL SURVEYS. Soil sampling within and without patches of diseased plants may reveal a correlation between a specific nematode and the outbreak of disease. This technique established a correlation between the distribution in raspberry and strawberry crops of *Arabis* mosaic virus with *X. diversicaudatum* and of tomato black-ring virus with that of *L. elongatus,* whereas neither virus related to the occurrence and

distribution of other plant parasitic nematodes (Harrison and Cadman, 1959; Harrison *et al.*, 1961). Similarly, soil sampling demonstrated the coincidence of *X. americanum* with patches of blueberry infected with necrotic ringspot disease (tobacco ringspot virus) (Griffin, Hoguelet, and Nelson, 1963) and with spearmint stunted by tobacco ringspot infection (Stone, Mink, and Bergeson, 1962). Conversely, Latta (1966) failed to relate nematodes to the distribution of lethal yellowing disease of coconut palms in Jamaica. Virus diseases may be established in a crop by planting material propagated vegetatively from diseased stock; diseased plants are then likely to be scattered at random and whether or not the spread of disease subsequently occurs from these infection foci depends on the presence or absence of a vector. Under these circumstances, a comparison of nematode populations in soils where the virus spreads with those where it does not can establish a satisfactory correlation with a vector. In this way, the association of *X. index* with grapevine fanleaf virus was found in California, U.S.A., and in many European countries (Amici, 1965; Cadman *et al.*, 1960; Hewitt *et al.*, 1958; Lamberti and Raski, 1964; Martelli and Raski, 1963; Raski and Amici, 1964; Vuittenez, 1961).

4. HANDPICKING SPECIMENS. If a soil survey fails to reveal a specific nematode as a likely vector, fractionation of infected soil by sieving can help to narrow the search. The usual procedure is to pass a suspension of the soil in water successively through a series of sieves diminishing in mesh size, for example, 240, 150, and 100 $\mu$, and then to test the infectivity of the debris remaining on each grade of sieve by adding it to bait plants grown in sterilized soil. Infectivity may then be associated with groups of organisms of different sizes (Cadman and Harrison, 1960; Sol *et al.*, 1960). A positive association between virus and vector can, however, only be established by transmission experiments using hand-picked nematodes. A typical experiment is that of Hewitt *et al.* (1958), who showed that fanleaf disease was transmitted from diseased to healthy grapevines grown in the same container only when *X. index* were added to the soil; no transmission took place in containers with healthy plants and *X. index,* or with healthy and diseased plants and no *X. index*. This double plant test has now largely been superseded by the technique in which infective nematodes, washed several times in distilled water, are added to bait plants growing in sterilized soil and the results assessed by inoculating sap from the roots or tops of the plants to suitable indicator plants.

Nematodes obtained directly from field soil may vary considerably in their capacity to transmit because of various factors such as poor growth of the infected crop plants, low concentration of virus in the roots, low soil temperatures depressing the activity of the nematode vector, and so on. The efficiency of transmission can often be increased by feeding the nematodes on suitable host plants before exposing them to bait plants. Walkinshaw *et al.* (1961) found that tobacco rattle virus was transmitted to tobacco bait plants more often when soil infested with *T. christiei* was first cropped with healthy maize than when the tobacco plants were planted directly in field soil; maize was a good host for both the nematode and virus, whereas tobacco was a poor host for the nematode. Harrison *et al.* (1961) used cucumber seedlings and potatoes inoculated with tomato black ring virus as "infector" plants in soil containing *L. elongatus* in order to maintain the nematode populations and to ensure a high level of infectivity.

### *B. Virus–Vector Associations*

The experimental approach and the terminology used in studies on the biological association between the virus and its nematode vector tend to parallel those used in work on insect-borne viruses, but any similarities between the two kinds of vectors are, on present evidence, superficial and fortuitous. The use of insect virus terminology such as persistent or circulative viruses and nonpersistent or mechanically transmitted viruses should be avoided as the evidence for nonpersistence of virus infection through the molt of nematode vectors does not appear to exclude the possibility of an intimate biological association with the vector (Taylor and Raski, 1964).

1. EFFICIENCY OF TRANSMISSION. Experimental evidence indicates that the adult and larval stages of most nematode vectors can transmit virus equally efficiently. All stages of *X. index* can transmit grapevine fanleaf virus (Raski and Hewitt, 1960) and all stages of *X. americanum* transmit tomato ringspot virus (Teliz *et al.*, 1966). Ayala and Allen (1966) showed that a single specimen of any stage of *T. allius,* except the first-stage larva which was not tested, could transmit tobacco rattle virus. *Arabis* mosaic and strawberry latent ringspot have been transmitted by adults and larvae of *X. diversicaudatum* (Jha and Posnette, 1961; Harrison, 1967), raspberry ringspot virus by females and larvae of *L. elongatus* (Taylor, 1962a), and tobacco rattle virus by females, males, and larvae of *T. pachydermus* (Gibbs and Harrison, 1964a; Sol, 1963), and by adults and larvae of *T. teres* (van Hoof, 1964). Harrison *et al.* (1961) reported that the larvae of *L. elongatus* transmitted tomato black ring virus whereas adults did not, but later work has shown that both adults and larvae transmit the virus equally well (C. E. Taylor, unpublished; Yassin, 1966b).

2. ACQUISITION AND INOCULATION. Few observations have been made

on the feeding behavior of nematode vectors, and techniques for determining the actual feeding time of a nematode on a host plant have yet to be developed. Various authors have recorded minimum periods for nematodes to acquire viruses from plants and then to transmit them to other plants but these are essentially access periods and not feeding periods. Teliz *et al.* (1966) use the term *acquisition access threshold* to express the minimum time that a nematode vector must have access to a virus source in order to become viruliferous, and *inoculation access threshold* for the minimum time that a viruliferous nematode must have access to a healthy plant for transmission to occur. In each definition there is no implication that the nematode feeds continuously during this time.

In some early experiments with *X. diversicaudatum, Arabis* mosaic virus (Jha and Posnette, 1961), *X. index,* and grapevine fanleaf virus (Hewitt and Raski, 1962), transmission was obtained after an acquisition access period of one day, the minimum time tried. Periods as short as 1 hr were shown to be sufficient for *X. americanum* to acquire tomato ringspot virus and transmit it to cucumber bait plants (Teliz *et al.,* 1966). With access periods up to and including 48 hr, the rate of transmission was low and variable and was not proportional to the length of the access period; with periods in excess of 48 hr, transmission occurred in all of the experimental pots. *Xiphinema americanum* acquired tobacco ringspot virus after 8–24 hr access to a virus-infected plant in experiments made by Bergeson and Athow (1963) and Bergeson *et al.* (1964). Using single *X. americanum,* McGuire (1964) obtained few transmissions of tobacco ringspot virus after access periods of 24 hr, but found that the rate of transmission increased proportionately with time up to 10 days, when the experiment ended. These results may, however, reflect the ability of the nematodes to survive the treatment or the ability of a single nematode to find the roots of a bait plant, rather than indicate an increase of infectivity of the vector with length of access period. On the other hand, Ayala and Allen (1966) found that *T. allius* could acquire tobacco rattle virus within an access period of 1 hr and that the number of transmissions increased proportionately with time up to 24 hr.

Experimental evidence indicates that the inoculation access threshold of most, if not all, of the nematode vectors is 24 hr or less. Results vary between workers using the same vector and virus, but there may be several reasons for this: the number of nematodes used per test, the length of the access period, the infector and bait plant used, temperature, and soil type all affect the result. Using tobacco ringspot-infected field soil containing large numbers of *X. americanum,* Bergeson *et al.*

(1964) obtained no transmissions in bait plants exposed for 4 or 8 hr, but 4 out of 5 transmissions after 24 hr and 5 out of 5 after 72 hr. The rate of transmission of tomato ringspot virus by *X. americanum* to 15-day-old cucumber plants was also found to be proportional to the length of the inoculation access time, few transmissions occurring after 1 hr inoculation access, but 100% transmission after 4 days (Teliz *et al.*, 1966). Similarly, efficiency of transmission of tobacco rattle virus by *T. allius* increased as the access time increased from 1 hr up to 48 hr (Ayala and Allen, 1966).

3. PERSISTENCE. Nematode-borne viruses usually persist in their vectors for long periods, much exceeding the *in vitro* longevity of the virus in plant sap (Table 2). In *Chenopodium* sap at 18–20°C, grapevine fanleaf virus retains infectivity for 10–28 days (Cadman *et al.*, 1960), but infected *X. index* held in moist soil without host plants retained the virus for up to 8 months (Taylor and Raski, 1964). *Arabis* mosaic was transmitted by *X. diversicaudatum* after a starvation period of 30 days (Jha and Posnette, 1961) and Harrison and Winslow (1961) found that populations of the nematode were still infective after being maintained for 8 months on Malling Jewel raspberry, a variety considered to be immune from the virus but a host for the nematode. Harrison (1967) found that strawberry latent ringspot virus also persisted in *X. diversicaudatum* adult females, kept in soil without plants, for at least 32 days. *Xiphinema americanum* containing tobacco ringspot virus and maintained in fallow soil at 10°C for 49 weeks transmitted the virus in 3 out of 5 experimental pots, an estimated 25% of the original nematode population remaining alive (Bergeson and Athow, 1963). *Longidorus* vectors may not retain infectivity for as long as *Xiphinema*. Raspberry ringspot and tomato black ring viruses persist in *L. elongatus* for a maximum of about 8 weeks, but the ability to transmit decreases rapidly after 2–3 weeks and it seems that a high level of transmission is maintained only if *L. elongatus* has continuous access to infected plants (Lister and Murant, 1962; Taylor, 1962b). Debrot (1964) found that *L. macrosoma* retained the English strain of raspberry ringspot virus for 34 days but did not experiment further. Sol (1963) found that the Dutch isolate of tobacco rattle virus was retained by *T. pachydermus* for at least 36 days, and Ayala and Allen (1966) found that *T. allius* transmitted the California isolate after 28 days of starvation or maintenance on virus-immune plants.

Despite the long retention of viruses in their nematode vectors, the little evidence available indicates that nematodes, unlike aphids, do not retain the virus through the molt. Using newly molted *X. diver-*

*sicaudatum,* Harrison and Winslow (1961) failed to transmit *Arabis* mosaic virus, but too few nematodes may have been used to exclude the possibility that infectivity is occasionally retained. Taylor and Raski (1964) transferred fourth instar larvae of *X. index* from populations raised on fanleaf-infected grapevines to fig, which is immune to the virus, and then failed to transmit the virus to grapevine with those that had molted to the adult stage. Similarly, larvae of *L. elongatus* populations from plants infected with raspberry ringspot and tomato black ring viruses failed to transmit either virus after they had molted to adult females when kept on virus-immune mint (*Mentha sativa*) (C. E. Taylor, unpublished; Yassin, 1966b). Experiments on transmission of virus through the egg stage have been made by maintaining virus-carrying gravid females on virus-immune hosts and using the resulting larvae in transmission tests; grapevine fanleaf virus in *X. index, Arabis* mosaic virus in *X. diversicaudatum,* and raspberry ringspot and tomato black ring viruses in *L. elongatus,* were not transmitted in such tests (Taylor and Raski, 1964; Taylor and Thomas, 1968; Yassin, 1966b). Ayala and Allen (1966) also failed to obtain any evidence for the transmission of tobacco rattle virus through the egg, or through the molt, of *T. allius.*

4. SPECIFICITY OF TRANSMISSION. Among the NEPO viruses, several are transmitted by one and the same vector species but serologically distinct strains of a particular virus appear to be transmitted by different, although closely related, species of the same genus (Table 1). Thus, whereas *X. diversicaudatum* transmits the three unrelated viruses—strawberry latent ringspot, *Arabis* mosaic, and cherry leaf roll—it does not transmit the grapevine fanleaf strain of *Arabis* mosaic virus, which has *X. index* as a vector (Dias and Harrison, 1963). *Longidorus elongatus* transmits the Scottish strains of raspberry ringspot and tomato black ring viruses, but the English strains which share only about half of their antigenic groups with the Scottish strains have different species of *Longidorus* as vectors (Table 1). The idea that serologically distinct forms of a virus have different specific vectors (Harrison, 1964b) is supported by the experimental evidence of specific transmission of strains of raspberry ringspot by *L. macrosoma* (Debrot, 1964; Harrison, 1964b), tomato black ring virus by *L. elongatus* (Yassin, 1966b) and *Arabis* mosaic virus by *X. diversicaudatum.* Harrison (1964b) suggests that the geographical distribution of each particular strain of virus reflects the different ecological requirements, such as soil type, moisture, and temperature, of the different nematode species.

On the other hand, variants of viruses distinguishable in host range and showing slight serological differences in gel diffusion tests, do not necessarily have different vectors. *Xiphinema americanum* can transmit tomato ringspot virus and the serologically distinct grape yellow-vein strain of the virus (Teliz *et al.,* 1966), and Sauer (1966) transmitted two serologically distinct strains of tobacco ringspot with a single specimen of *X. americanum* and found no evidence of protection of one strain against another in the nematode. Murant *et al.* (1968) found a variant of raspberry ringspot virus occurring with the normal Scottish strain and showed that both were transmitted by *L. elongatus.* They suggested that the normal form remains dominant in nature because it is transmitted through the seeds of infected weeds and thus provides an overwintering reservoir of infection, whereas this rarely occurs with the variant form, which therefore rarely survives.

American and European cultures of tobacco rattle virus are serologically distinguishable (Cadman, 1963; Oswald and Bowman, 1958) and are transmitted by different *Trichodorus* species. However, each isolate of the virus can be transmitted by more than one vector species, and conversely, a single *Trichodorus* species can transmit more than one isolate (Table 1; Ayala and Allen, 1966; van Hoof, Maat, and Seinhorst, 1966). The European and British isolates of pea early browning virus are also serologically distinguishable and are transmitted by several species of *Trichodorus; T. pachydermus* and *T. teres* apparently transmit both Dutch and British isolates of the virus (Gibbs and Harrison, 1964a; van Hoof, 1962) but Harrison (1967) found that *T. anemones* transmitted two variants of a British isolate but did not transmit a Dutch isolate. There is little evidence of specificity of transmission among NETU viruses but in general there is much less information about the details of their transmission than those of the NEPO viruses. Until experiments have been done in which specific isolates of NETU viruses are transmitted first by one *Trichodorus* species and then by another, the idea of specificity of transmission cannot be entirely dismissed.

5. MECHANISM OF TRANSMISSION. There is no evidence to indicate in what way virus is retained in the nematode or how it is transmitted, but with some nematodes it has at least been possible to demonstrate the presence of the virus in the vector. Tobacco rattle virus was detected in *Trichodorus* sp. by cutting infective nematodes in a drop of phosphate buffer solution and using this suspension to inoculate *C. amaranticolor* assay plants (Sänger, Allen, and Gold, 1962). Raski and Hewitt (1963) detected grapevine fanleaf virus in *X. index* by this technique, but transmissions were obtained only within the first few days with the nematodes feeding on young, infected roots and not at a

later time, although the nematodes remain infective for long periods (Taylor and Raski, 1964; D. J. Raski, private communication; Taylor, unpublished). Tomato black ring and raspberry ringspot viruses are readily detected in *L. elongatus* simply by cutting up infective nematodes in a drop of water and inoculating *C. quinoa* with the suspension (Taylor, 1964), but Debrot (1964) claimed little success using this technique with *L. macrosoma* and the English strain of raspberry ringspot virus.

The virus detected in nematodes in this way is probably that in the intestine, and because of the valvular apparatus between intestine and esophagus and the peristaltic feeding action of the nematode, it seems unlikely to play any part in the transmission process. Nevertheless, it remains an assumption that transmissible virus is associated with the esophageal region of the vector. Taylor and Raski (1964) pointed to the apparent differences in the retention of virus by *X. index* and *L. elongatus* and suggested these were indicative of a fundamental difference in the association of virus with the vector in the two genera, the virus having a close biological association with *Xiphinema* species; in *Longidorus* the retention could be mechanical or by contamination of the feeding apparatus and buccal capsule. Electron microscope examination of ultrathin sections of *L. elongatus* has revealed that the stylet is not only a hollow tube but has a slit 34–80 m$\mu$ wide running along its whole length (Taylor *et al.,* 1966b and unpublished). Conceivably virus particles could pass through this slit into the buccal capsule and guiding sheath of the stylet, where they would lodge until the stylet was again exerted and some particles passed into a plant with the salivary secretions of the nematode (Taylor, unpublished). In *X. index,* Roggen (1966) detected a difference in the morphology of the lateral chords in healthy and virus-infected nematodes, which lends some support to the idea of a biological association between virus and vector. However, there is no reason to exclude the idea of mechanical transmission of grapevine fanleaf virus as a contaminant of the stylet, as postulated for *L. elongatus*. NEPO viruses are readily transmitted by mechanical inoculation of sap from plant to plant and nematodes may be as efficient as humans in this technique.

## V. ECOLOGICAL ASSOCIATIONS

Viruses may be disseminated either through the distribution and use of infected planting material or, in some cases, through infected weed seeds, or more rarely through the transportation of nematode vectors in soil. The occurrence of grapevine fanleaf virus in California is almost certainly the result of importation of infected grapevines from Europe—either by the Spanish missionaries or by General Harazthy's massive importation of plant material from Europe; *X. index* and grapevine fanleaf virus were both probably imported into Victoria, Australia about 50 years ago with *Phylloxera*-resistant rootstocks from France. Such records as there are of *Arabis* mosaic virus in Scotland are attributable largely to the importation of infected planting material imported from England (Cadman, Harrison, and Lister, 1958) because the occurrence of *X. diversicaudatum* in Scotland is apparently very localized and few endemic outbreaks of disease have been identified (Osborne, 1964; Taylor *et al.,* 1966a). Flower bulbs are an effective way of disseminating tobacco rattle virus, and doubtless also tomato black ring, raspberry ringspot, and *Arabis* mosaic viruses since, for example, many commercial *Narcissus* stocks appear to be infected with these viruses (Brunt, 1966). Probably other kinds of ornamental plants are equally efficient. Weed seeds provide an effective means of dispersal for tomato black ring and raspberry ringspot viruses, but to what extent this is effective in nature is a matter for conjecture. Although Murant and Taylor (1965) conclude that weed seeds undoubtedly provide a reservoir of infection in an outbreak area and are the most important, if not the only, overwintering sources of infection, infected weed seeds appear to play no comparable role in the survival and dissemination of *Xiphinema*-transmitted viruses. Nevertheless, infection remains in the soil for long periods. During limited periods of fallow, infection may persist in the long-lived adults of *X. index* (Taylor and Raski, 1964) but over long periods infection may be maintained in pieces of grapevine roots which survive in the soil after the crop has been removed. Not only do the root pieces act as a reservoir of infection; they also help to maintain populations of *X. index* in the absence of suitable host plants (Raski *et al.,* 1965a; Raski *et al.,* 1965b). Root pieces may also play a part in the survival of other *Xiphinema* vector species and the viruses they transmit.

In the final analysis, distribution and spread of the viruses depends on the presence of the specific nematode vectors. Except for *X. index,* which appears to thrive only on grapevine and fig, the host range is unlikely to be a limiting factor in the distribution of nematode vectors, although cultivations associated with annual cropping may adversely affect species having a low rate of reproduction. Hence, *X. diversicaudatum* and *X. americanum* are usually associated with woody perennial plants and clover/grass leys (Griffin and Darling, 1964; Morgan, 1964; Pitcher and Jha, 1961; Taylor *et al.,* 1966a) in conditions where they are undisturbed for long periods, and not with herbaceous crops, although they are able to feed on them (Harrison and Winslow, 1961; McGuire, 1964). Soil type, possibly by its effect on soil water level, limits the distribution of many vector species. *Xiphinema diversicaudatum* is associated with heavy, clay soils and more particularly with moist and shady sites in the vicinity of water or with medium soils with a high soil moisture level (Fritzsche, 1966). *Longidorus elongatus* is found in light to medium soils, *L. macrosoma* in clays and heavy loams, and *L. attenuatus* in light, sandy soils (Harrison, 1964b). Among *Trichodorus* species, *T. teres* is found in marine, sandy soils, whereas *T. pachydermus* is generally widespread, but usually in lighter rather than heavy soils (van Hoof, 1962). With the possible exception of the egg stage, none of the nematode vectors is known to have a resistant stage such as the cyst of *Heterodera rostochiensis* or the fourth-stage larva of *Ditylenchus dipsaci*. In *X. americanum* the egg stage may be a means of circumventing adverse conditions (Norton, 1963). There is also no evidence to indicate that the nematodes become dispersed in soil by adhering to implements or tractor wheels such as occurs with *H. rostochiensis;* indeed the spread of *Xiphinema* and *Longidorus* species has been shown to be a slow process. It has been estimated that the rate of spread of *X. diversicaudatum* under natural conditions is about 1 ft per year (Harrison and Winslow, 1961) and from evidence of reinfestation of healthy plots of strawberry *L. elongatus* also seems to move at about this rate (Murant and Taylor, 1965; Taylor and Murant, 1968). In the laboratory *X. americanum* was found to move laterally about 1 ft in 10 weeks, but virus infection spread at a slightly greater rate because the roots of healthy plants entered the area of infective nematodes more quickly than the nematodes moved to them (Bergeson *et al.*, 1964).

Most of the preferred hosts of *L. elongatus* are surface rooting and consequently the bulk of the populations are found in the top 8 in of soil (Taylor, 1967). *Xiphinema index* and *X. diversicaudatum* prefer woody hosts and the vertical distribution of the nematodes in the soil is related to the distribution of the young roots on which they feed. In the stony, shallow soils of the Cote d'Or vineyards, Vuittenez (1961) found the largest populations of *X. index* at 30–60 cm, and in three regions of Switzerland, Savary (1968) found *X. index, X. vuittenezi,* and *X. diversicaudatum* all within the 20–40 cm zone, where the strongly branched, turgid roots of the vines were growing. In the more fertile and deeper soils of the California Napa Valley, *X. index* has been found at 3.6 m depth, but in irrigated crops the nematodes were found to a depth of only 1.9 m, about the limit of root development (Hewitt *et al.* 1962; Raski *et al.* 1965a). *Xiphinema diversicaudatum* were usually found at a 15–20 cm depth under strawberry crops, with very few below 100 cm (Harrison and Winslow, 1961). Both the horizontal and vertical distribution of *Trichodorus* species in the soil may be more variable than that of *Xiphinema* and *Longidorus,* because many of them are capable of increasing rapidly if suitable host plants are present (Kuiper and Loof, 1962; Rohde and Jenkins, 1957a); this may give rise to relatively rapid fluctuations in population. At one site *T. primitivus* and *T. pachydermus* were found mainly in the 20–40 cm zone, with very few above and none below (Taconis and Kuiper, 1963), but elsewhere *T. pachydermus* was found at 50 cm and tobacco rattle virus was detectable in soil at 80–100 cm depths (Sol, 1963). In 13 out of 16 samples taken from different places in the Netherlands, *T. teres* was found below 20 cm in the spring, but in summer and autumn large populations built up in the upper layers of the soil (Hijink and Kuiper, 1966).

## VI. METHODS OF CONTROL

Much can be done to prevent virus and vectors reaching new sites. The soil should be removed from transplants at planting time and, if feasible, the roots of the transplants should be dipped in a suitable chemical to kill any vectors that may be present. Only virus-free planting material should be used and growers should be particularly careful not to take rootings or runner material from plants growing in or near infected sites, even if the material looks healthy.

Most viruses and vectors have such a wide host range that the rotation of crops would have to be highly selective to eliminate the chances of infection. *Xiphinema index* populations decrease during periods of fallow or when non-host crops are grown, but to ensure the decay of viable root pieces remaining after the removal of the grapevine crop, a minimum break of 5 years is advocated before replanting with grapevines (Hewitt *et al.,* 1962; Raski *et al.,* 1965a). With *L. elongatus,* fallowing or the use of non-host crops are difficult to implement as a means of control because of the importance of weeks as hosts for the nematode and the viruses it transmits. In a crop of Redgauntlet strawberry, the amount of virus infection decreased and the yields increased as a result of chemical control of weeds but the yield increments compared unfavorably with those attainable by nematicidal treatment of the soil (Taylor and Murant, 1968). Weed control may, however, effectively prevent the spread of virus infection in a susceptible crop when the crop itself—raspberry, for example—is a poor host for the vector (Taylor, 1967). *Longidorus elongatus* not only fails to

*Fig. 5.* (top and bottom). An outbreak of raspberry ringspot virus in Malling Jewel raspberry at Blairgowrie, Scotland.

reproduce on raspberry, but the populations rapidly decrease if the crop residues of canes and roots are ploughed into the soil, apparently because these release a nematicidal chemical which kills the nematode (Taylor and Murant, 1966b).

In long-term crops such as grapevine, stone fruits, and raspberry, the planting of disease-escaping or virus-immune crop varieties is an attractive solution to the problem of nematode-transmitted virus infection. Varietal resistance against grapevine fanleaf virus is known in grapevine (Petri, 1937) and tolerance to infection by tobacco rattle virus is known in potato (Noordam, 1956), as is immunity to infection against *Arabis* mosaic, tomato black ring, and raspberry ringspot viruses in raspberry (Cadman, 1961). There are, however, two dangers in the use of virus-resistant varieties. First, although the current crop remains immune to infection, the nematode vector populations may continue to multiply on the host crop and may increase the chances of infection in subsequent susceptible crops once a virus source has been reintroduced; moreover, the nematodes cause direct damage to crops, so that an increase in population is to be avoided. Second, variants of a virus that infect varieties previously immune may evolve; all raspberry varieties are apparently susceptible to the variant of raspberry ringspot virus found in Lloyd George raspberry (Murant *et al.*, 1968) and varieties that have proved immune from an English isolate of *Arabis* mosaic virus in graft tests became infected in Scottish outbreaks (Taylor *et al.*, 1966a).

Treatment of the soil with nematicidal chemicals has so far proved the most satisfactory way of controlling the spread of nematode-borne viruses. Among the earliest experiments are those of Ravez (1930) who reported good control of court noué (grapevine fanleaf virus) with carbon disulfide, formaldehyde, and heavy doses of lime! Later, Vuittenez (1957, 1960, 1961) obtained good control of *X. index* in French vineyards with D–D (dichloropropane–dichloropropene), carbon bisulfide, and methyl bromide. In California, *X. index* survived soil treatment with D–D, EDB (ethylene dibromide), chloropicrin, and methyl bromide, but this was mainly because some of the nematodes survived at depths below 4 ft where the chemicals did not penetrate (Hewitt *et al.*, 1962). In most instances, however, D–D and methyl bromide are generally effective in preventing virus infection to newly planted grapevines although the complete elimination of the vector is rarely accomplished (Boubals, 1965), Raski and Schmitt (1964) obtained significant increases in yields of Tokay grapevines by applicants of DBCP (dichlorobromopropane) metered into irrigation water; the effect of the treatment was apparent for up to

3 years in reduction of *X. index* populations, but the control of root knot nematode (*Meloidogyne* sp.) was effective for one year only.

In glasshouses, D–D applied at 800 lb per acre controlled *X. diversicaudatum* (Peachey and Brown, 1965), but in other experiments DBCP proved more effective than D–D in reducing galling on roses and decreased *X. diversicaudatum* populations by 60% at 3 weeks and 90% at 14 weeks from application (Peachey, Green, and Greet, 1965). In the field, the spread of *Arabis* mosaic virus in strawberry crops was prevented by applications of D–D or methyl bromide at 2 lb per 100 sq ft (Harrison, Peachey, and Winslow, 1963). Both chemicals were effective to a depth of 28 in and killed about 99% of *X. diversicaudatum;* dazomet, tetramethyl thiuram disulfide, methyl isothiocyanate, metham sodium, and DBCP were ineffective. Dichloropropane–dichloropropene and quintozene gave good control of *L. elongatus* and prevented the spread of raspberry ringspot and tomato black ring viruses in strawberry crops in eastern Scotland (Murant and Taylor, 1965). Quintozene is normally used as a fungicide for the control of *Rhizoctonia* but proved as effective as D–D in

*Fig. 6.* Strawberry var. Redgauntlet growing in experimental plots after the soil was treated with D-D (left center) and quintozene (left foreground), showing the improvement in growth compared with the untreated plot (right center) in which the plants are infected with raspberry ringspot virus.

controlling *L. elongatus*. Although it does not act as quickly as D–D in the soil, it persists and its nematicidal effects may be apparent for 6 years or more (Taylor and Murant, 1966a; Taylor and Murant, 1968).

The problem of nematode-borne viruses can, of course, be completely avoided by not planting susceptible crops where nematode vectors and virus are present. Soil samples can be used to detect the presence of the vector by sieving or other methods of extraction, and the extent of virus infection can be assayed by bait testing. If, however, virus-infected land has to be used, then the results of such tests can help to decide whether chemical treatment of the soil need be undertaken. Chemical soil sterilization rarely leads to the complete elimination of a nematode population, but for most of the virus–vector species the effect is long lasting due to their low rate of multiplication. An increasing knowledge of the ecology of the nematode vectors and their association with the viruses they transmit will help to refine the control measures currently used and may avoid, or at least reduce, the need for somewhat costly chemical treatments.

## BIBLIOGRAPHY

Alhassan, S. A., and Hollis, J. 1966. Parasitism of *Trichodorus christiei* on cotton seedlings. Phytopathology 56:573–574.

Allen, M. W. 1948. Relation of soil fumigation, nematodes and inoculation techniques to big vein disease of lettuce. Phytopathology 38:612–627.

———. 1957. A review of the nematode genus *Trichodorus* with descriptions of ten new species. Nematologica 2:32–62.

Amici, A. 1965. Ulteriori ricerche sulla diffusione di *Xiphinema index* e sulla presenza di altri nematodi nei vigneti italiani. Riv. Pat. Veg. Ser. IV, 1:109–128.

Amici, A., Baldacci, E., Belli, G., Betto, E., Raski, D. J., and Refatti, E. 1964. Ricerche con *Xiphinema index* Thorne & Allen isolate in vigneti italiani, quale vettore di viras delta vite. Riv. Pat. Pavia Ser. 3, 4:1–11.

Athow, K. L., and Bancroft, J. B., 1959. Development and transmission of tobacco ringspot virus in soybean. Phytopathology 49:697–701.

Ayala, A., and Allen, M. W. 1966. Transmission of the California tobacco rattle virus by three species of the nematode genus *Trichodorus*. Nematologica 12:87. (Abstr.)

Behrens, J. 1899. Weiter Beiträge zur Kenntnis der Tabakpflanze. XIV. Die Mauche (Mauke) des Tabaks. Landwirtsch. Vers. Sta. 52:442–447.

Beijerinck, M. W. 1898. Über ein Contagium vivum fluidum als Ursache der Flecken-Krankheit der Tabaksblätter. Verhandl. Kon. Akad. Wetenschap. Amsterdam II. 6:1–22.

Bergeson, G. B., and Athow, K. L. 1963. Vector relationship of tobacco ringspot virus and *Xiphinema americanum* and the importance of this vector in TRSV infection of soybean. Phytopathology 53:871. (Abstr.)

Bergeson, G. B., Athow, K. L., Laviolette, F. A., and Thomasine, Sister Mary. 1964. Transmission, movement and vector relationships of tobacco ringspot virus in soybean. Phytopathology 54:723–728.

Blumer, S. 1954, 1955. Viruskrankheiten an Obstbäumen. Schweiz Z. Obst.-Weinbau, 63, 64:516–519, 525–529, 2–11.

Bock, K. R. 1966. *Arabis* mosaic and *Prunus* necrotic ringspot viruses in hop (*Humulus lupulus* L.). Ann. Appl. Biol. 57:131–140.

Bos, L., and van der Want, J. P. H. 1962. Early browning of pea, a disease caused by a soil- and seed-borne virus. Tijdschr. Plantenziekten 68:368–390.

Boubals, D. 1965. Résultats de traitements du sol contre les nématodes vecteurs de la dégénérescence inféctieuse, virose de la vigne. Compt. Rend. 8th Symp. Int. Nématol., Antibes 1965, p. 112–113.

Breece, J. R., and Hart, W. H. 1959. A possible association of nematodes with the spread of peach yellow bud mosaic virus. Plant Disease Reporter 43:989–990.

Brunt, A. A. 1966. The occurrence of cucumber mosaic virus and four nematode-transmitted viruses in British narcissus crops. Plant Pathol. 15:157–160.

Cadman, C. H. 1956. Studies on the etiology and mode of spread of Scottish raspberry leaf curl disease. J. Hort. Sci. 31:111–118.

——— 1959. Some properties of an inhibitor of virus infection from leaves of raspberry. J. Gen. Microbiol. 20:113–128.

——— 1960. Studies on the relationships between soil-borne viruses of the ringspot type occurring in Britain and continental Europe. Virology 11:653–664.

——— 1961. Raspberry viruses and virus diseases in Britain. Hort. Res. 1:47–61.

——— 1962. Evidence for the association of tobacco rattle viruses with a cell component. Nature 193:49–52.

——— 1963. Biology of soil-borne viruses. Ann. Rev. Phytopathol. 1:143–172.

Cadman, C. H., Dias, H. F., and Harrison, B. D. 1960. Sap-transmissible viruses associated with diseases of grapevine in Europe and North America. Nature 187:577–579.

Cadman, C. H., and Harris, R. V. 1952. Leaf curl, a virus disease of raspberries in Scotland. J. Hort. Sci. 27:201–211.

Cadman, C. H., and Harrison, B. D., 1959. Studies on the properties of soil-borne viruses of the tobacco rattle type occurring in Scotland. Ann. Appl. Biol. 47:542–556.

——— 1960. Studies on the behaviour in soils of tomato black ring, raspberry ringspot and arabis mosaic viruses. Virology 10:1–20.

Cadman, C. H., Harrison, B. D., and Lister, R. M. 1958. And now—soil-borne viruses! Comml. Grow. 14, Feb. 1958.

Cadman, C. H., and Lister, R. M. 1961. Relationship between tomato ringspot and peach yellow bud mosaic viruses. Phytopathology 51:29–31.

Chen, T. A., and Mai, W. F. 1965. The feeding of *Trichodorus christiei* on individually isolated corn root cells. Phytopathology 55:128. (Abstr.)

Christie, J. R. 1952. Some new nematode species of critical importance to Florida growers. Proc. Soil Sci. Soc. Florida 12:30–39.

——— 1959. Plant Nematodes. Their bionomics and control. Agr. Exp. Sta. Univ. Florida, Gainesville, 1959. 256 p.

Christie, J. R., and Perry, V. G. 1951. A root disease of plants caused by a nematode of the genus *Trichodorus*. Science 113:491–493.

Cremer, M. G., and Kooistra, G. 1964. Investigations on notched leaf ("kartelblad") of *Gladiolus* and its relation to rattle virus. Nematologica 10:69–70. (Abstr.)

Cropley, R. 1951a. Cherry leaf-roll virus. Ann. Appl. Biol. 49:524–529.

——— 1961b. Viruses causing rasp-leaf and similar diseases of sweet cherry. Ann. Appl. Biol. 49:530–534.

Crowley, N. C. 1959. Studies on the time of embryo infection by seed-transmitted viruses. Virology 8:116–123.

Dalmasso, A., and Caubel, G. 1966. Répartition des espèces des genres *Xiphinema* et *Longidorus* trouvées en France. Paris Acad. Agr. France Compt. Rend. Hebd. Séances 52:440–446.

Davis, R. A., and Jenkins, W. R. 1960. Nematodes associated with roses and the root injury caused by *Meloidogyne hapla* Chitwood 1949, *Xiphinema diversicaudatum* (Micoletzky, 1927) Thorne, 1939, and *Helicotylenchus nannus* Steiner, 1945. Bull. Maryland Agr. Exp. Sta. No. A–106. 16 pp.

Debrot, E. A. 1964. Studies on a strain of raspberry ringspot virus occurring in England. Ann. Appl. Biol. 54:183–191.

Dias, H. F. 1963. Host range and properties of grapevine fanleaf and grapevine yellow mosaic viruses. Ann. Appl. Biol. 51:85–95.

Dias, H. F., and Harrison, B. D. 1963. The relationship between grapevine fanleaf, grapevine yellow mosaic and arabis mosaic viruses. Ann. Appl. Biol. 51:97–105.

Diener, T. O., and Weaver, M. L. 1959. A caffeine additive to aid mechanical transmission of necrotic ring spot virus from fruit trees to cucumber. Phytopathology 49:321–322.

Duffield, C. A. W. 1925. Nettlehead in hops. Ann. Appl. Biol. 12:536–543.

Epstein, A. H., and Barker, K. R. 1966. Pathogenicity of *Xiphinema* americanum on *Vinca minor*. Plant Disease Reporter 50:420–422.

Flegg, J. J. M. 1965. The plant-pathogen relationship (Blackman Essay). Rep. East Malling Res. Sta., England, 1964, p. 62–70.

———. 1966. Once-yearly reproduction in *Xiphinema vuittenezi*. Nature 212:741.

Fritzsche, R. 1964. Untersuchungen über dei Virusübertragung durch Nematoden. *Wiss. Z. Univ. Rostock* 13:343–347.

———. 1966. Beitrag zur Ökologie von *Xiphinema diversicaudatum* (Micoletzky) Thorne. Nachrbl. Deut. Pflanzenschutzdienst 20:8–11.

Fritzsche, R., and Kegler, H. 1964. Die Ubertragung des Blattrolvirus der Kirsche (cherry leaf-roll virus) durch Nematoden. Inst. Phytopathol. Acad. Landwirt. Berlin Buchbesprechungen 12:299.

Fritzsche, R., and Schmidt, H. B. 1963. *Xiphinema paraelongatum* Altherr und *Xiphinema* n. sp. zwei Vektoren des Arabis-Mosaik virus. Naturwissenschaften 50:163.

Frost, R. R., Harrison, B. D., and Woods, R. D. 1967. Apparent symbiotic interaction between particles of tobacco rattle virus. J. Gen. Virol. 1:57–70.

Fulton, J. R. 1962. Transmission of tobacco ringspot virus by *Xiphinema americanum*. Phytopathology 52:375.

Gibbs, A. J., and Harrison, B. D. 1963. Plant Pathology Department. Rept. Rothamsted Exp. Sta. Harpenden, England, 1962, p. 113.

———, 1964a. Nematode-transmitted viruses in sugar beet in East Anglia. Plant Pathol. 13:144–150.

———. 1964b. A form of pea early-browning virus found in Great Britain. Ann. Appl. Biol. 54:1–11.

Gooding, G. V. 1963. Purification and serology of a virus associated with the grape yellow vein disease. Phytopathology 53:475–480.

Gooding, G. V., and Hewitt, W. B. 1962. Grape yellow vein symptomatology, identification, and the association of a mechanically transmissible virus with the disease. Amer. J. Enol. Viticulture 13:196–203.

Griffin, G. D., and Darling, H. M. 1964. An ecological study of *Xiphinema americanum* Cobb in an ornamental spruce nursery. Nematologica 10:471–479

Griffin, G. D., and Epstein, A. H. 1964. Association of dagger nematode, *Xiphinema americanum* with stunting and winter kill of ornamental spruce. Phytopathology 54:177–180.

Griffin, G. D., Hoguelet, J. E., and Nelson, J. W. 1963. *Xiphinema americanum* as a vector of necrotic ringspot virus of blueberry. Plant Disease Reporter 47:703–704.

Harrison, B. D. 1957. Studies on the host range, properties and mode of transmission of beet ringspot virus. Ann. Appl. Biol. 45:462–472.

———. 1958. Raspberry yellow dwarf, a soil-borne virus. Ann. Appl. Biol. 46: 221–229.

———. 1960. The biology of soil-borne plant viruses. *Advances Virus Res.* 7: 131–161.

———. 1951a. Identity of red currant spoon leaf virus. Tijdschr. Plantenziekten 67:562–565.

———. 1961b. Plant Pathology Department, Rep. Rothamsted Exp. Sta. Harpenden, England, 1960, p. 118.

———. 1962. Plant Pathology Department, Rep. Rothamsted Exp. Sta. Harpenden, England, 1961, p. 105.

———. 1964a. The transmission of plant viruses in soil. *In* M. K. Corbett and H. D. Sisler (ed.) *Plant virology*. Univ. Florida Press, 1964. 118–117.

———. 1964b. Specific nematode vectors for serologically distinctive forms of raspberry ringspot and tomato black ring viruses. Virology 22:544–550.

———. 1967. The transmission of strawberry latent ringspot virus by *Xiphinema diversicaudatum* (Nematoda). Ann. Appl. Biol. 60:405–409.

Harrison, B. D., and Cadman, C. H. 1959. Role of a dagger nematode (*Xiphinema* sp.) in outbreaks of plant diseases caused by *Arabis* mosaic virus. Nature 184:1624–1626.

Harrison, B. D., Mowat, W. P., and Taylor, C. E. 1961. Transmission of a strain of tomato black ring virus by *Longidorus elongatus* (Nematoda). Virology 14:480–485.

Harrison, B. D., and Nixon, H. L. 1959. Separation and properties of particles of tobacco rattle virus with different lengths. J. Gen. Microbiol. 21:569–581.

Harrison, B. D., Peachey, J. E., and Winslow, R. D. 1963. The use of nematicides to control the spread of *Arabis* mosaic virus by *Xiphinema diversicaudatum* (Micol.) Ann. Appl. Biol. 52:243–255.

Harrison, B. D., and Winslow, R. D. 1961. Laboratory and field studies on the relation of *Arabis* mosaic virus to its nematode vectors *Xiphinema diversicaudatum* (Micoletzky). Ann. Appl. Biol. 49:621–633.

Harrison, B. D., and Woods, R. D. 1966. Serotypes and particle dimensions of tobacco rattle viruses from Europe and America. Virology 28:610–620.

Hewitt, W. B. 1950. Fanleaf—another vine disease found in California. California State Dep. Agr. Bull. 39:62–63.

Hewitt, W. B., Goheen, A. C., Raski, D. J., and Gooding, G. W. 1962. Studies on virus diseases of the grapevine in California. Vitis 3:57–83.

Hewitt, W. B., and Raski, D. J. 1962. Nematode vectors of plants viruses. *In* K. Maramorosch (ed.) *Biological transmission of disease agents,* Academic Press, New York, 1962. p. 63–72.

Hewitt, W. B., Raski, D. J., and Goheen, A. C. 1958. Nematode vector of soil-borne fanleaf virus of grapevines. Phytopathology 48:586–595.

Hijink, M. J., and Kuiper, K. 1966. Waarnemingen over de Verdeling van Aaltjes in de Grond. Mededel. Rijksfaculteit Landbouwwetensschappen Gent 31:558–571.

Hoff, J. K., and Mai, W. F. 1962. Pathogenicity of the stubby root nematode to onion. Plant Disease Reporter 46:24–25.

Hollings, M. 1965. Anemone necrosis, a disease caused by a strain of tobacco ringspot virus. Ann. Appl. Biol. 55:447–457.

Hoof, H. A. van. 1962. *Trichodorus pachydermus* and *T. teres,* vectors of the early browning virus of peas. Tijdschr. Plantenziekten 68:391–396.

———. 1964. *Trichodorus teres* a vector of rattle virus browning virus. Netherlands J. Plant Pathol. 70:187.

Hoof, H. A. van, Maat, D. Z., and Seinhorst, J. W. 1966. Viruses of the tobacco rattle virus group in northern Italy: their vectors and serological relationships. Netherlands J. Plant Pathol. 72:253–258.

Hooper, D. J. 1961. A redescription of *Longidorus elongatus* (De Man, 1876) Thorne and Swanger, 1936 (Nematoda, Dorylaimidae) and descriptions of five new species of *Longidorus* from Great Britain. Nematologica 6:237–257.

———. 1962. Three new species of *Trichodorus* (Nematoda: Dorylaimoidea) and observations on *T. minor* Colbran, 1956. Nematologica 7:273–280.

———. 1963. *Trichodorus viruliferus* n.sp. (Nematoda: Dorylaimida). Nematologica 9:200–204.

Horner, C. E., and Jensen, H. J. 1954. Nematodes associated with mints in Oregon. Plant Disease Reporter. 38:39–41.

Jensen, H. J. 1961. Nematodes affecting Oregon agriculture. Bull. Agr. Exp. Sta. Oregon State Univ., Corvallis, 579, 34 p.

Jensen, H. J., and Allen, T. C. 1964. Transmission of tobacco rattle virus by a stubby-root nematode, *Trichodorus allius.* Plant Disease Reporter. 48:333–334.

Jha, A., and Posnette, A. F. 1959. Transmission of a virus to strawberry plants by a nematode. Nature 184:962–963.

———. 1961. Transmission of *Arabis* mosaic virus by the nematode *Xiphinema diversicaudatum* (Micol.) Virology 13:119–123.

Johnson, F. 1945. The effect of chemical soil treatments on the development of wheat mosaic. Ohio J. Sci. 45:125–128.

Kuiper, K., and Loof, P. A. A. 1962. *Trichodorus flavensis* n. sp. (Nematoda: Enoplida), a plant nematode from new polder soil. Verslag Plantenziekten Dienst Wageningen 136:193–200.

Lamberti, F., and Raski, D. J. 1964. Transmissione di 'malformazione infettiva' della vite attraverso terreno da *Xiphinema index* Thorne and Allen, in Apulia. Phytopathol. Mediterranea 3:41–43.

Latta, R. 1966. Attempt to relate nematodes to lethal yellowing disease of coconut palms in Jamaica. Trop. Agr. (Trinidad) 43:59–68.

Lima, M. B. 1968. A numerical approach to the *Xiphinema americanum* complex. Compt. Rend. 8th Symp. Int. Nématol., Antibes 1965, p. 30.

Lister, R. M. 1962. Nematode-borne viruses. Scottish Hort. Res. Inst. Rep. Dundee, 1961–1962, p. 66.

———. 1964. Strawberry latent ringspot: a new nematode-borne virus. Ann. Appl. Biol. 54:167–176.

———. 1966. Nematode-borne viruses. Scottish Hort. Res. Inst. Rep. Dundee, 1964 and 1965, p. 45–46.

Lister, R. M., and Murant, A. F. 1961. Soil-borne viruses. Scottish Hort. Res. Inst. Rep., Dundee, 1960–1961, p. 56.

———. 1962. Nematode-borne viruses. Scottish Hort. Res. Inst. Rep., Dundee, 1961–1962, p. 67–68.

———. 1967. Seed-transmission of nematode-borne viruses. Ann. Appl. Biol. 59: 49–62.

Lister, R. M., Raniere, L. C., and Varney, E. A. 1963. Relationships of viruses associated with ringspot disease of blueberry. Phytopathology 53:1031–1035.

Lownsberry, B. F., and Maggenti, A. R. 1963. Some effects of soil temperature and soil moisture on population levels of *Xiphinema americanum*. Phytopathology 53:667–668.

Maat, D. Z. 1965. Serological differences between red currant spoonleaf virus isolated from Eckelrade-diseased cherry trees and the Scottish raspberry ringspot virus. Netherlands J. Plant Pathol. 71: 47-53.

Maat, D. Z., van der Meer, F. A., and Pfaeltzer, H. J. 1962. Serological identification of some soil-borne viruses causing diseases in fruit crops in the Netherlands. Tijdschr. Plantenziekten 68:120–122.

Martelli, G. P., and Raski, D. J. 1963. Osservazioni su *X. index* (Thorne and Allen). Informatore Fitopatologico 13:416–420.

McGuire, J. M. 1964. Efficiency of *Xiphinema americanum* as a vector of tobacco ringspot virus. Phytopathology 54:799–801.

McKinney, H. H., Webb, R. W., and Dungan, G. H. 1957. Studies on chemical control, and overseasoning of, and natural inoculation with, the soil-borne viruses of wheat and oats. Plant Disease Reporter 41:256–66.

McLean, D. M. 1962. Differentiation of tobacco ringspot and tomato ringspot viruses by experimental host reactions. Plant Disease Reporter 46:877–881.

Meer, F. A. van der. 1965. Investigations of currant viruses in the Netherlands. II. Further observations on spoonleaf virus, a soil-borne virus transmitted by the nematode *Longidorus elongatus*. Netherlands J. Plant Pathol. 71: 33–46.

Milbrath, J. A., and Reynolds, J. E. 1961. Tomato ringspot virus isolated from *Eola* rasp leaf of cherry in Oregon. Plant Disease Reporter 45:520–521.

Morgan, H. G. 1964. *Xiphinema diversicaudatum* and *Arabis* mosaic virus in south-west England. Nematologica 10:70. (Abstr.)

Mowat, W. P., and Taylor, C. E. 1962. Nematode-borne viruses. Scottish Hort. Res. Inst. Rep., Dundee, 1961–1962, p. 68.

Murant, A. F., and Lister, R. M. 1967. Seed-transmission in the ecology of nematode-borne viruses. Ann. Appl. Biol. 59:63–70.

Murant, A. F., and Taylor, C. E. 1965. Treatment of soil with chemicals to prevent transmission of tomato black ring and raspberry ringspot viruses by *Longidorus elongatus* (de Man). Ann. Appl. Biol. 55:227–237.

Murant, A. F., Taylor, C. E., and Chambers, J. 1968. Properties, relationships and transmission of a strain of raspberry ringspot virus infecting raspberry cultivars immune to the common Scottish strain. Am. Appl. Biol. 61:175–186.

Noordam, D. 1956. Waardplanten en toetsplanten van het ratelvirus van de tabak. Tijdschr. Plantenziekten 62:219–225.

Norton, D. C. 1963. Population fluctuations of *Xiphinema americanum* in Iowa. Phytopathology 53:66–68.

Osborne, P. 1964. *Xiphinema diversicaudatum* (Micoletzky) Thorne in rose beds. Plant Pathol. 13:184.

Oswald, J. W., and Bowman, T. 1958. Studies on a soil-borne potato virus disease in California. Phytopathology 48:396.

Peachey, J. E., and Brown, E. B. 1965. Ridding glasshouse soil of dagger nematode before planting with rose. Exp. Hort. 13:45–48.

Peachey, J. E., Green, C. G., and Greet, D. N. 1965. Effects of D-D and Nemagon on eight-year-old rose bushes infested with dagger nematodes. Exp. Hort. 13:49–50.

Percival, J. 1895. The eelworm disease of hops. 'Nettle-headed' or 'skinkly' plants. J. South East Agr. College, Wye 1:5–9.

Perry, V. G. 1958. Parasitism of two species of dagger nematodes (*Xiphinema americanum* and *X. chambersi*) to strawberry. Phytopathology 48:420–423.

Petri, L. 1937. Rassegna dei casi fitopatologici osservati nel 1936. Boll. Staz. Patol. Veg. Roma (NS) 17:1–78.

Pitcher, R. S., and Flegg, J. J. M. 1965. Observations of root feeding by the nematode *Trichodorus viruliferus*. Nature 207:317.

Pitcher, R. S., and Jha, A. 1961. On the distribution and infectivity with *Arabis* mosaic virus of a dagger nematode. Plant Pathol. 10:67–71.

Pitcher, R. S., and Posnette, A. F. 1963. Vascular feeding by *Xiphinema diversicaudatum*. Nematologica 9:301.

Radewald, J. D. 1962. The biology of *Xiphinema index* and the pathological effect of the species on grape. Ph.D. thesis, University of California, Davis.

Radewald, J. D., and Raski, D. J. 1962a. A study of the life cycle of *Xiphinema index*. Phytopathology 52:748. (Abstr.)

———. 1962b. Studies on the host range and pathogenicity of *Xiphinema index*. Phytopathology 52:748. (Abstr.)

Raski, D. J., and Amici, A. 1964. Richerche sulla diffusione di *X. index* Thorne and Allen et sulla presenza di altri nematode fitoparasite nei vegreti italiani. Riv. Patol. Veg. Ser. III. 4:41–78.

Raski, D. J., and Hewitt, W. B. 1960. Experiments with *Xiphinema index* as a vector of fanleaf of grapevines. Nematologica 5:166–170.

———. 1963. Plant parasitic nematodes as vectors of plant viruses. Phytopathology 53:39–47.

Raski, D. J., Hewitt, W. B., Goheen, A. C., Taylor, C. E., and Taylor, R. H. 1965a. Survival of *Xiphinema index* and reservoirs of fanleaf virus in fallowed vineyard soil. Nematologica 11:349–352.

———. 1965b. Survival of *Xiphinema index* and fanleaf virus in fallowed vineyard soil. Nematologica 11:44–45. (Abstr.)

Raski, D. J., and Lider, L. 1959. Nematodes in grape production. California Agr. 13:13–15.

Raski, D. J., and Radewald, J. D. 1958. Reproduction and symptomatology of certain ectoparasitic nematodes on roots of Thompson seedless grapes. Plant Disease Reporter 42:941–943.

Raski, D. J., and Schmitt, R. V. 1964. Grapevine responses to chemical control of nematodes. Amer. J. Enol. Viticulture 15:199–203.

Ravez, L. 1930. Sur le court-noué. Ann. Ecole Nat. Agr. Montpellier NS 20: 110–125.

Roggen, D. R. 1966. On the morphology of *Xiphinema index* reared on grape fanleaf virus infected grapes. Nematologica 12:287–296.

Rohde, R. A., and Jenkins, W. R. 1957a. Host range of a species of *Trichodorus* and its host–parasite relationships on tomato. Phytopathology 47:295–298.

Rohde, R. A., and Jenkins, W. R. 1957b. Effect of temperature on the life cycle of stubby-root nematodes. Phytopathology 47:29. (Abstr.)

Sänger, H. L. 1961. Untersuchungen über schwer übertragbare Formen des Rattle-virus. Proc. conf. Potato virus diseases, 4th, Brunswick, Braunschweig, Germany, 1960, p. 22–28.

Sänger, H. L., and Brandenburg. E. 1961. Über die Gewinnung von infektiösen Pressaft aus "Wintertyp" Pflanzen das Tabak-Rattle-virus durch Phenol extraktion. Naturwissenschaften 48:391.

Sänger, H. L., Allen, M. W., and Gold, A. H. 1962. Direct recovery of tobacco rattle virus from its nematode vector. Phytopathology 52:750. (Abstr.)

Sauer, N. I. 1966. Simultaneous association of strains of tobacco ringspot virus within *Xiphinema americanum*. Phytopathology 56:862–863.

Savary, A. 1968. Les nématodes vecteurs du virus de la dégénérescence inféctieuse dans les vignobles du Bassin Lemanique. Compt. Rend. 8th Symph. Int. Nématol, Antibes 1965, p. 108.

Schindler, A. F. 1957. Parasitism and pathogenicity of *Xiphinema diversicaudatum* on ectoparasitic nematode. Nematologica 2:25–31.

——— 1958. Attempts to demonstrate the transmission of plant viruses by plant parasitic nematodes. Plant Disease Reporter 42:1348–1350.

Schindler, A. F., and Braun, A. J. 1957. Pathogenicity of an ectoparasitic nematode *Xiphinema diversicaudatum,* on strawberries. Nematologica II Suppl.: 91–93.

Seinhorst, J. W. 1963. A redescription of the male of *Trichodorus primitivus* (de Man) and the description of a new species, *T. similis*. Nematologica 9:125–130.

——— 1966. *Longidorus elongatus* on *Fragaria vesca*. Nematologica 12:275–279.

Sharma, R. D. 1965. Direct damage to strawberry by *Longidorus elongatus* (de Man, 1876) Thorne and Swanger, 1936. Mededel. Langbhogesch. OpzoekStns. Gent 30:1437–1443.

Smith, K. M., and Markham, R. 1944. Two new viruses affecting tobacco and other plants. Phytopathology 34:324–329.

Sol. H. H. 1963. Some data on the occurrence of rattle virus at various depths in the soil and on its transmission. Netherlands J. Plant Pathol. 69:208–214.

Sol. H. H., van Heuven, J. C., and Seinhorst, J. W. 1960. Transmission of rattle virus and *Atropa belladonna* mosaic virus by nematodes. Tijdschift. Plantenziekten 66:228–231.

Sol. H. H., and Seinhorst, J. W. 1961. The transmission of rattle virus by *Trichodorus pachydermus*. Tijdschrift. Plantenziekten 67:307–311.

Stace-Smith, R. 1962. Studies on *Rubus* virus diseases in British Columbia. IX. Ringspot disease of red raspberry. Can. J. Bot. 40:905–912.

Stone, W. J., Mink, G. I., and Bergeson, G. B. 1962. A new disease of American spearmint caused by tobacco ringspot virus. Plant Disease Reporter 46:623–624.

Sturhan, D. 1966. Vorkommen virusübertragender Nematoden in Bayern. Gesunde Pflanzen 18:93–96.

Taconis, P. J., and Kuiper, K. 1963. Overdracht van het Nicotiana virus 5 door aaltjes van het geslacht *Trichodorus* in zaailingen van 5 geswassen. Verslag Plantenziekten Dienst. 141:177–178.

Tarjan, A. C. 1968. Variability in *Xiphinema americanum*. Compt. Rend. 8th Symp. Int. Nématol, Antibes 1965, p. 29.

Taylor, C. E. 1962a. Transmission of raspberry ringspot virus by *Longidorus elongatus* (de Man) (Nematoda: Dorylaimoidea). Virology 17:493–494.

——— 1962b. Nematode-borne viruses. Scottish Hort. Res. Inst. Rep., Dundee, 1961–1962, p. 67.

———. 1964. Nematode-borne viruses. Scottish Hort. Res. Inst. Rep. Dundee, 1963–1964. p. 65.

——— 1967. The multiplication of *Longidorus elongatus* (de Man) on different host plants with reference to virus transmission. Ann. Appl. Biol. 59:275–281.

Taylor, C. E., and Murant, A. F. 1966a. The use of quintozene (PCNB) as a nematicide. Proc. Brit. Insect. Fungicide Conf., 3rd, 1965, p. 514–520.

——— 1966b. Nematicidal activity of aqueous extracts from raspberry cane and roots. Nematologica 12:488–494.

——— 1968. Chemical control of raspberry ringspots and tomato black ring viruses in strawberry, Plant Pathol. 17 (in press).

Taylor, C. E., and Thomas, P. R. 1968. The association of *Xiphinema diversicaudatum* (Micoletzky) with strawberry latent ringspot and *Arabis* mosaic viruses in a raspberry plantation. Ann. Appl. Biol. 62 (in press).

Taylor, C. E., and Raski, D. J. 1964. On transmission of grape fanleaf by *Xiphinema index*. Nematologica 10:489–495.

Taylor, C. E., Thomas, P. R., Cathro, J., and Roberts, I. A. 1966b. Nematode-borne viruses. Scottish Hort. Res. Inst. Rep., Dundee, 1964 and 1965, p. 47.

Taylor, C. E., Thomas, P. R., and Converse, R. H. 1966a. An outbreak of *Arabis* mosaic virus and *Xiphinema diversicaudatum* (Micoletzky) in Scotland. Plant Pathol. 15:170–174.

Teliz, D., Grogan, R. G., and Lownsbery, B. F. 1966. Transmission of tomato ringspot, peach yellow bud mosaic and grape yellow vein viruses by *Xiphinema americanum*. Phytopathology 56:658–663.

Vuittenez, A. 1957. Lutte préventive contre le court-noué de la vigne par la désinfection chimique du sol avant la plantation. Compt. Rend. Acad. Agr. France 43:185–196.

——— 1960. Nouvelles observations sur l'activité des treatments chimique du sol pour l'éridication des virus de la dégénérescence infectieuse de la vigne. Compt. Rend. Acad. Agr. France 46:89–99.

——— 1961. Les nématodes vecteurs de virus et le probleme de la dégénérescence inféctieuse de la vigne. "Les Nématodes". C.N.R.A. Versailles, 1961, p. 55–78.

Vuittenez, A., and Legin, R. 1964. Confirmation de la transmission du virus de la dégénérescence inféctieuse par *Xiphinema index* (Thorne et Allen) et de l'activité parasitaire de ce nématode sur la vigne. Etudes Virol. Appl. 5:59–62.

Wagnon, H. K., and Breece, J. R. 1955. Evidence of retention of peach yellow bud mosaic virus in soil. Phytopathology 45:696. (Abstr.)

Walkinshaw, C. H., Griffen, G. D., and Larson, R. H. 1961. *Trichodorus christiei* as a vector of potato corky ringspot (tobacco rattle) virus. Phytopathology 51:806–808.

Weiner, A., and Raski, D. J. 1961. New host records for *Xiphinema index* Thorne and Allen, 1959. Plant Disease Reporter 50:27–28.

Weischer, B. 1966. Contribution on the geographical distribution and biology of species of the genera *Xiphinema* and *Longidorus*. Mitt. Biol. Bundesanstalt Landwirt. Forstwirt. Berlin-Dahlem 118:101–106.

Whitehead, A. G. 1965. Nematodes associated with "Docking Disorders" of sugar beet. Brit. Sugar Beet Rev. 34:77–78, 83–84, 92.

Yassin, A. M. 1966a. Nematode-borne viruses. Scottish Hort. Res. Inst. Rep., Dundee, 1964 and 1965, p. 46.

——— 1966b. The biology of *Longidorus elongatus* de Man in relation to the transmission of tomato black ring virus. M.Sc. Thesis, University of St. Andrews.

Zuckerman, B. M. 1961. Parasitism and pathogenesis of the cultivated cranberry by some nematodes. Nematologica 6:135–143.

——— 1962. Parasitism and pathogenesis of the cultivated highbush blueberry by the stubby root nematode. Phytopathology 52:1017–1019.

# White Flies As Virus Vectors

A. S. Costa

*Instituto Agronômico, Campinas, Brazil*

## I. INTRODUCTION

Whitefly-transmitted viruses infect plants from many families, including representatives of Compositae, Convolvulaceae, Cruciferae, Cucurbitaceae, Euphorbiaceae, Geraniaceae, Labiatae, Leguminosae, Linaceae, Malvaceae, Pedaliaceae, Solanaceae, Scrophulariaceae, Tiliaceae, Urticaceae, Verbenaceae, and others. A review of 28 diseases of this group was made by Varma (1963). The relationship between them has not always been well established and it is possible that in some cases they represent diseases from different species caused by the same virus or by variants of the same virus complex that became better adapted to certain host plants.

Diseases transmitted by whiteflies are encountered mostly in tropical areas, but occur also under subtropical and temperate conditions such as in Israel, and California, Florida, Georgia and Maryland in the United States. For *Abutilon* infectious variegation,* Silberschmidt and Tommasi (1955) consider that the 30° parallel north and south of the equator limits the area in the American continent where the disease occurs.

Although they are not as important vectors as the aphids, whiteflies are responsible for the natural spread of diseases of economic importance on various crops: cotton leaf curl in Sudan and Nigeria (Kirkpatrick, 1931; Tarr, 1964); tobacco leaf curl in India (Pal and Tandon, 1937), Africa (McClean, 1940), and Java (Thung, 1932); cassava mosaic in various parts of Africa (Storey and Nichols, 1938), and others.

* This disease and the causal virus will henceforth be represented by the initials AIV and AIVV.

## II. TYPES OF VIRUS DISEASES TRANSMITTED BY WHITEFLIES

Virus diseases transmitted by whiteflies may be placed in two main groups:

(*1*) Those recognised by leaf symptoms such as clearing or yellowing of the veins, yellow net, or leaf crinkling and curling resulting from unequal growth of normal and mosaic areas. Mosaic symptoms are of the green or more frequently of the yellow type and may be assumed to be associated with parenchyma tissue.

(*2*) The curl type of disease in which mosaic symptoms are not evident, but invaded plants may show stunting, leaves with diffuse yellowing, crinkling, vein clearing or vein thickening, and leaf enations. In many diseases of the latter type the virus may be restricted to the phloem tissues.

Some diseases might induce symptoms of one type on a host plant and of the other type on another. It would be of interest to determine whether the occurrence of the virus in parenchyma tissues or in the phloem would parallel the symptomatology.

*Abutilon* infectious variegation from Brazil and some of the diseases that affect Malvaceae and Leguminosae in India are good examples of the first or mosaic group. Tobacco leaf curl from Africa (McClean, 1940) and India (Pal and Tandon, 1937) and cotton leaf curl (Kirkpatrick, 1931) are representative of the second group. Tobacco leaf curl as described from Venezuela (Wolf *et al.*, 1949) Puerto Rico (Bird, 1958), and Brazil (Costa and Carvalho, 1960b) are associated with mosaic symptoms and may differ from the disease present in India and Africa.

## III. PROPERTIES OF WHITEFLY-TRANSMITTED VIRUSES

### *A. Particle Morphology*

Whitefly-transmitted viruses have received little attention as to particle morphology. Tobacco leaf-curl virus in Venezuela was first reported as rod shaped (Sharp and Wolf, 1949), but later shown to be a soft, spherical particle of approximately 39 m$\mu$ in diameter (Sharp and Wolf, 1951).

In electronmicroscope studies of ultrathin sections of *Abutilon striatum* var. *thompsonii* leaves infected with AIVV, Sun (1964) described the presence of spheroidal particles about 80 m$\mu$ in diameter in the cytoplasm. These were considered to be the virus and consisted of an inner core about 16 m$\mu$ in diameter, surrounded by an outer envelope.

Several whitefly-transmitted viruses that occur in Brazil, such as AIVV in species of *Sida, Euphorbia* mosaic virus in *E. prunifolia* and *Datura stramonium,* bean golden mosaic virus, and the tomato leaf-curl virus from the Pesqueira (Pe.) area (Costa, unpublished) have been examined by Kitajima (unpublished) in dip preparations. No elongated or large spheroidal particle was found associated with them. In studies of ultrathin sections of lima bean (*Phaseolus lunatus*) infected with the golden mosaic virus* and *Sida rhombifolia, Abutilon striatum,* and cotton plants infected with AIVV from *Sida,* Kitajima (unpublished) noticed the presence of large spheroidal particles (*ca.* 100 mμ) in the cytoplasm of degenerating infected cells, but did not consider them to be the virus.

It is not definitely established whether cassava brown streak virus is transmitted by whiteflies. The morphology of this virus was studied by Kitajima and Costa (1964) on drief leaf samples received from Africa. Dip preparations showed the presence of elongated virus-like particles extracted from leaf tissue infected with cassava brown streak, but none from leaf tissue infected with the African cassava mosaic. The frequency distribution of particle length showed peaks at 300 and 600 mμ. These authors suggested that this last class might represent the normal length of the virus. It was also pointed out (Kitajima and Costa, 1964) that apparently most of the viruses transmitted by whiteflies that have been examined are expected to be generally spherical (based on negative evidence resulting from not finding elongated particles in dip preparations) and the cassava brown streak virus, if definitely confirmed to be transmitted by the whitefly, would be the first known case of an elongated virus having an insect of this group as a vector.

### *B. Inclusion Bodies Associated with Whitefly-Transmitted Viruses*

The presence of intranuclear inclusion bodies in the leaves of cotton plants (*Gossypium hirsutum*) infected with leaf crumple was reported by Tsao (1963). These inclusions were found inside the nucleus, but not in the cytoplasm or in the nucleoli. They are spheroidal in shape, and measure 2–4 μ in diameter.

In plants of several species infected with sweet potato virus B, transmitted by *Bemisia tabaci,* Sheffield (1958) did not detect any virus inclusion bodies.

* Probably identical to yellow mosaic of *Phaseolus lunatus* in India (Capoor and Varma, 1948).

### *C. Mechanical Transmission*

The viruses of tobacco leaf curl (McClean, 1940), cotton leaf curl (Kirkpatrick, 1931), and tomato leaf curl (Costa, unpublished) were not transmitted mechanically. This would not be surprising, if they prove to be restricted to phloem tissues. It is difficult to understand why only a few of the whitefly-vectored viruses that induce diseases of the mosaic type, associated with parenchyma tissues, have been transmitted by mechanical methods. The *in vitro* properties of viruses of this group (Costa and Carvalho, 1960a; Cohen and Nitzany, 1960) indicate that they are very labile, but this fact alone is not enough to explain all the negative results.

Storey and Nichols (1938) reported complete failure in their attempts to transmit the cassava mosaic virus in Tanganyika by mechanical inoculation, in spite of previous positive reports. Bird (1957) was unable to transmit mosaic of *Jatrophas gossypifolia* by mechanical inoculation in Puerto Rico. In India, mechanical transmission failed for the viruses that induce the following diseases: bhendi yellow vein mosaic and a disease of *Dolichos lablab* (Capoor and Varma, 1950a, 1950b); yellow vein mosaic of *Ageratum conyzoides;* yellow mosaic of *Corchorus trilocularis;* yellow mosaic of *Croton;* yellow vein mosaic of *Leucas;* yellow mosaic of *Phaseolus aureaus;* pumpkin yellow mosaic; and tobacco yellow net (Varma, 1963). Rao and Varma (1964) had negative results with the yellow-vein mosaic virus from *Malvastrum coromandelianum.*

Except for a virus from sweet potatoes in East Africa, mentioned later, the following whitefly-transmitted viruses that affect this plant do not pass by mechanical inoculation: vein clearing virus from Israel (Loebenstein and Harpaz, 1960) and from Ghana (Clerk, 1960); sweet potato mosaic virus from Georgia, U.S.A. (Girardeau, 1958); sweet potato yellow-dwarf virus, one component of the feathery mottle complex and probably identical to New Jersey yellow-dwarf virus; and Georgia mosaic virus (Hildebrand, 1960).

Other whitefly-transmitted viruses that failed to pass mechanically are those responsible for the following diseases: cotton leaf crumple (Dickson, Johnson, and Laird, 1954; Erwin and Meyer, 1961); mosaic of *Wissadula amplissima* (Schuster, 1964); golden mosaic of beans and lima beans (Costa, 1965); beet pseudo-yellows (Duffus, 1965); and tomato yellow leaf curl (Cohen and Nitzany, 1966).

Negative results were also experienced by early workers (Baur, 1904; Lindemuth, 1907; Hertzsch, 1928; Kunkel, 1930; Klebahn, 1932) and some more recent ones (Costa, 1937; Silberschmidt, 1943, 1948; Flores and Silberschmidt, 1958; Hollings, 1959; Flores, Silberschmidt, and Kramer, 1960) that attempted to transmit AIVV by mechanical inoculation.

It appears that the first record for the mechanical transmission of a whitefly-transmitted virus is that by Costa and Bennett (1950), who infected low percentages of *Euphorbia prunifolia* and *Datura stramonium* seedlings rubbed with juice containing the *Euphorbia* mosaic virus.

In experiments with a sweet potato virus transmitted by *Bemisia tabaci* in East Africa, Sheffield (1957, 1958) did not transmit it mechanically from sweet potato to sweet potato, but succeeded in the transmission from sweet potato to a species of *Ipomoea* and *Petunia* sp.; transmission was also accomplished when *Petunia* and *Ipomoea* were used as virus sources and test seedlings.

Crandall (1954) reported infection of cotton plants rubbed with juice from *Stizolobium, Mucuna,* and *Malvastrum* infected with a virus that possibly belongs to the AIVV complex. Costa (1955) questioned these results on the basis of experiments with the AIVV complex in Brazil.

Costa and Carvalho (1960a) mechanically transmitted many isolates of the AIVV complex that occur throughout Brazil—mainly when species of *Malva,* and particularly *M. parviflora,* were used as test seedlings. Yellow mosaic on *Leonurus sibiricus,* first described in Brazil by Silberschmidt, Flores, and Tommasi (1957) and reported to be caused by a virus transmitted by *Bemisia tabaci* (Costa and Carvalho, 1960b), was transmitted by mechanical inoculation with relative ease.

A virus that affects the cucumber plant and is transmitted by *Bemisia tabaci* in Israel (Cohen and Nitzany, 1960) passes easily by mechanical inoculation to many Cucurbitaceae. This virus is apparently different from the one described by Varma (1955b) from Cucurbitaceae in India, which is not transmitted mechanically.

As it was postulated before (Costa, 1955), it is expected that many whitefly-transmitted viruses found not transmissible mechanically thus far, especially those associated with mosaic of parenchyma cells, will be transmitted by modified techniques or the use of suitable test plants.

1. PHYSICAL PROPERTIES IN VITRO. Only a few whitefly-transmitted viruses have had their properties investigated *in vitro*. The properties of AIVV were determined by Costa and Carvalho (1960a). The virus had a low thermal inactivation point, losing most activity at 50°C for 10 min, but occasionally retaining a small fraction of activity after treatment at 55°C. The dilution end point was low, some transmission being obtained with the juice diluted to 1:625. Most of the activity was

lost on aging for 24 hr when the inoculum was extracted in the presence of water; however, some activity was demonstrable in preparations aged for 48 hr when extracted in phosphate buffer plus sodium sulfite. Costa and Carvalho (1960b) reported that the physical properties of the *Euphorbia* mosaic virus are rather similar to those of AIVV. The determination was by local lesions on *Datura stramonium*.

The physical properties of a cucurbit virus from Israel, transmissible by *Bemisia tabaci,* were determined by Cohen and Nitzany (1960) on samples of crude sap. The thermal inactivation point lies between 50 and 52°C; the dilution endpoint, between $10^{-2}$ and $10^{-3}$; and the resistance to aging *in vitro* between 21 and 27 hr. Virus activity was still present in leaves dried over calcium chloride after 66 days of storage but not in samples aged 75 days.

### *D. Seed Transmission*

Keur (1933, 1934) reported transmission of AIVV in the seed of the $F_1$ hybrid between *Abutilon mulleri* and *A. thompsonii*. Out of 271 seedlings, 204 were green and 67 showed variegated characteristics. He reported transmission of the variegated condition to *A. regnelli* scions by grafting in 4 cases out of 67 grafts. The symptoms were milder than in the hybrid. Keur's results need confirmation before they can be accepted because of the ample negative evidence obtained by many other investigators on seed transmission of identical or related viruses.

Silberschmidt (1945) studied the progeny of a plant of *Abutilon striatum* var. *spurium* infected with AIVV. Some seedlings in the progeny presented chlorotic symptoms in the leaves and could be thought to be infected. When tested for viruses, they gave negative results. The *Abutilon* infectious variegation virus in Brazil was not transmitted through the seeds collected from infected cotton plants (Costa, 1937); several species of *Sida* (Silberschmidt, 1943; Costa, 1955); *Phenax sonneratii* (Silberschmidt, 1948); and *Nicandra physaloides* (Costa, 1955). The virus that induces bean golden mosaic was not transmitted through the seeds (Costa, 1965). Tests on transmission of the *Euphorbia* mosaic virus through the seed from *E. prunifolia* (Costa and Bennett, 1950) and beans (Costa, 1965) were negative.

Negative seed transmission was recorded for the following whitefly-vectored viruses occurring in India: a virus that affects *Dolichos lablab* (Capoor and Varma, 1950b), yellow-vein mosaic viruses of bhendi (Capoor and Varma, 1950a); *Ageratum conyzoides* (Varma, 1963); and *Malvastrum coromandelianum* (Rao and Varma, 1964).

In Puerto Rico, Bird (1957) reported a lack of seed transmission of the virus that causes mosaic on *Jatrophas gossypifolia* and of the infectious chlorosis virus of *Sida carpinifolia* (Bird, 1958). Seed transmission was negative for the cotton leaf-crumple virus in California (Laird and Dickson, 1959). A mosaic virus vectored by *Bemisia tabaci* is not transmitted through the seeds of *Wissadula amplissima* or *Hibiscus cannabinus* (Schuster, 1964).

Cotton leaf-curl virus was not transmitted through the seeds of the Sakel variety in Africa (Kirkpatrick, 1931). Negative seed transmission resulted from tests with tobacco or *Datura stramonium* infected with the tobacco leaf-curl virus (McClean, 1940). Cassava mosaic virus was not seed transmitted (Chant, 1958).

### *E. Dodder Transmission*

Several investigators tried dodder in the transmission of whitefly-transmitted viruses, but the results have been negative (Capor and Varma, 1950b; Bird, 1957; Clerk, 1960; Erwin and Meyer, 1961; Rao and Varma, 1964).

### *F. Serological Properties*

The writer is not aware of any paper indicating successful serological testing of viruses transmitted by whiteflies. This line of investigation would be of great value in establishing relationships between the various groups of these viruses.

## IV. SPECIES OF WHITEFLY VECTOR

Within the aleyrodid vectors of plant viruses, *Bemisia tabaci* parallels *Myzus persicae* in the aphid group. It is by far the most important whitefly vector and responsible for the transmission of over 25 different diseases distributed throughout the world (Fig. 1 and 2).

Bay and Bennetts (1954) and Heinze (1959) list the whitefly species that have been reported to be vectors of plant viruses. Eliminating the synonymy as shown by Russell (1957), the remaining species can be listed as follows: *Aleurotrachelus socialis* Bondar, *Aleurothrixus floccosus* (Mask), *Bemisia manihotis* Frappa, *B. tabaci* Genn., *B. tuberculata* Bondar, *B. vayassieri* Frappa, *Trialeurodes abutilonea* Hald., *T. natalensis* Corb., and *T. vaporariorum* (Westw.).

It is possible the some of the *Bemisia* species above are synonyms of *B. tabaci* and some of the reports concerning species from other genera need confirmation before they can be accepted.

Wolf, Whitcomb, and Mooney (1949) reported that the vector of tobacco leaf curl in Venezuela is *Bemisia tabaci,* but that *B. tuber-*

*Fig. 1.* Whiteflies, *Bemisia tabaci,* feeding on a bean leaf (magnification ×6).

*culata* and *Aleurotrachelus socialis* may act as vectors. The results concerning these two species were obtained in tests with adult whiteflies collected from a field of yucca. The technique described did not preclude the possibility of the insects having been mixed with *B. tabaci,* since this species feeds on yucca.

*Aleurothrixus floccosus* is quoted by Heinze (1959) in association

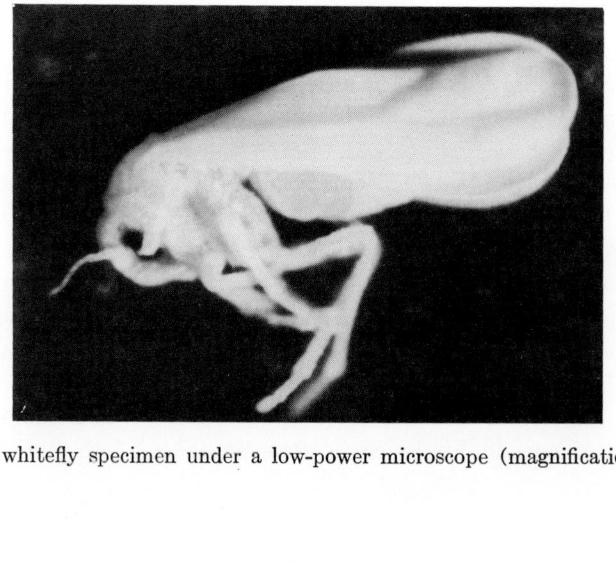

*Fig. 2.* A whitefly specimen under a low-power microscope (magnification ×80).

with the transmission of citrus tristeza virus, but he questions the record.

Transmission of tobacco leaf-curl virus by *Trialeurodes natalensis* in South Africa seems open to question. McClean (1940) reports that whiteflies from one of the colonies of this species did not transmit the virus. Another colony received originally from Rustemburg contained specimens of *T. natalensis* and transmitted the virus, but he acknowledged that possibly two species were present, suggesting in a footnote that *Bemisia gossypiperda* might be the other.

*Trialeurodes vaporariorum* was found to be the vector of beet pseudo-yellows virus in California and some of the virus–vector relationships were determined (Duffus, 1965). The same species was reported as a vector of a sunflower virus in Argentina (Traversi, 1949). This is a peculiar virus that is also transmitted by *Thrips tabaci* and *Myzus persicae*. This record needs confirmation.

*Trialeurodes abutilonea* is a vector of the sweet potato yellow-dwarf virus in Maryland and also of the sweet potato mosaic virus received from Georgia, which is considered identical to the first (Hildebrand, 1959, 1960). Sweet potato mosaic virus in Georgia is transmitted by *Bemisia tabaci* (Girardeau, 1958). This is apparently a case where two species of whiteflies transmit the same virus.

### *A. Races of Bemisia Tabaci*

Bird (1957) reared colonies of *Bemisia tabaci* from *Jatrophas* and determined that the insects would not feed or breed on *Sida carpinifolia*. Conversely, colonies of the same species from *Sida* would not feed or breed on *J. gossypifolia*. This host preference he attributed to physiological specialization.

*Bemisia tabaci* in Brazil feeds and breeds on a great number of host plants. Costa and Bennett (1950) list a number of species on which the insect was bred. Flores and Silberschmidt (1958) reared *B. tabaci* on several different species. They admit the existence of ecological biotypes of the species and explain Bird's results on this basis.

A list of many cultivated and wild plants in India that serve as food and breeding hosts for *Bemisia tabaci* is given by Pruthi and Samuel (1939, 1942).

## V. VIRUS–VECTOR–HOST PLANT RELATIONSHIPS

### *A. Feeding Habits of Whiteflies*

A discussion of feeding habits of whiteflies was presented by Varma (1963) in his review paper and will not be repeated. Nymphal instars of whiteflies are firmly attached to the host plant except for a short period in instar I and play no role in the field transmission of plant viruses. Adult whiteflies are phloem feeders, and the penetration of their stylets generally follows an intercellular course, except when it reaches the phloem. Capoor (1949) thinks that *Bemisia tabaci* might need more than 15–30 min to reach the phloem and Varma's data (1952) suggest that fasting reduces the time the insect needs to reach the phloem.

Whiteflies can introduce varying amounts of saliva, which may affect the tissues of the host plant. Costa and Bennett (1950) noticed a vein clearing induced by nonviruliferous *Bemisia tabaci* on several plants and attributed this effect to toxins introduced while feeding. Pollard (1955) reported that nymphs of *B. tabaci* induced a yellow spotting on cotton leaves caused by the saliva introduced in the stylet punctures. He also noticed an increase in anthocyanin associated with whitefly feeding. Bird (1957) also noticed nonvirus vein clearing induced as a result of adult *B. tabaci* feeding on *Jatrophas gossypifolia.* In addition, he reported that chlorotic spots appear on the upper leaf surface at the places where nymphs or pupae are located on the underside.

Varma (1963) points out that whiteflies travel only short distances from one plant to another, but it is possible that they can be transported over long distances by the wind. This agrees with the observations made by the writer on the spread of cotton common mosaic in Brazil. *Bemisia tabaci* can infect cotton only when acquiring a virus on infected malvaceous weeds. In extensive cotton fields in Brazil where virus sources occur only at the borders, mosaic infected cotton plants can be found several hundred meters from the nearest virus reservoir.

### *B. Manner of Virus Transmission by Whiteflies*

The transmission of plant viruses by whiteflies has not been so intensively studied as that by some other groups. Existing data permit the following assertions: (*1*) whitefly-transmitted plant viruses are not acquired as rapidly as in the case of mechanically transmitted aphid viruses; (*2*) the transmission efficiency of the vector increases with longer feeding periods of up to several hours on virus sources; (*3*) in most cases there is a definite, but relatively short, incubation period; (*4*) all viruses are retained in their vectors for periods of a few days to up to 20 days, but serial transmission is generally intermittent, it is not highly efficient, and sometimes there is loss of infectivity. These facts led most of the investigators to consider that the relationship of the whitefly-transmitted viruses to their vectors is of the circulative type as defined by Black (1959). The writer does not consider the evidence conclusive against a possible virus increase in whiteflies. Potato leaf-roll virus is accepted as capable of increase in *Myzus persicae* (Heinze, 1955; Stegwee and Ponsen, 1958) and its relationship to the vector as determined on test plants is not different from that of certain whitefly-transmitted viruses. Virus increase in the whitefly could occur early and cease or its reduction or depletion in the vector as the time since acquisition increases could be explained on the basis of partial or clean recovery as it occurs in host plants (Costa *et al.,* 1959).

### *C. Virus Acquisition by Whiteflies in the Preadult Stages*

The problem of determining whether immature forms can acquire virus was first investigated by Kirkpatrick (1931). He found it difficult to remove whitefly larvae or pupae from the leaves without injuring the insect. To test adult whiteflies that had no acquisition feeding on the virus source after emergence, he caged diseased leaves on which many pupae were attached and caught the adults as soon as they emerged. With the whiteflies thus collected, caged in numbers of 15, with 20 or more insects per cotton plant, he obtained infection on 9 out of 11 seedlings.

Costa and Bennett (1950) obtained pupae detached from diseased *Euphorbia prunifolia* leaves and placed them in petri dishes on healthy leaves. Adult whiteflies emerged 24–48 hr after the transfer and were caged on test seedlings for virus determination. Some had deformed wings, but most appeared to be normal and were able to feed. Caged on test seedlings at the rate of 1, 2, and 5 insects per plant, the transmission results were, respectively, 0/24, 0/24, and 9/48.

Laird and Dickson (1959) used a similar technique to test the acquisition of cotton leaf crumple virus by immature forms of *Bemisia tabaci.* Out of 603 pupae from diseased plants caged on 16 test plants, 120 adults emerged. None of the plants became diseased.

Hildebrand (1959) found that adult whiteflies, *Trialeurodes abutilonea,* emerging from pupal cases developed on sweet potato plants infected with feathery mottle, transmitted virus acquired in the preadult stages. He emphasized later (1965) that adult *T. abutilonea* is so short lived (3 days or less) that it cannot acquire and transmit the sweet potato yellow-dwarf virus because the incubation period in the vector is equal to its life span. Hildebrand (1960) had previously reported that sweet potato yellow dwarf is the whitefly-transmitted component of sweet potato feathery mottle.

Varma (1963) reported that the ability of adult whiteflies to transmit when only the immature stages had access to virus sources has been verified for several viruses at Poona; however, not all adults thus bred will be infective. A virus that induces *Wissadula amplissima* mosaic was transmitted through the pupal instar of *Bemisia tabaci* (Schuster, 1964).

Cohen and Nitzany (1966) transferred pupae from *Datura stramonium* infected with tomato yellow leaf-curl virus onto cotton leaves in petri dishes. More than half of 98 females tested singly onto *D. stramonium* seedlings survived for 48 hr and, from these, 28% transmitted the virus.

### *D. Virus Acquisition by Adult Whiteflies*

Whiteflies, as most insect vectors, can in most cases acquire viruses after reaching the adult stage. This has been shown for practically all the diseases they transmit.

Cohen and Harpaz (1964) reported what they called periodic acquisition for the relationship between *Bemisia tabaci* and tomato yellow leaf-curl virus. Whiteflies that had an acquisition feeding for 24 hr retained the virus for 10–12 days, but could not reacquire it until ceasing to transmit from the previous acquisition.

1. MINIMUM ACQUISITION PERIOD AND FULL CHARGE ACQUISITION. *Bemisia tabaci* requires the following minimum acquisition periods to become infective: 3 hr for cotton leaf-curl virus (Kirkpatrick, 1931); 5 hr for tobacco leaf-curl virus, but shorter periods were not tried (Pruthi and Samuel, 1939); 30 min for *Euphorbia* mosaic virus (Costa and Bennett, 1950); 1 hr for the bhendi yellow-vein mosaic virus (Varma, 1952), but this period could be reduced to 30 min if the whiteflies were given a fasting period of 1–6 hr prior to acquisition; 2 hr for the mosaic virus of *Jatrophas gossypifolia* (Bird, 1957); 15 min for the infectious chlorosis of *Sida carpinifolia* (Bird, 1958); 4 hr for cassava mosaic virus (Chant, 1958); 4–8 hours for cotton leaf-crumple virus (Laird and Dickson, 1959); 30 min for tomato yellow leaf-curl virus (Cohen and Nitzany, 1966). Beet pseudo-yellows virus is acquired by *Trialeurodes vaporariorum* in 1 hr (Duffus, 1965). Varma (1963) pointed out that whiteflies require a comparatively longer acquisition period that aphids. This is rather doubtful if the comparison is made among circulative viruses transmitted by the two groups of insects.

Whiteflies require a period several times that needed for minimum acquisition to become fully charged. This has been determined in several cases for *Bemisia tabaci:* 3 hr for the bhendi yellow-vein mosaic virus (Varma, 1952); 2–6 hr for *Jatrophas* mosaic virus (Bird, 1957); 4 hr for tomato yellow leaf-curl virus (Cohen and Nitzany, 1966). *Trialeurodes vaporariorum* acquires a full charge of the beet pseudo-yellows virus in 24 hr (Duffus, 1965).

2. INFLUENCE OF THE HOST PLANT AS VIRUS SOURCES. The ability of whiteflies to acquire the same virus from different host plants seems to be more at variance than in case of other groups of insect vectors. Although whitefly-transmitted viruses may have a wide host range, indicating that viruliferous vectors can infect a great number of species, the recovery of the virus from some is often poor or completely negative.

Kirkpatrick (1931) reported that cotton leaf-curl virus could be transmitted by the whitefly vector to *Hibiscus esculentum,* or *H. cannabinus,* inducing symptoms similar to those on cotton. Back-transmission from *H. esculentum* to cotton was negative, but it was readily transmitted from *H. cannabinus* to cotton. Also, the virus could be transmitted from *H. esculentum* to *H. cannabinus* and then back to cotton.

Pruthi and Samuel (1939) determined that tobacco leaf curl in northern India is generally transmitted from alternate hosts, *Crotalaria juncea* and *Ageratum conyzoides* to tobacco by the vector, and rarely from tobacco to tobacco. Back inoculation from tobacco to *C. juncea* is difficult, but *A. conyzoides* is easily infected. The disease was also transmitted by the vector more readily from *C. juncea* to *A. conyzoides* than vice versa.

Infection of cotton and other susceptible crops with AIVV in Brazil generally depends on the transmission of the virus from the malvaceous weeds to the crop plants. Spread within the crops is negligible. Costa (1954, 1955) compared infected cotton plants and *Sida micrantha* as virus sources for *Bemisia tabaci.* Whiteflies that fed on *S. micrantha* infected 50% of the cotton test plants while those feeding from the infected cotton source, infected none, even with 100 insects per plant or more.

Silberschmidt *et al.* (1957) obtained only low percentages of infection in attempts to transmit AIVV from *Abutilon thompsonii* to *Sida rhombifolia* by means of *Bemisia tabaci.* After the virus from *A. thompsonii* was passed to *S. rhombifolia* by grafting, nonviruliferous insects could acquire and transmit it more easily. A still better transmission was obtained with vectors which acquired the virus of *A. thompsonii* from *S. rhombifolia* infected by the vector. Recovery of the virus from infected *Malope trifida* back to *S. rhombifolia* also resulted in a low percentage of success, 5.4%. Using whiteflies bred for a few generations on *M. trifida* Silberschmidt *et al.* increased the transmission from *M. trifida* to *S. rhombifolia* to 26.3%. Flores and Silberschmidt (1958) infected *Datura stramonium*, tobacco, tomato, and *Cyphomandra betacea* with AIVV. Virus recovery from infected plants back to *S. rhombifolia* was generally difficult. Recovery from infected *C. betacea* plants improved when the vectors were fed for a few days prior to acquisition on a healthy *C. betacea* plant.

Silberschmidt *et al.* (1957) discussed factors that may influence the vector in the acquisition from certain virus sources and mentioned: (*1*) the palatability of the host plant for the vector, as in the case of *A. thompsonii;* (*2*) a low virus concentration or unequal distribution in the host tissues; and (*3*) the presence of virus inhibitors. A low concentration or the disappearance of virus from phloem tissues seems to the writer the best explanation for some of the known facts concerning the poor virus acquisition by whiteflies.

3. INFLUENCE OF THE FEEDING SITE ON VIRUS ACQUISITION. Varma (1952) inoculated okra plants on the youngest leaf and on the third below the youngest by means of the vector and then determined the time interval necessary for the virus to become recoverable from the four leaves, first to fourth. Regardless of whether the virus had been introduced in the youngest or fourth leaf, the vector was able to acquire virus from the younger leaves before it could do so from the older leaves. Testing the four leaves at 15-day intervals, he found that the virus disappeared from some of the older leaves. He explained the disappearance of the virus as being due to its moving out of the leaf.

Laird and Dickson (1959) compared recently infected plants and stub cotton as sources of leaf crumple virus for *Bemisia tabaci*. Transmission efficiency was slightly higher for whiteflies that acquired virus from the newly infected source, but not significantly different from that found for the insects that acquired virus from the long-infected source.

4. INFLUENCE OF PREACQUISITION FASTING. Varma (1952) determined that fasting periods of up to 4 hr increased the efficiency of *Bemisia tabaci* in the acquisition of bhendi yellow-vein mosaic virus.

5. ARTIFICIAL FEEDING OF WHITEFLIES. Only a few of the whitefly-transmitted viruses have been transmitted by mechanical inoculation, thus permitting their physical properties to be determined. Another way to determine them is by feeding the vectors on virus preparations.

The first indication that whiteflies might be fed artificially was furnished by Kirkpatrick (1931). He reported that *Bemisia tabaci* deprived of food and water will remain alive for only a few hours in the dry atmosphere of Sudan. When water is supplied on wet fileter paper, 50% of the insects will survive for 24 hr and a few for 36 hr.

Costa and Bennett (1950) tried feeding *Bemisia tabaci* in cages, one end of which was covered with Baudruche capping skin. The end with capping skin was turned up and received the drops of feeding solutions. Whiteflies in cages that received preparations containing juice from diseased *Euphorbia prunifolia* collected below the drops and remained there for considerable periods, apparently feeding. Their stylets could be seen projecting through the membrane into the liquid. When tested for virus infectivity, they gave negative results. Although transmission results were negative, *B. tabaci* apparently feeds well through a membrane. Artificial feeding of whiteflies deserves more attention.

### *E. Incubation Period in the Whitefly Vector*

Whitefly-transmitted viruses have a definite incubation period in the vector. This varies from 4 to 8 hr for most viruses transmitted by *Bemisia tabaci.* Data on this relationship are given and discussed in Varma's review paper (1963). Laird and Dickson (1959) suggested that the cotton leaf-crumple virus might have an incubation period in *B. tabaci* of 20 hr or longer. The incubation period was also 21 hr or more for the tomato yellow leaf-curl virus in the same whitefly (Cohen and Nitzany, 1966). Duffus (1965) determined that the incubation period is less than 6 hr for the beet pseudo-yellows virus in *Trialeurodes vaporariorum.*

### *F. Inoculation Feeding*

Introduction of virus by whiteflies into a suitable site of the host plant to permit infection requires from 10 min to 1–2 hr. Varma (1963) summarized this relationship for several viruses. Duffus (1965) determined that groups of *Trialeurodes vaporariorum* can inoculate test plants in less than an hour. Groups of 20–25 viruliferous female *Bemisia tabaci* need from 15 to 30 min to inoculate the tomato yellow leaf-curl virus (Cohen and Nitzany, 1966).

Varma (1963) pointed out that whiteflies generally require a shorter feeding period to infect the plant than to acquire the virus. He interpreted this fact as an indication that acquisition depends on the vector reaching the phloem tissues, whereas the introduction of the virus into the plant might occur in other kinds of tissues. The range from 10 min to 1–2 hr for inoculation feeding of the same whitefly species with different viruses indicates that other factors also are operative. Moreover, since penetration of the vector stylet is accepted as being intercellular (Hargreaves, 1915; Smith, 1926; Capoor, 1949; Pollard, 1955) until it reaches the phloem tissues, the introduction of virus in parenchyma cells, as would be necessary to infect, would be rare.

A whitefly-transmitted virus of cucurbits in Israel is easily passed by mechanical inoculation (Cohen and Nitzany, 1960). The vector–virus–host plant relationships were not worked out, but one might expect that its minimum inoculation feeding period should be very short if Varma's view is correct. The whitefly vector might even transmit it in a mechanical manner.

1. INFLUENCE OF FASTING PRIOR TO INOCULATION FEEDING. Varma (1952) found that periods of fasting for up to 3–4 hr prior to inoculation feeding had a favorable effect on the transmission efficiency of *Bemisia tabaci* for the bhendi yellow-vein mosaic virus when tested with single insects per plant or in groups of 20. Also, fasting reduced the minimum infection feeding period from 30 to 10 min.

2. INFLUENCE OF NUMBER OF INSECTS ON INFECTION. Single-insect transmission with *Bemisia tabaci* has been obtained with the viruses responsible for the following diseases: cotton leaf curl (Kirkpatrick, 1931); tobacco leaf curl (Pruthi and Samuel, 1939); AIV (Orlando and Silberschmidt, 1946); *Euphorbia* mosaic (Costa and Bennett, 1950); bhendi yellow-vein mosaic (Varma, 1952); *Jatrophas* mosaic (Bird, 1957); cassava mosaic (Chant, 1958); *Wissadula* mosaic (Schuster, 1964); and tomato yellow leaf curl (Cohen and Nitzany, 1966). Duffus (1965) had transmission of beet pseudo-yellows virus with single *Trialeurodes vaporariorum.*

Ten viruliferous *Bemisia tabaci* per plant will insure nearly 100% infection with the following viruses: AIVV (Orlando and Silberschmidt, 1946); *Euphorbia* mosaic virus (Costa and Bennett, 1950); bhendi yellow-vein mosaic virus (Varma, 1952). Cohen and Nitzany (1966) needed 15 vectors per plant to obtain 100% transmission of the tomato yellow leaf-curl virus.

The presence of nonviruliferous vectors in cages with the viruliferous insects did not affect the transmission of cassava mosaic virus (Chant, 1958).

3. THE INFLUENCE OF THE FEEDING SITE ON INOCULATION. That the plant part on which viruliferous whiteflies feed plays a role on transmission was first demonstrated by Storey and Nichols (1938) in their study of cassava mosaic. Viruliferous *Bemisia tabaci* transmitted the virus only to those plants on which the vectors had been caged

on the third and fourth leaves above the youngest mature leaf at the top.

A similar relationship was reported by Bird (1957), who found that viruliferous whiteflies did not infect *Jatrophas gossypifolia* test plants with the mosaic virus if the insects were fed on old leaves; a single insect, feeding on immature leaves, did.

Costa and Bennett (1950) found that viruliferous *Bemisia tabaci* insects were more efficient vectors of the *Euphorbia* mosaic virus when caged on young true leaves of *E. prunifolia* (64.2% transmission) than when fed on the cotyledons of comparable seedlings (42.4% transmission). Marked differences were obtained by Costa (1955) in the transmission of AIVV from *Sida* to cotton by *B. tabaci.* Caging 10–25 viruliferous insects per plant on the young true leaves infected 43 out of 60 inoculated plants; none were infected out of 60 when the insects were caged on the cotyledons.

Laird and Dickson (1959) also found that in cotton leaf crumple, the virus could not invade the plant systemically if viruliferous vectors were caged on the cotyledons. Kirkpatrick (1931) reported an infection with the leaf-curl virus when viruliferous vectors were caged on cotton seedlings at the cotyledon stage. He did not mention whether the whiteflies were removed from the plants before they formed true leaves. Silberschmidt and Ulson (1954) were able to infect *Sida rhombifolia* through the flowers.

In discussing the results with AIVV, *Euphorbia* mosaic virus, and cassava mosaic virus in Africa, Costa (1955) pointed out that the establishment and subsequent invasion of host plants by whitefly-transmitted viruses seems easier when the vectors are fed on young true leaves. Bird (1957) shares the same view.

### *G. Virus Retention in the Whitefly Vector*

Kirkpatrick (1931) tested the retention of cotton leaf-curl virus in *Bemisia tabaci* in three ways: (*1*) caging viruliferous whiteflies on immune *Dolichos lablab* and testing the insects after 1-, 2-, and 3-day intervals on cotton seedlings; (*2*) viruliferous whiteflies were tested after feeding on a healthy cotton plant for 7 days; (*3*) viruliferous insects were starved for 1 day and tested for infectivity. He obtained positive transmission in all cases.

Costa and Bennett (1950) investigated retention of the *Euphorbia* mosaic virus by serial transfer of single, viruliferous *Bemisia tabaci* onto healthy test seedlings. The virus was retained in the vector for 20 days, but distribution of the infected plants in the series was more erratic than in the case of leafhopper vectors.

The retention of the bhendi yellow-vein mosaic virus in *Bemisia tabaci* was studied by Varma (1952). Vectors fed for 12–24 hr on the virus sources retained infectivity for life. In another paper (Varma, 1955a) he compared virus retention in vectors fed on the virus source for periods of ½, 1, 2, 4, and 6 hr. Whiteflies fed for 4 and 6 hr retained the virus for life, but there was virus depletion in insects fed for shorter periods. His results with fully charged whiteflies were very consistent and approached those given by some leafhopper vectors in a comparable series.

Bird (1957, 1958) studied the retention of the viruses inducing mosaic on *Jatrophas gossypifolia* and infectious chlorosis of *Sida carpinifolia*. Virus was retained in the vector, *Bemisia tabaci* for 4 and 7 days, respectively.

Flores and Silberschmidt (1958) found that AIVV from *Sida* in Brazil could be retained in *Bemisia tabaci* for periods of 20 days. These investigators tried yellow cages to improve the intermittent manner of transmission, but the results were no better than those obtained with transparent ones. In serial transfer tests, Cohen and Nitzany (1966) determined that the tomato yellow leaf-curl virus may be retained in female vectors for 20 days, but not for life. Most insects stopped transmitting after the tenth day.

*Trialeurodes vaporariorum* retains the beet pseudo-yellows virus for 6–7 days (Duffus, 1965).

### *H. Ability of the Whitefly to Carry More Than One Virus Simultaneously*

Whiteflies, like leafhoppers (Giddings, 1950; Kunkel, 1955; Maramorosch, 1958) and aphids (Harrison, 1958; Costa *et al.*, 1959) can carry two or more viruses simultaneously. Varma (1955b) fed virus-free *Bemisia tabaci* for 24 hr on bhendi plants infected with yellow-vein mosaic and for the next 24 hr on pumpkin plants infected with a different yellow-vein mosaic virus. Other insects were fed in the reverse order. Samples of insects from the two lots were daily transferred to healthy okra and pumpkin plants, alternately, for 9 days. They transmitted each of the two viruses to their respective host plant. Varma (1963) reported the simultaneous transmission by *Bemisia tabaci* of yellow-vein mosaic virus of bhendi and yellow mosaic virus of double beans. He also obtained evidence on the simultaneous transmission of three viruses by the same vector.

Working with AIV and *Euphorbia* mosaic viruses, Costa and Carvalho (1960b) showed that individual *Bemisia tabaci* could carry the two viruses simultaneously. The whiteflies used in the tests were bred on a diseased source for the first virus, given a 48-hr acquisition feeding for the second, transferred to a test seedling for the second virus for 24 hr, and then to a test seedling for the first for 48 hr. Out of 160 insects bred on diseased *E. prunifolia,* fed subsequently on *Sida micrantha* infected with AIVV, and then caged singly on the differential test seedlings, 22 transmitted both viruses; 14, only AIVV; 65, the *Euphorbia* virus; and 59, neither. Out of 80 insects bred on infected *S. micrantha* and given the 48-hr-acquisition period on infected *E. prunifolia,* and subsequently tested as single insect per plant, 8 individuals transmitted the two viruses; 6, the AIVV; 23, the *Euphorbia* mosaic virus; and 43, neither.

Flores and Silberschmidt (1963) tested the simultaneous transmission of AIV and *Leonurus* mosaic viruses by *Bemisia tabaci.* They used a similar technique for acquisition on two different virus sources in succession and then tested the vectors singly on differential test plants, alternately, for a few days. In addition they tried having the whiteflies feed on a single plant infected with the two viruses. Single *Bemisia tabaci* transmitted the two viruses regardless of the feeding sequence utilized and also when the insects acquired the two viruses from the same infected plant. When the whiteflies fed on two different sources in succession, they transmitted the first virus more efficiently. This result differs from that obtained by Costa and Carvalho (1960b) in which one virus of the pair was more efficiently transmitted than the other in both sequences. Flores and Silberschmidt's results might be interpreted as interference in the vector, as verified for leafhoppers (Kunkel, 1955; Maramorosch, 1958), indicating a relationship between AIV and *Leonurus* mosaic viruses.

### *I. Female versus Male Ability to Transmit*

Orlando and Silberschmidt (1946) determined that single female *Bemisia tabaci* whiteflies were much more efficient vectors of AIVV from *Sida* (43.5%) than male insects (20%). Costa and Bennett (1950) also verified that single female *B. tabaci* were more efficient vectors of the *Euphorbia* mosaic virus (44.0%) than the males (24.7%). These investigators suggested that the greater ability of the female whiteflies to transmit viruses might result from a greater metabolic activity due to egg production, which would lead to acquisition and the release of larger amounts of virus during the feeding processes in the plant.

When comparing female and male whiteflies for the transmission of bhendi yellow-vein mosaic virus in India, Varma (1952) tested 250 insects of both sexes and determined that 77.6% of the females and 56.4% of the males transmitted the virus. Cohen and Nitzany (1966) reported that the female *Bemisia tabaci* were six times more efficient vectors of tomato yellow leaf-curl virus than males. This marked difference in efficiency led these investigators to carry out all their vector tests with females.

### *J. Tests for the Transovarial Passage of Virus to Offspring*

In investigating transovarial passage of the cotton leaf-curl virus, Kirkpatrick (1931) caged viruliferous female *Bemisia tabaci* on immune *Dolichos lablab* and other plants. The adult whiteflies resulting from the eggs were tested for infectivity, but the results were negative.

Costa and Bennett (1950) tested the progeny of fully charged *Bemisia tabaci* whiteflies (*Euphorbia* mosaic virus) from eggs laid on immune cotton plants. The progeny were fed in groups of 5–10 insects per plant on 120 *E. prunifolia* seedlings, but no seedlings became infected.

Capoor and Varma (1950b) tested the progeny from whiteflies carrying the bhendi yellow vein mosaic virus in an interesting manner. They allowed the viruliferous whiteflies to lay eggs on a healthy bhendi plant for 1 hr and then killed them. When the nymphs hatched, they were transferred carefully to healthy bhendi seedlings and permitted to feed for 7 days or through the nymphal and pupal stages. The immature forms of the vectors did not infect the test seedlings. The same authors had no transmission of bhendi yellow-vein mosaic virus with adult whiteflies that resulted from eggs laid by viruliferous females on the immune cotton plant.

Costa (1955) tested the offspring of viruliferous *Bemisia tabaci* bred on *Sida* plants infected with AIVV. The 600 insects resulting from eggs laid on an immune plant were tested on 60 *S. micrantha*, but no plants showed symptoms.

Bird (1957) tested the progeny of viruliferous whiteflies bred on a *Jatrophas gossypifolia* plant infected with mosaic. Since old leaves from the diseased plant contain practically no virus, he selected some senescent ones on which many whitefly pupae were present and placed them in a petri dish on wet filter paper. Whiteflies emerged from the pupal cases and were placed on healthy *J. gossypifolia* seedlings. No seedlings became diseased. In studies on the infectious chlorosis of *Sida carpinifolia*, Bird (1958) tested adult whiteflies from eggs laid by viruliferous females on virus-immune *Helianthus annuus*. Testing was done on *S. carpinifolia* seedlings, but no infection resulted.

Cohen and Nitzany (1966) tested the transovarial passage of tomato yellow leaf-curl virus to offspring. Out of 360 *Bemisia tabaci* females from eggs laid on cotton plants and tested on *Datura stramonium* seedlings, none transmitted the virus.

Day and Bennetts (1954) mention the transovarial passage of AIVV, but the original source of this information could not be located.

### *K. Interference Between Whitefly-Transmitted Viruses*

1. IN THE PLANT. Attempts to determine the relationship between AIVV and *Euphorbia* mosaic virus by cross protection were made by Costa and Carvalho (1960b), but the results were inconclusive. Erwin and Meyer (1961) showed that a mild strain of cotton leaf-crumple virus protected the infected plant against a severe strain.

Flores and Silberschmidt (1961/62) studied the relationship between AIVV from *Sida* and the *Leonurus* mosaic virus by cross protection on *Nicandra physaloides* and *S. rhombifolia*. The *Leonurus* virus that causes milder symptoms did not protect the plants against AIVV that superimposed its more severe symptoms. Protection was negative, but these authors considered the two viruses to be related. Silberschmidt and Flores (1962) also studied the interference between AIVV and three unrelated viruses: potato virus X, TMV, and cucumber mosaic virus. There was no intereference in the first case, the symptoms of AIVV and potato virus X being merely cumulative. The other two cases of mixed infection resulted in a complex disease.

Cross protection between tobacco yellow net and tobacco leaf-curl viruses was tried by Dinghra and Náriani (1962), but they found them to be distinct.

2. INTERFERENCE IN THE VECTOR. *Bemisia tabaci* is the vector of the majority of the whitefly-transmitted plant viruses. This permits testing the acquisition of more than one virus by single insects and thus study the interference that the one acquired first might have on the acquisition of the second. As was mentioned, Flores and Silberschmidt (1963) noted a slight interference in the acquisition of AIVV and *Leonurus* mosaic virus when *B. tabaci,* which had acquired one virus, was fed on a source of the second virus. Kunkel (1955) considered an interference in the acquisition of a second virus by a vector to be indicative of a relationship between the two viruses if there was an increase in the vector.

## BIBLIOGRAPHY

Baur, E. 1904. Zur Aetiologie der infektiösen Panaschierung. Ber. Deut. Bot. Ges. 22:453–460.

Bird, J. 1957. A white-fly transmitted mosaic of *Jatropha gossypifolia*. Agr. Exp. Sta. Univ. Puerto Rico Tech. Paper 22:1–35.

——— 1958. Infectious chlorosis of *Sida carpinifolia* in Puerto Rico. Agr. Exp. Sta., Univ. Puerto Rico Tech. Paper 26:1–23.

Black, L. M. 1959. Cycles of plant viruses in insect vectors. p. 157–185. *In* F. M. Burnet and W. M. Stanley [ed.] The viruses. Academic Press, New York.

Capoor, S. P. 1949. Feeding methods of the white-fly. Current Sci. 18:82–83.

Capoor, S. P., and Varma, P. M. 1948. Yellow mosaic of *Phaseolus lunatus* L. Current Sci. 17:152–153.

——— 1950a. Yellow vein mosaic of *Hibiscus esculentus* L. Indian J. Agr. Sci. 20:217–230.

——— 1950b. A new virus disease of *Dolichos lablab*. Current Sci. 19:248–249.

Chant, S. R. 1958. Studies on the transmission of cassava mosaic virus by *Bemisia* spp. (Aleyrodidae). Ann. Appl. Biol. 46:210–215.

Clerk, G. C. 1960. A vein clearing virus of sweet potato in Ghana. Plant Disease Reporter 44:931–933.

Cohen, S., and Harpaz, I. 1964. Periodic, rather than continual acquisition of a new tomato virus by its vector, the tobacco whitefly (*Bemisia tabaci* Gennadius). Entomol. Exp. Appl. 7:155–166. Seen only in Review Applied Mycology. 44:161. 1965.

Cohen, S., and Nitzany, F. E. 1960. A white-fly transmitted virus of Cucurbits in Israel. Phytopathol. Mediterranea 1:44–46.

——— 1966. Transmission and host range of the tomato yellow leaf curl virus. Phytopathology 56:1127–1131.

Costa, A. S. 1937. Nota sôbre o mosaico do algodoeiro. Rev. Agr. Piracicaba 12: 453–470.

——— 1954. Identidade entre o mosaico comum do algodoeiro e a clorose infecciosa das malváceas. Bragantia 13:XXIII–XXVII.

——— 1955. Studies on *Abutilon* mosaic in Brazil. Phytopathol. Z. 24:97–112.

——— 1965. Three whitefly-transmitted virus diseases of beans in São Paulo, Brazil. FAO Plant Protection Bull. 13:1–12.

——— Occurrence of a type of tomato leaf curl in the Pesqueira, Pe. area of Brazil. (Unpublished.)

Costa, A. S., and Bennett, C. W. 1950. White-fly transmitted mosaic of *Euphorbia* prunifolia. Phytopathology 40:266–283.

Costa, A. S., and Carvalho, A. M. B. 1960a. Mechanical transmission and properties of the *Abutilon* mosaic virus. Phytopathol. Z. 37:259–272.

——— 1960b. Comparative studies between *Abutilon* and *Euphorbia* mosaic viruses. Phytopathol. Z. 38:129–152.

Costa, A. S., Duffus, J. E., and Bardin, R., 1959. Malva yellows, an aphid-transmitted virus disease. J. Amer. Soc. Sugar Beet Technol. 10:371–393.

Crandall, B. S. 1954. Additions to the host and geographic range of *Abutilon* mosaic. Plant Disease Reporter. 38:574.

Day, M. F., and Bennetts, M. J. 1954. A review of problems of specificity in arthropod vectors of plant and animal viruses. Council Sci. Ind. Res. Organ. Canberra, p. 1–172.

Dickson, R. C., Johnson, M. McD., and Laird, E. F. 1954. Leaf crumple, a virus disease of cotton. Phytopathology 44:479–480.

Dinghra, K. L., and Nariani, T. K. 1962. Tobacco yellow net virus Indian J. Microbiol. 2:21–24.

Duffus, J. E. 1965. Beet pseudo-yellows virus, transmitted by the greenhouse whitefly (*Trialeurodes vaporariorum*). Phytopathology 55:450–453.

Erwin, D. C., and Meyer, R. 1961. Symptomatology of the leaf-crumple disease in several species and varieties of *Gossypium* and variation of the causal virus. Phytopathology 51:472–477.

Flores, E., and Silberschmidt, K. 1958. Relations between insect and host plant in transmission experiments with infectious chlorosis of Malvaceae. An. Acad. Brasileira Ci 50:535–560.

——— 1961/62. Observations on a mosaic disease of *Leonurus sibiricus* occurring spontaneously in São Paulo. Phytopathol. Z. 43:221–233.

——— 1963. Ability of single whiteflies to transmit concomitantly a strain of infectious chlorosis of Malvaceae and of *Leonurus* mosaic virus. Phytopathology 53:238.

Flores, E., Silberschmidt, K., and Kramer, M. 1960. Observações de clorose infecciosa das malváceas em tomateiros do campo. Biológico 26:65–69.

Giddings, N. J. 1950. Some interrelationships of virus strains in sugar-beet curly top. Phytopathology 40:377–388.

Girardèau, J. H. 1958. The sweet potato white-fly *Bemisia inconspicua* (Q) as a vector of sweet potato mosaic in South Georgia. Plant Disease Reporter 42:819.

Hargreaves, E. 1915. The life history and habits of greenhouse white-fly (*Aleyrodes vaporariorum* West). Ann. Appl. Biol. 1:303–304.

Harrison, B. D. 1958. Ability of single aphids to transmit both avirulent and virulent stains of potato leaf roll virus. Virology 6:278–286.

Heinze, K. 1955. Versuche zur Übertragung des Blattrollvirus der Kartoffel in den Überträger (*Myzodes persicae* Sulz) mit Injektionsverfahren. Phytopathol. Z. 25:103–108.

——— 1959. Phytopathogene Viren und ihre Überträger. Ducker and Humblot, Berlin. 290 pp.

Hertzsch, W. 1928. Beiträge zur infektiösen Chlorose. Z. Bot. 20:65–85.

Hildebrand, E. M. 1959. A white-fly *Trialeurodes abutilonea* (Hald) as an insect vector of sweet potato feathery mottle in Maryland. Plant Disease Reporter 43:712–714.

——— 1960. The feathery mottle virus complex of sweet potato. Phytopathology 50:751–756.

——— 1965. Adaptation of the *Abutilon* whitefly for laboratory use. Plant Disease Reporter 49:429–432.

Hollings, M. 1959. Host range studies with fifty-two plant viruses. Ann. Appl. Biol. 47:98–108.

Keur, J. Y. 1933. Seed transmission of virus causing variegation of *Abutilon*. Phytopathology 23:20.

——— 1934. Studies on the occurrence and transmission of virus diseases in the genus *Abutilon*. Bull. Torrey Bot. Club 61:53–76.

Kirkpatrick, T. W. 1931. Further studies on leaf curl of cotton in the Sudan. Bull. Entomol. Res. 22:323–363.

Kitajima, E. W. Electronmicroscopical studies of whitefly-transmitted viruses. (Unpublished.)

Kitajima, E. W., and Costa, A. S. 1964. Elongated particles found associated with cassava brown streak. East African Agr. Forestry J. 30:28–30.

Klebahn, H. 1932. Fortsetzung der experimentellen Untersuchungen über Alliophyllie und Viruskrankheiten. Phytopathol. Z. 4:1–36.

Kunkel, L. O. 1930. Transmission of *Sida* mosaic by grafting. Phytopathology 20:129–130.

——— 1955. Cross protection between strains of yellows-type viruses. Advances Virus Res. 3:251–273.

Laird, E. F., and Dickson, R. C. 1959. Insect transmission of the leaf-crumple virus of cotton. Phytopathology 49:324–327.

Lindemuth, H. 1907. Studien über die sogennante Panaschüre und über einige begleitende Erscheinungen. Landwirt Jahrbuch 36:807–861.

Loebenstein, G., and Harpaz, I. 1960. Virus diseases of sweet potatoes in Israel. Phytopathology 50:100–104.

McClean, A. P. D. 1940. Some leaf curl diseases in South Africa. Sci. Bull. S. Africa Dep. Agr. 225:1–70.

Maramorosch, K. 1958. Cross protection between two strains of corn stunt virus in an insect vector. Virology 6:448–459.

Orlando, A., and Silberschmidt, K. 1946. Estudos sôbre a disseminação natural do vírus da clorose infecciosa das mulváceas (*Abutilon* vírus 1 Baur) e a sua relação com o inseto–vector *Bemisia tabaci* (Genn.) (Homoptera-Aleyrodidae). Arq. Inst. Biol. São Paulo 17:1–36.

Pal, B. P., and Tandon, R. N. 1937. Types of tobacco leaf curl in Northern India. Indian J. Agr. Sci. 7:363–393.

Pollard, D. G. 1955. Feeding habits of the cotton white-fly. Ann. Appl. Biol. 43: 664–671.

Pruthi, H. S., and Samuel, C. K. 1939. Entomological investigations on the leaf curl disease of tobacco in Northern India. III. The transmission of leaf curl by white-fly *Bemisia gossypiperda* to tobacco, sannhemp and a new alternative host of the curl virus. Indian J. Agr. Sci. 9:223–275.

——— 1942. Entomological investigations on the leaf-curl disease of tobacco in Northern India. V. Biology and population of the white-fly vector (*Bemisia tabaci* Gen.) in relation to the incidence of the disease. Indian J. Agr. Sci. 12:35–57.

Rao, D. G., and Varma, P. M. 1964. Studies on yellow vein mosaic of *Malvastrum coromandelianum* Garcke in India. An. Acad. Brasileira Ci. Rio de Janeiro 36:207–215.

Russel, L. M. 1957. Synonyms of *Bemisia tabaci* (Gennadius) (Homoptera-Aleyrodidae). Bull. Brooklyn Entomol. Soc. 52:122–123.

Schuster, M. F. 1964. A whitefly-transmitted mosaic of *Wissadula amplissima*. Plant Disease Reporter 48:902–905.

Sharp, D. G., and Wolf, F. A. 1949. The virus of tobacco leaf curl. Phytopathology 39:225–230.

——— The virus of tobacco leaf curl. II. Phytopathology 41:94–98.

Sheffield, F. M. L. 1957. Virus disease of sweet potato in East Africa. I. Identification of the viruses and their insect vectors. Phytopathology 47:582–590.

——— 1958. Virus diseases of sweet potato in East Africa. II. Transmission to alternative hosts. Phytopathology 48:1–6.

Silberschmidt, K. 1943. Estudos sôbre a transmissão experimental da "clorose infecciosa" das Malváceas. Arq. Inst. Biol. São Paulo 14:105–156.

——— 1945. Observações suplementares sôbre a transmissão experimental da "clorose infecciosa" das Malváceas. Arq. Inst. Biol. São Paulo 16:49–64.

——— 1948. Infectious chlorosis of *Phenax soneratii*. Phytopathology 38:395–398.

Silberschmidt, K., and Flores, E. 1962. A interação do vírus causador da clorose infecciosa has malváceas com o vírus X da batatinha, o vírus do mosaico do fumo ou o vírus do mosaico do pepino, em tomateiros. An. Acad. Brasileira Ci. 34:125–141.

Silberschmidt, K., Flores, E., and Tommasi, C. R. 1957. Further studies on the experimental transmission of infectious chlorosis of Malvaceae. Phytopathol. Z. 30:387–414.

Silberschmidt, K., and Tommasi, L. R. 1955. Observações e estudos sôbre espécies de plantas suscetíveis a clorose infecciosa das malváceas. An. Acad. Brasileira Ci. 27:195–214.

Silberschmidt, K., and Ulson, C. M. 1954. The transmission of "infectious chlorosis" of malvaceae by grafting and insect vector. 8th Congr. Int. Botanique, Paris, 1954. Rapp. Commun. Sect. 18, 19, and 20. p. 223.

Smith, K. M. 1926. A comparative study of the feeding methods of certain Hemiptera and the resulting effects upon the plant tissue with special reference to the potato plant. Ann. Appl. Biol. 13:109–130.

Stegwee, D., and Ponsen, M. B. 1958. Multiplication of potato leafroll virus in the aphid *Myzus persicae* (Sulz.). Entomol. Exp. Appl. 1:291–300.

Storey, H. H., and Nichols, R. F. W. 1938. Studies on the mosaic diseases of cassava. Ann. Appl. Biol. 25:790–806.

Sun, C. N. 1964. Das Auftreten von Viruspartikeln in *Abutilon*-Chloroplasten. Experientia 20:497.

Tarr, S. A. J. 1964. Virus diseases of cotton. Commonwealth Mycological Institute, Kew. 23 pp.

Thung, T. H. 1932. De krul en kroepoek-ziekten van tabak en de oorzaken van hare verbreiding. Proefstation Vorstenlandsche Tabak. Meded. 72:1–54.

Traversi, B. A. 1949. Estudio inicial sobre una enfermedad del girasol (*Helianthus annuus* L.) en Argentina. Rev. Invest. Agr. Buenos Aires 3:345–351.

Tsao, P. W. 1963. Interanuclear inclusion bodies in the leaves of cotton plants infected with leaf crumple virus. Phytopathology 53:243–244.

Varma, P. M. 1952. Studies on the relationship of the bhendi yellow-vein mosaic virus and its vector, the white-fly (*Bemisia tabaci* Gen.). Indian J. Agr. Sci. 22:75–91.

——— 1955a. Persistence of yellow vein mosaic virus of *Abelmoschus esculentus* (L.) Moench in its vector (*Bemisia tabaci* Gen.). Indian J. Agr. Sci. 25:293–302.

——— 1955b. Ability of the white-fly to carry more than one virus simultaneously. Current Sci. 24:317–318.

——— 1963. Transmission of plant viruses by whiteflies. Nat. Inst. Sci. India Bull. 24:11–33.

Wolf, F. A., Whitcomb, W. H., and Mooney, W. C. 1949. Leaf curl of tobacco in Venezuela. J. Elisha Mitchell Sci. Soc. 65:38–47.

# Mites as Vectors of Plant Viruses

J. T. SLYKHUIS

*Cell Biology Research Institute,**
*Research Branch,*
*Canada Department of Agriculture,*
*Ottawa, Canada*

## I. INTRODUCTION

Mites have been suspected of being vectors of plant viruses since Amos *et al.* (1927) reported a correlation between the presence of the big bud or black currant gall mite *Phytoptus* (*Eriophyes*) *ribis* (Westw.) Nal. and the development of the reversion disease of black currants. Experiments by Massee (1952) confirmed that mites from diseased bushes caused the disease in test bushes. However, the mites caused toxic effects that were often difficult to distinguish from the effects of reversion disease.

The case for eriophyid mites as vectors of plant viruses was convincingly strengthened by the demonstration that wheat streak mosaic virus (WSMV) was transmitted by *Aceria tulipae* K. (Slykhuis, 1953, 1955). This virus had attracted much attention as the cause of a rapidly developing, destructive disease of an important annual crop, wheat, in the great plains area of North America. Once suspected, the mites were readily found on diseased plants and easily tested as vectors. Their role in the transmission of the virus was soon confirmed by many investigators, and this unquestionable example of the mite transmission of a virus encouraged other workers to take a closer look at the possibilities of mites as vectors of other viruses. Now there is published evidence that eight plant viruses are transmitted and others are suspected to be transmitted by mites of the family Eriophyidae. One virus has been reported to be transmitted by a species of the Tetranychidae.

The subject of mite transmission of plant viruses has recently been reviewed in detail (Slykhuis, 1965). Here, the evidence for mite

* Contribution No. 613.

transmission of each virus is discussed along with new evidence on some mite–vector relationships.

## II. ERIOPHYIDAE AS VECTORS

The tiny (0.2 mm long) four-legged, worm-like eriophyids have piercing, sucking mouth parts and usually feed on young tissues in buds or on leaves of their specific host plants (Keifer, 1952). Some species cause leaf discolorations, malformations, galls, bud blasting, swelling, and other injuries that may be confused with symptoms caused by some viruses and which may have hampered earlier recognition of the role of some of these mites as vectors of viruses.

Four principal tests have been used to prove that symptoms associated with mites on plants were caused by viruses (Table 1), but the usability and reliability of each test has differed with the nature of the host, the virus, and the mites.

*1. Manual transmission* by the leaf rub method is reliable for differentiating the symptoms of wheat streak mosaic, ryegrass mosaic, and *Agropyron* mosaic from the toxic effects of mites. The viruses can be transmitted by rubbing juice from diseased plants on the leaves of test plants with the complete assurance that no mites are carried over

TABLE 1
EVIDENCE THAT DISEASE SYMPTOMS ASSOCIATED WITH ERIOPHYID MITES ARE CAUSED BY VIRUSES

| Disease | Mite | Symptoms persist after mites killed | Other transmission: Sap | Other transmission: Graft | noninfective mites from eggs |
|---|---|---|---|---|---|
| Currant reversion | *Phytoptus ribis* | + | – | + | 0 |
| Wheat streak mosaic | *Aceria tulipae* | + | + | 0 | + |
| Wheat spot mosaic | *Aceria tulipae* | + | – | 0 | + |
| Fig mosaic | *Aceria ficus* | + | – | + | + |
| Peach mosaic | *Eriophyes insidiosus* | + | – | + | + |
| Ryegrass mosaic | *Abacarus hystrix* | + | + | 0 | + |
| *Agropyron* mosaic | *Abacarus hystrix* | + | + | 0 | + |
| Pigeon pea sterility | *Aceria cajani* | + | – | + | 0 |
| Latent virus of plum | *Vasates fockeui* | 0 | + | 0 | 0 |

[a] + = tests with positive results; – = tests with negative results; 0 = no satisfactory tests.

from the diseased plants. Unfortunately, not all viruses can be transmitted in this way.

*2. Graft transmission* has been used to prove that currant reversion, fig mosaic, peach mosaic, and pigeon pea sterility, which affect woody plants, are caused by viruses, but special care is necessary to assure the complete absence of mites on the scion material. Treatment with miticides can seldom be relied upon to eliminate all mites in buds and leaf whorls. However, for graft transmission tests, the absence of mites can be assured by grafting patches of bark on the stems of test plants or by using lengths of stem, stripped of all buds that could harbor mites, as intermediate grafts between a healthy root stock and healthy scion.

*3. Tests with nonviruliferous mites.* Since the capacity to induce toxic symptoms is inherent (Carter, 1962), the development of non-symptom-inducing colonies of eriophyids, by hatching eggs on healthy plants, can be used as evidence that the symptoms are caused by a virus. The evidence can be strengthened if the mites become able to induce symptoms only after feeding on diseased plants. Experiments with infective and noninfective mites have provided the primary evidence that wheat spot mosaic is caused by a virus. However, some leafhopper-transmitted viruses are transovarially transmitted so there could also be viruses that are transmitted through eggs of mites. If so, some of the symptoms now attributed to toxic feeding may be caused by transovarially transmitted viruses. Other criteria will be required to prove that such symptoms on plants are caused by virus.

*4. The continued development of symptoms after the elimination of mites* has been used as evidence that symptoms on plants were caused by a virus rather than by the toxic effects of associated mites. This method cannot be considered reliable but may be used as a supplement to other methods. As with insect toxins (Carter, 1962), there is no assurance that some of the toxic effects of eriophyids will not continue long after the mites have been removed and plant growth has continued. The uncertainties are accentuated with the eriophyids because these mites are so tiny that some of them may survive unobserved in buds and leaf whorls and continue to induce toxic symptoms even after plants have been treated with miticides.

### A. *Viruses that Infect Grasses*

Two grass viruses, wheat streak mosaic virus (WSMV) and wheat spot mosaic virus (WSpMV), are transmitted by one eriophyid mite, *Aceria tulipae* K. Two others, ryegrass mosaic virus (RMV) and *Agropyron* mosaic virus (AMV), are transmitted by another mite,

*Abacarus hystrix* Nal. None of these viruses has been found to be transmissible by both species of mites. Three of these viruses, WSMV, RMV, and AMV, are readily transmitted manually; they have flexuous, rod-shaped particles between 700 mμ and 710 mμ long (Brandes, 1964; Staples and Brakke, 1963); and they were recognized before they were found to be transmitted by mites. The other, WSpMV, has not been transmitted manually; no particles have been recognized; and it was not recognized as a virus until after the symptoms it causes were found to be induced by mites (Slykhuis, 1956).

*Wheat streak mosaic virus* was probably first collected in Nebraska in 1922 but was identified from collections in Kansas in 1932 (McKinney, 1937). It caused chlorotic dashes and streaks, stunting, and early death of wheat plants. It has caused serious losses of wheat in the central and northern great plains and western states and in Alberta, Canada (King and Sill, 1959; Staples and Allington, 1956; Slykhuis, 1955). More recently, it has been found in wheat and maize in Ontario (Slykhuis, 1964 Paliwal, Slykhis, and Wall, 1966) and Ohio (McKinney *et al.*, 1966). It has also been recognized in Jordan (Slykhuis and Bell, 1963), Rumania (Pop, 1962), the U.S.S.R. (Razvyazkina, Kopkova, and Belyanchikova, 1963) and Yugoslavia (Šutié and Tošié, 1964).

Wheat streak mosaic virus may infect spring wheat, maize, and other spring-sown crops, but it has been reported only in or near areas where winter wheat is grown. Being readily transmitted manually, the virus was recognized long before its natural means of spread was known. It was differentiated from other viruses that infected wheat by differences in symptoms on wheat, by its host range, and by transmission characteristics (McKinney, 1937, 1944). There was no evidence that it was seed or soil borne. It spread rapidly during the summer and although there was circumstantial evidence that it was spread by insects, there was no correlation between its transmission and the presence of any specific insect. Cages with screens sufficiently fine to prevent the passage of thrips did not protect plants from infection with WSMV. Attention was therefore focused on the tiny eriophyid mite *Aceria tulipae* K., and it was found to be an efficient vector (Slykhuis, 1953, 1955; Connin, 1953; Staples and Allington, 1956). The same mite has been found to transmit WSMV in Europe (Pop, 1962; Razvyazkina *et al.*, 1963).

About 30–50% of the mites reared on diseased wheat transmitted the virus when tested singly on wheat seedlings. Mites of all stages except the eggs carried the virus. Noninfective mites were therefore obtained by transferring eggs to healthy wheat to hatch. The mites multiplied parthenogenetically and completed the life cycle from egg to egg in 7–10 days at 24–27°C. Although the mites caused an inrolling of leaves of wheat and other graminaceous hosts, and sometimes caused a chlorotic mottling on barley, they did not cause chlorotic symptoms on wheat that could be confused with wheat streak mosaic.

The epidemiology and control of wheat streak mosaic is closely correlated with the history of the mites in the field. There are a number of annual grasses that can be infected with the virus and with the mites. The virus is probably harbored on one or more species of perennial grasses but no perennial has been found naturally infected with the virus and also infested with mites. Wheat is a most favorable host for both the virus and the mites, and although it is an annual, it is of predominant importance in perpetuating the virus and vector throughout the year (Slykhuis, 1955; Staples and Allington, 1956; Slykhuis, Andrews, and Pittman, 1957). Winter wheat infested with viruliferous mites in the fall harbors the disease over the winter. During the warm weather of the next spring and summer the mites multiply and are dispersed by wind. Any immature spring or volunteer wheat plants, or regrowth from hailed wheat, can perpetuate the virus and the mites after the winter wheat crop matures, and if still present, these plants may provide sources of infection for fall-sown crops. A break in the continuity of immature wheat practically eliminates the disease; therefore, wheat streak mosaic is not a problem in spring wheat areas where there is no winter wheat to harbor the virus and vector during the winter, or in winter wheat areas such as Oklahoma or the eastern United States and eastern Canada where immature wheat is absent during most of the summer. Exceptions occur in southwestern Ontario (Paliwal *et al.*, 1966) and Ohio (McKinney *et al.*, 1966), where maize appears to function as an oversummering host of the virus and the mites. Practical control of the disease can be achieved easily in most areas simply by eliminating oversummering hosts of the disease before sowing winter wheat in the vicinity.

The relationship of WSMV to its vector appears to be unique. Like most of the stylet-borne (nonpersistent), aphid-transmitted viruses, WSMV has flexuous, rod-shaped particles and is readily sap transmissible. Otherwise, it resembles the circulative (persistent) viruses in that its vector must feed for long periods (15 min or longer) to acquire or transmit it. The virus persists in its vector for long periods (up to 9 days) and is retained through the molts (Slykhuis, 1955; Staples and Allington, 1956; Orlob, 1966). An additional interesting fact is that nymphs readily acquire the virus, but mites that are not given access to diseased plants until they are adults do not become infective (Slykhuis, 1955; Del Rosario, 1959; Orlob, 1966).

Wheat streak mosaic virus has been detected in extracts of crushed viruliferous *A. tulipae* by manual inoculation of wheat and by examination with the electron microscope (Orlob, 1966). In addition to these methods of detection, Paliwal and Slykhuis (1967) used the serological precipitin ring test and also examined ultrathin sections with the electron microscope. Large concentrations of flexuous, rod-shaped particles characteristic of particles of WSMV were found in the midgut and the rectum-like sac of the hindgut of mites that acquired WSMV from manually inoculated wheat (Figs. 1–4). Most of the particles occurred in bundles in the midgut, indicating that the virus was either taken in from the plant in bundles or multiplied elsewhere and accumulated in this region. The particles were densely packed, but a few discrete bundles were evident in the hindgut, indicating that they had become dissociated during passage from the midgut. No particles were seen outside the intestinal region of the viruliferous mites or in any portion of the nonviruliferous mites. It was suggested that plant infection could occur either from virus particles that moved forward toward the mouth parts while the mites were feeding or from particles excreted by the mites. Excreted particles could be introduced into plant cells through subsequent feeding punctures or by the rubbing action of the caudal suckers and setae which aid the mites to cling to the leaf surface. The inability of mites to transmit if they did not feed on diseased plants until after they became adults may be attributed to an apparent degeneration of the connection between the midgut and hindgut, thus preventing the excretion of virus particles acquired by the feeding of adults.

Stein-Margolina, Cherin, and Razvyazkina (1966) also found filiform particles (16 $m\mu \times 670$–$720$ $m\mu$) in suspensions from leaves of wheat infected with WSMV and in infective mites, but when examining ultrathin sections, they observed polygonal particles in the cytoplasm of cells of the mesophyl of leaves of diseased wheat and in cells of the mites. The particles were 16–18 $m\mu$ in diameter and had light-colored centers 40–60 Å in diameter. In the early stages of plant infection the particles in the cytoplasm were scattered, but later they appeared in orderly groups. Sometimes they formed chains of filaments. In the cells of mites, conglomerates of particles were often surrounded by a membrane. The authors concluded that the observations indicate that WSMV multiples in the vector. They did not discuss the significance of the absence of flexuous, rod-shaped particles in ultrathin sections of diseased wheat and infective mites, and they did not suggest

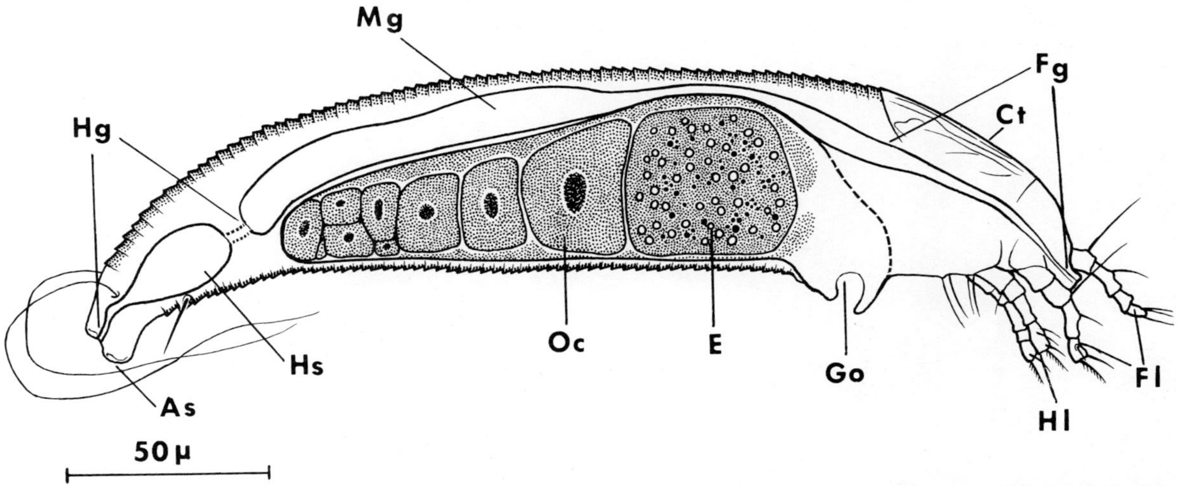


*Fig. 1.* Diagram of a young adult *Aceria tulipae* showing the alimentary canal and the ovary: Fg, foregut; Mg, midgut; Hg, hindgut; Ct, cephalothorax; Fl, forelegs, Hl, hind legs; Go, genital openings; E, egg; Oc, oocyte; Hs, rectum-like sac of the hindgut; As, anal sucker.

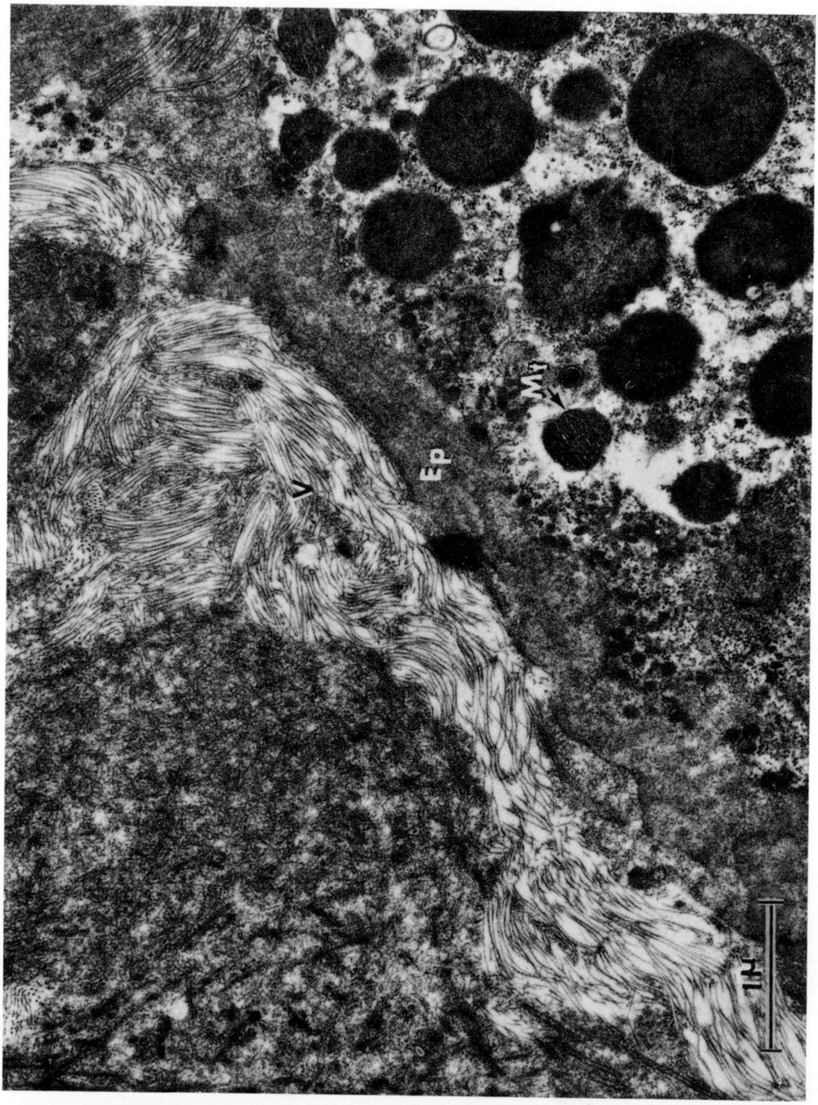


*Fig. 2.* Ultrathin section through the midgut of *A. tulipae* showing an accumulation of WSMV particles lining the epithelium (Ep) of the wall of the posterior part of the midgut: Mt, mitochondrion; V virus. Note the lack of distinct cell walls in the epithelium.

*Fig. 3.* Wheat streak mosaic virus particles in the lumen of the midgut of *A. tulipae*. Note the long bands and discrete bundles of particles.

whether or not the polygonal particles represent sectional components of the rod-shaped particles, different particles of the virus, or a different virus.

*Wheat spot mosaic virus* (WSpMV) was not detected until *A. tulipae* was recognized as a vector of WSMV in Alberta (Slykhuis, 1953, 1955, 1956). When mites from naturally diseased wheat were tested singly on wheat for the transmission of WSMV, some of the test plants did not develop streak mosaic symptoms, but instead, developed chlorotic spots, severe chlorosis, stunting, and necrosis, even though no sap-transmissible virus was present. The symptoms were at first attributed to the toxic effects of the mites, but they continued to develop even after the plants were freed of mites. Furthermore, mites hatched from

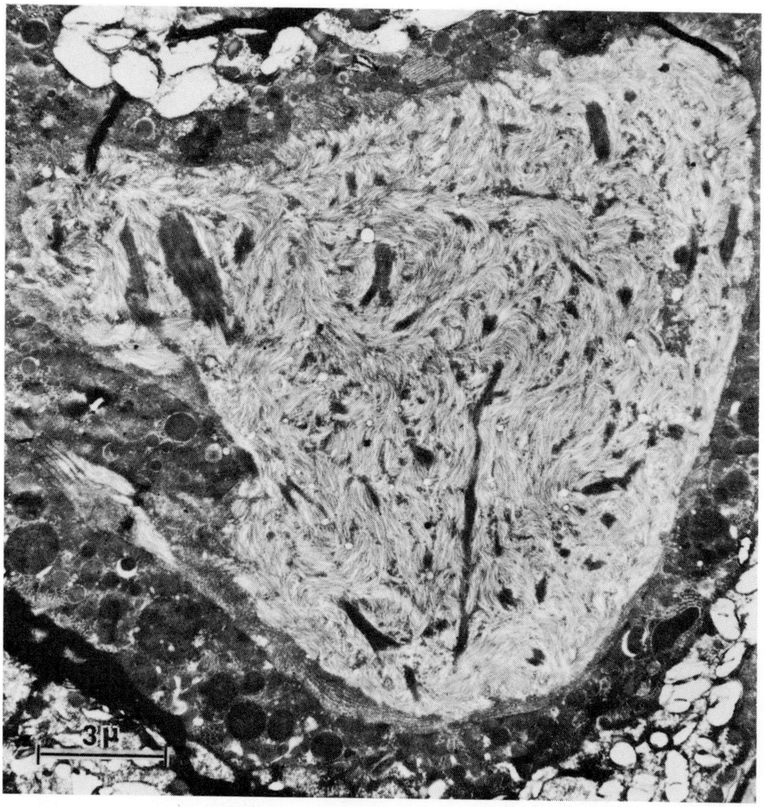


*Fig. 4.* Part of an ultrathin section through *A. tulipae* showing the rectum-like sac (hindgut) packed with virus particles.

eggs transferred to healthy wheat did not produce such symptoms unless they had first fed on diseased wheat. The symptoms were therefore attributed to a virus. Different isolates of the virus caused symptoms that differed greatly in severity. Some caused extreme leaf chlorosis, death of the emerging leaves, and they quickly killed the plants. Others caused milder effects. Plants infected with both WSMV and WSpMV were more severely diseased than plants infected with either virus alone.

Like WSMV, WSpMV was carried by all stages of *A. tulipae* except the eggs, but unlike WSMV, it was acquired by adults as well as nymphs. Some of the mites from plants infected with both WSMV and WSpMV carried both viruses, but a higher percentage of the mites transmitted WSpMV (65%) than WSMV (34%). Viruliferous mites placed on an immune host remained infective up to the twelfth day.

Although WSpMV has not been detected elsewhere, with the possible exception of a very mild chlorosis factor transmitted by *A. tulipae* in Ontario (Slykhuis, 1961), it has been isolated repeatedly since 1952 from diseased wheat from southern Alberta. It causes a very distinct mottling and spotting on a number of varieties of corn as well as on barley, oats, rye, and a number of annual grasses. Since it has the same vector as WSMV, its epidemiology and the requirements for control are believed to be similar, and thus it is considered with WSMV as part of a wheat mosaic complex.

*Ryegrass mosaic virus* (RMV) is a sap-transmissible virus first isolated in England in 1956 (Slykhuis, 1958). It has a flexuous, rod-shaped particle (Mulligan, 1960), which Brandes (1964) reported to have a normal length of 703 m$\mu$, almost identical with the particle for WSMV. It causes light-green to yellow mottling and streaking of ryegrass leaves, and infects Italian ryegrass (*Lolium multiflorum* L.), perennial ryegrass (*L. perenne* L.), *Dactylis glomerata* L., *Festuca* sp., rice, and oats, but not wheat or other hosts of WSMV.

Mulligan (1960) found that of three species of eriophyid mites on ryegrass only *Abacarus hystrix* (Nal.) transmitted the virus. When reared on diseased ryegrass, all stages of the mite except the eggs carried the virus. Noninfective mites acquired the virus during 2–12-hr periods on diseased ryegrass. In the absence of a virus source, viruliferous mites lost infectivity in less than 24 hr; hence, RMV resembles the stylet-borne (nonpersistent), aphid-transmitted viruses.

Viruses isolated from ryegrass in Washington, U.S.A. (Bruehl, Toko, and McKinney, 1957), and British Columbia, Canada (Slykhuis, 1964), have particles similar to RMV from England, but attempts to transmit them with *A. hystrix* collected in Ontario have failed.

*Agropyron mosaic virus* (AMV) has been found on *Agropyron repens* (L.) Beauv. in Virginia, Iowa, South Dakota, Saskatchewan, Ontario, and other eastern provinces in Canada (McKinney, 1937; Slykhuis, 1962). In Ontario it can be found in winter wheat each year, and causes mosaic symptoms similar to wheat streak mosaic. The virus is readily sap transmitted and has particles similar to WSMV but is only distantly related to WSMV serologically; and the two viruses are synergistic in cross-inoculation tests (Slykhuis and Bell, 1966).

*Agropyron* mosaic virus spreads from infected *A. repens* to wheat during the warmest summer weather in Ontario. Wheat test plants placed in the field near diseased *A. repens* or naturally diseased wheat during the summer may become infected unless protected with screens sufficiently fine to prevent the passage of eriophyid mites. Although *A. tulipae,* the vector of WSMV, and *Aculus mckenziei* (K.) are sometimes present, all attempts to transmit AMV with these mites have failed. *Abacarus hystrix* has been constantly found associated with the natural spread of AMV, and although many of the attempts to transmit the virus with this mite have failed, other tests have indicated that it is a very inefficient vector. Unpublished evidence that *A. hystrix* transmits the virus has been obtained in growth cabinets in which fans have been used to blow mites from source plants to test plants. The source plants were *A. repens* or wheat infected with AMV by manual inoculation. These were colonized with pure cultures of the species of mite to be tested. Wheat test plants were placed in each growth cabinet beyond contact with the source plants. A fan was placed so that it blew across the source plants toward the test plants and it was run daily during the light periods for 7–10 days. By this procedure many mites were transferred in a natural manner from the source plants to the test plants. In most tests with *A. hystrix,* 1% or more of the test plants became infected with AMV, but no test plants became infected in any of the tests with either *A. tulipae* or *A. mckenziei,* or with no mites on the source plants.

### B. *Viruses that Infect Trees*

Fig mosaic and peach mosaic, which are induced by eriophyid mites, are considered to be caused by viruses rather than by feeding toxins for three reasons: their causes are graft transmissible in the absence of mites; test plants infected by mites continue to develop the symptoms even after the elimination of the mites; and mites do not induce symptoms on healthy seedlings unless they have fed on diseased trees. Two other viruses, cherry mottle leaf and a latent virus of plum, are also suspected to be transmitted by mites.

*Fig mosaic* was the first tree disease reported to be caused by a mite-transmitted virus (Flock and Wallace, 1955). The disease, first reported in California in 1933 (Condit and Horne, 1933), occurs in widely scattered locations on at least four continents (Smith, 1957; Slykhuis, 1965). The symptoms, which include mottling, spotting, distortion of leaves, and sometimes leaf and fruit drop, may be confused with leaf spotting, chlorosis, and russetting caused by the fig mite, *Aceria ficus* (Cotte). When fig mosaic was first described as a virus disease (Condit and Horne, 1933), the fig mite was suspected as a vector. Healthy fig seedlings developed symptoms after mite-infected leaves and bud scales were placed on them. The cause was transmitted by grafting, but precautions were not taken to assure the elimination of mites that could have been transferred on the scion material and may have caused toxic effects. Flock and Wallace (1955) showed that mosaic symptoms continued to develop on fig cuttings dusted with sulfur to kill the mites and grown in a mite-free environment. Also graft transmission was successful in the absence of mites. Mite transmission of the virus was proved by transferring 1–200 mites from diseased trees to healthy test seedlings, then after 3–5 days, killing the mites with sulfur. Some mosaic-like symptoms developed in less than 10 days, but similar symptoms were also caused by nonviruliferous mites from eggs hatched on disease-free seedlings. The symptoms caused by fig mosaic virus required 10 or more days to appear. In tests with one viruliferous mite per test plant, 7 out of 10 plants developed mosaic symptoms.

With such an efficient and abundant vector, and the host being a perennial, it is not surprising that fig mosaic appears to have infected all fig trees in California. However, the varieties of figs in commercial production are sufficiently tolerant for satisfactory production to continue.

*Peach mosaic* was first recognized as a virus disease in Texas in 1932 (Hutchins, 1932) and has since been reported in many regions in the United States and Europe. The virus has been transmitted by grafting or budding to peach, plum, apricot, and almond (Smith, 1957). The symptoms on peach include a mosaic pattern on leaves, and the leaves are often small, narrow, crinkled, and irregular in shape. Internodes become shortened and there is a profuse growth of leaf axil buds.

Vectors of the virus were sought for many years without success. At one time the peach silver mite, *Vasates cornutus* (Banks), was suspected as a vector (Wilson and Cochran, 1952). It caused a silvering of leaves and the "yellow spot" disease, which included vein-associated chlorosis, longitudinal rolling, and spotting of peach leaves. A 24-hr feeding period by the mites was sufficient to produce the symptoms, and there was a 10-day delay in their appearance, which suggested that these symptoms could be caused by a virus, but patch bark grafting with precautions to assure the absence of mites failed to show that a virus was involved.

After more than 8000 tests with more than 150 species of arthropods, Wilson, Jones, and Cochran (1955) found that the virus was transmitted by an eriophyid mite later named *Eriophyes insidiosus* (Keifer and Wilson, 1956). The mite occurred beneath bud scales, and when varying numbers from diseased trees were transferred to the buds of 65 potted Rio Oso Gem peach seedlings, 17 seedlings developed symptoms in 14–100 days. The presence of the virus was confirmed by patch bark graft transmission to other healthy seedlings. Check plants that received equal numbers of *E. insidiosus* mites from healthy peach and plum did not develop symptoms. Also non-viruliferous mites were obtained from eggs laid in buds of diseased trees but hatched on healthy seedlings. Jones reported that in tests with one mite per test plant, 2.5% of the mites from diseased peach transmitted the virus (Slykhuis, 1965). With 30–50 mites per test plant, 67–75% of the test seedlings developed symptoms. Transmission resulted only when mites were transferred to buds rather than to leaf blades. The mites retained infectivity for at least 2 days on glass slides in the absence of a virus source.

Despite the common occurrence of the mites, not all susceptible peach and plum trees in the field are diseased. Removal of infected trees, nursery inspection, and quarantine procedures have been effective in limiting the spread of the virus (List, Landblom, and Sisson, 1956).

*Cherry mottle leaf virus,* as reported by Jones (Slykhuis, 1965), was suspected to be transmitted by an eriophyid mite in California, but no confirmation has been reported.

*A plum latent virus* in Germany was recently reported to be transmitted by an eriophyid mite (Proeseler and Kegler, 1966). The mites, (*Vasates fockeui* Nal.) were transferred from different plum varieties with and without symptoms of virus disease to *Chenopodium foetidum* Schrad., on which ochre colored local lesions surrounded by necrotic rings developed after 7–10 days. Similar local symptoms were produced by mechanical inoculation of *C. foetidum* with sap from plum leaves, and gray necrotic lesions developed on *C. quinoa* similarly inoculated. No systemic symptoms developed on either host. Particles 750 mμ long were found in both hosts. It was "assumed" that the symptoms on the *Chenopodium* sp. were caused by a virus latent in plum and transmitted both manually and by the mites.

### C. *Viruses that Infect Shrubs*

Currant reversion and pigeon pea sterility, which are induced by mites, are attributed to viruses principally on the basis of graft transmission and the observation that diseased bushes continue to develop the symptoms even after treatments to eliminate the mites.

*Currant reversion* was the first plant disease suspected to be caused by a mite-transmitted virus. It is called "reversion" because the character of the leaves of affected black currants (*Ribes nigrum*) changes so that the bushes appear to have reverted to the wild type. Affected bushes eventually cease to bear fruit. The disease is widespread in the British Isles and Europe (Smith, 1957).

The black currant gall mite, *Phytoptus* (*Eriophyes*) *ribis* (Westw.) Nalepa, was recognized as a serious pest of black currants long before the reversion disease was recognized (Massee, 1928). The mite caused the swelling of buds known as "big bud." Slightly swollen buds sometimes produced small, crumpled leaves, but most swollen buds dried up in early summer. The symptoms now attributed to the reversion disease were at first suspected to be caused by the mites. They included narrower, smaller than normal leaves with bases flattened by a reduction in the depth of the basal sinus. There were fewer subsidiary veins. The leaf margins were coarsely rugose. Careful observation was required to distinguish reversion and "false" reversion that may develop after physical injury or adverse environmental effects (Amos and Hatton, 1927; Thresh, 1964b). Certain "false" reversion symptoms are also attributed to the feeding effects of *P. ribis* because the plants recover after the mites are eliminated, but true reversion symptoms persist (Thresh, 1963).

The first evidence that reversion was caused by a virus was obtained when the disease developed on normal plants grafted to or inarched with reverted bushes. Some of the grafts were done by using lengths of stem stripped of buds to eliminate mites, as intermediate grafts between healthy root stock and scion (Amos *et al.*, 1927). However, there was still doubt that the mites had been eliminated. Massee (1952) infected 24 healthy plants with large numbers of mites from reverted bushes while keeping 6 control bushes free from mites by repeatedly spraying with lime–sulfur and dusting with sulfur. Reversion symptoms were evident on 10 bushes in the first year, 9 in the second, and on all 24 infected bushes by the third year, but none of the 6 control plants developed the symptoms. This experiment confirmed that the mites induced the symptoms but did not strengthen the evidence that the disease was caused by a virus. Smith (1962) achieved similar results when he successfully transferred single mites to the axils of new shoots of black currant seedlings during the normal period of mite migration in spring and early summer. He found that a single mite could start an infestation leading to bud galling and the development of reversion.

Since no successful tests have been done with colonies of non-infective mites obtained by hatching eggs from diseased bushes on healthy plants, there is still no conclusive evidence of the effects of non-viruliferous mites. However, Cropley, Posnette, and Thresh (1964) induced the disease symptoms on healthy bushes that received patch bark grafts from reverted bushes. Also, Thresh (1964b) infected small currant seedlings with mites from diseased bushes and after 4 days dipped them in 0.05% Endrin solution to destroy the mites. The seedlings did not develop any recognized mite injury or bud galls, but some developed reversion symptoms, and the presence of reversion virus was confirmed by patch bark grafting to healthy bushes. These results confirm that reversion symptoms are caused by a virus transmitted by *P. ribis*.

Despite the abundance of mites and their persistence on infected currant bushes, neither the mites nor the reversion disease appears to spread rapidly to new bushes. Healthy black currant bushes are not as readily infected with mites as reverted bushes, evidently because healthy bushes have a greater hairiness on flowers and vegetative parts, which impedes the movement of mites into susceptible buds (Thresh, 1964a). The greater susceptibility of reverted bushes indicates the importance of removing the diseased bushes as an effective control measure. The spread of mites and reversion can be prevented with 10 weekly sprays with a miticide such as endrin, endosulfan, or lime–sulfur during the period of mite migration (Smith 1961; Thresh, 1965).

*Pigeon pea sterility,* which is characterized by mosaic symptoms on the leaves and partial sterility of the flowers of the pigeon pea (*Cajanus cajan* (Linn) Millsp.) in India, has been reported to be caused by a virus transmitted by an eriophyid mite, *Aceria cajani* Channa Bosovanna (Seth, 1962, 1965). The virus was transmitted by wedge grafting to 35 of 99 plants of the pigeon pea variety N.P. 69. Attempts to transmit it by sap inoculation, with dodder (*Cuscuta reflexa* Roxb.) and with leafhoppers, aphids, whiteflies, thrips, and two species of mites found on pigeon pea failed. However, the symptoms developed in 3–5 weeks on 31 of 136 pigeon pea plants on which 5–20 eriophyid mites (*A. cajani*) from diseased pigeon pea were placed. After 4 days the plants were regularly sprayed with 0.1% Ekatox to kill the mites. Periodic examination of the plants under a binocular microscope did not reveal any surviving mites. Infection was also obtained by placing diseased, mite-infested leaves on the test plants. Mites from healthy plants did not cause the symptoms. Efforts to colonize the mite on pigeon pea and several other crop plants under controlled conditions were not successful; hence, it has not been possible to test the effects of noninfective mites from eggs transferred from diseased plants to hatch on healthy plants.

The mite and the virus affected all tested varieties of *C. cajan,* the only species in the genus *Cajanus*. No other hosts of either the mite or virus have been found. Possibly disease resistance can be obtained in intergeneric crosses with *Atylosia* sp. Since pigeon pea is grown both as an annual crop and as a perennial, especially along hedges, the perennial plants provide reservoirs of the virus and mites. It may be possible to control the disease in annual crops by removing all old plants in the vicinity well before sowing the new crops.

## III. TETRANYCHIDAE AS VECTORS

The Tetranychidae or spider mites are pests of a wide range of plants (Metcalf, Flint, and Metcalf, 1962). They feed by piercing and sucking and cause pale to reddish-brown specks to blotches on leaves. As they can move freely from plant to plant and can be dispersed widely on infected plant material, they are common pests in both the greenhouse and the field and have been suspected as vectors of plant viruses.

*"Cotton curlines"* disease in Azerbaijan was reported to be transmitted by a tetranychid, *Epitetranychus althaeae* to 7.7–10% of the test plants, as well as manually and by aphids including *Aphis gossypii* (85–100%), *A. laburni* (6.1–16.6%), and *Myzus persicae* (12.5–25%) (Moskovets, 1941). However, the incubation period of the virus in the plants varied from 35 to 56 days, and thus it is possible that infection attributed to the mites could have been caused by one of the alternative means of transmission. There has been no confirmation of the report.

*Potato virus Y* is the only virus for which proof of transmission by a tetranychid has been reported (Schulz, 1963). The two-spotted spider mite *Tetranychus telarius* (L.) transmitted two isolates of the virus from potato and *Nicotiana glutinosa* to Red Pontiac potato. The highest transmission was obtained with acquisition and inoculation feeding periods of 5 min, which resulted in the infection of 18 out of 47 potato test plants. The plants that developed symptoms were tested for the presence of virus Y by spot necrosis tests on tobacco. The mite *T. telarius,* like aphid vectors of the same virus, transmits the virus in a manner characteristic for stylet-borne (nonpersistent) viruses.

## IV. CONCLUSIONS

Many of the plant-infesting mites cause toxic feeding symptoms that may be confused with virus diseases, but six species of the Eriophyidae and one of the Tetranychidae have been shown to transmit plant viruses, and other mites are suspected.

Each of the eriophyid vectors is the only known vector of the virus or viruses it transmits. Of the four known eriophyid-transmitted grass viruses three of them, WSMV, RMV, and AMV, are sap transmissible and have flexuous, rod-shaped particles about 700 mμ long. Of these, RMV is retained less than 24 hr by its vector, *Abacarus hystrix,* and hence may be stylet borne. *Agropyron* mosaic virus is very inefficiently transmitted by *A. hystrix.* Wheat streak mosaic virus and the non-sap transmissible WSpMV are both persistent in their vector *Aceria tulipae,* being carried through molts and retained up to 9–12 days, respectively. Wheat spot mosaic virus can be acquired by adults as well as by nymphs, but WSMV is not acquired by adults. Electron microscopic studies revealed large concentrations of WSMV particles in the midgut and hindgut of viruliferous *A. tulipae,* indicating that the virus either accumulates or multiplies in the mite. The tree diseases, fig mosaic and peach mosaic associated with *Aceria ficus* and *Eriophyes insidiosus,* respectively are attributed to viruses rather than to mite feeding toxins because they are graft transmitted in the absence of mites and are not induced by mites from eggs hatched on healthy plants. Peach mosaic virus is retained at least 2 days by mites in the absence of a virus source. There is evidence that eriophyid mites also transmit a latent virus of plums.

Currant reversion and pigeon pea sterility, induced by *Phytoptus ribis* and *Aceria cajani,* respectively, are graft transmissible in the absence of mites, but the mites have not been cultured under controlled conditions so there have been no tests with noninfective mites from eggs hatched on healthy plants.

Potato virus Y, the only virus reported to be transmitted by a tetranychid mite (*Tetranychus telarius*), is readily transmitted manually and by aphids. It has a nonpersistent relationship with the mite as with the aphid vectors.

## BIBLIOGRAPHY

Amos, J., and Hatton, R. G. 1927. Reversion of black currants I. Symptoms and diagnosis of the disease. J. Pomol. Hort. Sci. 6:167–183.

Amos, J., Hatton, R. G., Knight, R. C., and Massee, A. M. 1927. Experiments in the transmission of reversion in black currants. Ann. Rep. East Malling Res. Sta. Kent, Suppl. II:126–150.

Brandes, J. 1964. Identifizierung von gestreckten pflanzenpathogenen Viren auf morphologischer Grundlage. Mitt. Biol. Bundesanstalt Land-Forstwirtsch. Berlin-Dahlem 110:1–130.

Bruehl, G. W., Toko, H., and McKinney, H. H. 1957. Mosaics of Italian ryegrass and orchard grass in western Washington. Phytopathology 47:517.

Carter, W. 1962. Insects in relation to plant disease. Wiley, New York. 705 p.

Condit, I. J., and Horne, W. T. 1933. A mosaic of the fig in California. Phytopathology 23:887–889.

Connin, R. V. 1953. Studies to determine the insect vectors of wheat streak-mosaic virus. U.S. Dept. Agr. Spec. Rep. W-13, 10 p.

Cropley, R., Posnette, A. F., and Thresh, J. M. 1964. The effects of black currant yellows virus and an atypical strain of reversion virus on yield. Ann. Appl. Biol. 54:177–182.

Del Rosario, M. S. E. 1959. Studies of *Aceria tulipae* Keifer (Eriophyidae) and other eriophyid mites in Kansas in relation to the transmission of wheat streak mosaic virus. Diss. Abstr. 20:863–864.

Flock, R. A., and Wallace, J. M. 1955. Transmission of fig mosaic by the eriophyid mite *Aceria ficus*. Phytopathology 45:52–54.

Hutchins, L. M. 1932. Peach mosaic—a new virus disease. Science 76:123.

Keifer, H. H. 1952. The eriophyid mites of California. Bull. California Insect Survey 2:123.

Keifer, H. H., and Wilson, N. S. 1956. A new species of eriophyid mite responsible for the vection of peach mosaic virus. Bull. California Dept. Agr. 44: 145–146.

King, C. L., and Sill, W. H., Jr. 1959. 1959 wheat streak mosaic epiphytotic in Kansas. Plant Disease Reporter 43:1256–1257.

List, G. M., Landblom, N. and Sisson, M. A. 1956. A study of records from the Colorado peach mosaic suppression program. Colorado Agr. Exp. Sta. Tech. Bull. 59. 28 p.

McKinney, H. H. 1937. Mosaic diseases of wheat and related cereals. U.S. Dept. Agr. Circ. 442. 23 p.

——— 1944. Descriptions and revisions of several species of viruses in the genera *Marmor, Fractilinia* and *Galla*. J. Washington Acad. Sci. 34:322–329.

McKinney, H. H., Brakke, M. K., Ball, E. M., and Staples, R. 1966. Wheat streak mosaic virus in the Ohio Valley. Plant Disease Reporter 50:951–953.

Massee, A. M. 1928. The life-history of the black currant gall mite, *Eriophyes ribis* (Westw.) Nal. Bull. Entomol. Res. 18:297–309.

——— 1952. Transmission of reversion of black currants. Ann. Rep. East Malling Res. Sta. Kent, 1951:102–165.

Metcalf, C. L., Flint, W. P., and Metcalf, R. L. 1962. Destructive and useful insects. 4th ed., McGraw-Hill, New York. 1087 p.

Moskovets, S. N. 1941. Plant virus diseases and their control. [In Russian]. Trans. Conf. Plant Virus Diseases, Moscow, 4–7/II 1940, Moscow-Leningrad Acad. Sci. USSR, p. 173–190.

Mulligan, T. 1960. The transmission by mites, host range and properties of ryegrass mosaic virus. Annu. Appl. Biol. 48:575–579.

Orlob, B. G. 1966. Feeding and transmission characteristics of *Aceria tulipae* Keifer as vector of wheat streak mosaic virus. Phytopathol. Z. 55:218–238.

Paliwal, Y. C., and Slykhuis, J. T. 1967. Localization of wheat streak mosaic virus in the alimentary canal of its vector *Aceria tulipae* Keifer. Virology 32: 344–353.

Paliwal, Y. C., Slykhuis, J. T., and Wall, R. E. 1966. Wheat streak mosaic virus in corn in Ontario. Can. Plant Disease Survey 46:8.

Pop, I. 1962. Die Strichelvirose des Weizens in der Rumanischen Volksrepublik. Phytopathol. Z. 43:325–336.

Proeseler, G., and Kegler, H. 1966. Ubertragung eines latenten Virus von Pflaume durch Gallmilben (Eriophyidae). Monatsberichte 8:472–476.

Razvyazkina, G. M., Kopkova, E. A., and Belyanchikova U. V. 1963. Wheat streak mosaic virus. [in Russian]. Zashchita Rast. Moskva 8:54–55.

Schulz, J. T. 1963. *Tetranychus telarius* (L.), new vector of virus Y. Plant Disease Reporter. 47:594–596.

Seth, M. L. 1962. Transmission of pigeon pea sterility by an eriophyid mite. Indian Phytopathol. 15:225–227.

———. 1965. Further observations and studies on pigeon pea sterility. Indian Phytopathol. 18:317–319.

Slykhuis, J. T. 1953. Wheat streak mosaic in Alberta and factors related to its spread. Can. J. Agr. Sci. 33:195–197.

———. 1955. *Aceria tulipae* Keifer (Acarina: Eriophyidae) in relation to the spread of wheat streak mosaic. Phytopathology 45:116–128.

———. 1956. Wheat spot mosaic, caused by a mite-transmitted virus associated with wheat streak-mosaic. Phytopathology 46:682–687.

———. 1958. A survey of virus diseases of grasses in northern Europe. FAO Plant Protection Bull. 6:129–134.

———. 1961. Eriophyid mites in relation to the spread of grass viruses in Ontario. Can. J. Plant Sci. 41:304–308.

———. 1962. *Agropyron* mosaic as a disease of wheat in Canada. Can. J. Bot. 40:1439–1447.

———. 1964. Noteworthy and new records of grass viruses in Canada in 1964. Can. Plant Disease Survey 44:242–243.

———. 1965. Mite transmission of plant viruses. Advances Virus Res. 11:97–137.

Slykhuis, J. T., Andrews, J. E., and Pittman, U. J. 1957. Relation of date of seeding winter wheat in southern Alberta to losses from wheat streak mosaic, root rot and rust. Can. J. Plant Sci. 37:113–127.

Slykhuis, J. T., and Bell, W. 1963. New evidence on the distribution of wheat streak mosaic virus and the relation of isolates from Rumania, Jordan and Canada. Phytopathology 53:236–237.

———. 1966. Differentiation of *Agropyron* mosaic, wheat streak mosaic, and a hitherto unrecognized *Hordeum* mosaic virus in Canada. Can. J. Botany 44:1191–1208.

Smith, B. D. 1961. The application of lime sulphur for the control of the black currant gall mite. Annu. Rep. Agr. Hort. Res. Sta. Long Ashton Bristol, 1960: 129–133.

———. 1962. Experiments in the transfer of the black currant gall mite (*Phytoptus ribis* Nal.) and of currant reversion. Annu. Rep. Agr. Hort. Res. Sta. Long Ashton Bristol, 1961:170–172.

Smith, K. M. 1957. Textbook of plant virus diseases. 2nd ed., Churchill, London. 652 p.

Staples, R., and Allington, W. B. 1956. Streak mosaic of wheat in Nebraska and its control. Univ. Nebraska Agr. Exp. Sta. Res. Bull. 178. 41 pp.

Staples, R., and Brakke, M. 1963. Relation of *Agropyron repens* mosaic and wheat streak mosaic viruses. Phytopathology 53:969–972.

Stein-Margolina, V. A., Cherin, N. E., and Razvyazkina, G. M. 1966. The wheat streak mosaic virus in plant cells and the mite vector [in Russian]. Dokl. Akad. Nauk SSSR, 169:1446–1448.

Šutić, D., and Tošić, M. 1964. Virus critivastog-mozaika Psenice u nasoj zemlji. Wheat streak mosaic virus in Yugoslavia. Zaštita Bilja 15:307–314.

Thresh, J. M. 1963. Abnormal black currant foliage caused by the gall mite *Phytoptus ribis* Nal. Annu. Rep. East Malling Res. Sta. Kent, 1962:99–100.

———. 1964a. Increased susceptibility to the mite vector (*Phytoptus ribis* Nal.) caused by infection with black currant reversion virus. Nature 202: 1028.

———. 1964b. Association between black currant reversion virus and its gall mite vector (*Phytoptus ribis* Nal.). Nature 202:1085–1087.

———. 1965. The chemical control of black currant reversion virus and its gall mite vector (*Phytoptus ribis* Nal.). Annu. Rep. East Malling Res. Sta. Kent, 1964.

Wilson, N. S., and Cochran, L. C. 1952. Yellow spot, an eriophyid mite injury on peach. Phytopathology 42:443–447.

Wilson, N. S., Jones, L. S., and Cochran, L. C. 1955. An eriophyid mite vector of peach-mosaic virus. Plant Disease Reporter 39:889–892.

# Plant Susceptibility to Virus Infection by Insect Transmission*†

K. G. Swenson

*Entomology Department, Oregon State University, Corvallis, Oregon*

## I. INTRODUCTION

Resistance is involved whenever an inherited characteristic of a plant limits damage by an insect or pathogen, according to recent usage (Painter, 1958; Kuc, 1966). Resistance and susceptibility are at the opposite ends of a spectrum (Kuc, 1966), with immunity as complete resistance and destruction of the plant as complete susceptibility. This obviously is not the only concept of susceptibility. Bawden and Kassanis (1950) pointed out that a distinction should be made between susceptibility to infection and sensitivity to the pathogen after infection (see also Baerecke, 1958; Bjorling, 1966). The process of infection and the effects on the plant after infection undoubtedly involve different plant responses. Thus, a plant may be relatively resistant to infection, but may suffer severely once infected. What we are concerned with in this chapter is the susceptibility to establishment of virus infection by insect transmission. The likelihood of infection may be determined, however, by plant characteristics other than those that govern susceptibility of plant cells to viruses. These other characteristics complicate the interpretation of experiments made to evaluate plant susceptibility.

## II. METHODS OF EVALUATING SUSCEPTIBILITY

### *A. Vector Host Status*

The status of a plant as a host of the vector may greatly influence the results of both field and laboratory tests. The likelihood of infection with circulative (persistent) viruses increases with increasing

* Approved as Oregon Agricultural Experiment Station Technical Paper 2239.

† The names of insects and viruses used herein are those used in the papers cited. The reader is referred to Kennedy, Day, and Eastop (1962), Nielson (1962), and Hopkins (1957) for more recently used names.

feeding time and vectors will feed more consistently on a host than on a nonhost plant. On the other hand, optimal transmission of stylet-borne (nonpersistent) viruses occurs with very short feeding periods, or probes. The aphids may make more of these on nonhosts than on host plants.

The most striking effect of vector resistance on virus spread occurs with the red raspberry mosaic viruses. Resistance of red raspberry varieties to the rubus aphids, *Amphorophora rubi* in Europe and *A. agathonica* in North America, gives very adequate control of these viruses. Field spread of the cranberry false-blossom virus in cranberry varieties was correlated with the suitability of the varieties as hosts of the leafhopper vector (Wilcox and Beckwith, 1933; Wilcox, 1951). Wilcoxon and Peterson (1960) found that virus spread in red clover varieties was correlated with susceptibility of the varieties to pea aphid and concluded that breeding pea aphid-resistant varieties was a more practical approach to reducing virus spread than developing varieties resistant to the several aphid-borne viruses in red clover.

But virus spread is not always less in vector-resistant varieties. Baerecke (1958) reviewed the literature relating to *Myzus persicae* resistance in potato varieties and the effect of resistance on leafroll incidence. She concluded that the use of aphid-resistant varieties is the wrong approach to leafroll control. Aphid-resistant varieties frequently have a high leafroll incidence and Folsom (1955) records 81% leafroll in one aphid-resistant variety. The mechanism of aphid resistance to *Amphorophora agathonica* and *A. rubi* in red raspberry must be quite different from the resistance so far known to *Myzus persicae* in potato. Apparently, the rubus aphids recognize the undesirable host status of resistant varieties so rapidly that they do not feed long enough to inoculate plants (Schwartze and Huber, 1937; Knight, 1962).

Quite different plant characteristics may be responsible for disease-escaping in the field. Comparisons of the susceptibility of potato varieties to potato virus Y gave different results in the field and in the greenhouse (Bagnall and Bradley, 1958). Spread within varieties in the field was related to their effectiveness as sources of virus for aphids. In the greenhouse tests, all aphids were fed on a common virus source and then placed on the different varieties.

### *B. Multiple Transfer Methods*

When evaluating plant susceptibility, more than one insect may be transferred to each healthy test plant, especially when transmission frequency is likely to be low and there is a need to conserve time and space. The results may not, however, accurately reflect the relative susceptibility of different groups of plants inoculated, so that an estimation of transmission by individual insects is required. The information available indicates that the probability of establishment of infection by one insect is independent of transmission by other insects present on the same plant (Bindra and Sylvester, 1961). This is in accord with the theory of independent action in infection by microorganisms (Meynell, 1957). The transfer of several insects to one plant is essentially a multiple-transfer method and the maximum likelihood estimator is the most satisfactory method of estimating the frequency of transmission by individual insects (Gibbs and Gower, 1957).

The usual method of using aphids to inoculate plants with stylet-borne viruses involves transferring one or more aphids to each test plant and leaving them there during their entire infective period. This is essentially a multiple-transfer method, even when only one aphid is transferred to each test plant, because an aphid may make several probes on a test plant. The probability of infection at the site of any one probe would be expected to be independent of events at other sites. If transmission is scored by systemic infection of plants, there is no way of knowing if a plant became infected from virus established at one site only or at more than one site. The probability of infection at more than one site would increase as plant susceptibility increased. As a result, the accuracy in measuring differences in susceptibility decreases as plant susceptibility increases (Swenson and Welton, 1966). An alternative is to allow each aphid one probe of limited duration on a test plant. Restriction of the period on the test plant to one probe helps also to separate aphid behavior from plant susceptibility in the infection of plants with viruses by aphid transmission.

### *C. Mechanical Inoculation*

Most research on treatments to alter plant susceptibility to virus infection by insect transmission has been done with aphid-borne viruses which are also mechanically transmissible. Studies of the susceptibility of plants to virus infection by mechanical inoculation have included some aphid-borne viruses. The response of plants to infection with different viruses, when inoculated mechanically, have been remarkably consistent for any particular environment or nutritional treatment, with the exception of post-inoculation temperature. This facilitates comparison with the limited amount of information concerning plant response to the same treatments following inoculation by aphids. So far, the same treatments either elicit no response or produce an entirely different effect when transmission is by aphids. Therefore, relative plant susceptibility to infection by insect transmission cannot be determined by mechanical inoculation.

A brief summary of the results obtained by mechanical inoculation follows: Data from similar experiments involving insect inoculation, presented later in this chapter, will show the lack of correspondence of the two methods. In general, susceptibility increases with increasing pre-inoculation temperature (Kassanis, 1957), although the time required to get the maximum effect may vary (Sinha, 1960). The effect of post-inoculation temperature varies among different viruses (Kassanis, 1957). Susceptibility increased with higher fertilizer levels that also increased plant growth (Bawden and Kassanis, 1950; Selman, 1945) and sometimes beyond that point (Selman and Grant, 1957), although the effect of fertilizers was sometimes dependent on water supply (Selman, 1945). Susceptibility was greater with an optimal rather than a minimal water supply (Selman, 1945; Tinsley, 1953). It is a general practice in many laboratories to keep assay plants in darkness for a period of time before inoculation because of the increased susceptibility which results from this treatment (Siegel and Zaitlin, 1964). Matthews (1953) found that during a natural 24-hr light-dark cycle susceptibility was greatest during the afternoon and least at the end of the night, although factors other than light were not controlled. Recently, Wilcoxon and El Kandelgy (1966) reported that post-inoculation darkness reduced the number of local lesions produced by red clover vein mosaic virus on *Gomphrena globosa,* as did pre-inoculation darkness to a lesser extent.

## III. INHERENT DIFFERENCES

Differences in susceptibility to infection by insect transmission have almost invariably been found among the genetically different host plants of a virus. Sometimes these differences are very large. They are of interest for two main reasons. Low susceptibility may be useful in developing crop plant varieties resistant to virus infection. On the other hand, high susceptibility of test plants is useful in transmission experiments, because the time and space requirements are reduced and, most importantly, so is the uncontrolled variation. Also, host plant of differential susceptibility may be useful in studying transmission mechanisms.

### A. Circulative Viruses

Potato varieties vary greatly in the readiness with which they are infected with leafroll virus by aphids (Bald *et al.,* 1946). Resistance to infection has been the main factor in breeding potato varieties resistant to this disease (Baerecke, 1958). Certain potato varieties were much less likely to be infected with potato yellow dwarf virus by the leafhopper, *Aceratagallis sanguinolenta,* than were others (Larson, 1945). No differences were observed among these varieties in their suitability as host plants of the vector. Differences in resistance to infection with aster yellows virus (AYV) have been found in several crops. Incidence of AYV in sunflower cultivars ranged from 0 to 100% in field plots (Putt and Sackston, 1960). Incidence of disease in single cross hybrids and their parents indicated that resistance was dominant. Flax varieties varied in susceptibility to AYV in greenhouse tests (Martin *et al.,* 1961). Selection through five generations resulted in a useful degree of field resistance. Comparison of a large number of carrot varieties in field plots showed differences in susceptibility to AYV (Hervey and Schroeder, 1949). Greenhouse tests indicated that field differences were more likely to be due to vector preferences than to plant resistance. Thompson (1949) observed considerable variation in the number of plants of different wild *Lactuca* species which were infected with AYV in field plots.

The search for better indicator plants for virus research has led to a number of comparisons of plant susceptibility among the host plants of particular viruses. *Datura stramonium* and *Physalis floridana* were about three times as susceptible as *P. angulata* to infection with potato leafroll virus by *Myzus persicae* inoculation (Kirkpatrick, 1948). Aster yellows virus was transmitted to *Plantago major* by *Macrosteles fascifrons* much more frequently than it was to aster in similar experiments (Freitag, 1968). Peach yellow leafroll virus was transmitted by the leafhopper *Colladonus montanus* to 28% of the celery plants inoculated, compared to less than 2% of peach seedlings (Jensen, 1957). However, the fact that peach was a poor food plant for the leafhopper was as likely an explanation of the difference in numbers of plants infected as was a difference in susceptibility of the two plants to the virus.

Sometimes one vector species may not be able to transmit a virus to a particular host plant at all. The typical short-winged strain of *Macrosteles fascifrons* transmitted the California aster yellows virus equally well to both celery and aster (Severin, 1947). A number of other leafhopper species transmitted the virus efficiently to celery but inefficiently, or not at all, to aster (Severin, 1947, 1948). *Euscelis plebejus* transmitted strawberry green-petal virus to and from a wide range of plants, but not to strawberry plants (Frazier and Posnette, 1957). Although strawberry is a poor host plant for this leafhopper, the authors did not believe the failure to infect strawberry could be explained on this basis alone. Later, *Aphrodes bicinctus* was found to transmit the virus and was considered the probable field vector to strawberry (Posnette and Ellenberger, 1963).

### B. *Stylet-Borne Viruses*

The proportion of potato plants infected with potato virus Y by *Myzus persicae* transmission varied from 6 to 70% among 13 varieties (Bawden and Kassanis, 1946). Similar results were obtained by Bagnall and Bradley (1958). The use of differences in resistance to infection has been an important factor in breeding resistant potato varieties (Ross, 1960). The Italian El variety of pepper was highly resistant to potato virus Y, both by aphid and mechanical inoculation (Simons, 1960). Once infected, however, Italian El developed a virus titer as high as that in the susceptible variety, California Wonder. The resistant variety contained an inhibitor not present in the susceptible variety (Simons and Moss, 1963). The potato variety, Katahdin, and other varieties, were highly resistant to infection with potato virus A by aphid transmission (Bagnall and MacKinnon, 1960). Resistance did not depend on hypersensitivity as in some other resistant varieties. When infected by grafting, the varieties resistant to infection showed symptoms and were good sources of virus for aphids.

The combined results for two aphid species used to inoculate eight crucifer species with *Brassica nigra* virus varied from 4.5 to 48% among eight plant species (Sylvester, 1953). Pepper was somewhat more susceptible than chard to infection with cucumber mosaic virus by aphid transmission (Simons, 1955). An interaction between aphid species and test plant species was demonstrated by Sylvester and Simons (1951). Less than 1% infection resulted from inoculation of *Brassica chinensis* with *Brassica nigra* virus by *Myzus persicae,* compared to about 42% infection of *Brassica juncea.* Use of the aphid, *Rhopalosiphum pseudobrassicae,* resulted, respectively, in 9 and 15% infection of the two species. The relative susceptibility of two *Phaseolus vulgaris* varieties to bean yellow mosaic virus by *Myzus persicae* transmission varied at different times of the year (Adlerz, 1959). Forty-three per cent of greenhouse-grown plants of each variety were infected when inoculated in August, whereas one variety was about twice as susceptible as the other when inoculations were made in November.

## IV. AGE AND PART OF PLANT

*Brassica juncea* plants varying from 1 to 5 weeks in age were about equally susceptible to *Brassica nigra* virus by *Myzus persicae* transmission (Sylvester, 1953a). The percentage of pineapple seedlings infected with *Commelina nudiflora* mosaic virus by aphid transmission decreased with the increasing age of the plants (Carter, 1937), as did the infection of sugar beet plants with beet yellows virus by aphid transmission (Watson and Healy, 1953). The susceptibility of potato plants to potato virus Y by *Myzus persicae* transmission decreased somewhat as plants aged from 1 to 8 weeks when leaves in the same position on plants of different ages were inoculated (Bagnall and Bradley, 1958). A much more striking effect was obtained with leaves of different ages. The upper leaves were about twice as susceptible to the establishment of virus infection as were the lower leaves. The number of lima bean plants infected with cucumber mosaic virus by aphids feeding on the primary leaves was about halved with a two-day increase in age (Stimmann and Swenson, 1967).

Sylvester (1953a) recorded the feeding sites of aphids on small *Brassica juncea* plants in transmission experiments with *Brassica nigra* virus. Swenson (1962) did the same with small pea plants in bean yellow mosaic virus experiments, as did Sylvester and Richardson (1963) with small *Brassica campestris* plants in cabbage mosaic virus experiments. The differences in transmission were difficult to evaluate in all three cases because equal numbers of aphids did not feed at all sites. The results did indicate, however, that the part of the plant inoculated by aphids was not an important factor in transmission in these experiments.

Storey (1928) found with maize streak virus transmission by the leafhopper, *Cicadulina mbila,* that the age of plants at the time of inoculation had a great influence on the number infected. Infection varied from 72% at 8 days of age to about 35% at 20 days. Young potato plants were more susceptible to leaf-roll virus by aphid transmission than were old plants (Knutson and Bishop, 1963). There was also marked decrease in susceptibility during bud formation and flowering. Sixty-eight per cent of inoculated pea plants were infected with pea enation mosaic virus when pea aphids fed in the veins as compared with 28% when aphids fed on interveinal areas (Nault and Gyrisco, 1966).

## V. ENVIRONMENT

### *A. Preinoculation Temperature*

More bean plants were infected with bean yellow mosaic virus by *Myzus persicae* transmission when kept for 2 days at 18°C than when kept for 2 days at 27°C (Swenson and Sohi, 1961). The susceptibility of pea plants to infection with bean yellow mosaic virus by aphid transmission decreased with increasing temperature between 15 and 33°C during the two days immediately before inoculation (Welton, Swenson, and Sohi, 1964). The change in susceptibility was greatest between 15 and 18°C, and was virtually linear between 18 and 33°C (Fig. 1). This effect of pre-inoculation temperature was not obtained with all bean yellow mosaic virus cultures (Swenson, 1968).

### *B. Postinoculation Temperature*

The number of pea plants infected with bean yellow mosaic virus by *M. persicae* transmission was greater when the plants were kept at 30°C for 2 days immediately after inoculation than when they were kept at 12–15°C for 2 days (Welton *et al.*, 1964). The greatest difference was between 24 and 30°C, twice as many plants being infected at 30 as at 24°C. Such an effect was almost always obtained in many experiments in which postinoculation temperatures of 15 and 30°C were compared (Welton *et al.*, 1964; Swenson, 1968). More lima bean plants were infected with cucumber mosaic virus by *M. periscae* when exposed to 30°C for 2 days immediately after inoculation than when exposed to 15°C for 2 days (Stimmann and Swenson, 1967). No difference in numbers of *Brassica campestris* plants infected with cabbage mosaic virus by *M. persicae* resulted when plants were placed at temperatures ranging from 5 to 30°C for 24 and 96 hr after inoculation (Sylvester, 1964).

### *C. Preinoculation Light*

The susceptibility of plants to virus infection by mechanical inoculation is usually increased by keeping the plants in the dark for a day or two before inoculation over that of plants kept in the light for the same time. Similar results have not been obtained by aphid inoculation. The number of *Nicotiana rustica* plants infected with henbane mosaic virus by *M. persicae* transmission was not affected by keeping plants in the darkness for two or four days before inoculation (Bradley, 1952). Similarly, no effect on number of *Brassica juncea* plants infected with *Brassica nigra* virus by *M. persicae* transmission occurred when plants were kept in darkness before inoculation for periods

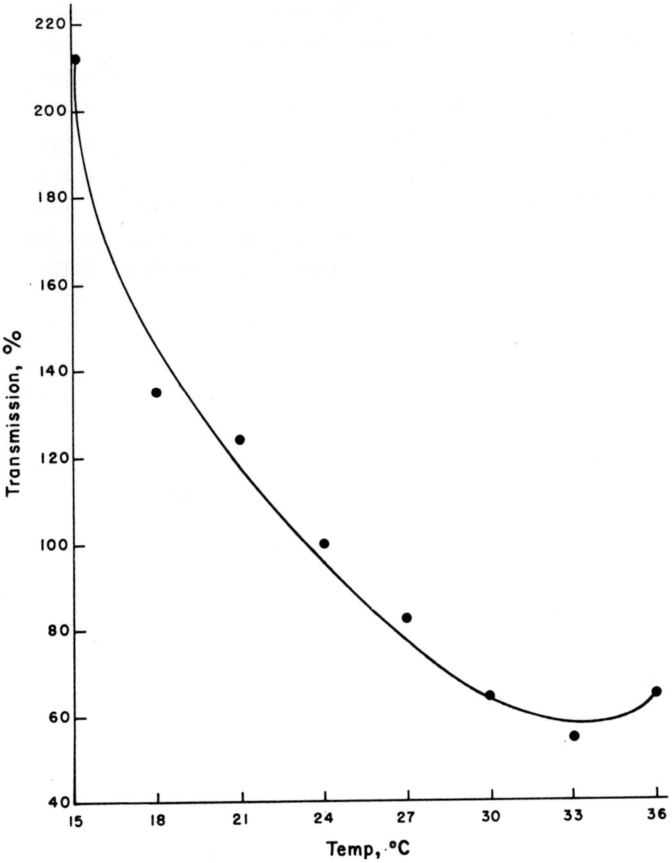


*Fig. 1.* Relation of preinoculation temperature to number of pea plants infected with bean yellow mosaic virus by aphid transmission.

varying from several hours to 48 hr (Sylvester, 1953). Exposure to 20 hr of darkness did not change the number of plants infected with bean yellow mosaic virus by *M. persicae* transmission (van Hoof, 1958).

Sylvester (1955) reported that lettuce plants kept in the dark for 24 and 48 hr before inoculation with lettuce mosaic virus by *M. persicae* were significantly more susceptible to lettuce mosaic virus than were plants kept under normal greenhouse conditions. The method of obtaining dark conditions might, however, have introduced temperature differences. No differences in susceptibility of pea and bean plants to bean yellow mosaic virus by aphid transmission occurred as a result of altering preinoculation light conditions when the temperature was controlled (Swenson, 1968). The experiments included the following comparisons: constant light with constant darkness for 2 days; 16-hr photoperiods of 550 ft-c, 1100 ft-c, and constant darkness at 15, 24, and 30°C for 2 days; 12-hr photoperiods at 500 ft-c with 16-hr photoperiods at 1000 ft-c, at 15 and 30°C. The susceptibility of lima bean plants to cucumber mosaic virus was not affected by keeping the plants in darkness for two days before inoculation by *M. persicae* (Stimmann and Swenson, 1967).

### *D. Postinoculation Light*

No effect on the number of *Brassica compestris* plants infected with cabbage mosaic virus by *M. persicae* transmission resulted when light intensities of about 1000 and 2000 ft-c for 24 and 96 hr were compared at temperatures ranging from 5 to 30°C (Sylvester, 1964). There were no significant differences between two days constant light and two days constant dark in the number of pea plants infected with bean yellow mosaic virus by *M. persicae* (Swenson, 1968). The effect of postinoculation light or dark was not affected by preinoculation dark treatments. Similar results were obtained in the transmission of cucumber mosaic virus to lima bean plants by *M. persicae* (Stimmann and Swenson, 1967).

## VI. PLANT NUTRITION

Fertilizer treatments that greatly influenced the growth of tobacco and potato plants in greenhouse trials had little effect on the number of plants infected with potato virus Y by *M. persicae* (Bawden and Kassanis, 1950). The susceptibility of bean plants to infection with bean yellow mosaic virus by *M. persicae* transmission was sometimes increased by raising the general (N, K, P) fertility level of soil in greenhouse trials (Swenson and Sohi, 1961; Swenson, 1968). In general, however, large differences in the fertility level of soil in which pea and bean plants were grown did not affect the number of plants infected. A comparison of pea plants grown in soil containing adequate amounts of nutrient elements with plants grown in an inert medium and irrigated with Hoagland's solution or with only water until two days after inoculation showed no differences in susceptibility to bean yellow mosaic virus by aphid transmission (Swenson, 1968).

## VII. WATER SUPPLY

Bean plants grown at two levels of soil moisture, field capacity and twice field capacity, did not differ in susceptibility to bean yellow mosaic virus by aphid transmission (Swenson, 1968). Similarly, pea plants did not differ in susceptibility to this virus when grown with adequate and markedly inadequate amounts of water (Welton, 1963). Storey (1928) found no difference in the numbers of plants infected with maize streak virus by *Cicadulina mbila* when watered heavily and lightly (just enough to prevent wilting).

## VIII. TRANSMISSION MECHANISMS

Many things affect the frequency of plant virus transmission by insects. These may be grouped into intrinsic and extrinsic factors. The former includes genetic factors determining the transmitting ability of the vector and the transmissibility of the virus by particular insects. Extrinsic factors include the susceptibility of plants to infection by insect transmission, availability of virus in diseased plants, and effect of vector host plants on the vector (Swenson, 1963).

The amount of virus acquired by vectors of circulative (persistent) viruses increases with increasing time on the diseased plants, so that low virus titer can be compensated for by longer feeding periods. The viruses are retained for long periods of time so that virus can be accumulated, stored, and, in some cases, increased by multiplication in the vector. Low probability of acquiring stylet-borne (nonpersistent) viruses cannot be offset by longer feeding on the diseased plants. Optimum acquisition periods are probes of 10–60 sec duration. As duration of the acquisition probe increases, the probability of acquiring virus decreases, *e.g.*, alfalfa mosaic virus transmission by aphids dropped from 28% with acquisition probes of 1 min to 2.5% with acquisition periods of 30 min (Swenson, 1952). So aphids cannot accumulate stylet-borne virus and the viruses are retained for only a short time by feeding aphids. Few aphids are infective after 15 min and many lose infectivity within 5 min after leaving the diseased plant.

Low plant susceptibility to circulative viruses may be offset by longer periods of vector feeding. Stylet-borne viruses, on the other hand, are lost so rapidly that an aphid must place virus in a site susceptible to infection during the brief period that it is infective, if successful transmission is to occur. For these reasons, transmission of stylet-borne viruses is much more dependent on the effects of extrinsic factors than is the transmission of circulative viruses, and differences in plant susceptibility might be expected to have greater effects on field spread.

## IX. CONCLUSIONS

The information available indicates that the host plants of an insect-borne virus will vary in susceptibility to infection by insect transmission. The efficiency of research involving virus transmission by insects will be greatly facilitated if, initially, some effort is given to selecting one of the more susceptible plants. Alternatively, the host plants most resistant to infection may be sources of field resistance in crop plants.

Greenhouse tests will suffice to find better host plants for virus studies. Both greenhouse and field tests are necessary when resistance in crop plants is sought. Some plant characteristics imparting field resistance will not be detected in greenhouse tests. This is likely to be particularly true of those involved in host plant selection by vectors. On the other hand, greenhouse tests are necessary to identify resistance mechanisms and may be used to supplement field tests.

It is quite clear that, among viruses transmitted both mechanically and by insects, the results of evaluations of susceptibility by mechanical inoculation have showed little similarity to the results obtained by insect inoculation. Among the stylet-borne viruses transmitted by aphids, temperature and plant age have so far proved to be the only factors which have an appreciable effect on plant susceptibility. Virtually nothing is known about plant susceptibility to circulative viruses.

## BIBLIOGRAPHY

Adlerz, W. C. 1959. Factors affecting transmission of bean yellow mosaic virus. J. Econ. Entomol. 52:260–262.

Baerecke, M. L. 1958. Blattrollresistenzzuchtung. p. 97–196. In: Handbuch der Pflanzenzuchtung, Bd. III. H. Kappert and W. Rudorf, eds. Paul Parey, Berlin.

Bagnall, R. H., and Bradley, R. H. E. 1958. Resistance to virus Y in the potato. Phytopathology 48:121–125.

Bagnall, R. N., and MacKinnon, J. P. 1960. Resistance to potato virus A in the Katahdin variety. Eur. Potato J. 3:331–336.

Bald, J. G., Norris, D. O., and Helson, G. A. H. 1946. Transmission of potato virus diseases. 5. Aphid populations, resistance, and tolerance of potato varieties to leaf roll. Australian Council Sci. Ind. Res. Bul. 196. 32 p.

Bawden, F. C. and Kassanis, B. 1947. Varietal differences in susceptibility to potato virus Y. Ann. Appl. Biol. 33:46–50.

———. 1950. Some effects of host nutrition on the susceptibility of plants to infection by certain viruses. Ann. Appl. Biol. 33:46–57.

Bindra, O. S., and Sylvester, E. S. 1961. Effect of insect numbers on aphid transmission of potato leafroll virus. Hilgardia 31:279–325.

Bjorling, K. 1966. Virus resistance problems in plant breeding. Acta. Agr. Scand., Suppl. 16:119–136.

Bradley, R. H. E. 1952. Studies on the aphid transmission of a strain of henbane mosaic virus. Ann. Appl. Biol. 39:78–97.

Carter, W. 1937. Aphid transmittal of *Commelina nudiflora* mosaic to pineapple. Ann. Entomol. Soc. Amer. 30:155–161.

Folsom, D. 1955. Testing potato seedling varieties in Maine for field resistance to leaf-roll and for desirable horticultural characteristics. Amer. Potato J. 32:372-385.

Frazier, N. W., and Posnette, A. F. 1957. Transmission and host-range studies of strawberry green-petal virus. Ann. Appl. Biol. 45:580–588.

Freitag, J. H. 1968. Chapter 14 of this volume.

Gibbs, H. J., and Gower, J. C. 1960. The use of a multiple-transfer method in plant virus transmission studies—some statistical points arising in the analysis of variance. Ann. Appl. Biol. 48:75–83.

Hervey, G. E. R., and Schroeder, W. T. 1949. The yellows disease of carrot. N. Y. Agr. Exp. Sta. Bul. 737, 29 p.

Hoof, H. A. van. 1958. Onderzoekingen over de biologische overdracht van een non-persistent virus. Mededel. Wageningen Inst. Plantenziekten. Onderzoekingstat. No. 161.

Hopkins, J. C. F. [ed.] 1957. Common names of plant viruses used in the Review of Applied Mycology. Rev. Appl. Mycol. 35, Suppl. 78 p.

Jensen, D. D. 1957. Differential transmission of peach yellow leaf roll virus to peach and celery by the leafhopper, *Colladonus montanus*. Phytopathology 47:575–578.

Kassanis, B. 1957. Effect of changing temperature on plant virus diseases. Advances Virus Res. 4:221–241.

Kennedy, J. S., Day, M. F., and Eastop, V. F. 1962. A conspectus of aphids as vectors of plant viruses. Commonwealth Inst. Entomol., London. 114 p.

Kirkpatrick, H. C. 1948. Indicator plants for studies with the leafroll virus of potatoes. Amer. Potato J. 25:283–290.

Knight, R. L. 1962. Fruit breeding. J. Royal Hort. Soc. 87:103–113.

Knutson, K. W., and Bishop, G. W. 1964. Potato leafroll virus—effect of date of inoculation on per cent infection and symptom expression. Amer. Potato J. 41:227–238.

Kuc, J. 1966. Resistance of plants to infectious agents. Annu. Rev. Microbiol. 20:337–370.

Larson, R. H. 1945. Resistance in potato varieties to yellow dwarf. J. Agr. Res. (U.S.) 71:441–451.

Martin, A. L. D., Frederiksen, R. A., and Westdahl, P. H. 1961. Aster yellows resistance in flax. Canad. J. Plant Sci. 41:316–319.

Matthews, R. E. F. 1953. Factors affecting the production of local lesions by plant viruses, I. The effect of time of day of inoculation. Ann. Appl. Biol. 40:377–383.

Meynell, G. G. 1957. Inherently low precision of infectivity titrations using a quantal response. Biometrics 13:149–163.

Nault, L. R., and Gyrisco, G. G. 1966. Relation of the feeding process of the pea aphid to the inoculation of pea enation mosaic virus. Ann. Entomol. Soc. 59:1185–1197.

Nielson, M. W. 1962. A synonymical list of leafhopper vectors of plant viruses (Homoptera, Cicadellidae). U.S. Dept. Agr. ARS-33-74. 12 p.

Painter ,R. H. 1958. The study of resistance to aphids in crop plants. Proc. Int. Congr. Entomol., 10th, London, 1956, 3:451–458.

Posnette, A. F., and Ellenberger, C. E. 1963. Further studies of green petal and other leafhopper-transmitted viruses infecting strawberry and clover. Ann. Appl. Biol. 51:69–83.

Putt, E. D., and Sackston, W. E. 1960. Resistance of inbred lines of sunflowers (*Helianthus annuus* L.) in a natural epidemic of aster yellows. Canad. J. Plant Sci. 40: 375–382.

Ross, H. 1960. Die Praxis der Zuchtung auf Infektionsresistenz und extremer Resistenz (Immunitat) gegen das Y-virus der Kartoffel. Eur. Potato J. 3: 296–306.

Schwartze, C. D., and Huber, G. A. 1937. Aphid resistance in breeding mosaic-escaping red raspberries. Science 86: 158–159.

Selman, I. W. 1945. The susceptibility of lettuce to mosaic virus in relation to nitrogen, phosphate and water supply. J. Pomol. Hort. Sci. 21:28–33.

Selman, I. W., and Grant, S. A. 1957. Some effects of nitrogen supply on the infection of tomato plants with tomato spotted wilt virus. Ann. Appl. Biol. 45:448–455.

Severin, H. H. P. 1947. Newly discovered leafhopper vectors of California aster-yellows virus. Hilgardia 17:511–519.

———. 1948. Transmission of California aster-yellows virus by leafhopper species in *Thamnotettix* group. Hilgardia 18:203–216.

Siegel, A., and Zaitlin, M. 1964. Infection process in plant virus diseases. Ann. Rev. Phytopathol. 2:179–202.

Simons, J. N. 1955. Some plant–vector–virus relationships of southern cucumber mosaic virus. Phytopathology 45:217–219.

Simons, J. N. 1960. Factors affecting field spread of potato virus Y in south Florida. Phytopathology 50:424–428.

Simons, J. N., and Moss, L. M. 1963. The mechanism of resistance to potato virus Y infection in *Capsicum annuum* var. Italian E. Phytopathology 53:684–691.

Sinha, R. C. 1960. Some effects of temperature and of virus inhibitors on infection of French-bean leaves by red clover mottle virus. Ann. Appl. Biol. 48:749–755.

Stimmann, M. W., and Swenson, K. G. 1967. Effects of temperature and light on plant susceptibility to cucumber mosaic virus by aphid transmission. Phytopathology 57:1072–1073.

Storey, H. H. 1928. Transmission studies of maize streak disease. Ann. Appl. Biol. 15:1–25.

Swenson, K. G. 1952. Aphid transmission of a strain of alfalfa mosaic virus. Phytopathology 42:261–262.

———. 1962. Bean yellow mosaic virus transmission by *Myzus persicae*. Australian J. Biol. Sci. 15:468–482.

———. 1968. Relation of environment and nutrition to plant susceptibility to bean yellow mosaic virus by aphid transmission. Oregon Agr. Exp. Sta. Tech. Bull. (in press).

Swenson, K. G., and Sohi, S. S. 1961. Factors determining the rate of bean yellow mosaic virus transmission by the aphid *Myzus persicae*. Phytopathology 51:67.

Swenson, K. G., and Welton, R. E. 1966. Evaluation of plant susceptibility to infection with bean yellow mosaic virus by aphid transmission. Phytopathology 56:269–271.

Sylvester, E. S. 1953a. *Brassica nigra* virus transmission: some vector-virus-host plant relationships. Phytopathology 43:209–214.

———. 1953b. Host range and properties of *Brassica nigra* virus. Phytopathology 43: 541–546.

———. 1955. Lettuce mosaic virus transmission by the green peach aphid. Phytopathology 45:357–370.

———. 1964. Some effects of temperature on the transmission of cabbage mosaic virus by *Myzus persicae*. J. Econ. Entomol. 57: 538–544.

Sylvester, E. S., and Richardson, J. 1963. Short-term probing behavior of *Myzus persicae* (Sulzer) as affected by fasting, anesthesia, and labial manipulation. Virology 20:302–309.

Sylvester, E. S., and Simons, J. N. 1951. Relation of plant species inoculated to efficiency of aphids in the transmission of *Brassica nigra* virus. Phytopathology 41:908–910.

Thompson, R. C. 1944. Reaction of *Lactuca* species to aster yellows virus under field conditions. J. Agr. Res. (U.S.) 69:119–125.

Tinsley, T. W. 1953. The effects of varying the water supply of plants on their susceptibility to infection with virus. Ann. Appl. Biol. 40:750–760.

Watson, M. A., and Healy, M. J. R. 1953. The spread of beet yellows and beet mosaic viruses in the sugar-beet crop. II. The effects of aphid numbers on disease incidence. Ann. Appl. Biol. 40:38–59.

Welton, R. E. 1963. Susceptibility of peas to aphid inoculation with bean yellow mosaic virus. Ph.D. thesis, Oregon State University, Corvallis. 130 p.

Welton, R. E., Swenson, K. G., and Sohi, S. S. 1964. Effect of temperature on plant susceptibility to infection with bean yellow mosaic virus by aphid transmission. Virology 23:504–510.

Wilcox, R. B. 1951. Tests of cranberry varieties and seedlings for resistance to the leafhopper vector of false-blossom disease. Phytopathology 41:722–735.

Wilcox, R. B., and Beckwith, C. S. 1933. A factor in the varietal resistance of cranberries to the false-blossom disease. J. Agr. Res. 47:583–590.

Wilcoxon, R. D., and El-Kandelgy, S. M. 1966. Effect of light on formation of lesions on *Gomphrena globosa* by red clover vein mosaic virus. Phytopathology 56:344.

Wilcoxon, R. D., and Peterson, A. G. 1960. Resistance of Dollard red clover to the pea aphid, *Macrosiphum pisi*. J. Econ. Entomol. 53:863–865.

# Virus Transmission by Aphids—A Viewpoint

EDWARD S. SYLVESTER

*Department of Entomology, University of California, Berkeley, California*

Until recently aphids were considered unique among the arthropod vectors of plant viruses in that they have two quite different mechanisms of transmission. The virus can be carried at the tips of the stylets (stylet-borne, Kennedy, Day, and Eastop, 1962) or the virus can be ingested, absorbed, translocated, and finally added to the salivary secretions from where it is put into the plant tissues during feeding (circulative viruses, Black, 1959).

Most plant viruses carried by aphids seem to be transmitted either one way or the other, but not both. However, both types of transmission occur in mosquito-borne eastern encephalitis virus, and there is evidence that at least one plant virus, cauliflower mosaic virus, can be transmitted as a stylet-borne virus and as a circulative virus by the cabbage aphid, *Brevicoryne brassicae* (Linn.). Why more examples of the dual method of transmission of plant viruses by aphids do not exist is not known, but perhaps can be explained in terms of the distribution of viruses in plant tissues, selective gut absorption, and inactivation in the vector.

The experimental demonstration of the type of transmission that occurs in any given situation can be outlined rather simply. With a stylet-borne virus, the inoculative potential of a vector is drastically reduced or eliminated if the tips of the stylets are treated with a viracidal agent such as ultraviolet irradiation (Bradley and Ganong, 1955). If, on the other hand, the inoculative capacity of the vector is transstadial, then the virus is considered to be circulative. However, there are experimental situations where application of these tests have been indecisive, and in such cases the mode of transmission remains obscured.

If there is no specific knowledge of the mechanism of transmission, then there is another classification of vector–virus relationships that has historical precedence, and which, particularly for aphid vectors,* has some advantages. In this classification scheme the transmission of plant viruses is divided into nonpersistent, semipersistent, and persistent relationhips † (Watson and Roberts, 1939; Sylvester, 1956). It is generally thought that the nonpersistent relationship involves stylet-borne virus, and that the persistent relationship occurs with a circulative virus. It has yet to be determined whether the semipersistent relationship involves stylet-borne or circulative virus, or both.

In addition to being circulative, a plant virus can also multiply, or become propagative, within the aphid host. This is a relatively new area of experimental emphasis with aphid-borne viruses, and to date the only example is that of the potato leaf-roll virus when transmitted by the green peach aphid, *Myzus persicae* (Sulzer) (Stegwee and Ponson, 1958).

It has been suggested (Duffus, 1963) that the aphid-transmitted sowthistle yellow-vein virus is propagative. In this case, however, the evidence is indirect, and consists mainly of the presence of a prolonged, temperature-sensitive, latent period. Strawberry crinkle virus (Prentice, 1949) is another candidate for a propagative aphid-borne virus on the basis of a prolonged latent period.

Three types of relationships are briefly discussed.

NONPERSISTENT. In the nonpersistent relationship, the acquisition and inoculation threshold periods are but a few seconds in duration (Sylvester, 1949), and the probability of acquisition increases if the aphids have fasted for a short time before making an acquisition probe into diseased tissue. This "preliminary fasting effect" (Watson, 1938), at least in part, results from an increased tendency for aphids that have been deprived of food to have a readiness to probe briefly into epidermal plant tissue (Bradley, 1952, 1961; Day and Irzykiewicz, 1954; Nault and Gyrisco, 1966). But this may not be the complete explanation (Day and Irzykiewicz, 1954).

As the duration of the acquisition probe, or feeding penetration, increases, there is a decrease in the probability of acquisition of virus. This phenomenon may reflect a decrease in accessible virus in subepidermal tissues (Bawden, Hamlyn, and Watson, 1954), or it could result from the continuous flushing and scouring activities that accompany the formation of the stylet sheath as a feeding probe is established

* Tungro virus of rice appears to have a semipersistent relationship with its leafhopper vector (Ling, 1966).

† Watson (1960) has outlined a modified grouping of vector-virus relationships in which the persistent category is subdivided into six groups, including all vectors of plant viruses except the fungi.

with deeper penetration of the stylets (McLean and Kinsey, 1965). During brief probes the salivary sheath may be represented simply by a small deposit of salivary material on the plant surface and between the epidermal cells (Nault and Gyrisco, 1966). As the stylet tips penetrate this material, it could readily become contaminated with virus, and then as the tips are withdrawn, a small amount of inoculum could be carried back to the tip of the labium (Sylvester, 1962). If the probe is not quickly terminated, the salivary sheath is more perfectly formed and the stylets are then being continuously bathed with a second liquid salivary component (Miles, 1959). Under such circumstances, the probability of carrying contaminated sheath material back to the labium upon withdrawal of the stylets is greatly reduced.

Once an aphid has acquired a "charge" of virus from a brief acquisition probe, the subsequent activities of the vector are critical if transmission is to occur. If the aphid has the opportunity to rapidly make a series of brief inoculation probes, successive transmissions of the virus, although somewhat limited in number, are possible; however, the probability of transmission tends to decrease with each successive probe. If, on the other hand, the inoculative aphid makes a prolonged feeding penetration, *i.e.*, a single penetration maintained for a time in excess of 15 min, the probability of transmission during that feeding is increased, but the chances that the aphid can inoculate a plant with a subsequent probe are quite low. When the inoculative aphid does not probe, or feed, on plant tissue, but simply remains away from a susceptible host, the probability that it will successfully inoculate any plant fed upon decreases rapidly with time. This is the "post-fasting effect" (Doolittle and Walker, 1928). The rate with which the inoculative capacity is lost differs somewhat with various viruses, and it can be varied, within limits, by controlling the ambient temperature. The rate of loss, however, tends to become somewhat exponential with time.

It is the relatively rapid loss of inoculative capacity by these vectors when transmitting stylet-borne viruses that gave rise to the term *nonpersistent* (Watson and Roberts, 1939).

This very briefly describes the transmission process in stylet-borne viruses. Most of these viruses have flexuous, rod-shaped particles and produce mosaic diseases. Transmission by juice inoculation is to be expected. The cucumber mosaic and alfalfa mosaic viruses are exceptions to the flexuous rod generalization, as is cauliflower mosaic virus, but the latter virus presents a more complicated pattern of transmission.

Some mention should be made of specificity and the nonpersistent relationship. In general, the stylet-borne viruses have a low level of vector specificity since they can be transmitted by several aphid species. However, the efficiency of transmission by different vectors can vary considerably, and in some instances of limited testing, no transmission has been obtained with specific vector–virus combinations (Doncaster and Kassanis, 1946). Little is known of the cause for such specificity, or variations in transmission efficiency, but it has been suggested that the salivary components that accompany the inoculation probe are involved (Day and Irzykiewicz, 1954; Sylvester, 1954). For example, there is evidence that the level of transmission can vary with the host plant species used, but as a function of inoculation, not acquisition (Sylvester and Simons, 1951). Furthermore, the work of Nishi (1963) strongly suggests that the salivary secretions of aphids have an inhibitory effect on the infectivity of certain viruses. There also is evidence that the host plant used to rear the vectors, as well as the species of plant used as a virus source, can influence vector efficiency (Simons, 1955). The entire question of virus inactivation as a function of association with vectors (Hoggan, 1933; Watson, 1936) awaits a final resolution.

In general terms the transmission might be thought of as being the product of the following factors: the probability of acquisition and of inoculation, the probability of virus inactivation during the interval between acquisition and inoculation, and finally the probability of salivary inhibition of virus activity at the site of inoculation. Specificity may be primarily a function of the latter.

Semipersistent. In the semipersistent relationship there is a positive correlation between the length of the acquisition feeding period and the probability of virus acquisition. The acquisition curve becomes asymptotic with time, but the maximum efficiency of acquisition is not realized until feeding has gone on for some hours. There also is a positive correlation between the duration of the inoculation feeding period and success of transmission, although the inoculation efficiency curve generally reaches a maximum within a shorter time interval than that of acquisition.

There does not appear to be any accumulation of inoculative capacity after an acquisition feeding period has ended. The greatest probability of transmission occurs upon leaving the virus source. The likelihood of transmission then gradually decreases to zero somewhat exponentially with time. The rate at which inoculativity decreases in either feeding or fasting aphids, while again influenced by the ambient temperature, is measurable in periods of hours, rather than in minutes as in the nonpersistent relationship, or days, as in the case of the persistent type of transmission. The acquisition of virus depends upon feeding, rather than probing; therefore, there is no benefit realized by using fasted aphids.

Such experimental results operationally define the semipersistent vector–virus relationship. Again this is a comparative term and must be considered in context with the nonpersistent and the persistent types. The transmission of beet yellows virus by the green peach aphid is considered to be typical of the semipersistent situation. Beet yellows is also a flexuous, rod-shaped virus. The semipersistent relationship seems to be common in the transmission of viruses infecting strawberry (Prentice and Harris, 1946; Mellor and Fitzpatrick, 1951; Prentice, 1952; Frazier and Posnette, 1958; Frazier, 1966a, 1966b) and *Rubus* (Cadman, 1952, 1954; Stace-Smith, 1955a, 1955b).

The "crucial" experimental tests to determine whether the virus is stylet borne or whether it is circulative have not been particularly fruitful in the case of semipersistent transmission. With beet yellows virus, both ultraviolet irradiation and formalin treatment of the stylets of inoculative aphids failed to give conclusive results (Bradley and Sylvester, 1962; Sylvester and Bradley, 1962). When transmission was reduced by such treatments, feeding was decreased on the test plants. Treated aphids that appeared to feed normally on test plants showed little, if any, reduction in transmission potential. Heinze (1963) reported reduction in the transmission of beet yellows following formalin treatment of the stylets, but no feeding behavior data were presented. Attempts to demonstrate transstadial passage of the beet yellows virus have been negative (Sylvester, 1962; Heinze, 1963), but those of the author were inconclusive because of the loss of inoculativity in the control insects. It remains unanswered as to whether the beet yellows virus is stylet borne or whether it is circulative. Frazier (1966b) has presented convincing evidence for the failure of either the strawberry mottle virus or the strawberry veinbanding virus to survive ecdysis by the aphid *Pentatrichopus jacobi* Hille Ris Lambers. However, he concluded that such negative evidence does not identify the mechanism of transmission.

The data on cauliflower mosaic virus transmission are contradictory and vector–virus relationships appear to be somewhat anomalous. The virus is, as mentioned before, a small polyhedron, and seems to be irregularly distributed in "packets" in host plant tissues (Day and Venables, 1961). Severin and Tompkins (1948) reported that the virus was juice inoculable to plants by using extracts from viruliferous aphids. The pattern of cauliflower mosaic virus transmission appears to vary with the species of aphid used. Preliminary fasting increases the acquisition efficiency by green peach aphids, but not by cabbage aphids (van Hoof, 1954). It has been reported that both stylet-borne virus and circulative virus are involved in transmission with the cabbage aphid (Chalfant and Chapman, 1962). The evidence given is fourfold: (*1*) there was a bimodal curve of acquisition efficiency, one peak occurring in the 2–5–min range, followed by a decline and then a second peak after about 8 hr of feeding; (*2*) fasted aphids tended to retain inoculativity for longer periods of time if they had acquired virus during long acquisition access periods rather than by brief probes; (*3*) serial transmission occurred more frequently if the acquisition access period was long rather than short; and (*4*) treatment of the stylets with formalin reduced or eliminated the inoculativity of aphids that acquired virus with short acquisition probes, but did not if the virus had been acquired during long acquisition access periods.

However, Orlob and Bradley (1961) reported that treatment of the stylets of cabbage aphids with formalin reduced or eliminated further inoculativity, even though the acquisition period was long. Limited trials in our laboratory in which the stylets of cabbage aphids, reared on plants infected with cauliflower mosaic virus, were treated with formalin gave results similar to those of Chalfant and Chapman (1962). Day and Venables (1961) failed to get transmission by injecting aphids when either purified virus or hemolymph from inoculative insects was used as inoculum, nor could they demonstrate transmission after the vector had molted.

The evidence thus far is variable and at times contradictory, and it would seem, at the very least, somewhat premature to accept the conclusion of Day and Venables (1961) that cauliflower mosaic transmission involves stylet-borne virus alone.

PERSISTENT. The persistent vector–virus relationship perhaps is the most interesting of all of the aphid vector–virus associations from a biological point of view. Here the transmission process is analogous to that found among the majority of the leafhopper-borne viruses. The singular exception to this generalization is that, to date, no case of transovarial passage of virus is known among the aphids.

Several of the leafhopper-borne plant viruses have been demonstrated to multiply in the vector as well as in the plant host, but, as previously noted, only in the case of potato leaf-roll virus has substantial evidence for multiplication in the body of the aphid vector been given (Stegwee and Ponson, 1958).*

* Potato leafroll virus recently has been reported to be transovarially passed to some of the offspring of the green peach aphid, *M. persicae* (Miyamoto and Miyamoto, 1966). Furthermore, this virus no longer is the only example of a

The following results are to be expected in transmission experiments with a virus that has the persistent type of relationship with its vector: (*1*) retention of inoculativity through a molt; (*2*) a positive correlation between the duration of both the acquisition and the inoculation access periods and the probability of virus transmission; (*3*) the presence of a latent period, although it may be brief and poorly defined in terms of a minimum; (*4*) a relatively long period of time (days) during which inoculativity is retained by vectors; and (*5*) the ability to transmit the virus is limited to a comparatively few aphid species. Two of these features are of special interest: the latent period and the long period of time over which inoculativity is retained.

If, as in the case of potato leaf roll, the virus is propagative, then the latent period and the prolonged period of transmission are perhaps to be expected. But if there is no direct evidence of a multiplication cycle within the aphid vector, can the latent period and the persistence of inoculativity be used as evidence for the virus being propagative? Apparently some authors (Day, 1955; Duffus, 1963) believe so, especially if the latent period is prolonged and if the efficiency of transmission remains high for a considerable portion of the insect's life. The difficulty with this approach lies in the fact that the time factor involved in both of these phenomena has yet to be defined in quantitative terms. For example, assume that one vector–virus combination has a latent period with an $LP_{50}$ (median latent period, Sylvester, 1965) value of 40 hr compared with another that has an $LP_{50}$ value of 100 hr, other conditions being equal. Does this indicate that the probability of the virus being propagative is less in the former instance than in the latter? A similar question can be asked concerning the time period during which inoculativity is retained. Or can these questions be answered in terms of virus concentration and stability?

To emphasize the significance of such questions, I would like to describe in some detail the present knowledge of the transmission of the pea enation mosaic virus by the pea aphid, *Acyrthosiphon pisum* (Harris). The vector–virus relationships found in this case are quite representative of those found in the persistent type of transmission. One exception is that the virus, or at least some of its isolates, is quite readily mechanically inoculable (Osborn, 1935, 1938). Apparently because of the susceptibility of the epidermal tissues to inoculation, the pea aphids have been found to be able to transmit virus by probing

---

propagative aphid-borne plant virus. There is now evidence from electron microscopy that both the lettuce necrotic yellows and the sowthistle yellow vein viruses cause systemic infections in an aphid vector, *Hyperomyzus lactucae* (O'Loughlin and Chambers, 1967; Richardson and Sylvester, 1968).

healthy plants for less than a minute (Nault, Gyrisco, and Rochow, 1964; Ehrhardt and Schmutterer, 1965).

Transstadial passage of inoculativity occurs (Osborn, 1935); therefore, the virus is circulative. Transmission is more efficient (more inoculative insects per unit of acquisition access time) by nymphs than by adults (Simons, 1954), suggesting perhaps a special relationship with immature forms. The virus occurs in the hemolymph of vectors fed on diseased plants, and this hemolymph can be used to inoculate a previously virus-free aphid, which, in turn, will then transmit the virus (Schmidt, 1959). Active virus occurs in the honeydew excreted by aphids feeding on diseased plants, but the alimentary tract can be cleared of detectable active virus by feeding the aphids on healthy plants for a day or so (Richardson and Sylvester, 1965).

When second to fourth instar nymphs were used to acquire virus, the following transmission characteristics were found in our laboratory at various controlled ambient temperature conditions (Sylvester and Richardson, 1966a). The acquisition threshold period was approximately 1.5 hr, independent of temperature, but the rate of acquisition (as measured by a gain in the proportion of aphids that acquired virus) varied with temperature. Thus, at 10°C the rate of acquisition was slower than at 20°C and some further acceleration in the rate occurred with a temperature of 30°C. The acquisition efficiency curve tended to reach its maximum after 12–24 hr of acquisition access time at 20 and 30°C. Aphids given varying inoculation access periods transmitted virus at a near maximum within a 30-min period at 20 and 30°C.

The latent period varied with the ambient temperature. Estimates for the value of the $LP_{50}$ were 70 hr at 10°C, and 25 and 14 hr at 20 and 30°C, respectively.

The net transmission rate (transmissions per aphid per generation, using 24-hr transfers to fresh plants) was approximately 33 plants at 10°C and 18 and 6 plants at 20 and 30°C, respectively. The mean weighted time for retention of inoculativity changed from approximately 30 days at 10°C to approximately 14 and 4 days at 20 and 30°C, respectively. These summarizing statistics were calculated by adapting the life table technique to transmission data and they apply to transmission from and to sweetpea, *Lathyrus odoratus,* in plant growth chambers with a constant light of 800–1000 ft-c.

At least an approximate twofold difference in both the latent period and in the persistence of inoculativity was found with 10°C changes in temperature.

The survival curves of aphids that lived to the adult stage indicated that 50% mortality was not reached until approximately 75, 40, and

15 days at 10, 20, and 30°C, respectively. Thus, there was a considerable loss of inoculativity before adult mortality became a significant factor. Results in our laboratory (Sylvester, 1967) indicate that the persistence of inoculativity varies with the isolate of virus being transmitted, and also that the loss of inoculativity by adult insects is associated with a decline in metabolic activity. For example, an isolate of the pea enation mosaic virus obtained from Professor R. J. Shepherd, University of California, Davis, had a mean weighted retention period of approximately 8.5 days while an isolate selected for work in our laboratory had a value of 13 days. Furthermore, studies indicate that the rate for natality and excretion also decline in a similar manner to the decline in transmission efficiency. Virus transmission by aging adult vectors apparently is influenced by many factors and may be a relatively poor indicator of virus content within the vectors.

Aphids that are gradually losing their capacity to transmit virus can be partially restored to potency if returned to a virus source for a second acquisition access period (Sylvester and Richardson, 1966b). The level of transmission in such "recharged" insects subsequently increased, indicating that increments of transmissible virus could be added to those already present. However, the recharging effect is quite temporary.

There is only a minor amount of information on the effect of dosage on the latent period and on the period of time that inoculativity persists. Ehrhardt and Schmutterer (1965) reported a decrease in the latent period and an increase in the persistence of inoculativity with increasing acquisition access periods. Preliminary trials in our laboratory in which nymphs were inoculated with varying dilutions of infectious honeydew suggested that as the dose of virus injected decreased, there was an increase in the length of the latent period and a decrease in the transmission efficiency of the individual aphids.

Finally, Shikata *et al.* (1966) reported virions of pea enation mosaic in the lumen of the gut and in the cell cytoplasm of the fat body of viruliferous aphids. This was considered direct evidence of infection of the aphid by a plant-pathogenic virus.

The following results in the transmission of the pea enation mosaic virus could be used as evidence for a hypothesis that the virus is propagative in the aphid vector:

*1.* Transstadial passage occurs, the virus can be recovered from the hemolymph, and virions have been reported in the fat body cytoplasm.

*2.* A latent period exists.

*3.* The average length of the latent period is a function of (*a*) the dose of inoculum, whether by feeding or by injection, (*b*) the ambient temperature, and (*c*) the age of the vector at the time of virus acquisition.

*4.* The inoculative capacity of the vector can persist for a relatively long period of time.

*5.* Retention of inoculativity is independent of the presence of detectable virus in the alimentary canal.

*6.* The effect of ambient temperature on both retention of inoculativity and the duration of the latent period is of the same order of magnitude expected with living systems.

The following experimental results can be used as evidence against a hypothesis that the virus is propagative in the vector.

*1.* The capacity of the vector to transmit virus gradually declines following acquisition.

*2.* The inoculative capacity of the vector is positively correlated with the dose of inoculum.

*3.* Vectors that are declining in their transmission rate can be "recharged" by using additional acquisition feedings.

*4.* The virus cannot be maintained within a vector population by the serial passage of hemolymph from one insect to another.

The failure to maintain transmission efficiency is a weak argument against virus multiplication since this also occurs with the propagative western X-disease virus (Whitcomb, Jensen, and Richardson, 1966). Furthermore, "blind" transfers occurred in the serial passage of aster yellows virus (Maramorosch, 1952) and potato leaf-roll virus (Stegwee and Ponson, 1958). Yet both of these viruses are considered to multiply within the body of their vectors. Blind transfers were cases of successful passage of the virus from one insect to another but the donor insect failed to transmit virus to the plant upon which it had fed. Transmission is not a simple expression of virus content.

The argument against multiplication that the inoculative capacity is a function of the dose of inoculum again is indecisive. Kirkpatrick and Ross (1952) and Heinze (1959) reported this to be the case in their work with potato leaf-roll virus, although Day (1955) did not get such results. The yield of propagative sigma virus of *Drosophila* has been reported to be a function of the dose of inoculum (L'Heritier, 1958). The yield would also be a function of dose if the multiplication of virus were limited or the infections were local (Black, 1953).

The ability of the vectors to be "recharged" has been used as evidence against multiplication in the transmission of curly-top virus (Freitag, 1936; Bennett and Wallace, 1938) and the tungro virus of rice (Ling, 1966). Recharging could be due to a temporary increase in the hemolymph titer from virus absorbed from the gut. This explanation has been used in connection with some anomalous results in the potato leaf-roll work (Stegwee, 1961).

The failure to maintain virus titer through its serial passage from insect to insect is perhaps the most critical argument against multiplication. All of our attempts to maintain inoculative insects by the serial injection method have failed beyond the first passage. Again, however, such negative evidence does not preclude multiplication.

The question of whether or not the pea enation mosaic virus is propagative in the pea aphid vector remains unanswered, but the evidence available to date would perhaps support limited multiplication as a tentative hypothesis. Such a hypothesis would provide the simplest explanation for the temperature-sensitive latent period, as well as for the relatively prolonged period of the retention of inoculativity.

Our failure to maintain transmission in a sequence of serial passages of hemolymph among aphid hosts may have been due to (*1*) an inefficient method of inoculation (infection via the alimentary canal may be most effective pathway), (*2*) the failure of the virus to become concentrated in the hemolymph (virus may be more closely associated with tissues other than blood cells), and (*3*) a rapid decrease in the amount and susceptibility of suitable substrate for virus multiplication during the maturation and wound healing of the vectors.

The site of virus increase, if it exists, is unknown, but the high efficiency of acquisition and the brevity of the latent period found with very young nymphs suggests a tissue system that is being developed and expanded during the nymphal instars. One such tissue system is the fat body, but there are others. Again, if limited multiplication does occur with pea enation mosaic virus, the process is evidently rapid and of a nondegenerative nature, for in our experiments, neither longevity nor the reproductive capacity of the inoculative aphids varied from that found with the noninoculative aphids of the same sample series (Sylvester and Richardson, 1966a).

Aphid transmission of plant viruses involves several levels of association of the viruses with their vectors. The location of the site of virus transport, *i.e.,* stylet borne or circulative, is only part of the information that is needed to describe or understand the transmission process. Additional information is implied when vector–virus relationships are described as nonpersistent, semipersistent, and persistent. Future research undoubtedly will lead to a more complete understanding of the transmission process in causal terms. Eventually we may be able to develop a more realistic and acceptable grouping of the various transmission patterns in evolutionary terms.

## BIBLIOGRAPHY


Bawden, F. C., Hamlyn, B. M. G., and Watson, M. A. 1954. The distribution of viruses in different leaf tissues and its influence on virus transmission by aphids. Ann. Appl. Biol. 41:229–239.

Bennett, C. W., and Wallace, H. E. 1938. Relation of the curly top virus to the vector, *Eutettix tenellus*. J. Agr. Res. 56:31–51.

Black, L. M. 1953. Transmission of plant viruses by cicadellids. Advances Virus Res. 1:69–89.

——— 1959. Biological cycles of plant viruses in insect vectors. *In* The Viruses 2. Burnet, F. M., and Stanley, W. M. (ed.), Academic Press, New York, p. 157–185.

Bradley, R. H. E. 1952. Studies on the aphid transmission of a strain of henbane mosaic virus. Annu. Appl. Biol. 39:78–97.

——— 1961. Our concepts; on rock or sand? Recent advances in botany, University of Toronto Press, Toronto. p. 528–533.

Bradley, R. H. E., and Ganong, R. Y. 1955. Evidence that potato virus Y is carried near the tip of the stylets of the aphid vector, *Myzus persicae* (Sulz.). Can. J. Microbiol. 1:775–782.

Bradley, R. H. E., and Sylvester, E. S. 1962. Do aphids carry transmissible sugar beet yellows virus via their stylets?—Evidence from ultraviolet irradiation. Virology 17:599–600.

Cadman, C. H. 1952. Studies in Rubus virus diseases III. A veinbanding disease of raspberries. Annu. Appl. Biol. 39:69–77.

——— 1954. Studies in Rubus virus diseases VI. Aphid transmission of raspberry leaf mottle virus. Annu. Appl. Biol. 41:207–214.

Chalfant, R. B., and Chapman, R. K. 1962. Transmission of cabbage viruses A and B by the cabbage aphid and the green peach aphid. J. Econ. Entomol. 55:584–590.

Day, M. F. 1955. The mechanism of the transmission of potato leaf roll virus by aphids. Australian J. Biol. Sci. 8:498–513.

Day, M. F., and Irzykiewicz, H. 1954. On the mechanism of transmission of non-persistent phytopathologic viruses by aphids. Australian J. Biol. Sci. 7:251–273.

Day, M. F., and Venables, D. G. 1961. The transmission of cauliflower mosaic by aphids. Australian J. Biol. Sci. 14:187–197.

Doncaster, J. P., and Kassanis, B. 1946. The shallot aphis, *Myzus ascalonicus* Doncaster, and its behaviour as a vector of plant viruses. Annu. Appl. Biol. 33:66–68.

Doolittle, S. P., and Walker, M. N. 1928. Aphis transmission of cucumber mosaic. Phytopathology 18:143, (Abstr.)

Duffus, J. E. 1963. Possible multiplication in the aphid vector of sowthistle yellow vein virus, a virus with an extremely long insect latent period. Virology 21:194–202.

Ehrhardt, P., and Schmutterer, H. 1965. Untersuchungen über die Beziehungen zwischen dem Enationenvirus der Erbse und seinen Vektoren III. Das Übertragungsvermögen verschiedener Entwicklungsstadien von *Acyrthosiphon pisum* (Harr.) nach unterschiedlichen Saugzeiten an virusinfizierten *Vicia faba*-Pflanzen. Phytopathol. Z. 52:73–88.

Frazier, N. W. 1966a. Pseudo mild yellow edge: a new strawberry virus disease in the Fragaria virginiana indicator clone $M_1$. Phytopathology 56: 571–572.

——— 1966b. Nonretention of two semipersistent strawberry viruses through ecdysis by their aphid vector. Phytopathology 56:1318–1319.

Frazier, N. W., and Posnette, A. F. 1958. Relationships of the strawberry viruses of England and California. Hilgardia 27:455–513.

Freitag, J. H. 1936. Negative evidence on multiplication of curly-top virus in the beet leafhopper, *Eutettix tenellus*. Hilgardia 10:305–342.

Heinze, K. 1959. Versuche zur Ermittlung der Haltbarkeit des Blattroll-Virus der Kartoffel und des Virus der Vergilbungskrankheit der Rübe im Überträger. Arch. Virusforsch. 9:396–410.

——— 1963. Über Virusübertragungen mit Blattläusen auf landwirtschaftliche Kulturpflanzen unter Berücksichtigung verschiedener Stadien des Entwicklungszyklus. Deut. Pflanzenschutzd. (Braunschweig) 15:5–9.

Hoggan, I. A. 1933. Some factors involved in aphid transmission of the cucumber-mosaic virus to tobacco. J. Agr. Res. 47:689–704.

Hoof, H. A. van. 1954. Differences in the transmission of the cauliflower mosaic virus by *Myzus persicae* (Sulz.) and *Brevicoryne brassicae* (L.) [in Dutch, with English summary]. Tijdschr. Plantenziekten 60:267–272.

Kennedy, J. S., Day, M. F., and Eastop, V. F. 1962. A conspectus of aphids as vectors of plant viruses. Commonwealth Inst. Entomol. London, p. 1–114.

Kirkpatrick, H. C., and Ross, A. F. 1952. Aphis-transmission of potato leafroll virus to solanaceous species. Phytopathology 42:540–546.

L'Heritier, P. 1958. The hereditary virus of *Drosophila*. Advances Virus Res. 5:195–245.

Ling, K. C. 1966. Nonpersistence of the tungro virus of rice in its leafhopper vector, *Nephotettix impicticeps*. Phytopathology 56:1252–1256.

McLean, D. L., and Kinsey, M. G. 1965. Identification of electrically recorded curve patterns associated with aphid salivation and ingestion. Nature 205: 1130–1131.

Maramorosch, K. 1952. Direct evidence for the multiplication of aster-yellows virus in its insect vector. Phytopathology 42:59–64.

Mellor, F. C., and Fitzpatrick, R. E. 1951. Studies of virus diseases of strawberries in British Columbia II. The separation of the component viruses of yellows. Can. J. Bot. 29:411–420.

Miles, P. W. 1959. Secretion of two types of saliva by an aphid. Nature 183: 756.

Miyamoto, S., and Miyamoto, Y. 1966. Notes on aphid-transmission of potato leafroll virus. Sci. Rptr. Hyogo Univ. Agric., Ser. Plant Protection 7:51–66.

Nault, L. R., and Gyrisco, G. G. 1966. Relation of the feeding process of the pea aphid to the inoculation of pea enation mosaic virus. Annu. Entomol. Soc. Amer. 59:1185–1197.

Nault, L. R., Gyrisco, G. G., and Rochow, W. F. 1964. Biological relationship between pea enation mosaic virus and its vector, the pea aphid. Phytopathology 54:1269–1272.

Nishi, Y. 1963. Studies on the transmission of plant viruses by aphids. Bull. Kyushu Agr. Exp. Sta. 8:351–408.

O'Loughlin, G. T., and Chambers, T. C. 1967. The systemic infection of an aphid by a plant virus. Virology 33:262–271.

Orlob, G. B., and Bradley, R. H. E. 1961. Where cabbage aphids carry transmissible virus B. Phytopathology 51:397–399.

Osborn, H. T. 1935. Incubation period of pea mosaic in the aphid, *Macrosiphum pisi*. Phytopathology 25:160–177.

——— 1938. Studies on pea virus 1. Phytopathology 28:923–934.

Prentice, I. W. 1949. Resolution of strawberry virus complexes III. The isolation and some properties of virus 3. Annu. Appl. Biol. 36:18–25.

——— 1952. Resolution of strawberry virus complexes V. Experiments with viruses 4 and 5. Ann. Appl. Biol. 39:487–494.

Prentice, I. W., and Harris, R. V. 1946. Resolution of strawberry virus complexes by means of the aphid vector *Capitophorus fragariae* Theob. Annu. Appl. Biol. 33:50–53.

Richardson, J., and Sylvester, E. S. 1968. Further evidence of multiplication of sowthistle yellow vein virus in its aphid vector, *Hyperomyzus lactucae*. Virology 35:347–355.

Richardson, J., and Sylvester, E. S. 1965. Aphid honeydew as inoculum for the injection of pea aphids with pea-enation mosaic virus. Virology 25:472–475.

Schmidt, H. B. 1959. Beiträge zur Kenntnis der Übertragung pflanzlicher Viren durch Aphiden. Biol. Zentralbl. 78:889–936.

Severin, H. H. P., and Tompkins, C. M. 1948. Aphid transmission of cauliflower-mosaic virus. Hilgardia 18:389–404.

Shikata, E., Maramorosch, K., and Granados, R. R. 1966. Electron microscopy of pea enation mosaic virus in plants and aphid vectors. Virology 29:426–436.

Simons, J. N. 1954. Vector–virus relationships of pea-enation mosaic and the pea aphid *Macrosiphum pisi* (Kalt.) Phytopathology 44:283–289.

——— 1955. Some plant–vector–virus relationships of southern cucumber mosaic virus. Phytopathology 45:217–219.

Stace-Smith, R. 1955a. Studies on Rubus virus diseases in British Columbia I. Rubus yellow-net. Can. J. Bot. 33:269–274.

——— 1955b. Studies on Rubus virus diseases in British Columbia II. Black raspberry necrosis. Can. J. Bot. 33:314–322.

Stegwee, D. 1961. Multiplication of plant viruses in their aphid vectors. Recent advances in botany. University of Toronto Press, Toronto. 533–535.

Stegwee, D., and Ponson, M. B. 1958. Multiplication of potato leaf roll virus in the aphid *Myzus persicae* (Sulz.) Entomol. Exp. Appl. 1:291–300.

Sylvester, E. S. 1949. Beet-mosaic virus—green peach aphid relationships. Phytopathology 39:417–424.

——— 1954. Aphid transmission of nonpersistent plant viruses with special reference to the *Brassica nigra* virus. Hilgardia 23:53–98.

——— 1956. Beet yellows virus transmission by the green peach aphid. J. Econ. Entomol. 49:789–800.

——— 1962. Mechanisms of plant virus transmission by aphids. *In* Biological transmission of disease agents. K. Maramorosch [ed.], Academic Press, New York. p. 11–31.

——— 1965. The latent period of pea-enation mosaic virus in the pea aphid, *Acyrthosiphon pisum* (Harris)—an approach to its estimation. Virology 25: 62–67.

——— 1967. Retention of inoculativity in the transmission of pea enation mosaic virus by pea aphids as associated with virus isolates, aphid reproduction and excretion. Virology 32:524–531.

Sylvester, E. S., and Bradley, R. H. E. 1962. Effect of treating stylets of aphids with formalin on the transmission of sugar beet yellows virus. Virology 17:381–386.

Sylvester, E. S., and Richardson, J. 1966a. Some effects of temperature on the transmission of pea enation mosaic virus and on the biology of the pea aphid vector. J. Econ. Entomol. 59:255–261.

——— 1966b. "Recharging" pea aphids with pea enation mosaic virus. Virology 30:592–597.

Sylvester, E. S., and Simons, J. N. 1951. Relation of plant species inoculated to efficiency of aphids in the transmission of Brassica nigra virus. Phytopathology 41:908–910.

Watson, M. A. 1936. Factors affecting the amount of infection obtained by aphid transmission of the virus Hy. III. Phil. Trans. B226:457–489.

——— 1938. Further studies on the relationship between Hyoscyamus virus 3 and the aphis *Myzus persicae* (Sulz.) with special reference to the effects of fasting. Proc. Roy. Soc. (London) B125:144–170.

——— 1960. The ways in which plant viruses are transmitted by vectors. Rep. Commonwealth Entomol. Conf., 7th, London, 1960:157–161.

Watson, M. A., and Roberts, F. M. 1939. A comparative study of the transmission of Hyoscyamus virus 3, potato virus Y and cucumber virus 1 by the vectors *Myzus persicae* (Sulz.), *M. circumflexus* (Buckton) and *Macrosiphum gei* (Koch). Proc. Roy. Soc. (London) B127:543–576.

Whitcomb, R. F., Jensen, D. D., and Richardson, J. 1966. The infection of leafhoppers by western X-disease virus II. Fluctuation of virus concentration in the hemolymph after injection. Virology 28:454–458.

# Specificity in Aphid Transmission of a Circulative Plant Virus*

W. F. Rochow

*Research Plant Pathologist, Crops Research Division, Agricultural Research Service, United States Department of Agriculture; and Professor of Plant Pathology, Cornell University, Ithaca, New York*

Less is known about circulative (persistent) aphid-transmitted plant viruses than about stylet-borne ones. This imbalance results from the facts that only a small proportion of aphid-transmitted viruses are in the circulative group and that circulative viruses are generally much harder to study than are stylet-borne viruses. On the basis of relationships with the host plant, circulative viruses are of two main types. Viruses of one type are also readily transmitted mechanically from plant to plant, apparently because many plant tissues, including epidermis, are susceptible. Pea enation mosaic virus, discussed by Sylvester in a previous chapter, is an example of a circulative virus of this type. Viruses of the second type are transmissible only by means of aphids, apparently because these viruses are restricted to the phloem tissue of their host plants. Potato leaf-roll and barley yellow-dwarf viruses are examples of this type, sometimes also referred to as yellows viruses. Although there are great differences between these two types of viruses in terms of the kinds of experiments that can be carried out to study them, the general patterns of circulative aphid–virus relationships of the two types appear to be similar.

Specificity in the relationship between a circulative plant virus and its aphid vector involves an interaction between the two biological systems. Since most experiments, and all outbreaks of disease in nature, also include transmission of the virus to a plant, a third biological

* Cooperative investigation, Crops Research Division, Agricultural Research Service, United States Department of Agriculture; and Cornell University Agricultural Experiment Station. Supported in part by Public Health Service Research Grant AI-02540 from the National Institute of Allergy and Infectious Diseases, and by Federal Regional Research funds from Project NE-23.

system is an integral part of the problem. Moreover, it is often difficult to eliminate factors that result from a fourth biological system, the investigator himself. This chapter will focus attention on some relationships between a vector and a virus, but it should be remembered that this interaction is only one component of the complete system. The spread of plant viruses by aphids in nature involves the simultaneous interaction of several biological systems with the environment; one should not forget what a small portion of the total picture can be studied in any one experiment in the laboratory.

The basis for this discussion will be barley yellow-dwarf virus (BYDV), a circulative aphid-transmitted plant virus that has not been transmitted mechanically from plant to plant, although its transmission by dodder has been recorded (Timian, 1964). Barley yellow-dwarf virus is a useful representative of the yellows type of circulative viruses about which little is known. Since the virus causes disease of considerable economic importance in many small grains and grasses throughout the world, much attention has been given to its control, its epidemiology, and to related practical problems that occur in nature. The virus also has been the subject of research on basic virus–vector relationships because (*1*) several kinds of specificity are known for BYDV, (*2*) two biological assay techniques not readily applicable to many other viruses of the same type can easily be applied to studies on BYDV (Rochow, 1963a), and (*3*) it is one of the relatively few such viruses that have been purified and whose properties have been investigated. Only certain aspects of BYDV that relate to specificity will be considered here. General treatments of barley yellow dwarf were prepared by Bruehl (1961) and Rochow (1961a).

A central theme in this discussion of vector specificity of BYDV will be that there are two distinct bases for the phenomenon. In some instances specificity results from variations among isolates of the virus; such cases will be emphasized here. In other instances, vector specificity results from variations based on differences among aphid species and especially on differences among clones of an aphid species. This viewpoint of a dual basis for specificity may be an oversimplification of the actual virus–vector interactions involved, but it has some advantages in an approach to understanding the many recorded examples of specificity.

## I. VECTOR SPECIFICITY OF BARLEY YELLOW-DWARF VIRUS

An unusual feature of the virus–vector relationship of BYDV is the range of specificity in the transmission of different virus isolates by different aphid species. Oswald and Houston (1953), and other early workers, generally found *Rhopalosiphum padi* (Linnaeus) to be the most efficient vector of the virus, but *R. maidis* (Fitch), *Macrosiphum avenae* (Fabricius), *Acyrthosiphon dirhodum* (Walker), and *Schizaphis graminum* (Rondani) all transmitted BYDV. (These names do not agree with many used originally because of various changes in names recognized by aphid taxonomists in the United States; see Hille Ris Lambers, 1960; Russell, 1963; Smith and Richards, 1963.) In tests on field-collected samples during 1957, however, Toko and Bruehl (1959) found that BYDV was transmitted from one sample only by *R. padi* and from another sample only by *M. avenae*. These two cases were exceptions; both aphid species transmitted virus from all other field samples collected in Washington. In New York during the same year, a contrasting situation was encountered because BYDV was transmitted from most field-collected samples only by *M. avenae* and not by *R. padi* (Rochow, 1958a). Similar variations have been encountered in other areas.

Four strains of BYDV that can be differentiated by means of three aphid species are currently under study at Cornell (Table 1). One virus strain (PAV) is transmitted nonspecifically by both *R. padi* and *M. avenae*, but transmissions by *R. maidis* are rare. A second virus strain (RPV) is transmitted specifically by *R. padi*. A third strain (MAV) is transmitted specifically by *M. avenae*. Purified preparations of all three of these virus strains have been prepared; comparisons between biological assays and electron micrographs have shown that infectivity of each virus strain is associated with a dense polyhedral particle about 30 mμ in diameter, as described for the MAV strain (Rochow and Brakke, 1964). The fourth virus strain (RMV), which is transmitted specifically by *R. maidis*, has not yet been used in membrane-feeding or injection assays and purified preparations have not been prepared.

TABLE 1

General Pattern of Transmission (+) or Nontransmission (−) of Four Strains of Barley Yellow-Dwarf Virus by Three Aphid Species

| Virus strain | *R. padi* | *M. avenae* | *R. maidis* |
|---|---|---|---|
| PAV | + | + | − |
| RPV | + | − | − |
| MAV | − | + | − |
| RMV | − | − | + |

These four strains of BYDV represent all of the major kinds of strains encountered to date in New York in tests of more than 600 field samples during a period of 10 years (Rochow, 1967b).

The patterns of transmission summarized in Table 1 raise a question about the relationship of the nonspecific virus strain (PAV) to the vector-specific ones. Is the PAV strain merely a mixture of the RPV and MAV virus strains? Two lines of study provide a negative answer to the question and support the view that PAV is a distinct virus strain. In one line of study, various attempts have been made to separate distinguishable virus isolates from plants infected by the PAV strain. Tests involving transmissions by different aphid species, the use of concentrated, partially purified inoculum injected into vectors, the feeding of vectors through membranes on concentrated inocula, and the fractionation of purified preparations of the PAV virus strain all failed to separate any distinct virus isolates from PAV (Rochow, unpublished data). The second line of study, which involves investigations on known mixtures of the RPV and MAV virus strains in oats, showed that mixed infections of the two vector specific virus strains differ from infections by the PAV virus strain in several respects (Rochow, 1965a). Experiences in other laboratories also support the view that virus isolates similar to the PAV strain are distinct and not the result of mixtures (Bruehl and Damsteegt, 1962; Timian and Jensen, 1964).

Since specificity in the relationship between BYDV and its vectors has been relative in our tests, results of any one comparison are not necessarily so clear-cut, as indicated in Table 1. Occasional transmissions of a virus strain by a "nonvector" aphid species do occur. For example, the routine method of making a comparative transmission at Cornell is to remove two adjacent leaves from a plant to be tested. Each leaf is then cut into two pieces to permit the simultaneous comparison of four aphid species. A portion of each leaf is infested with one of the four aphid species for an acquisition feeding period of 2 days at 15°C. Aphids from each leaf section are then transferred to seedlings of Coast Black oats (*Avena byzantina* C. Koch) for an inoculation test feeding period of 5 days. For each aphid species 10 aphids are usually placed on each of 3 oat seedlings. In any one test all 3 seedlings essentially always become infected when infested with aphids that transmit the virus strain in question. Occasionally, one or two of the seedlings in a pot may become infected following feeding by an aphid species that does not regularly transmit the virus strain.

The occasional transmissions of a virus strain by a "nonvector" aphid species represent transmission of the virus strain in question and not the selection of a mutant or different virus strain from a mixture. For example, when plants that became infected by means of *M. avenae* in tests with the RPV virus strain were tested, subsequent virus transmissions were only by *R. padi* and not by *M. avenae* (Rochow, 1959a, 1961b). Similar results in other laboratories have shown that the ability of a given species to transmit a strain of BYDV is not affected by previous transmissions by other vector species (Watson and Mulligan, 1960; Timian and Jensen, 1964).

Although the vector specificity of strains of BYDV is relative and can be influenced by various factors, this variation among virus strains is a stable, consistent, and important property of the virus. An illustration of the stability of vector specificity, of the striking differences among the virus strains, and of the relative nature of the specificity is given by results of parallel comparative transmission tests made during the past four years (Table 2). These data are results of 23 recent serial transfers involved in maintenance of the strains of BYDV under study in New York; the results are in general agreement with earlier comparative transfers of the same virus strains (Rochow, 1959a, 1961b; Rochow and Eastop, 1966). Data are given for a single isolate of each of the three vector-specific virus strains and for two isolates of the nonspecific PAV strain. One PAV isolate had been recovered initially and has been maintained solely by means of *R. padi*;

TABLE 2
TRANSMISSION OF FOUR STRAINS OF BARLEY YELLOW-DWARF VIRUS IN 23 SERIAL TRANSFERS BY MEANS OF FOUR APHID SPECIES

| Virus strain | Serial transfers from oats infected by means of aphid species shown | Transmission[a] in parallel tests with *R. padi* (RP), *M. avenae* (MA), *R. maidis* (RM), and *S. graminum* (SG) | | | |
|---|---|---|---|---|---|
| | | RP | MA | RM | SG |
| PAV | RP | 69/69 | 51/67 | 2/69 | 13/68 |
| PAV | MA | 68/69 | 58/69 | 1/69 | 20/67 |
| RPV | RP | 69/69 | 0/68 | 0/69 | 22/69 |
| MAV | MA | 2/69 | 69/69 | 0/69 | 0/69 |
| RMV | RM | 2/69 | 3/67 | 60/69 | 1/68 |
| None | (Aphid controls) | 0/69 | 0/69 | 0/69 | 0/69 |

[a] The numerator is the number of plants that became infected; the denominator is the total number of plants infested with about 10 aphids for an inoculation test feeding of 5 days following an acquisition feeding of two days at 15°C on a portion of a detached leaf.

the other PAV isolate, which originated from the opposite half of the same oat leaf, was recovered initially and has been maintained by means of *M. avenae.* Data for both PAV isolates are in agreement with observations from many laboratories that show *R. padi* to be the most efficient vector of nonspecific BYDV strains, such as PAV. The data also support the observation mentioned above that the PAV strain is distinct and not a mixture of vector specific strains. Although *S. graminum* has been included in most tests, this aphid species has not aided characterization of virus isolates in New York because *S. graminum* generally transmitted the same virus isolates as did *R. padi,* but did so more erratically than *R. padi* (Rochow and Eastop, 1966).

An important feature of the vector specificity is that the same pattern occurs when concentrated, purified virus preparations are used as inocula. In one series of experiments, for example, purified preparations of the RPV or MAV strains concentrated 200- to 500-fold were used to inject *R. padi* and *M. avenae* (Muller, 1965). Five injected aphids were allowed to feed on each oat test plant. When RPV was used as inoculum, 14 of 15 plants infested with injected *R. padi* became infected; none of 42 plants infested with injected *M. avenae* became infected. In parallel tests with the MAV virus strain, 15 of 15 plants infested with injected *M. avenae* became infected; none of 54 plants infested with *R. padi* became infected. None of 68 plants infested with 10 aphids as controls became infected. Similar results have been obtained when aphids were allowed to acquire virus by feeding through membranes on concentrated preparations of the virus strains. The same specificity also occurs when the inoculum for injection is obtained from viruliferous aphids (Rochow, 1963a).

## II. FACTORS AFFECTING VECTOR SPECIFICITY

Since the vector specificity of strains of BYDV tested in our laboratory is relative, it seems likely that a study of factors affecting the specificity will contribute to an understanding of the phenomenon. Some of the major factors now known to influence results of comparative transmission studies include the length of time aphids feed on plants, temperature, variation among clones of an aphid species, and the presence of more than one virus strain.

The length of the acquisition feeding period was an important variable recognized in initial studies. In general, the longer the acquisition feeding period, the better are chances for virus transmission by a "nonvector." For example, in one series of experiments (Rochow, 1961b) none of 72 plants became infected when *R. padi* had fed for 2 days on leaves infected by the MAV virus strain. When *R. padi* were permitted acquisition feedings of 3–5 days, however, 2 of 27 plants became infected by the MAV virus strain transmitted by *R. padi*. When aphids were reared on infected plants, specificity still prevailed but transmissions by "nonvectors" were more frequent than for 1- to 2-day acquisition feedings (Rochow, 1959a).

Another factor influencing vector specificity of strains of BYDV was temperature during aphid feeding. In a recent study (Rochow, 1967a) the temperature had a marked effect on virus transmission to oats in tests with the RMV virus strain. Transmission of the RMV strain by groups of *R. padi* or *M. avenae* was rare in all tests when acquisition or test feedings were at 15 or 20°C. This agrees with results from earlier greenhouse studies (Table 2). The transmission of the RMV virus strain by both *R. padi* and *M. avenae* increased with increasing temperatures and reached a maximum (70 and 83% of plants became infected, respectively) when both acquisition (2 days) and inoculation test feedings (5 days) were at 30°C. The effect of increasing temperature was not a general one, however, because the transmission of the RPV strain by *M. avenae* or the transmission of the MAV strain by *R. padi* was not increased by temperatures ranging from 15 to 30°C. Moreover, the relative efficiency of single aphids of either species in transmitting the PAV virus strain was not altered by the temperature treatments.

Variation among clones of aphid species has received considerable attention since the importance of this variable was emphasized by Stubbs (1955). Several kinds of variation among clones have been noted for vectors of BYDV. Three clones of *S. graminum*, each of which originated from a different area in the United States, were compared as vectors of four strains of BYDV in one study (Rochow, 1960). None of the clones transmitted the MAV strain of virus, but two of the clones did transmit three other strains of BYDV, which were essentially never transmitted by the third aphid clone. Thus, clones of *S. graminum* appeared to be either active or inactive as vectors of BYDV.

Saksena, Singh, and Sill (1964c) described a quantitative variation among clones of one aphid species. They compared four biotypes of *R. maidis* as vectors of BYDV. All of the biotypes, which had previously been identified on the basis of other physiological differences, transmitted BYDV, but one biotype (KS-2) was found to be a highly efficient vector (87% of the individual aphids tested transmitted virus). Two of the biotypes were intermediate in efficiency (44 and 46% transmitted virus, respectively) and one of the biotypes effected an average of 28% transmission. In a direct comparison of the KS-2 biotype with the New York clone of *R. maidis,* however, Rochow and Eastop (1966) did not find the KS-2 biotype to be an efficient vector.

A more specific variation was recently described for two clones of *R. padi.* A clone of *R. padi* from Kansas, which also had abnormal wing venation, regularly transmitted the RMV strain of virus that was rarely transmitted in parallel tests with the New York clone of *R. padi* (Rochow and Eastop, 1966). Differences between the clones of *R. padi* were pronounced only when tests were carried out in the greenhouse or under controlled conditions at 15 or 20°C. When the aphids fed at 30°, the differences between the clones were less pronounced than those at lower temperatures because the New York clone of *R. padi* also transmitted the RMV strain at 30°C. The unusual aspect of this variation between clones of *R. padi* was that the difference between aphid clones was pronounced only in tests with one of four strains of BYDV.

Although differences have been detected among clones of some aphid species, this does not mean that such differences always occur. Early attempts to detect possible differences among clones of *R. padi,* for example, were negative (Bruehl, 1958; Rochow, 1958a, 1958b). Gill (1967) found no differences among clones of five aphid species in the transmission of BYDV. Several other comparisons between clones of *M. avenae* and other aphid species also revealed no marked differences (Bruehl, 1958; Rochow, 1958b; Rochow and Eastop, 1966; Smith, 1963b).

In addition to variation among clones of an aphid species, variations among forms of a single aphid species also might be important. One example of such variation for a vector of BYDV was described by Orlob and Arny (1960). Oviparae, which normally do not feed on barley, and alatoid fundatrigeniae of *R. fitchii* were capable of transmitting BYDV, but four other forms of the same aphid species failed to transmit BYDV. When Orlob (1962) compared forms of *M. avenae,* oviparae of the nonmigratory forms reared in the greenhouse transmitted BYDV as efficiently as did apterous viviparous females.

All of the variations within a single aphid species are distinct from the variations in specificity based on differences among strains of the virus. As already emphasized, the specificity can result from variation in the vector as well as from variation in the virus. This dual basis for specificity sometimes causes confusion because it is difficult to tell which basis for specificity is operative in any one experiment unless comparative information is available on the aphid clones and virus strains involved. Bruehl (1958) and Rochow (1958b) exchanged aphids and thus were able to show that the vector specificity being studied rested with the virus and not with the aphid. Gill (1967) came to a similar conclusion by using more than one clone of each aphid species.

The presence of more than one strain of BYDV in a plant influences vector specificity in a variety of ways comparable to the range in the vector specificity phenomenon itself. Interactions of BYDV strains range from situations in which each virus in a mixed infection remains unaffected, through situations where one virus strain protects a plant from infection by a second strain, to situations where the viruses interact to produce an increase in the severity of disease and an alteration in vector specificity.

In many cases the presence of more than one strain of BYDV has no effect on vector specificity. Toko and Bruehl (1959) found that strains similar to RPV and MAV were transmitted selectively by *R. padi* or *M. avenae* both from single and double infections. Similarly, in New York tests on field-collected samples, situations have been encountered in which plants appear to have been infected by more than one virus strain and each of the strains was recovered selectively by the appropriate aphid species (Rochow, 1963b, 1967b).

Smith (1963a) noted that plants infected by a mild strain of BYDV appeared to be protected from later infection by a more virulent strain of the virus. Attempts in other laboratories to demonstrate such cross-protection among strains of BYDV, a well-established phenomenon for many plant viruses, had failed (Allen, 1957; Lindsten, 1964; Rochow, 1959a; Toko and Bruehl, 1959; Watson and Mulligan, 1960). Jedlinski and Brown (1965) also described an experiment in which oats inoculated with a mild strain of BYDV (MAV) were protected against subsequent invasion by either of two virulent strains (similar to PAV) under field conditions. Moreover, when oats were simultaneously inoculated with the three virus strains, the plants showed only mild symptoms, recovered completely, and no virus could be transmitted from the apparently healthy plants. This unique case of mutual exclusion, which was compared with similar interactions among certain viruses of the yellows group transmitted by leafhoppers, was considered to be additional evidence for the close relationship of the BYDV isolates studied (Jedlinski and Brown, 1965).

A still different interaction is under study at Cornell. When oat seedlings are doubly inoculated with both the RPV and MAV strains of BYDV, the infected plants usually develop symptoms more severe than those caused by either virus strain alone. Subsequent comparative transmissions from leaves of the doubly infected plants show that vector specificity still prevails in tests with *M. avenae,* but not in parallel tests with *R. padi* (Fig. 1). In many experiments over a period of six years, virus transmitted from one-half of a doubly infected leaf by groups of *M. avenae* was subsequently transmitted by *M. avenae* but not by *R. padi.* In contrast, virus transmitted from the opposite half of the same doubly infected leaf by groups of *R. padi* was subsequently transmitted by both *R. padi* and *M. avenae* in most cases. This interaction, which involves the apparent loss of vector specificity following double infection by two strains of BYDV, will be considered in some detail because it offers a useful approach to a study of the mechanism of vector specificity.

Details on the apparent loss of vector specificity have been published (Rochow, 1965a). The effect occurs for virus transmissions from doubly infected oats that had been inoculated simultaneously with the two virus strains or from plants inoculated with one of the strains before the other. Virus transmitted from doubly infected plants by *R. padi* continues to be transmitted by both *R. padi* and *M. avenae* in subsequent serial transfers, but virus transmitted each time by *M. avenae* is subsequently transmissible only by *M. avenae* (Fig. 1). When concentrated, partially purified inocula prepared from doubly infected plants are injected into aphids or when aphids feed through membranes on such inocula, the apparent loss of vector specificity occurs in tests with *R. padi,* but not in parallel tests with *M. avenae.* A difference between *R. padi* and *M. avenae* was also noted by Lindsten (1964) in transmission tests from plants infected by a mixture of BYDV strains.

Perhaps the simplest explanation for the apparent loss of vector specificity is that from doubly infected leaves *R. padi* transmits unaltered MAV together with RPV, perhaps as some aggregated complex

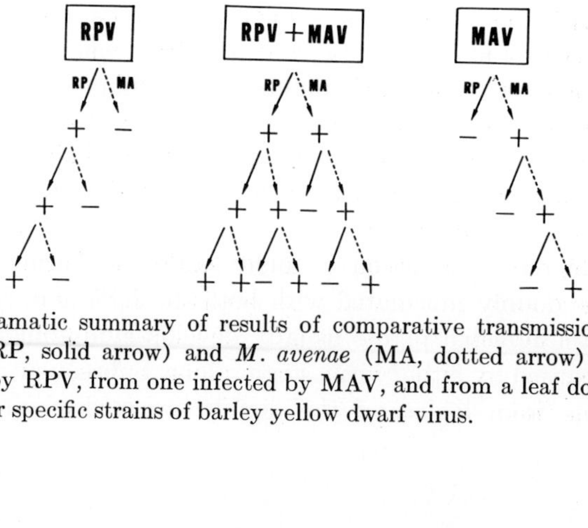


*Fig. 1.* Diagramatic summary of results of comparative transmissions by means of *R. padi* (RP, solid arrow) and *M. avenae* (MA, dotted arrow) from an oat leaf infected by RPV, from one infected by MAV, and from a leaf doubly infected by both vector specific strains of barley yellow dwarf virus.

of the two viruses. *Rhopalosiphum padi* might thus transmit MAV in the presence of RPV despite the general inability of *R. padi* to transmit the MAV strain alone. Other examples are known in which aphid transmission of one virus depends on the presence of a second virus. Smith (1965), who recently reviewed some examples of this phenomenon, suggested the term "helper virus" for the second virus. The "helper virus" idea suggests that *R. padi,* after becoming viruliferous for RPV, might then also be able to acquire and transmit the MAV virus strain, or that, in some way, the loss of specificity might occur if the two virus strains interact in the vector.

Many tests have been carried out to determine whether a loss of vector specificity occurs following alternate acquisition of the separate BYDV virus strains by *R. padi.* Aphids were allowed to feed on plants infected by one virus strain and then were exposed to the other virus strain by feeding on infected leaves, by the injection of concentrated inoculum, or by feeding through membranes on concentrated inoculum. In related experiments, aphids were injected with, or allowed to feed through membranes on, inocula made by mixing concentrated preparations of each of the separate virus strains. In one series of experiments, involving 265 separate tests, a loss of vector specificity did not occur, except in two tests (Rochow, 1965b). At the time, these two exceptions were not considered significant.

Results of more recent experiments on the possible interaction of the two virus strains within *R. padi* generally agree with those of earlier tests. A loss of vector specificity has occurred rarely in tests involving the alternate exposure of *R. padi* to the two virus strains. It may be significant, however, that in every case when a loss of vector specificity did occur, *R. padi* had been exposed to the MAV virus strain before being exposed to the RPV strain. In some recent experiments, for example, *R. padi* were reared on oats infected by the MAV strain or on oats infected by the RPV strain. Aphids were then permitted a second acquisition feeding, either by feeding through membranes on concentrated inoculum of the other virus strain or by feeding on leaves from singly infected plants. At least one case of an apparent loss of vector specificity occurred in each of four recent experiments. In every instance the aphids had been exposed to the MAV strain before RPV; the phenomenon did not occur in any parallel tests in which *R. padi* had been exposed to RPV before MAV.

Tests on the possible interaction of the virus strains within *R. padi* have thus provided little support for the idea that the apparent loss of vector specificity is based on the interaction of the virus strains within the aphid. Except for the cases just cited, all attempts to detect the apparent loss of vector specificity following possible interaction of the separate virus strains within the vector have been negative. To date, the apparent loss of vector specificity has been consistent only following transmissions from plants doubly infected by both virus strains. Since the synthesis of both virus strains occurs simultaneously in such doubly infected plants, the explanation could result from events that take place during this simultaneous virus synthesis. It is possible that the apparent loss of vector specificity is not really a loss of specificity of the MAV strain, but instead involves transmission by *R. padi* of a "different" virus isolate, one that is produced as a result of simultaneous synthesis of the two virus strains in the doubly infected plant. This possibility will be discussed further in the next section.

The wide range of interactions among strains of BYDV focuses attention on two problems basic for plant virology. One concerns the relationship of the different virus isolates. The characterized isolates of BYDV discussed here have been referred to as strains mainly for simplicity and convenience. The term "strain" generally indicates a degree of relatedness not clearly established for the BYDV strains. The positive cross-protection tests support the use of "strain" because such tests usually indicate a close relationship among virus isolates. On the other hand, the interaction of the RPV and MAV strains, which is accompanied by symptoms more severe than those caused by either virus alone, suggests a relationship more commonly associated with unrelated plant viruses than with related ones. The application of chemical, physical, and serological criteria will eventually help to resolve this question.

The second problem concerns the lack of basic knowledge on interactions among viruses such as BYDV that are restricted to phloem tissue of plants. Many studies on the interaction of plant viruses have been carried out with mechanically transmissible viruses of the mosaic type, viruses that generally infect most tissues of the plant. It seems reasonable to expect that interactions of phloem-limited viruses of the yellows type might be quite different from those of viruses not limited to specific plant tissues. Thus, the range in kinds of interactions among strains of BYDV probably is a reflection of a lack of knowledge about the system and about factors that influence it.

## III. MECHANISM OF VECTOR SPECIFICITY

Although data are available for some aspects of vector specificity of BYDV, an understanding of the phenomenon is not possible at present. Enough is known to suggest that it might now be possible to develop a theoretical model of the mechanism of BYDV vector specificity. I have attempted to do this. The model appears to be consistent with experimental evidence, but it should be emphasized that the following comments are largely speculative and intended only to represent a simple working hypothesis that might aid thought and experimentation on the mechanism of specificity. In other words, it might be helpful to have a definite picture of the mechanism as long as one does not become convinced prematurely that the picture is the correct one.

The first assumption is that composition and/or structure of the protein capsid of one BYDV strain is different from that of another. Although this is an assumption, some recent serological data suggest that it might be a reasonable one. When a new technique recently described by Gold and Duffus (1967) was used, antisera prepared for the MAV strain of BYDV were found to be active in tests with the MAV strain, but not in some parallel tests with the RPV strain (W. F. Rochow and E. M. Ball, unpublished data). Differences in the amino acid composition of strains are well established for several other plant viruses.

A second assumption is that the major barrier controlling specificity is the entrance of virus into the salivary glands of the aphid. This assumption seems reasonable in view of two lines of evidence that the gut wall is not a major barrier to the circulation of two strains of BYDV within the vector. First, specificity occurs when concentrated preparations of the RPV or MAV strains are injected into aphid hemolymph, just as it does when virus enters aphids through the alimentary tract. Second, some virus strains not regularly transmitted by *R. padi* or *M. avenae* are acquired and often can be detected in hemolymph (Rochow and Pang, 1961). This evidence suggests that BYDV acquired from phloem by feeding aphids passes through the gut wall and reaches the hemolymph, but that only certain virus strains are able to enter salivary glands in sufficient amounts to be inoculated to plants during subsequent feeding. There is no direct evidence for the central role of salivary glands in vector specificity, but Nault, Gyrisco, and Rochow (1964) have called attention to the fact that some workers believe aphid salivary glands function, at least in part, as excretory organs because aphids have no malpighian tubules, the excretory organ of most insects. In aphids salivary glands might play a more important role in the transmission of circulative viruses than they do in other vectors such as leafhoppers.

The suggested hypothesis is that vector specificity results from the interaction of the protein capsid of a virus strain with the membranes surrounding aphid salivary glands. Since BYDV purified from aphids is the same as that purified from plants (Rochow and Brakke, 1964), it is at least known that complete virus particles occur in the vector. In view of evidence that adsorptive complementarity between the virus capsid protein and protein receptors on plasma membranes is important in the specificity of animal viruses (Cords and Holland, 1964), it does not seem unreasonable to postulate that BYDV capsid protein could interact with aphid membranes. Perhaps the adsorption of virus to membranes is involved in specificity, possibly as a first step in the penetration of virus into salivary glands by pinocytosis. Thus, *R. padi* would transmit the RPV strain because the structure of the virus capsid protein confers on RPV a basis for entrance into salivary glands. *Rhopalosiphum padi* would fail to transmit the MAV virus strain because the capsid protein of MAV normally results in the exclusion of the virus from salivary glands of *R. padi,* but not from those of *M. avenae.* Changes in specificity could result either from differences in virus capsid protein, as might occur among strains of virus, or from physiological or morphological differences in salivary glands of individual aphids, as might occur among the clones of one aphid species.

The theoretical model also provides a possible basis for explaining the apparent loss of vector specificity following double infection by the RPV and MAV strains. During the simultaneous synthesis of the RPV and MAV strains in the same plant, a type of interaction similar to the phenotypic mixing of animal viruses could occur. Thus, MAV nucleic acid might be incorporated into some particles that have an RPV protein capsid, just as polio virus RNA becomes incorporated within the protein capsid of coxsackie B 1 virus (Cords and Holland, 1964). Such atypical virus particles might then function as does the RPV strain when they reach the salivary glands of *R. padi,* and enter the glands together with typical particles of the RPV strain, which would also be present. Both kinds of virus particles would later by inoculated into plants when *R. padi* feeds. Since the atypical particles contain MAV nucleic acid, virus synthesis in the infected plant would be comparable to an initial double infection because both RPV- and MAV-directed virus synthesis would occur. This possibility emphasizes the importance of simultaneous synthesis of the two virus strains in the infected plant, an observation consistent at present with most experimental evidence on the apparent loss of BYDV vector specificity in double infections.

A final consideration of the theoretical model is directed to the infrequent cases of loss of vector specificity when *R. padi* is permitted access to the MAV strain before being exposed to RPV. When *R. padi* feeds on MAV-infected plants, virus particles pass through the digestive tract, through the gut wall, and circulate in the hemolymph. Normally, *R. padi* does not transmit the MAV strain because MAV particles do not enter salivary glands. The method of entrance of virus particles into the cells of salivary glands is unknown, but Forbes (1964) has pointed out that pinocytosis occurs in aphids and that it is a reasonable mechanism to explain the passage of virus particles through aphid cells. Regardless of the mechanism of penetration through the membrane, the process could involve at least two critical steps. First, virus might be adsorbed to the salivary cell membranes, perhaps at specific sites on the membrane. Evidence for adsorption of proteins as a first step in pinocytosis has been presented (Brandt, 1958; Holter, 1959). Second, adsorbed virus, if its protein capsid is complementary, might initiate pinocytosis or some other process and enter the salivary gland. Perhaps MAV in *R. padi* can initiate only the first step of this process—adsorption to the salivary membranes. The MAV particles circulating in *R. padi* hemolymph might then tend to accumulate on the surface of the salivary glands as the aphid continues to feed on MAV-infected plants. When the aphid is later exposed to the second virus strain, RPV particles not only adsorb to the membrane but also readily initiate pinocytosis, in some cases incorporating one or more MAV particles into the same vesicle. Thus, the RPV strain might "carry" adsorbed MAV with it into the salivary gland in a manner somewhat similar to the ingestion and utilization of some carbohydrates by the amoeba only if the carbohydrates are supplied together with a pinocytosis-inducing protein (Holter, 1959).

This concept seems consistent with the current evidence that the apparent loss of vector specificity following the alternate exposure of *R. padi* to both virus strains occurs only if aphids are exposed to the MAV strain first. Presumably the incorporation of MAV into a pinocytosis vesicle with RPV rarely occurs, except when there is a previous accumulation of MAV on the membrane. Previous feeding on MAV-infected plants is the only experimental condition used thus far to bring about such an accumulation. MAV would very rarely be incorporated into pinocytosis vesicles together with RPV when *R. padi* is exposed to the RPV strain first, or when the two virus strains are injected simultaneously into aphid hemolymph, because the amount of MAV adsorbed on the membrane would be insufficient to result in more than a very low probability of its incorporation into a pinocytosis vesicle with RPV. If adsorbed MAV is "carried" into salivary glands by the RPV strain, then both virus strains could be inoculated into a plant when the aphid feeds and the apparent loss of specificity would occur in tests on such infected plants because they would be doubly infected by both RPV and MAV.

The multiplication of virus in the vector has not been incorporated into this theoretical model for the sake of simplicity and because it is not known whether such a multiplication is common in the circulative group of viruses transmitted by aphids. The only direct evidence for the multiplication of a circulative virus in its aphid vector is that for potato leaf-roll virus (see the chapter by Sylvester). Attempts by Irmgard Muller and me to transmit BYDV serially from aphid to aphid have been unsuccessful, as were similar attempts by Sylvester and Richardson (1966) for pea enation mosaic virus. Although BYDV can readily be inoculated into vectors, we have not succeeded in transmitting it to subsequent insects when injected aphids were maintained on plants immune from BYDV or on artificial diet. There are many possible explanations for the failure to demonstrate virus multiplication in aphids, but one related to the key role of salivary glands in the theoretical model just described is often overlooked. Perhaps plant virus synthesis in aphids occurs in salivary glands. If so, synthesized virus would probably be secreted during aphid feeding and might never accumulate to detectable levels within the aphid vector. This concept is not supported by recent evidence on the location of pea enation mosaic virus in the gut lumen and in the fat body of aphid vectors (Shikata, Maramorosch, and Granados, 1966), but the observations do not disprove the concept because virus synthesis and virus accumulation at specific sites within the vector are not necessarily related.

## IV. RELEVANCE OF VECTOR SPECIFICITY IN NATURE

The vector specificity of BYDV is not merely a laboratory curiosity providing an approach to studies on aphid–virus relationships. The specificity is important in nature. An appreciation of vector specificity of BYDV has helped to clarify several epidemiological aspects of the disease, such as variations in the disease from one location to another, and from one year to another at the same location. The specificity has also focused attention on certain practical problems. For example, it is clearly important to use more than one aphid species in attempts to confirm the diagnosis of the disease by means of aphid transmission tests. The differential disease reaction of varieties of small grains at different locations underscores the importance of variation among virus strains in evaluating tolerant plant varieties (Arny and Jedlinski,

1966; Bruehl *et al.,* 1962). Although vector specificity is only one of several kinds of variation known for isolates of BYDV, its relevance to various observations on the disease in nature, its potential role in variation and changes in the disease, and its simultaneous involvement of both vector and virus all suggest that the phenomenon is a key to understanding the disease.

Tests on field-collected samples made in different laboratories have revealed a great range in the vector specificity among isolates of BYDV. It is difficult to compare the results from all laboratories because of differences in the methods, plants, and aphids used in transmission tests. It is often not possible to determine whether the variation encountered in one location resulted from properties of strains of the virus or from characteristics of clones of the aphid species used in transmission tests. Experiments have not always been carried out in a manner to detect the presence of mixed infections by more than one strain of BYDV. The range of specificity in the transmission of different virus isolates from field-collected samples can probably best be summarized by arranging the virus isolates into four main groups.

The first group includes all BYDV isolates transmitted most efficiently by *R. padi.* Isolates in this group vary greatly in the efficiency by which they are also transmitted by other aphid species. At one end of a spectrum are virus isolates transmitted nonspecifically by *R. padi* and many other aphid species; at the other extreme are isolates transmitted specifically by *R. padi.* In Kansas, for instance, virus isolates were transmitted by all four aphid species tested. Saksena, Sill, and Kainski (1964b) found the relative transmission efficiencies of the aphid species to be as follows: *R. padi,* 93%; *S. graminum,* 77%; *M. avenae,* 74%; and *R. maidis,* 37%. Many virus isolates in this group are transmitted efficiently by *R. padi* and *M. avenae,* but other aphid species may be less efficient vectors. Other isolates appear similar to the PAV strain described above because efficient transmission occurs in tests with both *R. padi* and *M. avenae,* but *S. graminum* is an erratic vector and *R. maidis* only rarely transmits the virus. In New York, for example, we have not yet found a nonspecific virus isolate transmitted efficiently by both *R. padi* and *R. maidis.* Although *R. padi, M. avenae, R. maidis,* and *S. graminum* have sometimes all transmitted virus from a single field sample, subsequent tests of the virus isolates transmitted from the field sample by each aphid species showed either that the field plant was infected by a mixture of the PAV and RMV strains, or that virus transmissions by *R. maidis* in the initial test were merely occasional transmissions of the PAV strain by *R. maidis* (Rochow, 1967b). On the other hand, virus isolates transmitted by both *R. padi* and *R. maidis* have been reported from other locations in the United States (Saksena *et al.*, 1964b; Tetrault, Schulz, and Timian, 1963) as well as from Canada (Smith, 1963b), and Australia (Butler, Grylls, and Goodchild, 1960).

Much of the early work on BYDV appears to have involved isolates of this first group that are transmitted nonspecifically by *R. padi* and other aphid species. These virus isolates seem to be the most common ones in nature. Nonspecific virus isolates appear to be the predominating ones in many areas of the United States, such as California (Allen, 1957), Washington (Bruehl and Damsteegt, 1962), Illinois (Rochow *et al.*, 1965), and North Dakota (Timian and Jensen, 1964). In recent years such isolates represented the predominating ones in New York (Rochow, 1967b). These nonspecific virus isolates occur in many parts of the world. They have been encountered in several provinces of Canada (Gill, 1967; Orlob, 1961; Smith, 1963b). Similar isolates appear to occur also in England (Watson and Mulligan, 1960), Wales (Catherall, 1963), Finland (Ikäheimo, 1960), Sweden (Lindsten, 1964), The Netherlands (Oswald and Thung, 1955), Germany (Rademacher and Schwarz, 1958), Czechoslovakia (Vacke, 1964), Australia (Butler *et al.*, 1960), and India (Nagaich and Vashisth, 1963).

At the other extreme in this first group are the virus isolates transmitted efficiently by *R. padi*, but not transmitted regularly by other aphid species, isolates similar to the RPV strain described above. These vector-specific virus isolates often have been encountered in areas where the nonspecific isolates are common. For example, virus isolates transmitted specifically by *R. padi* have been isolated in Washington (Toko and Bruehl, 1959), in New York (Rochow, 1959a), and in Canada (Gill, 1967; Orlob, 1961). In Wales (Catherall, 1963) and in Sweden (Lindsten, 1964) such vector-specific isolates were encountered in addition to the nonspecific ones, but in Israel (Harpaz and Klein, 1965) only vector-specific BYDV was detected.

An important feature of all virus isolates in this first group is variation in virulence. Allen (1957) grouped 43 virus isolates transmitted by *R. padi* into 16 strains, which could be arranged into seven types. Endo and Brown (1963) noted a differential response among three oat varieties to a mild and a moderately virulent virus strain. Physiological differences also have been detected between plants infected by virulent and avirulent strains (Goodman, Watson, and Hill, 1965). The variation in virulence among virus isolates often has been associated with vector specificity. In general, the virus isolates transmitted nonspecifically by *R. padi* cause more severe symptoms in plants than do virus isolates transmitted specifically by one aphid species (Catherall, 1963; Gill, 1967; Jedlinski and Brown, 1965; Rochow *et al.*, 1965; Smith, 1963b).

The second main group of virus isolates includes those transmitted most efficiently by *M. avenae*. The vector specificity of this second group is much more pronounced than that of the first group. Many of the virus isolates, which appear similar to the MAV strain described above are also transmitted, often with low efficiency, by *Macrosiphum (Sitobion) fragariae* (Walker) and *Acyrthosiphon (Metopolophium) dirhodum* (Walker) (Gill, 1967; Smith, 1963b; Watson and Mulligan, 1960). The distinguishing feature of isolates in this group is that transmissions by *R. padi, S. graminum,* and *R. maidis* either do not occur or are rare. Isolates of this second group have been detected in the United States in Washington (Toko and Bruehl, 1959), were common for several years in New York (Rochow, 1959a, 1967b), and occur in other parts of the country (Rochow, 1959b). Similar virus isolates have been described from several provinces in Canada (Gill, 1967; Slykhuis *et al.*, 1959; Smith, 1963b), from England (Watson and Mulligan, 1960), from Scotland (Anonymous, 1961), from Wales (Catherall, 1963), and from Belgium (Roland, 1962).

The third group of BYDV isolates includes those transmitted most efficiently by *R. maidis*. These virus isolates appear to be less common in nature than those of the first two groups. Many isolates in this third group appear to be vector specific because transmissions by *R. padi, M. avenae,* and *S. graminum* are infrequent (Rochow, 1961b). Virus isolates in this group, which are similar to the RMV strain described above, were detected in most years in a low proportion of field samples tested in New York (Rochow, 1967b). Similar vector specific isolates also have been encountered in Manitoba (Gill, 1967). Gill (1967) described some additional isolates from Canada that were transmitted with rather low efficiency by five aphid species, the most efficient of which was *R. maidis*. These isolates differ from the usual nonspecifically transmitted ones because *R. padi* was not the most efficient vector. Since the vector specificity of the RMV strain is so dependent on temperature and the aphid clone (Rochow and Eastop, 1966), virus isolates in this third group might have been considered similar to nonspecific isolates of the first group in many tests on field samples.

The fourth group of BYDV isolates includes those transmitted most efficiently by *S. graminum*. Virus isolates in this group apparently are rare. Gill (1967) recovered one such virus isolate from oats; 13% of individual *S. graminum* tested transmitted the virus, but only 2% of individual *R. padi* transmitted virus, and no transmission was obtained by *M. avenae, A. dirhodum,* or *R. maidis. Schizaphis graminum* generally appears to transmit the same virus isolates as does *R. padi*. In some instances the efficiency of *S. graminum* was comparable to that of *R. padi* (Saksena, Dody, and Sill, 1964a; Saskena *et al.,* 1964b; Timian and Jensen, 1964). In other instances *S. graminum* was a less efficient vector than *R. padi* (Orlob, 1961; Oswald and Houston, 1953; Rochow and Eastop, 1966; Smith, 1963b). The variation among clones of *S. graminum* in their ability to transmit BYDV can be considerable (Rochow, 1960). This variation in the aphid probably is a main factor in the differences observed among laboratories in the transmission efficiency of *S. graminum,* although direct comparisons of New York and Kansas clones of *S. graminum* gave no support for this explanation (Rochow and Eastop, 1966).

An understanding of barley yellow dwarf must not only include a knowledge of the kinds of virus isolates in nature but also information on the virus isolates that predominate in any one year. Recent studies suggest that dramatic changes can occur in the strains of BYDV that predominate from year to year in one region. In tests with 5 aphid species, Gill (1967) found the transmission patterns of virus isolates recovered in Manitoba in 1964 to be completely different from transmission patterns obtained in similar tests in 1965. Results of tests in New York over a 10-year period indicated a change in the strain of BYDV predominating in one location. Although the MAV strain was recovered from most samples tested in early years, the PAV and RPV strains were the ones predominating in recent years (Rochow, 1967b).

Changes in strains of BYDV that predominate from year to year in one location are important in epidemiology mainly as they relate to corresponding populations of aphid vectors. Fluctuations in the populations of different aphid species on oats (Adams and Drew, 1964; Forbes, 1962) illustrate the complex nature of the epidemiology of barley yellow dwarf. Because of changes in both strains of virus and in aphid species or even in aphid clones that predominate—coupled with interactions among host plants for both virus and aphids, with the weather, and with the age and variety of crop plants—variations in the disease as it is observed in different locations are not surprising.

A specific example of the possible application of knowledge about vector specificity of BYDV might illustrate the relevance of vector specificity in nature (Rochow, 1967b). During late May and early June of 1966, populations of *M. avenae* on spring oats near Ithaca, New York, were higher than those observed in most previous seasons. Despite these high aphid populations, however, the percentage of oats infected by BYDV was much lower than had been observed in earlier years. A plausible explanation for this apparent discrepancy was suggested by data on changes in the strains of BYDV predominating during the different growing seasons. Since *M. avenae* is known to be a less efficient vector of the PAV strain, which predominated in 1966, than of the MAV strain, which predominated in earlier years, the discrepancy between aphid populations and the percentage of plants infected by BYDV in 1966 may have been simply a reflection of inefficient transmission of the predominating virus strain by the predominating aphid species.

An attempt has recently been made to relate observations made in the field to current knowledge about some of the interactions among plant hosts, virus, and vectors (Rochow *et al.*, 1965). Several differences in the BYDV disease of oats in different locations were noted. In one case distribution of infected plants along borders and in circular areas within a field and relatively severe outbreaks of disease were associated with the PAV strain and virus transmission by *R. padi.* In another case, a random pattern of distribution of infected plants throughout fields and relatively mild symptoms of the disease were associated with the MAV strain of virus being transmitted by *M. avenae.* The two general patterns, considered extremes in a range of variation, suggest some ways in which basic studies on virus–vector relationships and field studies on epidemiology can complement each other.

## BIBLIOGRAPHY


Adams, J. B., and Drew, M. E. 1964. Grain aphids in New Brunswick. I. Field development on oats. Can. J. Zool. 42:735–740.

Allen, T. C., Jr. 1957. Strains of the barley yellow-dwarf virus. Phytopathology 47:481–490.

Anonymous. 1961. Report of Scottish Plant Breeding Station. 78 p.

Arny, D. C., and Jedlinski, H. 1966. Resistance to the yellow dwarf virus in selected barley varieties. Plant Disease Reporter 50:380–381.

Brandt, P. W. 1958. A study of the mechanism of pinocytosis. Exp. Cell Res. 15:300–313.

Bruehl, G. W. 1958. Comparison of eastern and western aphids in the transmission of barley yellow dwarf virus. Plant Disease Reporter 42:909–911.

——— 1961. Barley yellow dwarf, a virus disease of cereals and grasses. Monogr. No. 1, Amer. Phytopathol. Soc. 52 p.

Bruehl, G. W., and Damsteegt, V. D. 1962. Re-examination of vector-specificity in the barley yellow dwarf virus in Washington. Phytopathology 52:1056–1060.

Bruehl, G. W., Damsteegt, V. D., Austenson, H. M., and Crandall, P. C. 1962. Resistance to yellow dwarf of oats in Washington. Plant Disease Reporter 46:579–582.

Butler, F. C., Grylls, N. E., and Goodchild, D. J. 1960. The occurrence of barley yellow dwarf virus in New South Wales. J. Australian Inst. Agr. Sci. 26:57–59.

Catherall, P. L. 1963. Transmission and effect of barley yellow-dwarf virus isolated from perennial ryegrass. Plant Pathol. 12:157–160.

Cords, C. E., and Holland, J. J. 1964. Alteration of the species and tissue specicity of poliovirus by enclosure of its RNA within the protein capsid of coxsackie B1 virus. Virology 24:492–495.

Endo, R. M., and Brown, C. M. 1963. Effects of barley yellow dwarf virus on yield of oats as influenced by variety, virus strain, and developmental stage of plants at inoculation. Phytopathology 53:965–968.

Forbes, A. R. 1962. Aphid populations and their damage to oats in British Columbia. Can. J. Plant Sci. 42:660–666.

——— 1964. The morphology, histology, and fine structure of the gut of the green peach aphid, *Myzus persicae* (Sulzer) (Homoptera: Aphididae). Mem. Entomol. Soc. Can. No. 36. 74 p.

Gill, C. C. 1967. Transmission of barley yellow dwarf virus isolates from Manitoba by five species of aphids. Phytopathology 57:713–718.

Gold, A. H., and Duffus, J. E. 1967. Infectivity neutralization—a serological method as applied to persistent viruses of beets. Virology 31:308–313.

Goodman, P. J., Watson, M. A., and Hill, A. R. C. 1965. Sugar and fructosan accumulation in virus-infected plants: rapid testing by circular-paper chromatography. Annu. Appl. Biol. 56:65–72.

Harpaz, I., and Klein, M. 1965. Occurrence of barley yellow dwarf virus (BYDV) in Israel. Plant Disease Reporter 49:34–35.

Hille Ris Lambers, D. 1960. The identity and name of a vector of barley yellow dwarf virus. Virology 12:487–488.

Holter, H. 1959. Pinocytosis. Int. Rev. Cytol. 8:481–504.

Ikäheimo, K. 1960. Two cereal virus diseases in Finland. Maataloustieteellinen Aikakauskirja 32:62–70.

Jedlinski, H., and Brown, C. M. 1965. Cross protection and mutual exclusion by three strains of barley yellow dwarf virus in *Avena sativa* L. Virology 26: 613–621.

Lindsten, K. 1964. Investigations on the ocurrence and heterogeneity of barley yellow dwarf virus in Sweden. Lantbrukshögskolans Annaler 30:581–600.

Muller, Irmgard. 1965. Aphid injection, an aid in the study of barley yellow dwarf virus. Cornell Plantations 20:68–71.

Nagaich, B. B., and Vashisth, K. S. 1963. Barley yellow dwarf: a new viral disease for India. Indian Phytopathol. 16:318–319.

Nault, L. R., Gyrisco, G. G., and Rochow, W. F. 1964. Biological relationship between pea enation mosaic virus and its vector, the pea aphid. Phytopathology 54:1269–1272.

Orlob, G. B. 1961. Aphids and the epidemiology of barley yellow dwarf virus in New Brunswick. Plant Disease Reporter 45:466–469.

——— 1962. Further studies on the transmission of plant viruses by different forms of aphids. Virology 16:301–304.

Orlob, G. B., and Arny, D. C. 1960. Transmission of barley yellow dwarf virus by different forms of the apple grain aphid, *Rhopalosiphum fitchii* (Sand.). Virology 10:273–274.

Oswald, J. W., and Houston, B. R. 1953. The yellow-dwarf virus disease of cereal crops. Phytopathology 43:128–136.

Oswald, J. W., and Thung, T. H. 1955. The barley yellow dwarf virus disease on cereal crops in The Netherlands. Phytopathology 45:695. (Abstr.)

Rademacher, B., and Schwarz, R. 1958. Die Rotblättrigkeit oder Blattröte des Hafers eine Viruskrankheit (*Hordeumvirus nanescens*). Z. Pflanzenkrankheiten 65:641–650.

Rochow, W. F. 1958a. Barley yellow dwarf virus disease of oats in New York. Plant Disease Reporter 42:36–41.

——— 1958b. The role of aphids in vector specificity of barley yellow dwarf virus. Plant Disease Reporter, 42:905–908.

——— 1959a. Transmission of strains of barley yellow dwarf virus by two aphid species. Phytopathology 49:744–748.

——— 1959b. Differential transmission of barley yellow dwarf virus from field samples by four aphid species. Plant Disease Reporter Suppl. 262:356–359.

——— 1960. Specialization among greenbugs in the transmission of barley yellow dwarf virus. Phytopathology 50:881–884.

——— 1961a. The barley yellow dwarf virus disease of small grains. Advances Agron. 13:217–248.

——— 1961b. A strain of barley yellow dwarf virus transmitted specifically by the corn leaf aphid. Phytopathology 51:809–810.

——— 1963a. Variation within and among aphid vectors of plant viruses. Annu. N.Y. Acad. Sci. 105:713–729.

——— 1963b. Recovery of barley yellow dwarf virus from field samples in 1961 and 1962. Plant Disease Reporter 47:139–143.

——— 1965a. Apparent loss of vector specificity following double infection by two strains of barley yellow dwarf virus. Phytopathology 55:62–68.

——— 1965b. Selective virus transmission by *Rhopalosiphum padi* exposed to two vector-specific strains of barley yellow dwarf virus. Phytopathology 55: 1284–1285. (Abstr.)

——— 1967a. Temperature alters vector specificity of one of four strains of barley yellow dwarf virus. Phytopathology 57:344–345. (Abstr.)

——— 1967b. Predominating strains of barley yellow dwarf virus in New York: changes during ten years. Plant Disease Reporter 51:195–199.

Rochow, W. F., and Brakke, M. K. 1964. Purification of barley yellow dwarf virus. Virology 24:310–322.

Rochow, W. F., and Eastop, V. F. 1966. Variation within *Rhopalosiphum padi* and transmission of barley yellow dwarf virus by clones of four aphid species. Virology 30:286–296.

Rochow, W. F., Jedlinski, H., Coon, B. F., and Murphy, H. C. 1965. Variation in barley yellow dwarf of oats in nature. Plant Disease Reporter 49:692–695.

Rochow, W. F., and Pang, E-Wa. 1961. Aphids can acquire strains of barley yellow dwarf virus they do not transmit. Virology 15:382–384.

Roland, G. 1962. Recherches effectuées sur le virus de la jaunisse de l'orge (barley yellow dwarf virus). Mededel. Landbouwhogeschool Opzoekingsstations Staat Gent 27:992–1009.

Russell, L. M. 1963. Changes in the scientific names of some common aphids. Coop. Econ. Insect Rept. 13:84.

Saksena, K. N., Dody, D. G., and Sill, W. H., Jr. 1964a. Importance of the greenbug, *Toxoptera graminum,* in field transmission of barley yellow dwarf virus. Plant Disease Reporter 48:127–130.

Saksena, K. N., Sill, W. H., Jr., and Kainski, J. M. 1964b. Relative efficiency of four aphid species in the transmission of Kansas isolates of barley yellow dwarf virus. Plant Disease Reporter 48:756–760.

Saksena, K. N., Singh, S. R., and Sill, W. H., Jr. 1964c. Transmission of barley yellow-dwarf virus by four biotypes of the corn leaf aphid, *Rhopalosiphum maidis*. J. Econ. Entomol. 57:569–571.

Shikata, E., Maramorosch, K., and Granados, R. R. 1966. Electron microscopy of pea enation mosaic virus in plants and aphid vectors. Virology 29:426–436.

Slykhuis, J. T., Zillinsky, F. J., Hannah, A. E., and Richards, W. R. 1959. Barley yellow dwarf virus on cereals in Ontario. Plant Disease Reporter 43:849–854.

Smith, H. C. 1963a. Interaction between isolates of barley yellow dwarf virus. New Zealand J. Agr. Res. 6:343–353.

——— 1963b. Aphid species in relation to the transmission of barley yellow dwarf virus in Canada. New Zealand J. Agr. Res. 6:1–12.

Smith, H. C., and Richards, W. R. 1963. A comparison of *Rhopalosiphum padi* (L) and *R. fitchii* (Sand.) as vectors of barley yellow dwarf virus. Can. Entomol. 95:537–547.

Smith, K. M. 1965. Plant virus–vector relationships. Advances Virus Res. 11: 61–96.

Stubbs, L. L. 1955. Strains of *Myzus persicae* (Sulz.) active and inactive with respect to virus transmission. Australian J. Biol. Sci. 8:68–74.

Sylvester, E. S., and Richardson, Jean. 1966. "Recharging" pea aphids with pea enation mosaic virus. Virology 30:592–597.

Tetrault, R. C., Schulz, J. T., and Timian, R. G. 1963. Effects of population levels of three aphid species on barley yellow dwarf transmission. Plant Disease Reporter 47:906–908.

Timian, R. G. 1964. Dodder transmission of barley yellow dwarf virus. Phytopathology 54:910. (Abstr.)

Timian, R. G., and Jensen, G. L. 1964. Absence of aphid species specificity for acquisition and transmission of a strain of barley yellow dwarf virus. Plant Disease Reporter 48:216–217.

Toko, H. V., and Bruehl, G. W. 1959. Some host and vector relationships of strains of the barley yellow-dwarf virus. Phytopathology 49:343–347.

Vacke, J. 1964. Žlutá zakrslost ječmene v ČSSR, barley yellow-dwarf virus disease in Czechoslovakia. Rostlinná Výroba 10:859–868.

Watson, M. A., and Mulligan, T. 1960. The manner of transmission of some barley yellow-dwarf viruses by different aphid species. Annu. Appl. Biol. 48: 711–720.

# Mechanism of Transmission of Stylet-Borne Viruses *

THOMAS P. PIRONE

*Department of Plant Pathology*
*Louisiana State University*
*Baton Rouge, Louisiana*

Stylet-borne virusese are, by definition, those carried at the tips of aphids' stylets (Kennedy, Day and Eastop, 1962). That so many enigmatic findings could result from such a seemingly simple virus–vector relationship is ample testimony to our lack of knowledge about other factors involved in the transmission of these viruses. Stylet-borne virus transmission has been the subject of several recent reviews (Sylvester, 1962; Maramorosch, 1963; Bradley, 1964; Smith, 1965), which cover the development of concepts regarding the mechanism of transmission of stylet-borne viruses from several different viewpoints. In this chapter an attempt will be made to discuss present concepts of stylet-borne virus transmission.

## I. EVIDENCE THAT THE VIRUSES ARE STYLET BORNE

The rapidity with which these viruses can be acquired and transmitted, and the fact that most of them can also be transmitted by mechanical inoculation, suggest that these viruses are carried by the sylets. Direct evidence for this was lacking however, until Bradley and Ganong (1955) showed, by exposing stylets of *Myzus persicae* to ultraviolet radiation, that potato virus Y was carried at the tips of the stylets. This finding was later extended to other viruses and other aphids (Bradley and Ganong, 1957; Orlob and Bradley, 1960; 1961).

Although most workers agree that the transmissible virus is located near the stylet tips, the exact site of such virus is uncertain. Van der Want (1954) suggested that virus could not enter the salivary duct, and that which entered the food duct could not be reinjected into the plant. Consequently, he assumed that virus must be carried on the outer

* Unpublished results cited herein were supported in part by grant GM-12389 from the Institute of General Medical Sciences, U.S. Public Health Service.

surface of the stylets. Van Hoof (1958) showed that there are ridges at the apices of the stylets of *Myzus persicae,* and suggested that transmissible virus might be carried in these ridges. Sylvester (1962) suggested that the transmissible virus is carried in saliva which adheres to the stylets. Sylvester and Richardson (1964) interpreted their findings as support for this hypothesis. Bradley (1966) has presented evidence that potato virus Y is carried on the maxillae, but the possibility that transmissible virus is also carried by the mandibles cannot be ruled out on the basis of his evidence.

## II. SITE OF VIRUS ACQUISITION

Several lines of evidence indicate that virus is acquired primarily from the epidermis. Bradley (1964) found that the stylets of an aphid do not extend even 5 $\mu$ beyond the end of the labium during the first 15 sec of probing. Nault and Gyrisco (1966) reported that the stylets of *Acyrthosiphon pisum* did not extend over 5 $\mu$ after a 10-sec probe, that the average extension after a 30-sec probe was 8 $\mu$, and that the depth of epidermal cells of herbaceous plants was about 10 $\mu$. Since the acquisition of stylet-borne viruses usually occurs within the first 15 sec of a probe (Bradley, 1964), in many cases the epidermis is the only tissue from which virus can be acquired. Bradley (1952) found that *Myzus persicae* required over 1 min to penetrate stripped tulip epidermis and van Hoof (1958) found that *M. persicae* and *Aphis fabae* required over 2 min to penetrate stripped hyacinth epidermis. These studies correlate well with those made on intact tissue and indicate that most virus uptake is from the epidermis.

Several workers (Bradley, 1956; van Hoof, 1958; Namba, 1962) have shown, by allowing aphids to probe leaves from which the epidermis had been removed, that aphids can also acquire virus from subepidermal tissues. Whether virus from this source plays a significant role in the transmission process if often difficult to assess, particularly for viruses with which the probability of uptake during the first 15 sec of probing is so high that further uptake might not be detected if it did occur (Bradley, 1964).

Bradley (1952) reported that aphids' stylets followed either an intercellular or intracellular path through stripped epidermis. However, van Hoof (1958) presented evidence that stylets usually followed an intercellular path through such tissue and suggested that virus was acquired intercellularly. His technique involved the observation of areas in which a salivary sheath was laid down, and the type of probes required to produce such a sheath might not be characteristic of those in which uptake usually occurs (Bradley, 1964). Swenson (1962) suggested that aphids locate the position of the cell walls by tapping the epidermis. Since most nontapping aphids (which would be more likely to probe into cells than between them, due to the relative areas involved) made probes of less than 45 sec, and probes of more than 45 sec resulted in reduced transmission, Swenson reasoned that most of the transmissible virus is acquired intracellulary. Others (Henning, 1963; Bradley, 1964; Nault and Gyrisco, 1966) have reported that aphids slide the rostrum over the leaf surface before probing, and suggest that this enables the aphid to locate intercellular areas.

Further evidence that aphids' stylets follow an intercellular path even in short probes has been presented by Nault and Gyrisco (1966), who found that the salivary flange produced by aphids (*A. pisum*) making short (7–26 sec) probes was located above an intercellular groove in 58 of 62 cases, and that in most cases sheath saliva was found between epidermal cells. Evidence of intracellular probing was found in 4 cases. Probes of longer duration were also nearly always intercellular.

On the basis of the probing behavior of certain aphids on certain plants, therefore, it would seem that if virus is acquired by these aphids from these hosts, it must be acquired intercellularly. The fact that Roberts (1940) showed that different species of aphids differed in the relationship of their stylets to tobacco cells may indicate the danger of generalizations based on observations of a few aphid–plant combinations, however. Whether virus can also be acquired intracellularly has not been determined.

Whether epidermal uptake occurs as the result of saliva-free probes has been the subject of some speculation, as it might explain the fact that these viruses are more readily acquired during probes of short duration. Bradley (1952) and van der Want (1954) suggested that saliva did not surround the stylets in the first brief probes. Hennig (1963) found that short probes by *Aphis fabae* may result in the production of salivary plugs and/or salivary sheaths; but in some cases no traces of saliva may be found. Nault and Gyrisco (1966) reported that sheath saliva was produced by most aphids (*A. pisum*) which made short (7–26 sec) probes into pea leaves. In several of the shorter probes, however, only flange saliva was produced. In some cases, therefore, salivary sheath production may occur within the short time in which virus acquisition can occur.

## III. METHOD OF VIRUS ACQUISITION

The exact way in which aphids acquire virus is not known, and it doubtless will require a better understanding of the nature of the events occurring at the stylet tips during probing. McLean and Kinsey (1964, 1965) have described a technique for recording and interpreting these events, and correlations between the behavior of various aphids on different hosts and their ability to acquire certain viruses may ultimately lead to a fuller understanding of how viruses are acquired. Hennig (1963), using plants labeled with $P^{32}$, found no evidence that *Aphis fabae* took up plant sap during short probes. Gamez and Watson (1964) did not obtain any transmission of henbane mosaic virus when the stylets of anesthetized aphids were inserted into cells of infected plants. They regarded this as evidence that active probing is required for virus acquisition. However, Barnett and Pirone (1966) found that anesthetized aphids, whose stylets were dipped into capillaries containing purified cucumber mosaic virus, acquired the virus and were able to transmit it after recovery from the anesthetic. Although this does not eliminate active probing as a means of virus acquisition, it establishes that virus can be acquired by passive contamination of the stylets.

## IV. FORM AND AMOUNT OF VIRUS ACQUIRED

There is no direct evidence of whether whole virus or viral nucleic acid is the form acquired by aphids from plants. Van Hoof (1958) suggested, on the basis of the inability of aphids to acquire virus from leaves killed by heat or freezing (from which virus could still be transmitted mechanically), that viral nucleic acid was the aphid transmissible form. This seems plausible in light of the evidence that some aphids probe intercellularly, and the fact that there is evidence that viruses can move from cell to cell in the nucleic acid form, via the plasmodesmata (see Schneider, 1965). However, Pirone and Megahed (1966) found that aphids could transmit virus acquired by making probes into purified preparations of intact cucumber mosaic virus but could not transmit viral RNA preparations of equal infectivity. Although this indicates that whole virus and not RNA is the transmissible form in this system, it does not eliminate viral RNA as the form transmitted from plants. Similarities between the transmission of purified virus and that acquired from plants led these authors to conclude that this experiment provided circumstantial evidence that whole virus is the transmissible form, however. Simons and Moss (1963) found that aphids were not made nonviruliferous by treatment of their stylets with ribonuclease. Although Bradley (1964) inferred that this was evidence against RNA being the transmissible form, the experiments were not done in such a way as to justify this conclusion.

There are as yet no quantitative data on the amount of virus acquired by probing aphids. Pirone and Megahed (1966) found that aphid transmission of purified alfalfa and cucumber mosaic viruses decreased with decreasing virus concentration, and that dilution of preparations containing these viruses resulted in the loss of aphid transmissibility before mechanical transmissibility was lost. No conclusions could be drawn as to the relative amounts of virus involved, however, since there were too many disparities between the systems. An approach toward determining the amount of virus carried by the stylets has been made by Pirone (1967), who allowed individual aphids to probe suspensions of purified tobacco mosaic virus (TMV) and to then probe buffer suspensions, which were assayed for the presence of virus. From 1 to 9 lesions were produced by 0.2 ml of the buffer, in the cases in which virus was recovered. Since a minimum of 50,000 virus particles are required to produce a single lesion (Steere, 1955), at least 450,000 particles must have been acquired by an aphid in some cases. Although this work is not directly applicable to the problem since TMV is not a stylet-borne virus, this approach should make it possible to do quantitative studies with these viruses.

## V. SITE OF VIRUS INOCULATION

The lines of evidence which indicate that virus is acquired intercellularly from the epidermis point to the same areas as the sites of inoculation. There is no evidence that stylet-borne viruses cannot also be inoculated into epidermal cells or subepidermal tissues, however.

## VI. SPECIFICITY OF TRANSMISSION

While the mechanics of the transmission process, as outlined above, would seem to indicate that transmission of stylet-borne viruses should be a fairly straightforward process, the fact that there are differences in the efficiency and specificity with which these viruses are transmitted indicate that there are numerous complexities involved.

Sylvester (1954) made a comprehensive review of the factors which may affect transmission of a particular virus. These include the effects due to differences in handling and behavior of the aphids; the species of aphid used; the species used as the virus source or test plant; the strain of virus used; and various environmental factors. The effects of these, as well as other factors, such as the clone of aphids used, the condition of the colony from which the aphids used in transmission experiments are taken, and the species of plant on which the aphids are reared, have also been reviewed by Swenson (1963). Even when other factors are kept constant, different viruses or strains of a virus may differ in the efficiency with which they are transmitted by a particular clone of aphids.

Numerous hypotheses have been proposed to explain these differences. All of the more recent ones suggest that the transmission process is essentially mechanical, with vector behavior and/or virus–vector–host interactions accounting for specificity. Day and Irzykiewicz (1954) proposed a *mechanical inactivator behavior* hypothesis, in which differential susceptibility of viruses to salivary components, as well as differences in the behavior of different aphid species, could account for specificity. Van der Want (1954) suggested a *mechanical surface adherence* hypothesis, in which differences in the surface structure of stylets account for differential adsorption to, and elution of virus from, the stylets. Sylvester (1954) proposed a *mechanical inactivator compatibility* hypothesis, in which specificity depended on compatibility of the combination of virus, saliva, and inoculated host cell. This was later modified to include aphid behavior (Sylvester, 1962).

The concept that salivary secretions of aphids may differentially inactivate certain viruses and thus play a role in vector specificity is an interesting and often proposed one, but there is at present no evidence to support it. Day and Irzykiewicz (1954) tested the effect of the saliva of a hemipterous insect (*Nezara viridula* L.) on the infectivity of potato virus Y and TMV in an attempt to show a differential effect of the saliva on the infectivity of these viruses. The saliva was less inhibitory to the TMV, however. They were also unable to show an inhibitory effect of saliva on TMV suspensions which had been fed on by *Myzus persicae*. We have had similar results when *M. persicae* were fed on purified TMV and cucumber mosaic virus suspensions (T. P. Pirone, unpublished). Nor is there any evidence as yet to support Sylvester's (1954) proposal that substances in aphid saliva may affect the susceptibility of the plant cell in a manner similar to that proposed for inhibitors of mechanical transmission.

The concept that the structure of the stylets may be a major basis for differential transmission is also difficult to reconcile with differences in the transmission of virus strains or of viruses with identical particle morphology. The possibility that transmissible viruses possess a special chemical group capable of specific combination with some component of an aphid's mouthparts, as proposed by Watson (1958), cannot be ruled out, however.

It would seem that any attempt to explain differences in specificity or efficiency of transmission must include the effects of the differences in probing behavior which may occur with particular aphid–plant combinations. In addition to readily observable differences, what may be needed is a fuller understanding of the subtle differences which may occur at the stylet tips during the probing process. The electronic recording technique described by McLean and Kinsey (1964, 1965) may be useful in obtaining this type of information. This might make it possible to determine differences between probes into preferred hosts and those made into less desirable species; or to determine the result of various types of pretreatments of aphids on the manner in which they probe.

Differences in transmissibility may also depend on differences in the chemical and physical properties of the viruses involved. Most research on stylet-borne virus transmission has been done with viruses acquired from plants, in which their form, condition, and location are not known with certainty. Recent work by Pirone and Megahed (1966) with purified cucumber and alfalfa mosaic viruses has shown the effects of pH, buffers, and virus concentration on the transmission of these viruses by *Myzus persicae*. The fact that viruses which cannot be transmitted from plant by certain aphids may be transmissible if acquired from purified preparations was shown for cucumber mosaic virus and *Toxoptera graminum* by Megahed and Pirone (1966). However, these workers also found that factors other than the virus–stylet relationship may play a role in acquisition, since *Myzus persicae* could not transmit purified cauliflower or turnip mosaic viruses, both of which are readily transmitted from plants by this aphid (Pirone and Megahed, 1966). A similar situation was found to occur with cucumber mosaic virus and *Aphis rumicis* (Megahed and Pirone, 1966), although the inability of the aphid to acquire this virus through a membrane was not ruled out in this case. This type of experiment may help to elucidate some of the mechanisms of vector specificity, although the obviously artificial method of virus acquisition raises other problems unless the experiments are carefully controlled.

Differences in the host–virus relationship may affect the ability of an aphid to acquire a particular virus, as well as the ability of a virus to infect a cell when placed at a given dosage at a particular location by the aphid, and would thus play an important role in determining transmissibility. Differences in transmission based on differences in the probing behavior of an aphid may, in part, be based on these factors.

Swenson (1963) has reviewed the subjects of virus availability in the source plant and the susceptibility of test plants. Most attempts to correlate such things as virus concentration and location with transmissibility have been made with whole leaves, or by comparing epidermal with subepidermal tissues. The applicability of studies made in such a manner to, for example, the concentration of virus available at the very limited and possibly very specific sites at which the aphid acquires the virus seems questionable.

The fact that the virus–host relationship may play an important role at the cellular, as well as the tissue, level in affecting virus transmissibility is indicated by the fact that the MCJ strain of cucumber mosaic virus cannot be transmitted from tobacco plants by *M. persicae*, even though local lesion assays indicate that it is present in epidermal tissues at the same concentration as transmissible strains of the virus. The fact that the virus itself is transmissible by *M. persicae* was established by tests with purified virus (R. A. Normand and T. P. Pirone, unpublished). Different viruses, as well as different strains, could differ in their location or condition in the cell, and thus contribute to differential transmissibility.

Another factor which may be of importance is that the conditions necessary for the initiation of infection after virus introduction into or around cells may differ with the virus and the intrinsic reaction of the host. It has been shown by Jedlinski (1964) that the infectible sites for mechanically transmitted TMV and tobacco necrosis virus are different on each of two hosts. Thus there may be specific receptor sites for the different viruses to which a plant is susceptible. If this were also true for stylet-borne viruses, it could obviously provide an explanation for some aspects of differential transmissibility.

## VII. FAILURE OF APHIDS TO TRANSMIT TMV AND OTHER MECHANICALLY TRANSMISSIBLE VIRUSES

The fact that TMV, potato virus X, southern bean mosaic, tobacco ringspot, turnip yellow mosaic, and a number of other viruses which are stable occur in high concentrations and are easily transmitted mechanically cannot be transmitted as stylet-borne viruses has long puzzled plant virologists. Of these, TMV has received by far the most attention in this respect. Although there have been reports of aphid transmission of TMV, it is now generally believed that this virus is not transmitted by aphids (Kennedy *et al.*, 1962; Orlob, 1963; Smith, 1965).

Several workers have attempted to determine whether TMV is acquired by aphids from infected plants, or from suspensions of purified virus. Matsui, Sasaki, and Kikumoto (1963) showed by electron microscopy that TMV particles can be acquired from infected plants by aphids and that they are released by these aphids into distilled water. The particles were not infectious, however. Bradley (1952) and Orlob (1963) dipped exposed stylets into purified TMV. In Bradley's experiments, the aphids were unable to transmit the virus. Orlob found that infectious virus could be recovered from the stylets before, but not after, ensheathment. As ensheathment must precede probing, presumably the aphids would not have been able to transmit the virus, although this was not tested. Others (Watson and Roberts, 1939; Sylvester, 1954; Day and Irzykiewicz, 1954; Schmidt, 1959) have suggested that substances present in insect saliva, or which are formed as a result of interaction between saliva and plant materials, might inhibit or inactivate TMV.

Recently, Pirone (1967) has shown that aphids can acquire TMV in short probes into purified virus suspensions, and that they release infectious virus when allowed to probe buffer solutions. Plants probed or fed on by such viruliferous aphids do not become infected, however. It would seem, therefore, that the reason for lack of TMV transmission by aphids must lie in the inoculation process. Possible reasons for lack of transmission would include: (*1*) the areas probed by the aphids are not susceptible to TMV, due either to intrinsic reasons or to the inhibitory activity of salivary components on plant susceptibility, as suggested by Sylvester (1954); (*2*) the TMV carried by the aphids is not released at susceptible sites; (*3*) the amount of TMV deposited at a given site may not be sufficient to cause infection, even though the amount of virus eluted by the buffer can cause infection if inoculated mechanically.

The work of Teakle and Sylvester (1962), showed that local lesions were produced when *Myzus persicae* probed and fed on *Nicotiana glutinosa* leaves which had been sprayed with TMV. This indicates that TMV can infect through wounds produced by probing or feeding aphids. Thus, the second or third explanation offered above might be more probable. Further investigation of this problem may increase our understanding of vector specificity.

## VIII. SUMMARY AND CONCLUSIONS

Although present evidence indicates that stylet-borne viruses are carried at the distal end of the stylets and that they are acquired from and transmitted to the epidermal tissues in brief probes by the aphid.

there are no satisfactory explanations for the observed differences in the efficiency or specificity of their transmission. To attribute these to differences in vector efficiency or specificity is an oversimplification, as the plant and virus involved must also play an important role in the process.

There is no reason to believe that the host–virus relationship should be the same for all viruses which happen to have a particular vector relationship. There is evidence (Jedlinski, 1964) that different plants have different infectible sites for different mechanically transmitted viruses. The fact that some circulative viruses may infect epidermal or subepidermal cells while others must be inoculated into the phloem (McEwen, Schroeder, and Davis, 1957; Nault, Gyrisco, and Rochow, 1964) shows that these viruses also differ in their host relationships. Subtle differences in virus form, concentration, or location in source plants and quantitative or qualitative differences in the susceptible sites in test plants could, when combined with differences in aphid behavior, provide an adequate explanation for differences in efficiency or specificity of transmission.

## BIBLIOGRAPHY

Barnett, C. B., Jr., and Pirone, T. P. 1966. Stylet-borne virus: active probing by aphids not required for acquisition. Science 154:291–292.

Bradley, R. H. E. 1952. Studies on the aphid transmission of a strain of henbane mosaic virus. Ann. Appl. Biol. 39:78–97.

———. 1956. Effects of depth of stylet penetration on aphid transmission of potato virus Y. Can. J. Microbiol. 2:539–547.

———. 1964. Stylet-borne viruses. p. 148–174. *In* M. K. Corbett and H. D. Sisler [ed.] Plant Virology. Univ. Florida Press, Gainesville.

———. 1966. Which of an aphid's stylets carry transmissible virus? Virology 29:396–401.

Bradley, R. H. E., and Ganong, R. Y. 1955. Evidence that potato virus Y is carried near the tip of the stylets of the aphid vector *Myzus persicae* (Sulz.). Can. J. Microbiol. 1:775–782.

———. 1957. Three more viruses borne at the stylet tips of the aphid *Myzus persicae* (Sulz.). Can. J. Microbiol. 3:669–670.

Day, M. F., and Irzykiewicz, H. 1954. On the mechanism of transmission of non-persistent phytopathogenic viruses by aphids. Australian J. Biol. Sci. 7:251–273.

Gamez, R., and Watson, M. 1964. Failure of anesthetized aphids to acquire or transmit henbane mosaic virus when their stylets were artifically inserted into leaves of infected or healthy tobacco plants. Virology 22:292–295.

Hennig, E. 1963. Zum Probieren oder sogenannten Probesaugen der schwarzen Bohenlaus (*Aphis fabae* Scop.). Entomol. Exp. Appl. 6:326–336.

Hoof, H. A., van. 1958. Onderzoekingen over de Biologische overdrach van een Non-persistent virus. Doctoral thesis, Wageningen Agr. Univ. Van Putten & Oortmeijer, Alkmaar, The Netherlands. 96 p.

Jedlinski, H. 1964. Initial infection processes by certain mechanically transmitted plant viruses. Virology 22:331–341.

Kennedy, J. S., Day, M. F., and Eastop, V. F. 1962. A conspectus of aphids as vectors of plant viruses. Commonwealth Inst. Entomol., London. 114 p.

McEwen, F. L., Schroeder, W. T., and Davis, A. C. 1957. Host range and transmission of the pea enation mosaic virus. J. Econ. Entomol. 50:770–775.

McLean, D. L., and Kinsey, M. G. 1964. A technique for electronically recording aphid feeding and salivation. Nature 202:1358–1359.

———. 1965. Identification of electrically recorded curve patterns associated with aphid salivation and ingestion. Nature 205:1130–1131.

Maramorosch, K. 1963. Arthropod transmission of plant viruses. Ann. Rev. Entomol. 8:369–414.

Matsui, C., Sasaki, T., and Kikumoto, T. 1963. Electron microscopy of tobacco mosaic virus particles from aphid stylets. Virology 19:411–412.

Megahed, E., and Pirone, T. P. 1966. Comparative transmission of cucumber mosaic virus acquired by aphids from plants or through a membrane. Phytopathology 56:1420–1421.

Namba, R. 1962. Aphid transmission of plant viruses from the epidermis and subepidermal tissue: *Myzus persicae* (Sulzer)–cucumber mosaic virus. Virology 16:267–271.

Nault, L. R., and Gyrisco, G. G. 1966. Relation of the feeding process of the pea aphid to the inoculation of pea enation mosaic virus. Ann. Entomol. Soc. Amer. 59:1185–1197.

Nault, L. R., Gyrisco, G. G., and Rochow, W. F. 1964. Biological relationship between pea enation mosaic virus and its vector, the pea aphid. Phytopathology 54:1269–1272.

Orlob, G. B. 1963. Reappraisal of transmission of tobacco mosaic virus by insects. Phytopathology 53:822–830.

Orlob, G. B., and Bradley, R. H. E. 1960. Drei weitere Blattlausarten, die das Y-virus der Kartoffel mit den Stechborstenspitzen übertragen. Z. Pflanzenkrankh. Pflanzenschutz 67:407–409.

———. 1961. Where cabbage aphids carry transmissible virus B. Phytopathology 51:397–399.

Pirone, T. P. 1967. Acquisition and release of infectious tobacco mosaic virus by aphids. Virology 31:569–571.

Pirone, T. P., and Megahed, E. 1966. Aphid transmissibility of some purified viruses and viral RNAs. Virology 30:631–637.

Roberts, F. M. 1940. Studies on the feeding methods and penetration rates of *Myzus persicae* Sulz., *Myzus circumflexis* Buckt., and *Macrosiphum gei*, Koch. Ann. Appl. Biol. 27:348–358.

Schmidt, H. B. 1959. Beiträge zur Kenntnis der Übertragung pflanzlicher Viren durch aphiden. Biol. Zentralbl. 78:889–936.

Schneider, I. R. 1965. Introduction, translocation, and distribution of viruses in plants. Advances Virus Res. 11:163–221.

Simons, J. N., and Moss, L. M. 1963. The mechanism of resistance to potato virus Y infection in *Capsicum annum* var. Italian El. Phytopathology 53:684–691.

Smith, K. M. 1965. Plant virus–vector relationships. Advances Virus Res. 11: 61–96.

Steere, R. 1955. Concepts and problems concerning the assay of plant viruses. Phytopathology 45:196–208.

Swenson, K. G. 1962. Bean yellow mosaic virus transmission by *Myzus persicae*. Australian J. Biol. Sci. 15:468–482.

———. 1963. Effects of insect and virus host plants on transmission of viruses by insects. Ann. N.Y. Acad. Sci. 105:730–740.

Sylvester, E. S. 1954. Aphid transmission of non-persistent plant viruses with special reference to *Brassica nigra* virus. Hilgardia 23:53–98.

———. 1962. Mechanisms of plant virus transmission by aphids. p. 11–31. *In* Karl Maramorosch [ed.] Biological transmission of disease agents. Academic Press, New York.

Sylvester, E. S., and Richardson, J. 1964. Transmission of cabbage mosaic virus by green peach aphids—stylet transmission studies. Virology 22:520–538.

Teakle, D. S., and Sylvester, E. S. 1962. Infection of plants by the feeding of aphids on leaves sprayed with tobacco mosaic virus. Virology 16: 363–365.

Want, J. P. H., van der. 1954. Onderzoekingen over Virusziekten van de Boon (*Phaseolus vulgaris* L.). 84 p. Doctoral Thesis, Wageningen Agr. Univ. A. Veenman & Zonen, Wageningen, The Netherlands.

Watson, M. A. 1958. The specificity of transmission of some non-persistent viruses. Proc. 10th Int. Congr. Entomol. 3:215–219.

Watson, M. A., and Roberts, F. M. 1939. A comparative study of the transmission of Hyoscyamus virus 3, potato virus Y and cucumber virus 1 by the vectors *Myzus persicae* (Sulz.), *M. circumflexus* (Buckton), and *Macrosiphum gei* (Koch). Proc. Roy. Soc. (London) B127:543–576.

# Morphology of the Homoptera, with Emphasis on Virus Vectors*

A. R. Forbes and H. R. MacCarthy

*Research Station,*
*Canada Department of Agriculture,*
*Vancouver, British Columbia*

More than 80% of the species of insect vectors of plant virus diseases are in the suborder Homoptera (Heinze, 1959; Ossiannilsson, 1966). Of these, about 40% occur in the Auchenorrhyncha and 60% in the Sternorrhyncha. In the Auchenorrhyncha 109 vector species are leafhoppers (Jassidae=Cicadellidae) (Nielson, 1962), 10 species are in the Cercopidae, 8 in the Delphacidae, 3 in the Cixiidae, and 1 in the Flatidae (Carter, 1962). In the Sternorrhyncha, more than 180 species are aphids, 17 are in the Coccoidea, 9 in the Aleyrodoidea, and 1 in the Psylloidea (Carter, 1962; Kennedy, Day, and Eastop, 1962; Ossiannilsson, 1966).

The plant viruses transmitted by auchenorrhynchous Homoptera are all circulative and several are propagative (Ossiannilsson, 1966). Transovarial transmission of virus has been demonstrated in some leafhoppers. Of the viruses transmitted by aphids 84 are stylet borne, 30 are circulative, and the status of about 45 is uncertain (Kennedy *et al.*, 1962). At least two viruses transmitted by aphids are probably propagative (Smith, 1965). Several viruses transmitted by whiteflies are circulative. The two best known viruses transmitted by coccids are stylet borne (Ossiannilsson, 1966). The virus transmitted by the psyllid is apparently circulative (Jensen, 1966).

Detailed information on the morphology, histology, and fine structure of insect vectors is prerequisite to understanding the mechanics and mechanisms of transmission. Data on the morphology of Homoptera are scattered and, with notable exceptions, few. Excellent general treatises by Weber (1930) and Pesson (1951) are standard references.

* Contribution No. 120, Research Station, Research Branch, Canada Department of Agriculture, 6660 N.W. Marine Drive, Vancouver 8, B.C.

A recent review by Auclair (1963, 1964), although not primarily concerned with virus transmission, has assembled useful morphological information on aphids.

Great gaps still exist in our knowledge of the morphology of vector insects. Until recently workers depended upon dissection, standard sectioning and staining, and light microscopy. Today new techniques are available, including electron, polarizing, phase, and interference microscopy; radioactive tracers; and microspectrophotometry. Histochemistry has been refined and makes fresh approaches possible. The resources of the new morphology are beginning to be applied to vector insects and important developments can be expected.

The emphasis in this review is perforce on aphids and leafhoppers. Special attention has been given to the cosmopolitan green peach aphid, *Myzus persicae* (Sulzer), the most important vector species and one of the best known. This aphid transmits 71 stylet-borne and 11 circulative viruses, one of which is propagative (Kennedy *et al.,* 1962).

This review is not exhaustive, but selected with the available space in mind, from published and other information about vectors and related plant sucking insects, and assembled to clarify problems associated with virus acquisition, transmission, and multiplication. In this large and complex group generalizations can be dangerous especially when the available information is scanty. However, in attempting to produce a unified picture of the structure of the vectors some generalizations were unavoidable and these should be taken as working concepts that may not hold in every instance.

## I. MOUTHPARTS

The mouthparts are the direct means of acquisition and transmission of both stylet-borne and circulative viruses. Bradley and Ganong (1955a, 1955b) showed that aphids carry stylet-borne viruses near the tip of the stylet bundle. Later Bradley (1966) was able to demonstrate the presence of transmissible virus on the maxillary stylets, but was unable to show whether or not the mandibular stylets also carry virus.

The mouthparts in Homoptera are remarkably alike, this fact being associated with their uniform piercing and sucking feeding habit and the fact that they are all phytophagous. The mouthparts consist of two pairs of needle-like stylets, with a labium and a labrum. Maxillary and labial palps are lacking, their sensory function probably taken over by tactile hairs and chemoreceptors (Pesson, 1951) at the tip of the labium. Associated with the mouthparts is a well-developed salivary syringe whose duct traverses the hypopharynx.

The stylets are admirably adapted for piercing plant tissue and extracting juices. They arise in the head in the bristle pouches, converge along the sides of the hypopharynx, and come tgether as they approach the labium (Fig. 1).* The outer pair of the stylet bundle or fascicle is the mandibular stylets; the inner is the maxillary stylets (Fig. 2). The maxillary stylets interlock by a series of ridges and grooves, the food and salivary canals lying between their apposed inner surfaces. The mandibles arise anteriorly in the head, the maxillae posteriorly. The relationships of the pouches and of the bases of the stylets to the sclerites of the head capsule are complex but have been excellently figured by Davidson (1914) for the woolly apple aphid, *Eriosoma lanigerum* (Hausmann), and by Weber (1928) for the black bean aphid, *Aphis fabae* Scop. The mandibular stylets are articulated with the head capsule by stout chitinous levers; the maxillary stylets are articulated by thinner levers. Each pair has protractor and retractor muscles.

In the aphid *E. lanigerum* (Davidson, 1914), the retractors of the mandibles are inserted on the enlarged base of the stylet and originate on the tentorium and the head capsule; the protractors are inserted on the mandibular lever and arise anteroventrally on the head capsule. These muscles would produce rapid oscillations in the plane of the longitudinal axis (Bradley, Sylvester, and Wade, 1962). The retractors of the maxillae are inserted on the base of the stylet and originate dorsally on the tentorium; the protractors also are inserted on the base of the stylet and originate ventrally on the walls of the maxillary plate. These muscles would produce an up and down movement of the maxillae together or sliding movements on each other. Another set of so-called retractor muscles originates laterally on the tentorium and is inserted medially on the maxillary lever. The origin and insertion of these muscles make it unlikely that they retract the maxillae; they almost certainly produce side-to-side movements, such as have been described in *M. persicae* (Bradley *et al.*, 1962).

The stylets of *M. persicae* have been studied in detail. The mean length of 75 single stylets dissected from adults was 496 $\mu$ with a stan-

* Abbreviations used in figures are as follows: BMdS, base of mandibular stylet (anterior); BMxS, base of maxillary stylet (posterior); C, cauda; CD, central duct; E, eye; Es, esophagus; FC, filter chamber; FdC, food canal; FM, functional mouth; Hg, hindgut; I, intestine; Lb, labium; LG, labial groove; MA, muscle attachment on midline of anterior wall of pump chamber; MdS, mandibular stylet; MT, malpighian tubule; MxS, maxillary stylet; R, rectum; SC, salivary canal; SP, sucking pump; SS, salivary syringe; St, stomach; TB, tentorial bar; THg, tubular hindgut; TS, tips of stylets; 1V, first ventriculus; 2V, second ventriculus; 3V, third ventriculus.

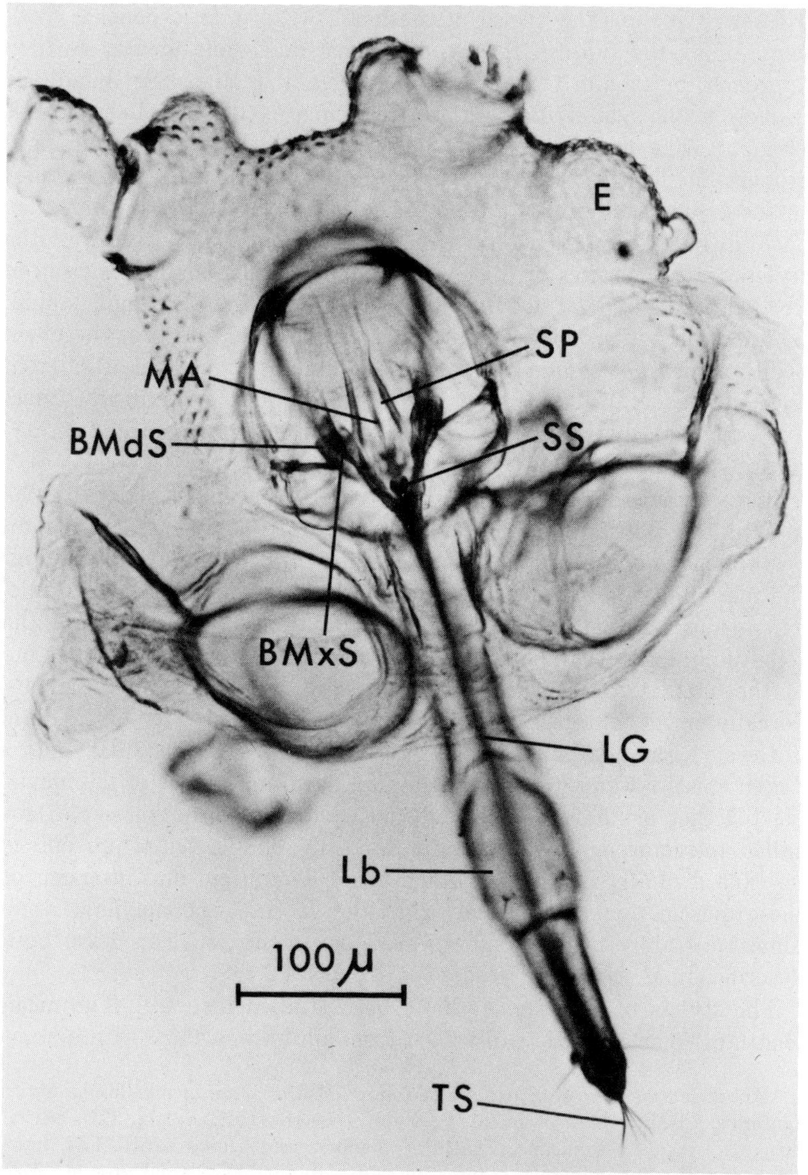


*Fig. 1.* Photomicrograph of an optical section from behind into the cleared head capsule of *Myzus persicae* (Sulzer) to show relative positions of stylets, labium, salivary syringe, and sucking pump. (Original.)

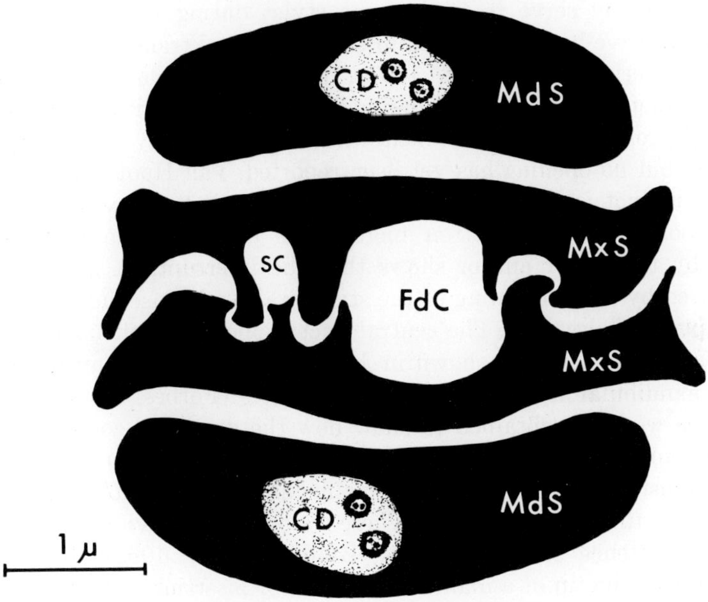


*Fig. 2.* Drawing from an electron micrograph of a cross section in the proximal part of the sylet bundle of *Myzus persicae* (Sulzer). (Original.)

dard deviation of 17 μ (A. R. Forbes, unpublished). Bradley (1962) reported the effective piercing organ of *M. persicae* to be roughly 300 μ long by 5 μ in diameter. The stylets are expanded within the head, especially so in the case of the mandibular pair (Fig. 1). The stylets taper toward the sharply pointed tips. Distally on the outer surface of the mandibular stylets is a series of more or less concentric, curved, and barb-like ridges (van Hoof, 1958) (Fig. 3). At a distance of 6 μ from the tip, the barbs project from the body of the stylet about 150 mμ, but the projection decreases towards the tip. These ridges extend over at least the distal 15 μ of the stylet but thereafter they become indistinct.

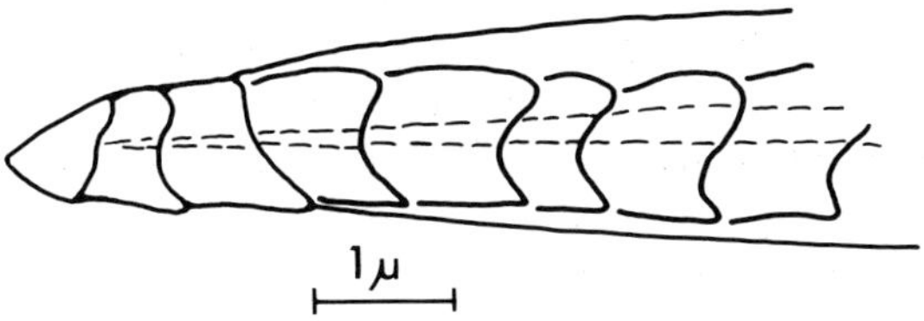


*Fig. 3.* Drawing from an electron micrograph of the outer surface of a mandibular stylet of *Myzus persicae* (Sulzer). The central duct is shown by dotted lines. (Original.)

They probably serve to anchor the stylet during insertion into the plant and have been implicated in the mechanical transfer of virus and plant material (van Hoof, 1957). Each mandibular stylet has a central duct (Figs. 2 and 3) running from its base to within 0.6 $\mu$ of the tip. Electron micrographs indicate that this canal may branch toward the tip, but no opening has yet been reported. Van Hoof (1958) states that this duct is interrupted along the length but an examination of van Hoof's published electron micrographs and many similar preparations by the senior author shows that the interruptions are artifacts produced by the thickening of the stylet wall and consequent complete absorption of electrons. The central duct is open throughout its length.

Recently, it has been shown in *M. persicae* that the central duct of each mandibular stylet contains two dendrites (Forbes, 1966) (Fig. 2). Sections were not obtained to show how the dendrites contact stimulating molecules, but it is suggested that mediation is by microvillus-like extensions of the dendrite plasma membrane at minute, thin-walled pores near the tip of the stylet. There were several previous indications of the existence of these nerves. Bradley (1960, 1962) found that amputating the tip of a mandibular stylet or inserting the intact stylet tip into various solutions prevented feeding but greatly increased larviposition. He suggested that this response demonstrated the presence of nerves in the stylets and observed that their central duct contained material that could be pulled as a thread from the cut end of the stylet. Wensler (1962) showed that the cabbage aphid, *Brevicoryne brassicae* (L.), perceives the specific feeding stimulus, sinigrin, with the stylets after they have penetrated the leaf surface rather than with sense organs on the antennae, labium, or tarsi. The existence of canals in the mandibular stylets of the Hemiptera–Homoptera has long been known, but nerves were not associated with them until Pinet (1963), using interference microscopy and whole mounts of stylets of *Rhodnius prolixus* Stål, traced nerves from bipolar neurons into the base of the stylets and via the central ducts to near the tip. The nerves arise from the anterior part of the subesophageal ganglion.

The inner, maxillary stylets of *M. persicae* (Fig. 2) are held closely together by interlocking ridges and grooves and between them form the food canal anteriorly and the smaller salivary canal posteriorly. The contours of their outer surface permit close apposition with the inner surface of the mandibular stylets, producing a compact stylet bundle. Three distinct ridges occur on the anterior surface close to the tips. The interlocking ridges and grooves permit movement of the maxillary stylets upon each other. An early investigator (Davidson, 1923) considered that the independent protraction and retraction of the maxillary stylets was probably slight, since such movements would dislocate the junctions of the salivary canal and duct, and of the food canal and sucking pump. Morphologically there is no reason to believe that the maxillary stylets cannot move independently and the frequency with which staggered tips have been observed (Bradley *et al.*, 1962; A. R. Forbes, unpublished; van Hoof, 1958) indicates that such movements do occur.

The food canal is considerably larger than the salivary canal (Fig. 2). In *M. persicae* the food canal tapers from about 1.5 $\mu$ in diameter near the head to 0.35 $\mu$ near the tip; the salivary canal tapers from 0.56 $\mu$ to 0.07 $\mu$ (A. R. Forbes, unpublished). The canals end short of the tips.

The tips of the maxillary stylets are sharply pointed and strengthened by the extension and expansion of the ridge that forms the posterior wall of the salivary canal. When the maxillary stylets are staggered, the openings of both the canals become open troughs, the length of which depends on the amount of displacement. No closed needle's eye such as that which has been described between the maxillary tips of *A. fabae* (Weber, 1928) has been seen in *M. persicae*.

The piercing mechanism of the Homoptera was beautifully described by Weber (1929) in a comprehensive study. On largely morphological evidence he deduced that penetration is by alternate protractions of the mandibular stylets, followed by protraction of the maxillary pair as a unit until their tips attain those of the mandibular stylets, the cycle being repeated several times to achieve penetration in depth. This mechanism has been confirmed for *M. persicae* (Bradley *et al.*, 1962). Barbs at the tips of the stylets and a clamping action by the end of the labium serve to anchor the stylets between thrusts.

Most Homoptera do not wholly depend on muscle action to reach their feeding site; salivary secretions injected during probing contain enzymes which facilitate entry and penetration. Pectinase (pectin polygalacturonase) was found in the saliva of *M. persicae* (Adams and McAllan, 1956) and 23 other species of aphids, in a vector leafhopper, *Dalbulus maidis* (DeLong and Wolcott), and in the balsam woolly aphid, *Adelges piceae* (Ratzelburg). In five species of aphids the enzyme was found in the apterae but not in the alatae (Adams and McAllan, 1958). Recently, a cellulose-hydrolyzing factor was detected in the saliva of 31 species of aphids and one species of psyllid (Adams and Drew, 1965).

A great mass of literature exists concerning the path of penetration of the stylets of aphids and the feeding sites reached. Smith (1926) found that the stylets of *M. persicae* usually entered potato leaves intercellularly, but sometimes through a stoma, to reach the phloem; Bradley (1952) observed intercellular penetration but Sukhov (1944) found mostly intracellular penetration of the stem of *Impatiens*. In general the sieve tubes of the phloem are the feeding goal of aphids and these are reached most commonly by intercellular penetration, less commonly by intracellular penetration, and sometimes by both (Mittler, 1957; Auclair, 1963; Sorin, 1966). Pectinase permits intercellular penetration by hydrolyzing the pectic middle lamella of plant tissues (McAllan and Adams, 1961).

The labrum is short and triangular, covering the base of the stylet bundle. Its inner surface has a longitudinal groove that positions and guides the stylet bundle, since the apposed labium is not usually grooved proximally.

The labium or rostrum in Homoptera, when well developed, is a segmented tube-like organ suspended from the head. A groove along its anterior surface contains the stylets. The complex musculature has been described and illustrated for *A. fabae* by Weber (1928). When not in the feeding position, the labium is flexed beneath the body, its tip directed posteriad. In aphids of the genus *Stomaphis* the labium exceeds the length of the body (Pesson, 1951); in *M. persicae* (Fig. 1) it is four-segmented, more than 300 $\mu$ long, and reaches the hind coxae when at rest. In aphids the labium telescopes to expose the stylets during feeding (Bradley, 1952; Sorin, 1966). Toward the tip of the labium of aphids, the edges of the labial groove are produced and apposed to form a more or less tight tube (Weber, 1928; Bradley and Ganong, 1955a). This feature presumably helps the stylets to penetrate into and withdraw from the plant.

The tip of the labium is usually equipped with sensory structures. In *M. persicae* the ultimate segment has several long setae and a row of much finer hairs around the opening of the labial groove. These are well innervated (A. R. Forbes, unpublished). Aphids are known to tap or otherwise explore leaf surfaces with the labium before probing, perhaps to locate the grooves between cells (Bradley, 1952; Nault and Gyrisco, 1966; Sukhov, 1944; Swenson, 1962).

Leafhoppers bring the labium and stylets into feeding position more rapidly than do aphids. The stylets usually enter the leaf at random and by pressure alone, piercing the cell walls with ease and following an intracellular track (Hagley and Blackman, 1966; Smith, 1926). The stylets differ from those of aphids only in degree. They are thicker, proportionately shorter, and stronger. The stylets are barbed, particularly in Cercopidae, *e.g.*, *Aphrophora* (Weber, 1930), but in spite of the barbs leafhoppers can pull out their stylets unharmed with great rapidity (Day and McKinnon, 1951).

Most leafhoppers feed mainly in phloem tissue, certain groups feed mainly in parenchyma and mesophyll, and most of the Tettigellinae or sharpshooters and some cercopids feed in the xylem. Certain groups have saliva toxic to the host plants. Toxic species are not known to transmit virus. After studying the efficient vector *Orosius argentatus* (Evans) and comparing it with other leafhoppers, Day, Irzykiewicz, and McKinnon (1952) concluded that jassids find the tissue they seek by random probing. In artificial media the four stylets extended about 70 $\mu$ into the fluid and made fairly rapid oscillating movements. If the insects settled to feed, normal homopteran salivary sheaths appeared. Leafhoppers fed as readily in the dark as in the light and appeared hardly to discriminate in the *p*H range between 4.2 and 9.0 (Day and McKinnon, 1951). The site of feeding was found to influence the acquisition of aster yellows virus by *Macrosteles fascifrons* (Stål) (Maramorosch, 1962).

## II. SALIVARY SYSTEM

The salivary system consists of a pair each of principal and accessory glands with associated ducts and a salivary syringe. In aphids the principal gland is large and bilobed but the accessory gland is smaller and single lobed (Fortin, 1958; Moericke and Wohlfarth-Bottermann, 1960; Sorin, 1961; Weber, 1928). The ducts from each gland fuse in the midline to form the common salivary duct, which leads to the salivary syringe. The histology of the salivary glands of *Macrosiphum euphorbiae* (Thomas) in the light microscope has been reported by Fortin (1958) and the detailed structure in *M. persicae* in the light and electron microscopes by Moericke and Wohlfarth-Bottermann (1960, 1963) and by Wohlfarth-Bottermann and Moericke (1960).

The salivary syringe in aphids (Davidson, 1914; Weber, 1928) is a cup-shaped, chitinous structure on the hypopharynx immediately posterior to the sucking pump (Fig. 1). Its distal end receives the common salivary duct; its proximal end is invaginated to form a piston. Inserted upon the piston are large muscles which retract it. Contraction of the pump muscles pulls the piston from the cylinder, drawing saliva into the chamber from the common salivary duct; when the muscles relax, the piston springs back, forcing the saliva through the outlet tube. Small muscles attached to the cylinder wall near the entry of the salivary duct apparently control the opening and provide valve action.

The syringe discharges into the salivary canal in the maxillary stylets by way of an outlet tube at the tip of the hypopharynx.

Most Homoptera, including aphids and leafhoppers, secrete a salivary sheath about their stylets when they penetrate plant tissue or an artificial substrate. Much has been written about its formation, supposed function, and possible effects on transmission of viruses. Recent contributions are by Miles (1959a, 1959b, 1965), Miles, McLean, and Kinsey (1964), Mittler and Dadd (1963), Naito (1965), Nault and Gyrisco (1966), and Sukhov (1944).

The salivary glands of leafhoppers have an underlying similarity. Their complexity and size proportional to the insect indicate their importance to the feeding process. In cercopids they are complex, but in jassids they are simpler, somewhat different among species but differing only in degree between nymphs and adults (Pesson, 1951), or even between individuals. Bilaterally symmetrical, the glands lie on each side of the esophagus in the head and thorax, close to the subesophageal ganglion. Dobroscky (1931a, 1931b) recorded their measurements in *M. fascifrons* and *Euscelis striatulus* Fallén as 360 $\mu$ long, 380 $\mu$ wide, and 270 $\mu$ deep. Each half consists of a principal gland, made up of large anterior and posterior lobes, with one or more small lobes beneath each of these, and a small and sometimes tubular accessory gland lying in front of the main mass (Sogawa, 1965). The principal gland delivers its secretion into a duct arising between the anterior and posterior lobes. The slender duct from the accessory gland joins it, and this duct runs down to join its opposite number, forming the common salivary duct, which passes to the salivary syringe and the mouthparts. The ducts are lined with spirally thickened intima (Dobroscky, 1931a).

According to species the lobes are made up of several follicles or acini, often arranged radially or in the upper part serially, on a single, small duct. The acini are large, binucleate, merocrine cells. There may be clusters of five, six, or more types of cells. Their canaliculi are intracellular so that the endoplasmic reticulum communicates with the extracellular ducteole (Herold and Munz, 1965; Sogawa, 1965). The serous and mucous secretions pass by the ducteoles to converge upon the main duct. Detailed descriptions of the types of acini have been given by Dobroscky (1931a) and Gil-Fernandez and Black (1965), with numerous illustrations and a histochemical study by Sogawa (1965).

Salivary glands are of prime interest in virology because circulative viruses are bound to pass through them in order to invade a fresh plant. Littau and Maramorosch (1960) showed that aster yellows virus can occur in the hemolymph of two nontransmitting vector species by injecting their hemolymph into aster leafhoppers, which subsequently transmitted the virus, proving that there was a barrier in the salivary glands of the nontransmitters. Aster yellows virus has been shown to occur in a higher concentration in the salivary glands than in other tissues of the aster leafhopper (Hirumi and Maramorosch, 1963), and the efficient vector of rice dwarf virus has large masses of particles in the salivary glands (Nasu, 1965; Shikata and Maramorosch, 1965). In the clover leafhopper, wound tumor-fluorescent antigens showed that virus was present in the salivary glands (Sinha, 1965a) but there was no accumulation of virus particles, which may explain the erratic and inefficient transmission record of this vector (Shikata and Maramorosch, 1965). The same situation may explain why *Peregrinus maidis* (Ashmead) was unable to transmit maize mosaic every day. The thick, rod-like virus particles were seen in midgut cells and in the ducteoles of salivary glands, but always in a low concentration (Herold and Munz, 1965).

Embryologically, the salivary syringe of leafhoppers is an enlarged and modified portion of the salivary duct (Muir and Kershaw, 1912). In the aster leafhopper it is barrel shaped, about 27 $\mu$ at the widest point by 39 $\mu$ at the longest. The piston, which fills the cylinder, is pyriform. The syringe is heavily chitinized, and the piston has considerable resilience to judge by the two strong, flat retractor muscles (Dobroscky, 1931a). The small efferent duct down which the secretion is forced, is an extension of the median duct, and it runs along the ventral surface of the hypopharynx, to open through a pointed process at the junction of the maxillary stylets, precluding the mixing of saliva and sap at this level (Gil-Fernandez and Black, 1965; Qadri, 1959).

## III. ALIMENTARY CANAL

The alimentary canal of Homoptera consists of the food canal in the maxillary stylets, the sucking pump, and the gut. On embryological and morphological grounds the gut is divided into: foregut, midgut, and hindgut (Snodgrass, 1935). Circulative viruses are ingested by way of the alimentary canal and presumably have to pass through the gut wall into the hemocoel.

The following discussion of the alimentary canal of *M. persicae* (Fig. 4) is based on Forbes (1964). The food canal within the maxillary stylets opens into a tubular functional mouth, which leads to the sucking or cibarial pump. The pump arcs dorsad to join the foregut at the transverse tentorial bar. The pump chamber is crescentic in cross section, its posterior wall is thick, and its anterior wall is thinner, flexi-

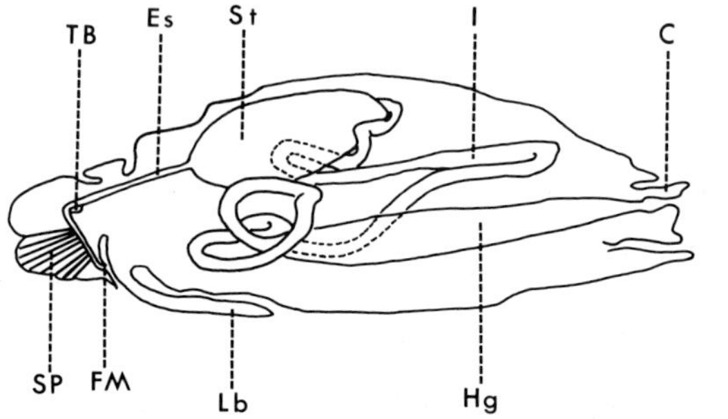


*Fig. 4.* Schematic diagram of a longitudinal section of an adult apterous *Myzus persicae* (Sulzer) to show the sucking pump and gut. (Redrawn from Forbes, 1964).

ble, and invaginated into the lumen. The dilator muscles originate on the clypeus and are inserted along the midline of the anterior wall of the chamber. The structure of the sucking pump is essentially the same in other aphids studied (Davidson, 1914; Sorin, 1961; Weber, 1928). When the dilators contract, the invaginated anterior wall is pulled forward, drawing plant sap into the chamber. When the dilators relax, the wall springs back, expelling the sap into the foregut. The role of suction in ingestion of sap has been much debated (Banks, 1964; Banks and Nixon, 1958; Davidson, 1914; Mittler, 1957; Zweigelt, 1914). Banks and Nixon (1958) suggested that the sucking pump probably acts as a valve which opens to allow sap under pressure to enter the gut. Morphology does not support this view and rapid pumping action has been observed in feeding aphids (Bradley, 1952; A. R. Forbes, unpublished).

Most Homoptera, including aphids, have on the epipharynx near the beginning of the sucking pump a specialized structure believed to be a gustatory organ (Pesson, 1951). In *E. lanigerum* (Davidson, 1914) and *A. fabae* (Weber, 1928), it consists of a group of large hypodermal cells, with darkly staining nuclei situated over the perforated epipharynx. These cells are supplied with nerves.

The gut of *M. persicae* (Forbes, 1964) appears to be typical. The unspecialized foregut consists of a very short pharynx and a longer esophagus. The esophagus is invaginated into the stomach, forming the esophageal or cardiac valve. The foregut intima extends into the stomach lumen and is reflexed to surround the valve. The stomach, or first section of the midgut, is large and sac-like and leads to the long, tubular intestine, or second section of the midgut. The intestine folds upon itself several times and expands to join the straight, thin-walled hindgut, which terminates at the anus at the base of the cauda. There is no filter chamber. A filter chamber has been described in the Lachnidae (Auclair, 1963; Börner, 1938), but there are no vectors in this group (Kennedy *et al.*, 1962). The absence of malpighian tubules is characteristic of aphids. Similar alimentary canals have been described for several other aphids including *A. fabae* (Weber, 1928), *E. lanigerum* (Davidson, 1913), and *M. euphorbiae* (Smith, 1939).

The gut wall consists of a layer of epithelial cells on a tunica propria, the latter being made up of the basement membranes of the epithelial cells, the muscle fibers, and the tracheoblasts, with connective tissue. Muscle fibers are applied to the tunica propria. The epithelial cells of the esophagus are thin and flattened, their free surfaces bearing sparse microvilli. The mitochondria and membranes of the endoplasmic reticulum are evenly but sparsely distributed in the cytoplasm. The epithelial cells of the stomach are mostly large and lobate. They have striated free borders, consisting of microvilli and microlabyrinths. The basal cell membranes are also elaborately infolded. Mitochondria are abundant at both the striated border and the basal cell border. The endoplasmic reticulum is well developed and in the light microscope this is associated with intense cytoplasmic basophilia attributable to the presence of RNA.

The epithelial cells of the intestine are also large. Their striated borders consist of microlabyrinths and microvilli, but here the border is sometimes almost entirely microvillar. The basal cell membranes are infolded. Mitochondria show the same concentration as in stomach cells, but the endoplasmic reticulum is less well developed. The epithelial cells of the hindgut are so thin that the nuclei characteristically project into the lumen. The free borders show a few cytoplasmic protrusions, and the basal cell membranes are mostly simple. The mitochondria and endoplasmic reticulum are relatively sparse and evenly distributed. The mitochondria, Golgi complex, endoplasmic reticulum, nuclei, intercellular membranes, and the basement membrane in the gut cells of *M. persicae* show no unusual fine-structural features. The complex elaboration of the membranes at the free and basal cell surfaces of the midgut cells and the presence there of endocytosis vesicles indicate active transport of material into and out of the cell, probably by membrane flow and vesiculation. Circulative plant viruses may be transported to the hemolymph by these routes.

The sucking pump in leafhoppers has strong musculature originating on the clypeus, with a long reach and evident power (Gil-Fernandez and Black, 1965) (Fig. 5). Leafhoppers are normally more precipitate in feeding than are aphids and their pump clearly works with more speed and force.

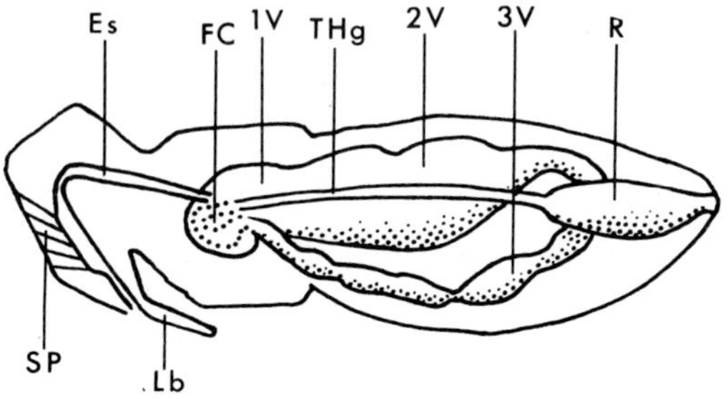


*Fig. 5.* Schematic diagram of a longitudinal section of a leafhopper to show the sucking pump, filter chamber, and gut. The malpighian tubules are omitted. (Original.)

In *Typhlocyba ulmi* (L.) the gustatory organ consists of two rows of five perforations in the midline of the epipharynx arranged alternately, then two more pores just before the perforated plate passes into the dorsal wall of the cibarial pump (Willis, 1949). Above the plate lies a mass of cells connected by nerve trunks with the frontal ganglion. The cells appear always to have intracytoplasmic ducts opening directly to the lumen of the epipharynx, and fibrillar terminals of the epipharyngeal nerve penetrate their cytoplasm (Pesson, 1951; Weber, 1930; Willis, 1949).

Past the sucking pump in leafhoppers the pharynx turns caudad over the tentorial bar. Beyond the bar and starting near the vertex of the head is the esophagus, an unspecialized tube that appears to have only a passive function in passing back ingested plant sap (Fig. 5). It has few muscle fibers, is narrow, straight, and thin-walled, running between the salivary glands and ending at the midgut or ventriculus. In those species studied closely, the lumen is narrow, with a folded cuticular intima. The walls are of syncitial tissue with small, elongate nuclei (Bharadwaj, Reddy, and Sinha, 1966; Dobroscky, 1931a; Gil-Fernandez and Black, 1965; Pesson, 1951; Willis, 1949). The esophagus enters the midgut as an invagination, comprising the esophageal or cardiac valve. This appears to function as a barrier to regurgitation but has no sphincter (Dobroscky, 1931a; Gil-Fernandez and Black, 1965; Willis, 1949). On the surface of the invagination, the large cells end abruptly as true midgut cells begin (Willis, 1949). In *M. fascifrons* the cells lining the invagination are typically elongate, with dense cytoplasm and large nuclei (Dobroscky, 1931a).

The midgut or ventriculus of those Homoptera having a filter chamber consists of three parts: the first ventriculus, receiving the invaginated esophageal valve and forming part of the filter chamber; the second ventriculus, a sac-like structure extending backward for about two thirds of the length of the abdomen, sometimes miscalled the crop; and the third ventriculus, a long, tubular portion which loops forward to enter the filter chamber and there join the hindgut (Snodgrass, 1935) (Fig. 5). The malpighian tubules usually emerge from the filter chamber with the tubular hindgut.

The second ventriculus is not a mere storage organ, as the term *crop* implies, but a region where digestion takes place. Its walls are pleated and thicker than those of the esophagus (Gil-Fernandez and Black, 1965). There is a well-marked, infolded basement membrane with an outer lattice of muscle fibers. The epithelial cells are large and cuboidal with two prominent, rounded nuclei and a syncitial tendency in the basal region. The cells, separated by narrow grooves, protrude into the lumen (Dobroscky, 1931a; Licent, 1912; Smith and Littau, 1960; Willis, 1949). There are vacuoles and round masses of lipoid material in the apical area, and the free border is striated (Willis, 1949). Near the border the cytoplasm contains numerous mitochondria, which typically are elongate and oriented parallel with the long axes of the striations or microvilli. The cytoplasm contains parallel arrays of the profiles of cisternae in the endoplasmic reticulum, which is more highly developed here than elsewhere in the midgut (Smith and Littau, 1960). At intervals are small replacement cells (Willis, 1949).

At the constriction where the third ventriculus begins there is in some species an abrupt change in the character of the cells. The cells are smaller and the cytoplasm appears sparse and ragged, with two very large nuclei per cell (Gil-Fernandez and Black, 1965; Dobroscky, 1931a). In *M. fascifrons,* Smith and Littau (1960) found little change. The cells continued with a well-developed, striated border having about 50 microvilli per square micron.

The filter chamber is a structure on which authors largely agree. Apparent disagreement arises from the definition of the term. In nearly all the leafhoppers studied the workers found the distal end of the third ventriculus and the proximal part of the hindgut insinuated under a membrane at the proximal end of the first ventriculus, or at least beneath an overlying web of muscle fibers, which always showed pulsing movements. Its accepted function is that of disposing of surplus water by passing it directly to the hindgut, leaving a more concentrated solution for the ventriculus (Licent, 1912; Pesson, 1951).

There is a developmental gradient in which the filter chamber ranges from simple apposition of the parts, as in *Typhlocyba* (Fig. 6*a*) (Willis, 1949); through types slightly less primitive (Figs. 6*b* and 6*c*) and as described by Smith and Littau (1960) in *M. fascifrons* and by Bharadwaj *et al.* (1966) in *Agallia constricta* Van Duzee; to that in

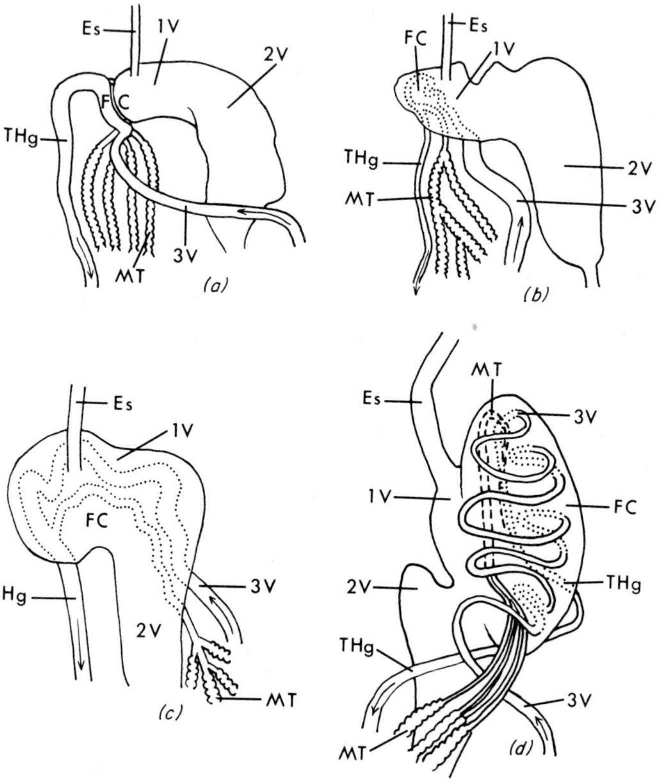


*Fig. 6.* Schematic diagrams of filter chambers of leafhoppers in order of increasing complexity, to show the relationship of the parts of the ventriculus with the malpighian tubules and tubular hindgut: (*a*) *Typhlocyba;* (*b*) *Idiocerus;* (*c*) *Athysanus;* (*d*) Cercopidae. (Redrawn from Licent, 1912.).

the Cercopidae (Fig. 6*d*), which must deal with very dilute xylem sap (Licent, 1912; Pesson, 1951). In *Typhlocyba,* there is no fusion of tissues within the chamber, but the walls of the distal part of the third ventriculus are much thinned with numerous vacuoles (Willis, 1949). Pesson (1951) and Licent (1912) consider this pattern, common in the Jassidae, to be rudimentary. In the Cercopidae, however, the approximating epithelia may become merged, extremely reduced and thin (Willis, 1949).

There is disagreement on the function of the looped third ventriculus. Evidence exists that it is excretory and secretory. Gil-Fernandez and Black (1965) confirmed Dobroscky's (1931a) finding of carbonates in the lumen and in cells. Smith and Littau (1960) point out that leafhoppers may receive into and eject from the hindgut clear fluid from the midgut or brochosomal material from the malpighian tubules.

The point of transition from midgut to hindgut may be beneath the membrane covering the filter chamber, as in the Cercopidae (Pesson, 1951) or immediately outside it as in some Jassidae (Willis, 1949) or Membracidae (Kershaw, 1913). In the Jassidae there appears to be no well-marked, functional pyloric valve or sphincter at the transition, but merely a simple enlargement of a few cells. The hindgut runs backward dorsally as a thin and fairly simple tube, expanding into a variously shaped rectum before the anus. There may be a small constriction with circular musculature between the tubular hindgut and the rectum (Gil-Fernandez and Black, 1965; Mukharji, 1959). Sometimes the closed distal ends of the malpighian tubules are attached on the forward part of the rectum (Bharadwaj *et al.*, 1966).

The tubular hindgut does not have well-defined cell walls, but consists of syncitial epithelium with large, elongate nuclei, thinly covered and protruding into the narrow lumen. The epithelium is much folded and thinner than that of the foregut, its inner surface lined with intima (Gil-Fernandez and Black, 1965; Willis, 1949). The folds or lamellae, which are about 2 $\mu$ long and richly supplied with mitochondria, greatly increase the area of contact with fluid in the lumen. Their role may be to complete the absorption of food constituents (Smith and Littau, 1960).

The walls of the rectum are very thin, with an outer layer of circular muscle overlaid by a few longitudinal muscle fibers. The cells are small and irregular with compact nuclei. Protuberances into the enlarged lumen give a corrugated appearance to the surface (Dobroscky, 1931a; Gil-Fernandez and Black, 1965; Mukharji, 1959).

## IV. HEMOCOEL

All the spaces in the body and appendages not occupied by organs or tissues constitute the hemocoel. The hemocoel is not an open blood system, for it is lined by a continuous, selectively permeable, connective tissue sheath (Pipa and Cook, 1958). The hemocoel is filled with blood, consisting of a liquid component or hemolymph and cellular components or hemocytes. Little research has been concentrated on homopteran blood, but recent reviews deal with hemolymph and blood cells generally (Wigglesworth, 1959; Wyatt, 1961). A special feature of the blood in aphids and coccids is the occurrence of wax cells, large hemocytes which are almost completely filled by vacuoles of lipid (Pesson, 1951).

The fat body consists of irregular masses of rounded or polygonal cells in the hemocoel. It is better described as diffuse tissue for storage of metabolic reserves than as a discrete organ. The fat body cells (trophocytes) of the English grain aphid, *Macrosiphum avenae* (F.) are large cells with central stellate or round nuclei and an abundance of vacuoles and lipid droplets. Conspicuous strands of cytoplasm radiate from the center of the cells to the periphery. The relative abundance of cells with stellate nuclei has been associated with the developmental stage of the aphid (Rutschky and Campbell, 1964).

In leafhoppers the fat body is found mostly in the abdomen and head, where the cell masses appear to fill up unoccupied spaces (Littau and Maramorosch, 1960). Colored whitish, pale yellow, or greenish, the large cells normally contain numerous fat-filled vacuoles, smooth, rounded nuclei, and homogenous, basophilic cytoplasm. Binucleate cells are common. The tissue is well supplied with tracheae and surrounded by many small cells that lack obvious accumulations of lipid. Fat is most abundant in females and juveniles, and lessens markedly with age. The fat body is in contact with the suspensory filament of the ovarioles, which circumstance may provide a source of nourishment and a route for it to the developing eggs (Gil-Fernandez and Black, 1965). It is a major site for propagation of some viruses (Fukushi and Shikata, 1963; Shikata and Maramorosch, 1965; Nasu, 1965; Sinha, 1965b).

Cicadellids, membracids, and fulgorids appear to have four malpighian tubules, but cercopids have up to 12. The tubules are often narrowed and fused near the point of origin, but their length lies in the hemocoel bathed in hemolymph. The pattern of thick and thin portions varies between groups. Histology indicates that they have more than an excretory function, and are almost certainly a site for increase of more than one virus (Fukushi and Shikata, 1963; Hirumi and Maramorosch, 1963; Nasu, 1965; Shikata and Maramorosch, 1965; Sinha, 1965b.)

Mycetomes are special organs in the hemocoel housing symbiotic microorganisms, which are often enclosed in special cells or mycetocytes. In *M. fascifrons* and *E. striatulus* the mycetomes typically are located on each side of the abdomen (Dobroscky, 1931b). In *A. constricta* they are similar, comprising two isolated masses, each having about 30 large cells, in the first to third abdominal segments on each side, held to the body wall by tracheae, covered by membrane and surrounded by fat body. Gil-Fernandez and Black (1965) saw only a few organisms resembling bacteria, and noted that the mycetomes degenerated as this insect aged. In *Nephotettix cincticeps* Uhler the mycetomes are already well defined in the first instar, consisting of many cells, in four layers. The innermost contains mycetocytes filled with bacteroids; the intermediate layers contain mycetocytes filled with larger symbiotes (Nasu, 1965).

Symbiotes are of proved benefit to cockroaches but their role in leafhoppers is less clear. Among Homoptera, only the Typhlocybinae appear to be without mycetocytes, which are thought to be derived from fat cells or leucocyte blood cells. The symbiotes of some Homoptera appear to be yeasts of which there may be as many as five types in a single insect (Pesson, 1951). The bacteroids are also polymorphic (Nasu, 1965). De Wilde states (1964) that infectious microorganisms in Homoptera are produced in the mycetomes, and carried free in the hemolymph or in loose mycetocytes. Pesson (1951) makes it clear that in spite of the mass of literature, the systematics of the symbiotes of Homoptera and their relationships with their hosts are comparatively little known and extraordinarily complex.

## V. OVARIES

Female Homoptera have eight or more telotrophic ovarioles, each of which is attached at the top by a filament to the fat body, and connected at the bottom by a short, hollow pedicel with the oviduct. Each tapering, tubular ovariole contains a succession of young egg cells or oocytes, and eggs in various stages of maturity. Interspersed between the oocytes and clustered near the apex are nurse or nutrient cells, from which nurse fibers lead into the developing oocytes. A mycetocyte at the entry to the pedicel has been demonstrated in *N. cincticeps* by Nasu (1965).

Nine or more plant viruses, some of them demonstrably propagative, are known to pass transovarially through the eggs of leafhoppers. The transmission may be highly efficient, as in *Delphacodes striatella* Fallén, in which Shikata carried the virus of rice stripe for 5 years through 40 generations, with no decline in titer (Maramorosch, 1960); or inefficient, as in *Euscelis plebejus* Fallén, in which the virus of clover witches' broom was transmitted occasionally through eggs laid in immune monocots (Posnette and Ellenberger, 1963). Most of the leafhopper-borne viruses are not yet known to pass transovarially, but undoubtedly more will be found. Leafhoppers which transmit two or more viruses are not necessarily those which transmit transovarially. There is evidence that transmissibility to eggs may depend on one or more of the following: permeability of membranes in the vector; genetic variability; some quality of the virus particle; long association between the virus and vector with special adaptations in the vector; the symbiotes.

Sinha (1960) found that only one female *Delphacodes pellucida* Fabricius transmitted to its progeny out of 20 that acquired wheat striate virus as adults. Since the virus must pass through the gut wall and into the ovariole and egg, the indication is that the permeability of membranes decreases with age (Sinha, 1963). Genetic variability in efficiency of transovarial transmission was shown by Yamada and Yamamoto (Maramorosch, 1960) with rice stripe virus in *D. striatella,* and with wound tumor in *A. constricta* by Sinha (1965a), who found virus fluorescent antigens in the ovarioles, ovaries, and oviducts of infectives. Nearly all leafhopper-borne viruses have spherical particles, the size of which makes it conceivable that they accompany the microorganisms into the developing ova. Indeed, within some aphids, cicadellids, membacids, fulgorids, psyllids, and coccids are follicles with so-called receptive cells, often near the caudal pole of the eggs, which appear to function solely to admit symbiotes from the mother (de Wilde, 1964; Pesson, 1951). Nasu (1965) demonstrated by electron microscopy an affinity between rice dwarf virus particles and the surface membrane of symbiotes in *N. cincticeps,* and deduced from his evidence that symbiotes free in the hemolymph migrated to the mycetocyte at the base of the developing egg. The symbiotes and their accompanying particles increased in numbers and then entered the oocyte at about the yolk-forming stage. By the time the eyes of the embryo were showing pigmentation, virus-like particles and symbiotes were numerous within the egg.

Black (1959) points out that there is no *a priori* reason why even a nonmultiplying virus should not pass readily through the mother's tissues to reappear in the progeny. There is, in fact, no reason to suppose that transovarial transmission cannot occur in aphids.

## BIBLIOGRAPHY


Adams, J. B., and Drew, M. E. 1965. A cellulose-hydrolyzing factor in aphid saliva. Can. J. Zool. 43:489–496.

Adams, J. B., and McAllan, J. W. 1956. Pectinase in the saliva of *Myzus persicae* (Sulz.). Can. J. Zool. 34:541–543.

———. 1958. Pectinase in certain insects. Can. J. Zool. 36:305–308.

Auclair, J. L. 1963. Aphid feeding and nutrition. Ann. Rev. Entomol. 8:439–490.

———. 1964. Recent advances in the feeding and nutrition of aphids. Can. Entomol. 96:241–249.

Banks, C. J. 1964. Aphid nutrition and reproduction. Rep. Rothamsted Exp. Sta., p. 299–309.

Banks, C. J., and Nixon, H. L. 1958. Effects of the ant, *Lasius niger* L., on the feeding and excretion of the bean aphid, *Aphis fabae* Scop. J. Exp. Biol. 35: 703–711.

Bharadwaj, R. K., Reddy, D. V., and Sinha, R. C. 1966. A reinvestigation of the alimentary canal in the leafhopper *Agallia constricta* (Homoptera: Cicadellidae). Ann. Entomol. Soc. Amer. 59:616–617.

Black, L. M. 1959. Biological cycles of plant viruses in insect vectors. p. 157–185. *In* The viruses. Vol. 2. F. M. Burnet and W. M. Stanley [ed.] Academic Press, New York.

Börner, C. 1938. Neuer Beitrag zur Systematik und Stammesgeschichte der Blattläuse. Abhandl. Naturwissenschaften Bremen 30:167–179.

Bradley, R. H. E. 1952. Studies on the aphid transmission of a strain of henbane mosaic virus. Ann. Appl. Biol. 39:78–97.

———. 1960. Effect of amputating stylets of mature apterous viviparae of *Myzus persicae*. Nature 188:337–338.

———. 1962. Response of the aphid *Myzus persicae* (Sulz.) to some fluids applied to the mouthparts. Can. Entomol. 94:707–722.

———. 1966. Which of an aphid's stylets carry transmissible virus? Virology 29:396–401.

Bradley, R. H. E., and Ganong, R. Y. 1955a. Evidence that potato virus Y is carried near the tip of the stylets of the aphid vector *Myzus persicae* (Sulz.). Can. J. Microbiol. 1:775–782.

———. 1955b. Some effects of formaldehyde on potato virus Y *in vitro,* and ability of aphids to transmit the virus when their stylets are treated with formaldehyde. Can. J. Microbiol. 1:783–793.

Bradley, R. H. E., Sylvester, E. S., and Wade, C. V. 1962. Note on the movements of the mandibular and maxillary stylets of the aphid, *Myzus persicae* (Sulzer). Can. Entomol. 94:653–654.

Carter, W. 1962. Insects in relation to plant disease. Wiley, New York. 705 p.

Davidson, J. 1913. The structure and biology of *Schizoneura lanigera,* Hausmann, or wooly aphis of the apple tree. 1. The apterous viviparous female. Quart. J. Microscopical Sci. 58:653–701.

———. 1914. On the mouth-parts and mechanism of suction in *Schizoneura lanigera* Hausmann. J. Linnean Soc. London Zool. 32:307–330.

———. 1923. Biological studies of *Aphis rumicis* Linn. The penetration of plant tissues and the source of the food supply of aphids. Ann. Appl. Biol. 10:35–54.

Day, M. F., Irzykiewicz, H., and McKinnon, A. 1952. Observation on the feeding of the virus vector *Orosius argentatus* (Evans), and comparisons with certain other jassids. Australian J. Sci. Res. Ser. B 5:128–142.

Day, M. F., and McKinnon, A. 1951. A study of some aspects of the feeding of the jassid *Orosius*. Australian J. Sci. Ser. B 4:125–135.

Dobroscky, I. D. 1931a. Morphological and cytological studies on the salivary glands and alimentary tract of *Cicadula sexnotata* (Fallen), the carrier of aster yellows virus. Contrib. Boyce Thompson Inst. 3:39–58.

———. 1931b. Studies on cranberry false blossom disease and its insect vector. Contrib. Boyce Thompson Inst. 3:59–83.

Forbes, A. R. 1964. The morphology, histology, and fine structure of the gut of the green peach aphid, *Myzus persicae* (Sulzer) (Homoptera: Aphididae). Mem. Entomol. Soc. Can. No. 36. 75 p.

———. 1966. Electron microscope evidence for nerves in the mandibular stylets of the green peach aphid. Nature 212:726.

Fortin, B. 1958. Étude préliminaire des glandes salivaires de *Macrosiphum solanifolii* Ashm. (Homoptera: Aphididae): Anatomie et histologie. Proc. Int. Cong. Entomol., 10th, Montreal 1:601–603.

Fukushi, T., and Shikata, E. 1963. Localization of rice dwarf virus in its insect vector. Virology 21:503–505.

Gil-Fernandez, C., and Black, L. M. 1965. Some aspects of the internal anatomy of the leafhopper *Agallia constricta* (Homoptera: Cicadellidae). Ann. Entomol. Soc. Amer. 58:275–284.

Hagley, E. A. C., and Blackman, J. A. 1966. Site of feeding of the sugarcane froghopper, *Aeneolamia varia saccharina* (Homoptera: Cercopidae). Ann. Entomol. Soc. Amer. 59:1289–1291.

Heinze, K. 1959. Phytopathogene Viren und ihre Überträger. Duncker & Humblot, Berlin. 291 p.

Herold, F., and Munz, K. 1965. Electron microscopic demonstration of viruslike particles in *Peregrinus maidis* following acquisition of maize mosaic virus. Virology 25:412–417.

Hirumi, H., and Maramorosch, K. 1963. Recovery of aster yellows virus from various organs of the insect vector, *Macrosteles fascifrons.* Contrib. Boyce Thompson Inst. 22:141–151.

Hoof, H. A. van. 1957. On the mechanism of transmission of some plant viruses. Koninkl. Ned. Akad. Wetenschap. Proc. Ser. C 60:314–317.

———. 1958. Onderzoekingen over de biologische overdracht van een non-persistent virus. Van Putten & Oortmijer, Alkmaar. 96 p.

Jensen, D. D. 1966. Pear psylla-pear decline virus relationships. Bull. Entomol. Soc. Amer. 12:299. (Abstr.)

Kennedy, J. S., Day, M. F., and Eastop, V. F. 1962. A conspectus of aphids as vectors of plant viruses. Commonwealth Inst. Entomol., London. 114 p.

Kershaw, J. C. 1913. The alimentary canal of Flata and other Homoptera. Psyche 20:175–188.

Licent, P. E. 1912. Recherches d'anatomie et de physiologie comparées sur le tube digestif des Homoptères supérieurs. Cellule Rec. Cytol. Histol. 28:5–164.

Littau, V. C., and Maramorosch, K. 1960. A study of the cytological effects of aster yellows virus on its insect vector. Virology 10:483–500.

McAllan, J. W., and Adams, J. B. 1961. The significance of pectinase in plant penetration by aphids. Can. J. Zool. 39:305–310.

Maramorosch, K. 1960. Leafhopper-transmitted plant viruses. Protoplasma 52: 457–466.

———. 1962. Acquisition and transmission of asters yellows virus. Phytopathology 52:1219. (Abstr.)

Miles, P. W. 1959a. Secretion of two types of saliva by an aphid. Nature 183:756.

———. 1959b. The salivary secretions of a plant-sucking bug, *Oncopeltus fasciatus* (Dall.) (Heteroptera: Lygaeidae)—I. The types of secretion and their roles during feeding. J. Insect Physiol. 3:243–255.

———. 1965. Studies on the salivary physiology of plant bugs: the salivary secretions of aphids. J. Insect Physiol. 11:1261–1268.

Miles, P. W., McLean, D. L., and Kinsey, M. G. 1964. Evidence that two species of aphid ingest food through an open stylet sheath. Experientia 20:582.

Mittler, T. E. 1957. Studies on the feeding and nutrition of *Tuberolachnus salignus* (Gmelin) (Homoptera, Aphididae) 1. The uptake of phloem sap. J. Exp. Biol. 34:334–341.

Mittler, T. E., and Dadd, R. H. 1963. Studies on the artificial feeding of the aphid *Myzus persicae* (Sulzer)—I. Relative uptake of water and sucrose solutions. J. Insect Physiol. 9:623–645.

Moericke, V., and Wohlfarth-Bottermann, K. E. 1960. Zur funktionellen Morphologie der Speicheldrüsen von Homopteren. 1. Die Hauptzellen der Hauptdrüse von *Myzus persicae* (Sulz.), Aphididae. Z. Zellforsch. 51:157–184.

———. 1963. Zur funktionellen Morphologie der Speicheldrüsen von Homopteren. II. Die Deck und die Zentralzellen der Speicheldrüse von *Myzus persicae* (Sulz.), Aphididae. Z. Zellforsch. 59:165–183.

Muir, F., and Kershaw, J. C. 1912. The development of the mouthparts in the Homoptera, with observations on the embryo of *Siphanta*. Psyche 19:77–89.

Mukharji, S. P. 1959. The structure and histology of the alimentary canal of Homoptera (Hemiptera). Ph.D. thesis, Banaras Hindu Univ., Varanasi 5, India. (Preface and summary.)

Naito, A. 1965. Collecting method of the salivary sheath material of leafhoppers and plant hoppers. Japan. J. Appl. Entomol. Zool. 9:142–144.

Nasu, S. 1965. Electron microscopic studies on transovarial passage of rice dwarf virus. Japan. J. Appl. Entomol. Zool. 9:225–237.

Nault, L. R., and Gyrisco, G. G. 1966. Relation of the feeding process of the pea aphid to the inoculation of pea enation mosaic virus. Ann. Entomol. Soc. Amer. 59:1185–1197.

Nielson, M. W. 1962. A synonymical list of leafhopper vectors of plant viruses (Homoptera, Cicadellidae). U.S. Dep. Agr. ARS-33-74. 12 p.

Ossiannilsson, F. 1966. Insects in the epidemiology of plant viruses. Ann. Rev. Entomol. 11:213–232.

Pesson, P. 1951. Ordre des homoptères. Traité de Zoologie 10:1390–1656.

Pinet, J. M. 1963. L'innervation sensorielle des stylets mandibulaires et maxillaires de *Rhodnius prolixus* Stål. (Insecte Hémiptère Hétéroptère). Compt. Rend. Acad. Sci. 257:3666–3668.

Pipa, R. L., and Cook, E. F. 1958. The structure and histochemistry of the connective tissue of the sucking lice. J. Morphol. 103:353–385.

Posnette, A. F., and Ellenberger, C. E. 1963. Further studies of green petal and other leafhopper-transmitted viruses infecting strawberry and clover. Ann. Appl. Biol. 51:69–83.

Qadri, M. A. H. 1959. Mechanism of feeding in Hemiptera. p. 237–245. *In* Studies in invertebrate morphology. Vol. 137. Smithsonian Inst. Pub. Misc. Collections. Washington, D.C.

Rutschky, C. W., and Campbell, R. L. 1964. Fat body morphology of English grain aphids infected with barley yellow dwarf virus. Ann. Entomol. Soc. Amer. 57:53–56.

Shikata, E., and Maramorosch, K. 1965. Electron microscopic evidence for the systemic invasion of an insect host by a plant pathogenic virus. Virology 27: 461–475.

Sinha, R. C. 1960. Comparison of the ability of nymph and adult *Delphacodes pellucida* Fabricius, to transmit European wheat striate mosaic virus. Virology 10:344–352.

———. 1963. Effect of age of vector and of abdomen punctures on virus transmission. Phytopathology 53:1170–1173.

———. 1965a. Sequential infection and distribution of wound-tumor virus in the internal organs of a vector after ingestion of virus. Virology 26:673–686.

———. 1965b. Recovery of potato yellow dwarf virus from hemolymph and internal organs of an insect vector. Virology 27:118–119.

Smith, C. F. 1939. The digestive system of *Macrosiphum solanifolii* (Ash.) (Aphidae: Homoptera). Ohio J. Sci. 39:57–59.

Smith, D. S., and Littau, V. C. 1960. Cellular specialization in the excretory epithelia of an insect, *Macrosteles fascifrons* Stål (Homoptera). J. Biophys. Biochem. Cytol. 8:103–133.

Smith, K. M. 1926. A comparative study of the feeding methods of certain Hemiptera and of the resulting effects upon the plant tissue, with special reference to the potato plant. Ann. Appl. Biol. 13:109–139.

———. 1965. Plant virus–vector relationships. Advances Virus Res. 11:61–96.

Snodgrass, R. E. 1935. Principles of insect morphology. McGraw-Hill, New York. 667 p.

Sogawa, K. 1965. Studies on the salivary glands of rice plant leafhoppers. I. Morphology and histology. Japan. J. Appl. Entomol. Zool. 9:275–290.

Sorin, M. 1961. The mouth parts of *Aphis craccivora* Koch and the penetration of stylets into host plants. Japan. J. Appl. Entomol. Zool. 5:217–224.

———. 1966. Physiological and morphological studies on the suction mechanism of plant juice by aphids. Bull. Univ. Osaka Pref. Ser. B. 18:95–138.

Sukhov, K. S. 1944. Salivary secret of the aphis *Myzus persicae* Sulz. and its ability to form a filtering apparatus. Compt. Rend. Acad. Sci. URSS 42: 226–228.

Swenson, K. G. 1962. Bean yellow mosaic virus transmission by *Myzus persicae*. Australian J. Biol. Sci. 15:468–482.

Weber, H. 1928. Skelett, Muskulatur und Darm der schwarzen Blattläus Aphis fabae Scop. Mit besonderer Berucksichtigung der Funktion der Mundwerkzeuge und des Darms. Zoologica 28:1–120.

———. 1929. Zur vergleichenden Physiologie der Saugorgane der Hemipteren. Mit besonderer Berücksichtigung der Pflanzenläuse. Z. Vergleich. Physiol. 8: 145–186.

———. 1930. Biologie der Hemipteren; eine naturgeschichte der Schnabelkerfe. Springer Verlag, Berlin. 543 p.

Wensler, R. J. D. 1962. Mode of host selection by an aphid. Nature 195:830–831.

Wigglesworth, V. B. 1959. Insect blood cells. Ann. Rev. Entomol. 4:1–16.

Wilde, J. de. 1964. Reproduction. p. 9–58. *In* The physiology of Insecta. Vol. 1. M. Rockstein [ed.] Academic Press, New York.

Willis, D. M. 1949. The anatomy and histology of the head, gut and associated structures of *Typhlocyba ulmi*. Proc. Zool. Soc. London 118:984–1001.

Wohlfarth-Bottermann, K. E., and Moericke, V. 1960. Zur funktionellen Morphologie der Speicheldrüsen von Homopteren. III. Die Nebendruse von *Myzus persicae* (Sulz.), Aphididae. Z. Zellforsch. 52:346–361.

Wyatt, G. R. 1961. The biochemistry of insect hemolymph. Ann. Rev. Entomol. 6:75–102.

Zweigelt, F. 1914. Beiträge zur Kenntnis des Saugphänomens der Blattläuse und der Reacktionen der Pflanzenzellen. Zentr. Bakteriol. Parasitenk. Abt. II. 42: 265–335.

# Families and Genera of Leafhopper Vectors

Tamotsu Ishihara

*College of Agriculture, Ehime University,*
*Matsuyama, Japan*

## I. INTRODUCTION

Leafhoppers are important vectors of plant-pathogenic viruses and the type of transmission is usually circulative or propagative. In some species, viruses are carried to the offspring by transovarial passage. The 131 species of leafhoppers treated in this paper belong to 14 families within 3 superfamilies:

Superfamily Cercopoidea
- Family 1. Cercopidae: 3 sp. of 2 genera (2 viruses)
- Family 2. Clastopteridae: 1 sp. (1 virus)

Superfamily Cicadelloidea
- Family 1. Ulopidae: 1 sp. (1 virus)
- Family 2. Agalliidae: 12 sp. of 6 genera (8 viruses)
- Family 3. Iassidae: 1 sp. (1 virus)
- Family 4. Macropsidae: 1 sp. (1 virus)
- Family 5. Gyponidae: 2 sp. of 1 genus (2 viruses)
- Family 6. Aphrodidae: 2 sp. of 1 genus (3 viruses)
- Family 7. Coelidiidae: 1 sp. (1 virus)
- Family 8. Tettigellidae: 28 sp. of 12 genera (4 viruses)
- Family 9. Cicadellidae: 4 sp. of 1 genus (5 viruses)
- Family 10. Deltocephalidae: 63 sp. of 27 genera (about 60 viruses)

Superfamily Fulgoroidea
- Family 1. Cixiidae: 2 sp. of 2 genera (2 viruses)
- Family 2. Delphacidae: 10 sp. of 9 genera (22 viruses)

In each family, the genera and species are arranged alphabetically. Because of limited space, some authorities have been abbreviated as follows:

Ashm. = Ashmead, Atk. = Atkinson, Bak = Baker, Beam. = Beamer, Brul. = Brullé, Brun. = Brunner, Curt. = Curtis, Dav. = Davidson, DeL.

=DeLong, Dist.=Distant, Edw.=Edwards, Fab.=Fabricius, Fall.= Fallén, Fieb.=Fieber, Fraz.=Frazier, Germ.=Germar, Horv.=Horváth, Ish.=Ishihara, Kirk.=Kirkaldy, Kirsch.=Kirschbaum, L.= Linnaeus, Lath.=Lathrop, Leth.=Lethierry, Linnav.=Linnavuori, Mats.=Matsumura, Mel.=Melichar, Metc.=Metcalf, Motsch.= Motschulsky, Msh.=Marshall, Muls. et R.=Mulsant et Rey, Nott.= Nottingham, Osb.=Osborn, Prov.=Provancher, Rib.=Ribaut, Sign.= Signoret, Uhl.=Uhler, VD.=Van Duzee, Wag.=Wagner, Walk.= Walker, Wolc.=Wolcott, Zak.=Zakhvatkin, Zett.=Zetterstedt.

## II. SUPERFAMILY CERCOPOIDEA

### *A. Family 1. Cercopidae*

GENUS 1. *Philaenus* Stål, 1864 (type, by subsequent designation: *Cicada leucophthalmus* L., 1758, which is a blackish colored "variety" of *Cicada spumarius* L., 1758). Members of this genus occur in the holarctic region.

*1. Philaenus spumarius* (L., 1758) [=*P. leucophthalmus* L. 1758]. Distribution: holarctic region. Viruses transmitted: peach yellows and lucerne dwarf.

GENUS 2. *Aphrophora* Germ., 1821 (type, by subsequent designation: *Aphrophora corticea* Germ., 1821). The distribution of this genus is almost worldwide.

*2. Aphrophora augulata* Ball, 1898. A nearctic species. A vector of lucerne dwarf.

*3. Aphrophora permutata* Uhl., 1872. A nearctic species. A vector of lucerne dwarf.

### *B. Family 2. Clastopteridae*

GENUS 1. *Clastoptera* Germ., 1838 (type, by subsequent designation: *Clastoptera xanthocephala* Germ., 1838). Distribution: new world.

*1. Clastoptera brunnea* Ball, 1919. A nearctic species. A vector of lucerne dwarf.

## III. CICADELLOIDEA

### *A. Family 1. Ulopidae*

GENUS 1. *Moonia* Dist., 1908 (type, by original designation: *Moonia sancita* Dist., 1908). This genus is oriental in distribution.

*1. Moonia albomaculata* Dist., 1916. An Indian species. A vector of sandal spike.

### *B. Family 2. Agalliidae*

GENUS 1. *Aceratagallia* Kirk., 1907 (type, by original designation: *Bythoscopus sanguinolentus* Prov., 1872). Distribution: new world.

*1. Aceratagallia curvata* Oman, 1933. A nearctic species. A vector of potato yellow dwarf (New York strain).

*2. Aceratagallia longula* (VD., 1894). A nearctic species. A vector of potato yellow dwarf (New York strain).

*3. Aceratagallia obscura* Oman, 1933. A nearctic species. A vector of potato yellow dwarf (New York strain).

*4. Aceratagallia sanguinolenta* (Prov., 1872). A nearctic species. A vector of potato yellow dwarf (New York strain).

GENUS 2. *Agallia* Curt., 1833 (type, by monotypy: *Agallia consobrina* Curt., 1833=*Jassus puncticeps* Germ. 1832, a palaearctic species). The distribution of this genus is almost worldwide.

*5. Agallia albidula* Uhl., 1895. Distribution: nearctic and neotropical. A vector of Brazilian beet curly-top.

*6. Agallia constricta* (VD., 1894). A nearctic species. Viruses transmitted: clover big vein (wound tumor) and potato yellow dwarf (New Jersey strain).

*7. Agallia quadripunctata* (Prov., 1872). A nearctic species. Viruses transmitted: clover big vein (wound tumor) and potato yellow dwarf (New Jersey strain).

*8. Agallia venosa* (Fall., 1806). A European species. A vector of leaf crinkle of tomato.

GENUS 3. *Agalliana* Oman, 1933 (type, by original designation: *Bythoscopus sticticollis* Stål, 1859, a neotropical species). Distribution of the genus: neotropical.

*9. Agalliana ensigera* Oman, 1934 [=*Aceratagallia sanguinolenta* Auct. (nec Prov., 1872); *A. sticticollis* Auct. (nec Stål, 1859)]. A neotropical species. Viruses transmitted: Argentinian beet curly-top and Brazilian tomato curly-top.

GENUS 4. *Agalliopsis* Kirk. 1907 (type, by original designation: *Jassus novellus* Say, 1830). Distribution of the genus: new world.

*10. Agalliopsis novella* (Say, 1830). A nearctic species. Viruses transmitted: clover club leaf and clover big vein (wound tumor).

GENUS 5. *Austroagallia* * Evans, 1936 (type, by original designation: *Austroagallia torrida* Evans, 1936). Distribution of the genus: Australian and Pacific regions.

* According to Evans' personal communication, he recently found that the genus *Austroagallia* is not congeneric with *Nehela* White, 1878 (type species, by monotypy: *Nehela vulturina* White, 1878, a species of St. Helena) which is a synonym of the type species shown below.

*11. Austroagallia torrida* Evans, 1936 [=*Peragallia lauensis* Linnav., 1960, from Fiji]. Distribution: Australia and Fiji. A vector of Datura rugose leaf curl.

GENUS 6. *Peragallia* † Rib., 1948 [=*Pachynus* Stål, 1866, *nom. praeocc.*] (type, by original designation: *Bythoscopus sinuatus* Muls. et R. 1855). Distribution: palaearctic and Ethiopian regions, and Oceania.

*12. Peragallia sinuata* (Muls. et R., 1855). A palaearctic species. Viruses transmitted: potato witches' broom and tomato big bud.

### *C. Family 3. Iassidae*

GENUS 1. *Batrachomorphus* Lewis, 1837 [=*Eurinoscopus* Kirk., 1906] (type, by monotypy: *Batrachomorphus irroratus* Lewis, 1834). The distribution of this genus is almost worldwide, except the Ethiopian region.

*1. Batrachomorphus punctatus* (Evans, 1940). An Australian species. A vector of Australian lucerne witches' broom.

### *D. Family 4. Macropsidae*

GENUS 1. *Macropsis* Lewis, 1834 (type, by subsequent designation: *Cicada virescens* Gmelin, 1789). Distribution: holarctic.

*1. Macropsis fuscula* (Zett., 1829). A palaearctic species. A vector of *Rubus* stunt.

### *E. Family 5. Gyponidae*

GENUS 1. *Gyponana* Ball, 1920 (type, by original designation: *Tettigonia octolineata* Say, 1829, a nearctic species). Distribution: new world.

*1. Gyponana hasta* DeL., 1942. A nearctic species. A vector of Californian aster yellows.

*2. Gyponana striata* (Burm., 1839). A nearctic species. A vector of eastern X disease of peach.

### *F. Family 6. Aphrodidae*

GENUS 1. *Aphrodes* Curt., 1832 [=*Acucephalus* Germ., 1833; *Acocephalus* Burm., 1835; *Phlotaera* Zett., 1840] (type by monotypy: *Aphrodes testudo* Curt., 1833=*Cicada albifrons* L., 1758). Distribution: holarctic.

*1. Aphrodes albifrons* (L., 1758) [=*Acocephalus fulginosa* Sign.,

† I am not familiar with the type species of this genus. It is possible that *Peragallia* is congeneric with *Austroagallia*.

1879, with many other synonyms]. A species of holarctic distribution. A vector of phyllody of clover.

*2. Aphrodes bicinctus* (von Schrank, 1776) [=*Acucephalus bicinctus* Curt., 1836, with many other synonyms]. Distribution: holarctic. Viruses transmitted: aster yellows and southern stolbur of tomato.

### *G. Family 7. Coelidiidae*

GENUS 1. *Coelidia* Germ., 1821 (type, by subsequent designation: *Coelidia venosa* Germ., 1821). Distribution of this genus is almost worldwide, though some recorded species may possibly be heterogenous.

*1. Coelidia indica* Walk., 1851. An Indian species. A vector of sandal spike.

### *H. Family 8. Tettigellidae*

GENUS 1. *Carneocephala* Ball, 1927 (type, by original designation: *Draeculacephala floridana* Ball, 1901, a nearctic species). Species of this genus are subtropical.

*1. Carneocephala flaviceps* (Riley, 1880) [=*Tettigonia flavicephalum* Riley, 1880]. A nearctic species. A vector of Pierce's disease of grapes.

*2. Carneocephala fulgida* Nott., 1932. A nearctic species. A vector of Pierce's disease of grapes.

*3. Carneocephala triguttata* Nott., 1932. A nearctic species. A vector of Pierce's disease of grapes.

GENUS 2. *Cuerna* Mel., 1924 (type, by subsequent designation: *Cercopis lateralis* Fab., 1798, which is a synonym of *Cercopis costalis* Fab., 1803). This genus is holarctic in distribution.

*4. Cuerna costalis* (Fab. 1803) [=*Cercopis lateralis* Fab., 1798, *nom. praeocc.; Ceropis marginella* Fab., 1803; *Tettigonia pyrrhotela* Walk., 1851]. A species of holarctic distribution. A vector of Pierce's disease of grapes.

*5. Cuerna yuccae* Oman et Beam., 1944. A nearctic species. A vector of Pierce's disease of grapes.

GENUS 3. *Draeculacephala* Ball, 1901 (type, by original designation: *Tettigonia mollipes* Say, 1851). Distribution of this genus: new world.

*6. Draeculacephala antica* (Walk., 1851). A nearctic species. A vector of Pierce's disease of grapes.

*7. Draeculacephala crassicornis* VD., 1915. A nearctic species. A vector of Pierce's disease of grapes.

*8. Draeculacephala inscripta* VD., 1915. A nearctic species. A vector of Pierces disease of grapes.

*9. Draeculacephala minerva* Ball, 1927. A nearctic species. A vector of Pierce's disease of grapes.

*10. Draeculacephala noveboracensis* (Fitch, 1851) [=*Tettigonia prasina* Walk., 1851]. A nearctic species. A vector of Pierce's disease of grapes.

*11. Draeculacephala portola* Ball, 1927 [=*D. californica* David. et Fraz., 1949; *D. cubana* Metc. et Brun., 1936; *D. paludosa* Ball et China, 1933]. A nearctic species. Viruses transmitted: Pierce's disease of grapes and sugarcane chlorotic streak.

Genus 4. *Friscanus* Oman, 1938 (type, by original designation: *Errhomenellus friscanus* Ball, 1909). This genus is represented by the following single Californian species of tautonymy.

*12. Friscanus friscanus* (Ball, 1909) [=*Memnonia simplex* VD., 1917]. A Californian species. A vector of Pierce's disease of grapes.

Genus 5. *Graphocephala* VD., 1916 (type, by original designation: *Cicada coccinea* Förster, 1771, a nearctic species). Distribution of the genus: new world.

*13. Graphocephala cynthura* (Bak., 1898). A nearctic species. A vector of Pierce's disease of grapes.

*14. Graphocephala versuta* (Say, 1830). A nearctic species. A vector of phony peach.

Genus 6. *Holochara* Fitch, 1851 (type, by monotypy: *Holochara communis* Fitch, 1851). Distribution: nearctic.

*15. Holochara delta* Oman, 1943. A Californian species. A vector of Pierce's disease of grapes.

Genus 7. *Homalodisca* Stål, 1869. (type, by subsequent designation: *Cicada triquetra* Fab., 1803). Distribution: new world, especially the neotropical region.

*16. Homalodisca insolita* (Walk., 1858). A nearctic species. Viruses transmitted: Pierce's disease of grapes and phony peach.

*17. Homalodisca liturata* Ball, 1901. A nearctic species. A vector of Pierce's disease of grapes.

*18. Homalodisca triquetra* (Fab., 1803) [=*Proconia admittens* Walk., 1858; *Proconia aurigena* Walk., 1858; *Tettigonia coagulata* Say, 1832; *Proconia excludens* Walk., 1858; *Tettigonia vitripennis* Germ., 1821]. A nearctic species. A vector of phony peach.

Genus 8. *Hordnia* Oman, 1949 (type, by original designation: *Tettigonia circellata* Bak., 1898). Distribution: nearctic region.

*19. Hordnia circellata* (Bak., 1898). A nearctic species. A vector of Pierce's disease of grapes.

Genus 9. *Keonolla* Oman, 1949 (type, by original designation: *Proconia confluens Uhl.,* 1861). Distribution: nearctic region.

*20. Keonolla confluens* (Uhl., 1861). A nearctic species. Viruses transmitted: Pierce's disease of grapes and western X disease of peach.

*21. Keonolla dolobrata* (Ball, 1901). A nearctic species. A vector of Pierce's disease of grapes.

GENUS 10. *Neokolla* Mel., 1926 (type, by subsequent designation: *Tettigonia hieroglyphica* Say, 1831). Distribution: nearctic region.

*22. Neokolla hieroglyphica* (Say, 1830) [=*Tettigonia gothica* Sign., 1854; *Tettigonia similis* Woodworth, 1890]. A nearctic species. A vector of Pierce's disease of grapes.

*23. Neokolla severini* DeL., 1948. A nearctic species. A vector of Pierce's disease of grapes.

GENUS 11. *Oncometopia* Stål, 1869 (type, by subsequent designation: *Cicada undata* Fab., 1794). Distribution: new world.

*24. Oncometopia undata* (Fab., 1794) [=*Proconia badia* Walk., 1851; *P. clarior* Walk., 1851; *P. lucernea* Walk., 1851; *P. marginata* Walk., 1851; *P. nigricans* Walk., 1851; *Cicada orbona* Fab., 1798, and 3 other synonyms]. A nearctic species. Viruses transmitted: Pierce's disease of grapes and phony peach.

GENUS 12. *Pagaronia* Ball, 1902 (type, by subsequent designation: *Pagaronia tredecimpunctata* Ball, 1902). This genus is distributed in California and North America. All four known species listed below transmit Pierce's disease of grapes.

*25. Pagaronia confusa* Oman, 1938.

*26. Pagaronia furcata* Oman, 1938.

*27. Pagaronia tredecimpunctata* Ball, 1902 [=*Pagaronia octopunctata* Kirk., 1909].

*28. Pagaronia triunata* Ball, 1902.

### *I. Family 9. Cicadellidae*

GENUS 1. *Empoasca* Walsh, 1862 (type, by subsequent designation: *Empoasca viridescens* Walsh, 1862). Distribution: almost worldwide.

*1. Empoasca devastans* Dist., 1918. An Indian species. Viruses transmitted: little leaf of eggplant, tomato big bud, and cotton leaf curl.

*2. Empoasca dilitare* DeL. et David., 1935. A nearctic species. A vector of bunchy top of papaya.

*3. Empoasca papayae* Oman, 1937. A species distributed in the West Indies. A vector of bunchy top of papaya.

*4. Empoasca fabae* (Harris, 1841). A nearctic species. A vector of potato yellow dwarf (New York strain).

### *J. Family 10. Deltocephalidae*

GENUS 1. *Acinopterus* VD., 1892 (type, by monotypy: *Acinopterus acuminatus* VD., 1892, a nearctic species). Distribution: new world.

*1. Acinopterus angulatus* VD., 1892. A nearctic species. A vector of Californian aster yellows.

GENUS 2. *Chlorotettix* VD., 1892 (type, by original designation: *Bythoscopus unicolor* Fitch, 1851, a nearctic species). Distribution: new world.

*2. Chlorotettix similis* DeL., 1918. A nearctic species. A vector of Californian aster yellows.

*3. Chlorotettix viridius* VD., 1892. A nearctic species. A vector of Californian aster yellows.

GENUS 3. *Cicadulina* China, 1926 (type, by original designation: *Cicadulina zeae* China, 1926, which is a subspecies of *C. bipunctella* (Mats., 1908). Distribution: tropical and subtropical.

*4. Cicadulina arachidis* China, 1928. An African species. A vector of groundnut rosette.

*5. Cicadulina bipunctella* (Mats., 1908). A species distributed in Japan (Kyushu), Formosa, Philippines, Micronesia, Australia, Africa, etc. and the type locality is Port Said, North Africa.

*5′. Cicadulina bipunctella bimaculata* (Evans, 1940). An Australian subspecies. A vector of maize wallaby ear.

*5″. Cicadulina bipunctella zeae* China, 1926. An East African subspecies, being found also in Cyprus and the Canary Islands. A vector of maize streak.

*6. Cicadulina latens* Fennah, 1960. An African species. A vector of maize streak.

*7. Cicadulina mbila* (Naudé, 1926). A species distributed in Africa and India. A vector of maize streak.

*8. Cicadulina parazeae* Ghauri, 1960. An African species. A vector of maize streak.

*9. Cicadulina similis* China, 1928. An African species. A vector of maize streak.

*10. Cicadulina storyi* China, 1936. An African species. A vector of maize streak.

GENUS 4. *Circulifer* Zak., 1935 [=*Distomotettix* Rib., 1938] (type, by original designation: *Jassus haematoceps* Muls. et R., 1855, a palaearctic species). This genus is primarily palaearctic and was introduced into North America.

*11. Circulifer tenellus* (Bak., 1896) [=*Thamnotettix rubicundula* VD., 1907; *T. ignavus* Mats., 1908; *T. indivisus* Haupt, 1927]. A species with holarctic distribution. Viruses transmitted; sugar beet curly-top and sowbane mosaic.

GENUS 5. *Colladonus* Ball, 1936 [=*Conodonus* Ball, 1936; *Friscanus* Ball, 1936, and *Hypospadianus* Rib., 1942] (type, by original designation: *Thamnotettix collaris* Ball, 1902, a nearctic species). Distribution: new world.

*12. Colladonus citellarius* (Say, 1830). A nearctic species. A vector of eastern X disease of peach.

*13. Colladonus commissus* (VD., 1917). A California species. A vector of Californian aster yellows.

*14. Colladonus flavocapitatus* (VD., 1890) [=*Colladonus eurekae* Bliven, 1954]. A nearctic species. A vector of Californian aster yellows.

*15. Colladonus germinatus* (VD., 1890). A nearctic species. Viruses transmitted: western X disease of peach, western X little cherry, yellow leaf role of peach, and Californian aster yellows.

*16. Colladonus intricatus* (Ball, 1911). A nearctic species. A vector of Californian aster yellows.

*17. Colladonus kirkaldyi* (Ball, 1911). A Californian species. A vector of Californian aster yellows.

*18. Colladonus montanus montanus* (VD., 1892). A nearctic species. This subspecies is closely related to the following subspecies, *C. montanus reductus,* but can be distinguished by its shorter pygofer spine and the presence of a distinct yellow or ivory spot on the clavii of the forewings [Nielson, M. W., 1957. Tech. Bull. U.S. Dep. Agr. No. 1156:35]. A vector of western X disease of peach.

*18′. Colladonus montanus reductus* (VD., 1917). Viruses transmitted by this subspecies: Californian aster yellows and yellow leaf roll of peach.

*19. Colladonus rupinatus* (Ball, 1911). A nearctic species. A vector of Californian aster yellows.

GENUS 6. *Dalbulus* DeL., 1950 (type, by original designation, *Deltocephalus elimatus* (Ball, 1900). Distribution: new world.

*20. Dalbulus elimatus* (Ball, 1900). A nearctic species. A vector of corn stunt.

*21. Dalbulus maidis* (DeL. et Wolc., 1923). A nearctic species. A vector of corn stunt.

GENUS 7. *Endria* Oman, 1949 (type, by original designation: *Jassus inimica* Say, 1830). Distribution: new world.

*22. Endria inimica* (Say, 1830) [=*Jassus sexpunctata* Prov., 1872]. A nearctic species. A vector of wheat striate mosaic.

GENUS 8. *Euscelidius* Rib., 1942 (type, by original designation: *Jassus* (*Athysanus*) *variegatus* Kirsch., 1858. Distribution: holarctic.

*23. Euscelidius variegatus* (Kirsch., 1858) [=*Athysanus irroratus* Scott, 1875, and 2 synonyms]. A European species, introduced into North America, where it is a vector of Californian aster yellows.

GENUS 9. *Euscelis* Brul., 1832 [=*Phrynomorphus* Curt., 1833; *Cono-*

*sanus* Osb. et Ball, 1902; *Metathysanus* Dahlbom, 1912] (type, by monotypy: *Euscelis lineolatus* Brul., 1832). Distribution: palaearctic region.

*24. Euscelis galiberti* Rib., 1952. A European species. A vector of green petal of strawberry.

*25. Euscelis lineolatus* Brul., 1832 [=*Amblycephalus nitidus* Curt., 1833; *A. irroratus* Curt., 1835; *A. agrestis* Msh., 1866; *Euscelis bilobatus* * Rib., 1952]. A palaearctic species. Viruses transmitted: clover witches' broom, phyllody of clover, and green petal of strawberry.

*26. Euscelis plebejus* (Fall., 1806) [=*Athysanus communis* Edw., 1888]. A palaearctic species. A vector of European clover stunt.

GENUS 10. *Exitianus* Ball, 1929 [=*Mimodrylix* Zak., 1935] (type, by original designation: *Cicadula extiosa* Uhl., 1880). Distribution: worldwide.

*27. Exitianus exitiosus* (Uhl. 1880). A nearctic species. A vector of potato curly dwarf.

GENUS 11. *Fieberiella* Sign., 1880 (type, by monotypy: *Selenocephalus florii* Stål, 1864). Distribution: holarctic.

*28. Fieberiella florii* (Stål, 1864). A European species, introduced into North America, where it transmits Californian aster yellows, eastern X disease of peach, and western X disease of peach.

GENUS 12. *Hishimonoides* Ish., 1965 (type, by original designation and monotypy: *Hishimonoides sellatiformis* Ish., 1965). Distribution: Japan.

*29. Hishimonoides sellatiformis* Ish., 1965. A Japanese species. A vector of mulberry dwarf.

GENUS 13. *Hishimonus* Ish., 1953 [type, by original designation: *Hishimonus disciguttus* Ish., (nec Walk. 1853) =*Thamnotettix sellatus* Uhl., 1896]. Members of this genus occur in Japan, the oriental region, Oceania, etc.

*30. Hishinonus sellatus* (Uhl., 1896). A Japanese species. A vector of mulberry dwarf.

*31. Hishimonus phycitis* (Dist., 1908). An Indian species. Viruses transmitted: little leaf of eggplant and tomato big bud.

GENUS 14. *Idiodonus* Ball., 1936 [=*Josanus* DeL., 1938; *Orolix* Rib., 1942] (type, by original designation: *Jassus kennicottii* Uhl., 1864, a nearctic species). Distribution: holarctic.

* According to Kramer (Oman, personal correspondence), this name was published by Ribaut in 1952 (Faun. de France, **57**:93) without having been described by Wagner in 1938 as shown by Ribaut and the species is nothing but a form of dimorphism of the present species.

*32. Idiodonus heidmanni* Ball, 1900. A nearctic species. A vector of Californian aster yellows.

GENUS 15. *Inazuma* * Ish., 1953 (type, by original designation and monotypy: *Deltocephalus dorsalis* Motsch., 1859). Distribution: palaearctic and oriental regions.

*33. Inazuma dorsalis* (Motsch., 1859) [=*Deltocephalus fulguralis* Mats., 1902]. A species distributed in Japan, Formosa, Malaya, India, Ceylon, Borneo, and Philippines. Viruses transmitted: rice dwarf and rice orange leaf.

GENUS 16. *Macrosteles* † Fieb., 1866. (type, by subsequent designation: *Cicada sexnotata* Fall., 1860). Distribution: worldwide.

*34. Macrosteles cristatus* (Rib., 1927). A European species. A vector of European aster yellows.

*35. Macrosteles fascifrons* (Stål, 1858) [=*M. divisus* Auct. (nec Uhl., 1877); *Cicadula pallida* Osb., 1915; *C. scripta* DeL., 1924; *C. warioni* Leth., 1878); *C. wilburni* Dorst, 1937, etc.]. A holarctic species. Viruses transmitted: aster yellows, flax crinkle, and oat blue dwarf.

*36. Macrosteles laevis* (Rib., 1927). A palaearctic species. A vector of European aster yellows.

*37. Macrosteles quadripunctulatus* (Kirsch., 1868) [=*Cicadula ramiger* Zak., 1933]. A palaearctic species. A vector of European aster yellows.

*38. Macrosteles sexnotatus* (Fall., 1806) [=*Jassus didymus* Mul. et R., 1855, etc.]. A holarctic species. A vector of European aster yellows.

*39. Macrosteles viridigriseus* (Edw., 1924). A palaearctic species. A vector of green petal of strawberry.

GENUS 17. *Nephotettix* Mats., 1902 (type, by subsequent designation: *Selenocephalus cincticeps* Uhl., 1896). Distribution: eastern areas of palaearctic region, oriental region, Ethiopian region, etc.

*40. Nephotettix apicalis* (Motsch., 1859) [=*Pediopsis nigromaculatus* Motsch., 1859; *Thamnotettix nigropicta* Stål, 1870]. A species distributed widely in Japan (Kyushu), Ryukyus, Formosa, China, Malaya, India, Ceylon, Philippines, Micronesia, Australia, and East and South Africa. Viruses transmitted: rice dwarf, rice yellow dwarf, rice transitory yellowing, and tungro of rice.

* The present genus is related to the genus *Recilia* Edw. 1932, but they are different in the shape of the crown and features of face. Whether they are congeneric or not will require further studies.

† In order to establish the actual vectors of plant viruses, the recorded species of the genus *Macrosteles* may have to be revised in the future.

*41. Nephotettix cincticeps* (Uhl., 1896). A species distributed in Japan, Korea, Manchuria, Ryukyus, and Formosa. Viruses transmitted: rice dwarf and rice yellow dwarf.

*42. Nephotettix impicticeps* Ish., 1964 [=*Cicada bipunctatus* Fab., 1802 (invalid preoccupied name)]. A species distributed in Japan (Shikoku and Kyushu), Ryukyus, Formosa, India, and Philippines. Viruses transmitted: rice yellow dwarf, rice transitory yellowing, and tungro of rice.

GENUS 18. *Nesophrosyne* Kirk., 1906 [*Orosius* Dist., 1918; *Nesaloha* Oman, 1943] (type, by original designation: *Eutettix perkinsi* Kirk., 1904 *). Distribution: oriental region, Australian, Polynesia, Micronesia, etc.

*43. Nesophrosyne lotophagorum* (Kirk., 1907) [=*Thamnotettix argentatus* Evans, 1938]. A species recorded from Malaya, Indonesia, Philippines, Australia, Polynesia, and Micronesia. Viruses transmitted: tomato big bud, tomato yellow dwarf, tobacco yellow dwarf, groundnut mosaic, and Australian lucerne witches' broom.

*44. Nesophrosyne orientalis* (Mats., 1914). A species recorded from Japan (Honshu), Ryukyus, and Formosa. A vector of pulse witches' broom.

*45. Nesophrosyne ryukyuensis* Ish., 1965. A Ryukyu species. A vector of sweet potato witches' broom.

GENUS 19. *Norvellina* Ball, 1931 (type, by original designation: *Eutettix mildradae* Ball, 1901). Distribution: nearctic region.

*46. Norvellina chenopodii* (Osb., 1923). A nearctic species. A vector of curly leaf of lambsquarters.

GENUS 20. *Osbornellus* Ball, 1932 (type, by original designation: *Scaphoideus auronitens* Prov., 1889, a nearctic species). Distribution: new world.

*47. Osbornellus borealis* DeL. et Mohr, 1936. A nearctic species. Viruses transmitted: western X disease of peach and yellow leaf roll of peach.

GENUS 21. *Paraphlepsius* Bak., 1897 [=*Pendarus* Ball, 1927] (type, by original designation: *Paraphlepsius ramosus* Bak., 1897, a nearctic species). Distribution: new world.

*48. Paraphlepsius apertinus* (Osb. et Lath., 1923). A nearctic species. A vector of Californian aster yellows.

GENUS 22. *Paratanus* Young, 1957 (type, by original designation: *Atanus exitiosus* Beam., 1943). Distribution: neotropical region.

* I am not convinced that the following 3 species are truly congeneric with this type species of the genus. It is possible that the genus is changed to *Orosius* treated here as a synonym.

*49. Paratanus exitiosus* (Beam., 1943). A neotropical species. A vector of beet yellow wilt.

GENUS 23. *Psammotettix* Haupt, 1929 [=*Ribautiellus* Zak., 1934] (type, by original designation: *Athysanus maritimus* Perris, 1857). Distribution: worldwide except Australian and the neotropical regions.

*50. Psammotettix striatus* (L., 1758) [=*Laevicephalus latipex* DeL. et David., 1935]. A holarctic species. A vector of winter wheat mosaic.

GENUS 24. *Scaphoideus* Uhl., 1889 (type, by subsequent designation: *Jassus immistus* Say, 1831, a nearctic species). Members of this genus occur in holarctic and oriental regions.

*51. Scaphoideus luteolus* VD., 1894 [=*S. baculus* DeL. et Mohr, 1936]. A nearctic species. A vector of elm phloem necrosis.

GENUS 25. *Scaphytopius* Ball, 1931 [=*Tumeus* DeL., 1943; *Hebenarus* DeL., 1944] (type, by original designation: *Platymetopius elegans* VD., 1890). Distribution: new world.

*52. Scaphytopius acutus* (Say, 1831) [=*Jassus modestus* Stål, 1854; *S. flamentus* DeL., 1945; *S. tenuis* DeL., 1945]. A nearctic species. Viruses transmitted: eastern X disease of peach and lucerne witches' broom.

*53. Scaphytopius delongi* Young, 1952. A nearctic species. A vector of Californian aster yellows.

*54. Scaphytopius irroratus* (VD., 1910). A nearctic species. A vector of Californian aster yellows.

*55. Scaphytopius magdalensis* (Prov., 1889) [=*Platymetopius carolinus* Lath., 1917; *P. obscurus* Osb., 1905; *P. vaccinii* DeL., 1945]. A nearctic species. A vector of blueberry stunt.

GENUS 26. *Scleroracus* VD., 1894 [=*Conogonus* VD., 1894; *Ophiola* Edw., 1922; *Omaniella* Ish., 1953] (type, by monotypy: *Athysanus anthracinus* 1894, a nearctic species). Distribution: holarctic region.

*56. Scleroracus flavopictus* (Ish., 1953). A Japanese species. Viruses transmitted: potato witches' broom and potato purple top.

*57. Scleroracus vaccinii* (VD., 1890) [=*Athysanus striatulus* Auct. (nec Fall. 1866)]. A nearctic species. A vector of cranberry false blossom.

GENUS 27. *Texananus* Ball, 1918 [type, by original designation: *Phlepsius* (*Texananus*) *mexicanus* Ball, 1918]. Distribution: new world.

In this genus, the following 6 species are known as vectors of Californian aster yellows:

*58. Texananus incurvatus* (Osb. et Lath., 1923).

*59. Texananus lathropi* (Bak., 1925) [=*Phlepsius annulatus* Osb. et Lath., 1923].

*60. Texananus latipex* DeL., 1943.
*61. Texananus oregonus* (Ball, 1931).
*62. Texananus pergradus* (DeL., 1938).
*63. Texananus spatulatus* (VD., 1892).

## IV. SUPERFAMILY FULGOROIDEA

### *A. Family 1. Cixiidae*

GENUS 1. *Hyalesthes* Sign., 1862 (type, by monotypy: *Hyalesthes obsoletus* Sign., 1862). Distribution: palaearctic region.

*1. Hyalesthes obsoletus* Sign., 1862. A European species. A vector of tomato big bud.

GENUS 2. *Oliarus* Stål, 1862 (type, by subsequent designation: *Cixius pallens* Germ. 1821, a palaearctic species). Members of this genus are almost worldwide in distribution.

*2. Oliarus atkinsoni* Myers, 1924. A species of New Zealand. A vector of *Phormium* yellow-leaf.

### *B. Family 2. Delphacidae*

GENUS 1. *Javesella* Fennah, 1963 (type, by original designation and monotypy: *Fulgora pellucida* Fab., 1794). Only the type species is known in this genus.

*1. Javesella pellucida* (Fab., 1794) [=*Fulgora marginata* Fab., 1794; *Delphax suturalis* Curt., 1837; *Delphax dubia* Kirsch., 1868, etc.]. A holarctic species. A vector of wheat streak.

GENUS 2. *Laodelphax* Fennah, 1963 [=*Callidelphax* Wagner, 1963] (type, by original designation and monotypy: *Delphax striatella* Fall., 1826). Only the type species is known in this genus.

*2. Laodelphax striatellus* (Fall., 1826) [=*Delphax notula* Stål, 1854; *Liburnia devastans* Mats., 1900; *Liburnia haupti* Lindberg, 1936, etc.]. A Papaearctic species, being distributed in some areas in tropics. Viruses transmitted: rice stripe, rice black-streaked dwarf, northern cereal mosaic, and oat pseudorosette.

GENUS 3. *Muellerianella* Wag., 1963 (type, by original designation: *Delphax fairmairei* Perris, 1857). Distribution: palaearctic region.

*3. Muellerianella fairmairei* (Perris, 1857). A palaearctic species. A vector of northern cereal mosaic.*

GENUS 4. *Nilaparvata* Dist., 1906 [=*Hikona* Mats., 1935; *Kalpa* Dist., 1906, etc.] (type, by original designation and monotypy:

* Ishii, T. (personal correspondence, 1966) states that this species, identified by me, is also a vector of northern cereal mosaic virus.

*Nilaparvata greeni* Dist., 1906, which is a synonym of *Delphax lugens* Stål, 1854). Distribution: almost worldwide.

*4. Nilaparvata lugens* (Stål, 1854) [=*Delphax oryzae* Mats., 1917; *Nilaparvata greeni* Dist., 1906, and 6 other synonyms]. A species distributed widely in Japan, the oriental region, Micronesia, Australia, etc. A vector of rice grassy stunt.

GENUS 5. *Peregrinus* Kirk., 1904 (type, by original designation and monotypy: *Delphax maidis* Ashm., 1890). Only the type species is known in this genus.

*5. Peregrinus maidis* (Ashm., 1890). Distribution: tropics and subtropics. A vector of maize mosaic virus.

GENUS 6. *Perkinsiella* Kirk., 1903 (type, by original designation: *Perkinsiella saccharicida* Kirk., 1903). Distribution: tropics and subtropics.

*6. Perkinsiella saccharicida* Kirk., 1903. A species distributed in Ryukyus, Formosa, Malaya, Java, Hawaii, Fiji, New Guinea, Australia, and South Africa. A vector of sugarcane Fiji disease virus.

*7. Perkinsiella vastatrix* Breddin, 1896. Distribution: Formosa, Malaya, Java, Borneo, Celebes, Philippines, New Guinea, etc. A vector of sugarcane Fiji disease virus.

GENUS 7. *Ribautodelphax* Wag., 1962 (type, by original designation: *Delphax collina* Boheman, 1849, a palaearctic species). Distribution: palaearctic region.

*8. Ribautodelphax albifascia* (Mats., 1900). Distribution: Japan and Siberia. A vector of northern cereal mosaic.

GENUS 8. *Sogatodes* Fennah, 1963 (type, by subsequent designation: *Sogatodes molinus* Fennah, 1963, a Mexican species). Distribution: new world.

*9. Sogatodes orizicola* * (Muir, 1926). Distribution: New world. A vector of hoja blanca of rice.

GENUS 9. *Unkanodes* Fennah, 1956 (type, by original designation

* Muir's trivial name, *orizicola,* was given apparently to describe a rice plant habitant (=*Oryza-cola*). I proposed to amend *orizicola* to *oryzicola* at the U.S.–Japan Cooperative Seminar (Tokyo, 1965). When there is clear evidence of an inadvertent error, the amendation is justified under Article 32 of the International Code of Zoological Nomenclature.

Recently Oman, having discussed the point with Sabrowsky, informed me of his opinion,

> "I believe most systematic biologists consider it to be the prerogative of the original authors of the names to depart from the strict rules for transliteration and name formation if they so desire. I believe Muir exercised that prerogative in proposing *orizicola*."

In this paper I followed Sabrowsky's and Oman's opinion.

and monotypy: *Unkana sapporona* Mats., 1935). Only the type species is known in this genus.

*10. Unkanodes sapporonus* (Mats., 1935). A species distributed in Japan, Formosa, and China. Viruses transmitted: rice stripe, rice black-streaked dwarf, and northern cereal mosaic.

## V. VIRUSES AND VECTORS

The plant-pathogenic viruses and their leafhopper vectors are listed below according to families of plants harboring the viruses.

### *A. Gramineae*

Corn stunt: *Dalbulus elimatus* (Ball); *D. maidis* (DeL et Wolc.)

Hoja blanca of rice: *Sogatodes orizicola* (Muir)

Maize mosaic: *Peregrinus maidis* (Ashm.)

Maize mottle: *cicadulina mbila* (Naudé)

Maize streak: *Cicadulina bipunctella zeae* (China); *C. latens* Fennah; *C. mbila* (Naudé); *C. parazeae* Ghauri; *C. storeyi* China.

Maize wallaby ear: *Cicadulina bipunctella* bimaculata (Evans)

Northern cereal mosaic: *Laodelphax striatellus* (Fall.); *Ribautodelphax albifascia* (Mats.); *Unkanodes sapporonus* (Mats.)

Oat blue dwarf: *Macrosteles fascifrons* (Stål)

Oat pseudorosette: *Laodelphax striatellus* (Fall.)

Rice black-streaked dwarf: *Laodelphax striatellus* (Fall.); *Unkanodes sapporonus* (Mats.)

Rice dwarf: *Inazuma dorsalis* (Motsch.); *Nephotettix apicalis* (Motsch.); *N. cincticeps* (Uhl.)

Rice grassy stunt: *Nilaparvata lugens* (Stål)

Rice orange leaf: *Inazuma dorsalis* (Motsch.)

Rice stripe: *Laodelphax striatellus* (Fall.); *Unkanodes sapporonus* (Mats.)

Rice transitory yellowing: *Nephotettix apicalis* (Motsch.); *N. cincticeps* (Uhl.)

Rice yellow dwarf: *Nephotettix apicalis* (Motsch.); *N. cincticeps* (Uhl.); *N. impicticeps* Ish.

Sugarcane chlorotic streak: *Draeculacephala portola* Ball

Sugarcane Fiji disease: *Perkinsiella saccharicida* Kirk.; *P. vastatrix* Breddin

Tungro of Rice: *Nephotettix apicalis* (Motsch.); *N. impicticeps* Ish.

Uba cane: *Cicadulina mbila* (Naudé)

Wheat streak: *Javesella pellucida* (Fab.)

Winter wheat mosaic: *Psammotettix striatus* (L.)

*B. Liliaceae*

Phormium yellow leaf: *Oliarus atkinsoni* Meyers

*C. Ulmaceae*

Elm phloem necrosis: *Scaphoideus luteolus* VD.

*D. Moraceae*

Mulberry dwarf: *Hishimonoides sellatiformis* Ish.; *Hishimonus sellatus* (Uhl.)

*E. Santalaceae*

Sandal spike disease: *Coelidia indica* Walk.; *Moonia albomaculata* Dist.

*F. Chenopodiaceae*

Argentine beet curly-top: *Agalliana ensigera* Oman
Beet curly-top: *Circulifer tenellus* (Bak.)
Beet yellow wilt: *Paratanus exitiosus* (Beam.)
Brazilian beet curly-top: *Agallia albidula* Uhl.
Curly leaf of lambsquarters: *Norvellina chenopodii* (Osb.)
Sowbane mosaic: *Circulifer tenellus* (Bak.)

*G. Rosaceae*

Eastern X disease of peach: *Colladonus citellarius* (Say); *C. germinatus* (VD.); *Fieberiella florii* (Stål); *Gyponana striata* (Burm.)
Green petal of strawberry: *Euscelis galiberti* Rib.; *E. lineolatus* Brul.; *Macrosteles viridigriseus* (Edw.)
Peach yellows: *Philaenus spumarius* (L.)
Phoney peach: *Graphocephala versuta* (Say); *Homalodisca insolita* (Walk.); *H. triquetra* (Fab.); *Oncometopia undata* (Fab.)
Rubus stunt: *Macropsis fuscula* (Zett.)
Western X disease of peach: *Colladonus germinatus* (VD.); *C. montanua montanus* (VD.); *Fieberiella florii* (Stål); *Keonolla confluens* (Uhl.); *Osbornellus borealis* DeL. et Mohr.
Western X little cherry: *Colladonus germinatus* (VD.)
Yellow leaf of peach: *Colladonus germinatus* (VD.); *C. montanus reductus* (VD.); *Osbornellus borealis* DeL. et Mohr

*H. Leguminosae*

Australian lucerne witches' broom: *Batrachomorphus punctatus* (Evans); *Nesophrosyne lotophagorum* (Kirk.)

Clover big vein (wound tumor): *Agallia constricta* (VD.); *A. quadripunctata* (Prov.); *Agalliopsis novella* (Say)

Clover club leaf: *Agalliopsis novella* (Say)

Clover witches' broom: *Euscelis lineolatus* Brul.

European clover stunt: *Euscelis plebejus* (Fall.)

Groundnut mosaic: *Nesophrosyne lotophagorum* (Kirk.)

Groundnut rosette: ? *Cicadulina arachidis* China; ? *C. similis* China.

Lucerne dwarf (=Pierce's disease of grapes)

Lucerne witches' broom: *Scaphytopius acutus* (Say)

Phyllody of clover: *Aphrodes albifrons* (L.); *Euscelis lineolatus* Brul.

Pulse witches' broom: *Nesophrosyne orientalis* (Mats.)

### *I. Linaceae*

Flax crinkle: *Macrosteles fascifrons* (Stål)

### *J. Vitaceae*

Pierce's disease of grapes (=lucerne dwarf): *Aphrophora angulata* Ball; *A. permutata* Uhl.; *Carneocephala flaviceps* (Riley); *C. fulgida* Nott.; *C. triguttata* Nott.; *Clastoptera brunnea* Ball; *Cuerna costalis* (Fab.); *C. yuccae* Oman et Beam.; *Draeculacephala crassicornis* VD.; *D. inscripta* VD.; *D. minerva* Ball; *D. noveborensis* (Fitch); *D. portola* Ball; *Graphocephala cythura* (Bak.); *Holochara delta* Oman; *Homalodisca liturata* (Ball); *Hordnia circellata* (Bak.); *Keonolla confluens* (Uhl.); *K. dolobrata* (Ball); *Neokolla hieroglyphica* (Say); *N. severini* DeL.; *Oncometopia undata* (Fab.); *Pagaronia confusa* Oman; *P. furcata* Oman; *P. tridecimpunctata* Ball; *P. triundata* Ball; *Philaenus spumarius* (L.)

### *K. Malvaceae*

Cotton leaf curl: *Empoasca devastans* Dist.

### *L. Caricaceae*

Papaya bunchy top: *Empoasca dilitare* DeL. et David.; *E. papayae* Oman

### *M. Ericaceae*

Blueberry stunt: *Scaphytopius magdalensis* (Prov.)

Cranberry false blossom: *Scleroracus vaccinii* (VD.)

### *N. Convolvulaceae*

Sweetpotato witches' broom: *Nesophrosyne ryukyuensis* Ish.

### *O. Solanaceae*

Brazilian tomato curly-top: *Agalliana ensigera* Oman
Datura rúgose leaf curl: *Austroagallia torrida* Evans
Leaf crinkle of tomato: *Agallia venosa* (Fall.)
Little leaf of eggplant: *Empoasca devastans* Dist.; *Hishimonus phycitis* (Dist.)
Potato curly dwarf: *Exitianus exitiosus* (Uhl.)
Potato purple top: *Scleroracus flavopictus* (Ish.)
Potato yellow dwarf (New Jersey strain): *Agallia constricta* VD.; *A. quadripunctata* (Prov.); *Agalliopsis novella* (Say)
Potato yellow drawf (New York strain): *Aceratagallia curvata* Oman; *A. longula* (VD.); *A. obscura* Oman; *A. sanguinolenta* (Prov.); *Agallia quadripunctata* (Prov.); *Agalliopsis novella* (Say); ? *Empoasca fabae* (Harris)
Potato witches' broom: *Peragallia sinuata* (Muls. et R.) [in Europe]; *Scleroracus flavopictus* (Ish.) [in Japan]
Southern stolbur of tomato: *Aphrodes bicinctus* (von Schrank)
Tomato big bud: *Empoasca devastans* Dist.; *Hishimonus phycitis* Dist.; *Hyalesthes obsoletus* Sign.; *Nesophrosyne lotophagorum* (Kirk.); *Peragallia sinuata* (Muls. et R.)
Tomato yellow dwarf: *Nesophrosyne lotophagorum* (Kirk.)

### *P. Compositae*

Aster stunt (yellow): *Aphrodes bicinctus* (von Schrank)
"Aster yellows": *Macrosteles fascifrons* (Stål)
Californian aster yellows: *Acinopterus angulatus* VD.; *Chlorotettix similis* DeL.; *Colladonus commissus* (VD.); *C. flavocapitatus* (VD.); *C. intricatus* (Ball); *C. kirkaldyi* (Ball); *C. montanus reductus* (Ball); *C. rupinatus* (Ball); *Euscelidius variegatus* (Kirsch.); *Fieberiella florii* (Stål); *Gyponana hastata* DeL.; *Idiodonus heidmanni* Ball; *Paraphlepsius apertinus* (Osb. et Lath.); *Scaphytopius delongi* Young; *S. irroratus* (VD.); *Texananus incurvatus* (Osb. et Lath.); *T. lathropi* (Bak.); *T. latipex* DeL.; *T. oregonus* (Ball); *T. pergradus* (DeL.); *T. spatulatus* (VD.)
European aster yellows: *Macrosteles cristatus* (Rib.); *M. laevis* (Rib.); *M. quadripunctatus* (Kirsch.); *M. sexnotatus* (Fall.)

## ACKNOWLEDGMENTS

I am indebted to Dr. Karl Maramorosch, Dr. Paul Oman, and Dr. J. P. Kramer of the United States and Dr. J. W. Evans of Australia, who helped me with references or gave valuable and friendly advice.

## BIBLIOGRAPHY *

Day, M. F., and Bennett, M. J. 1954. A review of problems of specificity in arthropod-vectors of plant and animal viruses. Commonwealth of Australia, Canberra. 172 p.

Heinze, K. 1951. Die Überträger pflanzlicher Viruskrankheiten. Mitt. Biologisch. Zentralanst. für Land-u. Forstwirtschaft, Heft 71, Berlin. 126 p.

Ishihara, T. 1965. Taxonomic position of some leafhoppers known as virus-vectors. Special Pub. Oct. 1965 for U.S.–Japan Cooperative Science Program Conference. 16 pp.

Ishihara, T., and Nasu, S. 1966. Leafhoppers transmitting plant-viruses in Japan and adjacent countries. Papers presented at the Div. Meeting on Plant Protection, 11th Pacific Sci. Congress, p. 159–170.

Nielson, M. W. 1962. A synonymical list of leafhopper vectors of plant viruses (Hemiptera: Cicadellidae). USDA, ARS-33-74. 12 p. Oman, P. 1949. The nearctic leafhoppers. Entomol. Soc. Washington. Washington, D.C. 253 p.

Ruppel, R. F. 1965. A revision of the genus *Cicadulina*. Pub. Museum Michigan State Univ. Biol. Ser. 8 (2):385–428.

* Excepting those concerned exclusively with taxonomy.

# Nonpropagative Leafhopper-Borne Viruses

K. C. Ling

*International Rice Research Institute,*
*Los Baños, Laguna, Philippines*

## I. INTRODUCTION

The nonpropagative leafhopper-borne viruses may be defined as those plant viruses that are transmitted by leafhopper vectors but the virus does not multiply in the vector, or even if the virus does multiply, it does not multiply sufficiently to maintain the infectivity of the insect. Few plant viruses of this group are known. However, their virus–vector relationship is different from that of the propagative leafhopper-borne viruses. Because information is limited, the present discussion is largely confined to curly-top virus and rice tungro virus. It may serve as a reminder of the statement (Black, 1959):

> "It now seems that the probability of virus multiplication in certain vectors of plant viruses is sometimes too readily accepted and that ideas that were advanced during the controversy over the problem are sometimes forgotten. Some of them may, however, merely be in eclipse and will prove eventually to have a place in the total virus picture."

## II. EVIDENCE FOR MULTIPLICATION OF VIRUS IN LEAFHOPPERS

Considering the evidence for multiplication of plant viruses in their leafhopper vectors is a necessary basis for dealing with the nonpropagative ones. The concept of plant viruses multiplying in their leafhopper vectors was initiated by Kunkel (1926), although Smith and Boncquet (1915) had mentioned earlier that some development or change of the curly-top virus took place within the body of beet leafhopper during the first few hours after it fed upon a diseased plant. The evidence for Kunkel's conclusion that, "it seems probable that the incubation period of aster yellows in *Cicadula sexonotata* (Fall.) is due to a development and multiplication of the causative agent in some tissue of the leafhopper," is (*1*) a period of at least 10 days is required for an aster leafhopper to become infective after feeding on a plant with yellows; (*2*) once the insect becomes infective, it remains so for the rest of its life; and (*3*) the insect does not lose its infectivity after molting. Direct and substantial evidence for the propagative relationship between some plant viruses and their leafhopper vectors has subsequently been accumulated. The multiplication of some plant viruses in their leafhopper vectors is now generally accepted. Comprehensive discussions and extensive reviews of this subject have been published by various authors (Bawden, 1964; Black, 1953a, 1953b, 1954, 1959, 1962; Carter, 1962; Leach, 1940; Maramorosch, 1955b, 1963, 1964; Smith and Brierley, 1956; Smith, 1958, 1965; Storey, 1939).

There are four approaches to the problem of multiplication of viruses in their leafhopper vectors. They are: (*1*) direct observation of clusters of virus particles in tissues of viruliferous insect; (*2*) determination of infectivity of the insect after the virus concentration is diluted beyond the maximum possible without multiplication in the insects; (*3*) measurement of the rate of increase of virus in the vector; and (*4*) examination of the length of incubation period in relation to various dilutions of virus or to different conditions for regulating their rate of multiplication. Transovarial passage and serial injection passage are two virus dilution methods.

In addition to the evidence mentioned above, *i.e.*, a definite long incubation period, a long virus retention period, and infectivity remaining after molting, there is some critical evidence for the multiplication of the virus in the leafhopper vector. First, there is transovarial passage of virus from generation to generation of the vectors (Fukushi, 1933, 1940; Black, 1950). Second, there is evidence that the virus is carried in serial passages from insect to insect with the inoculum adequately diluted at each passage (Maramorosch, 1952; Black and Brakke, 1952). Third, there is the presence of clusters of virus particles in tissues of viruliferous insects demonstrated by examination of ultrathin sections of insects under the electron microscope (Fukushi *et al.*, 1960; Shikata, 1966; Shikata *et al.*, 1964). Furthermore, Whitcomb and Black (1961) measured the rate of virus increase in the insect by assaying the soluble antigen. Other evidence, such as cross protection between strains of virus in the vector (Kunkel, 1955), effect of heat on incubation period (Kunkel, 1937, 1941), influence of temperature on incubation period (Maramorosch, 1950), effect of dosage (Maramorosch, 1950), effect of volume on incubation (Maramorosch, 1953), etc. also support the probability that multiplication occurs.

The only evidence from the literature in support of the curly-top virus multiplying in *Circulifer tenellus* is a longer incubation period in the insect being associated with a lower concentration of virus injected mechanically into the vector (Maramorosch, 1955a). However, this might be due to the mechanical movement of the virus through intervening fluids and tissues and be unrelated to virus increase in the vector (Bennett, 1962). It is obvious that the curly-top virus must remain in the nonpropagative category unless serial transfer of the virus from insect to insect can be achieved (Maramorosch, 1955b, 1964; Bawden, 1964).

## III. EVIDENCE FOR ABSENCE OF VIRUS MULTIPLICATION IN LEAFHOPPERS

To prove that a virus does not multiply in a leafhopper vector is more difficult than to demonstrate the opposite, especially when the virus passes internally through the insect body or if it is unstable. As Black (1953a) stated: "there may always be an element of uncertainty about those viruses which the evidence indicates do not multiply in the vector, because such evidence is, of necessity, negative." However, the evidence opposing the assumption of multiplication of virus in the leafhopper could be considered as a possible indication of the absence of multiplication. Some of the evidence for propagative viruses may not prove multiplication unquestionably unless it is certain that the evidence is not subject to the interpretations. Nevertheless, there are certain criteria which any evidence against the multiplication of the virus in the vector must meet. Black (1959) pointed out that the most critical are (*1*) the virus content decreases progressively and consistently, following termination of the acquisition feeding period; (*2*) transmission by single vectors consistently decreases, following the termination of certain appropriate acquisition feeding periods; and (*3*) transmission ability of individual leafhoppers is roughly proportional to the length of the acquisition feeding period. In addition, there is a short latent period of a few hours, in striking contrast to a long incubation period of several days. However, whether short and long incubation periods are indicative of the absence and presence of multiplication, respectively, cannot be determined until critical evidence on more cases has been accumulated (Black, 1954). A short latent period would neither prove nor disprove the existence of multiplication unless the speed of multiplication in the insect vector were known. On the other hand, after the insect has acquired the virus, a long incubation before the insect becomes infective would not be necessary unless the virus had first to be multiplied in the vector.

There is therefore a greater likelihood of nonpropagative viruses having a short incubation period.

Loss of infectivity following a molt has been considered a critical criterion for nonpersistent aphid-borne viruses (Day and Venables, 1961), probably because if the insect loses infectivity after molting, the virus seems most unlikely to multiply inside the insect. The phenomenon indicates lack of multiplication unless the loss of infectivity is assumed to be a reflection of the insect itself becoming incapable of transmitting the virus after molting. Consequently, the loss of infectivity after molting may be regarded as a criterion for nonpropagative leafhopper-borne viruses.

## IV. FEATURES OF CURLY-TOP AND TUNGRO VIRUSES AND THEIR RESPECTIVE VECTORS

### *A. Transmission*

Under natural conditions nonpropagative leafhopper-borne viruses are transmitted mainly by leafhopper vectors. The sugar beet curly-top virus is transmitted by *Circulifer tenellus* (Baker) (Smith and Boncquet, 1915) and the rice tungro virus by *Nephotettix impicticeps* Ishihara (International Rice Research Institute, 1963; Fajardo *et al.*, 1964; Rivera and Ou, 1965). But neither one is transmitted transovarially. This is probably why Stahl and Carsner (1918) were able to develop a method of obtaining noninfective beet leafhoppers by transferring newly hatched nymphs with a camel's hair brush from diseased to healthy beet plants. There is no direct evidence thus far to indicate that either one of them is transmitted through the soil (Severin, 1924; Fajardo *et al.*, 1964) or through seeds (Fajardo *et al.*, 1964). Severin (1921), for example, grew a total of 22,738 beets from the seeds of diseased plants but not a single case of disease developed. However, typical curly top has been readily produced in healthy plants by grafting into the top of the root pieces of an infected plant with the growing buds attached (Smith and Boncquet, 1915). The virus has also been transmitted to healthy beet via three species of dodder, *Cuscuta subinclusa*, *C. campestris*, and *C. californica* (Johnson, 1941; Bennett, 1944). Severin (1924) demonstrated that juice, pressed from the leaves and roots of curly-top beets and inoculated into the crown of healthy beets by repeated punctures with insects pins, caused typical curly-top symptoms in 9 out of 100 beets. Bennett (1934) also succeeded in transmitting curly-top virus mechanically by using as inoculum phloem exudate from the cut surfaces of diseased beets.

### B. *Latent or Incubation Period in the Vector*

A latent or incubation period in the insect refers to the time elapsing between the insect acquiring the virus and the insect becoming infective. This period has been known since 1915 when Smith and Boncquet (1915) pointed out that the beet leafhopper did not transfer the pathogenic factor mechanically and that an incubation period in the insect body was necessary. The existence of an incubation period has stimulated the many investigations which have been conducted into the possibility of virus multiplication in the vector. This period has been explained for circulative virus as being the time required for circulation or passage of virus inside the insect body. However, there is practically no latent period required by most stylet-borne aphid-transmitted viruses.

An incubation period of no more than 48 and possibly no more than 24 hr for curly-top virus in the beet leafhopper was estimated by Smith and Boncquet (1915). Later, Carsner and Stahl (1924) demonstrated that a single insect was able to transmit curly top within a period of $21\frac{3}{4}$ hr. However, they pointed out that the number of insects requiring longer than 24 hr was greater than the number requiring a shorter period to become infective. Severin (1931) showed that a single leafhopper could transmit curly top in 7 hr (6 on diseased beet and 1 on a healthy plant). Bennett and Wallace (1938), however, were able to demonstrate transmittal by an individual leafhopper in 4 hr. They achieved this by transferring insects successively and individually at 1-hr intervals after allowing an acquisition feeding period of 1 hr. They found also that beet leafhoppers could acquire the virus from a diseased plant in 1 min and viruliferous insects could infect healthy plants during a feeding period of 1 min.

By confining a group of insects to each seedling, Severin (1921) originally reported a minimum latent period of 4–6 hr. Later he (Severin, 1931) succeeded in transmitting curly top within 20 min to 1 of 41 healthy seedlings with a group of 40 previously noninfective leafhoppers. Results were negative, however, if beet seedlings were inoculated with 6–21 insects which had acquisition and inoculating feeding periods of 5 and 10 min, respectively (Severin, 1921).

Swezy (1930) proposed that two conditions might be responsible for the occasional transmission of the curly-top disease at short time intervals: (*1*) an abnormal state of the alimentary canal, consisting of a clump of bacteria in the lumen of the esophagus anterior to the esophageal valve, might hinder the free passage of food which would consequently be ejected shortly after consumption; and (*2*) the infective principle may, in rare instances, pass so promptly through the body of the insect that it reaches the salivary glands in an unchanged condition and is immediately ready to infect new plants on ejection with the saliva.

Rivera and Ou (1965) reported that a 24-hr incubation period appeared to be necessary for insects to transmit the tungro virus to an appreciable extent. However, they succeeded in transmitting the tungro virus by insects, following an acquisition feeding period of 30 min, and concluded that the minimum inoculation feeding period was 15 min. Later, Ling (1966) transmitted tungro virus within a period of 2 hr, using individual insects that were originally virus free (Table 1). No definite incubation period of the tungro virus in its vector could be shown because once the insects had become infective, they transmitted the disease agent immediately after acquisition. When the acquisition feeding period was shortened, the insects never become infective. However, one of 5 infective insects was noninfective during the first 1-hr period after the end of acquisition feeding but transmitted the virus during the next hourly interval (Table 1). This failure to transmit the disease during the first hour could be interpreted in terms of insect feeding behavior, variation in susceptibility of the host plant, etc. and is unlikely to indicate a definite incubation period since the other four insects transmitted the disease agent during the first hour. Nevertheless, if there was a definite latent period in the insect, it could not have been longer than 2 hr. No positive results

TABLE 1

SERIAL TRANSMISSION OF THE TUNGRO VIRUS OF RICE BY INDIVIDUAL LEAFHOPPERS OF *Nephotettix impicticeps* GIVEN AN ACQUISITION FEEDING PERIOD OF 1 HR (after Ling, 1966)

| Hours after acquisition feeding | Insect number [a] | | | | | |
|---|---|---|---|---|---|---|
| | 1 | 2 | 3 | 4 | 5 | 6–30 |
| 0–1 | + | + | + | + | – | – (25) |
| 1–2 | – | – | – | + | + | – (25) |
| 2–3 | D | – | – | + | + | – (25) |
| 3–4 | | – | – | + | + | – (25) |
| 4–24 | | – | – | – | – | – (25) |
| 24–48 | | – | – | – | – | – (23) |

[a] +, positive transmission; —, negative transmission; D, insect died. Numbers in parenthesis are numbers of insects tested.

demonstrated the successful transmission by a single insect, or a group of 10 insects, within a period of 2 min.

The incubation periods of viruses in leafhoppers are often longer than 1 day. However, Storey (1928) reported that the minimum incubation period for maize streak virus in *Cicadulina mbila* was 6 hr. Severin (1949, 1950) demonstrated that, by using 5–20 insects per plant, the minimum latent periods for the virus of Pierce's disease were 2, 2, 7, and 2–6 hours in red-headed sharpshooter (*Carneocephala fulgida*), blue-green sharpshooter (*Neokolla circellata*), green sharpshooter (*Draeculacephala minerva*), and four varieties of meadow spittle insect (*Philaenus leucophthalmus*), respectively. Further evidence is essential before we can conclude whether these two viruses do or do not multiply in the leafhoppers. However, Bawden (1964) pointed out that all the evidence from work with maize streak is against virus multiplication in its vectors.

### *C. Gradual Decrease in Vector Infectivity with Time*

If the virus does not multiply in the insect, the gradual decrease of infectivity of the insect with time after acquisition feeding should be explicable on the basis of the virus supply in the insect available for infecting plants being progressively exhausted after a single acquisition feeding on a virus source. But if the virus particles are unchanged in the insect body, the decreasing rate may be varied, depending upon the amount of virus picked up by the insect and the amount of virus released from it. To determine the gradual decrease of infectivity and the retention period of the virus in the vector, the necessity should be considered of preventing the insect reacquiring the virus from susceptible plants exposed to the viruliferous insect. Two techniques have generally been applied. One is to rear the infective insect on plants immune to the virus; the other is to transfer the insect to a fresh healthy seedling before the insect can pick up the virus from the plant after infecting. Transferring the insect at a 1-day interval is commonly practiced.

In the case of beet curly top, Freitag (1936) found that the average percentages of beet infected were 26.5, 15.0, 5.6, 3.9, 0.5, and 0.0%, respectively for each successive 30-day period from 1 to 180 days after acquisition feeding by adult insects. The corresponding figures were 31.9, 14.8, 5.2, 1.2, 0.0, and 0.8% for insects fed during nymphal stages on a diseased beet. Consequently, he concluded that there was a gradual decrease in the percentage of beets infected by adult leafhoppers when the insects were transferred daily to a healthy beet.

Similar results have also been demonstrated by Bennett and Wallace (1938).

In addition to examining the gradual decrease of infectivity, Bennett and Wallace (1938) applied the bioassay technique for determining the relative concentration of the curly-top virus in the insect. They concluded that the virus content gradually decreased over a period of 8–10 weeks in viruliferous leafhoppers when they were confined to a very resistant or immune plant after acquisition feeding. This was true, regardless of the size of the initial charge of virus in the leafhoppers. This seems to be the most significant evidence against multiplication of curly-top virus in its vector. Even if multiplication does occur, it is insufficient to maintain the original content.

The gradual decrease of infectivity after acquisition feeding by leafhoppers transmitting the tungro virus is more striking than in the case of curly top. The percentages of infective insects were 46.6, 21.4, 7.0, 1.9, 0.2, and 0.0% for 1, 2, 3, 4, 5, and 6 days after the termination of acquisition feeding. Ling (1966) concluded from the following evidence that the gradual decrease of infectivity was associated with time. (*1*) Although a pattern of intermittent transmission occurred, if the insects were transferred consecutively at an interval of 1 hr, there was a tendency for infectivity to gradually decrease with time in hours after acquisition feeding (Fig. 1). (*2*) Insects were able to transmit virus to seven seedlings during a 24-hr period and still remain as active as those transferred to only one seedling. (*3*) The vectors lost

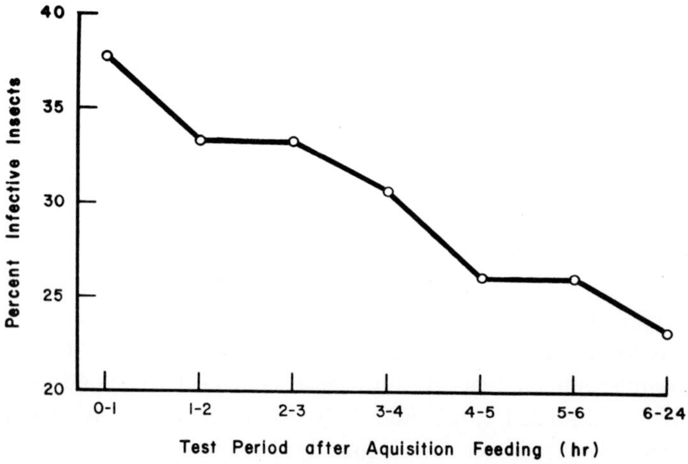


*Fig. 1.* Hourly loss of infectivity of a group of 69 *Nephotettix impicticeps* transmitting the tungro virus of rice following acquisition feeding.

infectivity gradually, even without feeding on seedlings. Therefore, the number of seedlings exposed to the insect was not the critical factor causing decrease of infectivity, the gradual loss of infectivity being related to the elapsed time from acquisition feeding.

### *D. Virus Retention in the Insect*

The length of time over which insects remain infective following a single feeding on a virus source varies among individuals and among viruses. Leafhopper-borne viruses frequently have long retention periods, often extending throughout their lives. Although long retention periods may be taken to indicate the multiplication of virus in the vector, they may also be explained by an ability of the virus to remain active in the body of the leafhopper for long periods and by a slow rate of virus loss from the vector (Bennett and Wallace, 1938). On the other hand, a short retention period may be taken as evidence against multiplication, assuming that the loss of infectivity is caused by an inadequate virus supply rather than some other factor.

The reported retention period of curly-top virus in beet leafhopper varies between authors. Boncquet and Stahl (1917) asserted that the retention period was 15–35 days. Other authors later demonstrated periods of 58–111 (Carsner, 1919), 97–104 (Severin, 1924), 1–161 (Freitag, 1936), and 92–121 days (Wallace and Murphy, 1938). However, most reports have indicated periods longer than 100 days.

The tungro virus retention period in the vector is much shorter than that of curly-top virus. Out of 243 infective rice green leafhoppers, one retained infectivity for 5 days after acquisition feeding, and this is the longest retention period of tungro virus thus far recorded (Ling, 1966).

### *E. Transmission Pattern*

Assuming that the susceptibility of individual plants is identical, a transmission pattern of a virus by its insect vector could be determined by transferring the infective insect successively to a series of healthy plants at a constant time interval. The transmission pattern of leafhopper-borne viruses is intermittent in most cases, so that some plants often remain uninfected after having been exposed to the insect. The uninfected plants are scattered irregularly in the series of plants inoculated by that insect. The transmission pattern by beet leafhopper of the curly-top virus, before the insect loses infectivity completely, is intermittent, regardless of whether the time interval is 1 hr (Table 4 of Bennett and Wallace, 1938) or 1 day (Figs. 3–5 and 7 of Freitag, 1936; Table 8 of Bennett and Wallace, 1938). However, the transmission pattern by rice green leafhopper of tungro virus depends upon the time interval. If it is 1 hr, the pattern is intermittent; if it is 1 day, the pattern is consecutive except for a few cases (4 out of more than 300 infective insects tested). The consecutive pattern indicates not only that the insect transmits the virus consecutively every day as long as it remains infective, but also that once the insect ceases to transmit, it remains noninfective until death (Ling, 1966).

### *F. Effect of Length of Acquisition Feeding Period on the Infective Capacity of the Vector*

If the virus does not multiply in the insect, the infective capacity of an insect is affected by the initial charge of the virus which can be at least experimentally regulated by the length of the acquisition feeding period. In other words, within the maximum charge of the virus, the infective capacity should be increased with the length of acquisition feeding period. Three kinds of data could be used for comparing the differences in infective capacity between insects after acquisition feeding on diseased plants for various lengths of time: (*1*) the number of infections produced by the insects if the transmission pattern is intermittent; (*2*) the number of infective insects if the transmission pattern is consecutive; and (*3*) the length of retention period. Among these, the difference in retention period seems to offer the most significant evidence.

Beet leafhoppers which had fed for only a short period on a curly-top beet were capable of producing an average of only 3.4 infections when transferred daily during adult life to successive healthy beets, whereas insects fed for long periods caused an average of 15.6 infections (Freitag, 1936). When adults or nymphs were fed for longer periods on diseased beets, they remained highly infective for a longer period than those that had been fed for short periods (Freitag, 1936). Bennett and Wallace (1938) came to the same conclusion that decreasing the size of the initial virus charge definitely shortens the time over which the leafhoppers are able to produce a maximum amount of infection.

Rivera and Ou (1965) concluded that for tungro virus the percentage of infective insects was increased by increasing the length of the acquisition feeding period. They demonstrated that the percentages of infective insects were 3, 4, 17, 48, 69, 81, and 83% for acquisition feeding periods of 0.5, 1, 3, 6, 12, 24, and 48 hr, respectively. Later, Ling (1966) confirmed their conclusion and pointed out that insects given an acquisition feeding period of 4 days retained their infectivity 1 or 2 days longer than those given an acquisition feeding period of 1 or 7 days

(Fig. 2). The infective capacity of insects given an acquisition feeding of 7 days was lower than that for 4 days feeding. This result was apparently caused by biological variations or because insects given 4-day acquisition feeding absorbed almost the maximum charge of the virus. However, none of the insects given a 2-hr acquisition feeding became infective (Fig. 2), and this contrasted with the positive results of insects given an acquisition of 1 hr (Table 1), possibly because the insects of the latter group were starved prior to the acquisition feeding. The retention period of insects given a 1-hr acquisition feeding was much shorter (Table 1).

It is obvious that these results demonstrate a positive effect of length of acquisition feeding period on the infective capacity of the insects transmitting either curly-top or rice tungro virus.

### *G. Vector Reinfected by Reacquisition Feeding*

There are few opportunities for studying the effect of reacquisition feeding on the vectors of propagative leafhopper-borne viruses. Such vectors remain infective without reacquiring the virus, and any result would therefore fail to elucidate the effect of reacquisition feeding

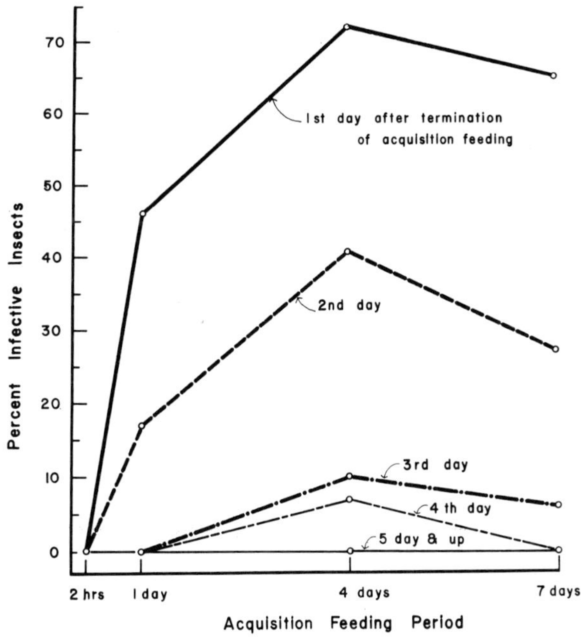


*Fig. 2.* Effect of length of acquisition feeding period on rice tungro virus transmission and retention by *Nephotettix impicticeps* (after Ling, 1966).

on infectivity, particularly for insects that, once infective, remain so for the rest of their lives. However, for nonpropagative viruses, a reacquisition feeding for the leafhopper vectors may be given to maintain insect infectivity for certain purposes and to eliminate the possibility of age being responsible for loss of infectivity with time. The fact that insects regain infectivity following a reacquisition feeding supports the view that the gradual loss of infectivity is caused by exhaustion of virus supply.

The percentage of beet leafhoppers able to acquire and transmit the curly-top virus decreases with age. Nevertheless, Freitag (1936) demonstrated that the insects which have lost infectivity become reinfective after reacquisition feeding on a diseased plant. This result has been confirmed by Bennett and Wallace (1938). Giddings (1950) reached the same conclusion that the curly-top strain 3 virus content of a leafhopper that had fed earlier on strains 2 and 3 was greatly increased by the reacquisition feeding on a strain 3-infected plant. In his experiments, the number of transfers for each leafhopper before and after reacquisition feeding was kept constant. Strain 3 symptoms developed in 13% of the test plants before reacquisition feeding and in 67% after the feeding.

For tungro virus the results are more significant since the virus retention period in the insect is shorter and the daily transmission pattern is consecutive. There were usually some insects that became infective immediately after reacquisition feeding, regardless of when it was provided after they lost infectivity (Fig. 3). Some of these insects definitely reacquired the virus while others, previously noninfective, became infective after the next acquisition feeding on diseased plants. There were no striking differences between infective insects that acquired virus only once and those that reacquired it by repeated feeding as far as the gradual decrease in infectivity and the virus retention in the vector were concerned. Furthermore, if a colony of rice green leafhoppers were provided with a daily reacquisition feeding of 15 hr/day, some insects could retain their infectivity for the length of their life span, although not all insects were able to transmit the tungro virus each day, and the percentage of infective insects fluctuated from time to time (Fig. 4). A daily reacquisition feeding for maintaining an infective colony has been used in testing rice varieties for tungro resistance (International Rice Research Institute, 1966).

### *H. Infective Insects Lost Infectivity after Molting*

The leafhopper-borne viruses are often retained by the vectors after they molt. This is also true for curly-top virus. Freitag (1936) and

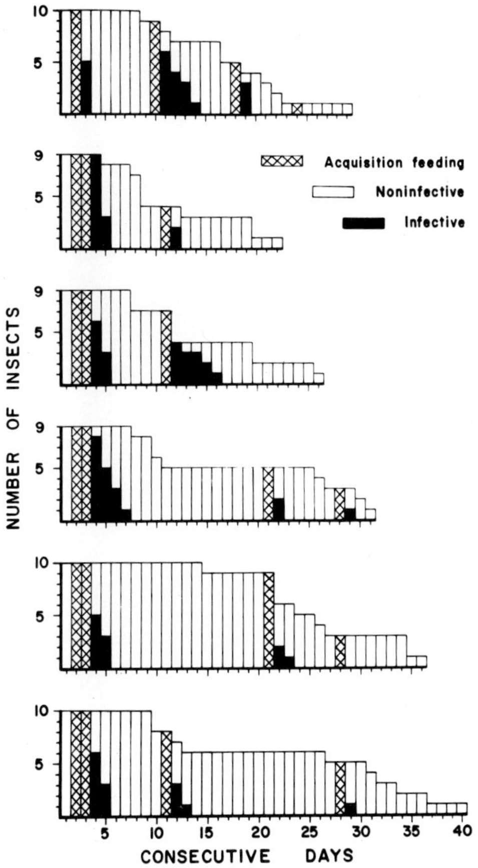


*Fig. 3.* Effect of reacquisition feedings on the infectivity of *Nephotettix impicticeps* transmitting the tungro virus of rice (after Ling, 1966).

Bennett and Wallace (1938) were able to obtain infective adults which had fed on diseased beet or the curly-top virus suspension only during the nymphal stages. Severin (1924) reported earlier that nymphs did not lose their infectivity during the process of molting. Infective nymphs of *N. impicticeps,* however, did not retain their infectivity after molting (Ling, 1966). This loss of infectivity after molting seems subject to the interpretation that the loss of infectivity was not due to molting but rather to the gradual decrease of infectivity in the insects over a passage of time because there must have been a time period required by insects to complete their molting and a time interval between two infectivity tests. Ling (1966) confined individual nymphs in the fifth instar to diseased leaves for acquisition feeding. Two days

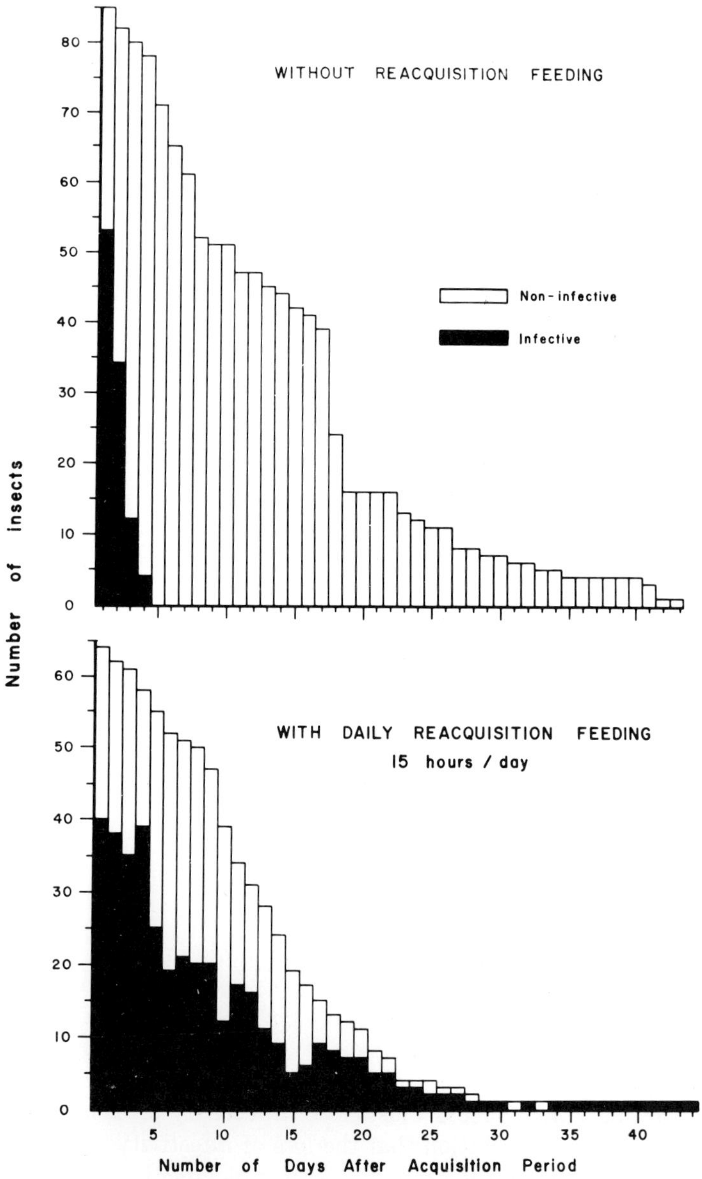


*Fig. 4.* Retention of infectivity of *Nephotettix impicticeps* transmitting the tungro virus of rice by daily reacquisition feeding (after Ling, 1966).

later, to test the nymphs for infectivity, healthy rice seedlings were exposed to the insects individually. After the inoculation feeding period of 5 hr, the insects were confined again to diseased leaves to prevent a decrease in infectivity over time. Since some insects which were not removed from diseased leaves immediately after molting became infective, the insects were observed at short time intervals. As soon as their molting was completed, they were removed and transferred individually from the diseased leaves to healthy rice seedlings to test their infectivity. The results in Table 2 indicate that, of the 45 insects removed from the diseased leaves immediately after molting, none became infective. Before molting, 25 had been infective and 20 noninfective. Although all the nymphs did not molt at the same time, 23 of the 25 infective nymphs molted within 24 hr after the termination of the 5-hr inoculation feeding period. In other words, these 23 insects had a maximum time interval of 24 hr between two infectivity tests, including time for the insects to complete the process of molting. In general, not every insect becomes noninfective 24 hr after acquisition feeding when the acquisition period is longer than two days. Consequently, it could be expected that at least a few of the 23 insects infective before molting would have remained infective after molting if the loss of infectivity were merely due to the gradual decrease over time. Molting *per se,* therefore, could not be eliminated completely from the factors contributing to the loss of infectivity of insects after molting.

TABLE 2

INFECTIVITY OF VIRULIFEROUS *Nephotettix impicticeps* TRANSMITTING RICE TUNGRO VIRUS BEFORE AND AFTER MOLTING

| | Before molting | | After molting | |
|---|---|---|---|---|
| Expt. | Number of infective | Number of non-infective | Number of infective | Number of non-infective |
| I | 3 | 3 | 0 | 6 |
| II | 5 | 3 | 0 | 8 |
| III | 17 | 14 | 0 | 31 |
| Total | 25 [a] | 20 | 0 | 45 |

[a] Twenty-three completed molting within 24 hr after infectivity test.

### *I. Vector Ability and Vector Specificity*

Variation in ability between individuals of beet leafhoppers to transmit the curly-top virus has been studied by Bennett and Wallace (1938) using successive selection and mating through several generations. Their results indicated that the degree of transmissive ability was heritable and that strains of leafhoppers below or above the average transmissive ability could be developed.

Giddings (1940, 1950) demonstrated that a beet leafhopper, after feeding on two or three distinct strains of curly-top virus and transferred daily to a susceptible sugar beet seedling, caused some plants to develop dominant symptoms characteristic of one strain while other plants developed symptoms characteristics of different strains. This result indicates that the insect is able to carry two or three distinct virus strains simultaneously. Consequently, there is no cross protection between strains of curly-top virus in the beet leafhopper as there is with aster yellows (Kunkel, 1955).

If the ability of an insect to transmit a virus depends merely on the ability of the insect to act as a host for the virus, vector specificity should occur only in the propagative viruses. However, the existence of aphid specificity of stylet-borne viruses (Bradley, 1964) indicates that vector specificity may not be limited to the propagative viruses, and the problem of vector specificity is not completely solved (Smith, 1965). For example, the curly-top virus prevalent in North America is transmitted by the only known vector, *C. tenellus*. One strain of the virus that occurs in Argentina is transmitted by *Agalliana ensigera* but not by *C. tenellus* (Bennett *et al.*, 1946). One of the Brazilian strains is transmitted by *Agallia albidula* but not by *Agalliana ensigera* (Bennett and Costa, 1949). Another strain in Brazil is transmitted by *Agalliana ensigera* and *A. stricticollis* but not by *Agallia albidula* (Costa, 1952). In Turkey, the virus not only is transmitted by *C. tenellus* (Bennett and Tanrisever, 1957) but also by *C. opacipennis* (Bennett and Tanrisever, 1958). All these facts indicate a high degree of vector specificity if all the curly-top viruses in various regions are different strains, but there is no adequate explanation for such complex virus–vector relationships.

Furthermore, the tungro virus is transmitted by *N. impicticeps* but positive evidence of transmission by nymphs and adults of *N. apicalis* has not yet been obtained. The reason for the inability of the latter species to act as a vector of the tungro virus remains obscure, but it could be attributed to vector specificity: *N. apicalis* but not *N. impicticeps* transmits rice dwarf virus (Nasu, 1963), but both of them are vectors of rice yellow-dwarf virus (Shinkai, 1965). Vector specificity for nonpropagative leafhopper-borne viruses therefore cannot be denied.

### *J. Bioassay Techniques and Distribution of Virus in the Insect*

Two bioassay techniques, injecting and feeding, have been developed for studying the curly-top virus. The injecting technique consists of mechanically transmitting the virus to the leafhopper by capillary glass needle inoculation. This technique has been used successfully by Maramorosch (1955a) to study the effect of dilution of the inoculum on the incubation period of curly-top virus in the insect.

Since Carter (1927) transmitted beet curly-top virus by feeding the juice of diseased beets to noninfective insects which, in turn, transmitted the virus to healthy plants, the feeding technique has been widely used by various investigators in studies of curly-top virus (Carter, 1928a, 1928b; Severin and Swezy, 1928; Severin, 1931; Severin and Freitag, 1933a; Bennett, 1935, 1962; Bennett and Wallace, 1938; Smith, 1941). The first method used a fishskin bag filled with a solution of the virus and suspended in the cage with the insects. This method was later improved by filling small dishes of various kinds with the feeding solution and covering them with a Baudruche transparent capping skin. The insects were confined on the top of the membrane. The feeding solution was prepared by extracting the virus from diseased beet plants or from viruliferous insects and adding sugar solution to reduce insect mortality. Noninfective insects were often starved for a couple of hours prior to feeding them the virus suspension through the membrane.

By using the feeding technique, the distribution of the curly-top virus in the insect body has been studied. Severin (1931) showed that the virus was present in the mouth parts of viruliferous insects, but he failed to obtain positive evidence of the virus in the blood and salivary glands by using the extract of insect heads after the mouth parts had been removed. He concluded, therefore, that the virus failed to pass through the wall of the midintestine into the blood and salivary glands. Later, Bennett and Wallace (1938) obtained positive evidence of the virus in the blood, salivary glands, alimentary tract, and feces of the insect. Smith (1941) demonstrated the presence of the virus in the secreting saliva. These results are significant evidence for the circulation or passage of the virus in the insect body rather than an indication of a general distribution of the virus in the insect body.

A bioassay technique for the rice tungro virus has not yet been developed.

### K. *Virus Properties and Host Range*

Some properties of the curly-top virus as determined by the membrane feeding technique have been reported by Severin and Freitag (1933a) and Bennett (1935). The curly-top virus has a very wide host range (Boncquet and Stahl, 1917; Carsner, 1919, 1925; Freitag and Severin, 1933; Lackey, 1932; Severin, 1927, 1929, 1934; Severin and Freitag, 1933b, 1934; Severin and Henderson, 1928; Starrett, 1929). All these studies have already been reviewed by Smith (1957).

According to Wathanakul (1964), *Eleusine indica, Echinochloa colonum,* and *Echinochloa crus-galli* are also host plants for the rice tungro virus. Although no symptoms have been observed on the last two species the virus has been recovered from them by insects.

### L. *Virus Attenuation and Strains*

In early experiments, the curly-top virus was attenuated by the passage of the virus through certain resistant host plants (Carsner, 1925; Carsner and Lackey, 1928; Lackey, 1929a, 1929b). The disease symptoms produced by attenuated virus were much milder than those induced by the original virus, and the attenuated virus could be reactivated almost to its original virulence by passage through the highly susceptible chickweed (*Stellaria media*) (Lackey, 1929b, 1931, 1932). However, both attenuation and restoration of virulence were probably due to a separation of virus strains (Giddings, 1940).

Strains of curly-top virus have probably been known since 1925 when Carsner reported variations in symptom severity, incited by different virus isolates. He found also that sugar beet plants infected by a less virulent form of the virus could be infected with a more virulent form, and would then develop severe symptoms. From 1938 to 1954, a total of 12 strains, differentiated by the reaction of a large number of host plants and varieties, have been described by Giddings (1938, 1940, 1944, 1950, 1954). Giddings (1959) also demonstrated that several strains of the virus were highly stable on common varieties of sugar beet and that no mutants were produced by successive selections over a period of 14 months or by transfers from newly infected young tissue.

## V. DISCUSSION

In the absence of a method to prove that multiplication of plant viruses does not occur in the vector, the elucidation of this question in relation to nonpropagative leafhopper-borne viruses must depend upon studies of virus concentration in the vector or on its infectivity following an appropriate single acquisition feeding. If the virus content or the infectivity of a large majority of individuals in a group of insects consistently, naturally, and gradually decreases with time after an acquisition feeding and can be restored by reacquisition feeding, the virus may be regarded as belonging to the nonpropagative group, unless there is direct evidence of multiplication in the insect.

Propagative and nonpropagative viruses sometimes have features in common, such as retention of infectivity after molting, vector specificity, lengths of incubation, and retention periods. These features, however, do not necessarily constitute critical evidence in support of multiplication in the insect. It appears that there is, in fact, no incontrovertible evidence from curly-top and tungro viruses interactions with their vectors that is opposed to the absence of multiplication in the insect. This may be taken as evidence for the existence of nonpropagative leafhopper-borne viruses, thus helping to complete the general picture of plant viruses transmitted by insects.

Most features of the relationships between curly-top and tungro viruses and their vectors are similar in general outline. However, they are distinguishable by certain features, as tabulated in Table 3. These features may serve as guides for separating virus–vector interactions among the nonpropagative group.

Circulative viruses, as described by Black (1959), are acquired through the mouth parts, accumulated internally without apparent multiplication, passed through the insect tissues, and introduced into plants again via the mouth parts of the carrier. The evidence available supports the inclusion of curly-top virus in this group (Maramorosch, 1964).

TABLE 3

DIFFERENCES IN RELATIONSHIPS BETWEEN CURLY-TOP AND TUNGRO VIRUSES AND THEIR RESPECTIVE LEAFHOPPER VOCTORS

| Virus | Curly top | Tungro |
|---|---|---|
| Minimum latent period in insect | 4 hr | Less than 2 hr, if any |
| Maximum virus retention period | Longer than 100 days | Shorter than 6 days |
| Daily transmission pattern | Intermittent | Consecutive |
| Infectivity after molting | Retained | Lost |

Much of the evidence seems to suggest that the tungro virus may not be circulative but stylet borne. This suggestion, however, needs to be verified by direct evidence. Since the term *nonpersistent* is generally applied when the virus retention in the vector is short and vectors lose infectivity after molting (Carter, 1962), the tungro virus may be regarded as nonpersistent in the vector if "short" refers to a duration of not longer than 1 week (Ling, 1966).

The differences between curly-top and tungro viruses suggest that there are various types of virus–vector relationships among the nonpropagative viruses.

## BIBLIOGRAPHY


Bawden, F. C. 1964. Plant viruses and virus diseases. 4th ed. Ronald Press, New York. 361 p.

Bennett, C. W. 1934. Plant-tissue relations of sugar-beet curly-top virus. J. Agr. Res. 48:665–701.

——— 1935. Studies on properties of the curly top virus. J. Agr. Res. 50:211–241.

——— 1944. Studies on dodder transmission of plant viruses. Phytopathology 34: 905–932.

——— 1962. Curly top virus content of the beet leafhopper influenced by virus concentration in diseased plants. Phytopathology 52:538–541.

Bennett, C. W., Carsner, E., Coons, G. H., and Brandes, E. W. 1946. The Argentine curly top of sugar beet. J. Agr. Res. 72:19–48.

Bennett, C. W., and Costa, A. S. 1949. The Brazilian curly top of tomato and tobacco resembling North American and Argentine curly top of sugar beet. J. Agr. Res. 78:625–693.

Bennett, C. W., and Tanrisever, A. 1957. Sugar beet curly top disease in Turkey. Plant Disease Reporter 41:721–725.

——— 1958. Curly top disease in Turkey and relationship to curly top in North America. J. Amer. Soc. Sugar Beet Technol. 10:189–211.

Bennett, C. W., and Wallace, H. E. 1938. Relations of the curly top virus to the vector, *Eutettix tenellus*. J. Agr. Res. 56:31–51.

Black, L. M. 1950. A plant virus that multiplies in its insect vector. Nature 166: 852–853.

——— 1953a. Transmission of plant viruses by Cicadellids. Advances Virus Res. 1:69–89.

——— 1953b. Viruses that reproduce in plants and insects. Ann. N.Y. Acad. Sci. 56:389–413.

——— 1954. Parasitological reviews. Arthropod transmission of plant viruses. Exp. Parasitol. 3:72–104.

——— 1959. Biological cycles of plant viruses in insect vectors. Vol. 2, p. 157–185. *In* F. M. Burnet and W. M. Stanley [ed.] The viruses. Academic Press, New York.

——— 1962. Some recent advances on leafhopper-borne virus. p. 1–9. *In* K. Maramorosch [ed.] Biological transmission of disease agents. Academic Press, New York.

Black, L. M., and Brakke, M. K. 1952. Multiplication of wound-tumor virus in an insect vector. Phytopathology 42:269–273.

Boncquet, P. A., and Stahl, C. F. 1917. Wild vegetation as a source of curly-top infection of sugar beets. J. Econ. Entomol. 10:392–397.

Bradley, R. H. E. 1964. Aphid transmission of stylet-borne viruses. p. 148–174. *In* M. K. Corbett and H. D. Sisler [ed.] Plant virology. Univ. Florida Press, Gainesville.

Carsner, E. 1919. Susceptibility of various plants to curly top of sugar beet. Phytopathology 9:413–421.

——— 1925. Attenuation of the virus of sugar beet curly-top. Phytopathology 15:745–758.

Carsner, E., and Lackey, C. F. 1928. Further studies on attenuation of sugar beet curly-top. Phytopathology 18:951.

Carsner, E., and Stahl, C. F. 1924. Studies on curly-top disease of sugar beet. J. Agr. Res. 28:297–320.

Carter, W. 1927. A technique for use with homopterous vectors of plant diseases with special reference to the sugar-beet leafhopper, *Eutettix tenellus* (Baker). J. Agr. Res. 34:449–451.

——— 1928a. An improvement in the technique for feeding homopterous insects. Phytopathology 18:246–247.

——— 1928b. Transmission of the virus of curly top of sugar beets through different solution. Phytopathology 18:675–679.

——— 1962. Insects in relation to plant disease. Wiley, New York. 705 p.

Costa, A. S. 1952. Further studies on tomato curly top in Brazil. Phytopathology 42:396–403.

Day, M. F., and Venables, D. G. 1961. The transmission of cauliflower mosaic virus by aphids. Australian J. Biol. Sci. 14:187–197.

Fajardo, T. G., Bergonia, H. T., Capule, N., and Novero, E. 1964. Studies on rice diseases in the Philippines. I. Progress report on "tungro" disease of rice. Paper presented in Meeting FAO-IRC Working Party on Rice Production and Protection, 10th, Manila, Philippines.

Freitag, J. H. 1936. Negative evidence on multiplication of curly-top virus in the beet leafhopper, *Eutettix tenellus*. Hilgardia 10:305–342.

Freitag, J. H., and Severin, H. H. P. 1933. List of ornamental flowering plants experimentally infected with curly top. Plant Disease Reporter 17:2–5.

Fukushi, T. 1933. Transmission of the virus through the eggs of an insect vector. Proc. Imp. Acad. (Tokyo) 9:457–460.

——— 1940. Further studies on the dwarf disease of rice plant. J. Faculty Agr. Hokkaido Imp. Univ. 45:83–154.

Fukushi, T., Shikata, E., Kimura, I., and Nemoto, M. 1960. Electron microscopic studies on the rice dwarf virus. Proc. Japan Acad. 36:352–357.

Giddings, N. J. 1938. Studies of selected strains of curly top virus. J. Agr. Res. 56: 883–894.

——— 1940. Curly-top virus strains. Phytopathology 30:786.

——— 1944. Additional strains of the curly top virus. J. Agr. Res. 69:140–157.

——— 1950. Some interrelationships of virus strains in sugar beet curly top. Phytopathology 40:377–388.

——— 1954. Two recently isolated strains of curly top virus. Phytopathology 44: 123–125.

——— 1959. The stability of sugarbeet curly top virus strains. J. Amer. Soc. Sugar Beet Technol. 10:359–363.

International Rice Research Institute. 1963. Annual Report 1963. p. 113–114.

——— 1966. Annual Report 1965. p. 118–120.

Johnson, F. 1941. Transmission of plant viruses by dodder. Phytopathology 31: 649–656.

Kunkel, L. O. 1926. Studies on aster yellows. Amer. J. Bot. 13:646–705.

——— 1937. Effect of heat on ability of *Cicadula sexnotata* (Fall.) to transmit aster yellows. Amer. J. Bot. 24:316–327.

——— 1941. Heat cure of aster yellows in periwinkles. Amer. J. Bot. 28:761–769.

——— 1955. Cross protection between strains of yellows-type viruses. Advances Virus Res. 3:251–273.

Lackey, C. F. 1929a. Attenuation of curly-top virus by resistant sugar beets which are symptomless carriers. Phytopathology 19:975–977.

——— 1929b. Further studies of the modification of sugar beet curly-top virus by its various host. Phytopathology 19:1141.

——— 1931. Virulence of curly-top virus restored by *Stellaria media.* Phytopathology 21:123–124.

——— 1932. Restoration of virulence of attenuated curly-top virus by passage through *Stellaria media.* J. Agr. Res. 44:755–765.

Leach, J. G. 1940. Insect transmission of plant diseases. McGraw-Hill, New York. 615 p.

Ling, K. C. 1966. Nonpersistence of the tungro virus of rice in its leafhopper vector, *Nephotettix impicticeps*. Phytopathology 56:1252–1256.

Maramorosch, K. 1950. Influence of temperature on incubation and transmission of the wound-tumor virus. Phytopathology 40:1071–1093.

——— 1952. Direct evidence for the multiplication of aster-yellows virus in its insect vector. Phytopathology 42:59–64.

——— 1953. Incubation period of aster yellows virus. Amer. J. Bot. 40:797–808.

——— 1955a. Mechanical transmission of curly top virus to its insect vector by needle inoculation. Virology 1:286–300.

——— 1955b. Multiplication of plant viruses in insect vectors. Advances Virus Res. 3:221–249.

——— 1963. Arthropod transmission of plant viruses. Ann. Rev. Entomol. 8:369–414.

——— 1964. Virus–vector relationship: Vectors of circulative and propagative viruses. p. 175–193. *In* M. K. Corbett and H. D. Sisler [ed.] Plant virology. Univ. Florida Press, Gainesville.

Nasu, S. 1963. Studies on some leafhoppers and planthoppers which transmit virus diseases of rice plant in Japan. Bull. Kyushu Agr. Exp. Sta. 8:153–149.

Rivera, C. T., and Ou, S. H. 1965. Leafhopper transmission of "tungro" disease of rice. Plant Disease Reporter 49:127–131.

Severin, H. H. P. 1921. Minimum incubation periods of causative agent of curly leaf in beet leafhopper and sugar beet. Phytopathology 11:424–429.

——— 1924. Curly-leaf transmission experiments. Phytopathology 14:80–93.

——— 1927. Crops naturally infected with sugar beet curly-top. Science 66:137–138.

——— 1929. Additional host plants of curly top. Hilgardia 3:595–637.

——— 1931. Modes of curly-top transmission by the beet leafhopper, *Eutettix tenellus* (Baker). Hilgardia 6:253–276.

——— 1934. Weed host range and overwintering of curly-top virus. Hilgardia 8: 263–280.

——— 1949. Transmission of the virus of Pierce's disease of grapevines by leafhoppers. Hilgardia 19:190–206.

——— 1950. Spittle-insect vectors of Pierce's disease virus. II. Life history and virus transmission. Hilgardia 19:357–382.

Severin, H. H. P., and Freitag, J. H. 1933a. Some properties of the curly-top virus. Hilgardia 8:1–48.

——— 1933b. List of ornamental flowering plants naturally infected with curly top or yellows diseases in California. Plant Disease Reporter 17:1–2.

——— 1934. Ornamental flowering plants naturally infected with curly-top and aster-yellows viruses. Hilgardia 8:233–260.

Severin, H. H. P., and Henderson, C. F. 1928. Some host plants of curly top. Hilgardia 3:339–392.

Severin, H. H. P., and Swezy, O. 1928. Filtration experiments on curly top of sugar beets. Phytopathology 18:681–690.

Shikata, E. 1966. Electron microscopic studies on plant viruses. J. Faculty Agr. Hokkaido Univ. 55:1–110.

Shikata, E., Orenski, S. W., Hirumi, H., Mitsuhashi, J., and Maramorosch, K. 1964. Electron micrographs of wound-tumor virus in an animal host and in plant-tumors. Virology 23:441–444.

Shinkai, A. 1965. Transmission of four rice viruses by leafhoppers [in Japanese]. Ann. Phytopathol. Soc. Japan 31:380–383.

Smith, F. F., and Brierley, P. 1956. Insect transmission of plant viruses. Ann. Rev. Entomol. 1:299–322.

Smith, K. M. 1941. Some notes on the relationship of plant viruses with vector and non-vector insects. Parasitology 33:110–116.

——— 1957. A textbook of plant virus diseases. 2nd ed. Little & Brown, Boston. 652 p.

——— 1958. Transmission of plant viruses by arthropods. Ann. Rev. Entomol. 3: 469–482.

——— 1965. Plant virus-vector relationships. Advances Virus Res. 11:61–96.

Smith, R. E., and Boncquet, P. A. 1915. Connection of a bacterial organism with curly leaf of the sugar beet. Phytopathology 5:335–342.

Stahl, C. F., and Carsner, E. 1918. Obtaining beet leaf-hoppers nonvirulent as to curly-top. J. Agr. Res. 14:393–394.

Starrett, R. C. 1929. A new host of sugar beet curly top. Phytopathology 19:1031–1035.

Storey, H. H. 1928. Transmission studies of maize streak disease. Ann. Appl. Biol. 15:1–25.

——— 1939. Transmission of plant viruses by insects. Bot. Rev. 5:240–272.

Swezy, O. 1930. Factors influencing the minimum incubation periods of curly-top in the beet leafhopper. Phytopathology 20:93–100.

Wallace, J. M., and Murphy, A. M. 1938. Studies on the epidemiology of curly top in Southern Idaho, with special reference to sugar beets and weed hosts of the vector *Eutettix tenellus*. U.S. Dep. Agr. Tech. Bull. 624:1–46.

Wathanakul, L. 1964. A study of the host range of tungro and orange leaf viruses of rice. M.S. Thesis. College Agr. Univ. Philippines. 35 p.

Whitcomb, R. F., and Black, L. M. 1961. Synthesis and assay of wound tumor soluble antigen in an insect vector. Virology 15:136–145.

# Relationships between Propagative Rice Viruses and Their Vectors

TEIKICHI FUKUSHI

*Department of Botany, Faculty of Agriculture, Hokkaido University, Sapporo, Japan*

The rice plant is extensively cultivated in India, China, Japan, Indonesia, and other oriental countries. In the western hemisphere rice culture has been carried on in the United States and various countries in Central and South America. Hoja blanca seems to be the only virus disease of rice in the western hemisphere while several leafhopper-borne virus diseases have been reported from the Orient. The viruses causing rice dwarf, stripe, yellows, and black streak in Japan have been shown to be propagative in their vectors and the virus of orange leaf reported from the Philippines and Thailand (Rivera *et al.*, 1963) probably belongs to this group, but as to "tungro" in the Philippines (Rivera and Ou, 1965), transitory yellowing in Taiwan (Chiu *et al.*, 1965), grassy stunt in the Philippines (John, 1965), and others, the problem still awaits more extended research.

As hoja blanca is mentioned elsewhere in this book, the relationships between leafhoppers and rice viruses in Japan are discussed here, laying stress upon the early history of leafhopper transmission of rice dwarf and the transovarial transmission of rice viruses. A brief account will also be given of the relationships between leafhoppers and orange leaf, "tungro," and others in Southeast Asia.

Four virus diseases of the rice plant have been recognized in Japan, *viz.*, dwarf (or stunt), stripe, yellows (or yellow dwarf), and black streak (or black streak dwarf). Dwarf is characterized by the stunting of the plant, excessive tillering, dark green foliage, and yellowish white specks or interrupted streaks along the veins of leaves (Fig. 1). The symptoms of stripe consist of abnormal elongation and the drooping of young leaves and one or more yellowish white stripes running parallel to the midrib. Yellows is manifested by the yellowing of leaves

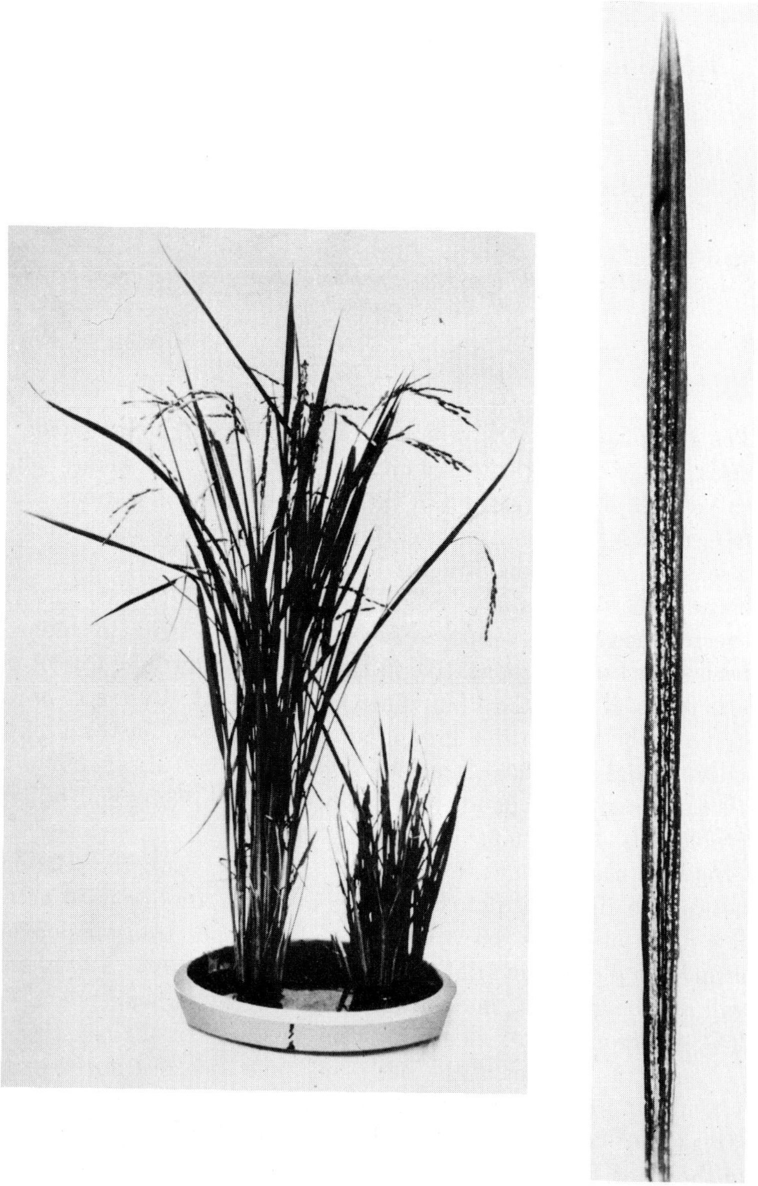

*Fig. 1.* Rice dwarf: (*A*) Healthy (left) and diseased rice plants. (*B*) Leaf of diseased rice plant, showing yellowish white specks and interrupted streaks along the veins.

and new shoots. The symptoms of black streak include stunting of the entire plant, diminished tillering, dark green foliage, and elongated swellings on the undersurface of leaves, as well as on leafsheaths and culms (Fig. 2). These swellings, which extend along the veins, are whitish in color at the early stage of development, but later turn brown or black. Dwarf and black streak seem to be endemic to Japan, stripe is known in Japan and Korea, while yellows* appears to be more or less widely distributed, its occurrence having been reported from Japan, Taiwan, and the Philippines.

* According to Hashioka (1964), rice yellows is known also in the Hainan Islands, South China, and Thailand.

*Fig. 2.* Rice black streak showing elongated swellings on culms. (After Kobi.)

## I. EARLY HISTORY OF THE LEAFHOPPER TRANSMISSION OF RICE DWARF DISEASE

It is well known that rice dwarf is of historical significance because it is the first plant virus disease shown to be transmitted by an insect (Kunkel, 1926; Hino, 1927; Fukushi, 1935; Katsura, 1936). However, there has been confusion in the literature (Leach, 1940; Cook, 1947; Smith, 1960; Carter, 1962; Bawden, 1964) on the early history of the leafhopper transmission of rice dwarf disease, and it will not be superfluous to explain it briefly here.

According to Ishikawa (1928), a rice grower in Shiga prefecture named Hashimoto was the first to prove experimentally the relationship of leafhoppers to dwarf disease of the rice plant. He noticed the disease in 1883 and subsequently suspected a relationship between leafhoppers and rice dwarf because of the frequent occurrence of the disease in rice fields where leafhoppers were abundant. In 1893, having realized the destructive nature of this trouble, the Government of Shiga Prefecture decided to investigate the disease in experimental plots located in various parts of the prefecture. Taking charge of the work in one of these experimental plots, Hashimoto planted young rice plants in a glass container and enclosed them in a cheesecloth cage, introducing numerous leafhoppers. Consequently, he discovered in 1894 the causal relation of leafhoppers to rice dwarf and confirmed it by the experiments carried out in the next year. This was the first case that proved experimentally the relationship between an insect and a virus disease. Hashimoto, however, failed to publish the results of his experiments and specify which leafhopper he dealt with. In 1895 Takata (1895–96) wrote in his paper that certain species of leafhopper which he named "mon-yokobai" was responsible for the disease. This leafhopper was subsequently renamed "inazuma-yokobai"* (Shiga Agricultural Experiment Station, 1898) and identified with *Deltocephalus dorsalis* Motsch. (Melichar, 1903) (Fig. 3). The investigations of rice dwarf disease in Shiga Prefecture mentioned above were carried out under the supervision of Takata and it is most likely that Takata's paper was written partly on the basis of Hashimoto's experimental results. As the writer has often pointed out, Takata was the first to state the causal relation of leafhopper to dwarf disease of rice plant. However, the leafhopper he mentioned was not *Nephotettix cincticeps*

* Inazuma-yokobai was described by Matsumura in 1902 as *Deltocephalus fulguralis* sp. n. but Melichar pointed out that it was identical with *D. dorsalis* Motsch. Ishihara (1953) transferred it to the genus *Inazuma* which he established in 1953.

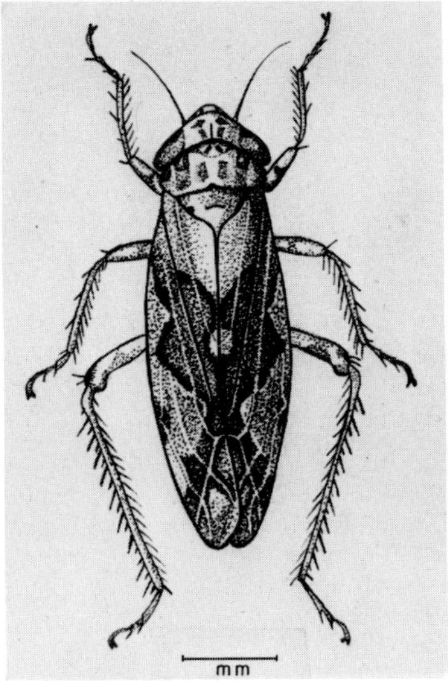


*Fig. 3.* *Inazuma dorsalis* (Motschulsky), the vector of rice dwarf and the first insect shown to induce a plant virus disease (approx. 4.5 mm in length).

(Uhl.) as has been referred to erroneously by several authors but *Inazuma* (*Deltocephalus*) *dorsalis,* as stated above.

Immediately after its establishment in 1895, the Shiga Agricultural Experiment Station undertook studies on this disease and published *Results of experiments with insect pests. Reports 1–8* in 1898–1908. Report 1 (1898) mentioned several species of leafhoppers including tsumaguro-,† inazuma-,* and futaten-yokobai‡ [*Nephotettix cincticeps* (Uhl.), *Inazuma dorsalis* (Motsch.) and *Macrosteles fascifrons* (Stal.), respectively] as causing the dwarf disease. In Report 2, which was published in 1900, however, only one species, tsumaguro-yokobai (*Nephotettix cincticeps*) (Fig. 4) was the true cause whereas several other species of leafhoppers, including inazuma-yokobai (*Inazuma dorsalis*), had no connection with the disease. This was confirmed in Reports 3–8 (1901–1908). It is evident that the disease was at that time entirely attributed to the leafhopper. Takami (1901), then the

† Tsumaguro-yokobai was first described by Uhler in 1896 as *Selenocephalus cincticeps.* Matsumura transferred it to the genus *Nephotettix* which he established in 1902 and later considered to be a variety of *Nephotettix apicalis* which had been described in India by Motschulsky (1859) as *Pediopsis apicalis* and accordingly designated it *N. apicalis* var. *cincticeps* (Matsumura, 1905).

‡ Futaten-yokobai was first named shiro-yokobai by the Shiga Agricultural Experiment Station.

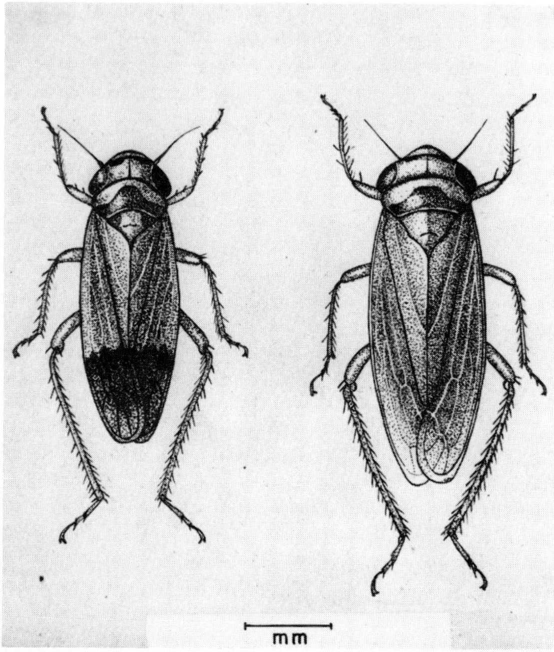


*Fig. 4.* *Nephotettix cincticeps* (Uhler), the vector of rice dwarf and the first insect shown to transmit a plant virus (male on the right, approx. 4.5 mm long, and female on the left, approx. 6 mm long).

principal agronomist at this station, outlined the early work of this experiment station and concluded that rice dwarf disease was due to the leafhopper tsumaguro-yokobai. Thus, Takata's claims that inazuma-yokobai was responsible for rice dwarf appeared to be incorrect.

Meanwhile, the entomologist in the Imperial Agricultural Experiment Station, predecessor of the present National Institute of Agricultural Sciences at Nishigahara, Tokyo, started experiments on the same subject. Ando (1910) briefly reported the results of experiments carried out in this station. According to him, the leafhoppers captured in the vicinity of Tokyo in 1905 were reared on dwarf-diseased rice plants taken from Shiga Prefecture. Their progeny of the second or third generation produced infections in healthy rice plants. He further stated that noninfective leafhoppers from Tokyo became infective after feeding upon diseased plants for about 15 days. Another experiment conducted at this station showed the exhaustion of the infectivity of the leafhopper. The infective leafhoppers collected in Shiga Prefecture were fed on healthy rice plants repeatedly replaced with fresh healthy ones at 5–7 day intervals to keep the leafhoppers from an access to infected plants. As a result of this experiment it was shown in 1906 that the leafhoppers of the first generation were capable of producing the disease in healthy rice plants but the progeny of the second and third generations were noninfective. Thus, he was led to the conclusion that rice dwarf was not caused by the leafhopper but by a certain unknown causative agent carried by the leafhopper. From the chronological point of view Ando's statement does not agree entirely with that of Murata as follows: According to Murata (1931), Onuki, entomologist at the Imperial Agricultural Experiment Station found in 1899 that the leafhoppers, tsumaguro-yokobai (*Nephotettix cincticeps*) captured in the vicinity of Tokyo were unable to produce rice dwarf in healthy rice plants. But through the experiments conducted at Ando's suggestion in 1902, he showed that these leafhoppers became infective if they had been fed on diseased plants. Thus, it became evident that the leafhopper was the carrier of the causal agent, the nature of which was unknown. This finding was confirmed by the Shiga Agricultural Experiment Station and written in Reports 7 and 8 in 1908. Accordingly, the Imperial Agricultural Experiment Station was the first to prove that fact, although the Shiga Agricultural Experiment Station reported the confirmed data in 1908. At any rate the true role of the leafhopper, *Nephotettix cincticeps* (Uhl.) as a vector of rice dwarf was established by two entomologists, N. Onuki at the Imperial Agricultural Experiment Station and T. Nishizawa at the Shiga Agricultural Experiment Station.

The Shiga Agricultural Experimental Station continued the work concerning insect transmission of rice dwarf for a period of 20 years and finally reached the same conclusion as mentioned above.

Consequently, it became the general belief that *Nephotettix cincticeps* was the sole agent transmitting rice dwarf and it was Takata who had demonstrated it. In 1937, however, the writer proved that *Inazuma dorsalis* was also a vector of rice dwarf virus and thus partly substantiated the conclusion reached by Takata (Fukushi, 1937). This leafhopper tends to kill rice plants before they begin to show symptoms even when they contract the disease. It is likely that the entomologist in the Shiga Agricultural Experiment Station overlooked this aspect. When Fukushi (1935) and Katsura (1936) gave a historical review of the leafhopper transmission of rice dwarf, *Nephotettix cincticeps* was believed to be the sole insect vector and Takata's paper was not adequately evaluated. Katsura in particular held Takami's paper in high esteem.

## II. TRANSOVARIAL TRANSMISSION OF RICE VIRUSES

### *A. Transovarial Passage of Rice Dwarf Virus*

After the Shiga Agricultural Experiment Station had reported that *Nephotettix cincticeps* caused rice dwarf, Onuki, entomologist at the Imperial Agricultural Experiment Station, attempted to verify this conclusion by using the leafhoppers of the same species captured in the vicinity of Tokyo, which, however, all failed to produce rice dwarf in healthy rice plants, as stated above. Consequently, he supposed that the leafhoppers in these areas might be different from each other. The idea prevailing at that time was that rice dwarf was entirely due to the infestation of leafhoppers and Onuki considered the capability of producing rice dwarf to be a characteristic specific to leafhoppers native to Shiga Prefecture. In order to clear up the question he collected the eggs of this leafhopper in Shiga Prefecture and brought them back to Tokyo. According to Murata (1931), Onuki and Murata discovered in 1902 that some leafhoppers hatched from these eggs produced the dwarf disease in healthy rice plants. Murata (1915) briefly wrote in his book that "the capability of producing dwarf disease" was transmitted from infective leafhoppers to the progeny through three or four generations but he gave no experimental data. He further stated that infective leafhoppers which had been reared on healthy rice plants, the food plants being renewed every 2 or 3 days, gradually lost their ability to transmit the disease. These entomologists undoubtedly did not understand that they were dealing with a virus disease in rice dwarf and that the virus might be transmitted through the egg of the leafhopper.

When the writer took up work on the leafhopper transmission of rice dwarf virus, he paid little attention to Murata's statement mentioned above because Murata had presented no reliable data; on the other hand, there was no record showing the congenital transmission of a plant virus in an insect vector. Before long the writer (Fukushi, 1934, 1940) found that not all individuals of *Nephotettix cincticeps* were capable of acting as the vector of this virus. As far as the leafhoppers of Tottori, Utsunomiya, and Shiraoka origin were concerned, only about 10% of them transmitted the disease, even when they had been reared on diseased plants. Thus, the writer experienced difficulties in obtaining enough potential transmitters to proceed with the work on the relationship between the leafhopper and the virus. Having supposed that there might be a strain of leafhopper which would readily transmit the virus, the writer introduced a few infective leafhoppers into an insect cage enclosing dwarf-diseased rice plants and tested the infective capability of individual leafhoppers of the progeny. About 90% of the leafhoppers tested produced infections in healthy rice plants. There are two interpretations: either the virus is congenitally transmitted through the eggs of the vector or the offspring from infective leafhoppers have the predisposition to carry the virus readily after feeding on diseased plants. As a result, a series of experiments were started to determine whether the virus was congenitally transmitted to the offspring from infective parents. It has been shown that the transovarial passage of virus occurs when the female leafhopper is infective (Fukushi, 1933, 1934). The majority (or 85%) of the offspring of the infective parents proved to be viruliferous; a relatively small percentage (60%) of the progeny were infective when infective females were crossed with noninfective males, whereas the virus was not transmitted to the progeny when noninfective females were mated with infective males. A period from 1 to 38 days with an average of about 15 days must elapse before most of the viruliferous nymphs become infective although a few individuals may transmit the disease immediately after their emergence from eggs. Most of these viruliferous nymphs retain their infectivity during all the nymphal stages and for considerably long periods, up to 88 days through their adult life, without renewed access to a source of virus. Considerable variation was observed in the infectivity of different leafhoppers. Some of them infected plants consistently on consecutive days while others did so only at great intervals. In some leafhoppers such noninfective periods lasted for about 15 days. In the progeny of infective females there were a few viruliferous female leafhoppers which proved to be noninfective during their entire life, but produced infective offspring. In such cases the virus presumably might have been suppressed from progagation because these leafhoppers were less susceptible to the virus and the congenitally transmitted virus might have localized by chance in the ovarioles to give rise to viruliferous eggs.

The progeny from the crosses between noninfective females and infective males are entirely free from virus, as stated above, but they have a greater ability to acquire and transmit the virus as compared with those derived from nonviruliferous parents. This can be readily explained if we assume that the virus is capable of multiplying only in susceptible leafhopper individuals and that this susceptibility to virus is a hereditary characteristic represented by a dominant factor or factors. On this assumption it is probable that the progeny from crosses between nonviruliferous females and infective males should have more susceptible insects in number than those derived from nonviruliferous parents because only a small proportion of ordinary nonviruliferous leafhoppers become infective after feeding on diseased plants, indicating that most of them are resistant to the virus. When the progeny from infective parents, from crosses between infective females and nonviruliferous males, and from nonviruliferous parents had been reared on diseased rice plants, 92, 68, and 12%, respectively, became infective. These facts can be also readily explained on the above assumption.

The writer attempted to determine to what extent the virus would be passed on to the offspring of leafhoppers. In a case shown in Fig. 5, starting with a single viruliferous female leafhopper, the writer demonstrated that the virus could be passed through the egg to 6 succeeding generations without replenishment from diseased plants. This experiment lasted longer than a year and during this time there was no evidence of a progressive decrease either in the percentage of infective leafhoppers or in their infectivity (Fukushi, 1939, 1940). On the basis of this evidence the writer decided that the rice dwarf virus multiplies in its insect vector. When Fukushi and Kimura (1959, 1960) successfully transmitted rice dwarf virus to virus-free leafhoppers by injecting extracts containing the virus using fine glass capillaries, they attempted to demonstrate the presence of the virus in the egg by this method. About 200 eggs laid by infective female leafhoppers were taken out before hatching, ground, and diluted with phosphate buffer to 1:100 or 1:1000. A small amount of the fluid was forced into the ventral side of the abdomen of a leafhopper. Four and 6 of each group of 49 leafhoppers inoculated with the respective extracts became infective. In order to obtain additional evidence of the transovarial passage of virus, Fukushi and Shikata (1963) attempted to show the virus in the egg of an insect vector with electron microscopy. They revealed virus particles in ultra-thin sections of the ovariole of an infective adult leafhopper, although they could not definitely demonstrate the presence of virus particles in egg cells. Quite recently Nasu (1965) discovered virus in the mycetomes of infective leafhoppers. The mycetome has been known to be present in every individual leafhopper and harbors symbiotic microorganisms which are transferred from generation to generation through the eggs of insects. This congenital transference of the mycetome and its accompanying symbionts may elucidate how the virus enters the developing egg. According to Nasu the virus particles were found to enter oocytes from the mycetocyte in the ovariole. The virus particles were seen in the cytoplasm of the germarium, mycetocytes, and pedicel cells of ovariole. When the symbionts in the hemolymph invade the mycetome in the ovariole, the virus particles may enter the mycetome at the same time. Nasu has shown that the mycetocytes with their

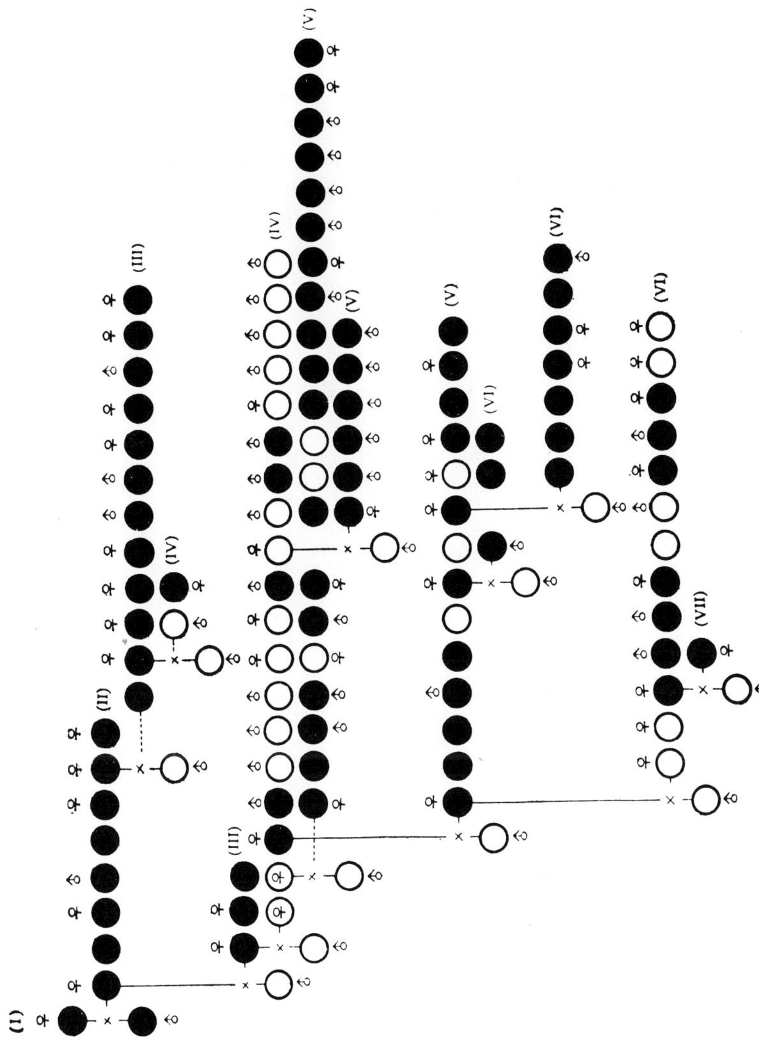


●○ indicate infective and noninfective leafhoppers, respectively

*Fig. 5.* Diagram showing the descent of infective and noninfective leafhoppers.

symbionts and virus particles entered oocytes at the yolk-formation stage. This is the only evidence to show the method of transovarial passage of rice dwarf virus in the leafhopper, *Nephotettix cincticeps*.

Shinkai (1958, 1962) has demonstrated the transovarial transmission of rice dwarf virus in *Inazuma dorsalis*. According to him, however, the rate of transovarial transmission in this case was lower than that in *Nephotettix cincticeps* and the number of viruliferous leafhoppers in the progeny decreased remarkably in successive generations. In his experiments the virus was transmitted through eggs to the progeny of the third generation, but not to that of the fourth generation. The offspring which had been congenitally viruliferous seemed to die prematurely. Recently Nasu (1963) reported kurosuji-tsumaguroyokobai [*N. apicalis* (Motsch.)] as the insect vector of rice dwarf. The virus is also transmitted through the egg of this leafhopper.

### *B. Transovarial Passage of Rice Stripe Virus*

Kuribayashi (1931a, b) was the first to demonstrate the transmission of rice stripe by *Delphacodes striatella* Fall* (Fig. 6). He found that

* Recently Fennah (1963) pointed out that this species did not belong to the genus *Delphacodes* and placed in a genus which he named *Laodelphax*.

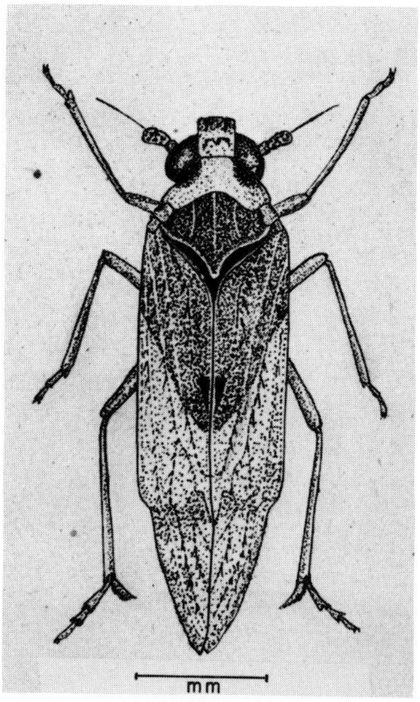


*Fig. 6. Laodelphax striatella* (Fallén), the vector of rice stripe and black streak (approx. 3.5 mm long).

some individuals of the progeny from infective female planthoppers produced the stripe disease in healthy rice plants, but he attributed it to experimental error because the nymphs newly emerged from eggs might feed on the plants which possibly contracted the disease during the feeding of the infective parents. More than 20 years later Yamada and Yamamoto (1954, 1955) and Shinkai (1954, 1962) definitely proved that the virus is passed through eggs of infective planthoppers to the progeny. They have shown that the virus from a single viruliferous female planthopper could be passed through eggs to a high proportion of the progeny in 23 (Yamada and Yamamoto) or 40 (Shinkai) succeeding generations. In the latter case the virus was retained for 6 years within the insect bodies and 95% of the individual planthoppers of the fortieth generation proved to be infective, showing that there was no progressive decline of the virus. These experiments provide additional evidence of the multiplication of virus in its insect vector. Most congenitally viruliferous nymphs began to transmit the virus to healthy plants on the day of hatching. They remained highly infective throughout the nymphal stage and for 2–3 weeks of early adult life. However, the infective capacity of some planthoppers apparently diminished in late adult life.

## III. RELATIONSHIPS BETWEEN RICE VIRUSES AND THEIR VECTORS IN JAPAN

### *A. Dwarf*

Since 1905(?) rice dwarf has been known to be transmitted by the leafhopper, *Nephotettix cincticeps*. Fukushi (1934, 1940) first discussed the relation of the virus to its vector in detail. He demonstrated that not all individuals of this leafhopper were capable of acting as the vector of the virus. Only a small number of nonviruliferous leafhoppers became infective after feeding on diseased rice plants, as has been already mentioned. The minimum acquisition feeding period was 3 days, but the leafhoppers acquired the virus more readily after a 10–50 day feeding period. [As will be mentioned later, the majority of leafhoppers of Shinkai's "active" (*or "susceptible" according to the writer's point of view*) race acquire the virus more readily.] Certain leafhoppers, however, failed to transmit the virus, even when the feeding period on diseased plants extended to 50–70 days. The incubation period of the virus within the insect varied from 10–25 to 60–73 days, generally being 30–45 days. The minimum inoculation feeding period was 5 min and a feeding period of 30 min was long enough to enable most of the infective leafhoppers to produce infections in healthy plants.

Most of the infective insects retained their infectivity until they died and some of them produced infections in every healthy plant on which an individual leafhopper was confined for 24 hr on consecutive days. In comparatively few instances the infectivity of the leafhopper was apparently reduced or lost even under favorable conditions. A viruliferous leafhopper may not produce infections in every healthy plant on which it feeds, even though the feeding period may be 24 hr or longer. It seems, therefore, that certain viruliferous leafhoppers sometimes fail to transmit the disease, presumably owing to the temporary exhaustion of the virus in the salivary glands. Recently, Shinkai (1956, 1962) found that the leafhoppers captured at different localities vary in their ability to transmit the virus. As for *Nephotettix cincticeps*, 69% of the leafhoppers, the progeny of those captured at Maruyama-machi, Chiba Prefecture, transmitted the virus after feeding on diseased plants whereas those from Yamagata Prefecture appeared to lack this ability. Kimura (1962) confirmed the above by demonstrating that the leafhoppers of Chiba origin transmitted rice dwarf virus more readily than those from Shiga or Akita Prefecture. Further, he showed that the latter leafhoppers became infective more readily when virus preparations were injected with a glass capillary into their bodies through a puncture made in the abdomen. However, the leafhoppers from Chiba still became infective more readily than those from Shiga or Akita when they were inoculated with rice dwarf virus by injection. In the case of *Inazuma dorsalis*, Shinkai reported that 43% of the leafhoppers from Odawara, Kanagawa Prefecture transmitted the virus after feeding on diseased plants whereas only 2% of those from Tokyo proved to be "active" (Storey, 1931). It is likely that the leafhoppers are either susceptible or resistant to the virus and that the former leafhoppers only enable the virus to multiply in their bodies.

According to Shinkai (1962) the majority of leafhoppers of the "active" or susceptible race of *Nephotettix cincticeps* of Chiba origin acquired the virus by feeding on diseased rice plants for 1 day, some nymphs in the first and second instars acquired the virus even by 1–3 min feeding on diseased plants. The latent period of the virus in this leafhopper ranged from 10 to 58 days, mostly 12–35 days. In *Inazuma dorsalis* also, almost 50% of the "active" or susceptible leafhoppers acquired the virus by feeding on diseased rice plants for 1 day. The minimum acquisition feeding period was 30 min and the incubation period of the virus in this insect was 9–42 days, mostly 10–15 days. Fukushi and Kimura (1959) succeeded in transmitting rice dwarf virus to virus-free leafhoppers, *Nephotettix cincticeps,* by injecting juice from viruliferous leafhoppers or affected rice plants using fine glass capillaries. Subsequently, Kimura (1962) found that the shortest incubation period of the virus in the insect vector to be 8–12 days and the longest, 33–37 days at a constant temperature of 18°C, whereas no leafhoppers became infective at 13°C. The virus concentration in the leafhopper was shown to increase most rapidly during a period of 15–20 days after infection. The virus was carried in the vectors through four serial passages. It was calculated that the quantity in the starting leafhopper would have undergone a dilution of $10^{-11}$ in the last group of the series if the virus had not multiplied. Since the dilution endpoint of the juice from infective insects was $10^{-4}$, Kimura came to the conclusion that the virus multiplied in the insect vector.

Through electron microscopic studies, Fukushi, Shikata, and Kimura (1962) and Shikata (1966) revealed rice dwarf virus in the bodies of infective leafhoppers *Nephotettix cincticeps.* The virus particles are spherical or polyhedral in shape, approximately 40–60 mμ in diameter and possess central dark areas which are surrounded by relatively transparent regions. These particles have been found in the cells of the fat body, blood, intestinal epithelium, salivary glands, Malpighian tubules, and ovarioles of infective leafhoppers. These particles are frequently seen in crystalline array in the cells of the infected insect tissues.

Similar particles have been shown to exist scattered or in clusters in the cytoplasm of cells in the chlorotic portions of diseased leaves of rice plants. They are occasionally seen in a crystal-like arrangement. The virus particles in a partially purified preparation extracted from the affected leaves of rice plants or bodies of infective leafhoppers are spherical or polyhedral in shape, are approximately 70 mμ in diameter and exhibit central electron-dense areas, 40–50 mμ in diameter, surrounded by less dense zones and outer membranes. These particles are covered with membranous envelopes, presumably of cytoplasmic origin. Toyoda *et al.* (1965) found that the above-mentioned envelopes could be removed by treating the virus preparation with phospholipase of snake venom or pancreatin. Then rice dwarf virus was eluted from *N,N*-diethylaminoethyl cellulose column with 0.2–0.25 M NaCl. The virus particles in the eluent were icosahedral in shape and provided with hollow capsomeres. The electrophoretic and sedimentation patterns of the effluent showed that the preparation consisted of homogeneous rice dwarf virus particles. The rice dwarf virus thus purified retained its infectivity to the rice plant when assayed through the insect vector *Nephotettix cincticeps.* Yoshii and Kiso (1959b) isolated ribonucleic acid from the tissues of affected rice plants and injected it into the abdomens of leafhoppers using glass capillaries. These leafhoppers, as well as those that ingested the RNA, became infective. Kimura (1962) extracted nucleic acid from a partially purified virus preparation. This nucleic acid was found to be infective when injected into the abdomens of nonviruliferous leafhoppers. The purified virus contains 11% RNA and no DNA (Toyoda *et al.,* 1964). According to Miura *et al.* (1966) this RNA has a complementary base composition, that is, both the mole ratios of adenine to uracil and guanine to cytosine are very close to unity. The optical density of the RNA in a 0.01 standard saline solution at 260 m$\mu$ shows a sharp increase at 80°C. This RNA did not react with formaldehyde at 37°C, but when heated at 100°C for 10 min and cooled quickly, it reacted with formaldehyde. In addition, it was more resistant to digestion by ribonuclease than was ribosomal RNA or transfer RNA. Thus, they concluded that ribonucleic acid from the rice dwarf virus is double stranded like reovirus RNA or wound-tumor virus RNA and consists of two complementary chains bound to each other through A–U and G–C pairings. Sato *et al.* (1966) also demonstrated the double-helix structure of this RNA with X-ray diffraction studies on the fibers of RNA that had been obtained from rice dwarf virus isolated from the infected rice leaves. Cytological changes in infective leafhopper tissues were found by Nasu (1963). In the viruliferous male *Nephotettix apicalis* (Motsch.), the fat body cell contained more numerous vacuoles, scanty cytoplasm, and a stellate nucleus which is similar to that described in *Macrosteles fascifrons* (Stal.) with aster yellows virus (Littau and Maramorosch, 1956, 1960). In some individuals such abnormalities were found in all fat bodies, but in others they were found only in several fat bodies in the abdomens. Seventy-four per cent of the infective adults of *Nephotettix apicalis,* 79% of those of *N. cincticeps,* and 67% of *Inazuma dorsalis* showed these changes in fat body cells. He further reported that the mycetome of vectors was also affected by rice dwarf virus. Viruliferous males carrying the virus showed shrunken mycetomes, in which the internal layers appeared to have hardened. This change occurred in 67% and 74% of infected males of *N. apicalis* and *N. cincticeps,* respectively.

The deleterious effect of rice dwarf virus on its vectors was also investigated by Nasu (1963). In congenitally viruliferous leafhoppers, *N. apicalis,* high mortality was caused by rice dwarf virus. The virus impaired the fecundity of the insect vectors. Infective females of *N. cincticeps* laid 28–68% fewer eggs than nonviruliferous leafhoppers. Likewise, 37–45% and 41% fewer eggs were laid by infective leafhoppers of *N. apicalis* and *I. dorsalis,* respectively. Among the infective females were frequently found those which laid no eggs throughout the whole oviposition period.

Yoshii (1959a) described metabolic abnormalities in *N. cincticeps* induced by rice dwarf virus. According to him oxygen consumption and the respiratory quotient were higher than normal in viruliferous leafhoppers; oxidative phosphorylation was also higher in the latter. The total phosphorus, inorganic, organic, lipid, nucleic acid, and protein phosphorus were found to increase in viruliferous leafhoppers. Yoshii came to the conclusion that the effect of the virus on the vector resulted in an increase of enzymic activities and in an accelerated metabolism. In his experiments, however, viruliferous leafhoppers were obtained by maintaining the insects for more than one month on diseased plants. Therefore, the changes found in the metabolism of viruliferous leafhoppers could have resulted from either the different diet or feeding on diseased plants.

### B. Stripe

Although Kuribayashi demonstrated as early as 1931 that rice stripe was readily transmitted by *Delphacodes striatella* Fall.,* the relationship between the vector and the virus was not studied closely until recently. Shinkai (1962) showed that about 20% of the planthoppers became infective after feeding on diseased plants. The shortest acquisition feeding period was 15 min; one day's feeding usually enabled the insects to acquire the virus. The latent period of the virus in insects ranged from 5 to 20 days, 5–10 days being most frequent. The infectivity of individual planthoppers appeared to be different, certain planthoppers having produced infections in plants on consecutive days while the others did so only at great intervals. The infective capacity of some planthoppers was apparently diminished in late adult life. The shortest inoculation feeding time was 10 min; one day's feeding enabled most of the infective planthoppers to produce infections in healthy plants.

Harmful and often lethal effect of this virus on its insect vectors was described by Nasu (1963). Fifty-two per cent of the eggs laid by the infective planthopper carrying rice stripe virus died prematurely. High mortality was observed in congenitally viruliferous plant hoppers; the mortality was especially high in nymphs of the first and second instars.

### C. Yellows and Black Streak

Rice yellows has been known in Kôchi Profecture at least since 1919. The experiments carried out in the Kôchi Agricultural Experiment

* Fennah (1963) placed this species in his new genus *Laodelphax*.

Station (1943) and those performed subsequently by Enjôji (1948) and Iida and Shinkai (1950) demonstrated that this virus was transmitted by *Nephotettix cincticeps*. According to Shinkai (1962), most leafhopper individuals acquired the virus by feeding on diseased rice plants for 1–3 hr; the minimum acquiistion feeding period was found to be 10 min. The incubation period of the virus in the insect was 20–39 days, generally 25–30 days. Recently *Nephotettix impicticeps* Ishiara* was also found to be a vector of the virus (Shinkai, 1959, 1962). Takahashi and Sekiya (1962) described cytological changes in *N. cincticeps* infected with rice yellows virus. Following infection, the nuclei of the fat body cells become enlarged and irregular in shape. Subsequently the nuclei shrink and vacuoles of varying size appear in the cytoplasm.

Kuribayashi and Shinkai (1952) were the first to distinguish black streak from rice dwarf for which the former must have been mistaken for many years. They discovered that black streak was transmitted by *Delphacodes striatella* Fall.† Shinkai (1962) showed that most individuals of the plant hopper could readily acquire the virus by feeding on diseased plants for one day, the minimum acquisition feeding period being 30 min. The latent period of the virus in the insect was 7–35 days, generally 7–12 days.

In conclusion, the relationships between insects and rice viruses in Japan are summarized as follows. Three species of leafhoppers, *Inazuma dorsalis* (Motsch.), *Nephotettix cincticeps* (Uhl.), and *N. apicalis* (Motsch.) transmit rice dwarf virus and *N. cincticeps* and *N. impicticeps* Ishihara are the vectors of rice yellows virus while the planthopper *Laodelphax striatella* (Fall.) transmits both rice stripe and black streak viruses. Among these viruses, the rice dwarf and stripe viruses have been demonstrated to pass through eggs of their insect vectors to the progeny. These insect vectors apparently are not only transmitters but also hosts of plant viruses.

## IV. LEAFHOPPER-BORNE VIRUS DISEASES OF THE RICE PLANT IN SOUTHEAST ASIA

Several leafhopper-borne virus diseases of the rice plant have been found in various countries in Southeast Asia. Among them the viruses

* Ishihara (1964) proposed this new name for *Nephotettix bipunctatus* (Fabricins) which is invalid.

† Recently Fennah (1963) created a new genus *Laodelphax* to which he assigned this species.

of yellows (or yellow dwarf) and orange leaf appear to be propagative in their insect vectors.

### *A. Rice Yellows*

Rice yellows is known in Taiwan (Kurosawa, 1940), Hainan Islands, South China, Thailand, and the Philippines (Hashioka, 1964). The relationship between this virus and its insect vector, *Nephotettix cincticeps,* has already been stated.

### *B. Orange Leaf*

The orange leaf disease of rice plant has been observed in Thailand since 1960 and found in the Philippines in 1962 (Rivera *et al.*, 1963). The symptom first appears on the leaves as a characteristic golden yellow to deep orange color when the plants are about one month old. As the disease advances, the affected leaves roll and die. The affected plants tend to develop few tillers. Rivera *et al.* (1963) have demonstrated that *Inazuma dorsalis* transmits this disease. About 14% of the virus-free population became infective after feeding on diseased plants. Adults became viruliferous after a minimum acquisition feeding period of 5 hr. These insects could infect healthy rice seedlings after a lapse of about 2–6 days and a minimum inoculation feeding of 6 hr was sufficient to cause the disease. Insects seemed to remain infective until their death. Symptoms first appeared approximately 13–15 days after the placement of the viruliferous leafhoppers. *Nephotettix apicalis,* which transmits rice dwarf, failed to transmit the orange-leaf disease.

### *C. "Tungro"*

"Tungro" disease (meaning degenerated growth) seems to be the most important virus disease of rice in the Philippines (Rivera *et al.*, 1965). The most conspicuous symptoms are the stunting of the plants and yellowing of the leaves. A yellowing of leaves may first appear as partial yellowing or striping with diffused chlorotic mottles, or with more or less defined yellowish to whitish spots on the younger leaves. In certain rice varieties the infected plants may show a few yellowish streaks along the veins of newly unfolded leaves. In other varieties the chlorotic leaves turn yellowish orange and frequently develop irregular dark brown blotches. Susceptible varieties infected at their early stages of growth produce few tillers, most of which develop symptoms. According to Rivera *et al.*, both the nymphs and adults of *Nephotettix impicticeps* acquired and transmitted the virus. The minimum acquisition feeding of 30 min on affected plants was sufficient to allow the transmission of disease, but greater success was obtained with longer feeding periods. The percentage of transmitters apparently increased from 3 to 83% when acquisition feeding periods were increased from 30 min to 48 hr. A 24-hr incubation period appeared to be necessary for adult insects to transmit the disease to an appreciable extent. One leafhopper was sufficient to infect healthy plants and the minimum inoculation feeding period was 15 min. Viruliferous insects remained infective until death, but transovarial transmission of the virus was not observed. As mentioned above, the incubation period of this virus within insect is extremely short and it has not been definitely shown that the virus multiplies in the bodies of the vector.

### *D. Transitory Yellowing*

The transitory yellowing of rice is of importance on the second crop in central and south Taiwan (Chiu *et al.*, 1965). Diseased plants show a yellowing of the leaves, starting with the older ones and extending from tip to base, and tillering is greatly reduced. It resembles "tungro" disease and is transmitted by *Nephotettix apicalis,* symptoms developing 2–4 weeks after feeding by infective leafhoppers. The varietal reactions to the two diseases differ.

### *E. "Penyakit merah"*

"Penyakit merah" (meaning red disease) of rice in Malaysia causes a marked stunting in plants, results in the discoloration of leaves in various shades from yellow to orange, and greatly reduces the yield. Young leaves show irregular areas of pale and dark green mottling, while more mature leaves develop yellow or yellow-orange tips and edges, the color persisting until they die. In plants infected early, tillering is limited and flowering delayed. This disease is transmitted by *Nephotettix impicticeps* (Ou *et al.*, 1965). From the symptoms, the insect vector and varietal reaction, the virus appears to be related to that of "tungro."

"Mentek" disease in Indonesia is also believed to be a virus disease related to "penyakit merah" or "tungro" (Ou *et al.*, 1965).

Serrano (1957) reported a virus disease, "accep na pula" (meaning red disease) which severely affected rice in the Philippines as well as in North Borneo, now the State of Sabah, Malaysia. This disease appears to be identical with "tungro" (Rivera *et al.*, 1965).

Grassy stunt is another virus disease of rice plant found in the Philippines. According to Iida (1965), Rivera and co-workers have shown that this virus is transmitted by *Nilaparvata lugens* Stal. and requires a 7-day incubation period within the insect.

## BIBLIOGRAPHY

Ando, H. 1910. On dwarf disease of rice plant [in Japanese]. J. Japan. Agr. Soc. No. 347, 1–3.

Bawden, F. C. 1964. Plant viruses and virus diseases. p. 20. 4th ed. Ronald Press, New York. 361 pp.

Carter, W. 1962. Insects in relation to plant disease. p. 270, 433. Interscience Pub., New York & London. 705 pp.

Chiu, R. J., Lo, T. C., Pi, C. L., and Chen, M. H. 1965. Transitory yellowing of rice and its transmission by the leafhopper *Nephotettix apicalis apicalis* (Motsch.). Bot. Bull. Acad. Sin. 6:1–18.

Cook, M. T. 1947. Viruses and virus diseases of plants. p. 156. Burgess. Minneapolis. 244 pp.

Fennah, R. G. 1963. New genera of Delphacidae (Homoptera: Fulgoroidea). Proc. Roy. Entomol. Soc. London Ser. B, 32:15–16.

Fukushi, T. 1933. Transmission of virus through the egg of an insect vector. Proc. Imp. Acad. 9:457–460.

———. 1934. Studies on the dwarf disease of rice plant. J. Fac. Agr. Hokkaido Univ. 37:41–164.

———. 1935. Early records of insect transmission of virus diseases [in Japanese]. Byôchûgai-Zasshi (J. Plant Protection) 22:38–46.

———. 1937. An insect vector of the dwarf disease of rice plant. Proc. Imp. Acad. 13:328–331.

———. 1939. Retention of virus by its insect vector through several generations. Proc. Imp. Acad. 15:142–145.

———. 1940. Further studies on the dwarf disease of rice plant. J. Fac. Agr. Hokkaido Univ. 45:83–154.

Fukushi, T., and Kimura, I. 1959. On some properties of the rice dwarf virus. Proc. Japan Acad. 35:482–484.

Fukushi, T., and Shikata, E. 1963. Localization of rice dwarf virus in its insect vector. Virology 21:503–505.

Fukushi, T., Shikata, E., and Kimura, I. 1962. Some morphological characters of rice dwarf virus. Virology 18:192–205.

Hashioka, Y. 1964. Virus diseases of rice in the world. Reprinted from Riso, Dec. 1964, 16 pp. (Rev. Appl. Mycol. 44:459. 1965).

Hino, I. 1927. Early records of great significance on phytopathological science in the Orient [in Japanese]. Agr. Hort. 2:1223–1232.

Iida, T. 1965. Geographical distribution of leafhopper-borne viruses of cereals. Paper presented at the Conference on Relationships Between Arthropods and Plantpathogenic Viruses, Tokyo, Oct. 25–28, 1965.

Iida, T., and Shinkai, A. 1950. Transmission of rice yellow dwarf by *Nephotettix cincticeps* (Uhl.) [in Japanese]. Annu. Phytopathol. Soc. Jap. 14:13–114. (Abstr.)

Ishihara, T. 1953. A tentative check list of the superfamily Cicadelloidea of Japan (Homoptera). Matsuyama Agr. Coll. Sci. Rep. No. 11. 1–72 p.

———. 1964. Revision of the genus *Nephotettix*. Trans. Shikoku Entomol. Soc. 8:39–44.

Ishikawa, R. 1928. The merit of Hatsuzo Hashimoto, the earliest investigator of dwarf disease of rice plant [in Japanese]. J. Plant Protection 15:218–222.

John, V. T. 1965. On the antigenicity of virus causing "tungro" disease of rice. Plant Disease Reporter 49:305–306.

Katsura, S. 1936. The stunt disease of Japanese rice, the first plant virosis shown to be transmitted by an insect vector. Phytopathology 26:887–895.

Kimura, I. 1962. Further studies on the rice dwarf virus. I [in Japanese]. Ann. Phytopathol. Soc. Japan 27:197–263; II:204–213.

Kimura, I., and Fukushi, T. 1960. Studies on the rice dwarf varus [in Japanese]. Annu. Phytopathol. Soc. Japan 25:131–135.

Kunkel, L. O. 1926. Studies on aster yellows. Amer. J. Bot. 13:646–705.

Kuribayashi, K. 1931a. Studies on the rice stripe disease [in Japanese]. Nagano Agr. Exp. Sta. Bull. 2:45–69.

———. 1931b. On the relationship between rice stripe disease and *Delphacodes striatella* Fall. [in Japanese]. J. Plant Protection 18:565–571, 636–640.

Kuribayashi, K., and Shinkai, A. 1952. Occurrence of rice black streak dwarf in Nagano and Tokyo. It is transmitted by *Delphacodes striatella* Fall. [in Japanese]. Ann. Phytopathol. Soc. Japan 16:41. (Abstr.)

Kurosawa, E. 1940. On rice yellows in Taiwan [in Japanese]. J. Plant Protection 27:161–166.

Leach, J. G. 1940. Insect transmission of plant diseases. p. 342. McGraw Hill, New York and London. 615 pp.

Littau, V. C., and Maramorosch, K. 1956. Cytological effects of aster yellows virus on its insect vector. Virology 2:128–130.

———. 1960. A study of the cytological effects of aster yellows on its insect vector. Virology 10:483–500.

Matsumura, S. 1902. Monographie der Jassinen Japans. Termes. Füzet, 25:353–404. p. 379–380, 391–392.

———. 1905. Die Hemipteren Fauna von Riukiu (Okinawa). Trans. Sapporo Nat. Hist. Soc. 1:15–38. p. 20.

Melichar, L. 1903. Homopteren-Fauna von Ceylon. 233 pp. (vid. p. 200)

Miura, K., Kimura, I., and Suzuki, N. 1966. Double-stranded ribonucleic acid from rice dwarf virus. Virology 28:571–579.

Motshulsky, V. 1859. Insectes des Indes orientales, et de contrées analogues. Etud. Ent. 8:25–118. (*Pediopsis apicalis*. p. 110)

Murata, T. 1915. Insect pests of rice and barley and their control [in Japanese]. p. 150–161. Kōzandō, Tokyo. 364 pp.

———. 1931. Dwarf disease of rice plant [in Japanese]. J. Japan. Agr. Soc. No. 604:47–50.

Nasu, S. 1963. Studies on some leafhoppers and planthoppers which transmit virus diseases of rice plant in Japan [in Japanese]. Kyushu Agr. Exp. Sta. Bull. 8:153–349.

———. 1965. Electron microscopic studies on transovarial passage of rice dwarf virus. Japan. J. Appl. Entomol. Zool. 9:225–237.

Ou, S. H., Rivera, C. T., Navaratnam, S. J., and Goh, K. G. 1965. Virus nature of "penyakit merah" disease of rice in Malaysia. Plant Disease Reporter 49: 778–782.

Rivera, C. T., and Ou, S. H. 1965. Leafhopper transmission of "tungro" disease of rice. Plant Disease Reporter 49:127–131.

Rivera, C. T., Ou, S. H., and Pathak, M. D. 1963. Transmission studies of orange-leaf disease of rice. Plant Disease Reporter 47:1045–1048.

Sato, T., Kyogoku, Y., Higuchi, S., Mitsui, Y. Iitaka, Y., Tsuboi, M., and Miura, K. 1966. A preliminary investigation on the molecular structure of rice dwarf virus ribonucleic acid. J. Mol. Biol. 16:180–190.

Serrano, F. B. 1957. Rice "accep na pula" or stunt disease—a serious menace to the Philippine rice industry. Philippine J. Sci. 86:203–230.

Shiga Agricultural Experiment Station. Results of experiments with insect pests [in Japanese]. Rep. 1:111–169 (1898); Rep. 2:1–26 (1900); Rep. 3:25–55 (1901); Rep. 4:19–65 (1903); Rep. 5:1–37, (1)–(36) (1904); Rep. 6:1–43 (1906); Rep. 7 and 8:1–43, (1)–(50) (1908).

Shikata, E. 1966. Electron microscopic studies on plant viruses. J. Fac. Agr. Hokkaido Univ. 55:1–110.

Shinkai, A. 1954. Transovarial transmission of rice stripe virus in *Delphacodes striatella* Fall. [in Japanese]. Annu. Phytopathol. Soc. Japan 18:169 (Abstr.)

———. 1956. Difference in rice dwarf virus transmitting ability of leafhoppers from various localities [in Japanese]. Annu. Phytopathol. Soc. Japan 21:127. (Abstr.)

———. 1958. Transovarial passage of rice dwarf virus in *Inazuma dorsalis* (Motsch.) [in Japanese]. Annu. Phytopathol. Soc. Japan 23:26. (Abstr.)

———. 1959. Transmission of rice yellow dwarf by taiwan-tsumaguro-yokobai [in Japanese]. Annu. Phytopathol. Soc. Japan 25:36. (Abstr.)

———. 1962. Studies on insect transmission of rice virus diseases in Japan [in Japanese]. Nat. Inst. Agr. Sci. Bull. C 14:1–112.

Smith, K. M. 1960. Plant viruses. 3rd ed. 209 pp. Methuen, London and Wiley & Sons, New York. (vid. p. 4–5)

Storey, H. H. 1931. The inheritance by a leafhopper of the ability to transmit a plant virus. Nature 127 (3216):928.

Takahashi, Y. and Sekiya, I. 1962. Adipose tissue of the green rice leafhopper, *Nephotettix cincticeps* (Uhler) infected with the virus of the yellow dwarf virus disease of the rice plant [in Japanese]. Japan. J. Appl. Entomol. Zool. 6:90–94.

Takami, N. 1901. On dwarf disease of rice plant and "tsumaguro-yokobai" [in Japanese]). J. Japan. Agr. Soc. No. 241:22–30.

Takata, K. 1895–1896. Results of experiments with dwarf disease of rice plant [in Japanese]. J. Japan. Agr. Soc. No. 171:1–4 (1895); No. 172:13–32 (1896).

Toyoda, S., Kimura, I., and Suzuki, N. 1964. Rice viruses with special reference to rice dwarf virus [in Japanese]. Protein Nucleic Acid Enzyme 9:861–867.

———. 1965. Purification of rice dwarf virus. Annu. Phytopathol. Soc. Japan 30: 225–230.

Yamada, W., and Yamamoto, H. 1954. Transmission of rice stripe virus through the eggs of *Delphacodes striatella* Fall. [in Japanese]. Ann. Phytopathol. Soc. Japan 18:169. (Abstr. 1)

———. 1955–1956. Studies on the stripe disease of rice plant [in Japanese]. I. Okayama Agr. Exp. Sta. Special Bull. 52:93–112 (1955); III. ibid. 55:35–56 (1956).

Yoshii, H. 1959a. Studies on the nature of insect transmission in plant viruses. V. On the abnormal metabolism of the virus transmitting green rice leafhopper, *Nephotettix bipunctatus cincticeps* (Uhler) as affected with the rice stunt virus [in Japanese]. Virus 9:415–422.

Yoshii, H., and Kiso, A. 1959b. Studies on the nature of insect transmission in plant viruses. IX. On the biological activity of ribonucleic acid from stunt virus-infected rice plant. 9:582–589.

# Interactions of Plant Viruses and Virus Strains In Their Insect Vectors

J. H. Freitag

*Department of Entomology and Parasitology,*
*University of California, Berkeley, California*

## I. INTRODUCTION

The transmission of the great majority of plant viruses by insects involves only the acquisition of a single plant virus from an infected plant and its inoculation into a healthy plant. Most plant viruses are transmitted independently and the presence of two or more viruses in the vector together results in no detectable interaction between them. Distinguished from this are a few viruses which are unique because they apparently cannot be transmitted alone by the vector. Insect transmission of these viruses is dependent on the presence of a second virus. A third type of relationship results in the interference or cross protection between viruses in the insect. In this instance the insect having acquired one strain is incapable of acquiring and transmitting a second related strain.

## II. INDEPENDENT TRANSMISSION

Unrelated viruses as well as related viruses are usually transmitted independently and simultaneously by a vector. The presence of two or more viruses in the vector results in no detectable interaction between them. Sylvester (1956) reported that two unrelated aphid-borne sugar beet viruses, yellows and yellow net, were independently transmitted at the same time. Similarly, aphids could transmit both potato leaf roll and the unrelated turnip latent virus simultaneously (MacKinnon, 1960). The acquisition of one of these two viruses by the aphids did not adversely affect their ability to acquire and transmit the other one.

Green peach aphids which acquired and transmitted an avirulent strain of potato leaf-roll virus were subsequently capable of acquiring and transmitting a virulent strain as readily as were previously virus-free aphids (Harrison, 1958). The lack of any interaction in the aphid vector in this instance is of special interest because of the evidence for multiplication of the leaf-roll virus in the green peach aphid presented by Stegwee and Ponsen (1958).

The autonomous concurrent aphid transmission of component viruses has been reported for complexes affecting turnips (MacKinnon and Lawson, 1966), citrus (Stubbs, 1964), strawberry (Mellor and Fitzpatrick, 1951), and raspberry (Stace-Smith, 1956). Although Allen (1957) failed to obtain any interactions between strains of barley yellow-dwarf virus in the aphid vector, Smith (1963) and Jedlinski and Brown (1965) reported interference interactions in the oat plant. The unusually high degree of vector specificity exhibited by the aphid vectors of different strains of barley yellow-dwarf virus have been noted several times (Rochow, 1961, 1963; Gill, 1967). The occurrence of vector-specific strains of barley yellow-dwarf virus and the fact that certain strains can be acquired from plants by aphids which are unable to transmit them to healthy plants are of special interest.

MacKinnon (1961, 1963) compared the efficiency of aphids as vectors of persistent viruses and found that it was important to consider not only the insects' stage of development and the host plant they were reared on, but also whether or not these hosts were infected with other unrelated viruses. He found that under certain conditions green peach aphids reared on leaf-roll-infected plants were better vectors of turnip latent virus than those reared on healthy plants.

Little is known about the simultaneous transmission of more than one stylet-borne virus. Since these viruses are carried on the tips of the aphids' stylets and do not multiply in the vector, one would not expect any interaction to occur. This fact was confirmed by the results obtained by Castillo and Orlob (1966), who showed that transmission of one strain of cucumber mosaic or alfalfa mosaic was independent of a second strain. Most aphids acquired and transmitted only one strain, while relatively few simultaneously transmitted two strains.

Dual transmission of plant viruses by whiteflies was first reported from India by Varma (1955) and later from Brazil by Flores and Silberschmidt (1963). In India the cotton whitefly was found to be capable of acquiring and transmitting both yellow-vein mosaic of pumpkin and yellow-vein mosaic virus of bhendi simultaneously. Single whiteflies transmitted the two viruses to healthy host plants susceptible to the respective viruses on the same day and would continue to do so for several days without access to a virus source plant. Flores and Silberschmidt (1963) reported the simultaneous transmission of both the infectious chlorosis of Malvaceae and the strain

*Leonurus* mosaic by single individuals of the whitefly *Bemesia tabaci* Genn. The percentage of insects which could acquire and transmit both virus strains was relatively low. Virus-free individuals acquired and transmitted the virus to a higher percentage of plants than those already carrying another virus. The presence of two virus strains in the vector reduced transmission of both virus strains.

There have been several reports of the simultaneous transmission of related as well as unrelated viruses by leafhoppers. Giddings (1950a,b) reported that beet leafhoppers which had acquired and transmitted one strain of the curly-top virus were readily capable of acquiring and transmitting a second strain. There was no evidence of any cross-protection interaction between strains, and individual leafhoppers transmitted as many as three strains simultaneously. Bennett (1967) confirmed the apparent absence of cross protection between strains of the curly-top virus in the beet leafhopper. He utilized certain host plants in his experiments that are immune or highly resistant to certain previously known virus strains, but are highly infectious to other recently discovered strains. He showed that a leafhopper fully charged with one strain of the virus was capable of acquiring and transmitting a second strain. There was no detectable interference with the acquisition and transmission of the second strain. The lack of cross protection between strains of curly-top virus in the leafhopper vector might be predicted on the basis that there is no evidence to indicate a multiplication of the virus in the insect. Transmission of curly-top virus over extended periods of time by individual leafhoppers has been accounted for by the acquisition of a full charge through virus accumulation during prolonged feeding on virus-infected source plants.

Nagaraj and Black (1962) found that the leafhopper *Agallia constricta* could simultaneously transmit two seriologically unrelated viruses, wound tumor and potato yellow dwarf. They could detect no interference between these two independent viruses although both are known to multiply in the body of the vector. Frederiksen (1964) in limited tests obtained an infrequent simultaneous transmission of both aster yellows and oat blue dwarf viruses by the leafhopper vector. The data were not adequate to permit detection of any inhibitory effects of one virus on the transmission of the other.

Rivera and Ou (1967) reported that two strains of the rice tungro virus protect against each other in plant hosts, but that there was no such cross protection of these strains in the insect vector. This virus is unique among leafhopper-borne viruses in that it has been shown to be nonpersistent in the leafhopper vector.

Sauer (1966) and Fulton (1967) found that single nematodes could acquire and transmit more than one virus at the same time. Sauer accomplished this for strains of tobacco ringspot while Fulton succeeded in showing that the same nematode could transmit both tobacco ringspot and tomato ringspot. There was no evidence that the transmission of one virus interfered with the transmission of another.

## III. DEPENDENT TRANSMISSION

A number of plant viruses are unique because they cannot be transmitted by insects when they occur alone in infected plants. Transmission of these viruses by insect vectors is dependent on the presence of another virus. Clinch *et al.* (1936) first called attention to this unusual interaction when they found that tuber blotch strain of potato aucuba mosaic virus could be transmitted by the green peach aphid only from plants that were also infected with potato virus A. Although both viruses were transmitted by the mechanical inoculation of sap to potato and a number of other solanaceous host plants, the aphid vector could not transmit the tuber blotch strain alone. Later, Kassanis (1961) found that the green peach aphid would not transmit any one of 12 strains of potato aucuba mosaic virus collected in various parts of Europe when the plants were infected only with a strain of this virus. The different strains, however, were found to be aphid transmissible if the plants were also infected with either potato virus A or Y.

Watson (1960) found that potato virus C was not aphid transmissible when potato plants were infected with it alone, but that the virus could be transmitted by aphids if the plants were also infected with the related potato virus Y. The close relationship of these two viruses suggested that the aphid-transmitted virus was produced through a process of genetic recombination. The experimental evidence, however, does not support such an explanation.

Tobacco rosette was first described from England by Smith (1946) and later by Vanderveken (1963) from Africa. The disease is caused by a complex of two viruses: (*1*) tobacco mottle which is sap transmissible, but cannot be transmitted by aphids except in the presence of tobacco vein distorting virus and (*2*) tobacco vein distorting virus, a persistent aphid-borne virus which enables tobacco mottle to be aphid transmitted. The component viruses persist for long periods in the body of the aphids, and as many as 20 successive tobacco plants have been infected in daily serial transfers without the insect having access to a fresh source of virus.

Watson (1962, 1964) described two other interactions in which she suggests that one virus is dependent on the presence of a second virus in order to be aphid transmitted. The first of these is sugar beet yellow net which may be transmitted by aphids only in the presence of another mild virus. Both viruses are of the persistent type and cannot be transmitted by a sap inoculation. Carrot motley dwarf is also reported to be caused by a complex of two viruses: carrot mottle which is sap transmissible, but not aphid transmitted except in the presence of the second component, red leaf, which is a persistent aphid-transmitted virus. It is claimed that the presence of red leaf in some way enables the aphid to transmit the mottle virus which the insect cannot transmit when it is present alone. More evidence is needed to substantiate such an interpretation.

Another unusual interaction between plant virus strains was reported by Bennett (1957). He described yellow vein as a unique mutant strain of curly-top virus that produces conspicuous symptoms in tobacco, but no symptoms on sugar beets. The yellow vein strain is dependent on the presence of typical curly-top virus for its transmission. The two strains have not been separated, but have been transmitted in combination by beet leafhoppers, dodder, and grafting to tobacco, tomato, and sugar beets. It is of interest that the presence of curly-top virus in tobacco partially protects against invasion by the yellow vein strain.

A number of suggestions have been made to explain the transmission of viruses by insects which are not transmissible when present alone in the plant, but which become insect transmissible in the presence of a second virus. These include the suggestion that the assisting virus in some way increases the multiplication of the assisted virus much as tobacco necrosis does for the satellite virus in plant tissues (Kassanis, 1962). This might result in a higher virus concentration in the whole plant or, more important, it might increase the virus titer in certain tissues such as epidermis or phloem and thus make the virus more available to the insect vector. Watson (1960) proposes that genetic recombination might occur between related viruses such as potato virus C and Y and that this genotypic exchange would result in strains that could be aphid transmitted. Kassanis (1961) indicates the possibility of phenotypic mixing between unrelated viruses resulting in the exchange of protein coats of the virion, thus providing the virus particle with the properties necessary for its aphid transmission. The aggregation of virus particles resulting from the presence of a second virus has also been suggested as enabling the attachment of the virus to the aphids' stylets in a manner required for transmission. It is evident that further work is necessary before this type of interaction between two plant viruses can be satisfactorily explained.

## IV. CROSS PROTECTION

The most interesting and provocative type of interaction occurs when the vector has acquired one strain and is protected or prevented from transmitting a second related strain. Cross protection has been demonstrated in the leafhopper vector for aster yellows (Kunkel, 1955, 1957; Freitag, 1967) and corn stunt (Maramorosch, 1958). The phenomenon is analogous to the cross protection shown by related virus strains in their plant hosts. Plant virus strains are usually antagonistic to one another and numerous examples of this have been described. This type of interaction has been used mainly to establish relationships and as an aid to classification. It is significant that plant viruses which multiply in their leafhopper vector, such as aster yellows, exhibit cross protection, but that ones which do not multiply, such as curly-top, fail to manifest this phenomenon.

Kunkel (1955, 1957) first demonstrated this type of interaction when he showed that aster yellows virus protected against a western strain of the virus in both plant and leafhopper host. Groups of 30 leafhoppers that had fed for 2 weeks on a plant infected with one strain of aster yellows virus and then 2 weeks on a plant infected with a second strain would later transmit only the first strain. The experiments showed that the strains protected against each other in the insect vector as effectively as they do in plants. Kunkel presumed infective leafhoppers to be infected and that when infected with one virus, they became immune to infection by another.

Maramorosch (1958) demonstrated unilateral cross protection of strains of corn stunt virus in individual *Dalbulus maidis* leafhoppers. When the leafhoppers first acquired the Rio Grande strain, then 2 weeks later the Mesa Central strain, they transmitted only the Rio Grande strain. If the insects, however, first acquired the Mesa Central strain, then 2 weeks later the Rio Grande strain, they transmitted the Mesa Central strain at first and the Rio Grande strain later. After one-day acquisition feeding periods on each of two strains, the initial transmissions were of either strain; later the Rio Grande virus was transmitted predominantly, and finally exclusively. Cross protection occurred only following certain favorably spaced or sufficiently long acquisition-feeding periods, but not following consecutive short one-day feeding periods on each of the two strains. In order for the cross-protection reaction to occur, it was necessary that the first strain be given adequate time to become well established before the leafhoppers were permitted to acquire the challenging strain. The reasons for the lack of cross protection in the individual leafhoppers tested when the

Mesa Central strain was acquired first are unknown, but it was suggested that the Rio Grande strain may multiply faster or may invade the leafhopper tissues more rapidly than the Mesa Central strain.

## V. CROSS PROTECTION BETWEEN STRAINS OF ASTER YELLOWS VIRUS IN THE LEAFHOPPER VECTOR

Three strains of aster yellows virus occurring in California were differentiated on the basis of symptoms produced on common plantain, *Plantago major* L., *Nicotiana rustica* L. var. *humilis,* China aster, *Callistephus chinensis* Nees, and periwinkle, *Vinca rosea* L. (Freitag 1964). The three strains of the virus were designated as the Severe, Dwarf, and Tulelake strains. Cross-protection tests were conducted in plant hosts such as common plantain and *N. rustica.* In common plantain, there was only partial cross protection between strains. The Dwarf and Severe strains protected *N. rustica* plants from infection by the Tulelake strain, while reciprocal cross protection resulted between the Dwarf and Severe strains. In *N. rustica* plants infected with Tulelake strain and challenged by Dwarf or Severe strains, an interaction occurred which resulted in the suppression of symptoms.

The present report provides additional evidence (Freitag, 1963, 1967) on the interference phenomenon between strains of aster yellows virus in the six-spotted leafhopper, *Macrosteles fascifrons* (Stål). Since the earlier work by Kunkel (1955, 1957) was done with other strains and mainly with groups of insects, these tests were conducted to determine if cross protection resulted with western strains of the virus in individual leafhoppers. The same three strains of aster yellows virus, Severe, Dwarf, and Tulelake, previously used (Freitag, 1964) to test for strain interference and cross protection in plant hosts were utilized. Various combinations of two of the three strains were tested in the leafhopper. The insects were given acquisition-feeding periods of varying sequences and durations. The interactions were determined on the basis of the symptoms developing on the test plants following transmission by the leafhopper vector.

## VI. INTERACTIONS OF SEVERE AND DWARF STRAINS IN THE VECTOR

Adult leafhoppers of approximately the same age were fed for 2 weeks on common plantain infected with one strain and then 2 weeks on plantain infected with a challenging strain. Twenty leafhoppers were then placed singly in 20 individual cages and transferred to a healthy seedling plantain daily for a period of 52 days when the test was terminated. Some insects died or were lost before the test was completed.

The results are shown in Table 1. When 10 female and 10 male leafhoppers fed for 2 weeks on plantain infected with Dwarf strain and then 2 weeks on Severe strain, 19 of 20 leafhoppers transmitted only the Dwarf virus. Only one leafhopper of the 20 tested simultaneously transmitted both strains. This insect transmitted Dwarf virus to 51 plants, and it transmitted Severe virus to only one plant. The evidence favors the hypothesis that the Dwarf strain provides strong protection against the Severe strain in the leafhopper vector.

When 10 female and 10 male leafhoppers were fed on plants in the reverse sequence, first 2 weeks on Severe strain followed by a similar feeding on a Dwarf-infected plantain, 17 of 20 transmitted only one or the other strain. Fifteen of 20 leafhoppers transmitted Severe strain only whereas 2 of 20 transmitted the Dwarf strain only. The two leafhoppers that transmitted only the Dwarf strain may have failed to acquire the Severe virus during the 2-week acquisition feeding period. However, failure to transmit the Severe strain may have resulted from an interaction between the two strains, in which case the Dwarf strain succeeded in being predominant.

TABLE 1

TRANSMISSION OF THREE STRAINS OF ASTER YELLOWS VIRUS BY LEAFHOPPERS GIVEN EQUAL ACQUISITION FEEDING PERIODS OF 14 DAYS ON EACH OF TWO STRAINS OR ALTERNATELY FED 2 DAYS ON EACH OF TWO STRAINS FOR A PERIOD OF 20 DAYS AND THEN TRANSFERRED DAILY TO HEALTHY PLANTAIN

| Virus sequence | Number of leafhoppers transmitting | | | | | |
|---|---|---|---|---|---|---|
| | One Strain | Dwarf only | Severe only | Tulelake only | Two strains | Neither strain |
| | *2-week acquisition feeding period/virus* | | | | | |
| Dwarf, Severe | 19/20 | 19/20 | | | 1/20 | 0/20 |
| Severe, Dwarf | 17/20 | 2/20 | 15/20 | | 1/20 | 2/20 |
| Dwarf, Tulelake | 24/24 | 22/24 | | 2/24 | 0/20 | 0/24 |
| Tulelake, Dwarf | 8/24 | | | 8/24 | 16/24 | 0/24 |
| | *2-day acquisition feeding period/virus*[a] | | | | | |
| Severe, Dwarf | 27/34 | 8/34 | 19/34 | | 3/34 | 4/34 |
| Dwarf, Severe | 11/11 | 8/11 | 3/11 | | | 0/11 |

[a] Total 10-day acquisition access time to each strain, alternating at 2-day intervals.

The majority of the individual leafhoppers proved to be extremely efficient vectors of both the Dwarf and Severe strains. They averaged as high as 0.853 transmissions/day with the Dwarf virus and 0.820/day with the Severe virus (Table 2). Several leafhoppers infected every plant or nearly every plant on which they fed. Two of these leafhoppers transmitted virus to 50 consecutive plants on which they fed. Once the latent period was completed, the leafhoppers transmitted virus almost daily. The reasons for the poor transmission by a few leafhoppers is not known. It may have been due to (*1*) limited virus multiplication in the vector, (*2*) inefficient transmission, or (*3*) some interaction of virus strains resulting in only intermittent transmission. The results of the present study with the Severe and Dwarf strains indicate reciprocal cross protection in the great majority of the leafhoppers. The simultaneous transmission of these two strains occurs in only a few individuals.

## VII. LEAFHOPPERS FED ALTERNATELY ON PLANTS INFECTED WITH DWARF AND SEVERE STRAINS

To determine if it were possible to increase the number of individual leafhoppers that would acquire and transmit two strains of the aster yellows virus simultaneously, the leafhoppers were given alternate 2-day feeding periods on plants infected with Dwarf or Severe strains for a period of 20 days. Thirty-four females fed first on plantain infected with Severe strain for 2 days and then alternately fed 2 days on plants infected with the Dwarf and then the Severe strain again for a period of 20 days (Table 1). Eleven male leafhoppers were similarly fed except that they were first fed on plantain infected with the Dwarf strain. Following the 20-day acquisition-feeding period, each leafhopper was caged individually and then transferred daily for the duration of its life to a fresh healthy plantain plant.

Nineteen of the 34 female leafhoppers transmitted only the Severe strain (Table 1), 8 transmitted only the Dwarf strain, 4 failed to transmit either strain, and 3 insects transmitted predominately Severe strain, but also transmitted Dwarf strain to a limited number of plants. The three leafhoppers that acquired and transmitted both strains were transferred to 62, 88, and 83 plantains and transmitted Severe strain to 51, 73, and 50 of these plants while also transmitting Dwarf strain to 2, 5, and 3 plantains, respectively. Two of these leafhoppers transmitted the Dwarf strain near the end of their lives whereas the third leafhopper transmitted the Dwarf strain to five plants before it transmitted the Severe strain to any plants. Most leafhoppers transmitted only one

TABLE 2

SUMMARY OF RESULTS OF TRANSMISSION OF THREE STRAINS OF ASTER YELLOWS VIRUS FOLLOWING VARIABLE SEQUENCES AND ACQUISITION ACCESS PERIODS

| Virus sequence | No. of leaf-hoppers | Virus trans-mitted | Fraction of plants infected | Plants infected per day | No. of days leafhoppers transferred to test plants, range | No. of transmissions | |
|---|---|---|---|---|---|---|---|
| | | | | | | range | mean |
| *2-week acquisition feeding period/virus* | | | | | | | |
| Dwarf, Severe | 19 | Dwarf | 624/731 | 0.853 | 14–52[a] | 4–51 | 35.6 |
| Severe, Dwarf | 15 | Severe | 452/551 | 0.820 | 5–52[a] | 3–51 | 30.1 |
| | 2 | Dwarf | 49/101 | 0.485 | 49–52[a] | 10–39 | 24.5 |
| Dwarf, Tulelake | 22 | Dwarf | 960/1162 | 0.826 | 6–86 | 5–68 | 43.6 |
| | 2 | Tulelake | 108/120 | 0.900 | 51–69 | 45–63 | 54.0 |
| Tulelake, Dwarf | 8 | Tulelake | 302/406 | 0.744 | 21–76 | 10–62 | 37.8 |
| *2-day acquisition feeding period/virus* | | | | | | | |
| Dwarf, Severe | 8 | Dwarf | 175/338 | 0.517 | 8–50[a] | 1–34 | 27.5 |
| | 3 | Severe | 86/150 | 0.573 | 50[a] | 25–31 | 28.7 |
| Severe, Dwarf | 19 | Severe | 813/1229 | 0.662 | 30–110 | 1–80 | 42.8 |
| | 8 | Dwarf | 256/542 | 0.472 | 23–108 | 2–88 | 32.0 |

[a] Daily transfers to fresh plantain test plants were discontinued at the number of days indicated.

strain even though they were given ample opportunity to acquire a second challenging strain. This result suggests that the first strain acquired by the leafhoppers preempts the multiplication sites and results in no multiplication of the second strain or, at most, a very restricted or limited virus increase.

It is difficult to explain certain virus transmission patterns by individual leafhoppers. Although most leafhoppers infected the great majority of the plants on which they fed, some transmitted virus only occasionally. One leafhopper, transferred daily to healthy plantain plants for 86 days, transmitted Severe virus to only one plant on the 68th day. Two others infected only four plants each, while feeding on 108 and 110 test plants, respectively. One of these leafhoppers infected four plants with the Severe strain; another infected four plants with the Dwarf strain.

The low efficiency of some individuals is contrasted to the high efficiency of others. Examples of a high rate of transmission can be illustrated by two leafhoppers that transmitted the Dwarf strain to 88 of 108 plants and 81 of 110 plants, respectively. Two other leafhoppers transmitted the Severe virus to 81 of 97 plants and 75 of 110 plants. One leafhopper infected 56 consecutive plants during its lifetime. The results indicate a multiplication of the two strains of the virus in the leafhopper with equally high transmission efficiency of both strains and a high degree of susceptibility of the plantain test plants. The average number of plants infected per day by individual leafhoppers following the alternate 2-day feeding periods on Dwarf and Severe strains are given in Table 2 and show a variation from 0.472 for 8 leafhoppers transmitting the Dwarf strain to 0.662 for 19 leafhoppers transmitting the Severe strain. Since these insects had all fed for a total of 20 days on the two strains, no time was allowed for an incubation period of the virus in the leafhopper in the calculation of these transmissions-per-day averages. If the averages had been figured from the time that the insects produced the first infection, they would have been higher.

Giving the leafhoppers alternate 2-day acquisition-access periods on each of two strains did not significantly increase the number of individual insects that transmitted both strains. Only three leafhoppers, while transmitting predominately Severe strain, also transmitted Dwarf strain to a limited extent. Furthermore, it is of interest that four other leafhoppers, following 10-day feeding periods on each of the two strains, failed to transmit virus to any of the test plants on which they fed. These individuals may either be resistant to virus multiplication or this might be the result of an interaction of the two strains resulting in an inactivation or a neutralizing of the infectivity of the two viruses.

Common plantain was found to be a more sensitive test plant than China aster (wilt-resistant crego mixed). In three instances when leafhoppers that had been transferred daily to plantain were transferred instead to asters daily, a greatly reduced number of plants developed aster yellows symptoms. One leafhopper transmitted Dwarf virus to 39 consecutive plantain plants. However, when transferred to asters daily this leafhopper infected only 11 of 22 plants. When returned to plantain, the leafhopper transmitted virus to 30 of 31 plants. Similarly a second individual, following 44 consecutive Dwarf infections of plantain, infected only 8 of 22 asters but transmitted the virus to 36 consecutive plants when again transferred daily to plantain. A third leafhopper that had transmitted Severe strain to 42 consecutive plantain plants, when transferred daily to aster plants, failed to infect 15 asters.

## VIII. INTERACTIONS RESULTING BETWEEN DWARF AND TULELAKE STRAINS IN THE LEAFHOPPER VECTOR

Evidence indicating that Dwarf and Severe strains gave a cross-protection reaction against one another raised the question whether a similar interaction occurs between other strain combinations. The interaction of Dwarf and Tulelake strains was tested rather than the interaction of Severe and Tulelake, because it is difficult to differentiate between the symptoms of the latter two strains, especially on plantain. However, on common plantain the symptoms of Dwarf are as readily distinguishable from Tulelake as they are from Severe.

Leafhoppers given acquisition-access periods of 2 weeks on Dwarf followed by 2 weeks on Tulelake transmitted only one virus strain. The results given in Table 1 show that 22 of 24 leafhoppers tested transmitted only Dwarf virus while 2 of 24 transmitted only Tulelake virus. The results indicate that Dwarf strain provides cross protection against the Tulelake strain.

When the feeding sequence was reversed and the insects first fed for 2 weeks on plants infected with the Tulelake strain and then for 2 weeks on plants infected with the Dwarf strain, it was found that the Tulelake strain did not provide protection against the Dwarf strain. The results indicate that only a unilateral protection occurs between these two strains. All 24 leafhoppers used in the test transmitted the Tulelake strain at first (Table 1), but later 16 of them abruptly changed and transmitted the Dwarf strain. Eight leafhoppers continued to transmit the Tulelake strain exclusively; however, two of these leafhoppers survived only 21 and 32 days, and it is not possible to predict what their transmission record would have been if they had lived longer.

The lack of cross protection in the leafhoppers resulted in a changeover from the transmission of Tulelake strain to the transmission of Dwarf strain that was variable and unpredictable in the individual insects. The strain interactions and the transmission behavior of the leafhoppers are examples of an extremely complex biological system. In four insects, the change from Tulelake transmission to that of Dwarf occurred without any interruption in transmission. The final Tulelake infection was followed immediately by Dwarf transmission. The last Tulelake infection by the group of 16 leafhoppers averaged 39.7 days, whereas the first Dwarf infection averaged 40.8 days, indicating that on the average there was only about a 1-day break. In six leafhoppers, there was an interruption of 1–13 days between the last Tulelake infection and the first Dwarf transmission. In six other leafhoppers, there was a period of overlapping when both strains were transmitted. There were no reversals following the changeover, indicating that once established, the Dwarf virus predominated and was transmitted exclusively for the remainder of the leafhopper's life. The 16 leafhoppers that transmitted Tulelake virus during the first part of their life and Dwarf during the latter part were transferred to 42–89 plants, with an average of 68.7 plants. They transmitted Tulelake strain to 19–56 plants, with an average of 37, and Dwarf strain to 1–46, with an average of 23.5 plants.

## IX. CROSS PROTECTION IN LEAFHOPPERS GIVEN UNEQUAL ACQUISITION FEEDING PERIODS ON DWARF AND TULELAKE STRAINS

Since leafhoppers given equal feeding periods on plants infected with Dwarf and Tulelake strains of the aster yellows virus showed a unilateral cross-protection reaction, it was considered important to determine the type of interaction resulting between these two strains when the insects were given unequal feeding periods. Leafhoppers were first given a 2-day feeding on a plant infected with one strain followed by a 14-day feeding period on a plant infected with the second strain. This was done to determine if the longer acquisition feeding on a second virus strain might somehow overwhelm a first strain acquired during a shorter feeding period. Following these two acquisition-feeding periods on infected plants, the individual leafhoppers were transferred to healthy plantain test plants daily in the first experiment, weekly in the second and third, and three times a week in the fourth. These transfers were continued for the duration of the leafhopper's life. To check the virus titer of plantain source plants and to determine the number of leafhoppers that acquired virus during a 2-day feeding period, leafhoppers were fed for only 2 days on plantain infected with either Dwarf or Tulelake strain and then transferred individually to healthy plantain at weekly intervals.

The results shown in Table 3 indicate that when individual leafhoppers were fed only 2 days on plants infected with either Tulelake or Dwarf strains, 57.7% acquired and transmitted the virus. Tests conducted with leafhoppers given 2 days' access to one strain followed by 14 days on a second strain resulted in 76.5% transmitting one strain, 4.8% transmitting two strains, and 18.7% failing to transmit either strain. The data provide strong support of the hypothesis that these two strains cross protect against each other and that the leafhopper can transmit only one strain even though given ample opportunity to acquire a second strain. If we eliminate the individuals that failed to transmit either virus, then the results show that 94.1% of the leafhoppers which transmitted virus transmitted only one strain, and only 5.9% transmitted both strains.

When the leafhoppers were given a 14-day feeding period on a plant infected with a second strain following a 2-day feeding period on a

TABLE 3

INTERACTIONS RESULTING BETWEEN DWARF AND TULELAKE STRAINS OF THE ASTER YELLOWS VIRUS WHEN INDIVIDUAL LEAFHOPPERS WERE GIVEN DIFFERENT ACQUISITION FEEDING PERIODS OF 2 DAYS ON PLANTS INFECTED WITH ONE STRAIN FOLLOWED BY 14 DAYS ON PLANTS INFECTED WITH A SECOND STRAIN AND THEN TRANSFERRED TO PLANTAIN TEST PLANTS [a]

| Sources and acquisition feeding periods | Number of leafhoppers transmitting | | | | |
|---|---|---|---|---|---|
| | One strain | Tulelake | Dwarf | Two strains | Neither strain |
| Tulelake 2 days | 49/100 | 49/100 | | | 51/100 |
| Dwarf 2 days | 74/113 | | 74/113 | | 39/113 |
| Tulelake 2 days× Dwarf 14 days | 88/115 | 60/115 | 28/115 | 8/115 | 19/115 |
| Dwarf 2 days× Tulelake 14 days | 88/115 | 49/115 | 39/115 | 3/115 | 24/115 |
| | 176/230 | 109/230 | 67/230 | 11/230 | 43/230 |

[a] Leafhoppers were transferred to healthy plantain test plants daily in Expt. 1, weekly in Expts. 2 and 3, and three times a week in Expt. 4.

plant infected with the first strain, there was a 23.6% increase (from 57.7 to 81.3%) in the percentage of leafhoppers transmitting virus above those which had fed for only 2 days on plants infected with one strain. This is less than one might expect and could be interpreted as favoring the occurrence of an interaction between the two virus strains in the vectors. In one experiment, there was actually a higher percentage of insects transmitting virus after they had fed for only 2 days on a plant infected with one strain than when they were given an additional 14-day feeding period on another plant infected with a second virus strain. The results in this instance suggest the possibility of an antagonistic interaction between the two strains resulting in a partial neutralizing effect of the two strains upon one another and a consequent failure of 21 of 60 leafhoppers to transmit either strain. The nature of this interaction is unknown, but it does appear to parallel a somewhat analogous interaction in *Nicotiana rustica,* when plants inoculated first with the Tulelake and then the Dwarf strain showed a strong tendency to recover from symptoms and resulted in a plant that appeared to be normal (Freitag, 1964).

An earlier experiment had indicated unilateral cross protection between Dwarf and Tulelake strains following 2-week access periods on each strain. The Dwarf strain gave cross protection against the Tulelake strain, but Tulelake failed to protect against Dwarf and was replaced by it in most leafhoppers. However, when the insects were given unequal access periods of 2 days on Tulelake and 2 weeks on Dwarf, no replacement of Tulelake virus resulted in the leafhoppers, as evidenced by no leafhoppers first transmitting Tulelake virus and then Dwarf. Hypothetically, a small charge of the Tulelake virus acquired in 2 days' feeding may have been displaced in these tests by a large charge of Dwarf virus acquired during a longer 2-week access period and before there was an opportunity for any transmission of the Tulelake strain. The leafhoppers usually transmitted the first strain to which they had access and once having established transmission of it, there was small likelihood that the strain would be replaced.

The replacement of one virus strain acquired during a short 2-day acquisition feeding period by a second strain acquired during a longer 14-day acquisition might be inferred from the data given in Table 3. The results show that 43.0% transmitted the first strain and 33.4% the second strain. Since 57.7% of the leafhoppers acquired and transmitted virus after only a 2-day acquisition period on one strain, there were 14.7% fewer leafhoppers transmitting the first strain when the insects were also given a 14-day period on the second strain than if they had been feeding on plants infected with only one strain.

The results of one experiment indicate that there may be some exceptions to cross protection and that it does not always occur. Apparently replacement of one strain by another can occur under some conditions. In one replication, only 4 of 30 leafhoppers transmitted Dwarf virus when fed for 2 days on Dwarf-infected plantain followed by 14 days on Tulelake-infected plantain, while a greater number transmitted the Tulelake strain. In contrast, 22 of 24 leafhoppers in this replication, which had fed during the same 2 days on the same Dwarf source plant but had not had access to the Tulelake source, transmitted Dwarf virus. These results could be interpreted as evidence that the Tulelake strain in this instance may have replaced the Dwarf strain before either strain was transmitted. Replacement of one strain by another appears, however, to be exceptional and the results of the majority of the tests conducted support the cross protection hypothesis.

While an interaction of the two strains might result in the failure of some leafhoppers to transmit either strain, limited evidence in a few individuals shows a greatly reduced rate of transmission of the Tulelake strain. As an example, one leafhopper following a 2-day acquisition feeding on Dwarf and a 14-day feeding on Tulelake transmitted the Tulelake virus to only 2 of 94 healthy plantain plants on which it fed. Two other leafhoppers, following similar acquisition feedings, transmitted the Tulelake strain to only 1 of 71 and 1 of 91 plants, respectively, when transferred daily to healthy plants. The results are too limited and do not provide enough evidence for any definite conclusions since the poor transmitters may merely be inefficient individuals.

Leafhoppers that successfully acquired the virus after feeding periods of 2 days on plantain infected with either the Dwarf or Tulelake strain were usually efficient vectors and transmitted the virus to more than 50% of the test plants on which they fed. The results of three such experiments indicate no consistent differences in the efficiency of the leafhoppers in transmitting these two strains of the aster yellows virus. When the leafhoppers were transferred weekly to plantain test plants, the average number of plants infected per week varied from 0.583 to 0.835. The mode of the number of transmissions following 12 weekly transfers varied from 7 to 9 and was 3–4 transmissions when only 4–5 weekly changes of the test plant were made.

Insects given the opportunity to acquire unequal charges of the Dwarf and Tulelake strains and then transferred daily to fresh plantain test plants transmitted virus to fewer plants than those given equal virus access periods (Table 2). The number of plants infected per day varied from 0.305 to 0.630 for those transmitting Dwarf and varied

interaction that significantly influenced the number of leafhoppers transmitting more than one strain.

Reciprocal cross protection was the rule between Severe and Dwarf strains and generally resulted in the exclusive transmission of only one of these two strains. The interaction between Dwarf and Tulelake was different and resulted in unilateral cross protection. The Dwarf strain protected the insect from the challenge of the Tulelake strain, but the Tulelake strain was replaced in some leafhoppers when challenged by the Dwarf strain. Such insects would at first exclusively transmit Tulelake strain and then later exclusively transmit the Dwarf strain. It is of interest that the Tulelake strain when challenged in a plant host, *Nicotiana rustica,* likewise failed to cross protect against Dwarf and Severe strains (Freitag, 1964). The results indicate that a somewhat similar interaction occurs in both the plant and insect host of the virus strains.

## BIBLIOGRAPHY


Allen, T. C. 1957. Strains of barley yellow dwarf virus. Phytopathology 47:481–490.

Bennett, C. W. 1957. Interactions of sugar-beet curly top virus and an unusual mutant. Virology 3:322–342.

———. 1967. Apparent absence of cross protection between strains of the curly top virus in the beet leafhopper, *Circulifer tenellus*. Phytopathology 57: 207–209.

Castillo, M. B., and Orlob, G. B. 1966. Transmission of two strains of cucumber and alfalfa mosaic viruses by single aphids of *Myzus persicae*. Phytopathology 56:1028–1030.

Clinch, P. E. M., Loughnane, J. B., and Murphy, P. A. 1936. A study of aucuba or yellow mosaics of the potato. Sci. Proc. Roy. Dublin Soc. 21:431–448.

Flores, E., and Silberschmidt, K. 1963. Ability of single whiteflies to transmit concomitantly a strain of infectious chlorosis of Malvaceae and of Leonurus mosaic virus. Phytopathology 53:238.

Frederiksen, R. A. 1964. Simultaneous infection and transmission of two viruses in flax by *Macrosteles fascifrons*. Phytopathology 54:1028–1030.

Freitag, J. H. 1963. Cross protection of three strains of the aster yellows virus in the leafhopper and in the plant. Netherlands J. Plant Pathol. 69:215.

———. 1964. Interaction and mutual suppression among three strains of aster yellows virus. Virology 24:401–413.

———. 1967. Interaction between strains of aster yellows virus in six-spotted leafhopper *Macrosteles fascifrons*. Phytopathology 57:1016–1024.

Fulton, J. P. 1967. Dual transmission of tobacco ringspot virus and tomato ringspot by *Xiphinema americanum*. Phytopathology 57:535–537.

Giddings, N. J. 1950a. Some interrelationships of virus strains in sugar-beet curly top. Phytopathology 40:377–388.

———. 1950b. Combination and separation of curly-top virus strains. Amer. Soc. Sugar Beet Technologists Proc. 6:502–507.

Gill, C. C. 1967. Transmission of barley yellow dwarf virus isolates from Manitoba by five species of aphids. Phytopathology 57:713–718.

Harrison, B. D. 1958. Ability of single aphids to transmit both avirulent and virulent strains of potato leafroll virus. Virology 6:278–286.

Jedlinski, H., and Brown, C. M. 1965. Cross protection and mutual exclusion by three strains of barley yellow dwarf virus in *Avena sativa* L. Virology 13:93–97.

Kassanis, B. 1961. The transmission of potato aucuba mosaic virus by aphids from plants also infected by potato viruses A and Y. Virology 13:93–97.

———. 1962. Properties and behaviour of a virus depending for its multiplication on another. Jour. Gen. Microbiol. 27:477–488.

Kunkel, L. O. 1955. Cross-protection between strains of yellows-type viruses. Advances Virus Res. 3:251–273.

———. 1957. Acquired immunity from infection by strains of aster-yellows virus in the aster leafhopper. Science 126:1233. (Abstr.)

MacKinnon, J. P. 1960. Combined transmission by single aphids of two viruses that persist in the vector. Virology 11:425–433.

———. 1961. Transmission of two viruses by aphids reared on different hosts. Virology 13:372–373.

———. 1963. Some factors that affect the aphid transmission of two viruses that persist in the vector. Virology 20:281–287.

MacKinnon, J. P., and Lawson, F. L. 1966. A second component found in turnip latent virus complex. Can. J. Bot. 44:795–801.

Maramorosch, K. 1958. Cross protection between two strains of corn stunt in an insect vector. Virology 6:448–459.

Mellor, F. C., and Fitzpatrick, R. E. 1951. Studies of virus diseases of strawberries in British Columbia. II. The separation of component viruses of yellows. Can. J. Bot. 29:411–420.

Nagaraj, A. N., and Black, L. M. 1962. Hereditary variation in ability of a leafhopper to transmit unrelated viruses. Virology 16:152–162.

Rivera, C. T., and Ou, S. H. 1967. Transmission studies of the two strains of rice tungro virus. Plant Disease Reporter 51:877–881.

Rochow, W. F. 1961. A strain of barley yellow dwarf virus transmitted specifically by the corn leaf aphid. Phytopathology 51:809–810.

———. 1963. Variation within and among aphid vectors of plant viruses. Ann. New York Acad. Sci. 105:713–729.

Sauer, N. I. 1966. Simultaneous association of strains of tobacco ringspot virus within *Xiphenema americanum*. Phytopathology 56:862–863.

Smith, H. C. 1963. Interaction between isolates of barley yellow dwarf virus. New Zealand J. Agr. Res. 6:343–353.

Smith, K. M. 1946. The transmission of a plant virus complex by aphids. Parasitology 37:131–134.

Stace-Smith, R. 1956. Studies on rubus virus diseases in British Columbia III. Separation of components of raspberry mosaic. Can. J. Bot. 34:435–442.

Stegwee, D., and Ponsen, M. B. 1958. Multiplication of potato leafroll virus in the aphid *Myzus persicae* (Sulz.) Entomol. Exp. Appl. 1:291–300.

Stubbs, L. L. 1964. Transmission and protective inoculation studies with viruses of the citrus tristeza complex. Australian J. Agr. Res. 15:752–770.

Sylvester, E. S. 1956. Beet yellows virus transmission by the green peach aphid. J. Econ. Entomol. 49:789–800.

Vanderveken, J. 1963. La rosette du tabac dans le homani (Katanga). Isolement des virus, symptomatologie et modes de transmission. Parasitica 19:65–80.

Varma, P. A. 1955. Ability of whitefly to carry more than one virus simultaneously. Current Sci. 24:317–318.

Watson, M. A. 1960. Evidence for interaction or genetic recombination between potato viruses Y and C in infected plants. Virology 10:211–232.

Watson, M. A., 1962. Yellow-net virus of sugar beet I. Transmission and some properties. Annu. Appl. Biol. 50:451–460.

Watson, M. A., Serjeant, E. P., and Lennon, E. A. 1964. Carrot motley dwarf and parsnip mottle viruses. Annu. Appl. Biol. 54:153–166.

# Maize Viruses and Vectors

Robert R. Granados

*Boyce Thompson Institute for Plant Research, Yonkers, New York*

## I. INTRODUCTION

In 1919 Brandes included corn (*Zea mays* L.) in the host range of sugarcane mosaic virus. This and subsequent virus diseases of corn or maize described prior to 1962 were considered to be of little economic importance. Losses were sometimes severe, but usually localized and large acreages of corn were never involved. During the past four years, there has been considerable interest in maize virus diseases in the United States and other countries of the world. In the United States much of the interest stemmed from the discovery of a new virus disease, maize dwarf mosaic virus, which caused severe losses in Ohio and other states from 1963 through 1965. The situation in the United States has become increasingly complex as a result of indistinct symptomatology and the occurrence of different viruses in the same areas. Most of the virus diseases described since 1962 have involved strains of known viruses such as sugarcane mosaic virus, corn stunt virus, and wheat streak mosaic virus.

Arthropod vectors are mainly responsible for the spread of maize viruses in the field. Several reviews on maize virus diseases have appeared in recent years, but most have been of a descriptive or historical nature, little or no attention being given to vector–virus relationships. In the present chapter, emphasis will be placed on the vectors of maize viruses. Due to their similarity to virus diseases, a section on arthropod-induced phytotoxemias of maize is included. For details concerning host range, distribution, symptomatology, etc. of maize viruses, the reader is referred to Smith (1957) or the reviews by Stoner (1964a, b, 1965a, c) Ullstrup (1966), Klinkowski and Kreutzberg (1958), and Stoner and Williams (1965).

## II. VIRUSES WITH APHID VECTORS

### *A. Sugarcane Mosaic Virus (SCMV)*

This virus primarily causes a disease of sugarcane although many grasses including corn may be infected (Brandes, 1919). Brandes (1920a) described a mosaic disease of corn and established its cause as sugarcane mosaic virus. Kunkel (1921) published a cytological study of a Hawaiian corn disease which he called "corn mosaic." Originally, it was believed that this disease was similar if not identical to the sugarcane mosaic virus disease of corn. Later Kunkel (1922) showed that the corn mosaic virus prevalent in Hawaii was transmitted from corn to corn by means of the delphacid, *Peregrinus maidis* (Ashmead), but not from corn to sugarcane. Subsequent studies by Kunkel (1927) and Hadden (1928) demonstrated the presence of two maize virus diseases in Hawaii. One virus, sugarcane mosaic, was transmitted by the cornleaf aphid, *Rhopalosiphum (Aphis) maidis* (Fitch), and the other, maize mosaic (corn stripe virus), by *P. maidis.*

Sugarcane mosaic virus is transmitted mechanically and is not seedborne (Brandes, 1919; Brandes and Klaphaak, 1923). Symptoms on corn appear as irregular, mottled, or mosaic areas on the leaves. These symptoms are most conspicuous on the newest growth. They consist of yellowish streaks, with irregular outlines, which extend between the veins for considerable distances. Early infection may cause a stunting of the plant and infected plants are generally lighter green than normal. The incubation period of the virus in corn is usually 5–6 days, but can vary from 3 to 30 days (Costa and Penteado, 1951; Holdeman and McCartney, 1965). Several strains of SCMV have been described (Summers, Brandes, and Rands, 1948; Abbott and Tippett, 1966). Sugarcane mosaic virus occurs in practically all the sugarcane growing countries of the world wherever susceptible sugarcane varieties are grown and corn in the vicinity could become infected.

Insect transmission of SCMV was first demonstrated by Brandes (1920a, b), with *R. maidis.* His results were confirmed by Kunkel (1924) in Hawaii, and later by many investigators. Eight other species of aphids are known to transmit SCMV. These include *Acyrthosiphon pisum* (Harris) (Abbott and Charpentier, 1962), *Amphorophora sonchi* (Oestl.) (Abbott and Charpentier, 1962), *Carolinaia cyperi* (Ainslie) (Tate and Vandenberg, 1939), *Dactynotus ambrosiae* (Thos.) (Abbott and Charpentier, 1962), *Hysteroneura setariae* (Thos.) (Ingram and Summers, 1936), *Myzus persicae* (Sulz.) (Anzalone and Pirone, 1964; Benada, Kvĭcala, and Schmidt, 1964), and *Schizaphis (Toxoptera) graminum* (Rond.) (Ingram and Summers, 1938). Lawas and Fernandez (1949) reported that *Aphis gossypii* Glover was a vector of corn mosaic in the Philippines. They considered the virus they worked with to be the same as that causing sugarcane mosaic. Kennedy, Day, and Eastop (1962) listed *Sipha flava* (Forbes) and *Aphis nerii* B. de F. as vectors of SCMV. Tate and Vandenberg (1939) showed that *S. flava* was not a vector and the transmission tests with *A. nerii* were inconclusive. Only *R. maidis* and *A. gossypii* have been shown to transmit SCMV to corn. The vectors *D. ambrosiae* and *H. setariae* acquired virus during a 5-min acquisition feeding on diseased sugarcane plants (Zummo and Charpentier, 1964, 1965). Virus transmission occurred within 5 min after the aphids were placed on healthy sugarcane plants and the aphids lost their ability to transmit SCMV 1–5 hr after being removed from a virus source plant. Anzalone and Pirone (1964) showed that 40–80 sec probes were sufficient to make *M. persicae* viruliferous. On the basis of these virus-vector relationships SCMV is classified as a stylet-borne (nonpersistent) virus.

Sugarcane mosaic virus is a rod-shaped particle, 12–15 mμ in diameter and about 750 mμ long (Brandes, 1964; Pirone and Anzalone, 1966; Herold and Weibel, 1963). Herold and Weibel (1963) demonstrated the presence of SCMV in ultrathin sections of diseased sugarcane and maize plants.

### B. *Maize Dwarf Mosaic Virus* (*MDMV*)

The maize dwarf mosaic disease was first observed in a corn field in Scioto County, Ohio in 1962 (Janson and Ellett, 1963). The following year an estimated 15,000 acres of corn were infected with MDMV and in 1964 the losses due to the virus were estimated at 5 million bushels (Janson *et al.*, 1965).

Initially, the disease was diagnosed as corn stunt, a disease which caused symptoms very similar to those induced by MDMV (Janson and Ellett, 1963; Stoner and Ullstrup, 1964.) Subsequent studies (Dale, 1964; Stoner, Williams, and Alexander, 1964; Williams and Alexander, 1964; Williams *et al.*, 1964) demonstrated that the maize dwarf mosaic disease was caused by an aphid-borne, mechanically transmissible virus which was similar in many respects to sugarcane mosaic virus. Serological tests showed that MDMV isolates from different areas of the United States were related (Bancroft *et al.*, 1966; Shepherd, 1965; MacKenzie, Wernham, and Ford, 1966; Wagner and Dale, 1966), and that MDMV was related to SCMV (Shepherd, 1965; Wagner and Dale, 1966; Williams and Alexander, 1965). There seems little doubt that MDMV is a strain of SCMV. It is similar to SCMV in the symptoms it induces on corn and other hosts and in its host range, properties *in vitro,* morphology, and transmissibility. In respect to host range, MDMV is different in that it readily infects Johnson grass, *Sorghum halepense* (L.) Pers., whereas previously known strains of SCMV rarely infected Johnson grass (Abbott and Tippett, 1964). Recently, Abbott and Tippett (1966) reported that Johnson grass plants inoculated with strain H of SCMV may not manifest any mosaic symptoms yet carry the virus in a latent form. Whether or not other strains of SCMV can infect Johnson grass is yet to be determined.

According to Shepherd and Holdeman (1965), a California isolate of MDMV is seed borne in both field and sweet corn to a low percentage (0.4%). Shepherd (1965) stated that seed transmission is probably responsible for the long-distance dispersal of the virus. Other investigators (Bancroft *et al.*, 1966; Sehgal, 1966) failed to demonstrate seed transmission with different isolates of MDMV. These differences in seed transmissibility could result from factors such as moisture content or stage of maturity of seed tested, the variety of corn used, and variation among isolates of MDMV. Wagner and Dale (1966) suggested that slight serological differences existed between some MDMV isolates. Two distinct strains of MDMV occurring in the northeastern United States were identified by host reaction and serological relationships (MacKenzie *et al.*, 1966). Maize dwarf mosaic virus has only been reported from the United States and it is predominant in the north-central region in states bordering the Ohio and Mississippi rivers (Stoner, 1965a).

Symptoms on corn first appear as a mosaic pattern of light and green areas at the base of young leaves. The mosaic may remain quite diffuse, but often the chlorotic areas may merge into narrow, continuous or broken streaks along the veins (Fig. 1*B*). In the field as diseased plants approach maturity, red or reddish purple spots and streaks develop in the leaves. When corn is infected early, severe stunting accompanies the above symptoms. The incubation period of the virus in maize seedlings is normally 5–7 days, but may vary from 3 to 14 days (Williams and Alexander, 1965; Stoner *et al.*, 1964).

Twelve species of aphids are known to transmit MDMV in a stylet-borne manner (Stoner *et al.*, 1964; Frazier, Freitag, and Gold, 1965; Bancroft *et al.*, 1966; Nault, personal communication). These include *Anuraphis maidiradicis* Forbes, *Aphis crassivora* Koch, *Aphis pisum* Harris, *A. gossypii, Brevicoryne brassicae* L., *M. persicae, Rhopalomyzus poae* Gill, *Rhopalosiphum padi* L., *R. maidis, S. graminum, Therioaphis maculata* Buckton, and *Dactynotus* sp. Comparative transmission tests with all vector species except *S. graminum* showed that the *Dactynotus* sp. was the most efficient vector (over 60% trans-

*Fig. 1.* Maize leaves showing symptoms induced by (*A*) Rio Grande corn stunt virus, (*B*) maize dwarf mosaic virus, and (*C*) Louisiana corn stunt virus.

mission) and *R. maidis* was least efficient (under 10% transmission) (Bancroft *et al.*, 1966). Individual *M. persicae* showed an acquisition threshold of 20 sec, and inoculation threshold of 1 min, and a retention period of 20 min.

Nault (personal communication) studied the acquisition of MDMV by a previously unreported aphid vector, *S. graminum*. A definite increase in rate of virus transmission occurred as the duration of acquisition probes increased from 15 to 60 sec. No appreciable increase occurred with probes from 60 to 300 sec and this indicated that virus uptake took place primarily in the first 60 sec of stylet penetration. It was concluded that *S. graminum* acquired MDMV intercellularly and primarily from the epidermis. MDMV is a rod-shaped particle 12–15 mμ in diameter and about 750 mμ in length (Bancroft *et al.*, 1966; Shepherd, 1965; Thornberry, Otterbacher, and Thompson, 1966).

In Europe, a virus disease of sorghum and maize has been known since 1938 (Goidanich, 1938). The disease is commonly known as sorghum red stripe (arrossamento striato del sorgo) or maize mosaic virus. On the basis of means of transmission, host range, and symptoms on host plants, it was concluded that sorghum red stripe virus (SRSV) and sugarcane mosaic virus (SCMV) were related (Grancini, 1957; Lovisolo, 1957). Serological and electron microscopic investigations demonstrated that SRSV and SCMV were closely related (Benada *et al.*, 1964; Dijkstra and Grancini, 1960). Sorghum red stripe virus or related strains have been reported throughout most of southern Europe (Pop and Tusa, 1966). Panjan (1966) isolated many strains of maize mosaic virus in Yugoslavia. The virus was serologically related to SCMV and SRSV. The maize mosaic disease in Europe is similar in many respects to the maize dwarf mosaic disease in the United States. Both are members of the sugarcane mosaic virus complex.

### *C. Abacá Mosaic Virus (AMV)*

Previous to 1963 it was believed that AMV was probably related to cucumber mosaic virus (Celino, 1940). Eloja and Tinsley (1963) presented evidence which suggested that AMV was a strain of sugarcane mosaic virus. Purified preparations of AMV contained flexuous rods 680 mμ long similar to those of sugarcane mosaic virus. Abacá mosaic virus antisera reacted specifically with sap from plants infected with sugarcane mosaic virus and failed to react with sap from plants infected with cucumber mosaic virus.

Abacá mosaic virus is mechanically transmitted and is stylet borne by seven aphid vectors (Kennedy *et al.*, 1962). Only *R. maidis* and *A. gossypii* have been shown to transmit AMV to corn (Celino and Ocfemia, 1941; Juliano, 1951).

Disease symptoms on corn are similar to those described for sugarcane mosaic. The incubation period in maize plants inoculated by aphids varies from 4 to 14 days. The disease occurs only in the Philippines and is not considered to be important.

### *D. Maize Leaf Fleck Virus (MLFV)*

The leaf fleck disease of maize was first observed in California in the summer of 1946 and later described as an aphid-borne virus by Stoner (1952). The virus is not soil or seed borne, nor is it mechanically transmissible. Several varieties of field, sweet, and pop corn have been successfully inoculated. Harding grass (*Phalaris tuberosa* L. var. *stenoptera* Hitch) is the only other known host plant. The disease is limited to the San Francisco Bay area of California.

The first disease symptoms in maize plants are expressed as small, round, pale areas at the tip of the older leaves. These circular spots enlarge to about 2 mm in diameter, then develop progressively toward the base of the leaf. When the distal part of the leaf is spotted, marginal and tip burning begins. The circular spots do not coalesce or produce any streaking in the leaf lamina. There is an advance of symptoms in alternate leaves up the plant until the entire syndrome is seen in the plant 2–3 weeks after the initiation of symptoms. At this stage of infection the plant has a speckled or flecked appearance. The incubation period of MLFV in maize ranges from 4 to 8 weeks, with an average of 6 weeks.

The virus is transmitted in a circulative (persistent) manner by the aphids *Rhopalosiphum prunifoliae* (Fitch), *R. maidis,* and *M. persicae.* For *R. prunifoliae* the acquisition threshold was between 5 and 10 min, the inoculation threshold was 5 min or less, and the transmission threshold was 19 min (Stoner, 1952). Consequently, due to the short time interval during which virus was transmitted, no measurable latent period was detected in *R. prunifoliae.* Infective *R. prunifoliae* and *M. persicae* retained the virus for 9 and 6 days, respectively. The relationship between MLFV and its aphid vector is unique since this is the only circulative, aphid-borne virus where a latent period has not been demonstrated in the vector.

Atanasoff (1965) recently described a seed borne virus disease of maize which he considered to be the same virus described by Stoner in 1952. The evidence given for the proof of seed transmission is questionable since the experiments were conducted under uncontrolled conditions in the field. Furthermore, the described disease symptoms varied and were different from those reported for the leaf fleck virus disease. Mechanical transmission tests were negative and no aphid transmission tests were reported. The papers by Jones (1945) and McKeen (1952) were cited as examples of earlier work with maize leaf fleck; this is purely speculative and unfounded. According to Atanasoff (1966) the "leaf fleck" disease of maize is universal and thus far it has been reported from Africa, Bulgaria, Czechoslovakia, The Netherlands, U.S.S.R., and Yugoslavia. It is presently considered as the most serious problem of maize in Bulgaria, Rumania, and Yugoslavia.

Recently, Panjan (1966) described a seed borne, mottle mosaic disease of corn in Yugoslavia. The symptoms on maize were similar to those described by Atanasoff (1965, 1966) and the disease was not transmitted mechanically. He concluded that the mosaic symptoms were caused by genetic factors.

### *E. Cucumber Mosaic Virus (CMV)*

Wellman (1934) showed that many plants of the Gramineae, among them *Z. mays,* were readily infected when inoculated with southern celery mosaic virus. Price (1935) later showed this virus was a strain of CMV. Several strains of CMV have been shown to infect corn (Stoner, 1964a; Harter, 1938; Bridgmon and Walker, 1952; Pound and Walker, 1948; Doolittle and Zaumeyer, 1953). Naturally infected corn has been reported from California and Florida.

Symptoms on corn vary considerably, depending on the virus strain and temperature (Bridgmon and Walker, 1952; Pound and Walker, 1948). The symptoms on corn described by Wellman (1934) consist of many light colored, intermittent, elliptical spots of various lengths and widths which form stripes parallel to the veins. There is a tendency for the leaf blades to split and the leaf tips to crumple. Severe infections may stunt plants and induce buff-colored necrotic areas on the leaves. The incubation period of the virus in corn varies from 3 to 20 days.

Cucumber mosaic virus is mechanically transmitted and numerous aphid vectors have been reported (Kennedy *et al.,* 1962). Transmission of CMV to corn was demonstrated with *A. gossypii* (Wellman, 1934) and *M. persicae* (Stoner, 1964a). The virus is stylet borne and the vector-virus relationship has been studied extensively by many workers. Cucumber mosaic virus is a spherical virus with an average diameter of 30 m$\mu$ (Murant, 1965).

The virus probably is distributed worldwide. In Yugoslavia the stripe mosaic disease of maize is caused by CMV (Panjan, 1966). It is not as widespread as maize mosaic (see maize dwarf mosaic virus).

### *F. Barley Yellow-Dwarf Virus (BYDV)*

Barley yellow-dwarf virus was first described and studied by Oswald and Houston (1951). Oats and barley are the small grains most severely affected by the disease; however, many other members of the grass family including corn also are susceptible. BYDV is present in many countries and is probably the most widely distributed virus affecting cereals (Rochow, 1961).

Oswald and Houston (1953a), using *R. padi* (*R. prunifoliae*) as a vector, reported that corn plants were immune to infection by BYDV. Allen (1957) induced symptoms in the sweet corn variety 'Golden Cross Bantam' with a BYDV strain isolated from Arizona. *Rhopalosiphum padi* was used as the vector. Allen's attempts to transfer the virus from corn to corn, barley, or oats were unsuccessful. In 1960, Watson and Mulligan (1960) transmitted a virulent strain of BYDV to the corn variety 'White Horse Tooth' by using infective *R. padi.* The virus was recovered from infected plants.

Previous to 1965 corn was considered only as an experimental host for BYDV since the virus had never been recovered from naturally infected plants. Stoner (1965b) isolated BYDV from infected oats obtained from a field in South Dakota. Infective *R. padi* transmitted the virus to five varieties of corn. Serial transmissions were made from barley to corn to corn, barley to corn to barley, and barley to corn to oats. Recovery trials were made from the test plants to determine the presence or absence of virus.

Symptoms on corn are first observed as dark red to purple colorations which develop in an irregular pattern at the apical portion of some of the oldest leaves (Stoner and Williams, 1965). After the appearance of initial symptoms on the lower leaves, similar symptoms develop in the next alternate older leaves. These symptoms may involve 4–6 of the older leaves. The center growth of the whorl and the upper half of the plant remain normal in appearance. Stunting of the affected plant is slight. The incubation period of BYDV in corn plants ranges from 3 to 6 weeks after inoculation by aphids.

Barley yellow-dwarf virus is not transmitted mechanically, nor is it seed or soil borne (Oswald and Houston, 1953b). Ten aphid species are reported as vectors of BYDV (Kennedy *et al.*, 1962), but only *R. padi* has been shown to transmit the virus to maize. The virus is circulative in its aphid vectors and once the insect acquires the virus, it may transmit virus as long as it lives. The virus can be retained by individual *R. padi* and *M. granarium* for 21 and 12 days, respectively (Rochow, 1959). A latent period in the vector was demonstrated by Rochow (1963b). In general, groups of aphids allowed acquisition feeding periods of 12 hr or less require about 5 days to reach maximum transmission ability. Many interesting virus–vector relationships have been reviewed by Rochow (1961, 1963a). Barley yellow-dwarf virus is a polyhedral particle 30 mμ in diameter (Rochow and Brakke, 1964). The virus was purified from infected oats and viruliferous *Macrosiphum avenae* (F.).

## III. VIRUSES WITH LEAFHOPPER AND PLANTHOPPER VECTORS

### *A. Corn Stunt Virus (CSV)*

The corn stunt disease, also known as maize stunt or achaparramiento, was first observed and described by Altstatt (1945) in Texas and possibly by Frazier 1945) in California. Kunkel (1946a, b; 1948) established the viral nature of the disease and showed that *Dalbulus maidis* DeL. & W. was a vector.

Corn and teosinte, *Euchlaena mexicana* Schrad., are the only known hosts of CSV (Kunkel, 1948). The syndrome of the corn stunt disease in corn plants depends on the age of the plants when infected, the variety, and the strains of virus involved. Maramorosch (1955a) described two strains of CSV and named them Rio Grande and Mesa Central. The strains are distinguished by symptoms in corn. Rio Grande CSV causes small chlorotic spots of various sizes at the bases of newly developed leaves. These spots become numerous and coalesce into stripes along or between the veins of succeeding leaves (Fig. 1*A*). Eventually fully chlorotic leaves develop. Numerous chlorotic secondary shoots appear at the axils of leaves and along ear-bearing branches. Mesa Central CSV is distinguished from the Rio Grande virus in chlorotic markings, severity of stunting, and amount of red coloration. The streaks are more continuous and there is no discoloration at the bases of leaves. Frequently the amount of reddening of leaves is very extensive and a deep purple color often appears. Affected plants are more stunted and in general the disease is more severe. Secondary axillary and basal shoots are produced.

An apparently new strain of CSV was described by Granados *et al.* (1966b). The most conspicuous disease symptom under greenhouse conditions was a tearing of the leaf lamina edges (Fig. 1*C*). Chlorotic stripes may occur on the leaves, but generally there was an absence of chlorotic tissue. Stunting of the plant and the proliferation of basal shoots were typical of the stunt disease. This isolate of CSV was obtained from Louisiana and was therefore named Louisiana CSV. In Louisiana corn fields the marginal tearing of the leaf lamina and the basal shoot proliferation were found only in about 1% of the plants. Stunting, chlorosis, and reddening, depending upon the corn variety, were the only consistent symptoms that were associated with the disease in the field.

What may be a fourth strain of CSV was described by Ancalmo and Davis (1961) from El Salvador. Diseased leaves showed a fine stipple striping along the blades that did not coalesce to form bands. The diseased plants were not stunted. This possible type of CSV was transmitted by *D. maidis*.

Corn stunt virus can be maintained in corn plant tissue culture and in submerged single-cell cultivation for several months (Maramorosch, 1958b). A low virus titer in the tissue culture was indicated by the fact that *D. maidis* injected with inoculum prepared from diseased tissue culture cells transmitted CSV only after a long incubation period.

Corn stunt virus cannot be transmitted mechanically nor is it carried in the seed (Kunkel, 1948). The virus can be transmitted mechanically from insect to insect or from diseased plans to healthy leafhoppers, but not from plant to plant (Maramorosch, 1951).

Corn stunt virus is transmitted by three leafhopper species belonging to the subfamily Deltocephalinae. These include *D. maidis, Dalbulus elimatus* (Ball) (Niederhauser and Cervantes, 1950), and *Graminella nigrifrons* (Forbes) (Granados *et al.*, 1966b). The virus is circulative in all three vectors and infective insects generally transmit CSV up to the time of their death. Corn stunt virus multiplies in *D. maidis* (Maramorosch, 1955b) and a similar virus–vector relationship probably exists with the other two vectors.

*Dalbulus maidis* is a highly efficient vector since individual viruliferous insects invariably transmitted CSV to every plant they were transferred to at daily intervals (Kunkel, 1948). The minimum incubation periods of CSV in *D. maidis, D. elimatus,* and *G. nigrifrons* were reported as 14 (Kunkel, 1948), 22 (Niederhauser and Cervantes, 1950), and 23 days (Granados *et al.,* 1966a, b), respectively. The differences in the minimum incubation periods were not necessarily meaningful since conditions during the tests may not have been the same. Granados and Maramorosch (1967) reported on comparative CSV transmission tests by three leafhopper species. The minimum incubation period of Louisiana CSV in colonies of *G. nigrifrons, D. maidis,* and *D. elimatus* ranged from 21 to 25, 17 to 22, and 14 to 16 days, respectively. These studies also showed that *G. nigrifrons* was a relatively inefficient vector of CSV since insects reared on diseased corn plants infected 8–25% of plants during 24–96 hr feeding periods, whereas *D. maidis* and *D. elimatus* infected 55–100% of the plants.

Maramorosch (1958a) studied the cross protection between two strains of CSV in individual *D. maidis* and demonstrated unilateral protection of a plant virus in an insect vector. Insects which acquired the Rio Grande virus first were immune to infection by the Mesa Central virus. When Mesa Central was acquired first, it was transmitted initially, but later was followed by transmission of Rio Grande virus. It was concluded that the Rio Grande strain either was more virulent, moved more rapidly, or multiplied faster than the Mesa Central strain.

The biology of *D. elimatus* and *D. maidis* was studied in detail by Barnes (1954). Granados and Maramorosch (1967) reported that *G. nigrifrons* had a longer nymphal instar duration than either *D.*

*maidis* or *D. elimatus*. The longevity of adult *G. nigrifrons* was shorter than that of the other two species. Mitsuhashi and Maramorosch (1963) reported on the biology of *D. maidis* reared on corn plants under aseptic conditions.

The corn stunt disease has never been of great economic importance in the United States and only sporadic localized losses have occurred in some southern states. Previous to 1962 the disease was known to occur only in Texas and in California, but it is now known to occur in several southern states (Granados *et al.*, 1966b).

It had long been assumed that the streak disease of corn in California described by Frazier in 1945 was corn stunt virus (Kunkel, 1946b, 1948). Kunkel (1946b) obtained diseased corn material from Texas and California and reported the transmission of a new virus, corn stunt, by *D. maidis*. Frazier, Freitag, and Holdeman (1966) believed that Kunkel's studies (1946b, 1948) gave no substantiating evidence of having transmitted CSV from the California material. The corn stunt disease has never again been reported from California and attempts to transmit CSV from streak-diseased corn by *D. maidis* failed in 1947, 1964, and 1965 (Frazier *et al.*, 1966). Frazier *et al.* (1965, 1966) concluded that the occurrence of CSV in California was doubtful and that CSV was not the cause of the streak disease of corn. They felt that the streak disease reported in 1945 (Frazier, 1945) was a Johnson grass-infecting strain of sugarcane mosaic virus which is now very common in California. On the other hand, Shepherd (1965) expressed doubt that the mosaic virus of corn and Johnson grass (maize dwarf mosaic virus) now present in California is the same as that described by Frazier in 1945.

### B. *Maize Mosaic Virus 1 (MMV)*

Maize mosaic virus 1, also referred to as corn stripe or enanismo rayado, has been known since 1922 when Kunkel (1922) succeeded in transmitting a mosaic disease of corn by *P. maidis*. However, it was not until 1927 that MMV was demonstrated to be distinct from the aphid-borne sugarcane mosaic virus (Kunkel, 1927; Stahl, 1927).

The symptoms on corn first appear as small, elongated white specks at the base of the youngest leaves. The specks may elongate and form fine stripes parallel to the leaf veins. These stripes may also coalesce and form broad yellow bands. The stripes generally extend the length of the leaf and may occur on the leaf sheath, earhusk, and stalk. The incubation period of the virus in corn seedlings ranges from 4 to 24 days (Carter, 1941).

Maize mosaic virus is not mechanically transmissible. It is persistent in the planthopper *P. maidis*. Carter (1941) found that following 1- or 2-day acquisition feeding periods insects could transmit MMV regularly during alternate 2-day feeding periods until shortly before death. Sometimes insects which initially transmitted virus suddenly lost their ability to transmit. Carter indicated that this might be due to an exhaustion of the virus content in the insect. Individual insects which were inefficient vectors of MMV were also observed in the vector population. The incubation period of the virus in single *P. maidis* ranged from 11 to 29 days and, in one case, a 4-day incubation period was recorded. The ability of *P. maidis* to transmit MMV after an abnormally short incubation period might be controlled by genetic factors (Carter, 1941).

McEwen and Kawanishi (1967) showed that the minimum acquisition feeding period was between 2 and 4 hr and small nymphs (first and second instars) were unable to acquire virus from diseased corn plants. They also showed that virus could be transmitted during 15-min inoculation feeding periods. However, virus transmission by single insects was erratic when they were transferred serially to several plants, especially during short feeding periods. The incubation period of MMV in single insects ranged from 14 to 31 days. These values were greater than those reported by Carter (1941) but less than the 9–58 day incubation period reported by Herold and Munz (1965). Unexpectedly, *M. persicae* aphids transmitted maize mosaic virus in one test but this could not be confirmed in subsequent trials. *Peregrinus maidis* recovered MMV from the corn plants inoculated by *M. persicae* (McEwen and Kawanishi, 1967).

Virus-like particles were found in MMV-infected corn plants (Fig. 2) and in salivary glands and intestinal epithelial cells of infective *P. maidis* (Herold, Bergold, and Weibel, 1960; Herold and Munz, 1965). The virus particles, 50 m$\mu$ in diameter and 224 m$\mu$ long, were found in low concentration in only 4 of 12 salivary glands of infective planthoppers. It was suggested that this might explain why certain infective *P. maidis* are poor vectors.

Direct experimental evidence for the multiplication of MMV in *P. maidis* is not available. Nevertheless, certain lines of evidence including the long incubation period in the vector and the presence of virus particles in certain tissues of infective insects strongly indicate that multiplication of virus does occur. Maize mosaic virus occurs in Hawaii and in several tropical countries.

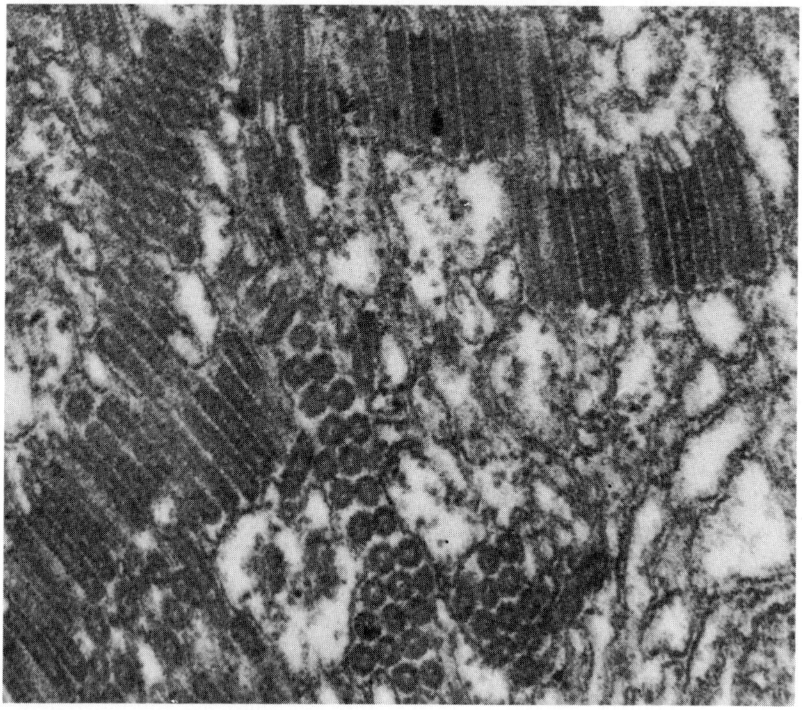

*Fig. 2.* Electron micrograph of a section through a cell of a diseased corn leaf showing maize mosaic virus particles in longitudinal and cross sections in the cytoplasm. × 75,000. (Courtesy of Dr. F. Herold, Instituto Venezolano de Investigaciones Científicas, Caracas, Venezuela, unpublished.)

### *C. Maize Streak Virus (MSV)*

Streak disease was the name given by Storey (1925a) to an aberrant chlorotic condition occurring in maize, sugarcane, and several wild grasses. The virus disease is characterized in maize by the development on the leaves of narrow, chlorotic streaks which parallel the veins. Several strains of the virus are known (Storey and McClean, 1930; McClean, 1947) and each is specialized to a particular host species. The incubation period of the virus in corn plants varies from 3 to 6 days.

Maize streak virus is not seed borne nor mechanically transmissible. It is transmitted by the Deltocephalinae leafhopper vectors *Cicadulina mbila* (Naude), *C. bipunctella zeae* China, *C. latens* Fennah, *C. parazeae* Ghauri, and *C. storeyi* China (Storey, 1925b, 1938; Ruppel, 1965).

The virus is circulative in its vectors and is not transmitted transovarially. Storey (1928, 1939b) indicated that the virus might be propagative in the vector, but definite experimental proof has not been obtained. The relationships existing between the virus and *C. mbila* were reported in the classical papers by Storey (1925b, 1928, 1932, 1933, 1938, 1939a).

Races of *C. mbila* were bred which were either able or unable to transmit the virus of streak disease in the natural process of feeding on maize plants (Storey, 1932). These races were termed "active" and "inactive," respectively. By crossing of the pure races Storey demonstrated that the ability to transmit was inherited as a simple dominant Mendelian factor, linked with sex. Inactive races of *C. mbila*, normally unable to transmit the virus, were made infective by needle inoculation with the streak virus. A simple puncture of the abdomen with a sterile needle, either following or followed by a feeding on a diseased plant, sometimes caused "inactive" *C. mbila* to become infective. The treatment was successful only if the needle penetrated some part of the intestine. These experiments showed that the "inactive" insects differed from the "active" ones primarily, but not entirely, in a property of the intestinal wall that resisted the passage of virus.

The minimum acquisition and inoculation feeding periods for "active" *C. mbila* were between 5 and 20 sec and 5 and 10 min, respectively (Storey, 1938). Maize streak virus must be inoculated directly into the phloem if infection is to follow. However, the virus is not confined to the phloem tissues and it may enter the mesophyll tissue from where insects may acquire it. The incubation period of MSV in *C. mbila* varied between 6 and 12 hr at 30°C (Storey, 1928). At temperatures above and below 30°C the incubation period was prolonged. All the nymphal instars as well as the adults of both sexes can acquire the virus. Females appear to be more efficient than males in transmitting virus.

Insects which acquire virus via the oral route usually continue to transmit virus up to the time of their death (Storey, 1928). In certain instances *C. mbila* may remain infective for as long as 9 weeks following an acquisition feeding of only 15 sec (Storey, 1939b). On the other hand, if insects obtain virus by abdominal injection, the insects, infective at first, become noninfective within 1 week and usually within 3 or 4 weeks (Storey, 1933, 1939b).

Storey (1937) described a new virus disease of maize which he named maize mottle virus. He concluded that the mottle virus and maize streak virus were different viruses since they produced different symptoms on maize and they failed to cross protect in plant and insect hosts.

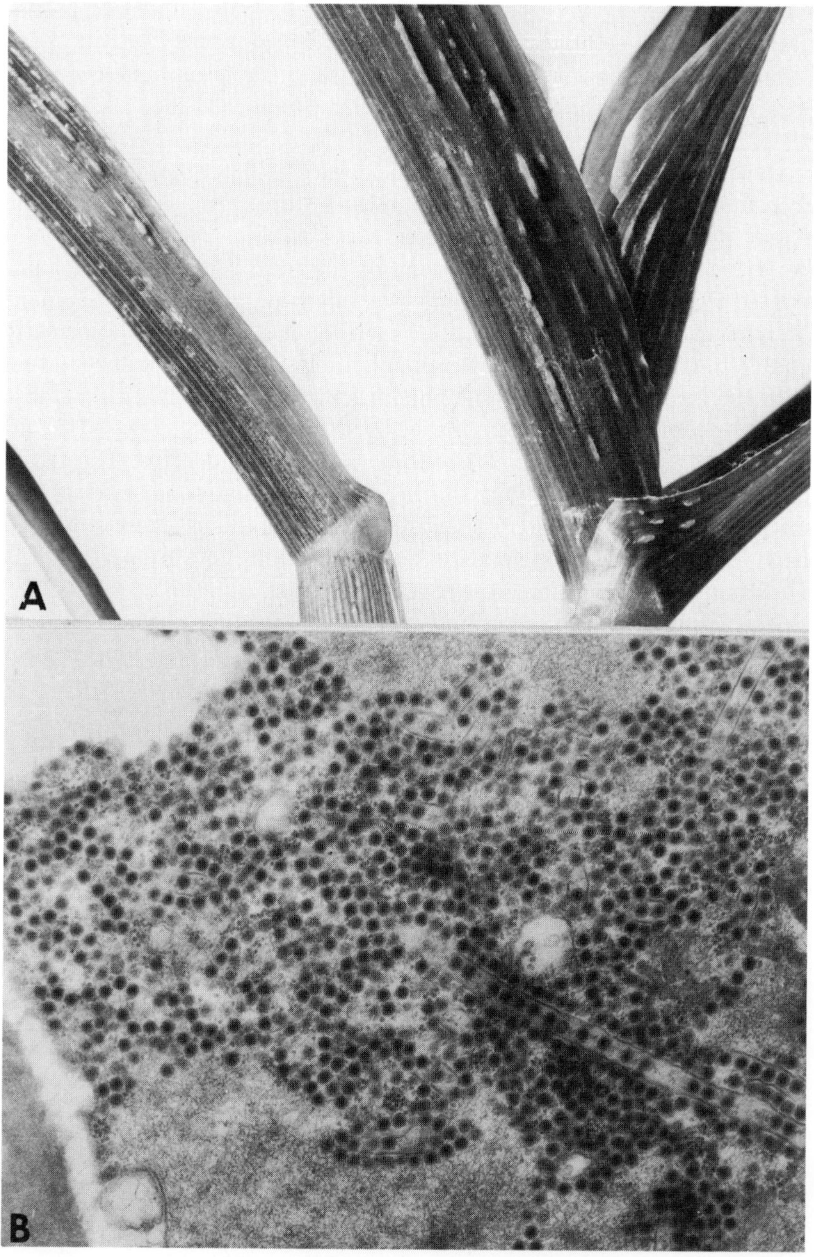
A
B

Either criterion by itself has long been shown to be unreliable as a basis for relating viruses. Smith (1957) classified this virus as the mottle strain of maize streak virus.

Maize streak virus is found only in Africa, where it is occasionally an economic problem on maize.

### *D. Maize Rough Dwarf Virus (MRDV)*

The rough dwarf disease of maize was first described from Italy in 1949 (Fenaroli, 1949) where it is known as nanismo ruvido. Harpaz, Minz, and Nitzani (1958) reported MRDV in Israel in 1958. Harpaz (1959) mechanically transmitted the virus from corn to corn by hypodermic injection of centrifuged sap extracted from diseased plants and established the viral nature of the disease. The maize rough dwarf disease is transmitted by planthoppers and is not seed borne (Harpaz, 1961; Harpaz *et al.*, 1965).

Infected maize plants become darker green than normal in the early stage of infection. Small galls described as veinal hyperplasia (Biraghi, 1952) develop on the undersurface of the leaves (Fig. 3*A*). Leaf symptoms continue to develop as interveinal chlorotic streaks, which may coalesce to form large whitish areas. A reddish discoloration may develop at the leaf margins and eventually involve the entire lamina. Infected plants are stunted and have a poor root system. The incubation period of MRDV in corn plants varies from 15 to 20 days. A detailed description of the disease was reported by Grancini (1958). The virus can infect other gramineous hosts (Vidano, Lovisolo, and Conti, 1966a, b; Lovisolo, Vidano, and Conti, 1966).

According to Harpaz (1961), Klinkowski and Kreutzberg (1958) erroneously inferred from a paper by Sibilia (1949) that the aphid *Tetraneura ulmi* (L.) was a vector of MRDV. Transmission of MRDV by *T. ulmi* has not been reported. The virus is transmitted by two planthopper species, *Laodelphax striatellus* Fallen (*Delphacodes striatella* Fallen) and *Javesella pellucida* (Fabricius) (*Calligypona pellucida* Fabricius) (Harpaz, 1961; Harpaz *et al.*, 1965). Comparative transmission experiments performed at temperatures ranging from 18 to 22°C showed that *L. striatellus* was a more efficient vector of MRDV. *Laodelphox striatellus* and *J. pellucida* transmitted MRDV

---

*Fig. 3.* Maize rough dwarf virus. (*A*) Symptoms in a maize plant experimentally infected by *Laodelphax striatellus* Fallén. (Courtesy of Dr. C. Vidano, Istituto di Entomologia della Universitá, Torino, Italy, unpublished). (*B*) Electron micrograph of virus particles in the cytoplasm of a neoplastic leaf cell. × 28,600. (Courtesy of O. Lovisolo and M. Conti, 1966.)

29% and 16% of the time, respectively. The minimum length of acquisition, incubation, and inoculation periods were equal for both vectors, being 1, 15, and 1 days, respectively (Harpaz *et al.*, 1965). In Italy, *J. pellucida* is found in alpine areas 1200–1500 m above sea level, where maize is not grown, and is not important in the epidemiology of MRDV. *Laodelphax striatellus* is common in areas where maize is grown and is responsible for the spread of MRDV. This planthopper does not breed on maize in nature and can survive on maize plants for only 7–17 days. This may account for the failure of this insect to transmit MRDV directly from one maize plant to another since the brief life span is not sufficient to complete the cycle of acquisition, multiplication, and inoculation of the virus (Harpaz, 1961, 1964).

Maize dwarf mosaic virus particles have been found in infected maize plants and in viruliferous *L. striatellus* (Gerola *et al.*, 1966; Gerola and Bassi, 1966; Lovisolo and Conti, 1966; Vidano, 1966; Vidano and Bassi, 1966). Virus particles about 70 mμ in diameter, possessing a darker central area of about 50 mμ surrounded by a clearer area, were found in the cytoplasm of cells of neoplastic tissues (Fig. 3*B*). Similar particles were found in the cell cytoplasm of fat body tissues, salivary glands, and mycetomes of viruliferous vectors. Negatively stained preparations of partially purified MRDV (Lovisolo *et al.*, 1967) showed particles about 60 mμ, in diameter, about 10 mμ smaller than those measured in ultrathin sections.

Blattný Pozděna, and Procházková (1965) described a rough dwarf and streak disease of maize which was transmitted by *J. pellucida*. Although the disease symptoms on maize were very similar to those induced by MRDV, it was believed that the virus was related to the European aster yellows virus group. Harpaz (1966) stated that the maize rough dwarf disease found in Italy and Israel was the same as the rough dwarf and streak disease in Czechoslovakia.

### *E. Rice Black-Streaked Dwarf Virus (RBSDV)*

In 1952 Kuribayashi and Shinkai (1952) described rice black-streaked dwarf virus and distinguished it from rice dwarf virus. Its host range includes many grass and cereal species.

The virus is transmitted only by the planthopper vectors *Laodelphax striatellus* (Kuribayashi and Shinkai, 1952) and *Unkanodes sapporona* (Iida, 1965). Shinkai (1962) found that the minimum acquisition feeding time for *L. striatellus* was 30 min. First instar nymphs acquired more virus than fifth instar nymphs during a 2-day feeding period. The incubation period of RBSDV in the vector was usually 7–12 days, but ranged from 7 to 35 days. Planthoppers transmitted virus during inoculation feeding periods of 3–5 min. There was considerable variation among individual insects in their ability to transmit RBSDV, but most remained viruliferous for long periods of time. The virus was not transmitted transovarially.

The virus overwinters in a high percentage of the leafhoppers and the overwintering viruliferous insects start transmitting virus early in the spring and remain viruliferous until early summer. The disease occurs only in Japan and can be an economic problem on maize (Hashioka, 1964).

## IV. VIRUSES WITH MITE VECTORS

### A. *Wheat Streak Mosaic Virus* (*WSMV*)

Wheat streak mosaic is a common disease of wheat and other cereals in the great plains area of the United States. The causal virus is transmitted mechanically and by the wheat curl mite, *Aceria tulipae* (Keifer) (Slykhuis, 1953). Corn has been reported as a host of WSMV under natural and experimental conditions (McKinney, 1949; Sill and Agusiobo, 1955; Sill and del Rosario, 1959; Slykhuis, 1955; Finley, 1957). Corn was also found to be a suitable host for the mite vector (Sill and del Rosario, 1959; How, 1963).

Maize mechanically inoculated with WSMV sometimes develops local lesions which are scattered and irregular (Sill and Agusiobo, 1955). Systemic symptoms consist of short streaks and dashes which disappear or develop into a mosaic pattern as the leaves mature. General yellowing and stunting sometimes occur.

Transmission of WSMV to corn by viruliferous *A. tulipae* has been reported by several investigators (Slykhuis, 1955; Sill and del Rosario, 1959; How, 1963). Slykhuis (1955) studied the feeding and transmission characteristics of *A. tulipae* and found that all stages of the vector except the egg could carry the virus, and that nymphs could acquire virus after 30 min. on diseased plants, but that adults were unable to do so. The virus persisted in the vector for at least 6 days.

Recently, Orlob (1966) reported that nymphal mites could acquire WSMV during a 15-min access feeding period on infected leaves. Similarly, the minimum inoculation feeding period was 15 min. Second instar mites were most efficient in transmitting WSMV but, although transmission efficiency decreased during adulthood, some adults remained infective for life. Wheat streak mosaic virus persisted in mites for at least 7 days at room temperature and for at least 61 days at 3°C.

Electron microscopic examination of homogenates from viruliferous

*A. tulipae* and of sap from infected plants revealed sinuous particles about 700 m$\mu$ in length (Orlob, 1966). These particles resembled those obtained from purified virus preparations (Gold, Houston, and Oswald, 1953; Brakke and Staples, 1958; Pop, 1961).

A disease causing red streaking of corn grain has been observed since 1963 in several midwestern states (Williams *et al.*, 1966). A virus was isolated from red-streaked corn grain and temporarily named 3A (Williams, Alexander, and Runnels, 1965). Subsequent tests showed that the 3A virus was a strain of WSMV (McKinney *et al.*, 1966; Williams *et al.*, 1967). The virus is transmitted mechanically and by *A. tulipae*. Other strains of WSMV have been described (McKinney, 1944; Lal *et al.*, 1957).

Atkinson (1952) reported that two species of aphids were vectors of western wheat mosaic virus (a strain of WSMV) in Colorado. These results have not been confirmed and Slykhuis (1965) suggested that either the aphids were transmitting barley yellow-dwarf virus or *A. tulipae* was present with the aphids that Atkinson tested.

## V. MAIZE VIRUSES WITH NO KNOWN VECTORS

Finley (1954) described a mosaic disease of sweet corn from naturally infected corn plants. Smith (1957) listed the causal virus as maize mosaic virus 2 to distinguish it from maize mosaic virus 1 which is transmitted by *P. maidis*. Maize mosaic virus 2 may be transmitted mechanically, but is not seed-borne. The symptom syndrome, host range, and physical–chemical characteristics of this virus suggest that it is different from other virus diseases of corn. No vector studies have been reported.

A seed-borne disease of corn was reported by McKeen (1952) in Canada. The disorder was observed in one of the corn inbred lines and it was assumed that the disease was either of viral or physiological origin. Mechanical transmission failed and *R. maidis* did not transmit the disease.

A disease of corn isolated from naturally infected plants in Iowa was reported by Lambe and Dunleavy (1965a). The disease was transmitted mechanically from corn to corn and it resembled the maize dwarf mosaic disease (Lambe and Dunleavy, 1965b). Although it was believed that a new virus disease was present in Iowa (Stoner, 1965a), subsequent studies (R. Ford, personal communication) showed that such a disease did not exist.

## VI. ARTHROPOD-INDUCED PHYTOTOXEMIAS OF MAIZE

### *A. Maize Wallaby Ear Disease*

This disease was first described by Schindler (1942) in Australia. The condition cannot be induced in test plants by rubbing with juice from diseased plants. The symptoms on maize are small, elongated swellings on secondary veins on the undersides of the top leaves. The foliage is darker green than normal and the leaves extend at a sharp angle from the stalk. The leaf edges are rolled inwards. If affected early, plants are dwarfed and growth is retarded. Affected plants sometimes recover. Since field-collected *Cicadulina bipunctella bimaculata* (Evans) (*Cicadula bimaculata* Evans) were capable of inducing disease symptoms on maize, Schindler assumed it to be caused by a virus. However, the viral nature of the disease has not been examined critically and it is now believed that the disease is toxin induced (Maramorosch, 1959; Maramorosch *et al.,* 1961; Holdeman and McCartney, 1965).

Field-collected *C. bipunctella bimaculata* and all their progeny reared on healthy plants are able to induce the wallaby ear disease (Maramorosch *et al.,* 1961). If a viral agent were involved, this would necessitate a 100% transovarial passage to the progeny, and this would be highly unusual. No "virus-free" insects are therefore available for control tests.

The swelling of the veins of the maize plant is the most characteristic symptom of the Wallaby Ear disease. Similar vein swellings have been associated with the feeding of *Cicadulina* sp. on maize in Africa (H. H. Storey, personal communication to R. F. Ruppel, 1965), on maize and rice in the Philippines (Maramorosch *et al.,* 1961), and with *Dalbulus elimatus* on maize in Mexico (Maramorosch, 1959). These reports strongly suggest that the leaf vein enations on wallaby ear-diseased plants are caused by insect toxins rather than by a virus.

### *B. Leaf Gall Disease of Rice and Maize*

This disease was first described from the Philippines by Agati and Calica (1949, 1950). The disease is neither seed borne nor mechanically transmissible. Symptoms on corn are very similar to those produced by the wallaby ear disease. Several gramineous species can be affected by this disease. On the basis of symptomatology and insect transmission tests with *Cicadulina bipunctella* (Matsumura), it was assumed that the disease was caused by a virus.

Maramorosch *et al.* (1961) showed that every insect collected in nature, as well as all insects raised on healthy maize, were able to produce leaf galls on maize. The vein swelling did not appear in the leaves on which the insect fed but on the next developing young leaf. The gall-inducing agent was localized since leaves subsequently formed remained free from galls. It is now believed that the leaf gall disease of rice and maize is caused by an insect toxin.

### *C. Kernel Red Streak* (*KRS*)

Red streaking of corn kernels was first observed in 1963 in parts of Ohio, Indiana, and Michigan (Williams *et al.,* 1966). Since then it has been reported in neighboring states and in Canada (Wall and Mortimore, 1965). Kernel red streak has been observed on many types of corn but it is most common on yellow dent. According to Nault *et al.* (1967) streaked kernels occur most commonly at the ear tips. The severity and incidence generally decrease from the apical to the basal portion of the ear (Fig. 4). Streak coloration ranges from deep red in yellow kernels to pink or purple in white kernels. The streaking is apparently formed by the deposition or formation of a red pigment within the pericarp tissues.

The KRS disease of maize was initially thought to be caused by

*Fig. 4.* Kernel red streak on yellow dent corn. Kernels showing severe streaking (top), mild streaking (middle), and no streaking (bottom). (Courtesy of Dr. L. R. Nault, Ohio Agricultural Research and Development Center, Wooster, Ohio.)

wheat streak mosaic virus. Although WSMV was isolated from red-streaked kernels, Paliwal, Slykhuis, and Wall (1966) reported that it could also be recovered from immature kernels without red-streaked pericarp from diseased plants. They found red-streaked pericarp on plants which were not infected with WSMV and in areas where WSMV did not occur. It was concluded that WSMV was not responsible for the red-streaked pericarp of maize kernels.

Nault *et al.* (1967) presented evidence that KRS was caused by a salivary phytotoxin produced by *A. tulipae,* the vector of WSMV. Mites were found on the silks, husks, and kernels of field-collected flint and yellow corn. The mite population on the kernels correlated with the amount and severity of KRS. Virus-free *A. tulipae* placed on kernels of sweet corn in the greenhouse produced KRS. Although a phytotoxin was suspected, it was not ruled out that a latent virus might be responsible for KRS.

Phytotoxin secretion by mites is not the only factor which can induce the red streaking of maize kernels. Ohio W49 yellow dent field corn, inoculated with maize dwarf mosaic virus and the 3A isolate of WSMV, produced on the average 55% ears with some degree of red-streaking kernels at maturity (Nault *et al.,* 1967). About 10% of the ears from uninoculated plants showed red-streaked kernels. Kernels with red coloration were also reported in an inbred line of maize (Jones, 1945). Atanasoff (1965, 1966) reported that the maize leaf fleck virus in Bulgaria induced a red-streaking of maize kernels.

## VII. VIRUSES WHICH HAVE BEEN SHOWN EXPERIMENTALLY TO INFECT CORN

Several viruses have been transmitted to maize under experimental conditions. They have been transmitted either by nematode or arthropod vectors or by mechanical means. These viruses are listed in Table 1. Canna mosaic virus and millet red-leaf virus are probably strains of known viruses. The remaining viruses appear to be distinct from others known to infect maize. The viral nature of the cadang-cadang (yellow mottle decline) disease is still questionable. The evidence in favor of a virus being involved is very strong, whereas the evidence against it is very weak (Price, 1958). The report on transmission of sunflower mosaic virus by thrips, aphids, and whiteflies seems doubtful and has never been confirmed.

Since these viruses are of no economic importance, there is very little information concerning the vector–virus relationship of those which are arthropod or nematode borne.

TABLE I.

VIRUSES WHICH DO NOT NATURALLY INFECT MAIZE IN THE FIELD

| Virus | Mechanical transmission | Vector [a] | References |
|---|---|---|---|
| Canna mosaic (sugarcane mosaic strain?) | + | *Aphis gossypii* *Rhopalosiphum maidis* | Castillo, *et al.*, 1956 |
| Millet red leaf (barley yellow-dwarf strain?) | − | Aphid vectors | Yu, *et al.*, 1957 |
| Filaree red leaf | − | Aphid vectors | Anderson, 1952 |
| Grassy shoot | + | Aphid vectors | Vasudeva, 1962 |
| Oat pseudo-rosette | − | *Laodelphax striatellus* (*Calligypona marginata*) | Soukhov and Vovk, 1938 |
| Oat sterile dwarf | − | *Javesella pellucida* (*Calligypona pellucida*) | Vacke and Průsă, 1962 |
| Winter wheat mosaic | − | *Psammotettix striatus* L. | Klinkowski and Kreutzberg, 1958 |
| Rice stripe | − | *Laodelphax striatellus* | Shinkai, 1962 |
| Wheat striate mosaic | − | *Endria inimica* (Say) | Slykhuis, 1962, 1963 |
| Wheat spot mosaic | − | *Aceria tulipae* | Slykhuis, 1956 |
| Sunflower mosaic | + | Thrips, aphid, and whitefly vectors (?) | Traversi, 1949 |
| Potato corky ringspot (tobacco rattle) | + | *Trichodorus christiei* Allen | Walkinshaw *et al.*, 1961 |
| Barley stripe mosaic | + | | McKinney, 1953 |
| Brome mosaic | + | | McKinney, 1944 |
| Cadang-cadang (virus?) | + | | del Rosario and Quiaoit, 1962 |
| Tobacco necrosis | + | | Fulton, 1950 |
| Tobacco ringspot | + | | Bridgmon and Walker, 1952 |
| Sugarcane ratoon stunting | + | | Steindl, 1955 Steib and Forbes, 1957 |

[a] Only the names of those vectors which transmitted the virus to corn are listed.

## BIBLIOGRAPHY

Abbott, E. V., and Charpentier, L. J. 1962. Additional insect vectors of sugarcane mosaic. Proc. Int. Soc. Sugar-Cane Technologists. 11:755–760.

Abbott, E. V., and Tippett, R. L. 1964. Additional hosts of sugarcane mosaic virus. Plant Disease Reporter 48:443–445.

———. 1966. Strains of sugarcane mosaic virus. U.S. Dep. Agr. Tech. Bull. 1340. 25 pp.

Agati, J. A., and Calcia, C. A. 1949. The leaf-gall disease of rice and corn in the Philippines. Philippine J. Agr. 14:31–38.

———. 1950. Studies on the host-range of the rice and corn leaf-gall virus. Philippine J. Agr. 15:249–259.

Allen, T. C., Jr. 1957. Symptoms of Golden Bantam corn inoculated with barley yellow-dwarf virus. Phytopathology 47:1.

Altstatt, G. E. 1945. A new corn disease in the Rio Grande Valley. Plant Disease Reporter 29:533–534.

Ancalmo, O., and Davis, W. C. 1961. Achaparramiento (corn stunt). Plant Disease Reporter 45:281.

Anderson, C. W. 1952. Additional notations on the red leaf virus of filaree. Phytopathology 42:110–111.

Anzalone, L., Jr., and Pirone, T. P. 1964. Transmission of sugarcane mosaic virus by *Myzus persicae*. Plant Disease Reporter 48:984–985.

Atanasoff, D. 1965. Leaf fleck disease of maize and its possible relation to cytoplasmic inheritance. Phytopathol. Z. 52:89–95.

———. 1966. Maize leaf fleck disease. Phytopathol. Z. 56:25–33.

Atkinson, R. E. 1952. Studies on viruses of cereal grains. Ph.D. thesis, Univ. Minnesota. (Diss. Abstr. 12:375.)

Bancroft, J. B., Ullstrup, A. J., Messieha, M., Bracker, C. E., and Snazelle, T. E. 1966. Some biological and physical properties of a midwestern isolate of maize dwarf mosaic virus. Phytopathology 56:474–478.

Barnes, D. 1954. Biología, ecología y distribucíon de las chicharritas, *Dalbulus elimatus* (Ball) y *Dalbulus maidis* (DeL. & W.). Mexican Secretar. Agr. Ganaderia, Ofic. Estud. Especiales, Folleto Tec. 11. 112 pp.

Benada, J., Kvīcala, B. A., and Schmidt, H. B. 1964. Die Rotstreifigkeit des Sorghum und das Streifenmosaik des Maises, eine für die ČSSR neue Viruskrankheit. Zentr. Bakteriol. Parasitenk. Abt. II. 117:683–691.

Biraghi, A. 1952. Ulteriore contributo alla conoscenza del nanismo ruvido del mais. Annu. Sper. Agr. (Rome) 6:1043–1053.

Blattný, C., Pozděna, J. and Procházková, Z. 1965. Virusbedingte Rauhverzwergung und Streifenkrankheit bei *Zea mays* L. Phytopathol. Z. 52:105–130.

Brakke, M. K., and Staples, R. 1958. Correlation of rod length with infectivity of wheat streak mosaic virus. Virology 6:14–26.

Brandes, E. W. 1919. The mosaic disease of sugar cane and other grasses. U.S. Dep. Agr. Dep. Bull. 829. 26 pp.

———. 1920a. Mosaic disease of corn. J. Agr. Res. 19:517–522.

———. 1920b. Artificial and insect transmission of sugar-cane mosaic. J. Agr. Res. 19:131–138.

Brandes, E. W., and Klaphaak, P. J. 1923. Cultivated and wild hosts of sugarcane or grass mosaic. J. Agr. Res. 24:247–262.

Brandes, J. 1964. Identifizierung von gestreckten pflanzenpathogenen Viren auf morphologischer Grundlage. Mitt. Biol. Bundesanstalt Land-Forstwirtsch. Berlin-Dahlem 110. 130 pp.

Bridgmon, G. H., and Walker, J. C. 1952. Gladiolus as a virus reservoir. Phytopathology 42:65–70.

Carter, W. 1941. *Peregrinus maidis* (Ashm.) and the transmission of corn mosaic. I. Incubation period and longevity of the virus in the insect. Annu. Entomol. Soc. Amer. 34:551–556.

Castillo, B. S., Yarwood, C. E., and Gold, A. H. 1956. Canna-mosaic virus. Plant Disease Reporter 40:169–172.

Celino, M. S. 1940. Experimental transmission of the mosaic of abacá, or Manila hemp plant (*Musa textilis* Née). Philippine Agriculturist 29:379–413.

Celino, M. S., and Ocfemia, G. O. 1941. Two additional insect vectors of mosaic of abacá, or Manila hemp plant, and transmission of its virus to corn. Philippine Agriculturist 30:70–79.

Costa, A. S., and Penteado, M. P. 1951. Corn seedlings as test plants for the sugar-cane mosaic virus. Phytopathology 41:758–763.

Dale, J. L. 1964. Isolation of a mechanically transmissible virus from corn in Arkansas. Plant Disease Reporter 48:661–663.

Dijkstra, J., and Grancini, P. 1960. Serological and electron microscopical investigations of the relationship between *Sorghum* red stripe virus and sugarcane mosaic virus. Tijdschr. Plantenziekten 66:295–300.

Doolittle, S. P., and Zaumeyer, W. J. 1953. A pepper ringspot caused by strains of cucumber mosaic virus from pepper and alfalfa. Phytopathology 43:333–337.

Eloja, A. L., and Tinsley, T. W. 1963. Abacá mosaic virus and its relationship to sugar-cane mosaic. Annu. Appl. Biol. 51:253–258.

Fenaroli, L. 1949. Il "nanismo" del mais. Notiz. Mal. Piante 3:38–39.

Finley, A. M. 1954. A mosaic disease of sweet corn. Phytopathology 44:488.

———. 1957. Wheat streak mosaic, a disease of sweet corn in Idaho. Plant Disease Reporter 41:589–591.

Frazier, N. W. 1945. A streak disease of corn in California. Plant Disease Reporter 29:212–213.

Frazier, N. W., Freitag, J. H., and Gold, A. H. 1965. Corn naturally infected by sugarcane mosaic virus in California. Plant Disease Reporter 49:204–206.

Frazier, N. W., Freitag, J. H., and Holdeman, Q. L. 1966. Evidence that corn stunt virus may not occur in California. Plant Disease Reporter 50:318–320.

Fulton, R. W. 1950. Variants of the tobacco necrosis virus in Wisconsin. Phytopathology 40:298-305.

Gerola, F. M., and Bassi, M. 1966. An electron microscopy study of leaf vein tumours from maize plants experimentally infected with maize rough dwarf virus. Caryologia 19:13–40.

Gerola, F. M., Bassi, M., Lovisolo, O., and Vidano, C. 1966. Virus-like particles in both maize plants infected with maize rough dwarf virus and the vector *Laodelphax striatellus* Fallén. Phytopathol. Z. 56:97–99.

Goidanich, G. 1938. Le malattie del sorgo succherino in Italia nelle ultime annate agrarie. Cellulosa 2:245–250.

Gold, A. H., Houston, B. R., and Oswald, J. W. 1953. Electron microscopy of elongated particles associated with wheat streak mosaic. Phytopathology 43:458–459.

Granados, R. R., and Maramorosch, K. 1967. Transmission of corn stunt virus by three leafhopper vectors. Phytopathology 57:340.

Granados, R. R., Maramorosch, K., Everett, T., and Pirone, T. P. 1966a. Leafhopper transmission of a corn stunt virus from Louisiana. Phytopathology 56:584.

———. 1966b. Transmission of corn stunt virus by a new leafhopper vector, *Graminella nigrifrons* (Forbes). Contrib. Boyce Thompson Inst. 23:275–280.

Grancini, P. 1957. Un mosaico del mais e del sorgo in Italia. Maydica 2:83–104.

———. 1958. I sintomi del "Nanismo ruvido" del mais. Maydica 3:67–79.

Hadden, F. C. 1928. Sugar cane mosaic and insects. Hawaiian Planters' Record 32:130–142.

Harpaz, I. 1959. Needle transmission of a new maize virus. Nature (London) 184: B.A. 77–78.

———. 1961. *Calligypona marginata,* the vector of maize rough dwarf virus. FAO Plant Protection Bull. 9:144–147.

———. 1964. Hostplant–vector and hostplant–virus relationships in the rough dwarf disease of maize. Maydica 9:16–20.

———. 1966. Further studies on the vector relations of the maize rough dwarf virus (MRDV). Maydica 11:18–23.

Harpaz, I., Minz, G. and Nitzani, F. 1958. Dwarf disease of maize. FAO Plant Protection Bull. 7:43.

Harpaz, I., Vidano, C., Lovisolo, O., and Conti, M. 1965. Indagini comparative su *Javesella pellucida* (Fabricius) e *Laodelphax striatellus* (Fallén) quali vettori del virus del nanismo ruvido del mais (maize rough dwarf virus). Atti. Accad. Sci. Torino. Classe Sci. Fis. Mat. Nat. 99:885–901.

Harter, L. L. 1938. Mosaic of lima beans (*Phaseolus lunatus macrocarpus*). J. Agr. Res. 56:895–906.

Hashioka, Y. 1964. Virus diseases of rice in the world. Reprinted from Riso, Dec. 1964. 16 pp. [Abstr. *in* Rev. Appl. Mycol. 44:2480. (1965).]

Herold, F., Bergold, G. H., and Weibel, J. 1960. Isolation and electron microscopic demonstration of a virus infecting corn (*Zea mays* L.). Virology 12:335–347.

Herold, F. and Munz, K. 1965. Electron microscopic demonstration of viruslike particles in *Peregrinus maidis* following acquisition of maize mosaic virus. Virology 25:412–417.

Herold, F., and Weibel, J. 1963. Electron microscopic demonstration of sugarcane mosaic virus particles in cells of *Saccharum officinarum* and *Zea mays*. Phytopathology 53:469–471.

Holdeman, Q. L., and McCartney, W. C. 1965. Virus diseases of corn, *Zea mays* L. A plant disease detection aid. California, Department of Agriculture, Bureau of Plant Pathology, Sacramento. 24 pp. (Mimeo.)

How, S. C. 1963. Wheat streak mosaic virus on corn in Nebraska. Phytopathology 53:279–280.

Iida, T. T. 1965. Geographical distribution of leafhopper-borne viruses of cereals. Papers, Conf. Relationships between Arthropods and Plant-Pathogenic Viruses, Tokyo, 1965. pp. 22–33.

Ingram, J. W., and Summers, E. M. 1936. Transmission of sugarcane mosaic by the rusty plum aphid, *Hysteroneura setariae*. J. Agr. Res. 52:879–887.

———. 1938. Transmission of sugarcane mosaic by the green bug (*Toxoptera graminum* Rond.). J. Agr. Res. 56:537–540.

Janson, B. F., and Ellett, C. W. 1963. A new corn disease in Ohio. Plant Disease Reporter 47:1107–1108.

Janson, B. F., Williams, L. E., Findley, W. R., Dollinger, E. J., and Ellett, C. W. 1965. Maize dwarf mosaic: New corn virus disease in Ohio. Ohio Agr. Exp. Sta. Res. Circ. 137. 16 pp.

Jones, D. F. 1945. Heterosis resulting from degenerative changes. Genetics 30: 527–542.

Juliano, J. P. 1951. A study of the mosaic disease of abacá, or Manila hemp plant (*Musa textilis* Neé), with special reference to sources of inoculum and possible transmission of the virus by mechanical means. Philippine Agriculturist 34:125–132.

Kennedy, J. S., Day, M. F., and Eastop, V. F. 1962. A conspectus of aphids as vectors of plant viruses. Commonwealth Institute of Entomology, London. 114 pp.

Klinkowski, M., and Kreutzberg, G. 1958. Vorkommen und Verbreitung von Gramineenvirosen in Europa. Phytopathol. Z. 32:1–24.

Kunkel, L. O. 1921. A possible causative agent for the mosaic disease of corn. Hawaiian Sugar Planters' Assoc. Exp. Sta. Bull. Bot. Ser. 3:44–58.

———. 1922. Insect transmission of yellow stripe disease. Hawaiian Planters' Record 26:58–64.

———. 1924. Studies on the mosaic of sugar cane. Hawaiian Sugar Planters' Assoc. Exp. Sta. Bull. Bot. Ser. 3:115–167.

———. 1927. The corn mosaic of Hawaii distinct from sugar cane mosaic. Phytopathology 17:41.

———. 1946a. Incubation period of corn stunt virus in the leafhopper *Baldulus maidis* (DeL. and W.). Amer. J. Botany 33:830–831.

———. 1946b. Leafhopper transmission of corn stunt. Proc. Nat. Acad. Sci. U.S. 32:246–247.

———. 1948. Studies on a new corn virus disease. Arch. Ges. Virusforsch. 4:24–46.

Kuribayashi, K., and Shinkai, A. 1952. Occurrence of rice black streak dwarf in Nagano and Tokyo. It is transmitted by *Delphacodes striatella* Fall [in Japanese]. Annu. Phytopathol. Soc. Japan 16:41. (Abstr.)

Lal, S. B., Sill, W. H., Jr., Rosario, M. S. del, and Kainski, J. M. 1957. Three new naturally occurring strains of wheat streak mosaic virus in Kansas and the Great Plains. Phytopathology 47:527.

Lambe, R. C., and Dunleavy, J. M. 1965a. A corn disease in Iowa. Plant Disease Reporter 49:339–341.

———. 1965b. A new disease in Iowa corn fields. Iowa Farm Sci. 19(9):3–4.

Lawas, O. M., and Fernandez, W. L. 1949. A study of the transmission of the corn mosaic and of some of the physical properties of its virus. Philippine Agriculturist 32:231–238.

Lovisolo, O. 1957. Contributo sperimentale alla conoscenza ed alla determinazione del virus agente dell'arrossamento striato del sorgo e di un mosaico del mais. Boll. Staz. Patol. Vegetale Roma Ser. 3 14:261–321.

Lovisolo, O., and Conti, E. M. 1966. Individuazione al microscopio elettronico del virus del nanismo ruvido del mais (MRDV) in piante di *Zea mays* L. sperimentalmente infettate. Atti. Accad. Sci. Torino Classe Sci. Fis. Mat. Nat. 100: 63–72.

Lovisolo, O., Luisoni, E., Conti, M., and Wetter, C. 1967. Partial purification of maize rough dwarf virus. Naturwissenschaften 54:73–74.

Lovisolo, O., Vidano, C., and Conti, E. M. 1966. Transmissione del virus del nanismo ruvido del mais ad *Avena sativa* L. con *Laodelphax striatellus* Fallén. Atti Accad. Sci. Torino Classe Sci. Fis. Mat. Nat. 100:351–358.

McClean, A. P. D. 1947. Some forms of streak virus occurring in maize, sugar-cane and wild grasses. Union S. Africa Dep. Agr. Sci. Bull. 265. 39 pp.

McEwen, F. L. and Kawanishi, C. Y. 1967. Insect transmission of corn mosaic; laboratory studies in Hawaii. J. Econ. Entomol. 60:1413–1417.

McKeen, W. E. 1952. A seed-borne disease of corn. Plant Disease Reporter 36: 145–146.

MacKenzie, D. R., Wernham, C. C., and Ford, R. E. 1966. Differences in maize dwarf mosaic virus isolates of the northeastern United States. Plant Disease Reporter 50: 814–818.

McKinney, H. H. 1944. Studies on the virus of brome-grass mosaic. Phytopathology 34:993.

———. 1949. Tests of varieties of wheat, barley, oats, and corn for reaction to wheat streak-mosaic viruses. Plant Disease Reporter 33:359–369.

———. 1953. New evidence on virus diseases in barley. Plant Disease Reporter 37:292–295.

McKinney, H. H., Brakke, M. K., Ball, E. M., and Staples, R. 1966. Wheat streak mosaic virus in the Ohio Valley. Plant Disease Reporter 50:951–953.

Maramorosch, K. 1951. Mechanical transmission of corn stunt virus to an insect vector. Phytopathology 41:833–838.

———. 1955a. The occurrence of two distinct types of corn stunt in Mexico. Plant Disease Reporter 39:896–898.

———. 1955b. Multiplication of plant viruses in insect vectors. Advances Virus Res. 3:221–249.

———. 1958a. Cross protection between two strains of corn stunt virus in an insect vector. Virology 6:448–459.

———. 1958b. Viruses that infect and multiply in both plants and insects. Trans. N.Y. Acad. Sci. 20:383–393.

———. 1959. An ephermeral disease of maize transmitted by *Dalbulus elimatus*. Entomol. Exp. Appl. 2:169–170.

Maramorosch, K., Calica, C. A., Agati, J. A., and Pableo, G. 1961. Further studies on the maize and rice leaf galls induced by *Cicadulina bipunctella*. Entomol. Appl. 4:86–89.

Mitsuhashi, J., and Maramorosch, K. 1963. Aseptic cultivation of four virus transmitting species of leafhoppers (Cicadellidae). Contrib. Boyce Thompson Inst. 22:165–173.

Murant, A. F. 1965. The morphology of cucumber mosaic virus. Virology 26:538–544.

Nault, L. R., Briones, M. L., Williams, L. E., and Barry, B. D. 1967. Relation of the wheat curl mite to kernel red streak of corn. Phytopathology 57:986–989.

Niederhauser, J. S., and Cervantes, J. 1950. Transmission of corn stunt in Mexico by a new insect vector, *Baldulus elimatus*. Phytopathology 40:20–21.

Orlob, G. B. 1966. Feeding and transmission characteristics of *Aceria tulipae* Keifer as vector of wheat streak mosaic virus. Phytopathol. Z. 55:218–238.

Oswald, J. W., and Houston, B. R. 1951. A new virus disease of cereals, transmissible by aphids. Plant Disease Reporter 35:471–475.

———. 1953a. Host range and epiphytology of the cereal yellow dwarf disease. Phytopathology 43:309–313.

———. 1953b. The yellow-dwarf virus disease of cereal crops. Phytopathology 43: 128–136.

Paliwal, Y. C., Slykhuis, J. T., and Wall, R. E. 1966. Wheat streak mosaic virus in corn in Ontario. Can. Plant Disease Survey 46:8.

Panjan, M. 1966. About some manifestations of mosaic on corn in Yugoslavia. Rev. Roumaine Biol. Ser. Bot. 11:159–162.

Pirone, T. P., and Anzalone, L., Jr. 1966. Purification and electron microscopy of sugarcane mosaic virus. Phytopathology 56:371–372.

Pop, I. 1961. Die Strichelvirose des Weizens in der Rumänischen Volksrepublik. Phytopathol. Z. 43:325–336.

Pop, I., and Tusa, C. 1966. Influence of maize mosaic on the growth and yield of some maize hybrids. p. 170–174. *In* A. B. R. Beemster and J. Dijkstra [eds.], Viruses of Plants: Proceedings of an International Conference on Plant Viruses, Wageningen, July, 1965. Wiley, New York.

Pound, G. S., and Walker, J. C. 1948. Strains of cucumber mosaic virus pathogenic on crucifers. J. Agr. Res. 77:1–12.

Price, W. C. 1935. Classification of southern celery-mosaic virus. Phytopathology 25:947–954.

———. 1958. Report to the government of the Philippines on the yellow mottle decline (Cadang-Cadang) of coconut. FAO Report No. 850. (Mimeo.)

Rochow, W. F. 1959. Transmission of strains of barley yellow dwarf virus by two aphid species. Phytopathology 49:744–748.

———. 1961. The barley yellow dwarf virus disease of small grains. Advances Agron. 13:217–248.

———. 1963a. Variation within and among aphid vectors of plant viruses. Ann. N.Y. Acad. Sci. 105:713–729.

———. 1963b. Latent periods in the aphid transmission of barley yellow dwarf virus. Phytopathology 53:355–356.

Rochow, W. F., and Brakke, M. K. 1964. Purification of barley yellow dwarf virus. Virology 24:310–322.

Rosario, M. S. del, and Quiaoit, A. R. 1962. Studies on the virus aspect of cadang-cadang disease of coconut. 1. Transmission and serology. Philippine Agriculturist 45:477–489. (Abstr. *in* Rev. Appl. Mycol. 41:734–735, 1962.)

Ruppel, R. F. 1965. A review of the genus *Cicadulina* (Hemiptera, Cicadellidae). Michigan State Univ. Mus. Biol. Ser. 2:385–428.

Schindler, A. J. 1942. Insect transmission of Wallaby Ear disease of maize. J. Australian Inst. Agr. Sci. 8:35–37.

Sehgal, O. P. 1966. Host range, properties and partial purification of a Missouri isolate of maize dwarf mosaic virus. Plant Disease Reporter 50:862–866.

Shepherd, R. J. 1965. Properties of a mosaic virus of corn and Johnson grass and its relation to the sugarcane mosaic virus. Phytopathology 55:1250–1256.

Shepherd, R. J., and Holdeman, Q. L. 1965. Seed transmission of the Johnson grass strain of the sugarcane mosaic virus in corn. Plant Disease Reporter 49: 468–469.

Shinkai, A. 1962. Studies on insect transmission of rice virus diseases in Japan [in Japanese, English summary]. Bull. Nat. Inst. Agr. Sci. Ser. C. No. 14:1–112.

Sibilia, C. 1949. Il nanismo del mais in Italia. Notiz. Mal. Piante 6:35–37.

Sill, W. H., Jr., and Agusiobo, P. C. 1955. Host range studies of the wheat streak mosaic virus. Plant Disease Reporter 39:633–642.

Sill, W. H., Jr., and Rosario, M. S. del. 1959. Transmission of wheat streak mosaic virus to corn by the eriophyid mite, *Aceria tulipae*. Phytopathology 49:396.

Slykhuis, J. T. 1953. Wheat streak mosaic in Alberta and factors related to its spread. Can. J. Agr. Sci. 33:195–197.

———. 1955. *Aceria tulipae* Keifer (Acarina: Eriophyidae) in relation to the spread of wheat streak mosaic. Phytopathology 45:116–128.

———. 1956. Wheat spot mosaic caused by a mite-transmitted virus associated with wheat streak mosaic. Phytopathology 46:682–687.

———. 1962. Wheat striate mosaic a virus disease to watch on the Prairies. Can. Plant Disease Survey 42:135–142.

———. 1963. Vector and host relations of North American wheat striate mosaic virus. Can. J. Bot. 41:1171–1185.

———. 1965. Mite transmission of plant viruses. Advances Virus Res. 11:97–137.

Smith, K. M. 1957. A textbook of plant virus diseases. 2nd ed. Little, Brown and Co., Boston. 652 p.

Soukhov, K. S. and Vovk, A. M. 1938. Mosaic of cultivated cereals and how it is communicated in nature. Compt. Rend. Acad. Sci. URSS NS 20:745–748.

Stahl, C. F. 1927. Corn stripe disease in Cuba not identical with sugar cane mosaic. Trop. Plant Res. Found. Bull. 7. 12 pp.

Steib, R. J., and Forbes, I. L. 1957. Johnson grass and corn as carriers of the virus of ratoon stunting disease of sugarcane. Sugar Bull. 35:375–379. (Abstr. *in* Rev. Appl. Mycol. 37:57, 1958.)

Steindl, D. R. L. 1955. p. 72–80. *in* Mungomery, R. W. 1955. Division of Entomology and Pathology. Rep. Bur. Sugar Exp. Sta. Queensland 55: 62–80, 1955. (Abstr. *in* Rev. Appl. Mycol. 35:327–328, 1956.)

Stoner, W. N. 1952. Leaf fleck, an aphid-borne persistent virus disease of maize. Phytopathology 42:683–689.

———. 1964a. Corn viruses in the United States through 1964. Proc. 19th Ann. Hybrid Corn Ind. Res. Conf.

———. 1964b. Corn stunt disease in the United States through 1963. Plant Disease Reporter 48:640–644.

———. 1965a. Some corn (maize) virus diseases in the United States in 1964. Plant Disease Reporter 49:918–922.

———. 1965b. Studies of transmission of barley yellow dwarf virus to corn (*Zea mays*). Phytopathology 55:1078.

———. 1965c. A review of corn stunt disease (Achapparramiento) and its insect vectors, with resumes of other virus diseases of maize. U.S. Dep. Agr. Spec. Rep. 33–99. 35 p.

Stoner, W. N., and Ullstrup, A. J. 1964. Corn stunt disease. Mississippi State Univ. Agr. Exp. Sta. Inform. Sheet 844.

Stoner, W. N., and Williams, L. E. 1965. Virus disease of maize in the continental United States. Southern Seedmen's Assoc. Annu. Rep. 47:67–76.

Stoner, W. N., Williams, L. E., and Alexander, L. J. 1964. Transmission by the corn leaf aphid, *Rhopalosiphum maidis* (Fitch), of a virus infecting corn in Ohio. Ohio Agr. Exp. Sta. Res. Circ. 136. 4 p.

Storey, H. H. 1925a. Streak disease, an infectious chlorosis of sugar-cane, not identical with mosaic disease. Imp. Botany Conf. 1924, Rept. Proc. 132–144.

———. 1925b. The transmission of streak disease of maize by the leafhopper *Balclutha mbila* Naude. Annu. Appl. Biol. 12:422–439.

———. 1928. Transmission studies of maize streak disease. Ann. Appl. Biol. 15:1–25.

———. 1932. The inheritance by an insect vector of the ability to transmit a plant virus. Proc. Roy. Soc. (London) Ser. B, 112:46–60.

———. 1933. Investigations of the mechanism of the transmission of plant viruses by insect vectors. I. Proc. Roy. Soc. (London) Ser. B 113:463–485.

———. 1937. A new virus of maize transmitted by *Cicadulina* spp. Annu. Appl. Biol. 24:87–94.

———. 1938. Investigations of the mechanism of the transmission of plant viruses by insect vectors. II. The part played by puncture in transmission. Proc. Roy. Soc. (London) Ser. B 125:455–477.

———. 1939a. Investigations of the mechanism of the transmission of plant viruses by insect vectors. III. The insect's saliva. Proc. Roy. Soc. (London) Ser. B 127:526–543.

———. 1939b. Transmission of plant viruses by insects. Bot. Rev. 5:240–272.

Storey, H. H., and McClean, A. P. D. 1930. The transmission of streak disease between maize, sugarcane and wild grasses. Annu. Appl. Biol. 17:691–719.

Summers, E. M., Brandes, E. W., and Rands, R. D. 1948. Mosaic of sugarcane in the United States, with special reference to strains of the virus. U.S. Dep. Agr. Tech. Bull. 955. 124 pp.

Tate, H. D., and Vandenberg, S. R. 1939. Transmission of sugarcane mosaic by aphids. J. Agr. Res. 59:73–79.

Thornberry, H. H., Otterbacher, A. G., and Thompson, M. R. 1966. Gamagrass, *Tripsacum dactyoides:* a new perennial host of maize dwarf mosaic virus. Plant Disease Reporter 50:65–68.

Traversi, B. A. 1949. Estudio inicial sobre una enfermedad del girasol ("Helianthus annuus" L.) en Argentina. Rev. Invest. Agr. (Buenos Aires) 3:345–351.

Ullstrup, A. J. 1966. Corn diseases in the United States and their control. Rev. U.S. Dep. Agr. Handbook 199. 44 pp.

Vacke, J., and Průsă, V. 1962. A study of the host range of oat sterile dwarf virus [in Russian]. Annu. Acad. Tchécosl. Agr. 35:463–474. (Abstr. *in* Rev. Appl. Mycol. 41:708, 1962.)

Vasudeva, R. S. 1962. Report of the Division of Mycology and Plant Pathology. Sci. Rep. Agr. Res. Inst. New Delhi, 1958–1959. p. 131–147. (Abstr. *in* Rev. Appl. Mycol. 42:593, 1963.)

Vidano, C. 1966. Il maize rough dwarf virus in ghiandole salivari e in micetoma di *Laodelphax striatellus* Fallén. Atti. Accad. Sci. Torino. Classe Sci. Fis. Mat. Nat. 100:731–748.

Vidano, C., and Bassi, M. 1966. Dimostrazione al microscopio elettronico di particelle del virus del nanismo ruvido del maiz (MRDV) nel vettore *Laodelphax striatellus* Fallén. Atti. Accad. Sci. Torino Classe Sci. Fis. Mat. Nat. 100:73–78.

Vidano, C., Lovisolo, O., and Conti, M. 1966a. Nuovi ospiti sperimentali del virus del nanismo ruvido del maiz (MRDV). Atti. Accad. Sci. Torino Classe Sci. Fis. Mat. Nat. 100:699–710.

———. 1966b. Tramissione del virus del nansimo ruvido del maiz (MRDV) a *Triticum vulgare* L. par mezzo di *Laodelphax striatellus* Fallen. Atti. Accad. Sci. Torino Classe Sci. Fis. Mat. Nat. 100:125–140.

Wagner, G. W., and Dale, J. L. 1966. A serological comparison of maize dwarf mosaic virus isolates. Phytopathology 56:1422–1423.

Walkinshaw, C. H., Griffin, G. D., and Larson, R. H. 1961. *Trichodorus christiei* as a vector of potato corky ringspot (tobacco rattle) virus. Phytopathology 51:806–808.

Wall, R. E., and Mortimore, G. G. 1965. Red striped pericarp of corn. Can. Plant Disease Survey 45:92–93.

Watson, M. A., and Mulligan, T. E. 1960. Comparison of two barley yellow-dwarf viruses in glasshouse and field experiments. Annu. Appl. Biol. 48:559–574.

Wellman, F. L. 1934. Infection of *Zea mays* and various other Gramineae by the celery virus in Florida. Phytopathology 24:1035–1037.

Williams, L. E., and Alexander, L. J. 1964. An unidentified virus isolated from corn in southern Ohio. Phytopathology 54:912.

———. 1965. Maize dwarf mosaic, a new corn disease. Phytopathology 55:802–804.

Williams, L. E., Alexander, L. J., Findley, W. R., Dollinger, E. J., Rings, R. W., and Treece, R. E. 1964. Maize dwarf mosaic. Ohio Rep. Res. Develop. 49:88–91.

Williams, E., Alexander, L. J., and Runnels, H. A. 1965. A virus isolated from red-streaked corn grain. Phytopathology 55:1083–1084.

Williams, L. E., Findley, W. R., Dollinger, E. J., Blair, B. D., and Spilker, O. W. 1966. Corn virus research in Ohio in 1965. Ohio Agr. Res. Develop. Center Res. Circ. 145. 22 pp.

Williams, L. E., Gordon, D. T., Nault, L. R., Alexander, L. J., Bratfute, O. E., and Findley, W. R. 1967. A virus of corn and small grains in Ohio and its relation to wheat streak mosaic virus. Plant Disease Reporter 51:207–211.

Yu, T. F., Pei, M. Y., and Hsu, H. K. 1957. Studies on the red-leaf disease of the foxtail millet (*Setaria italica* L. Beauv.). I. Red-leaf a new virus disease of the foxtail millet, transmissible by aphids. [in Chinese, English summary]. Acta Phytopathol. Sinica 3:1–18.

Zummo, N., and Charpentier, L. J. 1964. Vector–virus relationship of sugarcane mosaic virus. I. Transmission of sugarcane mosaic virus by the brick-red sowthistle aphid (*Dactynotus ambrosiae* Thos.). Plant Disease Reporter 48: 636–639.

———. 1965. Vector–virus relationship of sugarcane mosaic virus. III. Transmission of sugarcane mosaic virus by the rusty plum aphid (*Hysteroneura setariae* Thos.). Plant Disease Reporter 49:827–829.

# Hoja Blanca

Travis R. Everett

*Louisiana State University,*
*Baton Rouge, Louisiana*

AND

H. A. Lamey

*Crops Research Division,*
*U.S. Department of Agriculture, Baton Rouge, Louisiana*

Hoja blanca is one of the most destructive diseases of rice in the western hemisphere. The causal agent appears to be a circulative virus transmitted by the planthoppers *Sogatodes oryzicola* (Muir) and *Sogatodes cubanus* (Crawf.) (Fulgoroidea: Delphacidae). The reasons for this assumption are: (*1*) the causal agent has not been transmitted mechanically from plant to plant; (*2*) long acquisition-feeding periods are more effective than short ones; (*3*) there is an incubation period in the insect vector; (*4*) long transmission-feeding periods are more effective than short ones; (*5*) the agent is persistent in the insect vector; and (*6*) the agent is transmitted transovarially.

## I. DISTRIBUTION

Except for one unconfirmed report from mainland China (Wang *et al.*, 1964) hoja blanca has been reported only in the western hemisphere. The disease was probably observed in Colombia, South America as early as 1935 (Garcés-Orejuela *et al.*, 1958) and Osorio (1956) reported that "cinta blanca" (hoja blanca of later works) appeared in Cuba in 1946. During this year the disease caused severe damages in Cuba, "disappearing afterward and never returning until last year." It was observed in Panama in 1952 (Cralley, 1957). However, little attention was given to it prior to the discovery of diseased plants in Cuba in 1954. By 1956, hoja blanca was recognized as "one of the more destructive rice diseases of Cuba" according to Atkins and McGuire (1958). Malaguti (1956) reported severe yield losses due to hoja blanca in Venezuela in 1956. Following the reports of serious yield losses in Cuba and Venezuela and detection of the vector and disease in the United States (Atkins and Adair, 1957), work on the problem was intensified. Rice research workers in Central and South America became aware of the potential threat and detection of both the disease and vectors was soon reported in all major rice producing areas.

At the present time, hoja blanca and/or its insect vectors have been reported from the following countries: Argentina, Brazil, British Guina, British Honduras, Colombia, Costa Rica, the Dominican Republic, El Salvador, Ecuador, Guatemala, Honduras, Mexico, Nicaragua, Panama, Peru, Puerto Rico, Surinam, Venezuela, and the United States. In the United States the disease and insect occur sporadically in the rice growing areas of Florida, Louisiana, and Mississippi. Infestations of the insect were found in 1957, 1958, 1959, and 1964 in Florida, in 1959 and 1962 in Louisiana, and in 1958 and 1959 in Mississippi. The insect infestations of 1959, 1962, and 1964 were apparently eradicated by the use of insecticides and mortality due to subfreezing winter weather. Subsequent infestations are believed to have been the result of dispersal from Cuba via cyclonic or hurricane winds.

## II. SYMPTOMS

The hoja blanca disease is similar in symptomology to the rice stripe disease found in Japan. Mukoo and Iida (1957), Japanese research workers retained by the Cuban government to study the hoja blanca (HB) problem, noted that while HB and stripe disease were remarkably similar, they differed as follows: (*1*) the young leaves of plants infected with stripe virus did not unroll upon emergence and curved downwards whereas in hoja blanca the leaves developed normally except for the characteristic chlorosis; (*2*) mortality was not as high in young plants infected with hoja blanca as with stripe. However, it has been noted by subsequent workers that plants infected with hoja blanca in the 1–2 leaf stage seldom survive.

Atkins *et al.* (1961) found that there was no correlation between varietal resistance to stripe disease and resistance to hoja blanca in the 91 varieties tested. A partial list of varieties tested is given in Table 1 to illustrate the lack of correlation.

Lamey *et al.* (1964b) observed symptoms of HB on several small grains and grasses that were similar to those on rice. Symptoms similar to HB on rice were observed on *Echinochloa colonum* by van Hoof

TABLE 1
REACTION OF RICE VARIETIES TO HOJA BLANCA AND RICE STRIPE (ATKINS *et al.*, 1961)

| FAO Stock No. | CI or PI[a] No. | Variety Name | Source | Reaction to | |
|---|---|---|---|---|---|
| | | | | Stripe[b] | Hoja blanca[c] |
| 223 | 202943 | Aichi-Asahi | Japan | SS | R |
| 889 | 224792 | Aimasari | Japan | M | R |
| 1012 | 8990 | Bluebonnet 50 | USA | M | S |
| 217 | 2128 | Blue Rose | USA | M | S |
| 220 | 1344 | Fortuna | USA | S | S |
| | 9416 | Gulfrose | USA | RR | R |
| 913 | 224829 | Kamenoo 1 | Japan | R | S |
| 216 | 8318 | Magnolia | USA | R | S |
| 289 | 224850 | Norin 1 | Japan | R | R |
| | 222501 | Norin 31 | Japan | S | S |
| 936 | 224874 | Norin 37 | Japan | SS | R |
| 263 | 226184 | Rikuto-Norin 12 | Japan | RR | S |
| 1134 | 9013 | Toro | USA | RR | S |

[a] Accession number, USDA.
[b] RR = very resistant; R = resistant; M = moderately resistant; S = susceptible; SS = very susceptible.
[c] R = resistant; S = susceptible.

(1959) and Galvez *et al.* (1960b). Additional references for symptoms on grasses were given by McGuire *et al.* (1960).

Symptoms of HB vary somewhat according to the rice variety, the age of the plant when infected, and the duration of infection. Young plants are more severely affected than those infected at a later stage of growth and some varieties appear to be resistant or moderately resistant.

The chlorosis characteristic of the disease is observed only on leaves emerging after inoculation. At first the chlorotic areas consist of a spotting or mottling of the leaf. Initially the spots are yellow to greenish-yellow, approaching white with increasing age. As the leaf or subsequent leaves emerge, the chlorotic areas become more extensive and coalesce into stripes; the entire leaf may be chlorotic. Portions of infected leaves of the Bluebonnet 50 variety of rice are shown in Figure 1. Leaves emerging prior to inoculation do not become chlorotic and there is no increase in the size of chlorotic areas once a leaf has emerged. The first leaf showing symptoms may be normal at the tip with only a few chlorotic spots at the base. Thus, it appears that the hoja blanca

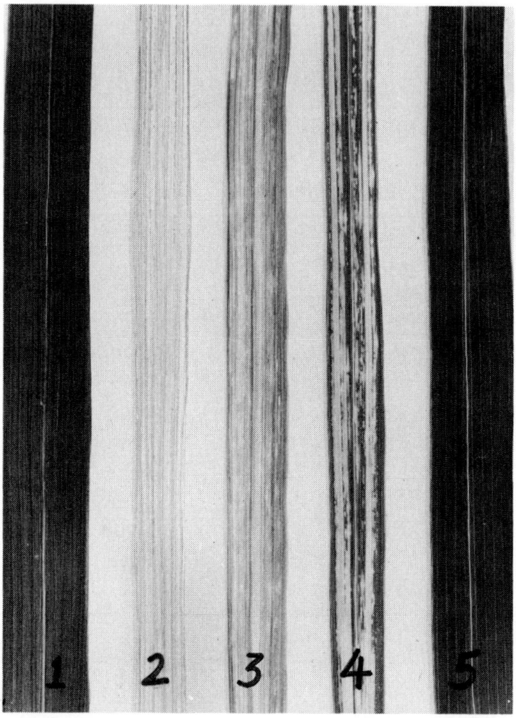


*Fig. 1.* Leaves of the Bluebonnet 50 variety of rice showing the symptoms of hoja blanca virus disease (2, 3, 4), as compared with healthy leaves (1 and 5).

virus (HBV) either inhibits chloroplast or chlorophyll formation or causes their premature destruction, rather than causing local destruction in mature cells. The mottling and progressive chlorosis serves to distinguish hoja blanca disease from genetic stripe or temporary physiological diseases. The former is a regular striping on all leaves while the latter is not found on succeeding leaves when the condition causing stress is removed.

If infection with HBV occurs in young plants, the leaves usually become totally chlorotic and begin to die back from the tip; very young plants ultimately die without flowering. Later infections may cause only chlorotic spots or streaks. The plants infected after flowering do not show symptoms. Atkins and McGuire (1958) reported both normal and diseased tillers on the same plant, the diseased tillers being reduced in height. Acuña (1957) found a 56% reduction in the height of plants infected at the five-leaf stage. Any panicles produced are reduced in size, the lemma and palea are distorted in shape and discolored, and most of the florets are sterile. Lamey *et al.* (1960) and Galvez *et al.* (1960a) reported that the virus was not transmitted in rice seeds.

Yield of infected plants is drastically reduced. McGuire *et al.* (1960) reported total loss of some plantings in Cuba. A yield loss of 95% was measured by Malaguti, Díaz, and Angeles (1957) in Venezuela. The reduction in yield is usually attributed to HBV infection. However, Osorio (1956) reported that the insect vector as well as the disease caused the reduction in yield. In 1964, Mario Sol * and Rufus Walker (personal communication) conducted experiments in El Salvador that showed yield increases of 100% when *S. oryzicola* was controlled with systemic insecticides. Their experiments included a variety (Nilo 10) that was moderately resistant to HBV and the increased yield in this variety could be attributed to vector control.

## III. INSECT VECTORS

Malaguti (1956) and Mukoo and Iida (1957) speculated that the hoja blanca virus was transmitted by insects, probably homoptera, owing to the similarity to rice stripe disease and presence of large numbers of leaf- and planthoppers in the rice fields of Cuba. A number of arthropods were tested as vectors by Acuña (1957) and McGuire *et al.* (1960). Negative results were obtained with *Hortensia similis* (Walker), *Draeculacephala portola* Ball, *Peregrinis maidis* (Ashm.), *Graminella nigrifrons* (Forbes), *Rhopalosiphon maidis* (Fitch), *Aphis maidis* Fitch, and *Tetranychus* sp. Two species of planthoppers have been confirmed as vectors of HBV: *Sogatodes oryzicola* (Muir) and *Sogatodes cubanus* (Crawford). *Hortensia similis* (Walker) and *Draeculacephala clypeata* Osb. have been reported as vectors. However, subsequent transmission tests gave negative results with these species.

*Sogatodes oryzicola* was described by Muir (1926) from specimens collected in British Guiana in 1923. *Sogatodes cubanus* was first described as *Dicranotropis cubanus* by Crawford (1914). Both species were included in the genus *Sogata* by Muir (1926). Recently Fennah (1963) transferred both species to his newly described genus *Sogatodes*. *Sogatodes oryzicola* was originally designated *S. orizicola*. The spelling of the trivium was corrected by Ishihara and Nasu (1966). *Sogatodes brasilensis* (Muir) was found by Fennah (1965) to be synonymous with *S. oryzicola* (Muir).

Adults of the two vector species are 2–4 mm long and cream to dark brown in color. Males are usually darker in color than females, but

* Mario A. Sol, Apartado 692, San Salvador, El Salvador.

light males and dark females may be observed. *Sogatodes oryzicola* is usually larger than *S. cubanus*. De Arévalo and Ruppel (1960) presented characters for separating *S. oryzicola, S. cubanus,* and *Sogata furcifera* (Horvath). (This species is probably *Sogatodes molinus* Fennah). The latter species is frequently collected from grasses in rice fields and may be mistaken for the vector species. A spot (stigma) on the wings of *S. cubanus* serves to separate that species from *S. oryzicola*. Males of *S. furcifera* also have a stigma and may be confused with *S. cubanus,* while females of the former lack a stigma and may be confused with *S. oryzicola*. The shape of the genital styles of the males and the second valvula of females are reliable characters for separating the three species (de Arévelo and Ruppel, 1960; McMillian, 1963).

Both alate and brachypterous forms of the vectors may be found. The proportion of short- to long-winged females may be quite high but in males brachyptery is rare.

The eggs of *S. oryzicola* are laid in clusters in the midvein of the leaf blade or in the leaf sheath. The number of eggs per cluster varies, but there is a tendency towards multiples of 7. A mated female lays an average of 161 eggs during her lifetime, at the rate of about 10 eggs per day (McMillian, 1963). Virgin females lay fewer eggs and these are laid in an erratic pattern. Parthenogenesis has not been observed.

The eggs hatch in about 8 days at 26.7°C. The first instar nymph emerges onto the adaxial surface of the leaf and begins feeding within 24 hr. Each of the 5 nymphal stadia occupies about 3 days. The first instar nymphs are pale cream to white in color, with faint longitudinal stripes. In each succeeding instar, the stripes become more pronounced and in the fifth stage they are dark brown to black in color.

McGuire *et al.* (1960) determined the thermal constant for nymphal development to be 257.6°C, with a developmental threshold of 8.2°C.

The adult stadium for *S. oryzicola* is 24–35 days. (McGuire *et al.*, 1960; Elías *et al.*, 1962). Mating occurs about 3 days after molting and a female needs to mate only once to insure fertility throughout her life span. McMillian (1963) observed abdominal vibrations in males and females just prior to copulation. These vibrations were assumed to be a mating signal. The male first vibrated his abdomen. If a receptive female was in the vicinity, she vibrated her abdomen. The male there scurried to her side and copulation ensued. Abdominal vibrations were not observed in mated females and it was assumed they mate only one time. Males mated repeatedly.

Little is known concerning the biology of *S. cubanus*. Elías *et al.* (1962) and Granados (1963) studied the life cycle of this species in an insectary at Cotaxtla, Vera Cruz, Mexico in 1960–1962. A summary of their data is presented in Table 2. With the exception of adult longevity, *S. cubanus* has a life history similar to *S. oryzicola*.

Infestations of *S. oryzicola* usually are found on the lower portion of the plant. McGuire *et al.* (1960) reported that

> "*S. orizicola* is of sedentary habits. The male is the most active of the three adult forms and will fly readily to other plants when disturbed. The alate femalea has a habit of flying to the ground or water rather than to other plants. All three forms prefer to move around the stem or to the other side of the leaf rather than to fly or jump."

The first and second instar nymphs will quickly drop from the plant when disturbed. The third to fifth instar nymphs are more apt to behave like the adults.

The insects migrate into rice fields soon after seedling emergence and immigration continues until about the time plants begin to form panicles. Soon after heading, the alate population leaves the field. McGuire *et al.* (1960) observed that the insects preferred young plants and migrated from older plantings to younger plants.

Following the immigration of insects into a field, there is a rapid buildup due to within-field reproduction. About 30 days are required for each generation. Thus, 3–4 generations may develop in a particular field. Initially, when the population is low, infestations occur in discrete patches. Dispersal is facilitated by wind and water currents, which spread nymphs and females that drop from the plant when disturbed.

TABLE 2
MEAN DURATION OF DEVELOPMENTAL STADIA FOR *Sogatodes cubanus* (CRAWF.) (ELIAS *et al.*, 1962; GRANADOS, 1963)

| Stage of development | Duration, days | | |
|---|---|---|---|
| | From Elías | From Granados | |
| | Nov.–Dec. | Nov.–Mar. | Sept.–Oct. |
| Egg | 7–8 | 8.25 | 6.42 |
| Nymph | | | |
| 1st instar | 2–3 | 2.82 | 2.1 |
| 2nd instar | 2–3 | 2.71 | 2.35 |
| 3rd instar | 1–2 | 2.28 | 2.35 |
| 4th instar | 2–2.5 | 2.73 | 2.39 |
| 5th instar | 4–4.5 | 3.98 | 3.04 |
| Preoviposition period | 3–4 | 4.81 | 5.3 |
| Adult longevity | 10–13 | 17.17 | 11.98 |

The proportion of alate females is greater in dense populations than in spare ones.

*Sogatodes oryzicola* requires high humidity and moderately high temperatures. McGuire *et al.* (1960) reported that the insect could survive 24 hr at temperatures of 37.8°C and −6.1°C. Insects survived temperatures that peaked at 46.7°C for 3 hr in the greenhouse at Baton Rouge, Louisiana. M. Yoshimeki and T. R. Everett (unpublished data) found that the $LT_{50}$ for insects caged with moist filter paper was 4.7 days at 26.7°C and 5.2 days at 15.6°C. Without the damp filter paper, the mean survival times were 3.5 and 8 hr, respectively. Both groups were maintained at 100% relative humidity.

The host range of *S. oryzicola* is restricted to the family Gramineae. Lamey *et al.* (1964b) tested grasses in 17 genera as host plants. Survival was good on *Oryza sativa* L., Rosen rye, Triumph wheat, *Oryza perennis* Moench, and domestic rye grass, but not on 21 other species of Gramineae. Egg deposition was less on Zayas Bazan and Dima varieties of rice than on Arkrose and Nato. Only *O. sativa* and *O. perennis* were satisfactory host plants for the completion of the life cycle. Some varieties of *O. sativa* were tested for preferential oviposition by the senior author. The variety Gulfrose was most preferred. Bluebonnet 50, Nilo 3, Chino, Dima, Palo Gordo 503, and Nato were intermediate; Nilo 1 was least preferred for egg deposition.

Cordero and Newsom (1962) found that *S. oryzicola* survived and oviposited as well on *Oryza perennis* var. *cubensis* (Moench), *O. perennis* var. *barthii* (Chevalier), and *O. balunga* Yeh and Henderson as on *O. sativa* L. var. Nato. Eight other species of rice and five species of grasses were not suitable hosts.

*Sogatodes cubanus* has been reared on *Echinochloa colonum* (L.), *Digitaria* sp., *Leptochloa scabra* Nees, and *Echinochloa polystachia* (Elías *et al.*, 1962; Granados, 1963; van Hoof, 1959). In tests reported by Granados (1963) adults survived for 10 days on *Panicum purpurascens,* 14 days on *Pennisetum ciliare,* 16 days on *Cynodon dactylon,* and 16 days on *Echinochloa polystachia*. Eggs were laid in *P. purpurescens* but the nymphs did not survive. No eggs were laid in *Pennisetum ciliare* and *Cynodon dactylon.*

Several parasites and predators appear to be important in checking infestations of both *S. oryzicola* and *S. cubanus*. McGuire *et al.* (1960) found two coccinellids, *Coleomegilla maculata cubensis* and *Cycloneda sanguinea limbifer* Cay. in Cuban rice fields. The larvae fed on nymphs of *Sogata sp.* A species of Dryinidae, *Anagarus sp.,* was found parasitizing eggs of *S. oryzicola* in Cuba (McGuire *et al.,* 1960) and both species of *Sogatodes* in Mexico (Elías *et al.,* 1962). McGuire *et al.*

(1960) also noted the parasitization of *S. oryzicola* by a species of Mymaridae.

A new genus and species of Streptiseptera was found as a parasite of *Sogatodes* at Cotaxtla, Mexico. Pierce (1961) described the species as *Sogatelenchus mexicanus* in the family Elenchidae. Parasitization may be as high as 75% (Elías *et al.*, 1962).

A mirid, *Tytthus parviceps* (Reuter), earwigs, empiid flies, mites, and nematodes have also beeen found in or on *S. oryzicola* and *S. cubanus* (Elias *et al.*, 1962; McGuire *et al.*, 1960; Showers, 1966).

*Sogatodes oryzicola* can be controlled effectively with several systemic insecticides. Bayer compound 39007 and Niagara 10242 applied to rice seeds at the rate of 1 lb/100 lb of seed controlled the insects feeding on the plants for about 30 days after planting.

Granular formulations of phorate, disulfoton, Niagara 10242, isolan, dimetilan, and Bayer 25141 applied at rates of 1–2 lb of toxicant per acre were all effective for at least 30 days in a test conducted in the greenhouse. Elías *et al.* (1962) found phorate and disulfoton granules applied at the rate of 1 kg of toxicant per hectare provided adequate control of *S. oryzicola* in rice fields at Cotaxtla, Mexico.

Sprays of malathion, carbaryl, parathion, or methyl parathion have been effective in reducing the population present in rice fields. However, there is little residual activity and the treated fields soon become reinfested. Sprays of Bidrin and phosphamidon at rates of 1 lb of toxicant per acre provide some residual activity. Mario Sol (personal communication) found that two sprays of Bidrin at approximately 30 and 60 days after planting protected the rice crop at Hacienda El Nilo, El Salvador.

## IV. VIRUS TRANSMISSION STUDIES

*Sogatodes oryzicola* was first confirmed as a vector of HBV by Acuña (1957). Galvez *et al.* (1960b) reported the transmission of HBV from rice to *Echinochloa colonum* with *S. cubanus*. Later, Galvez* (personal communication) was able to effect transmission from rice to rice and from *E. colonum* to rice. This confirmed *S. cubanus* as a vector of HBV.

Hoja blanca virus is a propagative or persistent type of virus owing to the incubation period in its insect vector and the fact that one acquisition feeding is sufficient to insure virus transmission throughout

* Guillermo E. Galvez, Plant Pathologist, Centro National de Investigaciones Agropecuarias Tibaitatá, Bogotá, Colombia.

the insect's life. McMillian *et al.* (1962) found that *S. oryzicola* could acquire HBV with an acquisition feeding of only 1 hr. However, 12 hr were needed to insure acquisition by all potential vectors. Their tests indicated "that insects began transmitting on the sixth day after virus acquisition. . . . Those insects that did not transmit by the sixth day never did. . . ." Galvez (personal communication) reported that he definitely had data that the incubation period of the virus in the insect was more than 28 days. In a later communication, he explained that virus-free insects were caged on HBV-diseased plants for 3 days and then transferred to young seedlings. The serial transfers were continued and "finally after 30–38 days" symptoms appeared. This difference in incubation period may be characteristic of the particular colonies tested, McMillian's colony being a more acceptable virus host strain than that of Galvez.

A normal population of *S. oryzicola* contains 7–15% potential vectors (Acuña *et al.*, 1958; Galvez *et al.*, 1961). The frequency of active vectors can be increased through selective breeding (McMillian *et al.*, 1960b). Granados (1963) found that 23% of a field population of *S. cubanus* collected at Cotaxtla, Mexico, transmitted virus to healthy seedlings of *E. colonum*. However, 80% of the insects transmitted virus after feeding on diseased plants. One would assume that *S. cubanus* is a better virus–host species than *S. oryzicola*.

Hendrick *et al.* (1965) selected active vectors and transferred them to a healthy rice seedling every day. From these serial transfers it was determined that an active vector does not necessarily transmit virus to every plant, even with a 24-hr transmission feeding.

Virus may be acquired by transovarial passage from active vector females of *S. oryzicola* through the egg to her offspring. The incidence of transovarial acquisition may approach 100% with selected lines of vectors. Showers (1966) observed transovarial acquisition from females that had failed to transmit virus to a series of six rice seedlings. Nymphs that acquire HBV transovarially may transmit virus within 24 hr of hatching.

McMillian *et al.* (1960a) showed that the latent period in the plant was directly proportional to plant age. Plants inoculated on the emerging leaf at 2 weeks of age developed symptoms in 7 days; those 1 month old developed symptoms in 18 days; and at 3 months of age the interval from inoculation to symptom expression was 26 days. The latent period was shorter when the vectors were fed on the emerging leaf than when fed on older leaves. Hendrick *et al.* (1965) found that plants inoculated 3 days after seedling emergence developed symptoms within 3 days.

The main objective of HBV transmission studies has been the evaluation of varieties for virus resistance and the establishment of the plant host range for the virus.

Lamey *et al.* (1964a) presented a method for screening rice varieties for HBV resistance in the greenhouse using mass insect transfers. Plants were grown in 1-gal crocks and confined periodically for 3–7 days in large cages containing many *S. oryzicola*. The plants were first confined when 5–8 weeks old and several weeks elapsed between the repeated confinements. Continuous confinement was not possible due to the damage inflicted by the dense insect populations. The results of their greenhouse tests, presented in Table 3, correlated with the results of field tests in Central and South America.

A similar technique was used to determine the plant host range of the virus (Lamey *et al.*, 1964b). The results of these experiments show that varietial resistance is present in some of the small grains (Table 4). This resistance may include resistance to attack by *S. oryzicola* since the insects were not force-fed on the plants. In tests with *Echinochloa colonum* (L.), the insects were confined on the plants, but no hoja blanca symptoms appeared. Galvez (personal communication) has succeeded in transmitting HBV from rice to *Echinochloa* with *S. cubanus*. It would appear, therefore, that *S. oryzicola* did not feed on the Echinochloa in the tests of Lamey *et al.* (1964b).

A technique for selecting active vectors of *S. oryzicola* for use in transmission studies was worked out by McMillian *et al.* (1962).

TABLE 3

Hoja Blanca Incidence in Rice Varieties after Confinement with Dense Populations of *Sogata orizicola* (Lamey *et al.*, 1964a)

| Rice variety | No. of tests | diseased plants,[a] mean % |
|---|---|---|
| Arkrose | 22 | 3.7 |
| Berlin (PI 202864) | 2 | 4.9 |
| Bluebonnet 50 | 51 | 68.0 |
| British Guiana No. 79 (PI 185800) | 2 | 66.7 |
| Dima | 2 | 66.3 |
| Gulfrose | 15 | 0.5 |
| Jojutla | 1 | 84.8 |
| Krakti | 2 | 59.3 |
| Nacaome (CI 2181) | 2 | 71.9 |
| Nato | 2 | 53.6 |

[a] Means of all tests, calculated by use of the arcsin conversion.

TABLE 4
HOJA BLANCA DISEASE DEVELOPMENT IN SMALL GRAINS AFTER PROLONGED CONFINEMENT WITH A LARGE POPULATION OF VIRULIFEROUS *Sogata orizicola* IN THE GREENHOUSE
(Lamey *et al.*, 1964b)

| Species | Plants diseased [a] |
|---|---|
| *Avena sativa* Albion | 0/34 |
| *A. sativa* Clintland | 0/23 |
| *A. sativa* Fayette | 0/20 |
| *A. sativa* Fulghum | 0/25 |
| *A. sativa* Newton | 0/24 |
| *A. sativa* Nortex | 0/24 |
| *A. sativa* Saia | 3/22 |
| *A. sativa* Victoria | 0/24 |
| *Hordeum vulgare* C.I. 3906-1 | 1/17 |
| *H. vulgare* C.I. 10572 | 0/24 |
| *H. vulgare* C.I. 10573 | 2/22 |
| *H. vulgare* Kindred | 0/21 |
| *H. vulgare* Pace | 0/22 |
| *H. vulgare* Valmore | 1/18 |
| *Secale cereale* Merced | 18/45 |
| *S. cereale* Petkus | 5/21 |
| *S. cereale* Prolific | 17/24 |
| *S. cereale* Rosen | 24/42 |
| *S. cereale* Tetra-Petkus | 0/2 |
| *Triticum aestivum* Genesee | 9/18 |
| *T. aestivum* Knox | 9/20 |
| *T. aestivum* Lee | 3/22 |
| *T. aestivum* Triumph | 14/21 |
| *T. aestivum* Wichita | 2/17 |
| *T. compactum* Omar | 4/20 |

[a] Numerator represents plants showing symptoms; denominator, plants tested.

However, the insects selected with their method had only a few days of their life span remaining when the potential for transmission was known. Hendrick *et al.* (1965) worked out a technique that produced active vectors, with most of the adult life span remaining. From a colony with 5% viruliferous individuals, active vectors were selected and mated to active vectors. The resulting progeny had a potential of 50–75% active vectors; the genes for virus transmission apparently are dominant. By continuous selective breeding of active vectors to active vectors a colony with 80–100% potential vectors was produced for studies of virus transmission. Techniques for the selection and maintenance of the colony are as follows: third to fifth instar nymphs that have been reared on HB diseased plants are caged singly on 1-day old rice seedlings. Within 10 days approximately 85% of the active vectors may be identified by the development of HB symptoms on the seedling. The insects are old enough to mate at that time. Individual pairs of active vector males and females are caged together for 24 hr. The females are then transferred to HBV-diseased plants for oviposition. Plants in the "oviposition cage" are changed every 4 days. This provides nymphs of fairly uniform age for a particular test. Several hundred active vectors may be produced each week with this procedure. This large stock of vectors has made it possible to test most of the rice varieties used in U.S. rice breeding programs.

The authors of the present paper have investigated the effect of plant age on susceptibility to HBV. The variety Gulfrose was found to be moderately susceptible and Arkrose susceptible in the one-leaf stage. Both varieties were resistant in the third- and fourth-leaf stage. Bluebonnet 50 and CI 6001 (Pandhori No. 4) were susceptible in the one- and two-leaf stage but the latter variety became more resistant in the third- and fourth-leaf stage.

A colony of nonviruliferous insects with a high incidence of potentially active vectors is necessary for many of the studies on HBV transmission. Selection of such a colony has been complicated by the transovarial acquisition of virus by *S. oryzicola*. Galvez (personal communication) developed a virus-free colony from a highly active one by selecting the individuals that did not transmit the virus and increasing their numbers. This was carried out for several generations. When reasonably increased, some of these insects were fed on diseased rice for 3 days and then transferred to young seedlings. An unspecified number transmitted HBV.

## V. INSECT BEHAVIOR AND VIRUS TRANSMISSION

Insects in general have specific behavorial patterns that govern their feeding activities. The mechanisms determining host plant acceptability or rejection may be classified into three broad categories: (*1*) attraction to the host plant, (*2*) an arrestant, and (*3*) feeding stimuli. The absence of any of these components (or presence of an objectionable component) results in rejection of a plant as a host.

A thorough understanding of vector feeding behavior is necessary in order to make comprehensive studies on plant–virus transmission. Any plant rejected as a host plant by the insect vector species would be assumed to be a nonhost plant of the virus. In actuality, the plant might well support the virus if the insect were induced to feed on it. Conversely, an insect species might be assumed to be a negative host for the virus if it did not feed on a viruliferous plant host. Thus, the acceptability of a plant as a host for insects may limit the host range for the virus to one or a few species of plants and insects. The host range might be extended by manipulating the feeding behavior of the insect or substituting exotic species as the insect vector.

Numerous examples of additions to the host plant range of insects are recorded in entomological literature. Indeed, some of the worst pests of agriculture result from introducing new plant species in the geographic range of a previously innocuous insect species or introducing the insect into a new region. A classic example of the first case is the spread of the Colorado potato beetle following the introduction of potatoes into the western United States. Prior to that time, the beetle had fed on a weed, *Solanum rostratum*. European corn borer, *Ostrinia nubilalis* (Hbn.), is an old world species that has become one of the major pests of maize since introduction of the crop into Europe and Asia and, subsequently, the introduction of the insect into the United States.

Additions and deletions in plant host range of insects may also come about through genetic or physiological changes. Genetic alterations are evident in the breeding programs, which are aimed at resistance to insect attack. An interesting alteration of the insect–host plant relationship as a result of physiological change in the plant was reported by Severin (1946). He found that an aster yellow virus infection in celery and aster plants made them more acceptable to nine species of leaf-hoppers that normally did not feed on either plant.

## VI. ORIGIN OF HOJA BLANCA DISEASE

Hoja blanca disease of rice has come into prominence rather recently. One naturally is curious as to the origin of the disease and the reasons for its sudden upsurge in prevalence.

The presence of wild species of *Oryza* in the epicenters of HBV leads one to suspect them as the original host plants from which the virus was transmitted to cultivated rice. The evaluations of Lamey, McMillian, and Hendrick (1964b) and Cordero and Newsom (1962) indicate that this is not the case. The most prevalent wild rice species, *Oryza alta* and *Oryza latifolia*, appear to be resistant to both HBV and *S. oryzicola*.

A more plausible hypothesis involves the virus disease present in *Echinochloa colonum* and transmitted by *S. cubanus.* Galvez (personal communication) has succeeded in transmitting HBV from *Echinochloa* sp. to rice with *S. cubanus.* This establishes the possibility of HBV being transmitted from jungle grass to rice. Once established in rice, the virus became available to *S. orizicola.* Thus, the original hosts of HBV are assumed to be *Echinochloa colonum* and *S. cubanus,* with *Oryza sativa* and *S. oryzicola* as adapted hosts. This hypothesis is further supported by the observations of Malaguti, Díaz, and Angeles (1957) where *E. colonum* appeared to be more susceptible to HBV than rice, and the report by Granados (1963) which indicated that *S. cubanus* had a higher incidence of active vectors than did *S. oryzicola.*

## BIBLIOGRAPHY

Acuña, J. 1957. Informaciones de interés general en relación con el arroz. Administración de Estabilización del Arroz, Bol. 4 and 5.

Acuña, J., Ramos, L., and Lopez, Y. 1958. *Sogata orizicola* Muir, vector de la enfermedad virosa hoja blanca del arroz en Cuba. Agrotecnia 1958, Sept.-Oct.: 23–34.

Arévalo, I. S. de, and Ruppel, R. F. 1960. La especie *Sogata orizicola* Muir, y otras allegadas, y la relacion que tienen con el virus causante de la enfermedad "Hoja blanca" del Arroz. Agr. Trop. 16(5) : 291–299.

Atkins, J. G., and Adair, C. R. 1957. Recent discovery of hoja blanca, a new rice disease in Florida, and varietal resistance tests in Cuba and Venezuela. Plant Disease Reporter 41:911–915.

Atkins, J. G., Goto, K. and Yasuo, S. 1961. Comparative reactions of rice varieties to the stripe and hoja blanca virus diseases. Int. Rice Comm. Newsletter 10(4):5–8.

Atkins, J. G., and McGuire, J. U. 1958. The hoja blanca disease of rice. FAO Plant Protection Bull. 6(11):161–166.

Cordero, A. D., and Newsom, L. D. 1962. Suitability of *Oryza* and other grasses as hosts of *Sogata orizicola* Muir. J. Econ. Entomol. 55:868–71.

Cralley, E. M. 1957. "Hoja blanca"—white leaf—A new disease of rice. Arkansas Farm Res. 6(5):9.

Crawford, D. L. 1914. A contribution towards a monograph of the homopterous insects of the family Delphacidae of North and South America. Proc. U.S. Nat. Mus. 46:557–640.

Elías, R., Granados, G., and Oretga, A. 1962. El estado actual de la hoja blanca en México. Agr. Técnica en México 2(1):2–7.

Fennah, R. G. 1963. The delphacid species-complex known as *Sogata furcifera* (Horvath) (Homoptera:Fulgoroidea). Bull. Entomol. Res. 54(1):45–79.

——— 1965. *Sogatodes brasilensis* (Muir). A new synonym of *Sogatodes orizicola* (Muir) (Fulgoroidea, Delphacidae) Bull. Entomol. Res. 56(2):215–7.

Galvez, G. E., Jennings, P. R., and Thurston, H. D. 1960a. Transmission studies of hoja blanca in Colombia. Plant Disease Reporter 44:80–81.

Galvez, G. E., Thurston, H. D. and Jennings, P. R. 1960b. Transmission of hoja blanca of rice by the planthopper, *Sogata cubana*. Plant Disease Reporter 44:394.

——— 1961. Host range and insect transmission of the hoja blanca disease of rice. Plant Disease Reporter 45:949–53.

Garcés-Orejuela, C., Jennings, P. R., and Skiles, R. L. 1958. Hoja blanca of rice and the history of the disease in Colombia. Plant Disease Reporter 42(6):750–751.

Granados, G. 1963. Biología, ecología, combate y pruebas de transmissión, con *Sogata orizicola* Muir y *Sogata cubana* (Crawf.) (Araepodidae-Homoptera), vectores del virus de la "hoja blanca" del arroz. Thesis, Escuela Nacional de Agricultura, Chapingo, Mexico. 57 p.

Hendrick, R. D., Everett, T. R., Lamey, H. A., and Showers, W. B. 1965. An improved method of selecting and breeding for active vectors of hoja blanca virus. J. Econ. Entomol. 58(3):539–42.

Ishihara, T., and Nasu, S. 1966. Leafhopper-transmitting plant viruses in Japan and adjacent countries. Paper presented at the divisional meeting on plant protection, 11th Pacific Science Congress, Tokyo, 1966. Japan Plant Protection Association, Tokyo. p. 159–170.

Lamey, H. A., Lindberg, G. D., and Brister, C. D. 1964a. A greenhouse testing method to determine hoja blanca reaction of rice selections. Plant Disease Reporter 48(3):176–9.

Lamey, H. A., McMillian, W. W., and Hendrick, R. D. 1964b. Host ranges of the hoja blanca virus and its insect vector. Phytopathology 54(5):536–41.

Lamey, H. A., McMillian, W. W., and McGuire, J. U. 1960. Transmission and host range studies on hoja blanca. Proc. Rice Tech. Working Group. p. 20–21.

McGuire, J. U., McMillian, W. W. and Lamey, H. A. 1960. Hoja blanca disease of rice and its insect vector. Rice J. 63(13):15–16, 20–24.

McMillian, W. W. 1963. Reproductive system and mating behavior of *Sogata orizicola* (Homoptera:Delphacidae). Annu. Entomol. Soc. Amer. 56(3):330–4.

McMillian, W. W., McGuire, J. U., and Lamey, H. A. 1960a. Relationship of hoja blanca to the inoculation point and to the age and yield of rice plants. Plant Disease Reporter 44(6):387–9.

——— 1960b. Hoja blanca studies at Camaguey, Cuba. Proc. Rice Tech. Working Group, p. 21.

——— 1962. Hoja blanca transmission studies on rice. J. Econ. Entomol. 55:796–7.

Malaguti, G. 1956. La "hoja blanca". Extrana enfermedad del arroz en Venezuela. Agron. Trop. 6(3):141–145.

Malaguti, G., Diaz, H., and Angeles, N. 1957. La virosis "hoja blanca" del arroz. Agron. Trop. Nogent-sur-Marne 6(4):157–63.

Mukoo, H., and Iida, T. 1957. Informe sobre la investigación de la hoja blanca del arroz en Cuba. Administración de Estabilización del Arroz. Bol. No. 1, p. 5–12.

Muir, F. 1926. Contributions to our knowledge of South American Fulgoroidea (Homoptera) Part I. The family Delphacidae. Bull. Exp. Sta. Hawaiian Sugar Planters Assoc. Entomol. Ser., Bull. 18. 51 p.

Osorio, J. M. 1956. Harmful insects of the rice crop in Cuba. Unpublished paper presented at the 10th International Congr. Entomol., Montreal, Canada.

Pierce, W. D. 1961. A new genus and species of Strepsiptera parasitic on a leafhopper vector of a virus disease of rice and other Gramineae. Annu. Entomol. Soc. Amer. 54(4):467–474.

Severin, H. H. P. 1946. Longevity, or life histories, of leafhopper species on virus infected and on healthy plants. Hilgardia 17:121–33.

Showers, W. B. 1966. Observable effects of hoja blanca virus of rice on its plant-hopper vector, *Sogata orizicola* Muir. M.S. thesis Louisiana State Univ. 86 p.

van Hoof, H. A. 1959. The delphacid *Sogata cubana* vector of a virus of *Echinochloa colonum*. Tijdschr. Plantenziekten 65:188–189.

Wang, F. M., Chen, Y. L., and Pai, K. C. 1964. A preliminary investigation on the transmission and hosts of the hoja blanca disease of rice (in Chinese). Plant Protection (mainland China) 2:9–10.

# Localization of Viruses in Vectors: Serology and Infectivity Tests*

R. C. SINHA

*Plant Research Institute, Canada Department of Agriculture, Ottawa, Ontario*

## I. INTRODUCTION

Most of the work on the localization of plant viruses in insect vectors has been with leafhopper-borne viruses. The discussion that follows, therefore, deals mainly with such viruses. The presence of certain viruses in the internal organs of vectors other than leafhoppers and in nonvector insects is also discussed.

For successful transmission of a virus by leafhoppers after they acquire it from infected plants the virus must pass through some part of the alimentary tract into the hemolymph, reach the salivary glands, and then be injected into the healthy plants with salivary secretions. Indications for this sequence of events first came from the early work on maize streak (Storey, 1933) and sugar beet curly-top (Bennett and Wallace, 1938) viruses. Certain features of the transmission of curly-top virus suggested that it may not multiply in its leafhopper vector (Freitag, 1936), and such viruses are now commonly referred to as "circulative" (see Black, 1959). Several other viruses have been shown to multiply in their leafhopper vectors and they are called "propagative" (see Black, 1953; Maramorosch, 1955).

The only virus that does not seem to circulate in its leafhopper vector is the tungro virus (Ling, 1966), transmitted by *Nephotettix impicticeps* (Ishihara). Ling reported that leafhoppers could transmit the virus after acquisition and inoculation feeding periods of only 1 hr each. Moreover, unlike other leafhopper-borne viruses, the infective nymphs failed to transmit the virus after molting into adults. Ling suggested that tungro virus may be stylet borne and "nonpersistant" in its leafhopper vector.

* Contribution No. 599 from Plant Research Institute, Canada Department of Agriculture, Ottawa, Ontario.

Several viruses have been shown to occur in the hemolymph and in certain internal organs of their insect vectors by means of the infectivity test, serology, or electron microscopy. I will only deal with the results reported by infectivity or serological tests. Site or sites of virus multiplication in an insect, wherever pertinent, will also be discussed.

## II. INFECTIVITY TEST

The following procedure has been used to demonstrate the presence of viruses in the hemolymph and in the extracts of internal organs of vectors carrying the virus. Nonviruliferous vectors are either fed on the extracts or are injected with them and the virus is then allowed to incubate in the treated insects before they are tested for their infectivity on plants susceptible to the virus. By this method several days of virus incubation in insect vectors and in plants are needed before the specific symptoms of disease permit identification of infective virus in the extract. Nevertheless, this test is one of the most sensitive methods to determine the presence of virus in an inoculum.

### *A. Maize Streak Virus*

Storey (1928) showed that the minimum incubation period of the virus in its vector *Cicadulina mbila* Naude was only 6–12 hr, but that once the leafhoppers became infective, they remained so throughout their lives. He believed that "the evidence clearly indicates the multiplication of the virus in the insect."

Storey (1933) was the first to show that a plant virus can be transmitted to its leafhopper vector by mechanical inoculation. He introduced the inoculum, prepared from viruliferous leafhoppers or from infected plants, into a puncture made in the abdomen or in the leg of insects with a finely pointed needle. By employing this technique, he showed that the virus was present in the general contents of the thorax, the abdomen, and in the hemolymph of leafhoppers carrying the virus. The virus in the hemolymph was detectable before the end of incubation period in insects.

### *B. Sugar Beet Curly-Top Virus*

Freitag (1936) found that after leafhoppers *Circulifer tenellus* (Baker) acquired the virus by feeding on infected plants and the insects were tested for their ability to transmit the virus each day, the percentage of plants infected by single leafhoppers decreased progressively. In some cases, the leafhoppers eventually lost their ability to infect plants. The ability of leafhoppers to infect plants was approximately proportional to the length of acquisition period. Freitag interpreted his data as evidence of the absence of virus multiplication in insects.

Bennett and Wallace (1938), using the artificial feeding technique, ascertained that the virus occurred in the hemolymph, alimentary tract, and salivary glands of viruliferous leafhoppers. The hemolymph was the richest source of virus inoculum. The infectivity of leafhoppers and their virus content gradually decreased over periods of 8–10 weeks, but the reduction of infectivity was slight if the original virus concentration was high. Bennett and Wallace also believed that their findings suggest the absence of virus multiplication in the leafhopper vector.

### *C. Aster yellow virus*

1. RECOVERY OF THE VIRUS FROM THE INTERNAL ORGANS OF A VECTOR. The virus has been shown to multiply in its leafhopper vector *Macrosteles fascifrons* (Stål) by several workers (Kunkel, 1937; Black, 1941; Maramorosch, 1952a). However, studies on the distribution of the virus in the vector have been reported only by Hirumi and Maramorosch (1963). Using the injection technique, the authors found that the salivary glands of leafhoppers removed 19 days after the start of a 5-day acquisition feeding, had a large concentration of the virus, but that no virus was recovered from intestines, Malpighian tubules, ovaries, and testes at this time. The salivary glands and intestines removed immediately after the acquisition feeding period showed a very low concentration of the virus. Hirumi and Maramorosch suggested that "some passageways seem to exist for the ingested virus, by which it can penetrate from the gut into the insects body cavity and accumulate in the salivary glands." The authors did not have enough data to draw any definite conclusions regarding the site or sites of virus multiplication in the insect.

In our laboratory we have studied the distribution of a celery strain of aster yellows virus in the vector *M. fascifrons.* Results (Sinha and Chiykowski, 1967a) showed (Table 1) that the virus was present in the intestine (midintestine and hindintestine), hemolymph, ovaries, and salivary glands of viruliferous leafhoppers. No virus was recovered from the mycetomes, Malpighian tubules, fat body tissue, testes, or brain. The age of insects at the time of acquisition feeding apparently had no effect on the subsequent distribution of the virus in leafhoppers. However, more virus was recovered from the ovaries of insects that had fed on infected plants as adults than as nymphs

TABLE 1

DISTRIBUTION OF ASTER YELLOWS VIRUS IN THE LEAFHOPPER VECTOR *Macrosteles Fascifrons*

| Source of extract | Transmission by insects injected with the extracts of tissues from leafhoppers that had acquisition feed as [a] | |
|---|---|---|
| | Nymphs | Adults |
| Intestine | 4/50 [b] (8%) | 4/100 [b] (4%) |
| Hemolymph | 72/86 (84%) | 29/35 (83%) |
| Salivary glands | 12/64 (19%) | 20/111 (18%) |
| Ovaries | 2/105 (2%) | 24/74 (32%) |
| Testes | 0/94 (0%) | 0/61 (0%) |
| Fat body tissue | 0/90 (0%) | 0/40 (0%) |
| Mycetomes | ND | 0/69 (0%) |
| Malpighian tubules | 0/80 (0%) | 0/98 (0%) |
| Brain | 0/89 (0%) | 0/71 (0%) |

[a] Combined results of two experiments (Sinha and Chiykowski, 1967a).

[b] Numerator is the number of injected insects that transmitted the virus; denominator is the number of insects tested; ND means treatment not done.

(Table 1). In both cases, hemolymph was the best source of virus inoculum.

In subsequent experiments leafhoppers were given an acquisition feeding of 3 days and then at various intervals from the start of acquisition their intestines were examined for the virus. The combined results of two such experiments showed that when extracts of intestines prepared on the third, sixth, twelfth, and twenty-fourth days were injected into the virus-free leafhoppers, the percentages of insects that became infective were about 2 ($1/53$), 14 ($9/55$), 43 ($19/44$), and 3% ($2/74$), respectively. These results suggest that the intestine, or some part of it, is the initial site of virus multiplication. The concentration of the virus in the intestine, however, increases by the twelfth day and then decreases sharply by the twenty-fourth day. The presence of virus in the hemolymph in large concentration indicates that hemocytes may be the main site of virus synthesis. It is not known if the virus multiplies or only accumulates in salivary glands and ovaries.

Transovarial transmission of aster yellows virus has never been shown. The successful transmission of viruses to progeny insects perhaps depends on the ability of virus to penetrate the eggs. How viruses penetrate the eggs is not known, but the electron microscopic work on rice dwarf virus suggests that the virus may be transported by the symbiotes of mycetocytes into the oocytes of the vector, resulting in the transovarial transmission of the virus (Nasu, 1965). It is possible that although some aster yellows virus occurs in the ovaries, the virus cannot reach the oocytes and, therefore, no transovarial transmission can occur.

2. RECOVERY OF THE VIRUS FROM NONVECTOR INSECTS. Maramorosch (1952b) demonstrated that aster yellows virus is not only acquired by the nonvector leafhopper *Dalbulus maidis* Del and Wol., but is also retained by it for several days. The presence of virus in the nonvector insects was demonstrated by injecting their extracts into healthy vector *M. fascifrons*, and then testing the injected insects for their infectivity. Maramorosch also found that no virus could be recovered from the nonvector insects immediately after the acquisition feeding of 2 days, but it was recovered after 21 days. No virus, however, was recovered from the nonvector insects injected with an infective dose of the virus. Maramorosch suggested that either the virus was unable to multiply in nonvector leafhoppers or it multiplied to a very limited extent.

Recovery of viruses from nonvector insects after they had fed on infected plants is not uncommon. As early as 1938, Bennett and Wallace demonstrated that several nonvector species of insects can acquire curly-top virus from infected plants. Tobacco mosaic and turnip yellow-mosaic viruses have been shown to be acquired by certain aphids and other insects (Ossiannilsson, 1958; Kikumoto and Matsui, 1962; Orlob, 1963; Hutchinson and Matthews, 1963). Orlob (1966) showed the presence of wheat streak mosaic virus particles in nonvector species of mites after they had fed on infected plants.

Recently, it was demonstrated that a celery strain of aster yellows virus can be recovered at various times from the nonvector leafhopper *Agallia quadripunctata* Prov. after they have acquired the virus from infected plants (Sinha and Chiykowski, 1967b). The virus increased in concentration by the twelfth day, then decreased by the twenty-fourth day. Subsequent experiments showed that the virus occurred in the intestine, but none was recovered from the hemolymph and the salivary glands. When extracts of intestines prepared on the sixth, twelfth, and twenty-fourth days after the start of the acquisition feeding period of 6 days by the nonvector insects, were injected into virus-free vector insects (*M. fascifrons*), the percentages of insects that became infective were 4, 21, and 17, respectively (Table 2). The results suggested that the virus had increased in concentration in the intestine. The

TABLE 2

RECOVERY OF ASTER YELLOWS VIRUS FROM INTESTINE OF A NONVECTOR AT VARIOUS TIMES AFTER THE START OF ACQUISITION FEEDING [a]

| Days from start of acquisition | Infectivity of extracts [b] (experiment number) | | | | Transmission, % |
|---|---|---|---|---|---|
| | 1 | 2 | 3 | 4 | |
| 6 | 1/18 [c] | 0/31 | 1/23 | 2/23 | 4 |
| 12 | 5/24 | 11/47 | 2/14 | — | 21 |
| 24 | 2/26 | 2/50 | 3/11 | 16/49 | 17 |

[a] Nonvector insects *Agallia quadripunctata,* in second or third instar were caged on infected aster plants for 6 days and then were maintained on healthy plants. In each treatment intestines of 30 exposed nonvector insects were ground in 0.1 ml of saline and their extract injected into healthy vector *M. fascifrons* (Sinha and Chiykowski, 1967b).

[b] Injected insects were maintained in groups for 3 weeks on healthy plants and then tested singly for 2 weeks on aster for their transmissibility.

[c] Numerator is the number of injected insects that transmitted the virus; denominator is the number of insects tested.

authors interpreted their data as evidence for virus multiplication in the nonvector insect.

Sinha and Chiykowski failed to recover any virus from nonvector insects that were injected with a massive dose of infective virus. They hypothesized that the virus in the injected nonvector insects either does not survive in the hemolymph or, if it does, it cannot pass through the gut wall into the susceptible cells of the intestine, which perhaps are the only cells that can support virus synthesis. By contrast, when the nonvector insects feed on infected plants, the acquired virus reaches the intestine and multiplies there. Vector specificity of aster yellows virus, therefore, cannot be explained solely on the basis of virus multiplication in an insect. As pointed out by Bawden (1964), apart from multiplication other limiting factors, such as the inability of the virus to pass through the gut wall into the body cavity or its inability to enter or survive in salivary glands, could prevent a leafhopper from acting as a vector.

### *D. Potato Yellow-Dwarf Virus*

This is the only leafhopper-borne virus that can be readily transmitted by sap inoculation. The long incubation period and persistance of the virus in leafhoppers suggest that it may multiply in the vectors (Black, 1959).

By means of the injection technique, the virus has been shown to occur in all the internal organs and in the hemolymph of viruliferous leafhoppers *Agallia constricta* Van Duzee (Sinha, 1965a). As sources of inoculum the hemolymph, intestine, brain, and fat body tissue were good; salivary glands and mycetomes were intermediate; and ovaries and testes were poor (Table 3). These results demonstrated that the virus is capable of systemically invading the leafhopper vector.

None of the viruses in leafhoppers, including the potato yellow dwarf, has been shown to be transmitted through the sperm of viruliferous males to the progeny insects. However, the above results show that some virus can occur in testes of insects.

### E. *Beet Crinkle Virus*

The virus is transmitted by the lace bug *Piesma quadratum* Fieb, and has been shown to multiply in its insect vector (Proeseler, 1966). This was demonstrated by the serial passage of the virus from insect to insect, until the dilution attained exceeded the dilution endpoint of the starting inoculum. This method, which provides direct evidence for

TABLE 3

RECOVERY OF POTATO YELLOW-DWARF VIRUS FROM HEMOLYMPH AND INTERNAL ORGANS OF AN INSECT VECTOR

| | Leafhoppers that transmitted the virus after being injected with the extract | | | |
|---|---|---|---|---|
| Source of extract | Expt. 1 | Expt. 2 | Total | Transmission, % |
| Hemolymph | 6/24 [a] | 15/20 | 21/44 | 48 |
| Intestines | 5/21 | 11/16 | 16/37 | 43 |
| Brain | 3/10 | 7/13 | 10/23 | 43 |
| Fat body tissue | 7/33 | 8/15 | 15/48 | 43 |
| Malpighian tubules | 1/19 | 7/22 | 8/41 | 20 |
| Salivary glands | 4/21 | 5/27 | 9/48 | 19 |
| Mycetomes | 0/36 | 7/23 | 7/59 | 12 |
| Ovaries | 1/25 | 2/18 | 3/43 | 7 |
| Testes | 1/30 | 1/18 | 2/48 | 4 |

[a] Numerator is the number of insects which transmitted the virus; denominator is the number of insects tested (Sinha, 1965a).

the multiplication of the virus in a vector, was first used with aster yellows virus in *M. fascifrons* (Maramorosch, 1952a).

Proeseler also demonstrated that the virus occurs in the intestine, hemolymph, and salivary glands of viruliferous insects. It is not known if the virus can invade other organs of the vector.

## III. SEROLOGICAL TEST

In the investigation of leafhopper-borne viruses, immunological techniques have been used in only a few cases. Antisera have been reported for at least 3 viruses: wound tumor (Black and Brakke, 1954), potato yellow dwarf (Wolcyrz and Black, 1956), and rice dwarf (Kimura, 1962). However, most extensive applications of serological techniques for detecting virus antigens in a vector have been made only with wound tumor virus. The discussion that follows, therefore, deals only with this virus.

### *A. Wound Tumor Virus*

The virus has been shown to multiply in the leafhopper vector *Agallia constricta* Van Duzee (Black and Brakke, 1952). Black and Brakke (1954) detected virus antigens in viruliferous leafhoppers, as well as in the plants infected with the virus. The authors also demonstrated that ultracentrifugal supernatants, obtained either from extracts of viruliferous leafhoppers or from infected plants, contained antigens smaller than the virus particles that reacted specifically with the antiserum prepared against the virus. These antigens were termed wound tumor-soluble antigens.

The distribution of virus antigens in the leafhopper vector *A. constricta* has been investigated by means of the precipitin ring test and the fluorescent antibody technique.

1. PRECIPITIN RING TEST. The precipitin ring test requires no more than 0.1 ml of each solution (antigen and antiserum) per tube, and even less can be used if desired. The antiserum, usually diluted in 10% glycerine saline, is placed at the bottom of the small tubes and the solutions to be assayed for virus antigens are layered on the top. After diffusion of the two solutions, a ring of precipitate is formed where optimum concentrations of antigen and of antibody are present. The precipitin ring test is very sensitive in being able to detect small amounts of virus antigens.

The soluble antigens of wound tumor virus were detected in the intestine, fat body, salivary glands, and ovaries of leafhoppers that had acquired the virus from infected plants (Sinha and Black, 1963), but the Malpighian tubules and testes failed to show a positive reaction (Table 4). By use of fluorescent antibodies, the virus antigens were also detected in the smears of hemolymph and in the fat body tissue. The results indicated that virus antigens were present in higher concentration in the fat body and in the hemolymph than in other organs. Sinha and Black hypothesized that the intestine, or some part of it, is the initial site of virus multiplication, and that as the incubation period progresses, the virus reaches other susceptible sites (such as hemocytes, cells of the fat body, and salivary glands) and may multiply there.

2. FLUORESCENT ANTIBODY TECHNIQUE. The virus antigens were detected in individual viruliferous leafhoppers by means of immunofluorescent technique (Nagaraj, Sinha, and Black, 1961; Sinha and Black, 1962). Sinha and Reddy (1964) improved the technique and studied its applications in detecting the virus antigens in individual leafhoppers. Sinha (1965b) studied the distribution of the virus antigens in viruliferous leafhoppers *A. constricta.* The virus antigens were detected in various parts of the following organs: the intestine, Malpighian tubules, female reproductive organs, brain, mycetomes, and salivary glands. The antigens were found to be localized in the cytoplasmic particulates of hemocytes and fat body cells but not in the nuclei. No antigens were detected in the male reproductive organs.

Fluorescent antibody technique was used to study the sequence of infection sites in leafhoppers that had ingested virus from infected plants (Sinha, 1965b). Leafhoppers were given an acquisition feeding of 1 day and then at various intervals, the intestine, hemolymph, fat body, brain, Malpighian tubules, and salivary glands were examined for the presence of virus antigens. The days on which the antigens could first be detected in different tissues were as follows: day 4—one corner of the filter chamber of the intestine; day 7—entire filter chamber and part of the ventriculus; day 12—hemolymph; day 14—fat body, brain, and Malpighian tubules; day 17—salivary glands.

These results suggest that the filter chamber of the intestine is the primary site of infection and virus multiplication. As the incubation period progresses, the infection spreads to other parts of the intestine and to other susceptible sites. The detection of virus antigens in hymocytes, fat body cells, and the acini of salivary glands may be taken as *prima facie* evidence that the virus multiplies in them too. That the virus systemically invades its leafhopper vector has also been shown by the presence of virus particles in various tissues (Shikata and Maramorosch, 1965).

## TABLE 4

DETECTION OF WOUND TUMOR-SOLUBLE ANTIGENS IN THE VARIOUS ORGANS OF LEAFHOPPERS AS TESTED BY PRECIPITIN RING TEST

| Expt. No. | Number, class, and sex of the leafhoppers dissected | Results of the ring test on various organs [a] | | | | | |
|---|---|---|---|---|---|---|---|
| | | Fat body | Intestine | Ovaries | Salivary Glands | Malpighian tubules | Testes |
| 1 | 20 exposed females | + + +[b] | + + | +[b] | | – | |
| 2 | 18 exposed females | + + + | + + | + | | – | |
| 3 | 18 exposed females | + + + | + + | | | – | |
| 4 | 11 exposed females | + + | + | | | – | |
| 5 | 17 exposed females | + | + + | | | – | – |
| 6 | 18 viruliferous males | + | + + | | + | – | – |
| 7 | 19 viruliferous females | + + + | + + | + | + | – | |

[a] Organs of the same kind from the number of leafhoppers indicated were pooled and ground in 0.1 ml of 0.85% saline; the extracts were tested for WTSA by the ring test (Sinha and Black, 1963).

[b] The greater the number of plus signs, the thicker the ring after 1 hr; a minus sign means that the results were negative; a blank space indicates that the test was not included.

## IV. SUMMARY

All the viruses discussed were shown to occur in the intestine, hemolymph, and salivary glands of their vectors. Wound tumor and potato yellow-dwarf viruses were shown to be capable of invading their leafhopper vectors systemically. Immunofluorescent studies on the sequential infection of the internal organs of a vector after ingestion of wound tumor virus by insects provided evidence that the filter chamber of the intestine was the primary focus of infection and multiplication. As the incubation period progressed, the virus spread to and infected other susceptible sites, such as hemocytes, fat body tissues, mycetomes, ovaries, and salivary glands.

Infectivity tests showed that the intestine of the leafhopper vector was the initial site of aster yellows virus multiplication. The virus was also recovered from the salivary glands and ovaries, and it was present in high concentration in the hemolymph, which suggested that hemocytes may be the main site of virus synthesis.

Aster yellows virus not only was acquired by nonvector leafhoppers after they had fed on infected plants, but also multiplied in them to a limited extent. The virus was recovered from the intestine but not from the hemolymph or the salivary glands. The virus also increased in concentration in the intestine. It is possible that the intestine is the only site of virus multiplication in the nonvector insects.

There is enough evidence to suggest that "propagative" viruses, after being acquired by their insect vectors, infect the intestine, are then released from the intestine into the hemolymph, are carried to salivary glands and infect them, and then are injected back into the healthy plants while the insect feeds. "Circulative" viruses follow the same sequence but they may not multiply in the vector.

## BIBLIOGRAPHY

Bawden, F. C. 1964. Plant viruses and virus diseases. 4th ed. Ronald Press, New York. 123 p.

Bennett, C. W., and Wallace, H. E. 1938. Relation of the curly-top virus to the vector, *Eutettix tenellus*. J. Agr. Res. 56:31–50.

Black, L. M. 1941. Further evidence for multiplication of aster-yellows virus in the aster leafhopper. Phytopathology 31:120–135.

——— 1953. Transmission of plant viruses by Cicadellids. Advances Virus Res. 1:69–89.

——— 1959. Biological cycles of plant viruses in insect vectors. p. 157–185. *In* F. M. Burnet and W. M. Stanley [ed.] The viruses II. Academic Press, New York.

Black, L. M., and Brakke, M. K. 1952. Multiplication of wound-tumor virus in an insect vector. Phytopathology 42:269–273.

——— 1954. Serological reactions of a plant virus transmitted by leafhoppers. Phytopathology 44:482. (Abst)

Freitag, J. H. 1936. Negative evidence on multiplication of curly-top virus in the beet leafhopper, *Eutettix tenullus*. Hilgardia 10:305–342.

Hirumi, H., and Maramorosch, K. 1963. Recovery of aster-yellows virus from various organs of the insect vector, *Macrosteles fascifrons*. Contrib. Boyce Thompson Inst. 22:141–152.

Hutchinson, P. B., and Matthews, R. E. F. 1963. The fate of turnip yellow mosaic virus in the nonvector *Hyadaphis brassicae* (L.). Virology 20:169–175.

Kikumoto, T., and Matsui, C. 1962. Electron microscopy of plant viruses in aphid midguts. Virology 16:509–510.

Kimura, I. 1962. Further studies on the rice dwarf virus—I, II. Ann. Japan. Phytopathol. Soc. 27:197–213.

Kunkel, L. O. 1937. Effect of heat on ability of *Cicadula sexnotata* (Fall.) to transmit aster yellows. Amer. J. Bot. 24:316–327.

Ling, K. C. 1966. Nonpersistence of the tungro virus of the rice in its leafhopper vector, *Nephotettix impicticeps*. Phytopathology 56:1252–1256.

Maramorosch, K. 1952a. Direct evidence for the multiplication of aster-yellows virus in its insect vector. Phytopathology 42:59–64.

——— 1952b. Studies on the nature of the specific transmission of aster-yellows and corn-stunt viruses. Phytopathology 42:663–668.

——— 1955. Multiplication of plant viruses in insect vectors. Advances Virus Res. 3:221–249.

Nagaraj, A. N., Sinha, R. C., and Black, L. M. 1961. A smear technique for detecting virus antigen in individual vectors by the use of fluorescent antibodies. Virology 15:205–208.

Nasu, S. 1965. Electron microscopic studies on transovarial passage of rice dwarf virus. J. Japan. Appl. Entomol. Zool. 9:225–237.

Orlob, G. B. 1963. Reappraisal of transmission of tobacco mosaic virus by insects. Phytopathology 53:822–830.

——— 1966. Feeding and transmission characteristics of *Aceria tulipae* Keifer as vector of wheat streak mosaic virus. Phytopathol. Z. 55:218–238.

Ossiannilsson, F. 1958. Is tobacco mosaic virus not imbibed by aphids and leafhoppers? Kgl. Lantbrukshogskolans Ann. 24:369–374.

Proeseler, G. 1966. Beziehungen Zwischen der Rübenblattwanze *Piesma quadratum* Fieb. und dem Rübenkräuse virus. Phytopathol. Z. 56:213–237.

Shikata, E., and Maramorosch, K. 1965. Electron microscopic evidence for the systemic invasion of an insect host by a plant pathogenic virus. Virology 27: 461–475.

Sinha, R. C. 1965a. Recovery of potato yellow dwarf virus from hemolymph and internal organs of an insect vector. Virology 27:118–119.

——— 1965b. Sequential infection and distribution of wound-tumor virus in the internal organs of a vector after ingestion of virus. Virology 26:673–686.

Sinha, R. C., and Black, L. M. 1962. Studies on the smear technique for detecting virus antigens in an insect vector by use of fluorescent antibodies. Virology 17:582–587.

——— 1963. Wound-tumor virus antigens in the internal organs of an insect vector. Virology 21:183–187.

Sinha, R. C., and Chiykowski, L. N. 1967a. Initial and subsequent sites of aster yellows virus infection in a leafhopper vector. Virology 33:702–708.

———. 1967b. Multiplication of aster yellows virus in a nonvector leafhopper. Virology 31:461–466.

Sinha, R. C., and Reddy, D. V. R. 1964. Improved fluorescent smear technique and its application in detecting virus antigens in an insect vector. Virology 24:626–634.

Storey, H. H. 1928. Transmission studies on maize streak disease. Ann. Appl. Biol. 15:1–25.

——— 1933. Investigation of the mechanism of the transmission of plant viruses by insect vectors. I. Proc. Roy. Soc. (London) B113:463–485.

Wolcyrz, S., and Black, L. M. 1956. Serology of potato yellow dwarf virus. Phytopathology 46:32. (abst.)

# Electron Microscopy of Insect-Borne Viruses *in Situ*

EISHIRO SHIKATA

*Department of Botany,*
*Faculty of Agriculture,*
*Hokkaido University,*
*Sapporo, Japan*

AND

KARL MARAMOROSCH

*Boyce Thompson Institute for Plant Research,*
*Yonkers, New York*

## I. INTRODUCTION

Electron microscopy of thin sections of plant and insect vector tissues provides a useful technique for the identification and localization of insect-borne plant-pathogenic viruses. The first electron micrographs of a circulative virus in plants and insects were by Fukushi *et al.* (1960) who demonstrated the presence of rice dwarf virus in the cytoplasm of virus-carrying *Nephotettix cincticeps* (Uhler) leafhoppers, as well as in diseased rice plants. The wound tumor virus was localized in plant tumors and in leafhopper vectors by Shikata *et al.* in 1964, and the invasion of vector tissues was studied in great detail (Shikata and Maramorosch, 1965a, 1965b, 1965c; 1967a, 1967b; Granados, Hirumi, and Maramorosch, 1967; Hirumi *et al.,* 1967). Several other viruses have been studied in thin sections of plant hosts. Corn mosaic virus was photographed in plant cells by Herold, Bergold, and Weibel (1960), Herold (1963); tomato spotted wilt virus by Martin (1964), Best and Palk (1964), Ie (1964), and Kitajima (1965b); wheat striate mosaic virus by Lee (1964, 1965); potato yellow-dwarf virus by MacLeod, Black, and Moyer (1966); maize rough dwarf virus by Bassi and Gerola (1965), Gerola *et al.* (1966), and Lovisolo and Conti (1966). Herold and Munz (1965) described corn mosaic virus in leafhopper cells, while maize rough dwarf virus was studied in vectors by Bassi and Gerola (1966), Vidano (1966), and Vidano and Bassi (1966). Two circulative aphid-borne viruses have also been localized in plants and in aphids. Pea enation mosaic virus has been photographed by Shikata and Maramorosch (1966b) and Shikata, Maramorosch, and Granados (1966), and lettuce necrotic yellows virus by Chambers, Crowley, and Francki (1965), and by O'Loughlin and Chambers (1967). Some of the above viruses are icosahedral and less than 30 mμ in diameter. Others, also icosahedral, are 60 or even 70 mμ in diameter and reminiscent of typical insect-pathogenic viruses and of viruses affecting higher animals. Potato yellow-dwarf, lettuce necrotic yellows, and wheat striate mosaic virus are bullet shaped and morphologically resemble the vesicular stomatitis group of animal viruses.

## II. COMPARATIVE STUDIES OF RICE DWARF VIRUS AND WOUND TUMOR VIRUS

This chapter will describe some of the salient findings in rice dwarf virus and wound tumor virus because both have been among the more thoroughly studied plant pathogenic viruses. The outstanding common features of the two viruses are: (*1*) transovarian passage to the progeny of insect vectors (Fukushi, 1933; Black, 1953); (*2*) multiplication in insect vectors (Fukushi, 1935, 1939, 1940; Black and Brakke, 1952); (*3*) morphologic similarity of virus particles isolated from plant and insect hosts (Brakke, Vatter, and Black, 1954; Fukushi *et al.*, 1960); (*4*) comparatively large size and icosahedron shape (Brakke *et al.*, 1954; Fukushi *et al.*, 1960); (*5*) electron microscopic localization in the cytoplasm of the plant and insect cells (Fukushi, Shikata, and Kimura, 1962; Shikata *et al.*, 1964); and (*6*) a double-stranded RNA core of the virions (Black and Markham, 1963; Gomatos and Tamm, 1963; Miura, Kimura, and Suzuki, 1966).

Wound tumor is a "laboratory disease," since it has never been reported to occur in nature, and it is therefore of no economic importance. In this respect rice dwarf and wound tumor differ strikingly, since rice dwarf virus causes serious damage to rice plants throughout Japan, with the exception of the northern island of Hokkaido. A summary of the main characteristics of the two viruses is given in Table 1.

The host range of rice dwarf virus is restricted to Gramineae, while wound tumor virus has a wide host range of at least 43 species in some 20 families of plants (Black, 1944). The extensive work carried out during the past years with wound tumor virus followed the discovery by Black that this virus has tumorigenic properties. It induces tumors on the stem and roots of sweet clover, *Melilotus alba* Desr. and *M. officinalis* L.; enlarged veins on leaflets of crimson clover, *Trifolium incarnatum* L.; wart-like enations on leaves and tumors on the roots of sorrel, *Rumex acetosa L.;* and systemic infection in these and many other plants. These signs of plant disease are different from those induced by rice dwarf virus which causes various degrees of chlorosis and chloroplast disintegration of infected leaves.

The morphology of wound tumor virus was first determined by Brakke *et al.* (1954) in purified virus preparations obtained from stem tumors of sweet clover plants as well as from extracts prepared from *Agallia constricta* Van Duzee, one of the insect vectors transmitting the virus. Negatively stained preparations revealed details of the fine structure of wound tumor virus and showed it to have a capsid 60 mμ in diameter with 92 capsomeres (Bils and Hall, 1962). The details of the fine structure of rice dwarf virus were recently studied. A purified preparation of the virus was negatively stained with phosphotungstic acid, revealing an icosahedral symmetry of the particles. In particles of 2– or 3–fold axis symmetry the diameter of the long axis was about 750 Å and of the short axis about 660 Å. The structural units composing the capsomeres consisted of hollow tubes, approximately 60 Å in diameter and 95 Å in length. According to Kimura and Shikata (1968) the virus particles have 32 capsomeres, each composed of 5 or 6 tubular structural units, that is, a total of 180 structural units on the surface of a particle.

There has been substantial evidence that rice dwarf virus, isolated by differential centrifugation, has an outer envelope (Fukushi *et al.*, 1962). Unshadowed particles, hexagonal in shape with a darker central area, are surrounded by relatively transparent regions and outer envelopes. Accumulations of isolated particles are accompanied by thin membranous structures which seem to be of cytoplasmic origin and derived from diseased cells. Particles that were only lightly shadowed, or were shadowed from two directions, clearly revealed these envelopes. A series of electron micrographs of particles was examined after shadowing with varying amounts of platinum. Envelopes were found to be composed of thin membranous structures, and the width of the shadow from particles within envelopes always appeared wider than the actual diameter of the particles themselves (Shikata) 1966). It has been found that the envelopes are composed of lipoprotein; the removal of the envelopes by the use of snake venom enhanced the purification of the virus (Toyoda, Kimura, and Suzuki, 1965).

Chlorotic portions of diseased rice leaves contain more virus than green portions (Kimura, 1962). Correspondingly, wound tumor virus

TABLE 1

Summary of Main Characteristics of Wound Tumor Virus and Rice Dwarf Virus

| | Wound tumor virus | Rice dwarf virus |
|---|---|---|
| Scientific name | *Reovirus neoformans* (Anonymous, 1965) | *Fractilinea oryzae* (Holmes, 1948) |
| Host plants | 43 species in 20 families. | 15 species in Gramineae. |
| Distribution | Vectors found near Washington, D.C. | In the rice fields of Japan and Philippines. |
| Symptoms on plant hosts | Systemic, enlarged vein, stem, and root tumors, enations. | Systemic, chlorotic specks along the vein, yellowish-white streaks, stunt. |
| Histological symptoms | The tumor on sweet clover results from periclinal division on the pericycle, hyperplastic, and hypertrophic cells. | Chlorotic modifications in the mesophyll cells adjacent to some of the vascular bundle. Chloroplasts in these cells are disintegrated. |
| Cytological symptoms | Spherules are found in the cytoplasm of the tumor cells, 1–4 $\mu$ diam, necrotic cells in hyperplastic cells. | Inclusion body in the cytoplasm of the infected plant cells, 3–10 × 2.5–8.5 $\mu$, with vacuoles. |
| Localization of virus in plants | Cytoplasm of tumor cells; comparatively small accumulation of virus particles. | Cytoplasm of the cells in chlorotic portions. |
| Insect vectors | *Agalliopsis novella* (Say)<br>*Agallia constricta* Van Duzee<br>*Agallia quadripunctata* (Provancher) | *Nephotettix cincticeps* (Uhler)<br>*Nephotettix apicalis* (Motsch.)<br>*Inazuma dorsalis* (Motsch.) |
| Incubation period, | 14–28 days (25–32°C) | 8–10 days |
| in plants | 13–15 days (minimum) | |
| in insects | 25 days (25–32°C) | 12–25 days |

| | | |
|---|---|---|
| Transovarial passage | 1.8% of the progeny.<br>70% of the progeny from selected insects. | 85% of the progeny of infective male and female.<br>60% of the progeny of healthy male and infective female. |
| Transmissibility by feeding | 4% by unselected insects.<br>84% by selected insects. | 69% by *N. cincticeps*.<br>90% by selected *N. cincticeps*.<br>43% by *I. dorsalis*. |
| Localization of virus in vectors | Fat body, gut, malpighian tubules, trachea, muscle, epidermal cells, mycetome, salivary gland, blood cells, central nervous system, ovaries. | Fat body, gut, malpighian tubules, trachea, muscle, epidermal cells, mycetome, salivary gland, blood cells, ovarian tubules. |
| Histological symptoms | Changes in the fine structure of nervous system, fat body, and gut cells. | Striated nuclei of fat body, vacuolated fat body, abnormal mycetome. |
| Properties of virus in crude sap | $10^4$ infective (Crimson clover).<br>$10^5$ infective (*Agallia constricta*).<br>100 times higher than nontumor segments.<br>12 months infective at −25°C.<br>50°C for 10 min infective. | $10^3$ infective (eggs).<br>$10^4$ infective (*N. cincticeps*).<br>$10^3$ infective (rice leaves).<br>$10^4$ infective (chlorotic portion of the leaves)<br>8 months infective at −30 to −35°C.<br>48 hr infective at 0 to 4°C.<br>40°C for 10 min infective. |
| Purification | Obtained from plant tumors and insect vectors | Obtained from rice plants, *Echinocloacrus-galli* Beauv. and insect vectors. |
| Morphology of virus | 60 mμ in diam, polyhedron with 92 capsomeres | 75 mμ in diam, icosahedron, 32 capsomeres, 180 structural units. |
| Nucleic acid | Double-stranded RNA | Double-stranded RNA. |

concentrations in stem tumors were found to be 100 times higher than those in nontumor segments of the same plants (Brakke *et al.*, 1954). Apparently rice dwarf and wound tumor virus particles become concentrated in specific areas. This has been confirmed by direct particle counting methods applied to wound tumor virus (Gámez and Black, 1967; Streissle *et al.*, 1968).

Accumulations of rice dwarf virus particles were found in chlorotic cells near vascular bundles, and sometimes also in the vicinity of nuclei, but never inside of nuclei, chloroplasts, or mitochondria. During early stages of infection, rice dwarf virus was found loosely arranged throughout the cell cytoplasm but, later, aggregates formed and the particles became regularly packed in a crystalline form.

Electron micrographs of stem and root tumors revealed the presence of large masses of wound tumor virus particles in necrotic and non-necrotic cells. No virus particles were found in epidermal cells, and they were always absent from nuclei, chloroplasts, and mitochondria. In some of the non-necrotic tumor cells particles of wound tumor virus were within membranes located near the cell wall. Scattered particles were found also in tracheid cells of root tumors. These findings confirmed those obtained by fluorescent antibody techniques (Nagaraj and Black, 1961), which established a concentration of virus antigen in the pseudophloem of root tumors and thick-walled cells in the xylem region, but the lack of virus antigen in the epidermis, cortex, most xylem cells, ray cells, and pith.

In crimson clover infected with wound tumor, electron micrographs revealed scattered virions in non-necrotic cells of vein enations as well as in necrotic and tracheid cells. Sections through the earliest detectable vein swellings, 14–17 days after virus inoculation, showed only a few particles, scattered or aggregated in the cytoplasm. After a few days, large virus clusters were found in sections of enations.

In order to determine the location of rice dwarf and wound tumor viruses in their respective insect vectors, ultrathin sections from infected leafhoppers were examined. Rice dwarf virus was found in the cells of the fat body, malpighian tubules, epithelial cells of the intestine, salivary gland, and ovarian tubules. Recently it was also detected in trachea, muscle, and epidermal cells (Shikata, unpublished). No virions were found in nuclei or mitochondria. Nasu (1965) also reported the presence of virions in the mycetome.

Wound tumor virus clusters were found in cells of the fat body, malpighian tubules, salivary gland, the nervous system, hemocytes, epidermis, trachea, muscle, and mycetome (Shikata and Maramorosch, 1965c, 1967a, 1967b; Hirumi, Granados, and Maramorosch, 1967;

Granados *et al.*, 1967; Granados, Ward, and Maramorosch, 1968). The above findings provided direct evidence for the systemic infection of the respective vectors by plant-pathogenic viruses. The detailed study of wound tumor virus revealed the presence of submicroscopic, cytopathic lesions in cells of the fat body and the nervous system. Finally, the role of the hemolymph in virus multiplication and movement to different sites in the body of the vector was clarified.

In the salivary gland of *A. constricta* wound tumor virions were found either scattered or, less frequently, in accumulations in defined structures (Fig. 1). Only one or two of the lobes contained virus particles. In this connection it is of interest to recall that Whitcomb and Black (1961), Sinha and Reddy (1964), and Sinha, Reddy, and Black (1964), reported that some insects with a positive wound tumor-soluble antigen reaction failed to transmit the virus to plants. Most likely the virus does not multiply sufficiently in the salivary gland. Otherwise, one would expect to find it in concentrations comparable to those in other organs. In a less efficient vector, *A. novella,* Granados *et al.* (1967) found a much lower concentration of virus and a lesser degree of systemic invasion, which further supports the contention that relative virus concentration is correlated with vector efficiency. In contrast, large masses of rice dwarf virions were found in the salivary gland of its vector (Fukushi and Shikata, 1963b; Shikata, 1962, 1966). This difference in salivary gland invasion is paralleled by differences in transmitting efficiency, since the rice leafhopper is an extremely efficient transmitter of the rice dwarf virus.

Electron microscopy of thin sections and serology has led to similar findings and conclusions. Whitcomb and Black (1961), Sinha and Reddy (1964), and Sinha *et al.* (1964) have reported that many *A. constricta* leafhoppers that reacted positively to wound tumor-soluble antigen failed to transmit the virus. When Sinha and Black (1963) first detected wound tumor virus antigen in various organs of vectors, none was found in the malpighian tubules. This could have been caused by a low concentration, which prevented serologic reaction.

Later Sinha (1965) found the virus soluble antigen also in malpighian tubules. We were able to detect virions quite frequently in malpighian tubules by electron microscopy (Shikata and Maramorosch, 1965c), a method that usually is considered less sensitive than serology.

Since a small percentage of viruliferous female leafhoppers transmit the virus transovarially to their offspring, it would seem possible to detect the virions in thin sections of ovaries, and perhaps also in the mycetome, as Nasu (1965) did in the case of rice dwarf virus.

Rice dwarf virions, engulfed in defined structures surrounded by

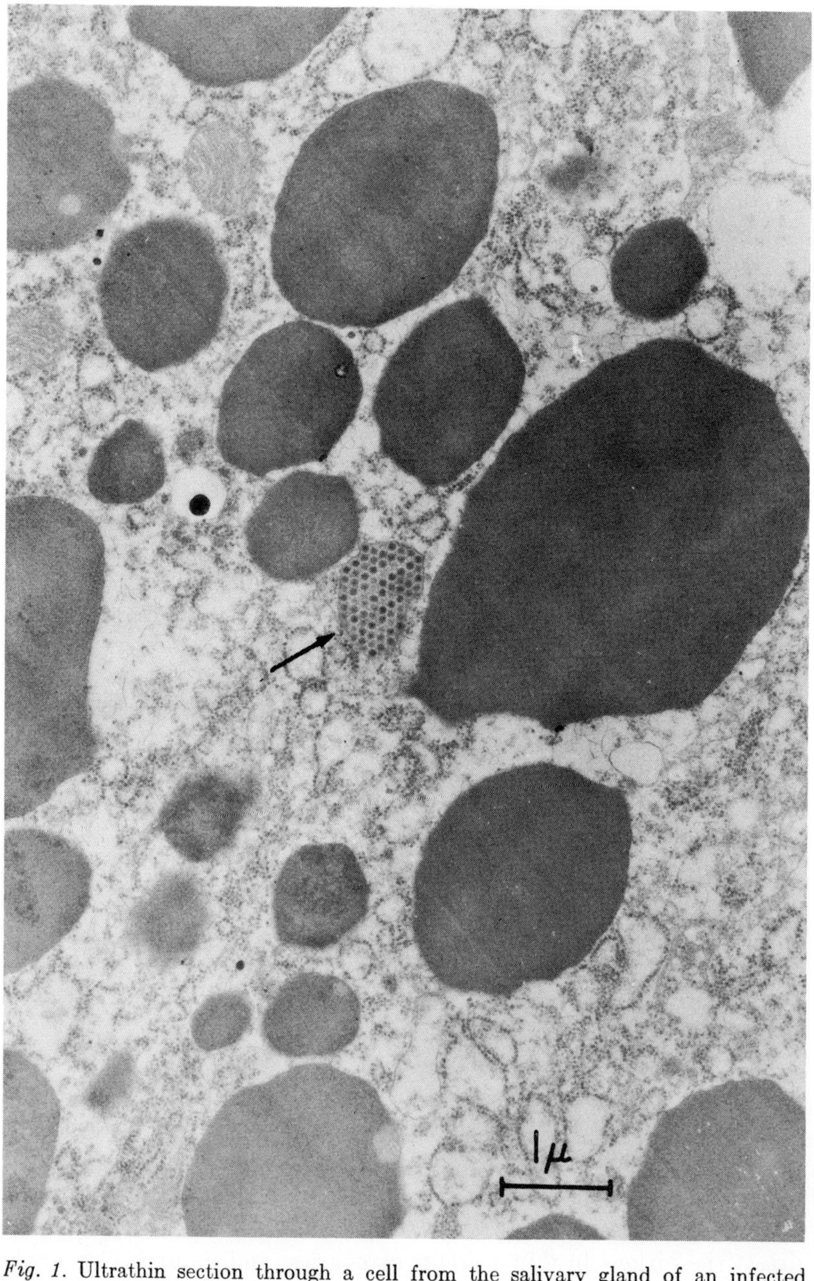


*Fig. 1.* Ultrathin section through a cell from the salivary gland of an infected leafhopper *Agallia constricta,* vector of wound tumor virus. Arrow points to a crystalline accumulation of wound tumor virions in a defined structure, enclosed by a membrane. Magnification, × 15,000.

several layers of membranes, were commonly found in thin sections of rice leafhopper vectors. Such myelin-like structures enclosing wound tumor virions were also common in the cell cytoplasm of infected organs of *A. constricta*.

Crystalline accumulations of virus particles were detected in sections of vectors of both viruses. A high magnification of the crystalline arrangement revealed the hexagonal structure of individual particles. The size of rice dwarf virions was measured as 40–60 m$\mu$ in diameter by center-to-center measurement and single-particle measurement of material embedded in methacrylate, which corresponded to measurements of purified virus pellets embedded in methacrylate. Accumulations of crystalline inclusions of wound tumor virus particles, as shown by Shikata and Maramorosch (1956a), were arranged in a three-dimensional fashion, and the diameter of virions, embedded in Epon, was 58–60 m$\mu$, which corresponds to the actual size determined from negatively stained preparations (Bils and Hall, 1962). Recently rice dwarf virions were also measured in Epon embedding and found to have a diameter of approximately 70 m$\mu$ (Shikata, unpublished).

Among the intriguing structures found in both rice dwarf and wound tumor virus infected cells were cisternae containing long strings of virus particles. These cisternae or tubules had both ends open. They were found in infected plants as well as in insect vectors of wound tumor and rice dwarf viruses. Their nature has not yet been determined but it is possible that the virions line along, or inside, some membranous structures occurring in the cytoplasm, such as the endoplasmic reticulum. Although microtubules or filamentous structures have been observed during the development of certain animal viruses (Fawcett, 1957; Chitwood and Bracken, 1964; Dales, 1963; Dales, Gomatos, and Hsu, 1965), the animal viruses were not arranged inside of tubular structures but rather on the outside. Maize rough-dwarf virus has been described in similar cisternae (Vidano, 1966) and the first report of a small RNA virus of animals in such structures appeared recently (Zee, Talens, and Hackett, 1967). Further investigations will be required to elucidate the function of the tubules.

As indicated in Table 1, one of the characteristic features of rice dwarf virus is the formation of inclusion bodies in infected rice leaves (Fukushi, 1931). Hirai *et al.* (1964) observed a specific reaction against fluorescent antibodies of rice dwarf virus antigen within the chlorotic portion of diseased leaves. These authors also found, by means of various staining reactions, that inclusion bodies were distributed not only in the chlorotic portion of diseased leaves, but also the phloem, vascular bundles, and cells adjacent to the vascular bundles. According to Suzuki, Kimura, and Toyoda (1964), inclusion bodies, initially scattered within infected cells, join together to form spherical or ellipsoidal structures. Such observations confirm that the inclusion bodies reported in rice dwarf-infected plants by Fukushi consist of virus particles (Shikata, 1962; Shikata and Kimura, 1965). After careful examination of hundreds of thin sections by electron microscopy, Shikata (1966) found only one type of virus accumulation that seems to correspond with the inclusion bodies described by Fukushi (1931). Perhaps with other embedding and fixing methods another type of inclusion body could be detected.

Littau and Black (1952) reported the occurrence of spherules in the cytoplasm of wound tumor-infected cells. These spherules were hyaline, homogenous, usually solid in appearance, and occasionally vacuolate. So far as our own observations are concerned, there exist in tumor cells of clover plants, apart from nuclei, two kinds of masses with electron dense structures. One of these consists of virus clusters enclosed within defined membranes near the cell wall. The other is an electron dense homogenous structure (Fig. 2) in the cytoplasm of tumor cells, containing virus particles around the periphery. We have identified the second type as viroplasms, the sites of virus assembly, and demonstrated their presence in plant cells and cells of infective leafhoppers (Fig. 3) (Shikata and Maramorosch, 1967a, 1967b). There is no information available at present as to whether the spherules are identical with the viroplasms. Nagaraj and Black (1961) detected an intense reaction of wound tumor virus antigen in the spherules, which would support the view that they are either the virus assembly sites, that is, the viroplasms, or that they represent virus storage locations, perhaps of the type described as microcrystals in electron micrographs.

## III. PEA ENATION MOSAIC VIRUS IN PLANTS AND APHIDS

The interest in pea enation mosaic virus has steadily increased in recent years, not only because of its economic importance as a plant disease agent, but also because of the fundamental implications in the study of another virus-induced neoplasia. Pea enation mosaic virus belongs to the comparatively small group of circulative aphid-borne viruses (Nault, Gyrisco, and Rochow, 1964). Although no direct evidence has been presented until now for the multiplication of this virus in its aphid vectors, there had been considerable speculation concerning the possible infection of the insects that carry the virus from plant to plant. This speculation began with the findings by

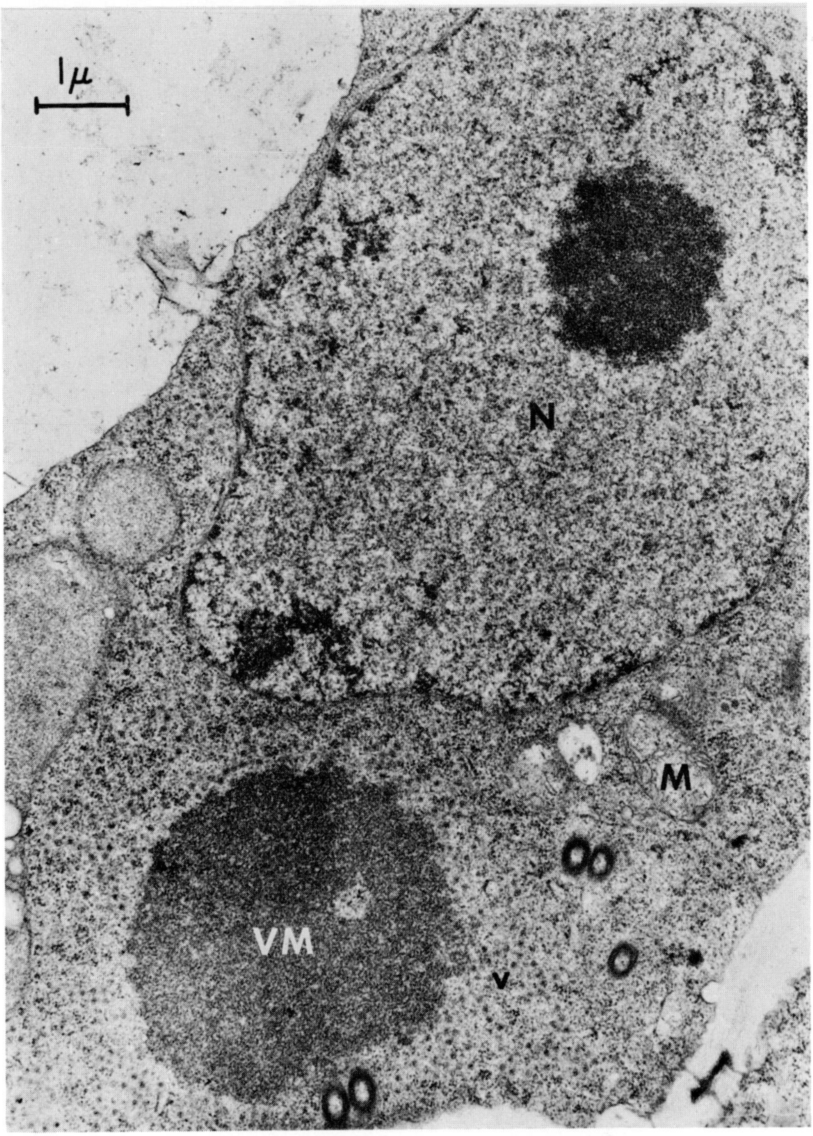


*Fig. 2.* An ultrathin section of a leaf enation from a crimson clover plant with wound tumor. Note viroplasm (VM) in the cytoplasm. Numerous virus particles are seen in the area (v) around the viroplasm. N indicates the nucleus and M mitochondria. Magnification, × 13,000.

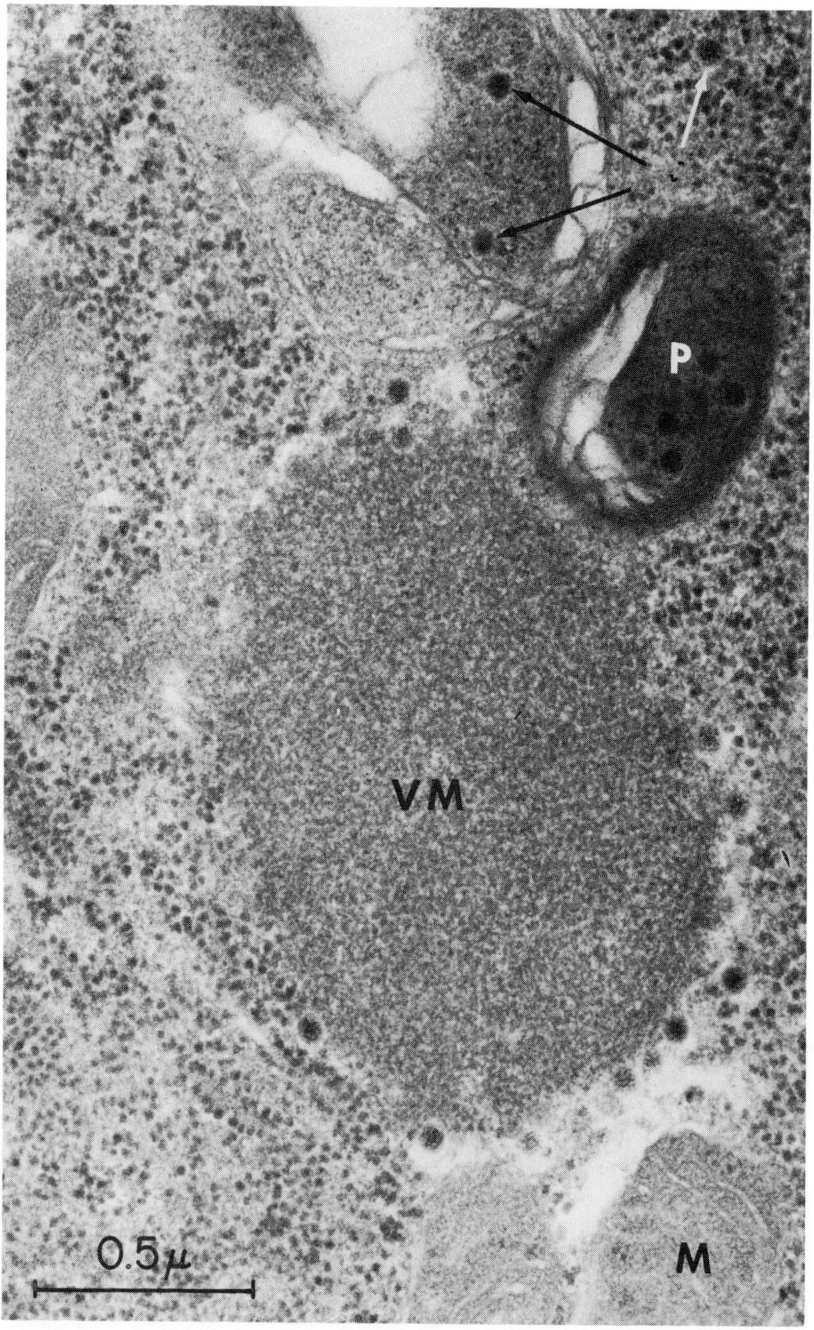


*Fig. 3.* An ultrathin section of a fat body cell from *A. constricta* carrying wound tumor virus. P represents a phagocytic structure enclosing several virus particles surrounded by membranes. VM represents a viroplasmic matrix with numerous wound tumor virions at the periphery. Arrows point to virus particles at some distance from the viroplasm. M indicates mitochondria. Magnification, × 60,000.

Osborn (1935) that the virus requires an incubation period of 9–48 hr in aphids. The first purification, electron microscopy, and serology of partially purified pea enation mosaic virus were done by Bustrillos (1964). Further purification of pea enation mosaic was carried out by Bozarth, Chow, and Gross (1965), Izadpanah and Shepherd (1966), Gibbs, Harrison, and Woods (1966), and by Bozarth and Chow (1966).

In ultrathin sections from chlorotic lesions of diseased pea leaves and pea pods with pea enation mosaic, large accumulations of virions were found in nuclei (Fig. 4) (Shikata and Maramorosch, 1966b). These accumulations were greater than those in the cytoplasm and were observed in cells obtained during earliest stages of disease. From these observations it was concluded that pea enation mosaic virus invades, and then multiplies in, cell nuclei, from which it penetrates into the cytoplasm of the infected cell. Pea enation virus was the first plant-pathogenic virus of cubical symmetry observed by electron microscopy in nuclei of sectioned plant cells.

Shikata (personal communication) calculated the particle size from several different types of preparations made by him. In ultrathin sections of leaves, the diameter of particles in the cytoplasm was 25–28m$\mu$; in the nucleus, 25–26 m$\mu$; in microcrystals of particles measured from center to center, as well as in individual particles, 24–27 m$\mu$; and in dip preparations, 28–30 m$\mu$.

Although the infection of aphids by circulative plant-pathogenic viruses has long been suspected, until recently no direct evidence supported this assumption. Ultrathin sections of pea aphids, *Acyrthosiphon pisum* Harris, were prepared from individuals that had been tested individually for their ability to transmit the virus. Large masses of virus particles were found in the lumen of the gut, and virus accumulations were also detected in the cytoplasm of fat body cells (Shikata *et al.*, 1966). Thus, pea enation mosaic virus became the first aphid-borne virus found by electron microscopy in plants as well as in aphid vectors *in situ.* It is possible and even likely that the virus will be detected in other organs, as has been the case with several leafhopper-borne viruses. The results presented the first direct evidence of the infection of an aphid vector by a plant pathogenic virus.

The second case of aphid infection by a plant-pathogenic virus has been reported recently by O'Loughlin and Chambers (1967). These workers found the large bacilliform virions of lettuce necrotic yellows virus in *Hyperomyzus lactucae* that fed on diseased plants and transmitted the virus. No particles were seen in aphids reared on virus-free plants or in embryos developing within infective vectors. Virions

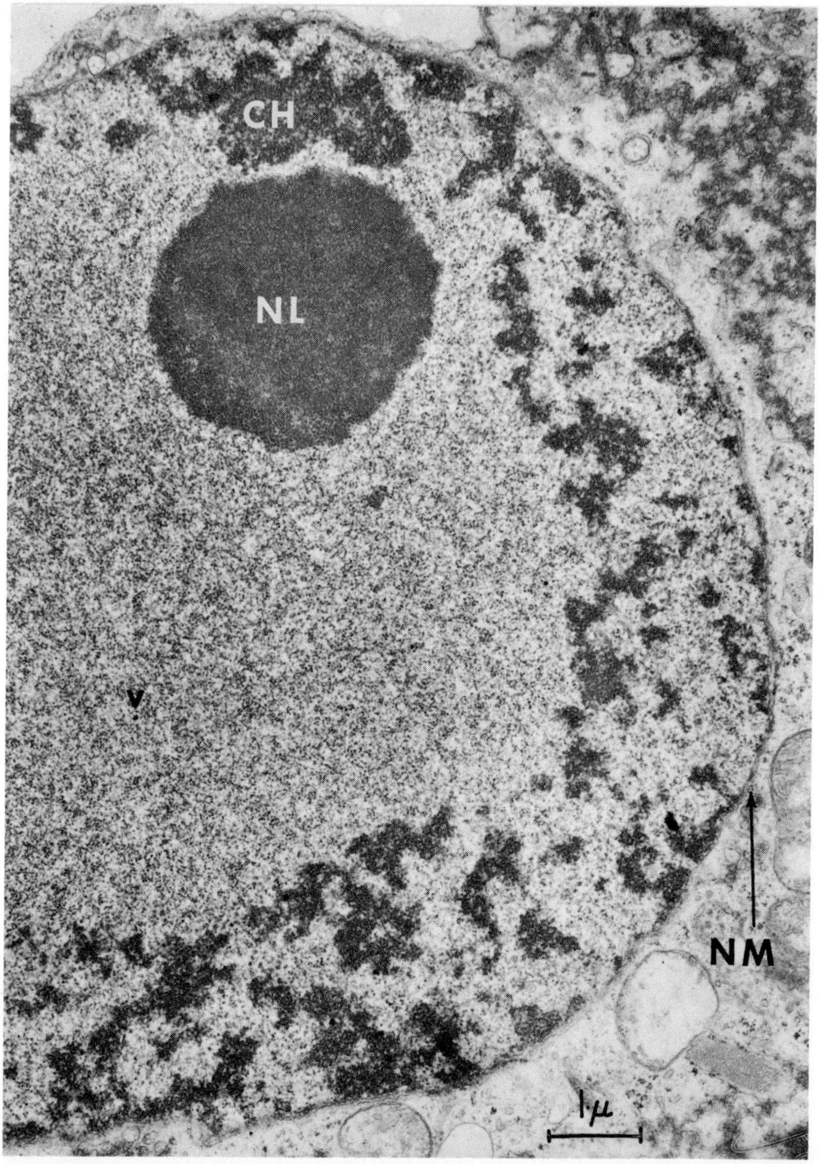


*Fig. 4.* An ultrathin section of a leaf enation from a chlorotic spot of a pea plant infected with pea enation mosaic virus. Note pea enation mosaic virus particles (v) in the central part of the nucleus. Chromatin substances (CH) are arranged at the periphery of the nucleus. NL indicates the nucleolus. NM the nuclear membrane. Magnification, × 13,000.

were detected in the muscles, fat body, brain, mycetome, tracheae, epidermis, salivary glands, and the cells of the alimentary canal. The authors concluded from their findings that the virus multiplied within the aphid.

## IV. CORN MOSAIC VIRUS AND OTHER BACILLIFORM OR BULLET-SHAPED VIRUSES

Among the important achievements obtained through electron microscopy has been the morphological characterization of the large, bacilliform, bullet-shaped plant-pathogenic viruses of corn mosaic, wheat striate mosaic, lettuce necrotic yellows, and potato yellow dwarf. These forms have now been detected in thin sections of plants as well as in those of their insect vectors.

Herold *et al.* (1960) and Herold (1963) found bacilliform particles of $242 \times 48$ m$\mu$ in corn plants infected with maize mosaic virus. Later Herold and Munz (1965) found similar particles within the cells of the salivary glands and intestines of infected insect vectors. Wheat striate mosaic virus was observed in wheat plant cells by Lee (1964). Harrison and Crowley (1965) and Crowley, Harrison, and Francki (1965) isolated and photographed the lettuce necrotic yellows virus. Observations of this virus in thin sections of host plants were carried out by Chambers *et al.* (1965) and in aphid vectors by O'Loughlin and Chambers (1967). The bacilliform particles were $277 \times 46$ m$\mu$. Earlier studies on potato yellow dwarf (Black, Moseley, and Wyckoff, 1948; Brakke, Black, and Wyckoff, 1951) revealed considerable variation in the shape of virus particles, including a spherical shape 110 m$\mu$ in diameter. Recently, MacLeod *et al.* (1966) clearly demonstrated that potato yellow-dwarf virus particles have a characteristic bullet-shaped appearance. Their dimensions were reported as $380 \times 75$ m$\mu$ in cells of diseased plants, quite different from that of the isolated virus. However, Black *et al.* (1948) had already suggested that the virus might possibly be rod shaped.

Purification of intact particles in this virus group involved considerable difficulties, particularly with lettuce necrotic yellows and potato yellow-dwarf viruses. The purified preparations were not homogenous, and small, as well as large, particles were found in final preparations. From recent studies of wheat striate mosaic virus by Lee (personal communication), it appears that the purified particles are not intact, and that this might explain their low infectivity. The complete virions seem to be composed of a bullet-shaped longer part and a much shorter cap, the latter having the same diameter as the bullet-shaped part. The cap easily detaches in purified preparations, exposing the nucleic acid to inactivation.

It is interesting to note that the vectors of bullet-shaped viruses belong to a wide range of insect species which include leafhoppers, planthoppers, and aphids. In ultrathin sections, the viruses usually appeared near the perinuclear region of infected plant cells, as was seen in corn mosaic, potato yellow dwarf, and in a *Gomphrena* virus studied by Kitajima and Costa (1966).

It should be pointed out that bullet-shaped plant-pathogenic viruses bear a striking similarity to the sigma virus that causes carbon dioxide sensitivity of fruitflies (Seecof, 1968) and, even more important, to the vesicular stomatitis group of animal-pathogenic viruses. Whether the morphological similarities are coincidental, or whether all these viruses belong to a closely related group, remains to be established by careful serological and chemical investigations.

## V. PRESENT AND FUTURE APPLICATIONS OF ELECTRON MICROSCOPY TO THE STUDY OF INSECT-BORNE PLANT-PATHOGENIC VIRUSES

Electron microscopy of insect-borne plant-pathogenic viruses has been used to a limited extent for the determination of virus morphology, for the observation of virus localization in host tissues and organs, and for the investigation of sites of virus assembly within infected cells of plants and insect vectors. Such investigations will no doubt be extended to many other arthropod-borne viruses affecting plants.

There are several techniques available at present for the determination of virus morphology by electron microscopy: shadow casting of isolated virus particles, replica techniques, positive staining, and the negative staining of preparations of crude "dip" material (Horne, 1967). Direct negative staining (Doi *et al.*, 1965a, 1965b), and the examination of viruses in thin sections of infected hosts (Morgan and Rose, 1967) are also quite useful for virus characterization.

Electron micrographs of shadowed, isolated virus particles, dried on grids, have been commonly used in many instances to determine virus morphology. In 1954, wound tumor virus was shown to have the same morphology, whether isolated from plants or insect vectors (Brakke *et al.*, 1954). Shadowed particles were, however, not very accurately measured and almost always the calculation of size resulted in an overestimation when compared with negatively stained virions. The discrepancies were due to several factors, mainly to the flattening of virions and their drying directly on grids. Certain embedding materials, such as methacrylate, shrink and cause distortion, while others, especially the harder epoxy resins, preserve the actual size and shape of particles.

A rapid and simple method of negative staining has been described by Doi *et al.* (1965b) and by Kitajima (1965a). Hitchborn and Hills (1965) used the dip method, which had earlier given good results with large, rod-shaped viruses, to measure small, icosahedral-shaped viruses.

The discovery of viroplasms, that is, sites of virus assembly, in cells of animals grown *in vitro* has been reported by Morgan *et al.* (1956), Morgan, Rifkind, and Rose (1962), Higashi *et al.* (1959) Higashi, Ozaki, and Fukada (1960), Dales (1963), Dales *et al.* (1965), Mattern and Daniel (1965), and others. The first descriptions of viroplasms of plant-pathogenic viruses were made by Shikata and Maramorosch (1967a, 1967b) for wound tumor virus and by Shikata (1968) for rice dwarf virus. In the course of these investigations it was also found that cytopathic changes in the fine structure, detectable by electron microscopy but not by light microscopy, are caused in cells of the fat body of insect vectors carrying wound tumor virus. Hirumi *et al.* (1967) discovered cytopathic lesions in the nervous system of leafhoppers carrying wound tumor virus. Lesions were also found in cells of the gut (Maramorosch *et al.*, 1968c). The finding of a cytopathic effect of a plant tumor virus on invertebrate animals has been made possible only in recent years with the advent of refined electron microscopy techniques.

At the moment one of the most important uses of electron microscopy in relation to plant virus studies is in the definitive establishment of the viral nature of etiologic agents. This has been demonstrated recently in yellows-type diseases. The etiologic agents of these diseases are transmitted by leafhoppers. Yellows diseases are characterized by chlorosis, stunting, proliferation of adventitious shoots, phyllody of flowers, and sterility of seed. Aster yellows, corn stunt, potato stolbur, mulberry dwarf, and many other diseases belong to this group. It had been generally assumed that yellows diseases were caused by viruses, since no bacteria or fungi have been detected in diseased plants and the transmission was effected only by specific leafhoppers or by grafting. However, all attempts to find virus-like particles in diseased plants and in leafhopper vectors have failed. Thin-section electron microscopy provided a clue to the possible nature of the etiologic agents of yellows diseases. Work by Doi *et al.* (1967), Shikata *et al.* (1968), and Maramorosch *et al.* (1968a, 1968b) revealed the presence of bodies resembling mycoplasma (PPLO) in the phloem cells of yellows-diseased plants. The likelihood that yellows diseases are actually caused by mycoplasma-like agents was further strengthened by the partial cure or suppression of several yellows diseases, and by the inhibition of insect transmission through treatment with tetracyclines (Ishiie *et al.*, 1967; Davis, Whitcomb, and Steere, 1968). Mycoplasma-like bodies were also found in leafhopper vectors, but not in controls (Maramorosch *et al.*, 1968a, 1968b). As a result of these findings it now appears that several diseases, discussed in this volume might eventually have to be reclassified (Ou *et al.*, 1968) if and when their nonviral etiology becomes established.

## BIBLIOGRAPHY


Anonymous. 1965. Proposals and recommendations of the Provisional Committee for Nomenclature of Viruses (P.C.N.V.). Ann. Inst. Pasteur **109**:625–637.

Bassi, M., and Gerola, F. M. 1965. Intracellular localization and ways of aggregation of some plant viruses. Atti Congr. Ital. Microscop. Elettron., **50**, Bologna: 125–128.

Best, R. J., and Palk, B. A. 1964. Electron microscopy of strain E of tomato spotted wilt virus and comments on its probable biosynthesis. Virology **23**: 445–460.

Bils, R. F., and Hall, C. E. 1962. Electron microscopy of wound-tumor virus. Virology **17**:123–130.

Black, L. M. 1944. Some viruses transmitted by agallian leafhoppers. Proc. Amer. Phil. Soc. **88**:132–144.

——— 1945. A virus tumor disease of plants. Amer. J. Botany **32**:408–415.

——— 1953. Occasional transmission of some plant viruses through the eggs of their insect vectors. Phytopathology **43**:9–10.

——— 1965. Physiology of virus-induced tumors in plants. Handbuch Pflanzenphysiol. XV/2:236–266.

Black, L. M., and Brakke, M. K. 1952. Multiplication of wound-tumor virus in an insect vector. Phytopathology **42**:269–273.

Black, L. M., and Markham, R. 1963. Base-pairing in the ribonucleic acid of wound-tumor virus. Netherlands J. Plant Pathol. **69**:215.

Black, L. M., Moseley, V. M., and Wyckoff, R. W. G. 1948. Electron microscopy of potato yellow-dwarf virus. Biochim. Biophys. Acta **2**:121–123.

Bozarth, R. F., and Chow, C. C. 1966. Pea enation mosaic virus: purification and properties. Contrib. Boyce Thompson Inst. **23**:301–309.

Bozarth, R. F., Chow, C. C., and Gross, S. 1965. Purification of pea enation mosaic virus. Phytopathology **55**:127.

Brakke, M. K., Black, L. M., and Wyckoff, R. W. G. 1951. The sedimentation rate of potato yellow-dwarf virus. Amer. J. Bot. **38**:332–342.

Brakke, M. K., Vatter, A. E., and Black, L. M. 1954. Size and shape of wound-tumor virus. Brookhaven Symp. Biol. No. **6**:137–156.

Bustrillos, A. D. 1964. Purification, serology and electron microscopy of pea enation mosaic virus. Ph.D. thesis, Michigan State University.

Chambers, T. C., Crowley, N. C., and Francki, R. I. B. 1965. Localization of lettuce necrotic yellows virus in host leaf tissue. Virology **27**:320–328.

Chitwood, L. A., and Bracken, E. C. 1964. Replication of *Herpes simplex* virus in a metabolically imbalanced system. Virology **24**:116–120.

Crowley, N. C., Harrison, B. D., and Francki, R. I. B. 1965. Partial purification of lettuce necrotic yellows virus. Virology 26:290–296.

Dales, S. 1963. The uptake and development of vaccinia virus in strain L cells followed with labeled viral deoxyribonucleic acid. J. Cell Biol. 18:51–72.

Dales, S., Gomatos, P. J., and Hsu, K. C. 1965. The uptake and development of reovirus in strain L cells followed with labeled viral ribonucleic acid and ferritin–antibody conjugates. Virology 25:193–211.

Davis, R. E., Whitcomb, R. F., and Steere, R. L. 1968. Chemotherapy of aster yellows disease. Phytopathology 58:884.

Doi, Y., Teranaka, M., Yora, K., and Asuyama, H. 1967. Mycoplasma- or PLT group-like microorganisms found in the phloem elements of plants infected with mulberry dwarf, potato witches' broom, aster yellows, or paulownia witches' broom. [in Japanese, English summary]. Ann. Phytopathol. Soc. Japan 33:259–266.

Doi, Y., Toriyama, S., and Asuyama, H. 1965a. A simple method for observation of viruses by means of direct negative staining [in Japanese]. Ann. Phytopathol. Soc. Japan 30:265.

Doi, Y., Yora, K., and Asuyama, H. 1965b. Observation of viruses by means of direct negative staining. II [in Japanese]. Ann. Phytopathol. Soc. Japan 30: 295.

Fawcett, D. W. 1957. Electron microscope observations on intracellular virus-like particles associated with the cells of the Lucké renal adenocarcinoma. J. Biophys. Biochem. Cytol. 2:725–742.

Fukushi, T. 1931. On the intracellular bodies associated with the dwarf disease of rice plant. Trans. Sapporo Natural History Soc. 12:35–41.

——— 1933. Transmission of the virus through the eggs of an insect vector. Proc. Imp. Acad. (Japan) 9:457–460.

——— 1935. Multiplication of virus in its insect vector. Proc. Imp. Acad. (Japan) 11:301–303.

——— 1939. Retention of virus by its insect vectors through several generations. Proc. Imp. Acad. (Japan) 15:142–145.

——— 1940. Further studies on the dwarf disease of rice plant. J. Faculty Agr. Hokkaido Imp. Univ. 45:83–154.

Fukushi, T., and Shikata, E. 1963a. Fine structure of rice dwarf virus. Virology 21:500–503.

——— 1963b. Localization of rice dwarf virus in its insect vector. Virology 21: 503–505.

Fukushi, T., Shikata, E., and Kimura, I. 1962. Some morphological characters of rice dwarf virus. Virology 18:192–205.

Fukushi, T., Shikata, E., Kimura, I., and Nemoto, M. 1960. Electron microscopic studies on the rice dwarf virus. Proc. Japan Acad. 36:352–357.

Gámez, R., and Black, L. M. 1967. Application of particle-counting to a leafhopper-borne virus. Nature 215:173–174.

Gerola, F. M., Bassi, M., Lovisolo, O., and Vidano, C. 1966. Virus-like particles in both maize plants infected with maize rough dwarf virus and the vector *Laodelphax striatellus* Fallén. Phytopathol. Z. 56:97–99.

Gibbs, A. J., Harrison, B. D., and Woods, R. D. 1966. Purification of pea enation mosaic virus. Virology 29:248–351.

Gomatos, P. J., and Tamm, I. 1963. Animal and plant viruses with double-helical RNA. Proc. Nat. Acad. Sci. U.S. 50:878–885.

Granados, R. R., Hirumi, H., and Maramorosch, K. 1967. Electron microscopic evidence for wound-tumor virus accumulation in various organs of an inefficient leafhopper vector, *Agalliopsis novella*. J. Invertebrate Pathol. 9:147–159.

Granados, R. R., Ward, L. S., and Maramorosch, K. 1968. Insect viremia caused by a plant pathogenic virus: electron microscopy of vector hemocytes. Virology 34:790–796.

Harrison, B. D., and Crowley, N. C. 1965. Properties and structure of lettuce necrotic yellows virus. Virology 26:297–310.

Herold, F. 1963. Estudios sobre dos enfermedades virales del maíz en Venezuela. Acta Cient. Venezolana Suppl. 1:221–227.

Herold, F., Bergold, G. H., and Weibel, J. 1960. Isolation and electron microscopic demonstration of a virus infecting corn (*Zea mays* L.). Virology 12:335–347.

Herold, F., and Munz, K. 1965. Electron microscopic demonstration of viruslike particles in *Peregrinus maidis* following acquisition of maize mosaic virus. Virology 25:412–417.

Higashi, H., Ozaki, Y., and Fukada, T. 1960. Electron microscopic studies on the growth of pox virus in monolayer culture of strain L cells and Hela cells. Inter. Kongr. Elektronmikroskopie, 4, Berlin. 1958. Verhandl. 2:573–576.

Higashi, N., Ozaki, Y., Notake, K., Ichinomiya, M., and Fukada, T. 1959. Virological and electron microscopic studies on the multiplication of pox virus group in vitro. Virus (Osaka) 9:165–190.

Hirai, T., Suzuki, N., Kimura, I., Nakazawa, M., and Kashiwagi, Y. 1964. Large inclusion bodies associated with virus diseases of rice. Phytopathology 54:367–368.

Hirumi, H., Granados, R. R., and Maramorosch, K. 1967. Electron microscopy of a plant-pathogenic virus in the nervous system of its insect vector. J. Virol. 1:430–444.

Hitchborn, J. H., and Hills, G. J. 1965. The use of negative staining in the electron microscopic examination of plant viruses in crude extracts. Virology 27:528–540.

Holmes, F. O. 1948. Order Virales. The filterable viruses. Suppl. No. 2, p. 1127–1296. *In* Bergey's Manual of determinative bacteriology. 6th ed. Williams & Wilkins, Baltimore.

Horne, R. W. 1967. Electron microscopy of isolated virus particles and their components. *In* Maramorosch, K. and Koprowski, H. [ed.] Methods in virology, Vol. 3 (4 vol.) Academic Press, New York, p. 521–574.

Hosaka, Y., Kokozeki, H., and Shikata, E. 1963. Morphology of rice dwarf virus. Abstr. Meeting Soc. Japanese Virologists, 11th, p. 20.

Ie, T. S. 1964. An electron microscope study of tomato spotted wilt virus in the plant cell. Netherlands J. Plant Pathol. 70:114–115.

Ishiie, T., Doi, Y., Yora, K., and Asuyama, H. 1967. Suppressive effects of antibiotics of tetracycline group on symptom development of mulberry dwarf disease [in Japanese, English summary]. Ann. Phytopathol. Soc. Japan 33:267–275.

Izadpanah, K., and Shepherd, R. J. 1966. Purification and properties of the pea enation mosaic virus. Virology 28:463–476.

Kimura, I. 1962. Further studies on the rice dwarf virus I. II. Ann. Phytopathol. Soc. Japan 27:197–213.

Kimura, I., and Shikata, E. 1968. Structural model of rice dwarf virus. Proc. Japan Acad. 44:538–543.

Kitajima, E. W. 1965a. A rapid method to detect particles of some spherical plant viruses in fresh preparations. J. Electron Microscopy (Tokyo) 14:119–121.

——— 1965b. Electron microscopy of vira-cabeça virus (Brazilian tomato spotted wilt virus) within the host cell. Virology 26:89–99.

Kitajima, E. W., and Costa, A. C. 1966. Morphology and developmental stages of *Gomphrena* virus. Virology 29:523–539.

Lee, P. E. 1964. Electron microscopy of inclusions in plants infected by wheat striate mosaic virus. Virology 23:145–151.

——— 1965. Electron microscopy of inclusions associated with wheat streak mosaic virus. J. Ultrastructure Res. 13:359–366.

Littau, V. C., and Black, L. M. 1952. Spherical inclusions in plant tumors caused by a virus. Am. J. Bot. 39:87–95.

Lovisolo, O., and Conti, M. 1966. Individuazione al microscopio elettronico del virus del nanismo ruvido del mais (MRDV) in piante di *Zea mays* L. sperimentalmente infettate. Atti Accad. Sci. Torino Classe Sci. Fis. Mat. Nat. 100:63–72.

MacLeod, R., Black, L. M., and Moyer, F. H. 1966. The fine structure and intracellular localization of potato yellow dwarf virus. Virology 29:540–552.

Maramorosch, K., Shikata, E., Granados, R. R., and Hirumi, H. 1968a. Structures resembling mycoplasma in insects and diseased plants. Trans. N.Y. Acad. Sci. 30(6): (in press).

——— 1968b. Mycoplasma-like bodies in leafhoppers and diseased plants. Phytopathology 58:886.

Maramorosch, K., Shikata, E., Hirumi, H., and Granados, R. R. 1968c. Multiplication and cytopathology of a plant tumor virus in insects. Abstr. Symp. Neoplasia Invertebrate Primitive Vertebrate Animals, Smithsonian Institution, Washington, D.C.

Martin, M. M. 1964. Purification and electron microscope studies of tomato spotted wilt virus (TSWV) from tomato roots. Virology 22:645–649.

Mattern, C. F. T., and Daniel, W. A. 1965. Replication of poliovirus in Hela cells: electron microscopic observations. Virology 26:646–663.

Miura, K.-I., Kimura, I., and Suzuki, N. 1966. Double-stranded ribonucleic acid from rice dwarf virus. Virology 28:571–579.

Morgan, C., Howe, C., Rose, H. M., and Moore, D. H. 1956. Structure and development of viruses observed in the electron microscope. IV. Viruses of the RI–APC group. J. Biophys. Biochem. Cytol. 2:351–360.

Morgan, C., Rifkind, R. A., and Rose, H. M. 1962. The use of ferritin-conjugated antibodies in electron microscopic studies of influenza and vaccinia viruses. Cold Spring Harbor Symp. Quant. Biol. 27:57–65.

Morgan, C., and H. M. Rose. 1967. The application of thin sectioning. *In* K. Maramorosch and H. Koprowski [ed.] Methods in virology, Vol. 3 (4 vol.) Academic Press, New York, p. 575–616.

Nagaraj, A. N., and Black, L. M. 1961. Localization of wound-tumor virus antigen in plant tumors by the use of fluorescent antibodies. Virology 15:289–294.

Nasu, S. 1965. Electron microscopic studies on transovarial passage of rice dwarf virus. Japan. J. Appl. Entomol. Zool. 9:225–237.

Nault, L. R., Gyrisco, G. G., and Rochow, W. F. 1964. Biological relationship between pea enation mosaic virus and its vector, the pea aphid. Phytopathology 54:1269–1272.

O'Loughlin, G. T., and Chambers, T. C. 1967. The systemic infection of an aphid by a plant virus. Virology 33:262–271.

Osborn, H. T. 1935. Incubation period of pea mosaic in the aphid, *Macrosiphum pisi*. Phytopathology 25:160–177.

Ou, S. H., Iida, T. T., Maramorosch, K., Raychaudhuri, S. P., and Suzuki, N. 1968. Report of the committee on nomenclature of rice virus diseases. *In* The virus diseases of the rice plant. Johns Hopkins University Press (in press).

Seecof, R. 1968. The Sigma virus infection of *Drosophila melanogaster*. Current topics Microbiol. 42:59–91.

Shikata, E. 1962. Observation of rice dwarf virus by means of an electron microscope. Plant Protection 16:313–318.

——— 1966. Electron microscopic studies on plant viruses. J. Faculty Agr. Hokkaido Univ. 55:1–110.

——— 1968. Electron microscopic studies of rice viruses. *In* The virus diseases of the rice plant. Johns Hopkins University Press (in press).

Shikata, E., and Kimura, I. 1965. Rice dwarf virus [in Japanese]. Ann. Phytopathol. Soc. Japan 31:125–132.

Shikata, E., and Maramorosch, K. 1965a. Plant tumour virus in arthropod host: microcrystal formation. Nature 208:507–508.

——— 1965b. Wound tumor virus in insect hosts inoculated parenterally. J. Appl. Phys. 36:2620.

——— 1965c. Electron microscopic evidence for the systemic invasion of an insect host by a plant pathogenic virus. Virology 27:461–475.

——— 1966a. An electron microscope study of plant neoplasia induced by wound tumor virus. J. Nat. Cancer Inst. 36:97–116.

——— 1966b. Electron microscopy of pea enation mosaic virus in plant cell nuclei. Virology 30:439–454.

——— 1967a. Electron microscopy of wound tumor virus assembly sites in insect vectors and plants. Virology 32:363–377.

——— 1967b. Electron microscopy of the formation of wound tumor virus in abdominally inoculated insect vectors. J. Virol. 1:1052–1073.

Shikata, E., Maramorosch, K., and Granados, R. R. 1966. Electron microscopy of pea enation mosaic virus in plants and aphid vectors. Virology 29:426–436.

Shikata, E., Maramorosch, K., Ling, K. C., and Matsumoto, T. 1968. On the mycoplasma like structures encountered in the phloem cells of American aster yellows, corn stunt, Philippine rice yellow dwarf, and Taiwan sugar cane white leaf diseased plants [in Japanese]. Ann. Phytopathol. Soc. Japan 34:208–209.

Shikata, E., Orenski, S. W., Hirumi, H., Mitsuhashi, J., and Maramorosch, K. 1964. Electron micrographs of wound-tumor virus in an animal host and a plant tumor. Virology 23:441–444.

Sinha, R. C. 1965. Sequential infection and distribution of wound-tumor virus in the internal organs of a vector after ingestion of virus. Virology 26:673–686.

Sinha, R. C., and Black, L. M. 1963. Wound-tumor virus antigens in the internal organs of an insect vector. Virology 21:183–187.

Sinha, R. C., and Reddy, D. V. R. 1964. Improved fluorescent smear technique and its application in detecting virus antigens in an insect vector. Virology 24:626–634.

Sinha, R. C., Reddy, D. V. R., and Black, L. M. 1964. Survival of insect vectors after examination of hemolymph to detect virus antigens with fluorescent antibody. Virology 24:666–667.

Streissle, G., Bystricky, V., Granados, R. R., and Strohmaier, K. 1968. Number of virus particles in insects and plants infected with wound tumor virus. 1968. J. Virol. 2:214–217.

Suzuki, N., Kimura, I., and Toyoda, S. 1964. Staining of inclusion bodies in the rice dwarf virus infected rice plants. Ann. Phytopathol. Soc. Japan 29:72.

Toyoda, S., Kimura, I., and Suzuki, N. 1965. Purification of rice dwarf virus. Ann. Phytopathol. Soc. Japan 30:225–230.

Vidano, C. 1966. Il maize rough dwarf virus in ghiandole salivari e in micetoma di *Laodelphax striatellus* Fallén. Atti Accad. Sci. Torino Classe Sci. Fis. Mat. Nat. 100:731–748.

Vidano, C., and Bassi, M. 1966. Dimonstrazione al microscopio elettronico di particelle del virus del nanismo ruvido del mais (MRDV) nel vettore *Laodelphax striatellus* Fallén. Atti Accad. Sci. Torino Classe Sci. Fis. Mat. Nat. 100:73–78.

Whitcomb, R. F., and Black, L. M. 1961. Synthesis and assay of wound-tumor soluble antigen in an insect vector. Virology 15:136–145.

Zee, Y. C., Talens, L., and Hackett, A. J. 1967. Localization of a small ribonucleic acid virus within cytoplasmic cisternae. J. Virol. 1:1271–1273.

# The Fate of Plant-Pathogenic Viruses in Insect Vectors: Electron Microscopy Observations

KARL MARAMOROSCH, EISHIRO SHIKATA,* AND ROBERT R. GRANADOS

*Boyce Thompson Institute for Plant Research, Yonkers, New York*

Although many aspects of interactions between viruses and vectors have been studied in great detail during the past decades, there is comparatively little information available on the fate of viruses acquired by vectors. Two major types of virus–vector interactions, classified according to actual routes of virus transport in vectors, have been defined by Kennedy, Day, and Eastop (1962). The first, and by far the most common, is stylet borne and comprises the viruses that are carried externally, as defined by Watson (1960). Only limited attempts have been made to detect viruses on vector stylets by electron microscopy (van Hoof, 1958), and no actual localization has been reported. Stylets of certain aphids were found to differ in fine surface structure and the differences in protrusions and grooves may account for the specificity of vectors of stylet-borne viruses. Descriptions of stylets in other groups of vectors of "nonpersistent" (Watson, 1960) viruses have not been published.

Recently attempts have been made to locate nematode-borne viruses on or in nematode vectors. It has not been established whether nematodes carry viruses externally or internally. Electron microscopy of stylets of *Longidorus* sp. (see Taylor and Cadman's chapter in this book), *Xiphinema* sp. (Raski, personal communication) and *Trichodorus* sp. (Chen, Hirumi, and Maramorosch, 1968; Hirumi, Chen, and Maramorosch, 1968) revealed the highly complicated morphology of such stylets. Thus far, no localization of nematode-borne viruses has been reported. The same lack of information applies to viruses transmitted by mealybugs, whiteflies, thrips, fungi, and mites.

The second type of virus–vector interaction comprises "circulative"

* Present address: Department of Botany, Faculty of Agriculture, Hokkaido University, Sapporo, Japan.

(Black, 1959) viruses. Several representatives of this type have been localized in aphid and leafhopper vectors by electron microscopy (see the chapter by Shikata and Maramorosch). From such investigations, as well as from serological studies (Sinha, 1965; Sinha, 1968; see also the chapter by Sinha in this book) and from additional sequential sectioning of insects that had acquired viruses artifically by abdominal injection (Maramorosch and Shikata, 1965), information has been gained to permit speculation on the fate of at least one virus in a leafhopper vector. That virus causes the wound tumor disease of plants. Although the disease is of no economical importance whatsoever and, in fact, causes only an "experimental" laboratory disease of plants, the plant tumor-inducing properties of the virus and its morphological and chemical resemblance to the widespread human-pathogenic reoviruses (Vasquez and Tournier, 1962; Gomatos and Tamm, 1963; Black and Markham, 1963; Bils and Hall, 1962; Streissle *et al.*, 1968; Rosen, 1968) made a detailed study seem worthwhile. The interpretation of data concerning the fate of wound tumor virus in leafhopper vectors is in part speculative and the findings are limited to a single virus. The fate of other viruses transmitted in the circulative manner by leafhoppers, or by other vectors, may be different.

## I. MECHANICAL INOCULATION OF INSECTS

Virus-free *Agallia constricta* insects, confined to wound tumor-diseased plants, seldom acquire the virus within a short feeding period. Usually oral acquisition is slow, and several days, or even weeks, may elapse before the virus is taken up by feeding leafhoppers. Since no precise timing of virus acquisition by the oral route is possible, it was decided to study the sequential stages in virus development by using insects that had acquired the virus by needle inoculation. Mechanical inoculation of wound tumor virus (Maramorosch, Brakke, and Black, 1949) is based on the classical technique developed by Storey (1933). To facilitate the tedious operation, a specially designed stage was employed (Maramorosch and Jernberg, 1964), so that 20 insects were lined up in a row and held in the proper position for injection (Fig. 1), carried out under a dissecting binocular microscope. This permitted the rapid injection of several hundred insects in a short period of time. Following the abdominal inoculation of stock insects with virus-containing extracts from diseased plants or from virus-carrying vectors, the leafhoppers were confined to plants. The search for virus was made in the fat body tissues, because this type of tissue in the past provided

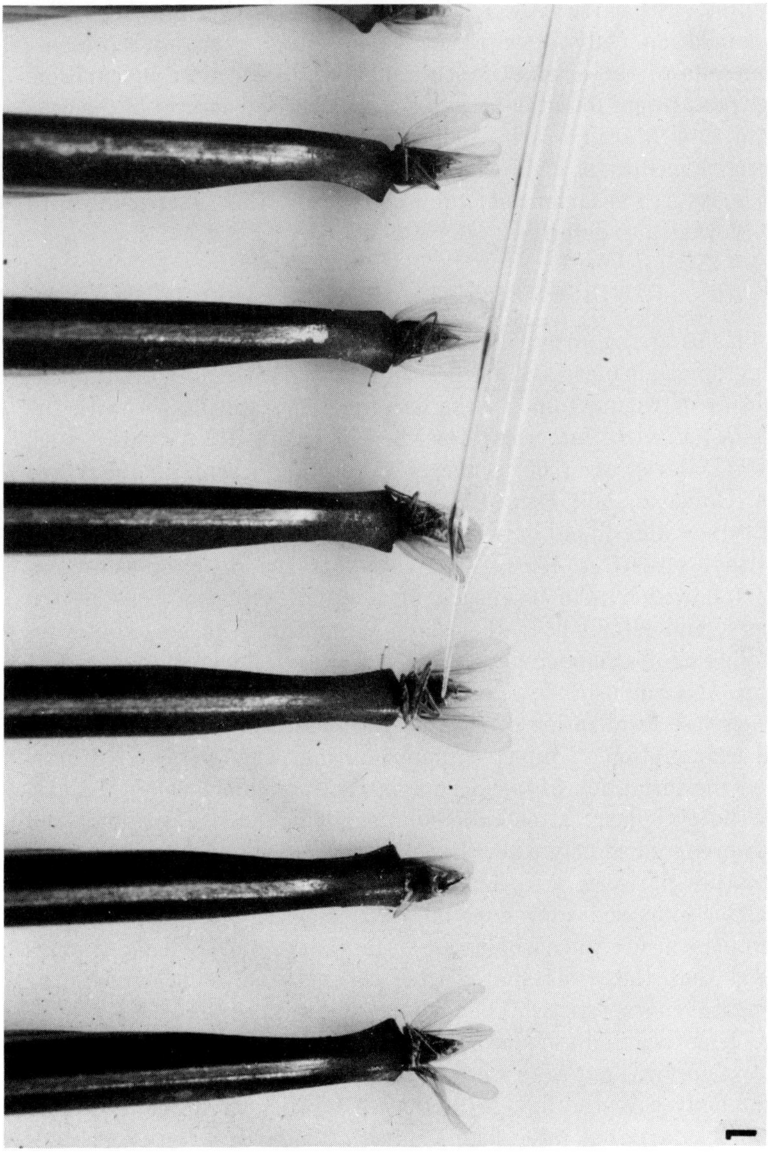

*Fig. 1.* Closeup of insects, immobilized in an insect holder. The third leafhopper from the left is being injected by means of a glass capillary.

the best material for wound tumor virus electron microscopy in ultrathin sections of leafhoppers.

The first virus particles were detected in the fat body 3 days after virus inoculation. Only a few particles were found, probably representing remnants of the original inoculum. Seven days after inoculation, virus particles were found in many electron dense structures of the cells (Shikata and Maramorosch, 1967b; Maramorosch, 1968). After 10 days pronounced cytopathic changes were noticed in a limited area of the fat body. At the same time, a few viroplasms, that is, areas of virus assembly, were also detected.

## II. VIRUS ASSEMBLY IN VIROPLASMS

Viroplasms are aggregates of finely textured, electron dense material in the cytoplasmic matrix of a cell, in which virus assembly takes place. Viroplasms of wound tumor virus correspond in appearance with the viroplasms or "virus factories" described in several RNA viruses, such as polio (Dales *et al.*, 1965a), mengo (Dales and Franklin, 1962), reo (Dales, Gomatos, and Hsu, 1965b), and others. The viroplasms of wound tumor were found only subsequent to virus inoculation, but not in healthy, virus-free controls. The appearance of viroplasms was always followed by the formation of complete virus particles at the periphery, and later within the whole viroplasm. The assembly sites were found in leafhopper vectors as well as in infected plants. The detection of viroplasms by electron microscopy provided a technique for the precise localization of sites of virus assembly within cells. The careful examination of infected plants during early stages of disease revealed the formation of viroplasms in the root tumor cell cytoplasm and in the cytoplasm of cells of enlarged leaf veinlets. In plant cell sections prepared shortly after the appearance of the first signs of disease, mature particles were initially formed at the periphery of the viroplasmic zone, and a few days later virions were detected in increasing numbers inside the viroplasmic area. These morphologic findings indicated that the virus shares with other cytoplasmic viruses the fundamental characteristic of developing within the cytoplasmic matrix proper. There was no evidence that virus assembly occurs in the nucleus, and no virus particles were detected in cell nuclei at any time.

In fat body cells the first formation of viroplasms probably occurs less than 10 days, but later than 7 days, after virus injection because on the tenth day several viroplasms were already seen with virions at their margin, while most other viroplasms had no discernible particles. A viroplasm with several mature virions at the periphery is shown in Figure 2. In addition to the virions, empty virus shells devoid of the

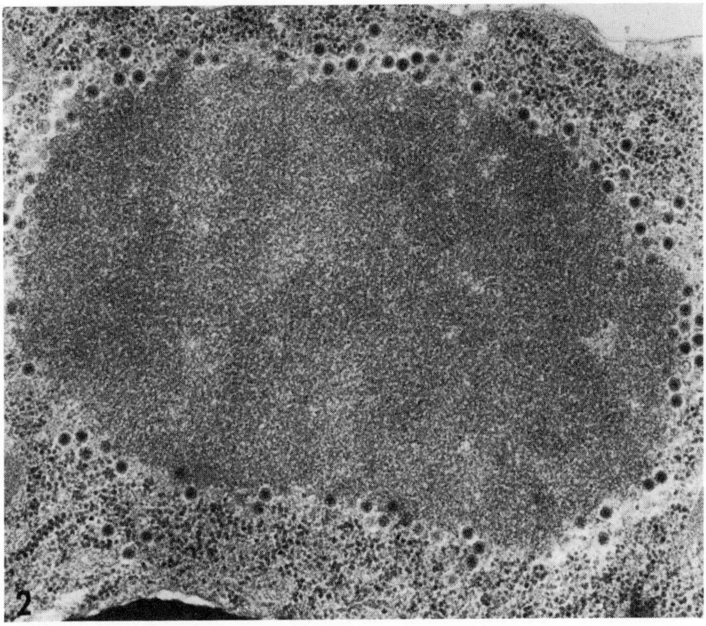

*Fig. 2.* Viroplasm with wound tumor virions at the periphery. Such viroplasms first appear in the fat body cells between 7 and 10 days after intrahemocoelic injection with virus. Later, viroplasms can be found in hemocytes and various organs and tissues of the vector. Magnification, × 29,440.

inner core were also detected. The viroplasms were always found in the proximity of ribosomes and polyribosomes that apparently came in contact with the viroplasmic matrix. In some fat body cells virus particles were observed within electron dense structures surrounded by several myelin-like layers. A week later wound tumor virus particles were found in practically all fat body areas. The cytoplasm at this stage was highly vacuolated, but, contrary to the fine structure changes observed on the tenth day, the vacuolation after 17 days was not specific for the injected insects since it also occurred in virus-free controls. As the number of viroplasms increased, virions were observed not only at the periphery but also within the viroplasms, as shown in Figure 3. The electron dense structures surrounded by myelin-like layers also increased in number. These structures now contained large masses of particles, often forming regular microcrystals. Apparently no microcrystals were formed at the sites of viroplasms but they appeared in close proximity in the same cells that actively assembled the virus.

Fat body cells are neither the only, nor the first, site of virus multiplication. Viroplasms were also observed in muscle, as shown in Figure

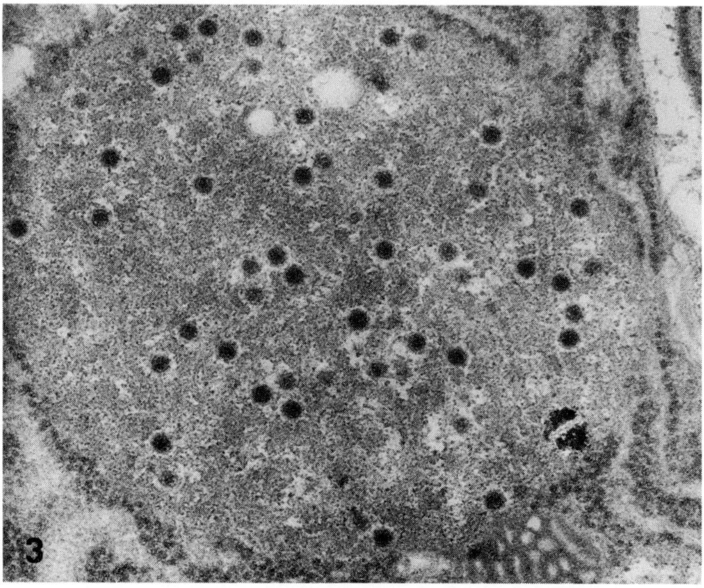

*Fig. 3.* Viroplasm with virus within the matrix. The formation of virions within the viroplasm in the fat body cells follows their formation at the periphery, illustrated in Figure 2. Magnification, × 48,000.

4, in malpighian tubules, trachea, salivary gland, epidermal cells, the nervous system (Hirumi, Granados, and Maramorosch, 1967), and hemocytes (Granados, Ward, and Maramorosch, 1967b, 1968).

## III. MOVEMENT OF INJECTED VIRUS

It appears likely that the original inoculum, injected into the abdominal cavity, is taken up by the hemolymyh of the insect and transported to various organs and tissues, among them the fat body cells (Maramorosch and Shikata, 1965). The hemolymph acts not only as a carrier, but also as the principal site of virus multiplication, as indicated by the finding of numerous viroplasms and of huge accumulations of virions in spherule cells and plasmatocytes (Granados *et al.*, 1968). A hemocyte containing wound tumor virus is shown in Figure 5. In the fat body, the original inoculum deposited by hemocytes apparently becomes phagocytized and engulfed in electron dense structures. Subsequently, the virus genome is probably released from the electron dense structures, which enables it to reach the sites of virus assembly, that is, the viroplasms. Virions are assembled first at the periphery, and

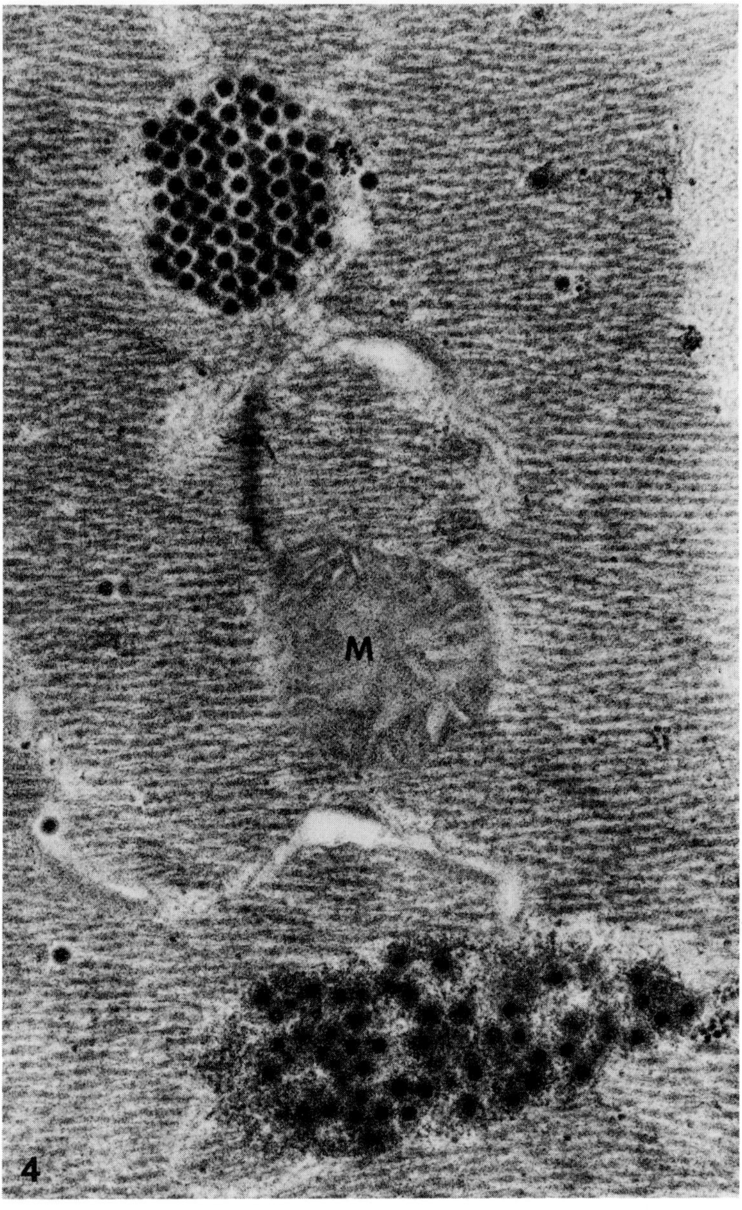


*Fig. 4.* Wound tumor virus in muscle of *Agallia constricta.* Note viroplasm at bottom, individual virus particles within the cell cytoplasm, and a virus microcrystal; M = mitochondrion. Magnification, × 44,000.

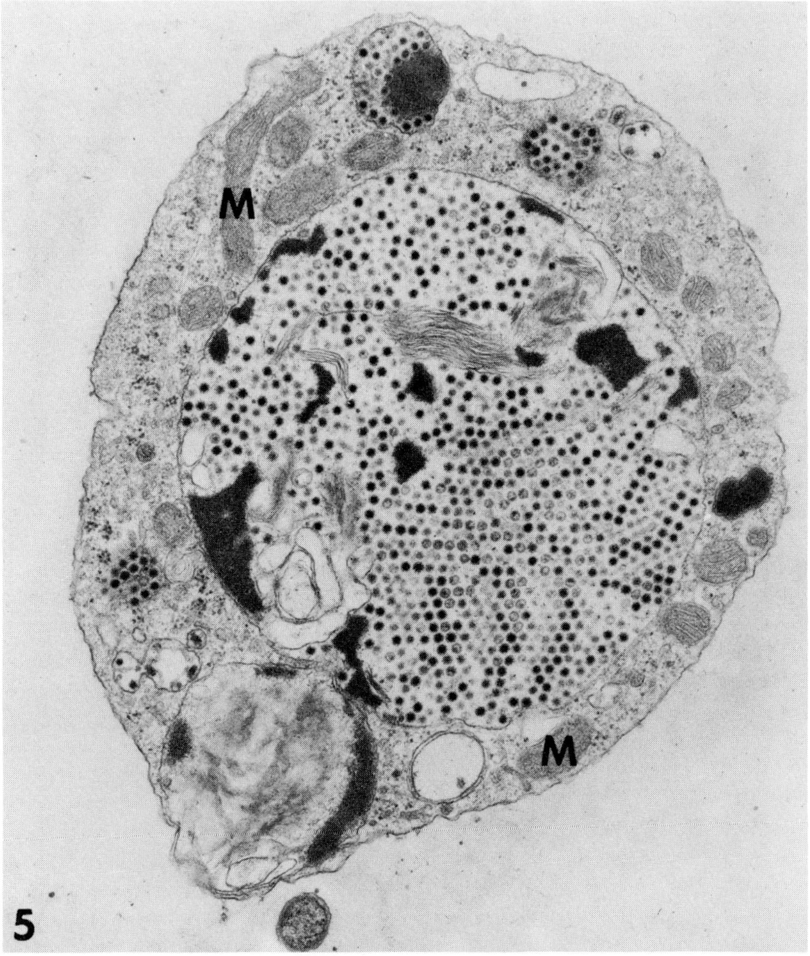


*Fig. 5.* Wound tumor virus in a hemocyte (plasmatocyte) of the leafhopper vector, *Agallia constricta.* The infection of the blood represents a viremia of the invertebrate animal; M = mitochondrion. Magnification, × 19,710. (Granados *et al.,* 1968.)

later throughout the viroplasm. Gradually, these new virions appear in the electron dense structures surrounded by myelin-like layers, where they become packed regularly into virus microcrystals, sometimes of considerable dimensions. This sequential process, reconstructed from electron micrographs, is strikingly reminiscent of the assembly of cytoplasmic animal viruses (Luria and Darnell, 1967).

## IV. ORAL ACQUISITION

Interpretation of the observations and suggested sequence in insects that acquired the virus by injection was supplemented by data obtained from insects inoculated by feeding on diseased plants (Shikata and Maramorosch, 1967a). Since the natural, oral acquisition is of greater interest than the artificial injection route, the sequential infection of insects that were rendered infective by feeding will be reconstructed from available findings.

One of the major differences between abdominally inoculated and orally inoculated leafhoppers was the absence of virus in the gut of injected individuals. Virus acquired orally enters the insect's gut lumen (Fig. 6). Individual virions were detected in, and around, the gut microvilli but the actual process of viral entry into the intestinal cells of the filter chamber has not been observed. The first accumulation of virus in the filter chamber of the gut was detected in electron micrographs.

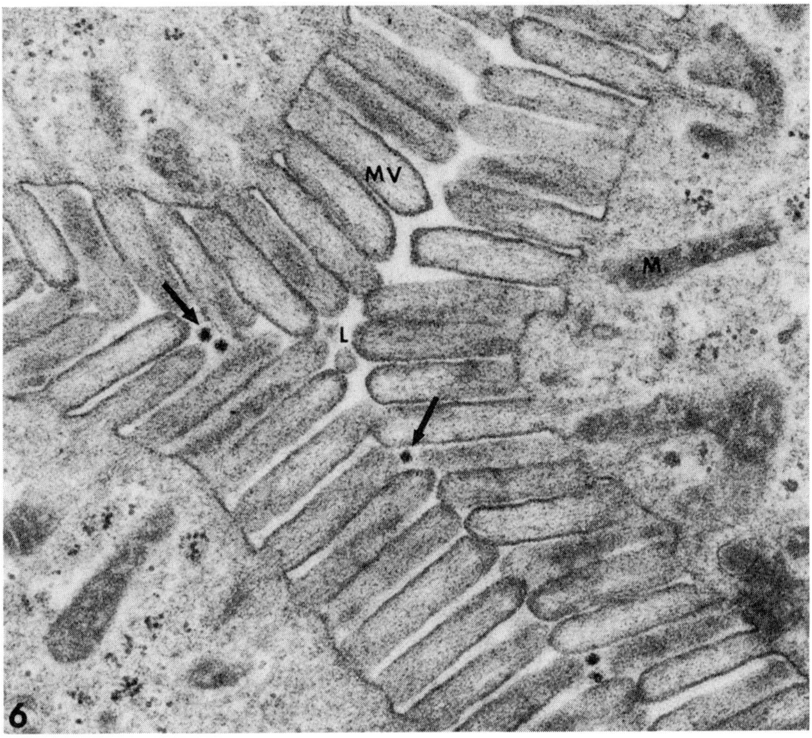


*Fig. 6.* Wound tumor virus (arrows) in the gut lumen: MV = microvilli; M = mitochondrion; L = lumen of gut. Magnification, × 38,160.

Wound tumor virus particles were found in electron dense phagocytic inclusions in the cell cytoplasm in the filter chamber 2 or 3 days following virus acquisition (Fig. 7). Numerous viroplasms with virions were located in the filter chamber cells 4 days after virus acquisition (Fig. 8). The finding of viroplasms proved that multiplication, not merely the accumulation, of virions took place. The abundance of viroplasms supports Sinha's (1965) hypothesis, based solely on the presence of virus antigen detected by fluorescent antibody tests. Unfortunately such tests fail to distinguish between virus accumulation and virus multiplication. The electron microscopy findings demonstrated that the filter chamber cells were, in fact, the primary foci of wound tumor virus infection.

After wound tumor virus has penetrated through the gut wall, it infects primarily or solely spherule cells and plasmatocytes of the hemolymph (Granados *et al.*, 1968). These hemocytes carry the virus to different sites in the insect body, as they do when virus is artifically introduced into the body cavity by needle inoculation. Since blood cells become infected, the plant-pathogenic virus causes viremia (blood infection) of the insect, which persists throughout the vector's life. The virus multiplies not only in the various locations at which it is deposited by the hemocytes, but also within the hemocytes, where it accumulates in large quantities. This finding explained the observed lifelong transmission once the individuals were rendered infective.

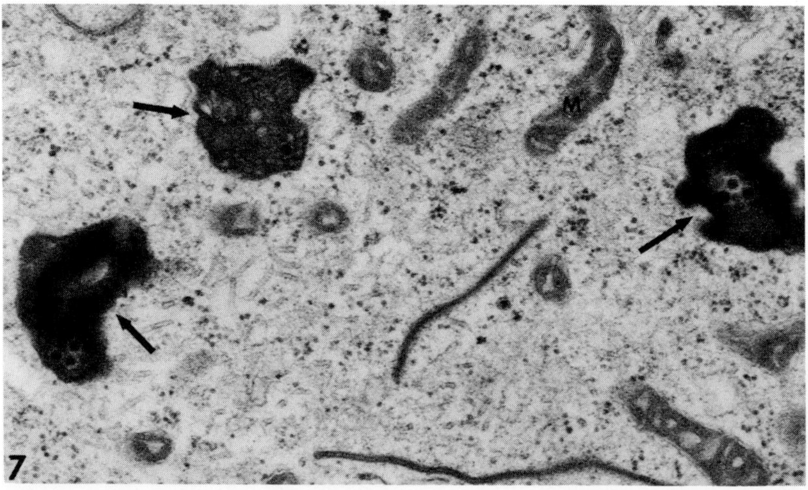


*Fig. 7.* Three electron dense areas (arrows) with wound tumor virus in gut cells, 2–3 days after oral acquisition of the virus; M = mitochondrion. Magnification, × 33,920.

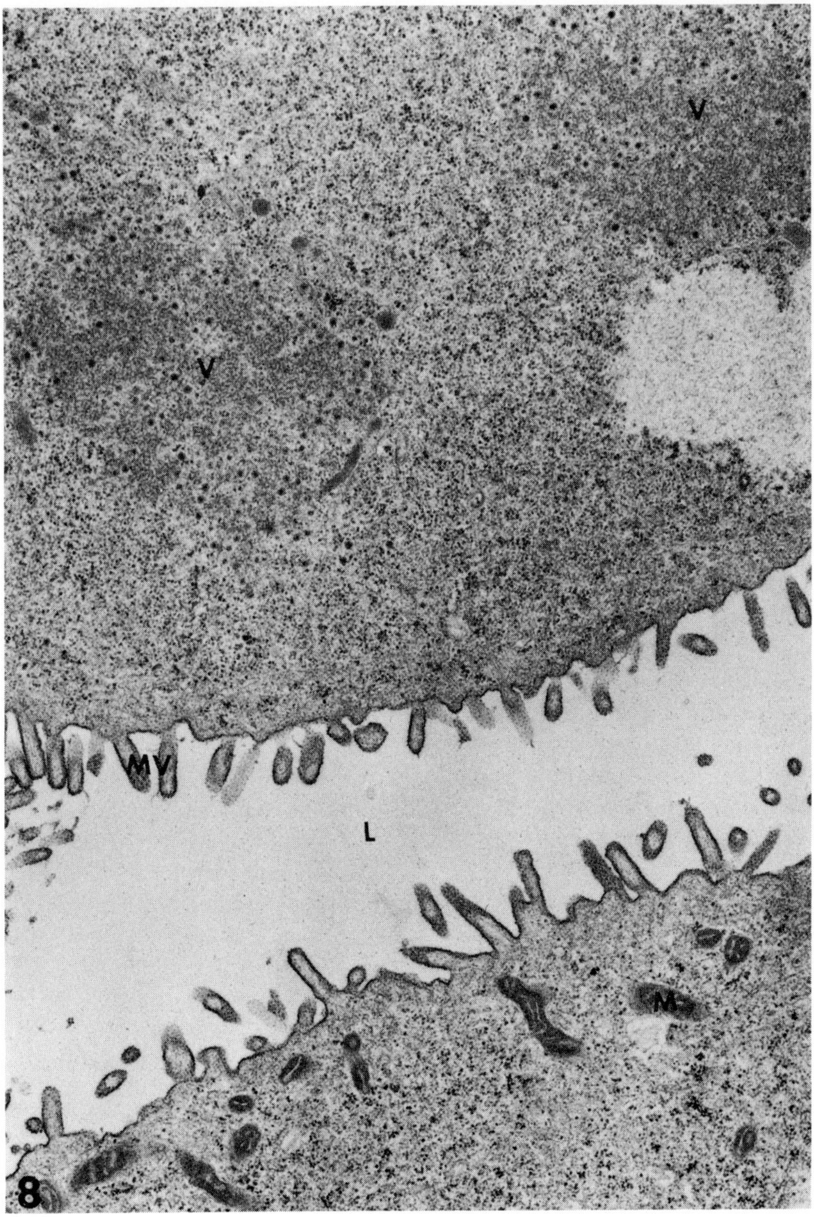


*Fig. 8.* Viroplasms in gut cells, 4 days after oral infection of the vector: V = viroplasm; M = mitochondrion; L = lumen of gut; MV = microvilli. Magnification, × 15,600.

The hemolymph reaches, among others, the salivary glands, where the virus may multiply and wherefrom it can enter the saliva, which acts as the vehicle for transmission into susceptible phloem cells of plants. The pathway and infection sites of wound tumor virus in the body of the leafhopper *A. constricta,* following virus acquisition through feeding, is diagrammatically presented (Fig. 9).

Sinha (1968) attempted to study wound tumor virus distribution in salivary glands by serological techniques and reported that virus antigens were located in all the lobes, but were concentrated in the anterior lobes. In other lobes only scattered fluorescent particles were observed. The disadvantage of the immuno-fluorescence technique for this kind of investigation is that entire organs are scanned for virus antigens, and small, precisely defined areas cannot be distinguished. Electron microscopy of leafhopper salivary glands indicated that usually only one of the several lobes contained wound tumor virus particles and, in most insects, the quantity of virus was low. Only in exceptionally efficient individuals that transmitted the virus to a large number of plants were virus accumulations detected in the form of microcrystals accompanied by viroplasmic matrixes. In most others, the virus appeared in the form of widely scattered particles. An abundance of empty shells was also found in less efficient individuals. Since the coats were found predominantly in poor transmitters, it seems reasonable to assume that they were incomplete, defective virus particles that formed in the salivary glands of the less efficient transmitters, and thus accounted for the erratic transmission obtained. This interpretation was further supported by the findings made in *Agalliopsis novella,* a nonefficient vector of wound tumor virus, where the total accumulation of virus was lower and the accumulation in the salivary glands was so low as to be inadequate for electron microscopical visualization (Granados, Hirumi, and Maramorosch, 1967a).

## V. CONCLUSIONS

The interpretations of the fate of wound tumor virus, outlined in this chapter, were based on the combined evidence obtained by us through electron microscopy, and by Sinha's (1965) use of fluorescent antibodies. In general, fluorescent antibody tests are much faster and simpler, and usually more sensitive for the detection of virus material than are electron micrographs of ultrathin sections. However, serological tests of wound tumor virus in vectors failed to distinguish between complete virions, incomplete virus particles, virus microcrystals, or virus assembly sites. Such distinctions were aided by the slower, more cumber-

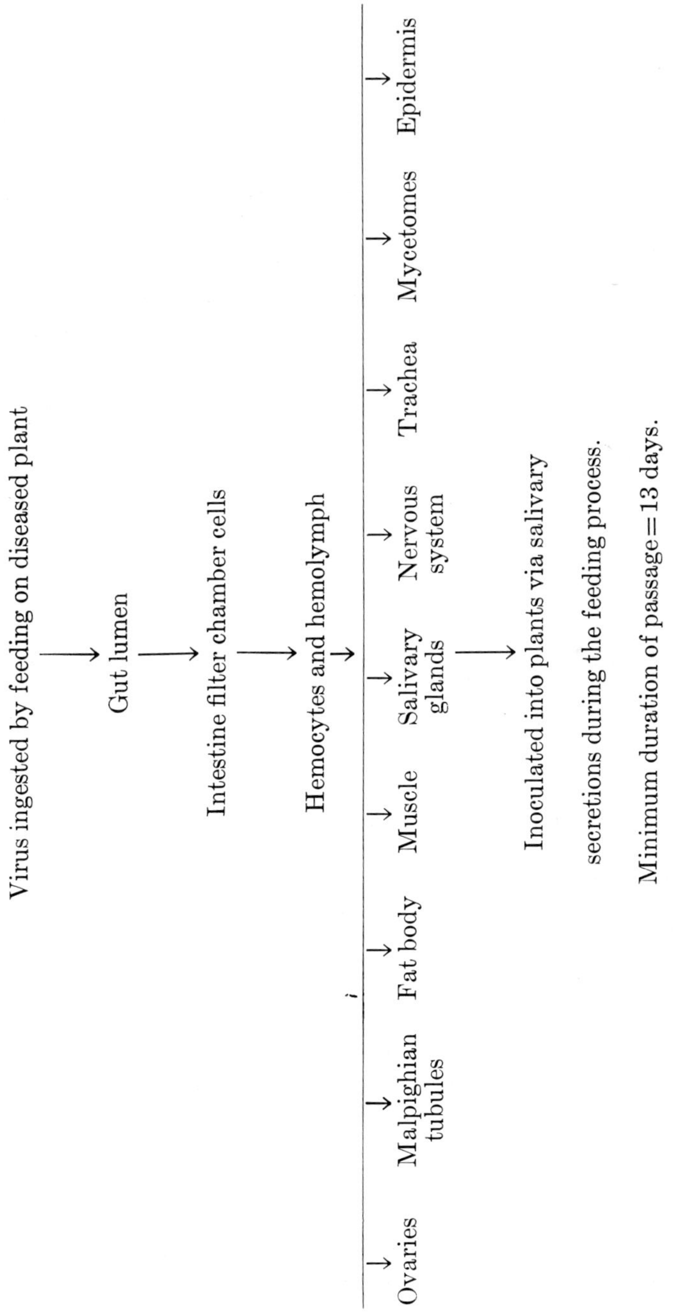


*Fig. 9.* Pathway and infection sites of wound tumor virus in the leafhopper vector *A. constricta*.

some, and difficult electron microscopy techniques that also helped in localizing wound tumor virus in cell structures with great precision. Virus ingested by a feeding leafhopper enters the gut lumen and the cells of the intestine filter chamber, where it accumulates and where initial multiplication takes place. After penetration through the gut wall, the virus is taken up by hemocytes in which further multiplication occurs. The hemolymph distributes the virus throughout the body of the invertebrate animal host, while viremia persists for the remainder of the insect's life. The passage and multiplication of virus require suitable, susceptible hemocytes, permeability of the gut and salivary gland, and an adequate level of sustained multiplication. Approximately 2 weeks are required for the virus to infect the insect, multiply, and contaminate the salivary secretions. Upon completion of the 2-week incubation period, the insect is able to inoculate the virus into the phloem of a susceptible plant.

## BIBLIOGRAPHY

Bils, R. F., and Hall, C. E. 1962. Electron microscopy of wound-tumor virus. Virology 17:123–130.

Black, L. M. 1959. Biological cycles of plant viruses in insect vectors. Vol. 2, p. 157–185. *In* The viruses. F. M. Burnet and W. M. Stanley [ed.] Academic Press, New York.

Black, L. M., and Markham, R. 1963. Base-pairing in the ribonucleic acid of wound-tumor virus. Netherlands J. Plant Pathol. 69:215.

Chen, T. S., Hirumi, H., and Maramorosch, K. 1968. Fine structure of *Trichodorus christiei.* Nematologica 14:4.

Dales, S., Eggers, H. J., Tamm, I., and Palade, G. E. 1965a. Electron microscopic study of the formation of poliovirus. Virology 26:379–389.

Dales, S., and Franklin, R. M. 1962. A comparison of the changes in fine structure of L cells during single cycles of viral multiplication, following their infection with the viruses of Mengo and encephalomyocarditis. J. Cell Biol. 14:281–302.

Dales, S., Gomatos, P. J., and Hsu, K. C. 1965b. The uptake and development of reovirus in strain L cells followed with labeled viral ribonucleic acid and ferritin–antibody conjugates. Virology 25:193–211.

Gomatos, P. J., and Tamm, I. 1963. Animal and plant viruses with double-helical RNA. Proc. Nat. Acad. Sci. U.S. 50:878–885.

Granados, R. R., Hirumi, H., and Maramorosch, K. 1967a. Electron microscopic evidence for wound-tumor virus accumulation in various organs of an inefficient leafhopper vector, *Agalliopsis novella.* J. Invertebrate Pathol. 9: 147–159.

Granados, R. R., Ward, L., and Maramorosch, K. 1967b. Electron microscopy of leafhopper blood cells infected with wound tumor virus. Bull. Entomol. Soc. Amer. 13:199.

Granados, R. R., Ward, L. S., and Maramorosch, K. 1968. Insect viremia caused by a plant-pathogenic virus: electron microscopy of vector hemocytes. Virology 34:790–796.

Hirumi, H., Chen, T. A., and Maramorosch, K. 1968. Feeding apparatus of *Trichodorus christiei:* three dimensional model of the ultrastructure. Phytopathology 58:400.

Hirumi, H., Granados, R. R., and Maramorosch, K. 1967. Electron microscopy of a plant-pathogenic virus in the nervous system of its insect vector. J. Virol. 1:430–444.

Hoof, H. A. van. 1958. Onderzoekingen over de biologische overdracht van een non-persistent virus. Doctoral thesis, Wageningen Agr. Univ. Van Putten & Oortmeijer, Alkmaar, The Netherlands. 96 p.

Kennedy, J. S., Day, M. F., and Eastop, V. F. 1962. A conspectus of aphids as vectors of plant viruses. Commonwealth Inst. Entomol., London. 114 p.

Luria, S. E., and Darnell, J. E., Jr. 1967. General virology. Wiley, New York. 512 p.

Maramorosch, K. 1968. Plant pathogenic viruses in insects. Current Topics Microbiol. 42:94–107.

Maramorosch, K., Brakke, M. K., and Black, L. M. 1949. Mechanical transmission of a plant tumor virus to an insect vector. Science 110:162–163.

Maramorosch, K., and Jernberg, N. 1964. A device for rapid inoculation of arthropods. Bull. Entomol. Soc. Amer. 10:171.

Maramorosch, K., and Shikata, E. 1965. The fate of wound tumor virus in vectors and non-vectors: an electron microscopy study. Papers Conf. Relationships Arthropod Plant-Pathogenic Viruses, Tokyo, p. 70–72.

Rosen, L. 1968. Reoviruses. Virology Monogr. 1:73–107.

Shikata, E., and Maramorosch, K. 1967a. Electron microscopy of wound tumor virus assembly sites in insect vectors and plants. Virology 32:363–377.

———. 1967b. Electron microscopy of the formation of wound tumor virus in abdominally inoculated insect vectors. J. Virol. 1:1052–1073.

Sinha, R. C. 1965. Sequential infection and distribution of wound-tumor virus in the internal organs of a vector after ingestion of virus. Virology 26:673–686.

———. 1968. Recent work on leafhopper-transmitted viruses. Advances Virus Res. 13:181–223.

Storey, H. H. 1933. Investigations of the mechanism of the transmission of plant viruses by insect vectors. I. Proc. Roy. Soc. (London) 113B:463–485.

Streissle, G., Bystricky, V., Granados, R. R., and Strohmaier, K. 1968. Number of virus particles in insects and plants infected with wound tumor virus. J. Virol. 2:214–217.

Vasquez, C., and Tournier, P. 1962. The morphology of reovirus. Virology 17: 503–510.

Watson, M. A. 1960. The ways in which plant viruses are transmitted by vectors. Rep. Commonwealth Entomol. Conf., 7th, London, 1960, p. 157–161.

# Electron Microscopy of the Transovarial Passage of Rice Dwarf Virus

Sôchô Nasu

*Division of Entomology*
*National Institute of Agricultural Sciences*
*Tokyo, Japan*

Several plant viruses such as rice dwarf virus, rice stripe virus, hoja blanca virus, clover club leaf virus, European wheat striate mosaic virus, wound tumor virus, and potato yellow-dwarf virus, have been reported to be transmitted transovarially (Fukushi, 1933; Shinkai, 1962; Administración de Estabilización del Arroz, 1959; Black, 1948, 1953; Slykhuis and Watson, 1958). Rice dwarf virus can be transmitted transovarially in *Nephotettix apicalis, Nephotettix cincticeps,* and *Inazuma dorsalis* (Fukushi, 1933, 1939; Nasu, 1963).

The particles of rice dwarf virus are about 70 m$\mu$ in diameter and icosahedron in shape (Fukushi *et al.*, 1960; Fukushi and Shikata, 1963a). The particles have been photographed in cells of the fat body, intestine, malpighian tubules, ovarioles, hemocoels, and salivary glands of the vectors (Fukushi *et al.*, 1960; Fukushi, Shikata, and Kimura, 1962; Fukushi and Shikata, 1963b). However, the processes of multiplication of this virus in the above-mentioned organs, and of transovarial passage of this virus have only recently been demonstrated by electron microscopy.

The author has studied the processes of transovarial passage as well as that of the multiplication of this virus in viruliferous individuals of the green rice leafhopper, *N. cincticeps* (Nasu, 1965).

## I. THREE POSSIBLE MECHANISMS FOR THE VIRUS INVASION OF OOCYTES

Based upon the experimental results, the three routes of virus invasion into the oocytes of *N. cincticeps* ovarioles have been postulated as working hypotheses (Fig. 1).

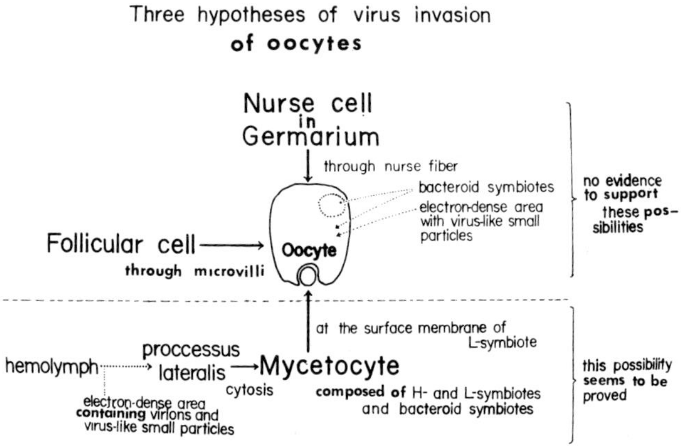


*Figure 1.*

(*1*) Viruses enter the oocytes accompanied by the cytoplasm and other materials that pass from the nurse cells of the germarium into the oocytes through the nurse fibers.

The first possibility has not been proved, although virus-like small particles were found in electron-dense areas of the cytoplasm in nurse cells and oocytes in viruliferous leafhoppers as shown in Figure 2, but not in nonviruliferous leafhoppers. The virus-like small particles are specific to the viruliferous leafhoppers and may have some relations to the virus disease.

(*2*) Viruses enter the oocytes accompanied by yolk and other substances supplied from the follicular cells through the microvilli.

No evidence has been obtained to support the second possibility in this study, and further work is needed to prove or disprove this possibility.

(*3*) Viruses enter the eggs accompanied by the mycetocyte of the ovariole.

The third possibility is supported by the evidence presented in Figures 4 and 5, which demonstrate the affinity of virus particles to the surface of L symbiotes, and suggest the transpotation of virus particles into oocytes by mycetocytes.

## II. VIRUS PARTICLES IN THE MYCETOCYTE OF AN OVARIOLE

One side of an ovary of *N. cincticeps* consists of about ten ovarioles (Nasu, 1963). Ovarioles of leafhoppers are telotrophic as shown in Figure 3 and can be divided histologically into three zones: the germarium, follicular cells, and pedicel. In the germarium many nurse

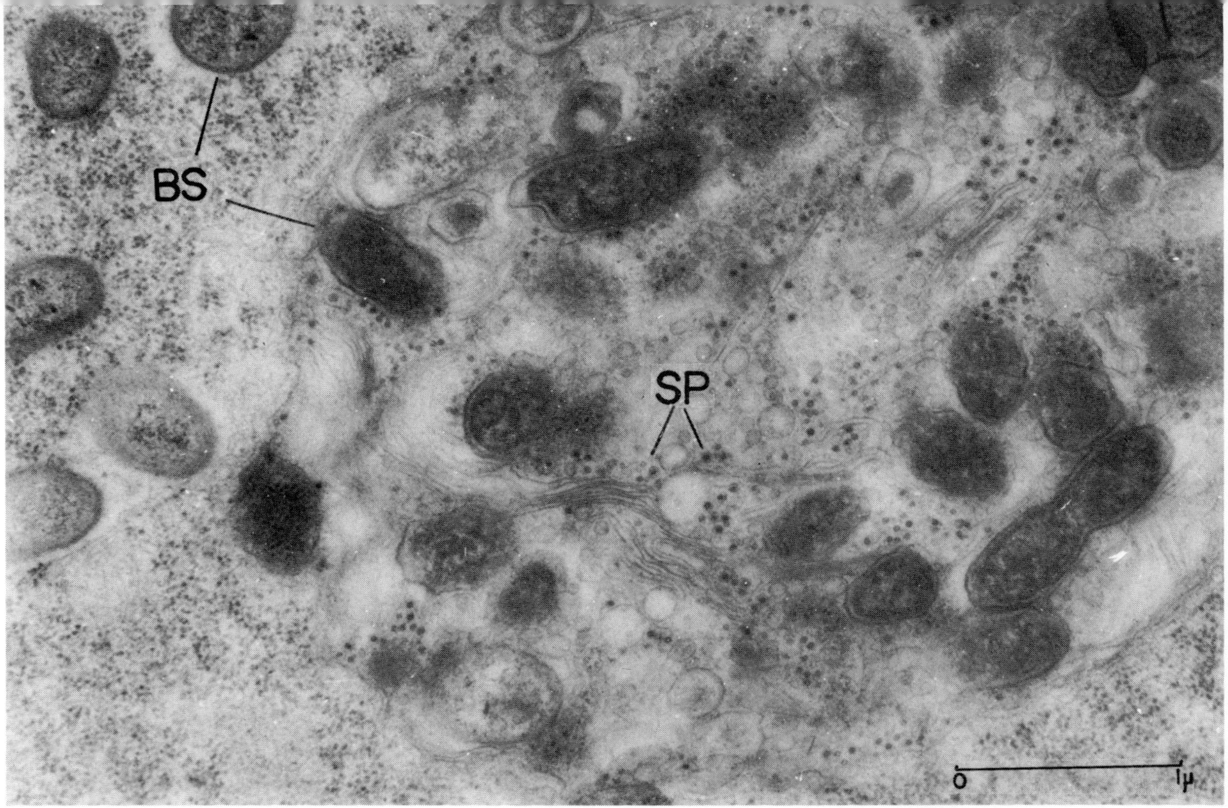


*Fig. 2.* Electron micrograph of the young oocyte (a zone which represents Part A of Fig. 3) in a section through the germarium of a viruliferous adult ovariole of *N. cincticets*. An electron-dense area is shown containing virus-like small particles (SP) : BS, bacteroid symbiotes.

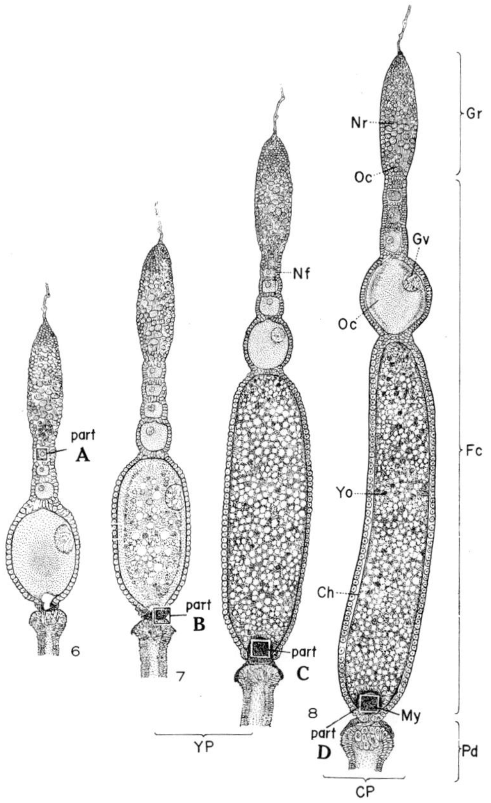


*Fig. 3.* Diagrammatic representation of an ovariole on *N. cincticeps* as seen in longitudinal section, illustrating the major zones. The ovariole was divided into several zones on the basis of its histological and cytological features according to Nasu (1963): Ch, chorion; CP, chorion-forming period; Fc, follicular cells; Gr, germarium; Gv, germinal vesicle; My, mycetocyte; Nr, nurse cells; Nf, nurse fiber; Oc, oocyte; Pd; pedicel; Yo, yolk; YP, yolk-forming period. Part A–D refer to electron micrographs. Numbers show the days elapsed after emergence at 25°C.

cells are found, and nurse fibers extend to oocytes from the nurse cells. The oocytes are surrounded by a layer of the follicular cells. In each ovariole mycetocytes are located at the place in which the layer of follicular cells comes into contact with the pedicel.

The cytoplasm of the mycetocyte contains two kinds of spheroids, approximately 5–6 $\mu$ in diameter and having different features (Fig. 4). One spheroid of low electron density is hereafter designated as the "L symbiote," whereas the other spheroid, high in electron density,

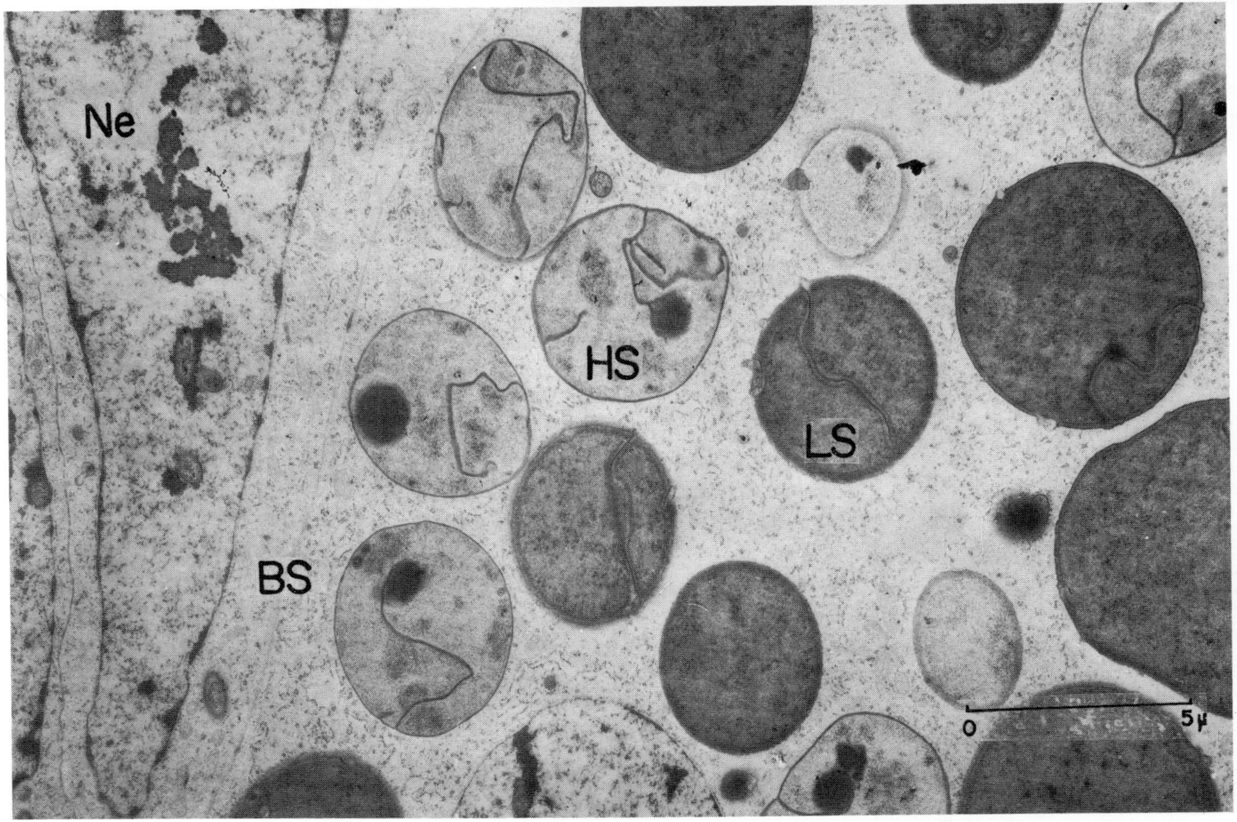


*Fig. 4.* Electron micrograph of mycetocyte (a zone which represents Part B of Fig. 3) in a section through the ovariole of the non-viruliferous adult of *N. cincticeps*: BS, bacteroid symbiotes; HS, H symbiotes; LS L symbiotes; Ne, nucleus.

is hereafter designated as the "H symbiote," although these spheroids have not yet been identified as symbiotes.

Each L symbiote, filled with numerous minute granules 10–15 mμ in diameter, usually has an electron-dense area of a round shape with no structures. The surface of each L symbiote is surrounded by two layers of membrane structures with minute fibers and a few big invaginations.

Each H symbiote also has numerous minute granules with some areas of condensation, and a membrane structure similar to the membrane of endoplasmic reticulum. The surface of H symbiotes consists of a double-layered membrane, covering another, limited membrane. The membranes have some invaginations.

Both the H and L symbiotes increase in number, and fill up the cytoplasm of the mycetocyte, which soon migrates into oocytes of the yolk-forming stage.

The viruliferous leafhoppers have electron-dense areas with virions and virus-like small particles in the cytoplasm of their mycetocytes (Fig. 5), whereas nonviruliferous ones have similar areas without virions and virus-like small particles. The areas seem to come together with H and L symbiotes, which are probably derived from the hemolymph in which they may exist independently.

Figure 6 shows a portion of the mycetocyte in an oocyte corresponding to C in Figure 3. As shown in that figure, virus particles are found on or between the membrane structures of L symbiotes. Electron micrographs of the viruliferous mycetocyte during the chorion-forming period are also shown in Figure 7. The rice dwarf virus particles can be found on the surface of the L symbiotes or between the neighboring symbiotes.

The method of Tawde and Sri Ram (1962) in immune electron microscopy was used to demonstrate the rice dwarf virus particles. Upon treating the sample of viruliferous oocytes with the 1:1000 complement fraction titer antiserum of rice dwarf virus conjugated with ferritin, the precipitated particles of ferritin were gathered at the virions on the surface of L symbiotes, as shown in the insert, Figure 7. This reaction showed that the virions were antigenic to the antiserum of rice dwarf virus, and this clearly proved that the virions on the surface of the L symbiotes were the rice dwarf virus particles.

## III. VIRUS PARTICLES ON THE DEVELOPING EMBRYOS

The duration of egg stage of *N. cincticeps* ranges from 240 to 285 hr at 25°C, having the mode value of 260 hr (Nasu, 1963).

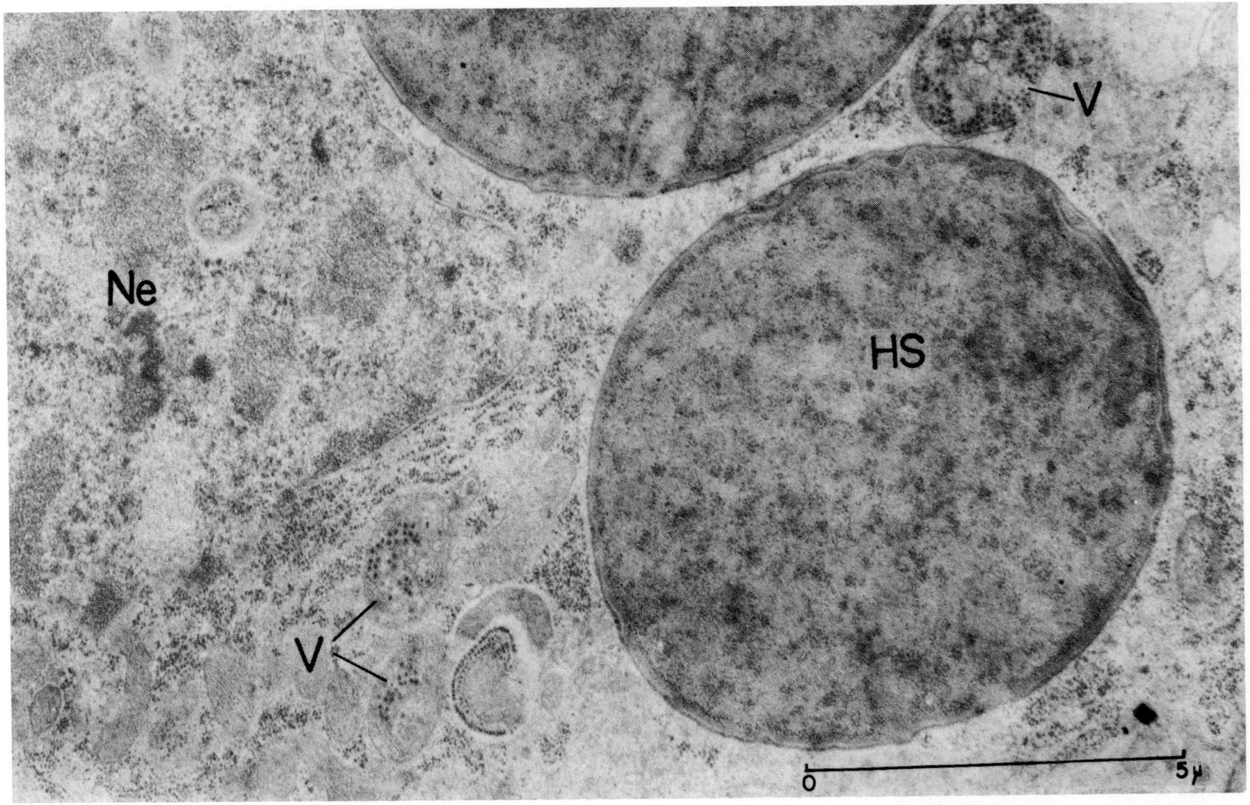


*Fig. 5.* Electron micrograph of mycetocyte (a zone which represents Part B of Fig. 3) in a section through the ovariole of a viruliferous adult of *N. cincticeps*: HS, H symbiotes; Ne, nucleus; V, virus particles.

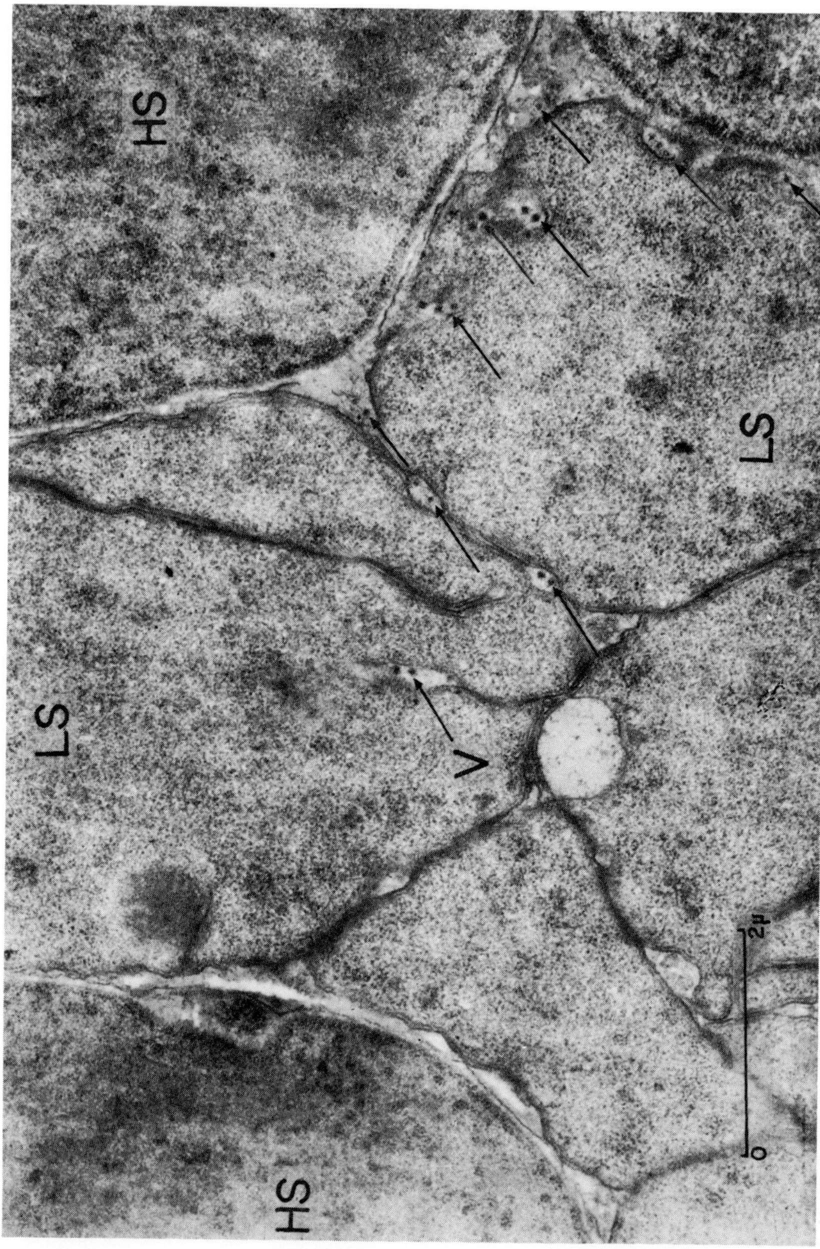


*Fig. 6*. Electron micrograph of the mycetocyte (a zone which represents Part C of Fig. 3) in a section through the ovariole of the yolk-forming stage in a viruliferous adult of *N. cincticeps* showing rice dwarf virus particles (V, arrows) in or between the surface structure of L symbiotes (LS): HS, H symbiotes.

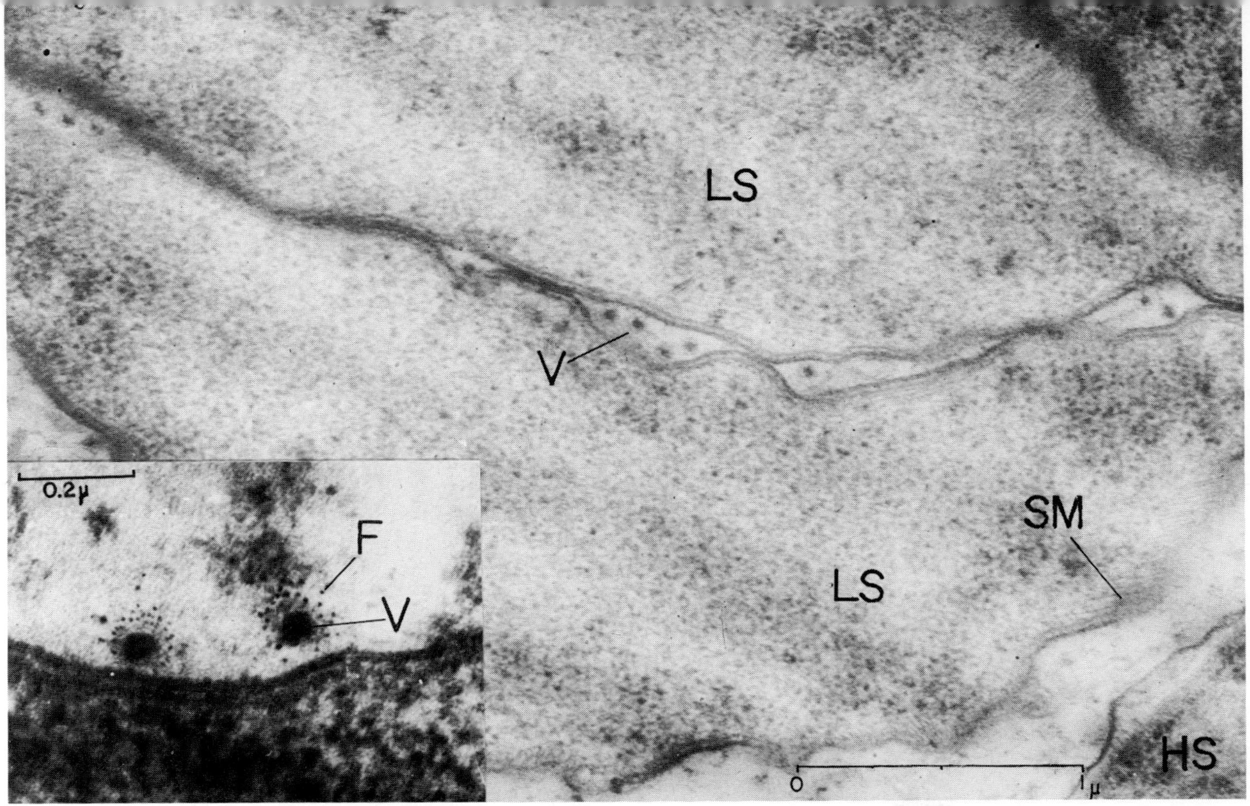


*Fig. 7.* Electron micrographs of the symbiotes in mycetocyte (a zone which represents Part D of Fig. 3) in a section through the ovariole of the chorion-forming stage in a viruliferous adult of *N. cincticeps:* V, virus particles; HS, H symbiote; SM, surface structure of L symbiote. Insert: a demonstration of rice dwarf virus particles (V) on the surface structure of an L symbiote by the ferritin (F) conjugated antibody technique.

In an area of the mycetome at the blastoderm stage embryo, L and H symbiotes are found simultaneously in each cell of the mycetome. The virions are found on the surface of L symbiotes. As shown in Figures 8 and 9, the virions are still on the surface membrane of L symbiotes in the mycetome at the blastokinetic stage, although L and H symbiotes are no longer found in the same cell at this stage.

In the mycetome of further developed embryos (metaphase of the eye pigmentation stage), a considerable number of virions are found also in cells which contain H symbiotes (Fig. 10).

The mycetome of the embryo after the eye pigmentation stage has an increased number of rice dwarf virus particles in electron-dense areas of the cytoplasm, along with other virus-like small particles (Fig. 11). There are still yolk bodies in the abdomen of the embryo at this age. Between the yolk bodies are many bacteroid symbiotes, which are increasing in number and forming electron-dense areas in which a number of virus-like small particles can be recognized.

## IV. VARIOUS PROBLEMS CONCERNING SYMBIOTES

*Nephotettix cincticeps* has two kinds of symbiotes, H symbiotes in the intermediate layer and L symbiotes in the internal one of the mycetome. They seem to be identical with symbiotes which were called *"Risensymbionten"* by Buchner (1953). They are not uniform in shape in the mycetomes of nymph and adult leafhoppers. As shown in Figures 4 and 5, in the mycetocyte of an ovariole, however, they are relatively uniform in shape and have a spherical outline. Their sizes are also uniform, 5–6 $\mu$ in diameter. The former, in the mycetomes of the abdomen, may be in the multiplication stage, and the latter, in the mycetocytes of the ovarioles, may be in their migrant form.

The above observations led to the conclusion that the multiplication stage of symbiotes may be transformed into the migrant form, which moves into the hemolymph in a free state and finally reaches the mycetocytes of the ovariole.

As pointed out, L and H symbiotes differ in electron density and in membrane structure. A limiting membrane is found in H symbiotes beneath two layers that surround both types of symbiotes. It is possible that the two membrane layers were originally derived from structures of host cells, since the structures are not so closely attached to the surface of the symbiotes as are the membranes of L symbiotes. Further studies will be required to elucidate this assumption.

Rice dwarf virus particles selectively attach to the surface membrane

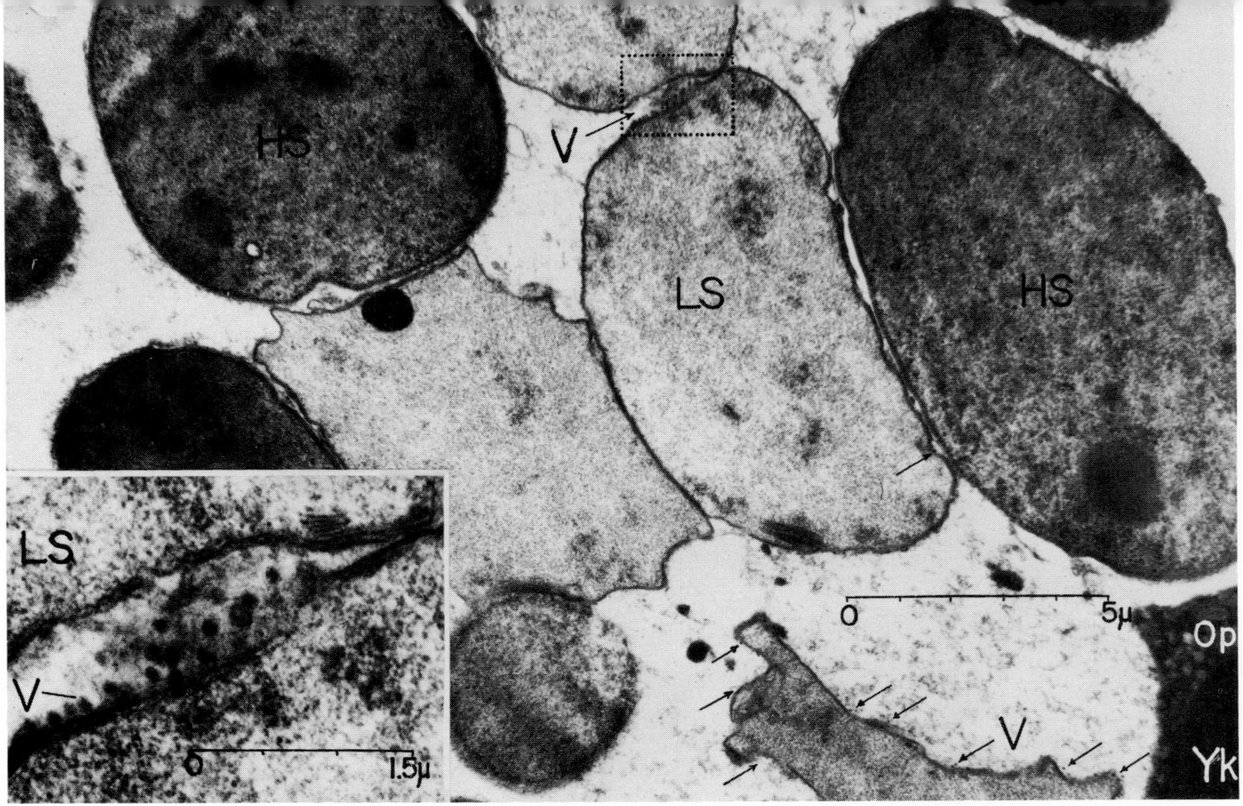


*Fig. 8.* Electron micrograph of a mycetome at the blastoderm stage of a viruliferous embryo of *N. cincticeps:* V, virus particles; HS, H symbiotes; Op, ooplasm; Yk, yolk. Insert: rice dwarf virus particles on the surface structure of an L symbiote (LS).

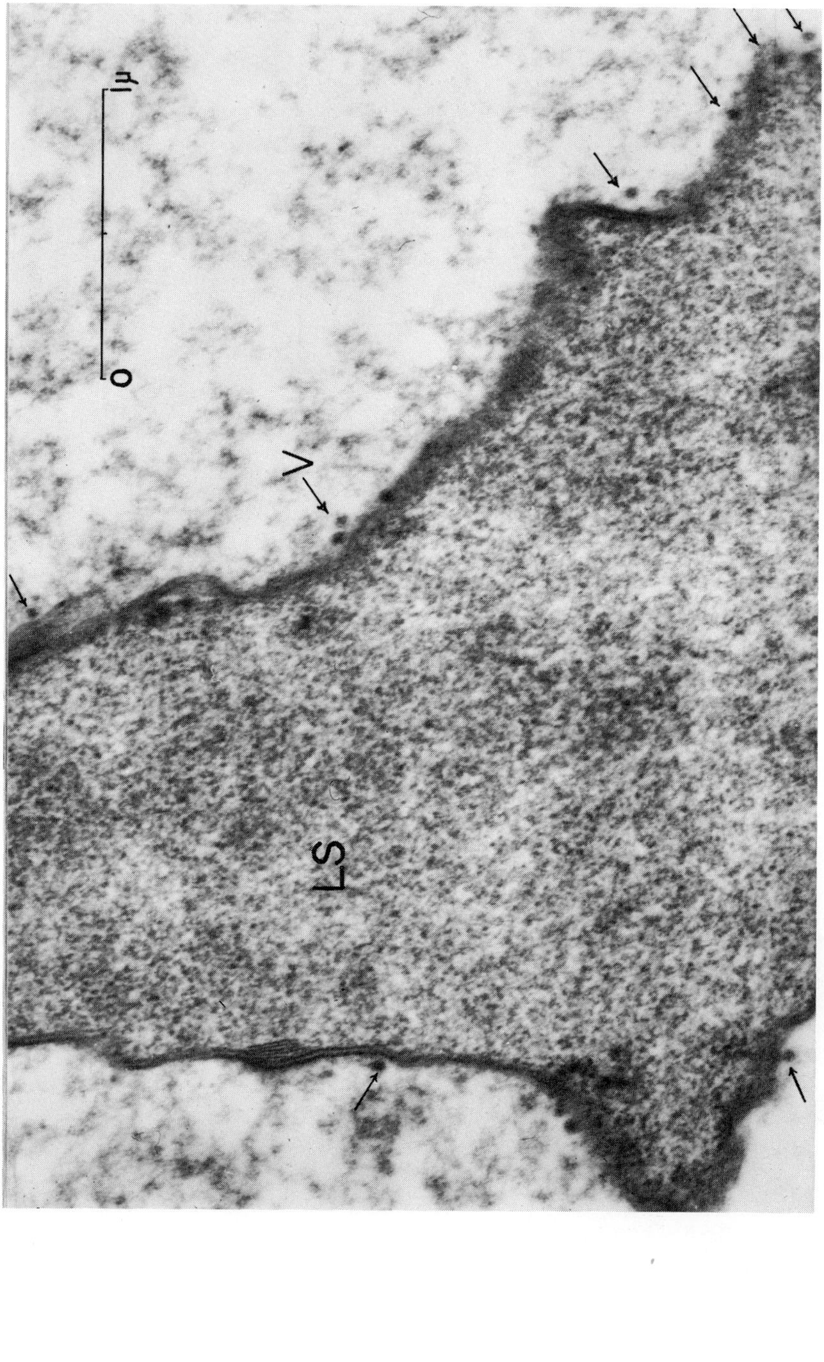


*Fig. 9*. Electron micrograph of mycetome at the blastoderm stage of a viruliferous embryo of *N. cincticeps*.

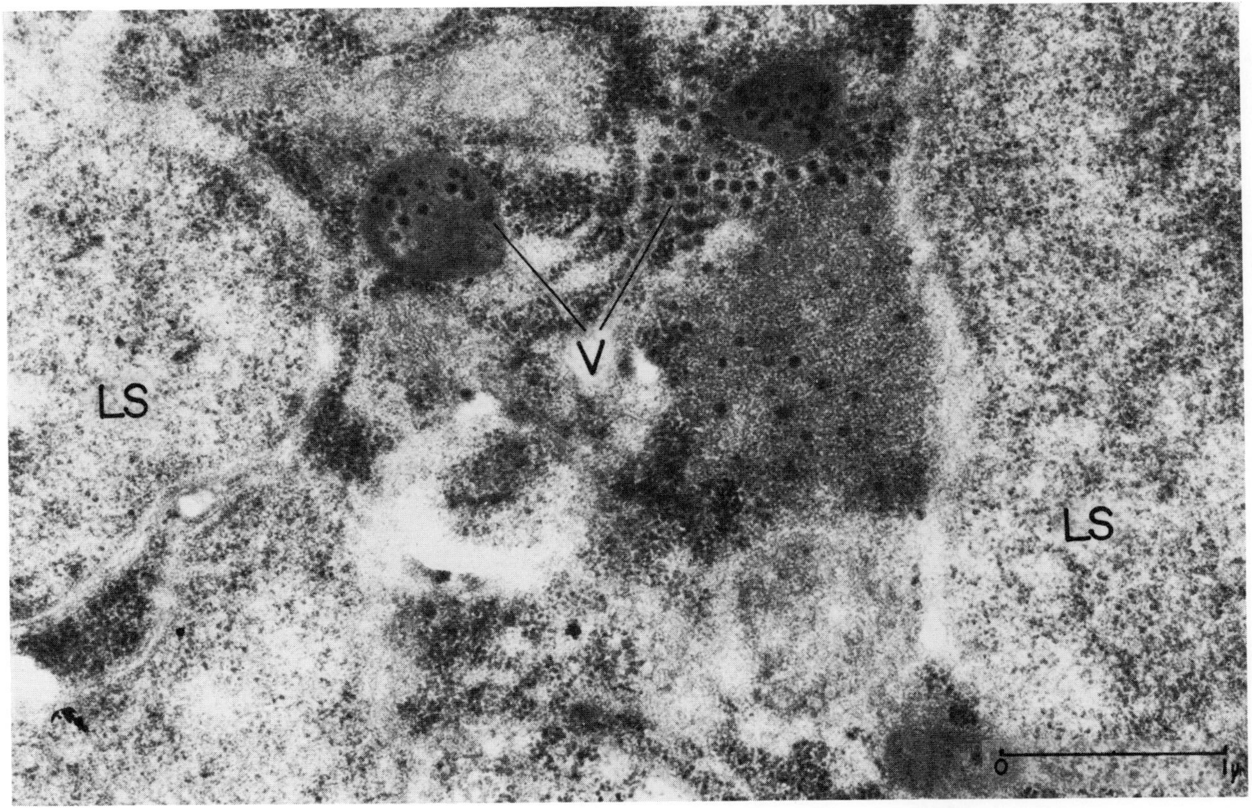


*Fig. 10.* Electron micrograph of a mycetome at prophase of the eye pigmentation stage of a viruliferous embryo of *N. cincticeps*. Note the rice dwarf virus particles (V) between L symbiotes (LS).

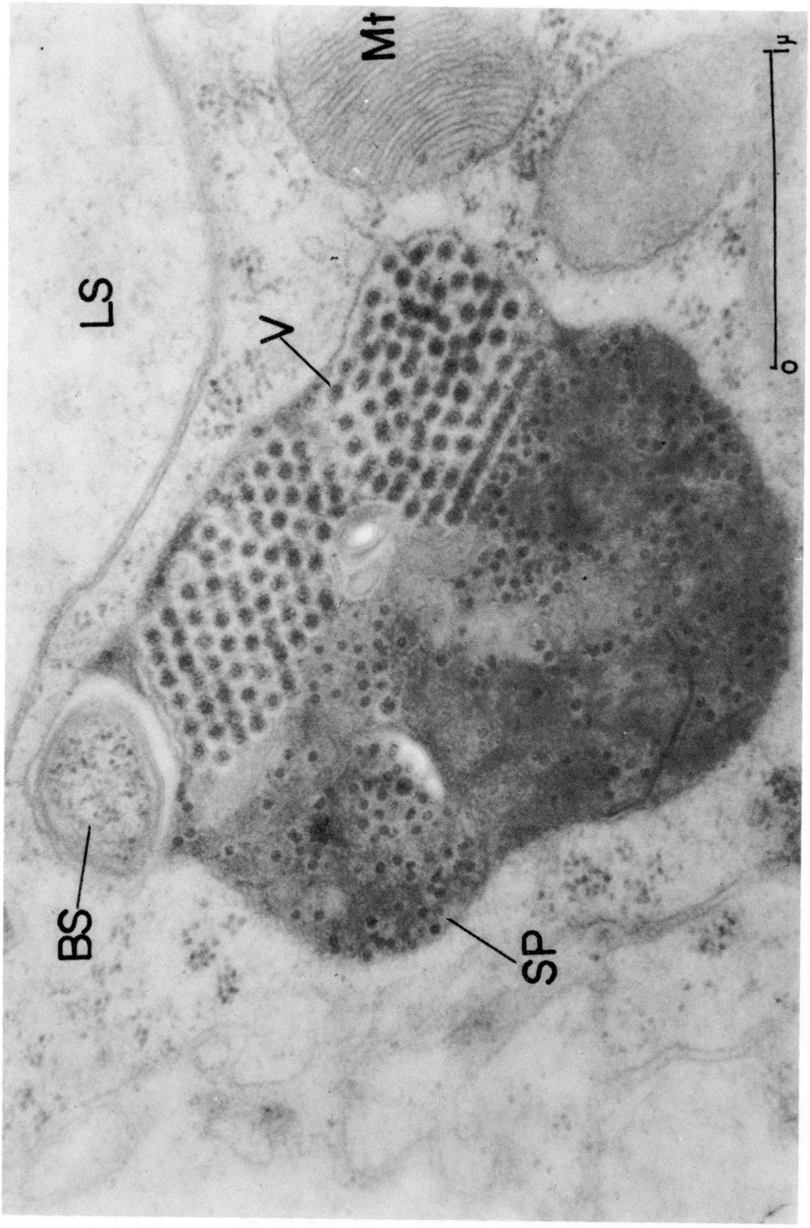


*Fig. 11.* Electron micrograph of the internal layers of the mycetome at anaphase during the eye pigmentation stage in a viruliferous embryo of *N. cincticeps*: BS, bacteroid symbiotes; LS, L symbiote; Mt, mitochondria, SP, virus-like and small particles; V, rice dwarf virus particles.

of L symbiotes in the mycetocyte of the ovariole, but not to H symbiotes. The membrane of L symbiotes may, therefore, act as a carrier of these virus particles. This affinity of rice dwarf virus particles to membrane structures of L symbiotes requires further studies. Such investigations might provide an explanation for the inability of this virus to pass transovarially in the related leafhopper vector *N. cincticeps.*

## V. CONCLUSIONS

As shown in Figures 4–7, the L and H symbiotes increase in numbers in the mycetocyte of the ovariole. Since L and H symbiotes gather around the *processus lateralis* of the mycetocyte, the increase seems to be due to the migration of the symbiotes that occur in a free state in the hemolymph and pass selectively into the mycetocyte. When L and H symbiotes increase in numbers, electron-dense areas containing the rice dwarf virus particles and the virus-like small particles appear in the mycetocyte simultaneously. The appearance of these areas may be linked to the migration of symbiotes, and this migration may play a role in the infection process of the mycetocyte by the virus.

When the mycetocyte becomes filled with L and H symbiotes, the rice dwarf particles attach to the surface of L symbiotes, and a part of the mycetocyte enters the neighboring oocyte. In this way oocytes of the progeny become infected with virus particles from the mother insects. The virus that entered oocytes multiplies in the cytoplasm of the mycetome during embryonic development. This seems to be the only feasible explanation for the transovarial passage of rice dwarf virus in *N. cincticeps*.

It is interesting that the incubation period of the virus in *N. cincticeps* leafhoppers that acquired the virus by feeding is of similar length to the period required for the migration of mycetomes into oocytes, and the transmission of virus by nymphs derived from viruliferous embryos.

## BIBLIOGRAPHY

Administración de Estabilización del Arroz. 1959. Informaciones de Interes General en Relacion con el Arroz. 12:20–21.

Black, L. M. 1948. Transmission of clover club-leaf virus through the egg of its insect vector. Phytopathology 38:2.

Black, L. M. 1953. Occasional transmission of some plant viruses through the eggs of their insect vectors. Phytopathology 43:9–10.

Buchner, P. 1953. Endosymbiose der Tiere mit Phlanzlichen Mikroorganismen. Verlag Birkhäuser, Basel. 771 p.

Fukushi, T. 1933. Transmission of virus through the egg of an insect vector. Proc. Imp. Acad. Japan. 9:457–460.

——— 1934. Studies on the dwarf disease of rice plant. J. Faculty Agr. Hokkaido Imp. Univ., 37:41–164.

——— 1935. Multiplication of virus in its insect vector. Proc. Imp. Acad. Japan. 11:301–303.

——— 1937. An insect vector of the dwarf disease of rice plant. Proc. Imp. Acad., Japan. 13:328–331.

——— 1939. Retention of virus by its insect vectors through several generations. Proc. Imp. Acad., Japan. 15:142–145.

——— 1940. Further studies on the dwarf disease of rice plant. J. Faculty Agr. Hokkaido Imp. Univ. 45:83–154.

Fukushi, T., and Kimura, I. 1959. On some properties of the rice dwarf virus. Proc. Imp. Acad., Japan. 35:482–483.

Fukushi, T., and Shikata, E. 1963a. Fine structure of rice dwarf virus. Virology 21:500–503.

——— 1963b. Localization of rice dwarf virus in its insect vector. Virology, 21: 503–505.

Fukushi, T., Shikata, E., and Kimura, I. 1962. Some morphological characters of rice dwarf virus. Virology, 18:192–205.

Fukushi, I., Shikata, E., Kimura, I., and Nemoto, M. 1960. Electron microscopic studies on the rice dwarf virus. Proc. Imp. Acad., Japan. 36:352–357.

Nasu, S. 1963. Studies on some leafhoppers and planthoppers which transmit virus diseases of rice plant in Japan. Bull. Kyushu Agr. Exp. Sta. 8:153–349.

——— 1965. Electron microscopic studies on transovarial passage of rice dwarf virus. Japan. J. Appl. Entomol. Zool. 9:225–237.

Shinkai, A. 1962. Studies on insect transmissions of rice virus diseases in Japan. Bull. Nat. Inst. Agr. Sci. C 14:1–112.

Slykhuis, J. T., and Watson, M. A. 1958. Striate mosaic of cereals in Europe and its transmission by *Delphacodes pellucida* (Fab.). Ann. Appl. Biol. 46:542–553.

Tawde, S. S., and Sri Ram, J. 1962. Conjugation of antibody to ferritin by means of *p,p′*-difluoro-*m,m′*-dinitrodiphenylsulphone. Arch. Biochem. Biophys. 97: 430.

# Bioassay of Plant Viruses Transmitted Persistently By Their Vectors

ROBERT F. WHITCOMB

*Plant Virus Vector Pioneering Research Laboratory,*
*Entomology Research Division,*
*Agricultural Research Service,*
*U.S. Department of Agriculture,*
*Beltsville, Maryland*

## I. INTRODUCTION

The bioassay of plant viruses borne persistently by their vectors is a subject which has received less attention than it deserves. In part, this has been due to the relatively small number of workers in the field. Also, most of these workers, including myself, have not been mathematically oriented. Finally, it happens that simple, qualitative, "common sense" interpretation of percentages or latent periods often suffices. However, it is inevitable that extensions of interpretative methodology are required as our experimental methodology becomes more complex. Thus, in recent years, various innovative analyses, mostly variants of the graded response type, have made it possible to answer questions which previously could not be asked. Recently, a fluorescent antibody tissue culture enumeration assay (Chiu, Reddy, and Black, 1966) was developed which should greatly enhance experimental possibilities, at least for viruses to which high-titer antisera can be prepared. Enumeration assays are not only more rapid than other methods, but their theory and analysis are relatively simple. Unfortunately, many workers will be forced to rely upon analysis of insect injection or membrane feeding experiments until tissue culture methods can be devised. This being the case, the most economical and theoretically sound interpretation of those assays is still of great interest. Details of their interpretation are stressed in this review with emphasis on present inadequacies.

## II. MECHANICAL INOCULATION TO PLANTS

Viruses which are closely restricted to phloem, such as wound tumor virus (WTV), may be inoculable by repeated pinwounding of crown tissue (Brakke, Vatter, and Black, 1954). Positive results from pinprick assays are infrequent and the method has not yet been applied to routine assay.

A quantal response assay for wheat streak mosaic virus was developed by Brakke (1958) which enabled an analysis of error to be made. His procedures could be used for most assays where the percentages of plants becoming infected are used for statistical analysis.

If a virus is capable of invading parenchyma, it may be assayed by counting the lesions produced by rubbing virus directly onto leaves (Holmes, 1929; Yarwood and Fulton, 1967). Persistently borne viruses for which local lesion enumeration assays have been found include potato yellow-dwarf virus (PYDV) (Black, 1938), tomato spotted wilt virus (TSWV) (Samuel and Bald, 1933), pea enation mosaic virus (PEMV) (Hagedorn, Layne, and Ruppel, 1964; Bozarth and Chow, 1965; Izadpanah and Shepherd, 1966), and lettuce necrotic yellows virus (LNYV) (Stubbs and Grogan, 1963; Crowley, 1967).

If local lesions are counted, a plethora of theoretical interpretations are available (Bald, 1937a, 1937b, 1937c; Youden, Beale, and Guthrie, 1935; Lauffer and Price, 1945; Kleczkowski, 1950; Furumoto and Mickey, 1967a, 1967b; Whitcomb, 1967). The diversity of interpretations need not prevent the pragmatic use of lesion numbers as follows: A dilution curve is constructed by inoculating dilutions of a known standard inoculum. The lesion counts from these dilutions are then plotted on graph paper on an appropriate scale, preferably one which linearizes the data. Lesion counts from samples of unknown concentration can then be converted to relative concentrations by referring to the standard curve. With mechanical inoculations, there is always danger of inhibition, as with insect extracts (Black, 1939), or enhancement, as with sucrose or normal plant components (Whitcomb and Sinha, 1964). Inhibition or enhancement can reach hundredfold proportions. Thus, the most careful controls are always necessary in experiments where virus concentration is to be measured.

## III. MECHANICAL TRANSMISSION TO VECTORS

Many viruses which are transmitted persistently by their vectors are transmissible by injection into their vector (Storey, 1933). In cases where subsequent virus multiplication ensures a high concentration of virus in the vector's body, injection is readily applicable.

For unknown reasons, some viruses, including PEMV (Schmidt, 1959; Nault, Gyrisco, and Rochow, 1964; Richardson and Sylvester, 1965), beet curly-top virus (CTV) (Maramorosch, 1955), and potato leafroll virus (PLRV) (Harrison, 1958; Stegwee and Ponsen, 1958), have proved to be poorly transmissible by injection. It is of interest that these three viruses have a number of characteristics in common—a very short latent period in the vector, small particle size, and relatively high stability, for example. Barley yellow-dwarf virus (BYDV), on the other hand, a virus with similar properties (Rochow and Brakke, 1964), has proved more amenable to injection. Even with BYDV, however, location of the virus zone by injection after density gradient centrifugation proved difficult. The location of WTV zones by injection, on the other hand (Brakke *et al.*, 1954) was possible. In general, agents which have a longer latent period—such as PYDV, WTV, the western X-disease agent, and the aster yellows disease agent —have proved readily amenable to assay by injection.

### A. *Techniques*

Adult insects, or more frequently, large nymphs, are used as test animals. They are stored in a moist chamber at a temperature slightly above freezing. Injection is most conveniently accomplished under a flow of moist carbon dioxide at room temperature under a dissecting microscope, after which the insects are returned to a plant, preferably immune. Much classical work has been done with needles pulled in gas flames (Storey, 1933; Black and Brakke, 1952; and many others), but many workers now use machine-pulled needles (Plus, 1954; Stegwee and Ponsen, 1958; Mueller and Rochow, 1961; Whitcomb, Jensen, and Richardson, 1966a, 1966b). The inocula are usually stored in an ice bath during injection. Small amounts of inoculum are drawn into the needle with a vacuum. After the insects are punctured abdominally with the needle, about 0.1 $\mu$liters of inoculum is expelled into the insect—preferably with compressed air, but with many workers, by blowing with the mouth. An "operating table" (Maramorosch and Jernberg, 1964) which holds a series of insects immobilized has the potential of greatly speeding the operation, but has not yet come into general use. Complex systems have been used for injection of aphids (Stegwee and Ponsen, 1958), but have not been widely adopted. Mortality of insects after injection may result from a number of causes (Jensen, Whitcomb, and Richardson, 1967), but microbial contamination of inocula and test insects is common (Whitcomb, Shapiro, and Richardson, 1966c).

### *B. Quantal Response*

With quantal assays an "all-or-none" response is recorded. Thus, after injection, an insect either transmits or fails to do so; it either dies from viral infection or lives to an age which approximates the normal life expectancy; it either develops a detectable antigenic titer or fails to do so. The fraction of insects responding in any chosen way is then used as a statistic, which is in some way related to the virus concentration employed in the injection. As in all bioassays, the *sine qua non* is the control dilution curve, which can be plotted on any scale, but preferably one which makes it approximately linear.

Brakke *et al.* (1954) plotted the logarithms of observed percentages of infected insects versus the logarithms of virus dilution, thereby obtaining a straight line in the range 0–50% of infected insects. It has turned out that this plot approximated the data much better than the plot

$$\log_{10} \log_{10} 1/Q = \log c$$

where $Q = 1 - P$, $P$ is the observed fraction of infected insects, and $c$ is virus concentration. This expression has theoretical merit (Plus, 1954) if the particles in the inoculum are evenly dispersed, if susceptibilities do not vary greatly from insect to insect, and if none of the background materials are inhibitory to the establishment of infection. The theoretical expectancy was, in fact, obtained with sigma virus of *Drosphila* (Plus, 1954). Analysis of published dilution curve data (Whitcomb, unpublished) has made it clear that conditions have not been found to date for WTV which approximate the theoretical expectation. Slopes of $\log_{10} \log_{10} 1/Q$ vs. concentration for WTV in crude extract were 0.48, 0.50, and 0.67 instead of 1.0. It is possible that reversible aggregation, variations in susceptibility among test insects, and interfering material, including inactive virus itself, may all have contributed to reduce the slope. Careful experiments with highly active, purified virus are needed to elucidate this issue.

Slopes with values other than one are not unique to assays of insect viruses. Assays of vertebrate viruses in whole animals frequently behave in this way, and the theoretical Poisson relationship is actually the exception rather than the rule (Dougherty, 1964). This has not prevented empirical analyses such as the Reed-Muench method (1938) or the Spearman-Karber method (Finney, 1952) from being widely and profitably used. For suggestions involving analysis of error, the reader is referred to the review of Dougherty, 1964, since the statistical analysis of a bioassay has not yet been applied to viruses borne persistently by their vectors.

### *C. Graded Response*

Whereas quantal response assays measure response as all or none, graded response assays measure the degree of response to the injected inocula. The *length of time* for transmission to begin, the *length of time* for viral mortality to ensue, and the *amount* of viral antigen present in infected insects are values which can be measured and plotted as a function of virus concentration. Viral mortality was used in both quantal and graded response assays of sigma virus (Plus, 1954; L'Heritier, 1958). It was less useful for assays of the western X-disease agent (Jensen *et al.*, 1967) in which incidental deaths obscured viral deaths. The most useful response has proved to be some variation of the "incubation," or "latent" period in the vector. This method was used by Maramorosch (1953), and later by Hirumi and Maramorosch (1963) in a semiquantitative way, to estimate amounts of the aster yellows agent in extracts.

An understanding of the graded response assays which depend on parameters of the transmission curve presupposes appreciation of the dynamics of viral infection of the insect. Assuming multiplication, following the injection of the infectious agent, there ensues a growth curve in the vector. This growth curve has an associated *transmission curve* which is similarly shaped in its ascending phase. Families of such curves are generated by the data from injection of dilutions (Fig. 1). The ascending phases of these curves probably correspond to a similar family of growth curves. In some insect virus systems the final plateau virus concentration depends upon the injected dose (Plus, 1954). With some plant viruses, higher transmission rates are recorded after inoculation with higher dosages, especially after longer feeding times (Kunkel, 1954; Day, 1955; Freitag, 1936). The details of this phenomenon should be thoroughly investigated since it seems to offer a parameter which might be useful for bioassay. Of no obvious value is the possibility of delayed or defective infection (reviewed by Fraser, 1967) which occurs after infection with high multiplicities of myxoviruses, and which may occur at high multiplicities of plant viruses in their vectors. This may be especially true of PYDV infection in its vectors, since its morphology resembles that of the myxoviruses and vesicular stomatitis groups (Black *et al.*, 1965; MacLeod, Black, and Moyer, 1966).

Any parameter of the transmission curve can be selected as a response statistic. For some purposes (Whitcomb, Jensen, and Richardson, 1966b) the time in days by which 50% of infected insects are transmitting ($T_{50}$) is suitable. For other purposes, $LP_{50}$ values (Sylvester, 1965) or average latent periods (Whitcomb, Jensen, and

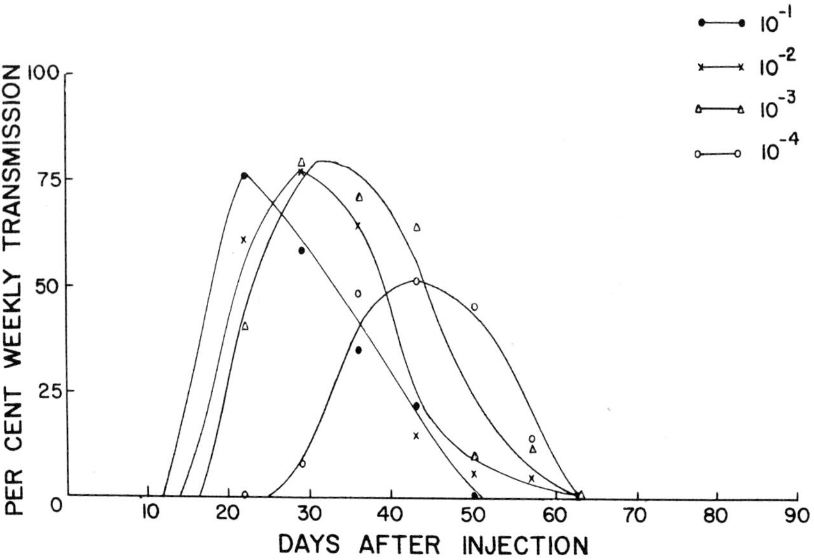


*Fig. 1.* Transmission of WXV in weekly inoculation access intervals on healthy celery following injection of serial tenfold dilutions of extract from infected vectors. Only insects which demonstrated inoculativity were used in the calculations. In some cases, the peak in inoculativity for $10^{-4}$ reached a level equal to that of other dilutions. In this case, it did not, making a $T_{50}$ analysis difficult. Variance in this peak makes an analysis such as $LP_{50}$ necessary.

Richardson, 1968) may be best. As an illustration of the graded response method, an array of average latent periods is presented in Figure 2. Of special note is the computer program of Sylvester and Richardson (1966), by which $LP_{50}$ values can be computed. A modification of such a program to permit computerized graded response assay is a desirable goal.

### *D. Total Latent Period*

With the aster yellows agent, we have recently begun using as a graded response statistic, the total latent period, $LP_T$, between injection and the appearance of symptoms in the plant:

$$LP_T = LP_I + LP_P$$

Latent periods in the plant, $LP_P$, are dosage sensitive (Kunkel, 1937; Duffus, 1963; etc.), as are latent periods in the insect, $LP_I$. However, the rise in inoculativity in the ascending phase of transmission curves is very steep, so that insects pass quickly through the period when virus dosages being inoculated into the plant are detectably different,

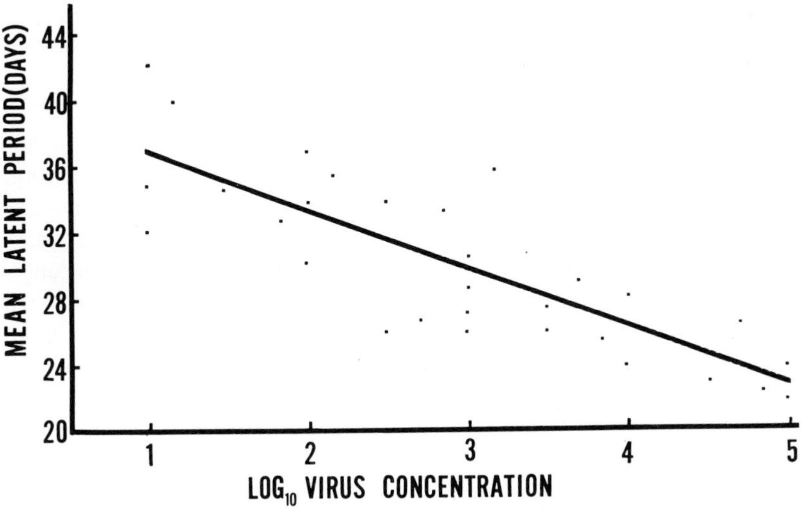


*Fig. 2.* Relationship between the mean latent period in the leafhopper and the relative virus concentration (based on tenfold dilution series). Test intervals were 3, 4, 5, or 6 days in duration. The median day of the test interval in which an insect first transmitted was used as an estimate of the WXV latent period. Points in the graph are means of latent periods of groups of insects, sample size varying from 4 to 54. The line was eye-fitted by inspection, but a rigorous analysis would require its calculation. Latent periods from insects injected with unknown concentrations of virus could be converted to relative virus concentrations by interpolation on this figure.

*i.e.*, the plant is quickly saturated with enough virus to produce the shortest possible response. (This results in the peculiar circumstance that "poor" virus–vector–host combinations, such as WXV–celery–*Colladonus montanus* (Van Duzee), in which transmission is relatively inefficient are the best for studying fluctuations in amounts of an agent actually transmitted.) Therefore, the latent period in the plant can be considered to be a variable with a constant mean, whereas the latent period in the insect varies with the injected dose. The variance of $LP_P$ will vary from system to system, but can reasonably be expected to increase with increasing mean. Therefore, the method should be best when $LP_P$ is short. It is also evident, that when $LP_I$ is very short (1–2 days with PLRV), differences due to the initial dosage would be obscured by the variability of $LP_P$ (more than 2 weeks in Chinese cabbage). Thus, the method would be expected to work best for viruses with latent periods of 8 or more days in their vectors. It can be seen that the aster yellows agent ($LP_I = 12+$ days; $LP_P = 6-18$

days) is ideally suited to this type of assay. There are, however, a number of other systems to which the method should be applicable.

### *E. Combined Quantal-Graded Response Assay*

If graded response data have been recorded for groups of injected test insects, and for a corresponding control curve, the percentage data will also be available from the same records. With sigma virus (Plus, 1954; L'Heritier, 1958) and with the western X-disease agent, it has been possible to use a quantal interpretation if the fraction of infected insects falls between 0 and 90%, and the graded response interpretation if the fraction is above 90%. This procedure has great theoretical merit, for examination of the Poisson probabilities for "hits" of infectious units indicates that there is a sizable range covering the lower percentages (about 0–40%), where most infections would result from a single particle, and should initiate the same $LP_I$. Thus, graded response interpretation is theoretically inefficient in that range.

An ingenious method (Reddy and Black, 1966) permitted the combination of quantal and graded response in a single analysis. The antigen titer 21 days after injection of groups of insects was measured and plotted as a function of virus concentration. Because the injection of diluted inocula resulted in less than 100% infection, the antigen titer depended not only on the time after injection, but also on the fraction of insects infected. This would be expected to magnify the difference in antigen concentration between treatments, and it is not surprising that the method proved to be more useful than previous bioassays for WTV.

## IV. MEMBRANE FEEDING TECHNIQUES

The infection of vectors by feeding on solutions through membranes (Carter, 1927, 1928) has been attempted for many viruses. The technique has proved most useful for the study of stable viruses such as CTV (Severin and Swezy, 1928; Severin and Freitag, 1933), BYDV (Rochow, 1960) and beet western yellows virus (Duffus and Gold, 1965). The technique was valuable (Gold and Duffus, 1967; Rochow and Ball, 1967) for a demonstration of the neutralization of infectivity by antiviral serum.

Two major drawbacks have prevented the widespread use of membrane techniques: (*1*) the instability of many viruses in feeding solutions and (*2*) variable uptake by the test insects, with the resultant variability in level of infection. Uptake of feeding solution has been measured by incorporation of radioactive substances in the medium (Hamilton, 1935; Day and McKinnon, 1951; Day and Irzykiewicz, 1953; Duffus and Gold, 1967). A number of factors including color and the chemical composition of the medium influence the rate of uptake. Materials which inhibit feeding may be difficult to fractionate from virus. Such materials, if present in the unknown samples, should also be present in the dilutions of standard inoculum. This will not always be easy to achieve. As a result of the above difficulties, workers have not attempted to construct dosage response curves. It is probably relevant to point out that injection of virus, while possessing various drawbacks, has not been shown to be greatly affected by "background materials"; volumes normally introduced do not vary greatly (Black and Brakke, 1952); volume can be controlled with even greater precision, if desired (Maramorosch, 1951); and the introduction of infectious agents in the proximity of susceptible sites is nearly instantaneous, so that very labile disease agents may be assayed. Ideally, a worker would wish to have both methods available (as with BYDV), to suit particular purposes.

## V. TISSUE CULTURE ENUMERATION ASSAY

It is now possible (Chiu and Black, 1966) to infect leafhopper cells in tissue culture with plant disease agents. This discovery culminated a series of progressive advances, largely on the part of two research groups. Maintenance of leafhopper tissues in media was accompanied by completion of the latent period of the aster yellows agent (Maramorosch, 1956). Recent results which suggest that the aster yellows agent might be a large, complex agent of the *Mycoplasma* or psittacosis type (Ishiie *et al.*, 1967; Doi *et al.*, 1967) enhance the interest in this observation. Ornithosis virus was shown (Crocker and Eastwood, 1963) to be capable of multiplication in nonnucleate cytoplasmic fragments; and it may well be that agents of high complexity require much less than total cellular function for multiplication. Later, dissected organs were preserved in media with the retention of the aster yellows agent (Hirumi and Maramorosch, 1963), or with demonstration of virus antigens (Sinha, 1965). Cultivation of leafhopper cells *in vitro* was then accomplished (Vago and Flandre, 1963; Mitsuhashi and Maramorosch, 1964; Hirumi and Maramorosch, 1964a, 1964b).

Tissue cultures were obtained by Chiu *et al.* (1966) from embryonic tissue sampled 6–7 days after oviposition (Hirumi and Maramorosch, 1964a). The dissected, trypsinized embryonic tissue was maintained on coverslips in the culture media of Mitsuhashi and Maramorosch

(1964), supplemented with serum and antibiotics. Ten-day-old cultures were inoculated with filtered WTV preparations. Multiplication of virus was evidenced by (*1*) specific staining with fluorescein labeled antibodies; (*2*) an increase in relative infectivity in the culture medium as measured by the assay of Reddy and Black (1966) described earlier in this review; and (*3*) the retention of viral titer through 7 serial passages. Careful experiments of this sort were in order since two earlier attempts to inoculate cells with viruses (aster yellows agent; Maramorosch *et al.*, 1965; rice dwarf virus; Mitsuhashi, 1965) were not accompanied with any evidence of viral multiplication. Because WTV produced no cytopathic effects on cells, it was necessary to use fluorescent antibody cell counts for this assay. Statistical aspects of this type of assay have been discussed in the review of Dougherty (1964). Frequently such tissue culture enumeration assays are found to follow simple proportionality, with variances closely approximating the theoretical Poisson values. It is to be hoped that in the near future, the exact statistical details will be available.

## VI. ABSOLUTE CONCENTRATION OF VIRIONS

All methods described in previous sections yield relative estimates of concentration. It has been possible with one virus (WTV) to determine actual numbers of particles. Gamez and Black (1967), employing conventional electron microscope particle counting techniques, estimated the minimum infective dose for successful injection of a leafhopper as $10^{2.6}$ virus particles. This value compared with a minimum infective dose of between $10^{4.4}$ and $10^{4.6}$ for inoculation of cells in tissue culture. Estimates of particle numbers in root tumors, in leafhoppers, fractions at various stages of purification, and at the serological endpoint were also made. It was possible to follow the fluctuation in particle numbers after infection of the leafhopper (Gamez and Black, 1968), and it was determined that at the sixth, thirtieth, and fortieth days, the average numbers of extractable virions were $10^{6.6}$, $10^{9.26}$, and $10^{8.7}$, respectively.

Development of a new $H_2O$–$D_2O$ gradient sedimentation technique (Strohmaier, 1967) for numbers of particles in crude preparations enabled Streissle *et al.* (1968) to obtain estimates of particle numbers in leafhoppers and tumors. Their counts for particle number were considerably lower than those of Gamez and Black (1967). Although they concluded that their lower counts probably reflected lower virus titer in original tissue, it is also possible that procedural differences accounted, in part, for the discrepancy. The requirement for centrifugation of virus in $D_2O$–$H_2O$ gradients without protective ions, plus the possible elution of virus from grids after sedimentation, are procedural differences which could account for some loss of particles.

## BIBLIOGRAPHY


Bald, J. G. 1937a. The use of numbers of infections for comparing the concentration of plant virus suspensions. I. Dilution experiments with purified suspensions. Ann. Appl. Biol. 24:33–55.

——— 1937b. The use of numbers of infections for comparing the concentration of plant virus suspensions. II. Distortion of the dilution series. Ann. Appl. Biol. 24:56–76.

——— 1937c. The use of numbers of lesions for comparing the concentration of plant virus suspensions. IV. Modification of the sample dilution equation. Australian J. Exp. Biol. Med. Sci. 15:211–220.

Black, L. M. 1938. Properties of the potato yellow-dwarf virus. Phytopathology 28:863–874.

—— 1939. Inhibition of virus activity by insect juices. Phytopathology 29:321–327.

Black, L. M., and Brakke, M. K. 1952. Multiplication of wound-tumor virus in an insect vector. Phytopathology 42:269–273.

Black, L. M., Smith, K. M., Hills, G. J., and Markham, R. 1965. Ultrastructure of potato yellow-dwarf virus. Virology 27:446–449.

Bozarth, R. F., and Chow, C. C. 1965. Pea enation mosaic virus: a new local lesion host. Phytopathology 55:492–493.

Brakke, M. K. 1958. Properties, assay, and purification of wheat streak mosaic virus. Phytopathology 48:439–445.

Brakke, M. K., Vatter, A. E., and Black, L. M. 1954. Size and shape of wound-tumor virus. Abnormal and Pathological Plant Growth. Brookhaven Symposia in Biology 6:137–156.

Carter, W. 1927. A technique for use with homopterous vectors of plant diseases, with special reference to the sugar-beet leafhopper, *Eutettix tenellus* (Baker). J. Agr. Res. 34:449–451.

——— 1928. An improvement in the technique for feeding homopterous insects. Phytopathology 18:246–247.

Chiu, R., Reddy, D. V. R., and Black, L. M. 1966. Inoculation and infection of leafhopper tissue cultures with a plant virus. Virology 30:562–566.

Crocker, T. T., and Eastwood, J. M. 1963. Subcellular cultivation of a virus: growth of ornithosis virus in nonnucleate cytoplasm. Virology 19:23–31.

Crowley, N. C. 1967. Factors affecting the local lesion response of *Nicotiana glutinosa* to lettuce necrotic yellows virus. Virology 31:107–113.

Day, M. F. 1955. The mechanism of the transmission of potato leaf roll virus by aphids. Australian J. Biol. Sci. 8:498–513.

Day, M. F., and Irzykiewicz, H. 1953. Feeding behaviour of the Aphides *Myzus persicae* and *Brevicoryne brassicae,* studied with radio-phosphorus. Australian J. Biol. Sci. 6:98–108.

Day, M. F., and McKinnon, A. 1951. A study of some aspects of the feeding of the Jassid *Orosius*. Australian J. Sci. Res. Ser. B. 4:125–135.

Doi, Y., Teranaka, M., Yora, K., and Asuyama, H. 1967. *Mycoplasma*- or PLT group-like microorganisms found in the phloem elements of plants infected with mulberry dwarf, potato witches' broom, aster yellows, or paulownia witches' broom. Ann. Phytopathol. Soc. Japan 33:259–266.

Dougherty, R. M. 1964. Animal virus titration techniques. *In* J. C. Harris [ed.] Techniques in Experimental Virology, Academic Press, New York. p. 169–223.

Duffus, J. E. 1963. Possible multiplication in the aphid vector of sowthistle yellow vein virus, a virus with an extremely long insect latent period. Virology 21: 194–202.

Duffus, J. E., and Gold, A. H. 1965. Transmission of beet western yellows virus by aphids feeding through a membrane. Virology 27:388–390.

——— 1967. Relationship of tracer-measured aphid feeding to acquisition of beet western yellows virus and to feeding inhibitors in plant extracts. Phytopathology 57:1237–1241.

Finney, D. J. 1952. Statistical methods in biological assay. Griffin, London.

Fraser, K. B. 1967. Defective and delayed myxovirus infections. Brit. Med. Bull. 23:178–184.

Freitag, J. H. 1936. Negative evidence on multiplication of curly-top virus in the beet leafhopper, *Eutettix tenellus*. Hilgardia 10:305–342.

Furumoto, W. A., and Mickey, M. R. 1967a. A mathematical model for the infectivity–dilution curve of tobacco mosaic virus: experimental tests. Virology 32:224–233.

——— 1967b. A mathematical model for the infectivity-dilution curve of tobacco mosaic virus: Theoretical considerations. Virology 32:216–223.

Gamez, R., and Black, L. M. 1967. Application of particle-counting to a leafhopper-borne virus. Nature 215:173–174.

——— 1968. Particle counts of wound tumor virus during its peak concentration in leafhoppers. Virology 34:444–451.

Gold, A. H., and Duffus, J. E. 1967. Infectivity neutralization—a serological method as applied to persistent viruses of beets. Virology 21:308–313.

Hagedorn, D. J., Layne, R. E. C., and Ruppel, E. G. 1964. Host range of pea enation mosaic virus and use of *Chenopodium album* as a local-lesion host. Phytopathology 54:832–848.

Hamilton, M. H. 1935. Further experiments on the artificial feeding of *Myzus persicae* (Sulz.). Ann. Appl. Biol. 22:243–258.

Harrison, B. D. 1958. Studies on the behavior of potato leaf roll and other viruses in the body of their aphid vector *Myzus persicae* (Sulz.). Virology 6:265–277.

Hirumi, H., and Maramorosch, K. 1963. Recovery of aster yellows virus from various organs of the insect vector, *Macrosteles fascifrons*. Contrib. Boyce Thompson Inst. 22:141–151.

Hirumi, H., and Maramorosch, K. 1964a. Insect tissue culture; use of blastokinetic stage of leafhopper embryo. *Science* 144:1465–1467.

——— 1964b. The *in vitro* cultivation of embryonic leafhopper tissues. Exp. Cell Res. 36:625–631.

Holmes, F. O. 1929. Local lesions in tobacco mosaic virus. Bot. Gaz. 87:39–55.

Ishiie, T., Doi, Y., Yora, K., and Asuyama, H. 1967. Suppressive effects of antibiotics of tetracycline group on symptom development of mulberry dwarf disease. Ann. Phytopathol. Soc. Japan 33:267–275.

Izadpanah, K., and Shepherd, R. J. 1966. *Galactia* sp. as a local lesion host for the pea enation mosaic virus. Phytopathology 56:458–459.

Jensen, D. D., Whitcomb, R. F., and Richardson, J. 1967. Lethality of injected peach western X-disease virus to its leafhopper vector. Virology 31:532–538.

Kleczkowski, A. 1950. Interpretating relationships between the concentrations of plant viruses and numbers of local lesions. J. Gen. Microbiol. 4:53–69.

Kunkel, L. O. 1937. Effect of heat on ability of *Cicadula sexnotata* (Fall.) to transmit aster yellows. Amer. J. Bot. 24:316–327.

——— 1954. Maintenance of yellows-type viruses in plant and insect reservoirs. p. 150–163. *In* F. W. Hartman, F. L. Horsfall, and J. G. Kidd [ed.] The Dynamics of virus and rickettsial infections. Blakiston, New York.

Lauffer, M. A., and Price, W. C. (1945). Infection by viruses. Arch. Biochem. 8: 449–468.

L'Heritier, P. 1958. The hereditary virus of *Drosophila.* Advances Virus Res. 5: 195–245.

MacLeod, R., Black, L. M., and Moyer, F. H. 1966. The fine structure and intracellular localization of potato yellow dwarf virus. Virology 29:540–552.

Maramorosch, K. 1953. Incubation period of aster-yellows virus. Amer. J. Bot. 40: 797–809.

——— 1953. Incubation period of aster-yellows virus. Amer. J. Bot. 40:797–809.

——— 1955. Mechanical transmission of curly top virus to its insect vector by needle inoculation. Virology 1:286–300.

——— 1956a. Multiplication of aster yellows virus in *in vitro* preparations of insect tissues. Virology 2:369–376.

——— 1956b. Semiautomatic equipment for injecting insects with measured amounts of liquids containing viruses or toxic substances. Phytopathology 46:188–190.

Maramorosch, K., and Jernberg, N. 1964. A device for rapid inoculation of arthropods. Bull. Entomol. Soc. Amer. 10:171.

Maramorosch, K., Mitsuhashi, J., Streissle, G., and Hirumi, H. 1965. Bacteriol. Proc., 1965, Abstr. 65th Ann. Meeting Amer. Soc. Microbiol., p. 120.

Mitsuhashi, J. 1965. Preliminary report on the plant virus multiplication in the leafhopper vector cells grown *in vitro.* Japan J. Appl. Entomol. Zool. 9:137–141.

Mitsuhashi, J., and Maramorosch, K. 1964. Leafhopper tissue culture: embryonic, nymphal, and imaginal tissues from aseptic insects. Contrib. Boyce Thompson Inst. 22:435–460.

Mueller, W. C., and Rochow, W. F. 1961. An aphid-injection method for the transmission of barley yellow dwarf virus. Virology 14:253–258.

Nault, L. R., Gyrisco, G. G., and Rochow, W. F. 1964. Biological relationship between pea enation mosaic virus and its vector, the pea aphid. Phytopathology 54:1269–1272.

Plus, N. 1954. Étude de la multiplication du virus de la sensibilité au gaz carbonique chez la drosophile. Bull. Biol. France Belg. 88:248–293.

Reddy, D. V. R., and Black, L. M. 1966. Production of wound-tumor virus and wound-tumor soluble antigen in the insect vector. Virology 30:551–561.

Reed, L. J., and Muench, H. 1938. A simple method of estimating fifty percent endpoints. Amer. J. Hyg. 27:493–497.

Richardson, J., and Sylvester, E. S. 1965. Aphid honeydew as inoculum for the injection of pea aphids with pea-enation mosaic virus. Virology 25:472–475.

Rochow, W. F. 1960. Transmission of barley yellow dwarf virus acquired from liquid extracts by aphids feeding through membranes. Virology 12:223–232.

Rochow, W. F., and Ball, M. 1967. Serological blocking of aphid transmission of barley yellow dwarf virus. Virology 33:359–362.

Rochow, W. F., and Brakke, M. K. 1964. Purification of barley yellow dwarf virus. Virology 24:310–322.

Samuel, G., and Bald, J. G. 1933. On the use of the primary lesions in quantitative work with two plant viruses. Ann. Appl. Biol. 20:70–99.

Schmidt, H. B. 1959. Beiträge zur kenntnis der übertragung pflanzlicher viren durch aphiden. Biol. Zentralbl. 78:889–936.

Severin, H. H. P., and Freitag, J. H. 1933. Some properties of the curly-top virus. Hilgardia 8:1–48.

Severin, H. H. P., and Swezy, O. 1928. Filtration experiments on curly top of sugar beets. Phytopathology 18:681–690.

Sinha, R. C. 1965. Sequential infection and distribution of wound-tumor virus in the internal organs of a vector after ingestion of virus. Virology 26:673–686.

Stegwee, D., and Ponsen, M. B. 1958. Multiplication of potato leafroll virus in the aphid *Myzus persicae*. Entomol. Exp. Appl. 1:291–300.

Storey, H. H. 1933. Investigations of the mechanism of the transmission of plant viruses by insect vectors. I. Proc. Roy. Soc. (London) B113:463–485.

Streissle, G., Bystricky, V., Granados, R. R., and Strohmaier, K. 1968. Number of virus particles in insects and plants infected with wound tumor virus J. Virol. 2:214–217.

Strohmaier, K. 1967. A new procedure for quantitative measurements of virus particles in crude preparations. J. Virology 1:1074–1081.

Stubbs, L. L., and Grogan, R. G. 1963. Necrotic yellows: a newly recognized virus disease of lettuce. Australian J. Agr. Res. 14:439–459.

Sylvester, E. S. 1965. The latent period of pea-enation mosaic virus in the pea aphid, *Acyrthosiphon pisum* (Harris)—an approach to its estimation. Virology 25:62–67.

Sylvester, E. S., and Richardson, J. 1966. "Recharging" pea aphids with pea enation mosaic virus. Virology 30:592–597.

Vago, C., and Flandre, O. 1963. Culture prolongée de tissus d'insects et de vecteurs de maladies en coagulum plasmatique. Ann. Epiphyties 14:127–139.

Whitcomb, R. F. 1967. Infectivity dilution curve of a leafhopper-borne virus on plant leaves. Phytopathology 57:836. (Abstr.)

Whitcomb, R. F., Jensen, D. D., and Richardson, J. 1966a. The infection of leafhoppers by western X-disease virus. I. Frequency of transmission after injection or acquisition feeding. Virology 28:448–453.

——— 1966b. The infection of leafhoppers by western X-disease virus. II. Fluctuation of virus concentration in the hemolymph after injection. Virology 28:454–458.

——— 1968. The infection of leafhoppers by western X-disease virus. V. Properties of the infectious agent. J. Invert. Pathol. (in press).

Whitcomb, R. F., Shapiro, M., and Richardson, J. 1966c. An *Erwinia*-like bacterium pathogenic to leafhoppers. J. Invert. Pathol. 8:229–307.

Whitcomb, R. F., and Sinha, R. C. 1964. Effect of host components and sucrose on infection by potato yellow dwarf virus. Phytopathology 54:142–146.

Yarwood, C. E., and Fulton, R. W. 1967. Mechanical transmission of plant viruses. Methods Virol. 1:237–266.

Youden, W. J., Beale, H. P., and Guthrie, J. C. 1935. Relation of virus concentration to the number of lesions produced. Contrib. Boyce Thompson Inst. 7: 37–53.

# Hemagglutination of Leafhopper-Borne Viruses

YASUO SAITO

*Institute for Plant Virus Research,*
*959 Aobacho, Chiba, Japan*

Sheep erythrocytes, when treated with a dilute solution of tannic acid, acquire the property of adsorbing protein. Such protein-coated red blood cells are agglutinated by specific antiserum directed against the protein used for adsorption. This procedure is called an indirect or a passive hemagglutination test.

Boyden (1951), for the first time, showed the adsorption of protein of *Mycobacterium tuberculosis* onto the surface of red blood cells which had been altered by tannic acid. Subsequently, this phenomenon has been studied by many investigators. The proteins proved to be adsorbed on tanned cells were crystalline ovalbumin, conalbumin, gamma globulin, human serum albumin, insulin, tuberculin proteins, diphtheria toxins and toxoid, various clostridal toxins, other bacterial proteins, and deoxyribonucleic proteins, etc. This technique also has been used to titrate the sera from patients of herpes simplex virus, adeno virus, and polio virus, etc. However, the titers obtained were relatively low and the reactions were rather labile. Therefore, this hemagglutination technique was not preferable to other serological methods in the field of animal virus research.

This technique was introduced into the field of plant virus research and modified by Saito (1961), Saito and Iwata (1964), and Saito, Takanashi, and Iwata (1964). The hemagglutination reaction is often more sensitive for the titration of antiplant virus antibodies than the other serological reactions. The method facilitates the detection of viruses in plant tissues or in viruliferous leafhoppers by the hemagglutination-inhibition test or by the hemagglutination test with antibody-sensitized cells.

## I. THE PRINCIPLE

By electron microscopic examination, virus particles were found attached to sensitized red blood cells. The amorphous material, which was thought to be antiserum covering the surface of virus particle, was found between two red blood cells. Thus, the virus particles, attached to the cells combining with the antiserum, constitute the connecting links between the agglutinated cells (Fig. 1).

In other serological reactions, such as the precipitation and gel-diffusion reactions, particles of the antigen are linked together by the antibody and these aggregates grow larger and larger until a visible precipitate is produced. This visible precipitate contains large numbers of virus particles. In a passive hemagglutination test, however, the antigen is adsorbed on the red blood cell. Since the relative proportion of a red blood cell to the virus particle is very large, it requires very small quantities of antigens and antisera to form a visible agglutination. This is the reason for the extreme sensitivity of this hemagglutination reaction. The red color of the erythrocytes is helpful in reading the results of the reaction.

In hemagglutination tests, barley stripe mosaic virus (BSMV) reacted specifically with anti-BSMV serum but did not react with antisera to any other plant viruses, plant-pathogenic bacteria, plant proteins, or egg albumin (Table 1). Rabbits were injected intravenously with purified BSMV at weekly intervals for 6 weeks, and the increase in antibody response was followed on sera sampled at each injection time. The results showed that the titer increased continuously, the hemagglutination titer being always about 10,000 times higher than complement fixation (CF) titer (Fig. 2). These facts seem to indicate

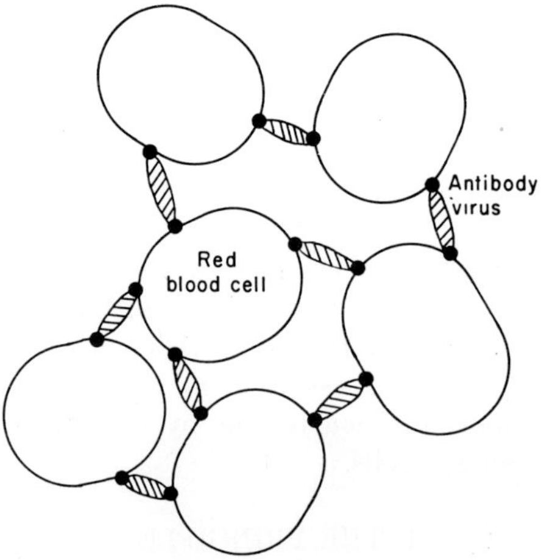


*Fig. 1.* A diagrammatic representation of hemagglutination with antibody-sensitized red blood cells.

TABLE 1

THE REACTION OF BSMV-SENSITIZED RED BLOOD CELLS WITH HOMOLOGOUS AND HETEROLOGOUS ANTISERA

| Antiserum | Reciprocal of highest dilution: With homologous antigen (precipitation) | Reciprocal of highest dilution: With BSMV-sensitized cells (hemagglutination test) |
|---|---|---|
| Wheat green-mosaic virus | 2000 | 0 |
| Cucumber mosaic virus | 200 | 0 |
| Turnip mosaic virus | 5000 | 0 |
| *Pseudomonas marginalis* | 25,000 | 0 |
| *Erwinia aroideae* | — | 0 |
| *Erwinia carotovora* | — | 0 |
| *Xanthomonas oryzae* | 6400 | 0 |
| Barley protein | 32 | 0 |
| Wheat protein | 32 | 0 |
| Egg albumin | — | 0 |
| Barley stripe mosaic virus | 1280 | 13,107,200 |

that the hemagglutination is primarily due to the combination of the virus with its homologous antiserum.

## II. THE PROCEDURE

### *A. Hemagglutination Test (HA) with Antigen-Sensitized Cells*

One volume of 3.5% sheep red blood cell suspension was mixed with an equal volume of 0.005% tannic acid saline, and kept in a 37°C water bath for 10 min. The mixture was centrifuged. The sedimented cells were washed with 1 volume of Veronal-buffered saline, at *p*H 7.2, and resuspended in 1 volume of Veronal-buffered saline. One volume of the 3.5% tanned cell suspension prepared by this procedure was mixed with 4 volumes of 0.1 M phosphate-buffered saline, at *p*H 5.0, and 1 volume of 0.05% purified virus suspension. This mixture was kept for 15 min at room temperature and then centrifuged. The sediment was washed and resuspended in 1 volume of serum diluent (0.5% N rabbit serum in Veronal-buffered saline). Twofold serial dilutions of the antiserum were made in 0.5-ml amounts in Veronal-buffered

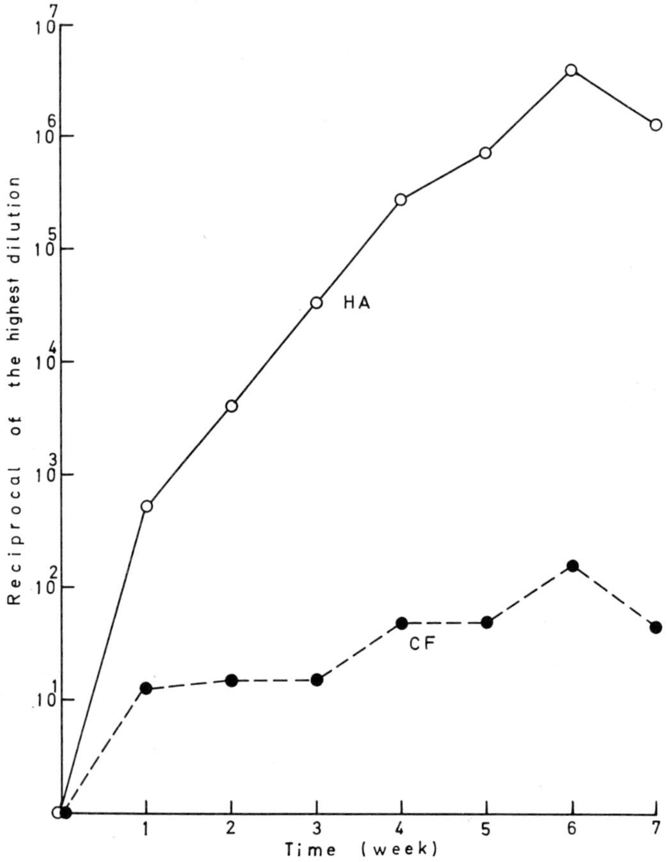


*Fig. 2.* Hemagglutination titers and CF titers of the sera sampled at each injection time. The rabbit was injected intravenously with purified BSMV at weekly intervals for 6 weeks.

saline, at *p*H 7.2, in $1 \times 10$-cm test tubes, and 0.05 ml of the tanned cells were sensitized with virus were added to each.

Readings were made after incubation for 3 hr at room temperature, or for one night at 5°C. A positive agglutination pattern consisted of a relatively uniform thin layer of red blood cells covering the bottom of the tube, whereas a negative pattern consisted of a more compact sedimentation of cells, sometimes forming a ring, at the bottom of the tube (Fig. 3).

Influences of several factors on this hemagglutination were studied, and the most important factor was found to be the *p*H value of diluent in which the tanned cells were sensitized with the virus. In the case

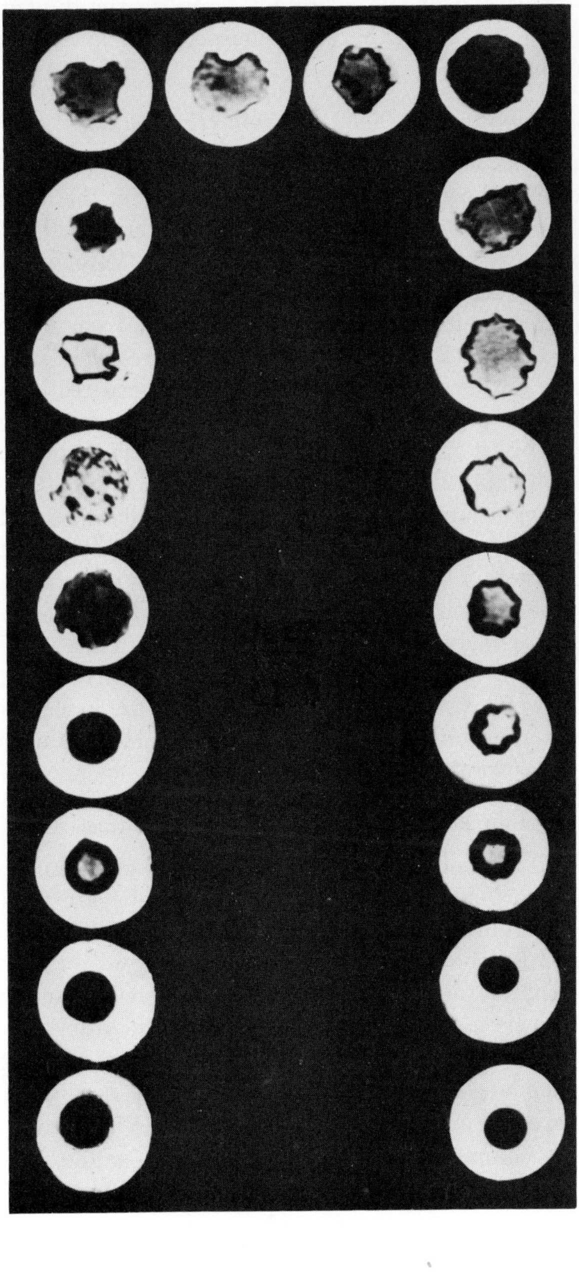

*Fig. 3.* Hemagglutination test with anti-BSMV serum and BSMV-sensitized sheep red blood cells. Serial twofold dilutions of the antiserum were made, the dilution of the first tube (upper left) being 1:100. The last tube (lower left) served as the control.

of BSMV, the optimum *p*H value was 5.0, and the optimum concentrations of tanned cells and of normal rabbit serum in the diluent were 3.5 and 0.5%, respectively. Otherwise, nonspecific hemagglutination or hemolysis might occur.

### *B. Hemagglutination-Inhibition (HAI) Test*

This technique has been devised to determine virus quantitatively with antigen-sensitized cells. There are two methods for the hemagglutination-inhibition test. One of these is as follows: serial twofold dilutions of test antigen were made in 0.2-ml amounts in Veronal-buffered saline, and 0.5-ml of serum, suitably diluted in Veronal-buffered saline to contain four hemagglutination doses (calculated from hemagglutination titration), was added to each. These mixtures were incubated for one night at 5°C. Then 0.05 ml of antigen-sensitized red blood cell suspension was added to each mixture. The antigen titer was expressed as the reciprocal of the highest dilution of the antigen showing no hemagglutination (Table 2).

The other hemagglutination-inhibition test is as follows: serial twofold dilutions of antiserum were made in 0.5-ml amounts in Veronal-buffered saline. Another twofold dilution was set up as a control. Suspensions of test antigen in 0.1-ml amounts were added to each tube. These mixtures were incubated for one night at 5°C. Then 0.05 ml of antigen-sensitized red blood cell suspension was added to each tube. The antigen titer was expressed in terms of the decrease in the reciprocal of the highest dilution of the antiserum giving hemagglutination, compared with the control titer (Table 3).

### *C. Hemagglutination Test with Antibody-Sensitized Cells*

In the rabbit, all antibodies are contained in the $\gamma_2$ globulin fraction. If $\gamma_2$ globulin is purified, and tanned red blood cells are sensitized with it, the cells agglutinate in the presence of virus. This reaction also could be used for the detection of specific virus and for quantitative determinations of the virus. The $\gamma$-globulin fraction or globulin fraction could also be used in place of the $\gamma_2$-globulin fraction.

The procedure is the same as that of hemagglutination with BSMV-sensitized red blood cells, except that the cells are sensitized with globulin in place of virus. The simplified method (Yanagita, 1964b) is as follows: one volume of 3.5% red blood cells in saline, 1 volume of 0.05% antiserum globulin in phosphate buffered saline, at *p*H 5.0, and 4 volumes of 0.005% tannic acid saline were prepared and these three solutions were mixed at the same time, and were kept at 37°C for 10 min. Then the mixture was centrifuged. The sedimented cells

TABLE 2

TITRATION OF VIRUS BY USE OF THE HAI TEST
(4 UNITS ANTIBODY METHOD)

| Concentration of antigen | Antigen dilution $2^1$ | $2^2$ | $2^3$ | $2^4$ | $2^5$ | $2^6$ | $2^7$ | $2^8$ | $2^9$ | $2^{10}$ | $2^{11}$ | $2^{12}$ | $2^{13}$ | $2^{14}$ | $2^{15}$ | $2^{16}$ | Saline |
|---|---|---|---|---|---|---|---|---|---|---|---|---|---|---|---|---|---|
| 1:100 | — | — | — | — | — | — | — | — | ++ | +++ | +++ | +++ | +++ | +++ | +++ | +++ | +++ |
| 1:1000 | — | — | — | — | ++ | ++ | +++ | +++ | +++ | +++ | +++ | +++ | +++ | +++ | +++ | +++ | +++ |
| 1:10,000 | — | ++ | +++ | +++ | +++ | +++ | +++ | +++ | +++ | +++ | +++ | +++ | +++ | +++ | +++ | +++ | +++ |
| 1:100,000 | +++ | +++ | +++ | +++ | +++ | +++ | +++ | +++ | +++ | +++ | +++ | +++ | +++ | +++ | +++ | +++ | +++ |
| 1:1,000,000 | +++ | +++ | +++ | +++ | +++ | +++ | +++ | +++ | +++ | +++ | +++ | +++ | +++ | +++ | +++ | +++ | +++ |
| Control | +++ | +++ | +++ | +++ | +++ | +++ | +++ | +++ | +++ | +++ | +++ | +++ | +++ | +++ | +++ | +++ | +++ |

TABLE 3

TITRATION OF VIRUS BY USE OF THE HAI TEST
(ANTIBODY DILUTION METHOD)

| Concentration of antigen | Antibody dilutim × 2500 | | | | | | | | | | | | | | | | |
|---|---|---|---|---|---|---|---|---|---|---|---|---|---|---|---|---|---|
| | $2^1$ | $2^2$ | $2^3$ | $2^4$ | $2^5$ | $2^6$ | $2^7$ | $2^8$ | $2^9$ | $2^{10}$ | $2^{11}$ | $2^{12}$ | $2^{13}$ | $2^{14}$ | $2^{15}$ | $2^{16}$ | Saline |
| 1:100 | — | — | — | — | — | — | — | — | — | — | — | — | — | — | — | — | — |
| 1:1,000 | +++ | +++ | +++ | +++ | +++ | +++ | — | — | — | — | — | — | — | — | — | — | — |
| 1:10,000 | +++ | +++ | +++ | +++ | +++ | +++ | +++ | ++ | — | — | — | — | — | — | — | — | — |
| 1:100,000 | +++ | +++ | +++ | +++ | +++ | +++ | +++ | +++ | ++ | + | — | — | — | — | — | — | — |
| 1:1,000,000 | +++ | +++ | +++ | +++ | +++ | +++ | +++ | +++ | ++ | + | — | — | — | — | — | — | — |
| Control | +++ | +++ | +++ | +++ | +++ | +++ | +++ | +++ | +++ | +++ | ++ | ++ | — | — | — | — | — |

were washed with one volume of Veronal-buffered saline, at *p*H 7.2, and resuspended in one volume of serum diluent. The globulin-sensitized tanned cells were used for the test.

### *D. Detection of Viruliferous Leafhoppers by the Hemagglutination Test*

In the case of rice stripe virus, one leafhopper was placed at the bottom of a 1×10-cm test tube. An aliquot of diluent was added to the test tube, and the leafhopper was macerated using a glass rod (3–4 mm in diameter, 16–17 cm in length). The mixture was centrifuged. Twofold serial dilutions of the supernatant fluid were made in 0.5-ml amounts in Veronal-buffered saline, at *p*H 7.2, in test tubes, and 0.05 ml of the tanned cells, sensitized with antiserum globulin, were added to each. If the only purpose of the test is to detect whether the leafhopper is viruliferous or not, this process of serial dilutions might be ommitted. In this case, a leafhopper was macerated with a glass rod and then 0.5 ml of the sensitized cell suspension, diluted with tenfold diluents, was added directly to the test tube (Yasuo and Yanagita, 1963b; Yanagita, 1964a).

## III. THE SENSITIVITY OF THE PASSIVE HEMAGGLUTINATION TEST

With anti-BSMV serum, the hemagglutination titer was 1:13,107,000, while the CF titer was 1:320. Several other plant viruses also reacted with homologous immune sera specifically, and with high sensitivity (Table 4).

The sensitivity of agglutination techniques with erythrocytes coated

TABLE 4

COMPARISON OF COMPLEMENT FIXTION AND HEMAGGLUTINATION TITERS WITH THE SAME SERA

| Virus | Reciprocals of antisera dilution endpoints | |
|---|---|---|
| | CF | HA |
| Barley stripe mosaic virus | 320 | 13,107,200 |
| Wheat mosaic-rosette virus | 160 | 200,000 |
| Rice dwarf virus | 1600 | 1,600,000 |
| Rice stripe disease virus | 320 | 50,000 |

with antibody is very high, too. Serial dilutions of BSMV in 0.5-ml amounts in Veronal-buffered saline, at *p*H 7.2, were made, the amount of virus in the first tube being 100 $\mu$g. After the addition of the 0.05 ml of sensitized cell suspension, the thirteenth tube was the highest dilution of antigen showing hemagglutination, indicating that about $10^{-8}$ g of virus was detectable.

## IV. APPLICATION OF THE PASSIVE HEMAGGLUTINATION TEST TO DETECT THE VIRUS IN VIRULIFEROUS LEAFHOPPERS AND DISEASED PLANT TISSUES

As shown in Table 4, rice dwarf virus and rice stripe virus could be detected by the HAI test with virus-sensitized red blood cells.

It is also possible to detect the virus in the viruliferous leafhoppers or in diseased plants by hemagglutination with antibody-sensitized cells.

Yasuo and Yanagita (1963) found that the rice stripe virus in plants and vectors was agglutinated by red blood cells which had been sensitized with partially purified antiserum globulin. Rice dwarf virus (Ishii, 1966) and rice yellow-dwarf virus (Seki and Onizuka, 1964) could also be detected in viruliferous leafhoppers by this technique. It is not necessary to purify the virus from viruliferous leafhoppers for this hemagglutination test. When the viruliferous leafhoppers which reacted positively in hemagglutination test where fed on healthy plant seedlings, 71% of the plants indicated infection by rice stripe virus. All controls reacted negatively (Yasuo and Yanagita, 1963a). But in rice yellow dwarf virus, Seki and Onizuka did not obtain definite correlation between the results of hemagglutination test and of infection test.

For detection of the virus in vectors or plants, this technique has the following merits: the method is simple, requires no expensive equipments, is time saving, and is of high sensitivity and specificity. Therefore, in spite of its weak points that random samples of the vector must be killed and the red blood cells are somewhat labile, this technique has been adopted by many investigators in Japan in the studies of rice virus diseases. For example, it was useful for determination of viruliferous individuals in the progeny of leafhoppers (Sakurai *et al.*, 1963; Yanagita and Ihsii, 1963; Ishii and Ono, 1964; Asaga and Ono, 1966), as well as those collected from paddy fields (Yanagita and Ishii, 1963; Seki and Onizuka, 1964; Uehara and Sato, 1964; Asaga, Yanagita, and Ono, 1965). It was also used for a study of the distribution and concentration of virus in rice plant tissues (Yanagita, 1964a;

Kitani and Kiso, 1965; Sonku and Sakurai, 1965). Recently, it is becoming one of the useful tools in the prediction of epidemics of rice virus diseases.

## BIBLIOGRAPHY

Asaga, K., and Ono, K. 1966. Clones of small brown planthopper with low ability to transmit rice stripe virus [in Japanese]. Ann. Phytopathol. Soc. Japan 32:90. (Abstr.)

Asaga, K., Yanagita, K., and Ono, K. 1965. Difference in the ability of *Laodelphax striatillus* from various localities in Japan to acquire rice stripe virus [in Japanese]. Ann. Phytopathol. Soc. Japan 30:83–84. (Abstr.)

Boyden, S. V. 1951. The adsorption of proteins on erythrocytes treated with tannic acid and subsequent hemagglutination by antiprotein serum. J. Exp. Med. 93: 107–124.

Ishii, M., and Ono, K. 1964. Effect of temperature and light on rice virus diseases. III. Effect of kinds and nutrition of plants on acquisition and transmission of rice stripe virus by its vector [in Japanese]. Ann. Phytopathol. Soc. Japan 29:264. (Abstr.)

——— 1966. Decrease of viruliferous individuals in *Nephotettix cincticeps* congenitally infected with rice dwarf virus [in Japanese]. Ann. Phytopathol. Soc. Japan 32:317. (Abstr.)

Saito, Y. 1961. Serological diagnosis of plant virus [in Japanese]. Plant Protection 15:531–534.

Saito, Y., and Iwata, Y. 1964. Hemagglutination test for titration of plant virus. Virology 22:426–428.

Saito, Y., Takanashi, K., and Iwata, Y. 1964. Studies on the soil-borne virus diseases of wheat and barley [in Japanese, English summary]. Bull. Nat. Inst. Agr. Sci. C-17:23–39.

Sakurai, Y., Ezuka, A., Yunoki, T., and Okamoto, H. 1963. Decrease of viruliferous individuals in *Laodelphax striatellus* congenitally infected with rice stripe virus [in Japanese]. Ann. Phytopathol. Soc. Japan 28: 294. (Abstr.)

Seki, M., and Onizuka, S. 1964. Detection of rice yellow dwarf virus in leafhopper vectors, by use of hemagglutination test [in Japanese]. Ann. Phytopathol. Soc. Japan 29:276. (Abstr.)

Sonku, Y., and Sakurai, W. 1965. Studies on rice stripe disease. VII. Translocation and multiplication of virus in the plant tissues. [in Japanese]. Ann. Phytopathol. Soc. Japan. (Abstr.)

Uehara, H., and Sato, Y. 1964. Sift of the rate of the leafhoppers infected with rice stripe virus in the field [in Japanese]. Ann. Phytopathol. Soc. Japan 29: 270–271. (Abstr.)

Yanagita, K. 1964a. Some improvements in preparation of antiserum to rice stripe virus [in Japanese]. Proc. Kanto-Tosan Plant Protection Soc. 11:16.

——— 1964b. Serological study on rice stripe and dwarf disease. III. Investigation on hemagglutination test [in Japanese]. Ann. Phytopathol. Soc. Japan 29: 73–74. (Abstr.)

Yanagita, K., and Ishii, M. 1963. Appearance of non-viruliferous progenies from viruriferous clones of *Laodelphax striatillus* with rice stripe virus [in Japanese]. Ann. Phytopathol. Soc. Japan 28:309. (Abstr.)

Yasuo, S., and Yanagita, K. 1963a. Serological study on rice stripe and dwarf disease. II. Hemagglutination test for rice stripe virus [in Japanese]. Ann. Phytopathol. Soc. Japan 28:84. (Abstr.)

——— 1963b. Serological diagnosis of rice stripe virus in insect vectors [in Japanese]. Plant Protection 17:215–218.

# Plant-Pathogenic Viruses in Insect Vector Tissue Culture

JUN MITSUHASHI

*Division of Entomology,*
*National Institute of Agricultural Sciences,*
*Nishigahara, Kita-ku, Tokyo, Japan*

## I. INTRODUCTION

Recently, insect tissue culture has been given special attention in various fields of biological studies because it is a very useful tool for studies at the tissue or cellular level. Insect tissue culture work was initiated in 1915 by Goldschmidt, and since that time little progress was made in this field for many years, despite a considerable number of investigations. The obstacles to achieving the long-term growth and survival of the cultivated insect tissues have been discussed in several reviews (Day and Grace, 1959; Martignoni, 1960, 1962; Jones, 1962; Mitsuhashi, 1965f, 1966d). Only during the last decade has insect tissue culture progressed. In 1962 Grace established four strains of cells from ovarioles of *Antheraea eucalypti* and very recently two other insect cell lines have been reported to be established (Grace, 1966; Horikawa, Ling, and Fox, 1966). Besides the establishment of cell lines, in recent years, many workers have reported to be able to cultivate the tissues of various insects for long periods.

The first attempt to use insect tissue culture for the study of viruses was made by Glaser (1917). Since then, insect-pathogenic viruses, as well as insect-borne animal viruses, have been studied in insect tissue cultures. It is also of some importance to both plant pathologists and insect virologists to cultivate the tissues of insect vectors of plant-pathogenic viruses. It is a great advantage of tissue cultures that the cytopathological effects of viruses on the vector cells can be observed directly under the phase-contrast microscope. Furthermore, the mode of virus multiplication can be studied by one-step growth experiments on the cultivated vector cells. If a cell line from vector insects is established, it may be used for quantitative studies of the plant virus. These experiments are possible only on a monolayer of cells, but the cultivated plant tissues usually do not form monolayer cell sheets. For this reason, tissue culture of vector insects, which produces monolayer cell sheets, seems to be very useful for the study of plant viruses. This chapter presents the techniques for the cultivation of some vector insects of plant viruses, and the inoculation experiments of the cultivated cells with the plant virus.

## II. CULTIVATION OF VECTOR TISSUES

In 1956, Maramorosch attempted to maintain small pieces of leafhopper tissues alive in a hanging drop preparation in order to prove conclusively that the aster yellows virus could multiply in its vector leafhopper, *Macrosteles fascifrons* Stål. Some tissue fragments survived for 10 days, but no outgrowth was obtained. Grace (1959) made several attempts to obtain the growth of cells and tissues of *M. fascifrons* and *Dalbulus maidis* DeL. & W., the latter being a vector of corn stunt virus. The gut tissue survived *in vitro* for several days, but no growth of cells or tissues was observed. Hirumi and Maramorosch (1963a, 1963b) maintained various organs of *M. fascifrons* and of *Agallia constricta* (van Duzee), both vectors of aster yellows and wound tumor virus, for several weeks *in vitro,* but no growing cells were obtained. The first successful cultivation of leafhopper cells was reported by Vago and Flandre (1963). They obtained cells from the gonads, gut, and hypodermis of *Philaenus spumarius* L., which is known to transmit lucerne dwarf virus or peach yellow virus according to its respective subspecies. They also obtained fibroblast-type and epithelial-type cells from *Macrosteles sexnotatus* Fall., which is a vector of tomato big-bud virus. Hirumi and Maramorosch (1964a, 1964b, 1964c) and Hirumi (1965) cultivated the embryonic tissues of *M. fascifrons,* and obtained cell growth solely from the embryos of the blastokinetic stage. Mitsuhashi and Maramorosch (1964) also cultivated tissues of *M. fascifrons* and obtained cell growth from embryos of various developmental stages, as well as growth from tissues of nymphs and adults. They also obtained similar cell growth in the cultivation of tissues of *D. maidis* and *A. constricta.* Furthermore, Mitsuhashi (1965b) successfully cultivated the embryonic cells of *Nephotettix cincticeps* Uhler, which is one of the vectors of rice dwarf virus and rice yellow-dwarf virus. The tissues of nymphs and adults of the same species have also been cultivated (Mitsuhashi, 1965e). The embryonic tissues of *Nephotettix apicalis* Motschulsky, which is a vector of rice dwarf virus and rice yellow-dwarf virus, and of *Inazuma dorsalis* Motschulsky, which is a vector of rice dwarf virus, have also been cultivated (Mitsuhashi, unpublished). Recently, Chiu, Reddy, and Black (1966) successfully cultivated embryonic tissues of *A. constricta* according to the culture method of Mitsuhashi and Maramorosch (1964).

The above-mentioned successful cases of cultivation were all primary cultures, and the cells survived for several months. However, very few attempts have been made to subculture the leafhopper cells grown in primary cultures (Vago and Flandre, 1963; Mitsuhashi and Maramorosch, 1964), and the results of subculturing were not encouraging. A cell strain or a cell line from leafhopper vectors has not been established thus far.

### *A. Methods for Setting Up Cultures*

Various methods have been used by different workers to culture leafhopper tissues. Hirumi and Maramorosch (1964a, 1964c) were the first to succeed in culturing embryonic tissues by dissociating them with 0.02% trypsin solution. They needed a minimum of 100 eggs for each experiment using 10 culture flasks, and when smaller numbers were used per flask, no growth could be obtained. Since they applied repeated centrifugation during the treatment, it is assumed that some fragments of embryos might have been lost before all the procedures could be completed. Mitsuhashi and Maramorosch (1964) and Mitsuhashi (1965b) employed a modified method for setting up cultures of embryonic tissues. About 5–10 eggs were required for setting up one culture, and the treated embryo fragments were forced to adhere to the glass surface, so as to minimize the fragments which fail to produce cell growth. This method is described in detail below.

Leafhoppers lay their eggs in plant tissues. In order to obtain the eggs of about the same developmental stage, mated female adults were confined to their host plants and were transferred every day. The eggs were excised from the leaf tissues under a binocular dissecting microscope, usually 3 days after oviposition at 25°C. At this time the embryos were at the blastokinetic stage, which is known to provide the most suitable materials for *in vitro* cultivation (Hirumi and Maramorosch, 1964a). However, embryos of other stages could also be cultivated (Mitsuhashi and Maramorosch, 1964; Mitsuhashi, 1964b). The excised eggs were collected in a glass tube filled with water, care being taken to sink all the eggs to the bottom of the tube. The eggs were then surface-sterilized by replacing the water with sterilizing agents, washed twice with sterilized distilled water. [The sterilizing agent should be determined in each species. Some leafhopper species, for instance *M. fascifrons,* can tolerate treatment with 0.1% mercuric chloride for 5 min (Mitsuhashi and Maramorosch, 1964), while others such as *N. cincticeps*, are seriously affected by this treatment (Mitsuhashi, 1965b). In the latter case treatment with 70% ethyl alcohol for 1 min can be used without deleterious effects on embryos.] The surface-sterilized eggs were then transferred into Rinaldini's salt solution placed on a Maximow slide by means of a pipet. The composition of Rinaldini's solution is as follows: 0.8 g of NaCl, 0.02 g of KCl, 0.005 g of $NaH_2PO_4 \cdot H_2O$, 0.1 g of glucose, 0.1 g $NaHCO_3$, and 0.0676 g sodium citrate—all made up to 100 ml with distilled water.

The tips of the eggs were cut off and the embryos were squeezed out of their chorions through the opening. This was carried out under a dissecting microscope, facilitated by a fine forceps and a knife. The chorions were removed from the solution and the yolk, which adhered to the embryos, was separated by the aid of fine needles. The embryos thus freed from yolk were then transferred to another Maximow slide with a few drops of Rinaldini's solution, and cut in half longitudinally to facilitate the following trypsinization.

The Rinaldini's solution was replaced with the solution of 0.1% trypsin (Trypsin 1:250, purchased from Difco, Detroit, Michigan) dissolved in Rinaldini's solution by means of pipets, care being taken not to expose the embryo fragments to air. The trypsinization was carried out for about 5 min at a room temperature of approximately 25°C. Agitation of the embryo fragments during trypsinization was avoided because it was found to cause a clumping of the tissues and the formation of very viscous, gel-like masses. When the surface of the tissues became sticky, the embryo fragments were transferred into the modified Ringer-Tyrode's salt solution (Carlson, 1946), which was placed on the bottom of the culture vessels. It was important not to allow the embryo fragments to come in contact with air during the transfer of the tissues. The tissues, released from a pipet into the Ringer-Tyrode's solution soon adhered to the glass surface of the bottom of the culture vessels. The modified Ringer-Tyrode's solution consisted of 0.7 g of NaCl, 0.02 g of KCl, 0.02 g of $CaCl_3 \cdot 2H_2O$, 0.01 g of $MgCl_2 \cdot 6H_2O$, 0.02 g of $NaH_2PO_4$, 0.012 g of $NaHCO_3$, 0.8 g of glucose, and distilled water added to make 100 ml.

The Ringer-Tyrode's solution was then replaced with culture medium, which was done by gently changing the medium five times. During these changes, care was taken to avoid exposure of the tissues to air. Fragments that accidentally became exposed to air detached themselves from the glass and spread on the surface of the liquid; such fragments or cells never attached themselves to the glass again, and soon deteriorated. It was not as important to replace the salt solution completely with the culture medium, and the small residue of Ringer-Tyrode's solution or even of the trypsin solution did not seem to have deleterious effects on the cell growth.

After the change of medium was completed, the tops of vessels were sealed. The used culture vessels consisted of microslide rings of 25-mm diameter and 10-mm height and cover glasses. The microslide rings and the cover glasses had been sterilized previously by dry heat at 180°C for 2 hr. The assembly of the culture vessels was made just before setting up the cultures. One side of the ring was fixed to a cover glass with melted paraffin–balsam (1:1) to make the bottom of the culture vessel; the opposite side was closed with another cover glass by means of grease after the culture was set up. All procedures were carried out under sterile conditions.

When nymphal or imaginal tissues are to be cultivated, the leafhoppers should be reared aseptically. It is extremely difficult to completely sterilize septically reared leafhoppers. The cultivation of alimentary canals may be carried out successfully only when the organs are derived from aseptic leafhoppers; otherwise, contamination by various microorganisms may disturb the cultivation.

To provide aseptic materials for tissue culture, various methods for aseptic rearing of leafhoppers have been developed (Vago and Flandre, 1963; Mitsuhashi and Maramorosch, 1963a, 1963b; Mitsuhashi, 1964a, 1965a, 1965d). Vago and Flandre (1963) were the first to cultivate nymphal or imaginal tissues of leafhoppers by surface-sterilizing septic leafhoppers, as well as by using aseptic leafhoppers. The aseptic rearing method, which has been routinely used by the author, is as follows:

The leafhopper eggs, just before hatching, were excised from the leaf tissues, and fastened onto strips of wax paper with egg albumin. The eggs were surface-sterilized by dipping the strips in the solution of sterilizing agent. The same sterilizing procedures could be applied here as with embryonic tissue culture. The sterilized eggs were placed in culture tubes containing seedlings of host plants grown on agar media from sterilized seeds. Hoagland and Knop's medium, modified by Chen, Kilpatrick, and Rich (1961), and modified Kimura's medium (1931, 1932) were used for the cultivation of aseptic plants.

The nymphs emerged within a few days and moved from the wax paper onto the sterile plants. Their rearing was carried out at 25°C, with 16 hr of light per day. Development from egg to adult took approximately the same length of time as it did under nonsterile conditions. Adults were put into new tubes with fresh plants, and allowed to mate and oviposit. Aseptically grown generations were reared continuously by transferring them into new tubes when the seedlings became yellowish.

For setting up the culture from nymphs or adult leafhoppers, Vago and Flandre (1963) treated organs excised from surface-sterilized leafhoppers by submersion in the mixture of antibiotics before dissociation of the organs. When aseptic leafhoppers were used, the leafhoppers were dissected directly in the culture medium. The dissociation of the organs to be cultivated was carried out mechanically and no trypsinization was employed. Mitsuhashi and Maramorosch (1964) applied essentially the same techniques here as with embryonic tissues to set up the cultures from aseptically reared leafhoppers. The nymphs or adults were first dipped into 70% ethyl alcohol. This prevented the formation of air bubbles when the leafhoppers were subsequently placed in the salt solution. The alcohol was removed by washing in sterilized, distilled water, and the leafhoppers were placed on a Maximow slide in a few drops of Rinaldini's solution. A few tiny air bubbles, still present, were removed and the wings of the adults were cut off. After transferring the leafhoppers to another Maximow slide containing Rinaldini's solution, they were either cut into small pieces, or dissected so as to provide various organs for cultivation. Then the substitution of Rinaldini's solution was made with 0.1% trypsin solution. Subsequent steps were the same as for the embryonic tissue culture.

Tokumitsu and Maramorosch (1966) cultivated tissues of *Acyrthosiphum pisum* (Harris), which is a vector of pea enation mosaic virus. The surface-sterilized or aseptically reared aphids were dissected in Ringer-Tyrode's solution and brought into culture vessels by the same procedure described by Mitsuhashi and Maramorosch (1964).

### *B. Medium*

Various culture media have been devised by different workers, some of which are listed in Table 1.

Vago and Flandre (1963) analyzed the hemolymph of leafhoppers belonging to genera *Macrosteles* and *Philaenus* and found a high potassium and magnesium content in the hemolymph, while that of sodium was low. In some insects, *Blattella germanica* Linnaeus, for example, the tissues require a definite sodium/potassium ratio in the culture medium for their growth *in vitro* (Ting and Brooks, 1965), but leafhopper tissues seem to proliferate in a wide range of salt compositions (Mitsuhashi and Maramorosch, 1964).

Glucose provides an adequate carbon and energy source for developing leafhopper tissues, and the substitution of a part of glucose with fructose, sucrose, or trehalose, or a combination of them did not improve the growth of cells (Mitsuhashi and Maramorosch, 1964).

Insect blood reportedly contains a higher concentration of amino acids than human blood (Buck, 1953; Martignoni, 1960; Wyatt, 1961). Wyatt, Loughheed, and Wyatt (1956) showed that the amino acid content of the blood plasma of two species of Lepidoptera and one species of Hymenoptera was almost 70 times as great as that of human plasma. Therefore, Wyatt's tissue culture medium for lepidopterious insects contained a tenfold increase in the concentration of amino acids over the TC-199 medium used routinely for vertebrate tissue culture (Wyatt, 1956). Surprisingly, it was found that leafhopper tissues did not require such a high concentration of amino acids and other protein derivatives and if TC-199 medium was supplemented with 20% fetal bovine serum, it proved satisfactory at half-strength (Mitsuhashi and Maramorosch, 1964).

The addition of a vitamin mixture did not improve the cell growth of leafhopper tissue, but the addition of TC-yeastolate did. The omission of TC-yeastolate still permitted limited growth if the medium was supplemented with fetal bovine serum, suggesting that the serum provided the minimum vitamin requirements for leafhopper tissues (Mitsuhashi and Maramorosch, 1964).

Most insect tissue culture workers prefer to incorporate insect blood in their culture media. Hirumi and Maramorosch (1963a) supplemented the culture medium with the serum of *Leucania unipuncta*, leafhopper extract, or crustacean hemolymph but the effects were not encouraging. Sera from vertebrates are sometimes incorporated in the insect tissue culture media. Mitsuhashi and Maramorosch (1964) reported that leafhopper tissues could not proliferate without fetal bovine serum, and only newborn calf serum could be substituted for fetal bovine serum, while normal calf, chicken, and horse sera were unsatisfactory.

The effects of some biologically active substances have been examined (Mitsuhashi, unpublished). The addition of ecdysone to the culture medium had no promoting effects on leafhopper cell growth. Phytohemagglutinin, which is known to initiate mitoses in human leucocytes (Nowell, 1960), showed no marked effects on the cell growth when it was incorporated in the culture medium for leafhopper embryonic cells. Agmatine sulfate, which is reported to accelerate mitoses in the culture of grasshopper neuroblasts (St. Amand, Anderson, and Gaulden, 1960), showed rather deleterious effects on the leafhopper cells.

Various modifications of the culture medium were attempted in the

TABLE 1

COMPOSITION OF CULTURE MEDIA USED FOR LEAFHOPPER TISSUE CULTURE

(all values in milligrams or milliliters per 100 milliliters)

| | Mitsuhashi and Maramorosch (1964) | | | | | | | Mitsuhashi (1965) | Mitsuhashi (unpublished) | | Hirumi and Maramorosch |
|---|---|---|---|---|---|---|---|---|---|---|---|
| | No. 1 | No. 2 | No. 3 | No. 4 | No. 5 | No. 6 | TC-199A | NCM-2B | NCM-4B | NCM-5A | (1964) |
| $NaH_2PO_4 \cdot H_2O$ | 20 | 48 | 60 | 8 | 10 | 20 | | 8 | 16 | 16 | 52 |
| $MgCl_2 \cdot 6H_2O$ | 10 | 120 | 150 | 4 | 5 | 10 | | 4 | 8 | 8 | 130 |
| $MgSO_4 \cdot 7H_20$ | | 160 | 200 | | | | | | | | 174 |
| KCl | 20 | 120 | 150 | 8 | 10 | 20 | | 8 | 16 | 16 | 130 |
| $CaCl_2 \cdot 2H_2O$ | 20 | 40 | 50 | 8 | 10 | 20 | | 8 | 16 | 16 | 43.5 |
| NaCl | 700 | | | 280 | 350 | 700 | | 280 | 560 | 560 | |
| $NaHCO_3$ | 12 | | | 4.8 | 6 | 12 | | 5 | 9.6 | 9.6 | |
| Glucose | 400 | 80 | 100 | 80 | 100 | 400 | | 160 | 400 | 800 | 30.4 |
| Sucrose | | | | | | | | | 400 | | 17.4 |
| Fructose | | 80 | 100 | 80 | 100 | | | | | | 17.4 |
| L-tryptophan | | | | | | | | | 6.4 | | |
| L-cystein | | | | | | | | | 1.6 | | |
| Lactalbumin hydrolisate [a] | 650 | 520 | 375 | 520 | 650 | 650 | | 520 | 640 | 800 | 435 |

| | | | | | | | | | | | |
|---|---|---|---|---|---|---|---|---|---|---|---|
| Bacto-peptone [b] | | 520 | 375 | 520 | 650 | | | 520 | | | |
| TC-yeastolate [b] | 500 | 200 | 250 | 200 | 250 | | | 200 | 160 | 160 | |
| Choline chloride | | 40 | 50 | 40 | 50 | | | | | | |
| TC-199 medium (100×) [c] | | 4 | | 4 | | | 5 | 2 | | | 4.3 |
| Essential amino acids (50×) [c] | | | 1 | | | 1 | | | | | |
| Nonessential amino acids (100×) [c] | | | 0.5 | | | | | | | | |
| Vitamins (100×) [c] | | | 0.5 | | | 0.5 | | | | | |
| L-Glutamic acid (100×) [c] | | | 0.5 | | | 0.5 | | | | | |
| Fetal bovine serum [c] | 20 | 18 | 20 | 18 | 20 | 20 | 20 | 20 | 20 | 20 | 12.9 |
| Antibiotics (penicillin and streptomycin) [c] | 2 | 2 | 2 | 2 | 2 | 2 | 2 | | | | 2 |
| Dihydrostreptomycin sulfate [d] | | | | | | | | 5 | 10 | 10 | |

[a] Nutritional Biochemicals Corporation, Cleveland, Ohio.
[b] Difco Laboratories, Detroit, Michigan.
[c] Microbiological Associates, Inc., Bethesda, Maryland.
[d] Takeda Chemical Industries Ltd., Osaka, Japan.

cultivation of leafhopper tissues (Mitsuhashi and Maramorosch, 1964; Mitsuhashi, unpublished). The growth of cells in the modified medium was about the same in the simple medium (No. 1 medium, Mitsuhashi and Maramorosch, 1964) previously described. Unfortunately, no medium has been found that would markedly improve cell growth.

Chiu *et al.* (1966) reported that in the cultivation of embryonic tissues of *A. constricta*, the basic culture medium of Mitsuhashi and Maramorosch (No. 1 medium in Table 1) gave more satisfactory results than several other media tried. Tokumitsu and Maramorosch used the modified Mitsuhashi and Maramorosch's medium (1964) for their aphid tissue culture.

### *C. Maintenance of the Culture*

It is one of the advantages that insect tissues or cells do not require a definite temperature for their growth. Accordingly, most workers did not seem to pay much attention to the culture conditions. However, both Grace and Vago (personal communications) consider temperature control important. It is generally considered that 25°C is suitable for insect tissue culture, but cultures may be maintained without deleterious effect at temperatures ranging from 20 to 35°C.

For long-term culture, a change of medium is necessary. Hirumi and Maramorosch (1964c) suggested frequent changes of the culture medium. Mitsuhashi and Maramorosch (1964) and Mitsuhashi (1965b) maintained the cultivated cells of leafhopper embryos for over 5 months by changing the culture medium once a week. One culture of *N. cincticeps* nymphal tissues was maintained for 11 months by changing the medium once a week (Mitsuhashi, unpublished).

When subculturing is attempted, it is necessary to detach the proliferated cells or the explants from the glass surface of the culture vessel. For this purpose, Mitsuhashi and Maramorosch (1964) employed the following procedures: the medium was first replaced with Rinaldini's solution and then replaced with 0.1% trypsin solution. The trypsinization was stopped when the surface of cells became somewhat viscous. The original explants and the cells that grew from them were detached from the glass surface by forcing the solution gently in and out of a small pipet. The detached tissues were then transferred into culture vessels containing Ringer-Tyrode's solution. The trypsinized explants, as well as the outgrowths, adhered to the bottom of the vessel as soon as they were released from a pipet into the Ringer-Tyrode's solution. The salt solution was finally replaced with the culture medium. Vago and Flandre (1963) employed the plasma clot method for the cultivation of leafhopper tissues, and when subculturing was performed, a part of the clot in which cells were actively multiplying was cut and transferred onto fresh clot.

### *D. Growth of Cells*

Vago and Flandre's cultures (1963) were considered to be a combination of monolayer culture and explant culture. In this case, the scattered individual cells began to multiply by mitosis, while the explanted tissue masses gave rise to some growing cells. In this way, fibroblast-type cells were obtained 4–5 days after the culture was set up. Epithelial-type cells were also obtained from the explanted gut wall.

Three types of cells have been reported in the embryonic tissue culture of *M. fascifrons* (Hirumi and Maramorosch, 1964b; Mitsuhashi and Maramorosch, 1964). According to Hirumi and Maramorosch, these types were epithelial cells, fibroblast-like cells, and phagocytes. Since some of the epithelial-type cells and fibroblast-type cells evidently perform phagocytosis, the classification of some cells as phagocytes seems inadequate. Mitsuhashi and Maramorosch recognized epithelial-type cells, fibroblast-type cells, and wandering cells. Similar cell types were also recognized in the embryonic tissue culture of *N. cincticeps* (Mitsuhashi, 1965b). These three types were further divided into several subtypes, but the classification of cell types was generally based on the morphological characteristics of cells, and the change of cells from one type to another type has not been clarified. Furthermore, the origin of cells of each type has not been determined. Although both Hirumi and Maramorosch (1964b) and Mitsuhashi and Maramorosch (1964) obtained epithelial-type cells and fibroblast-type cells, their subtypes were not identical, and a comparison of observed cell types is difficult. Since the ratio of each cell type is not always equal, probably depending on the culture conditions, the differences in cell types may also be due to the differences in culture methods employed.

In the embryonic tissue culture of leafhoppers, the most predominant cells are epithelial-type cells, designated as type-A epithelial cells by Mitsuhashi and Maramorosch (1964) or smaller epithelial cells by Hirumi and Maramorosch (1964b). Similar cells have been obtained in the embryonic tissue culture of *D. maidis, A. constricta, N. cincticeps, N. apicalis,* and *I. dorsalis.* These cells usually began to migrate within 24 hr, and multiplied rapidly, forming a typical cell sheet (Fig. 1). Very frequent mitoses were seen on the periphery of the cell sheet. Epithelial cells of other subtypes often appeared near the outer margin of a growing cell sheet. Some of them seemed to derive sec-

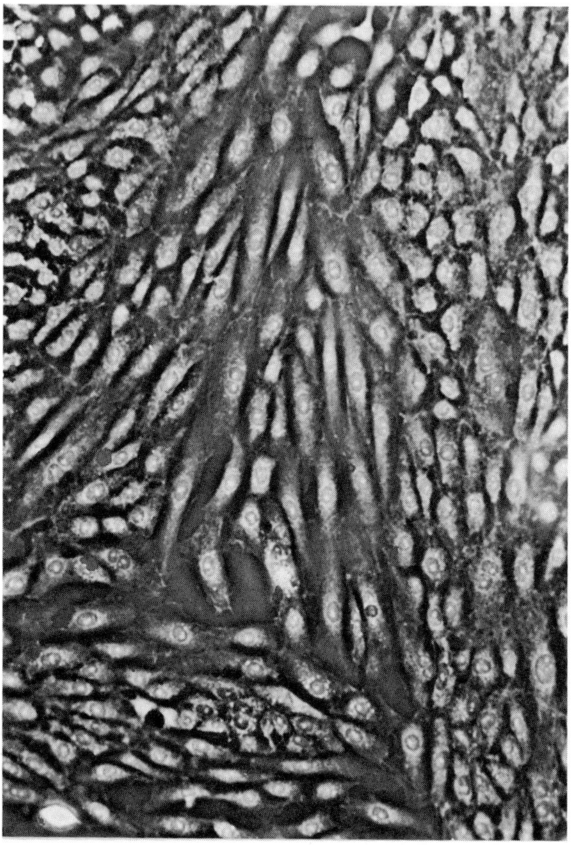

*Fig. 1.* A typical cell sheet formed by epithelial cells migrated from embryonic tissues of *Nephotettix cincticeps* (× 120).

ondarily from the type-A epithelial cells. Among them, giant cells and multinucleate cells were often observed.

Fibroblast-type cells, consisting of slender spindle-shaped cells, multiplied, forming a network (Fig. 2). This type of cell has been obtained in all the leafhopper species cultivated. Mitosis occurred frequently in these cells. In old cultures, very large fibroblast-like cells were sometimes obtained (Mitsuhashi and Maramorosch, 1964). These cells occasionally agglomerated and formed a strand that began to contract independently from the contraction movement of the original explant.

Wandering cells are characterized by their active amoeboid movement. For their active movement, these cells do not form a cell sheet

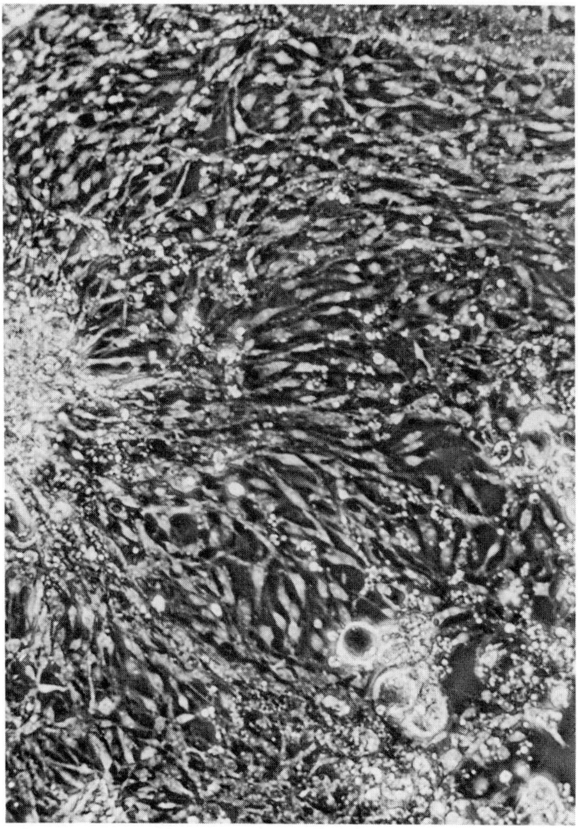

*Fig. 2.* Network of fibroblasts migrated from embryonic tissues of *Nephotettix cincticeps* (× 120).

or a network, and are usually distributed outside the growing cell sheets (Fig. 3). Some subtypes of wandering cells multiplied by mitosis; Binucleate cells were found among wandering cells. Two subtypes of wandering cells have been reported in the tissue cultures of *M. fascifrons* (Mitsuhashi and Maramorosch, 1964), and three in *N. cincticeps* (Mitsuhashi, 1965b), but no identical subtypes of wandering cells were found in both species.

Tokumitsu and Maramorosch (1966) obtained two types of cells from the explanted nymphal tissues of aphids. Both of them were reported to be fibroblast-type cells, and survived for up to 20 days, but no cell division was observed.

Most of the migrating cells in leafhopper tissue cultures multiply by mitosis, although mitotic figures have not been obtained in some

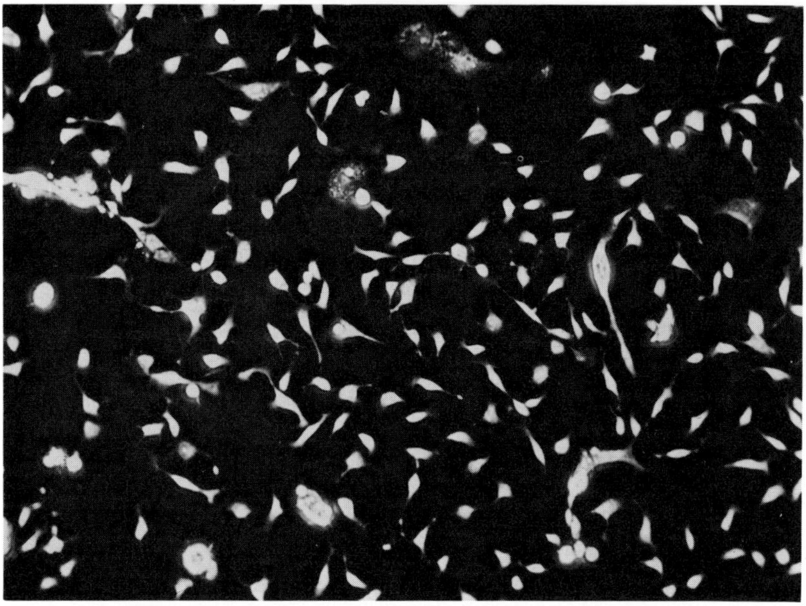

*Fig. 3.* Wandering cells migrated from embryonic tissues of *Nephotettix cincticeps* (× 100).

types of cells. When observed with a phase-contrast microscope, mitosis can be easily followed (Fig. 4). Besides normal bipolar mitoses, multipolar mitoses most often, tripolar mitoses (1) have been observed (Hirumi, 1965; Mitsuhashi, 1965b). Multipolar mitoses uusally seem to occur in polyploid cells. An example of tripolar mitosis, observed in an embryonic tissue culture of *N. cincticeps,* is shown in Fig. 5. At the late prophase (Fig. 5*a*) individual chromosomes can be counted. Since diploid numbers of chromosomes in *N. cincticeps* have been reported to be 15 in males and 16 in females (Mitsuhashi, 1966b), this cell might be octoploid. In mammalian tissue cultures, it has been reported that multipolar mitoses are sporadically observed in the culture of cancer cells (Nakahara, 1953; Makino and Nakahara, 1953; Půža, 1963). The occurrence of multipolar mitoses in leafhopper tissue cultures does not give sufficient evidence that the cells underwent a malignant transformation. Multipolar mitoses have also been reported in cultivated, normal embryonic chick cells (Stillwell, 1943, 1947) and in cultivated hemocytes of *Chilo suppressalis* (Mitsuhashi, 1966a). Other types of unusual mitoses, such as the one which resulted in the formation of a binucleate cell, or the mitosis of a binucleate cell which resulted in the formation of two mononucleate daughter cells, have

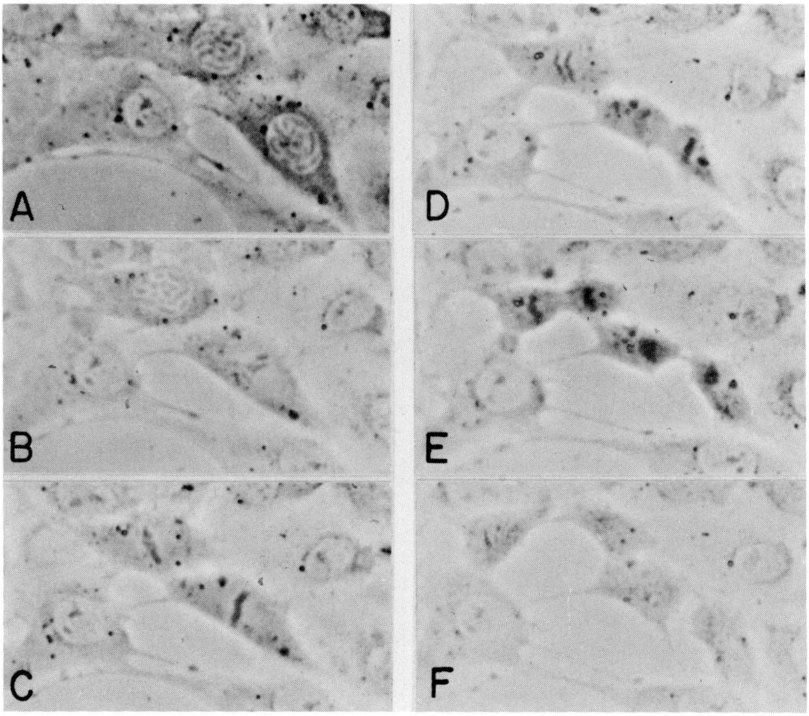


*Fig. 4.* Process of normal bipolar mitosis observed in the epithelial cell migrated from embryonic tissues of *Nephotettix cincticeps* (×540). (*A*) Two cells are at late prophase. Individual chromosomes are visible. (*B*) Nuclear membrane is about to disappear in the left cell, and chromosomes are gathering onto the equatorial plane in the right cell. (*C*) Metaphase: the chromosomes are arranged on the equatorial plane in both cells. (*D*) Early anaphase in the left cell, and early telophase in the right cell. (*E*) Early telophase in the left cell, and late telophase in the right cell. (*F*) Interphase: nuclei are formed in the four daughter cells.

been observed in the embryonic tissue culture of *N. cincticeps* (Mitsuhashi, 1965b).

When tissues of nymphs or adults of *M. fascifrons* were cultured, the same types of cells as those obtained in the embryonic tissue cultures were obtained (Mitsuhashi and Maramorosch, 1964). If the leafhoppers were cut into several parts and all of the fragments were brought into one culture vessel, the cut openings of the fragments adhered to the glass surface, and the cells began to proliferate (Fig. 6). Growth of the cells occurred in the same way as in the embryonic tissue cultures. Some explanted organs kept their functions for more than 2 months. Contractions of the leg muscles were very common

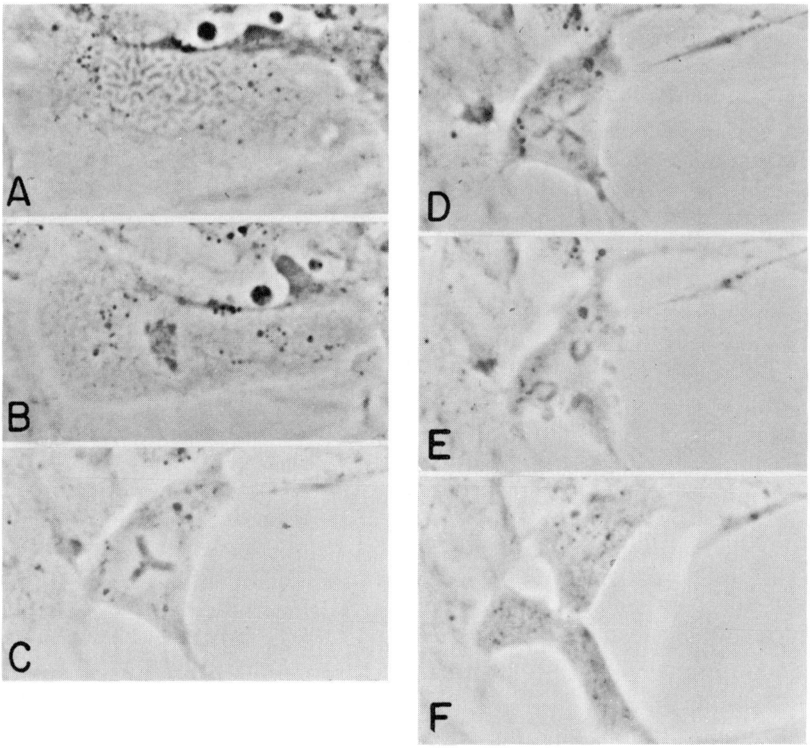


*Fig. 5.* Process of tripolar mitosis observed in the giant epithelial cell in the embryonic tissue culture of *Nephotettix cincticeps* (× 540). (*A*) Late prophase: individual chromosomes are visible. (*B*) Prometaphase: chromosomes are gathering onto the equatorial plane. (*C*) Metaphase: chromosomes are arranged in Y shape. (*D*) Early anaphase: chromosomes have started to separate. (*E*) Early telophase. (*F*) Late telophase.

and the legs themselves continued an extension and contraction movement, while numerous cells proliferated from their cut openings. The alimentary tracts and oviducts retained peristaltic movement for over 2 months. When the organs were cultured separately, the stomach adhered to the glass surface most easily, and very large cells proliferated from them. Epithelial-type cells proliferated from the ovarioles of young adults. No growing cells were obtained from the nervous system, salivary glands, muscles, fat bodies, testes, Malpighian tubules, mycetomes, or epidermis. Similar results were obtained in the cultivation of *N. cincticeps* nymphs or adults (Mitsuhashi, 1965e).

Subcultures of leafhopper tissues, either embryonic or nymphal and imaginal, were successful only with original explants but not with

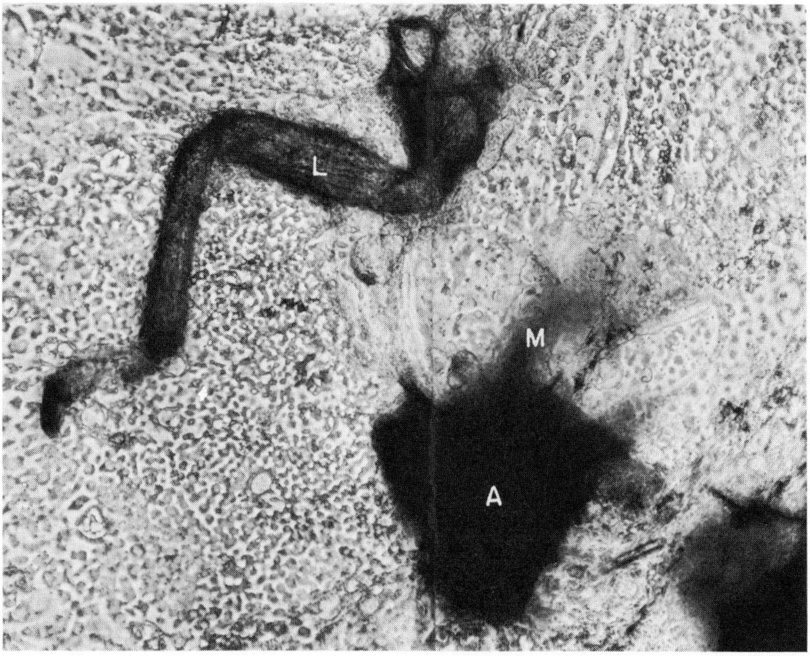


*Fig. 6.* Cell multiplication in the nymphal tissue culture of *Macrosteles fascifrons* (× 65) : A, a part of abdomen; L, hind leg; M, midgut.

cells that had proliferated from explants (Mitsuhashi and Maramorosch, 1964). After subculturing, cell growth occurred in the same way as it did in the primary cultures. Cell migration from the subcultured original explants occurred so fast as to give the impression that the original explants were stimulated by the subculturing. The same explants were subcultured several times at intervals of about 1 month and, whenever they were subcultured, they showed the same increased activity in cell production. On the other hand, cells that proliferated from original explants and were subcultured separately grew very poorly, in spite of their almost immediate attachment to the glass surface. Some of the subcultured cells showed mitosis but most cells soon stopped growing and began to degenerate after a week.

## III. CULTIVATION OF VIRULIFEROUS VECTOR TISSUES *IN VITRO*

Some of the leafhopper-borne, persistent plant viruses have been reported to exert cytopathological effects on vector cells *in vivo*. These viruses multiply in the vector cells, but do not kill their host cells. Therefore, attempts were made to cultivate virus-containing tissues from viruliferous leafhoppers. Maramorosch (1956) cultivated small pieces of *M. fascifrons* which had previously been allowed to feed on plants infected with aster yellows virus. After 10 days, no growth of cells was obtained, but some pieces were found to be still alive, and eventually it became evident that aster yellows virus completed its incubation in the vector tissues *in vitro*.

Hirumi and Maramorosch (1963b) maintained various organs of *M. fascifrons,* excised after a 5-day acquisition feeding, for 14 days *in vitro,* and showed that aster yellows virus was maintained in the Malpighian tubules *in vitro*. No virus was recovered from homogenates of the gut, salivary glands, ovaries, and testes, which were treated in the same way as the Malpighian tubules. It has been reported that in some leafhopper species, the fat bodies and mycetomes showed histological changes when plant viruses were harbored in them (see Maramorosch and Jensen, 1963). Unfortunately, no experiments have been conducted to examine whether or not plant viruses can multiply in fat bodies or mycetomes maintained *in vitro*.

Rice dwarf virus has been known to be transmitted from female adults of *N. cincticeps* to their progeny through ovaries. Mitsuhashi (1965c, 1965g) cultivated viruliferous embryonic tissues in the same way as virus-free embryonic tissues. Cell migration and cell multiplication occurred normally in explanted viruliferous tissues. Epithelial-type cells formed cell sheets in a large area around explants. These cells contained more granules compared with the identical type of cells obtained from healthy embryonic tissues. A considerable histolysis of the explants occurred simultaneously with the development of cell sheets. Part of the developed cell sheets began to degenerate 40 days after the culture was set up. The initiation of the cell degeneration seemed to occur somewhat earlier in the viruliferous tissue cultures than in the healthy tissue cultures. When the cultivated cells were examined with an electron microscope on the twenty-seventh day of cultivation, ultrathin sections revealed the presence of rice dwarf virus particles in the developed epithelial cells (Fig. 7). The virus particles were found in abundance in almost all of the cells examined. Most of the rice dwarf virus particles were arranged in lines and were accompanied by a sheath-like substructure. Since only small numbers of rice dwarf virus particles are found in the limited areas of viruliferous embryos (Nasu, 1965) it is evident that rice dwarf virus multiplied in the cells grown *in vitro*.

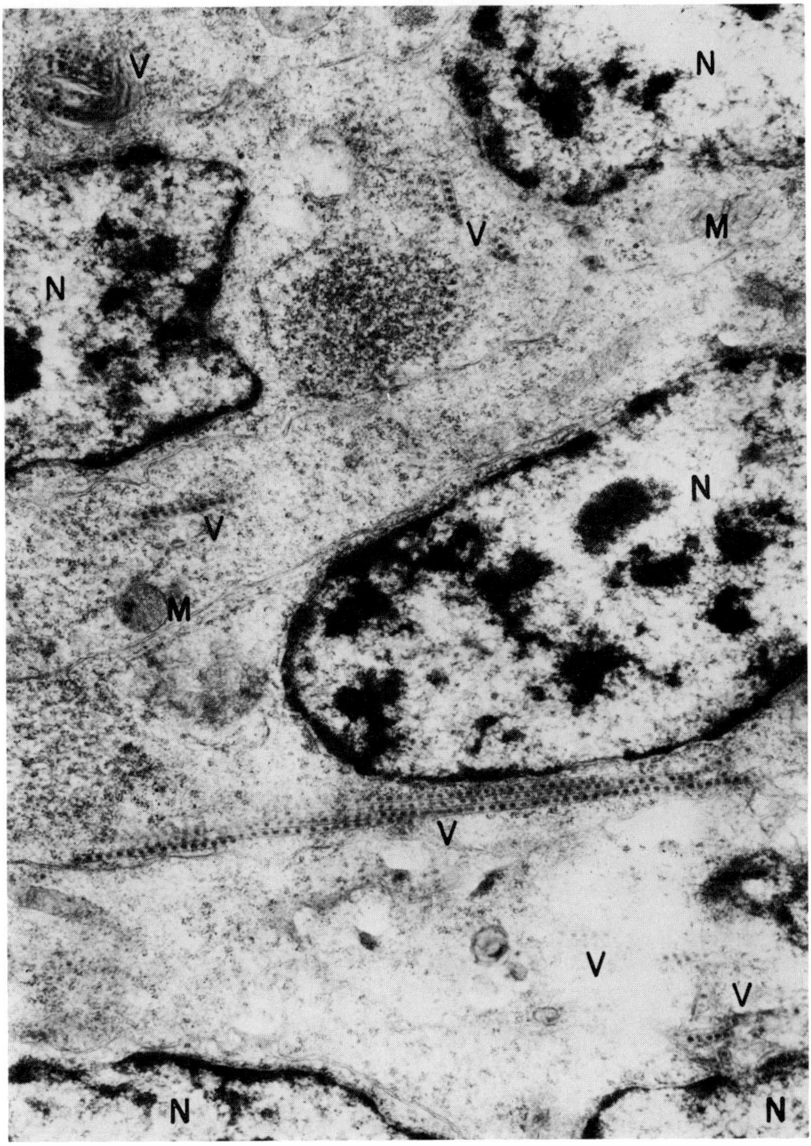


*Fig. 7.* Ultrathin section of the cultivated cells from viruliferous embryos of *Nephotettix cincticeps* (× 10,880): M, mitochondrion; N, nucleus; V, rice dwarf virus particles.

## IV. INOCULATION OF THE CULTIVATED VECTOR TISSUES WITH PLANT VIRUSES

Many experiments have been carried out on the inoculation of primary cultures of lepidopterous insects with insect viruses, mostly polyhedrosis viruses, but similar experiments with plant viruses have been reported only recently.

Mitsuhashi (1965c, 1965g) attempted to inoculate a healthy embryonic tissue culture of *N. cincticeps* with rice dwarf virus by bringing the inoculant into the culture medium. The inoculant was prepared from viruliferous female adults. The fat bodies, which were aseptically excised from 5 viruliferous female adults, were homogenized in 0.5 ml of the culture medium, and the homogenates were centrifuged at 3000 rpm for 5 min. The clear fluid between the lipid layer and sediment was used as the inoculant. The inoculation of virus was carried out by simply replacing the culture medium with the inoculant. The inoculated culture was maintained at 25°C for 24 hr with the inoculant, and then the inoculant was thoroughly washed away by changing the culture medium several times.

When a two-week-old culture, in which epithelial cell sheets were well formed, was inoculated with rice dwarf virus by the above mentioned method, the cells underwent some cytological changes. Granules began to appear around the nuclei within 24 hr. From time-lapse cinematographic analysis, it appeared as if the granules consisted of substances which were taken into the cells from the environment by active phagocytosis of the cells. After a histochemical examination of the granulated cells, the granules were found to be lipid in nature. The granules then increased in number day after day, and finally filled the whole area of the cytoplasm (Fig. 8). This was followed by the destruction of cells. This granulation was quite marked in the epithelial cells. The granulation proceeded from the periphery of the cell sheets to the inner parts. During the first 3 days following inoculation, mitoses could be seen even in granulated cells at the periphery of cell sheets. With the advance of granulation, the cells shrank and the cell sheets were destroyed. The explants themselves were also affected by the inoculation of virus. At the time of inoculation, they were contracting and this movement continued for a week after inoculation and then ceased. The cells of the explants were also filled with granules at the end of the first week following inoculation.

In the control experiments, in which the inoculant was prepared from 5 healthy female adults, the cells and explants did not show marked changes, although some cells were slightly granulated.

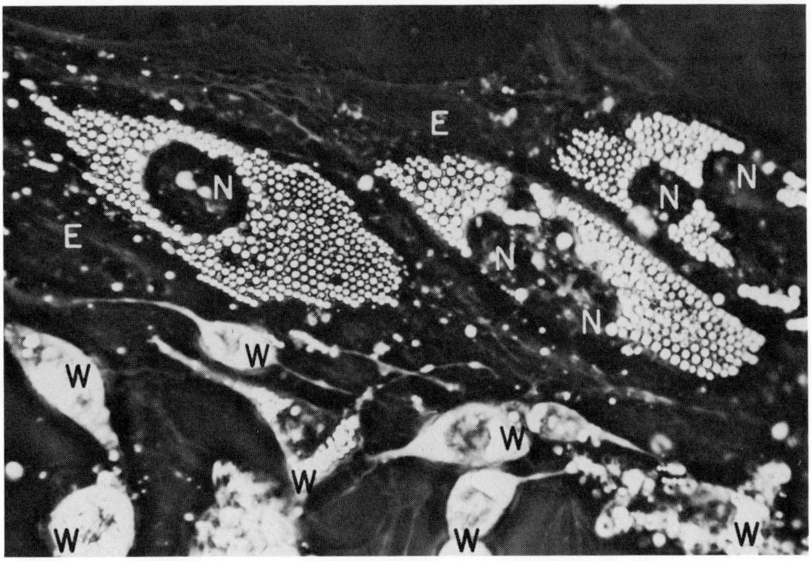


*Fig. 8.* Granulated epithelial cells in the embryonic tissue culture of *Nephotettix cincticeps* after inoculation of rice dwarf virus (× 320); E, epithelial cell; N, nucleus; W, wandering cell.

Similar granulation of cells was also observed when the embryonic tissue cultures of *M. fascifrons* and *A. constricta* were inoculated with aster yellows virus and wound tumor virus, respectively, although in both cases the infection of cells with respective virus was not evident (Maramorosch *et al.*, 1965).

Granules were also seen when nonviruliferous cells began to degenerate. The granulation of cells, which was observed in the inoculation experiments, might therefore present a secondary effect of virus multiplication in the cells or be caused by substances other than the virus. To clarify this point, purified virus should be used as the inoculant.

Electron microscopic examination was carried out on an ultrathin section of the inoculated culture on the ninth day after inoculation. Many small particles, resembling rice dwarf virus particles in shape, but far smaller (30 m$\mu$ in diameter) than the rice dwarf virus particles (70 m$\mu$ in diameter), were seen in abundance in almost all of the cells (Fig. 9), while a few normal rice dwarf virus particles were found in some cells. The geometrical arrangement of the small particles suggests that these particles are most likely virus particles. Since polymorphism or developmental stages of rice dwarf virus particles are not known, there is a possibility that the small particles represent some

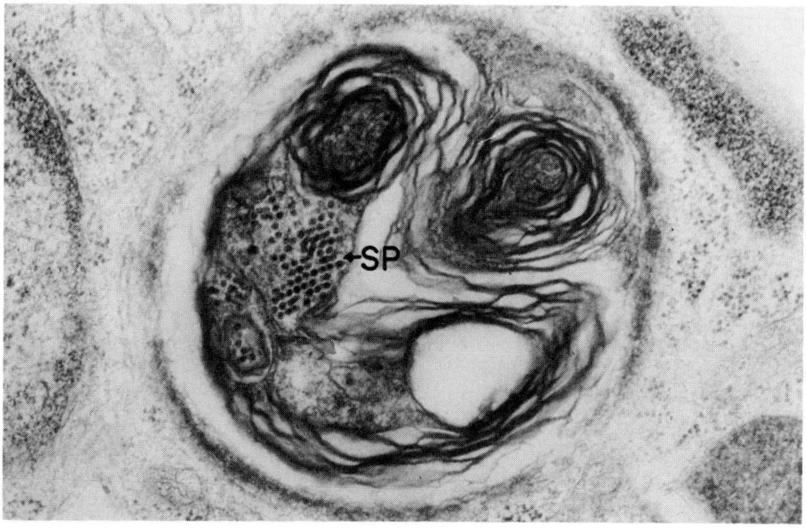


*Fig. 9.* Small particles (SP) appeared in the cells of *Nephotettix cincticeps* inoculated with rice dwarf virus *in vitro* (× 12,600).

unknown virus contaminant. Similar small particles have also been found in the cells of intact viruliferous leafhoppers (Nasu, 1965), but their infectivity has not been determined. Further investigations are required to clarify the nature of these small particles.

Recent electron microscopic studies demonstrated the multiplication of rice dwarf virus in the inoculated embryonic cells of *N. cincticeps*. In some sections viral matrix was seen in abundance and rich dwarf virus particles were often found at the periphery of the matrix (Fig. 10). Rice dwarf virus particles were also found in crystalline arrangement (Fig. 11), as well as in linear arrangement accompanied by a sheath-like substructure (Fig. 10). It therefore became evident that rice dwarf virus can multiply in the cultivated vector cells *in vitro* when artificially inoculated.

Recently Chiu *et al.* (1966) succeeded to inoculate and infect embryonic tissue cultures of *A. constricta* with wound tumor virus by the same technique as that of Mitsuhashi (1965c). The inoculant was prepared by homogenizing 30–60 viruliferous insects in 2 ml of the culture medium. This inoculant seemed to contain much more virus than the inoculant used by Mitsuhashi. The inoculation was carried out by replacing the culture medium with the inoculant and allowing an adsorption period of 2 or 3 hr at room temperature. No cytopathic effect of the virus infection, including granulation, was observed under

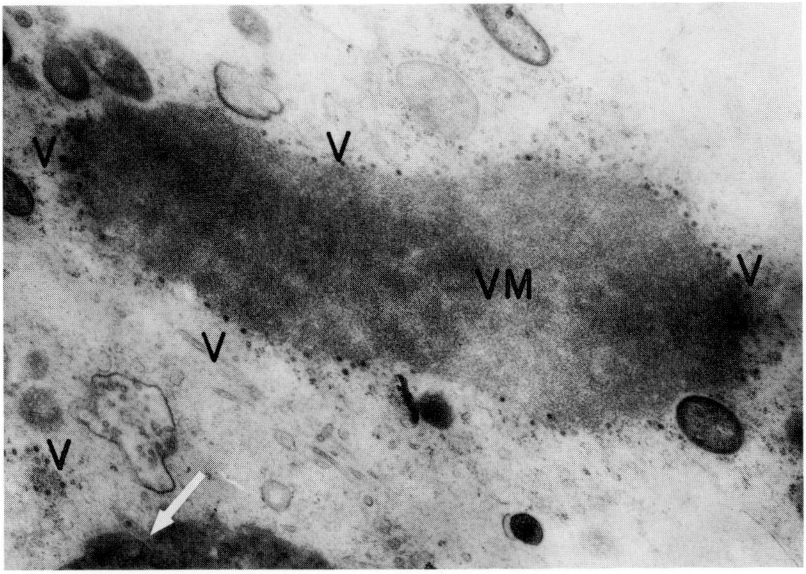


*Fig. 10.* Viral matrix (VM) and rice dwarf virus particles (V) appear in the cells of *Nephotettix cincticeps* inoculated with rice dwarf virus *in vitro.* Arrow shows rice dwarf virus particles accompanied by sheath-like substructure ($\times$ 7,900).

the light microscope. Evidence of virus multiplication in the inoculated cultures was obtained in three ways. On the third day after inoculation, the first evidence of viral antigens was noted in a few cells in the inoculated cultures by fluorescent antibody staining. The intensity of staining and the number of stained cells increased on later days. The viral antigens appeared only in the cytoplasm and not in nuclei. The relative concentration of infective virus was determined from the results of precipitin ring time tests on extract from the insects which were injected with the homogenate of inoculated tissues or the medium. Although the relative virus concentration detected in the cells did not exceed the level in the original inoculant, there was an increase in infectivity as the inoculated cultures were incubated. Serial passages of the virus in cultures were carried out at 8- to 11-day intervals. At each passage the virus was recovered at about the same concentration from the cell samples. Besides these evidences of virus multiplication, it was shown that the subcultured original explants, which were inoculated with the virus at the primary cultures, continued to yield cell sheets which took the specific staining with conjugated antiserum.

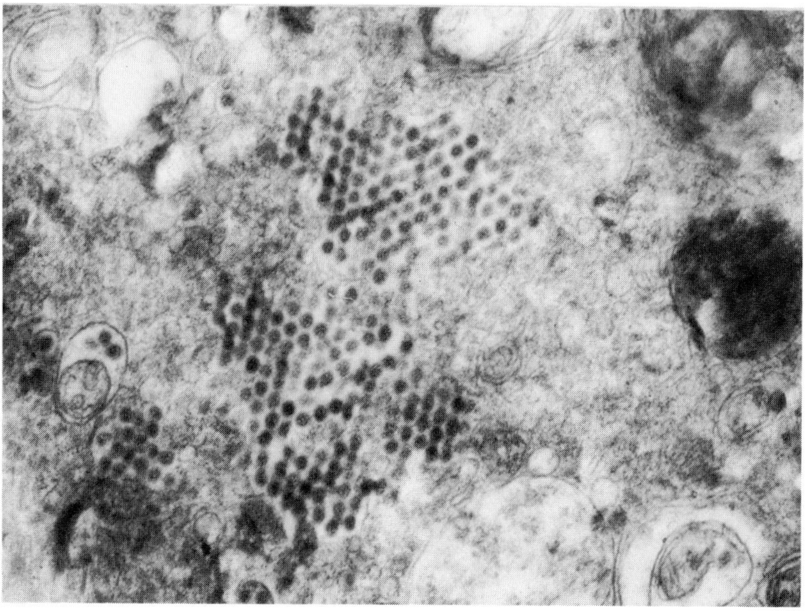

*Fig. 11.* Rice dwarf virus particles in a crystalline arrangement which appeared in the cells of *Nephotettix cincticeps* inoculated with rice dwarf virus *in vitro* (× 20,280).

## V. CONCLUSIONS

At present, we can obtain some growing cells from primary cultures of leafhopper vector tissues. Reproducibility of embryonic tissue cultures, especially, is very high. Such cells as those observed in the primary cultures can be used for plant virus research. However, for the titration of plant viruses by means of plaque formation or lesion formation, a uniform, monolayer cell sheet which covers the entire area of the glass surface may be necessary. Therefore, the establishment of a vector cell line is the ultimate goal of vector tissue culture. In studies of insect viruses, inoculation experiments have already been carried out on some established cell lines or cell strains (Bellett and Mercer, 1964; Bellett, 1965a, 1965b; Mitsuhashi, 1966c). Furthermore, Grace (1966) has established the cell lines of *Aedes aegypti,* which is a vector of yellow fever virus. These cell lines can be used in the studies of animal viruses transmitted by *A. aegypti.* It is hoped that the cell lines of vector insects which transmit plant viruses will be established in the near future.

Tissue cultures of several leafhopper vector species have now become possible, at least as primary cultures. However, more leafhopper vector species should be examined for their ability to produce growing cells *in vitro,* since, the same culture techniques might not necessarily be applicable to all leafhopper species, and each plant virus may require its own vector cells when it multiplies *in vitro*. The tissue cultures of an aphid (Tokumitsu and Maramorosch, 1966) or planthoppers (Mitsuhashi, unpublished) are still unsatisfactory and need more improvement, and, furthermore, tissue cultures of vector insects other than homopterous insects should also be studied in the future. Besides, mite- and nematode-borne viruses should be investigated by means of tisue cultures. With some animal viruses, for instance, eastern encephalomyelitis virus, the tissue culture of tick vectors has been used for *in vitro* inoculation experiments (Řeháček, 1963, 1965; Řeháček and Pešek, 1960).

The provide the most suitable insect materials for the vector tissue culture, the establishment of aseptic rearing methods of vectors is very important. There are various methods of rearing insects under aseptic conditions, and improvements and modifications may be necessary for diverse species of insects. The rearing of insects on synthetic diets under aseptic conditions is preferable. The same is true in cases of mites or plant parasitic nematodes. Recently, a mite, *Tetranychus ulticae,* was reared on a chemically defined diet under aseptic conditions (Rodriguez, 1966). A method to obtain and store an aseptic plant-parasitic nematode, *Pratylenchus penetrans,* has been established, and some special storage trap systems have been devised in order to provide aseptic nematodes (Tiner, 1966). These techniques may be applied to the rearing of mite vectors and nematode vectors of plant viruses, respectively.

Recently, some intracellular symbiotes of *N. cincticeps* were reported to play a role in the transovarial transmission of rice dwarf virus (Nasu, 1965). Moreover, the intracellular bacteroid symbiotes of *N. cincticeps* have been suspected to have some relationship to virus multiplication in the vector cells (Nasu, 1965). Generally, it is known that most homopterous insects have more than one species of intracellular symbiotes (Buchner, 1953). Therefore, interactions between intracellular symbiotes and plant viruses might play a role in homopterous insect vectors other than *N. cincticeps.* Fortunately, some intracellular symbiotes are known to be able to multiply in the cultivated host cells *in vitro* (Mitsuhashi and Maramorosch, 1964), and also in artificial culture media (Orenski *et al.,* 1965). The cultivation of vector cells which contain intracellular symbiotes, as well as cultivation of the symbiotes themselves, may be used as tools for the studies of the relationship between these symbiotes and plant viruses.

Finally, the present author would like to emphasize again the urgent need for the establishment of vector cell lines for the progress of the studies in this particular field.

## ADDENDUM

Recently establishment of cell lines from leafhopper vectors was reported (Chiu and Black, 1967). The cell lines derived from the embryos of *Agallia constricta* were comprised mostly of epithelial like cells. The cells formed monolayer cell sheets, and were susceptible to wound tumor virus. The infected cells did not show any cytopathic effect, and grew well even after infection. The authors suggested that their method for establishing leafhopper cell lines would be applicable to other leafhopper species.

## BIBLIOGRAPHY

Bellett, A. J. D. 1965a. The multiplication of *Sericesthis* iridescent virus in cell cultures from *Antheraea eucalypti* Scott. II. An *in vitro* assay for the virus. Virology 26:127–131.

——— 1965b. The multiplication of *Sericesthis* iridescent virus in cell cultures from *Antheraea eucalypti* Scott. III. Quantitative experiments. Virology 26: 132–141.

Bellett, A. J. D., and Mercer, E. H. 1964. The multiplication of *Sericesthis* iridescent virus in cell cultures from *Antheraea eucalypti* Scott. I. Qualitative experiments. Virology 24:645–653.

Buchner, P. 1953. Endosymbiose der Tiere mit Pflanzlichen Mikroorganismen. Verlag Birkhäuser, Basel. 771 p.

Buck, J. B. 1953. Physical properties and chemical composition of insect blood. p. 147–190. *In* K. D. Roeder [ed.] Insect physiology. Wiley, New York.

Carlson, J. G. 1946. Protoplasmic viscosity changes in different regions of the grasshopper neuroblast during mitosis. Biol. Bull. 90:109–121.

Chen, T. A., Kilpatrick, R. A., and Rich, A. E. 1961. Sterile culture techniques as tools in plant nematology research. Phytopathology 51:799–800.

Chiu, R-J., and Black, L. M. 1967. Monolayer cultures of insect cell lines and their inoculation with a plant virus. Nature 215:1076–1078.

Chiu, R-J., Reddy, D. V. R., and Black, L. M. 1966. Inoculation and infection of leafhopper tissue cultures with a plant virus. Virology, 30:562–566.

Day, M. F., and Grace, T. D. C. 1959. Culture of insect tissues. Ann. Rev. Entomol. 4:17–38.

Glaser, R. W. 1917. The growth of insect blood cells *in vitro*. Psyche 24:1–7.

Goldschmidt, R. 1915. Some experiments on spermatogenesis *in vitro*. Proc. Nat. Acad. Sci. U.S. 1:220–222.

Grace, T. D. C. 1959. Tissue culture for arthropod viruses. Trans. N.Y. Acad. Sci. Ser. II. 21:237–241.

——— 1962. Establishment of four strains of cells from insect tissues grown *in vitro*. Nature 195:788–789.

——— 1966. Establishment of a line of mosquito (*Aedes aegypti* L.) cells grown *in vitro*. Nature 211:366–367.

Hirumi, H. 1965. Study on insect tissue culture. Cultivation of embryonic leafhopper tissues *in vitro* [in Japanese with English summary]. Wakayama Igaku 15:325–334.

Hirumi, H., and Maramorosch, K. 1963a. Cultivation of leafhopper (Cicadellidae) tissues and organs *in vitro*. Ann. Epiphyties 14:77–79.

———. 1963b. Recovery of aster yellows virus from various organs of the insect vector, *Macrosteles fascifrons*. Contrib. Boyce Thompson Inst. 22:141–152.

——— 1964a. Insect tissue culture: Use of blastokinetic stage of leafhopper embryo. Science 144:1465–1467.

——— 1964b. Insect tissue culture: Further studies on the cultivation of embryonic leafhopper tissues *in vitro*. Contrib. Boyce Thompson Inst. 22:343–352.

——— 1964c. The *in vitro* cultivation of embryonic leafhopper tissues. Exp. Cell Res. 36:625–631.

Horikawa, M., Ling, L-N. and Fox. A. S. 1966. Long-term culture of embryonic cells of *Drosophila melanogaster*. Nature 210:183–185.

Jones, B. M. 1962. The cultivation of insect cells and tissues. Biol. Rev. Cambridge Phil. Soc. 37:512–536.

Kimura, J. 1931. On the behavior of rice to mineral nutrients in solution-culture, especially compared with those of barley and wheat [in Japanese with English summary]. J. Imp. Agr. Exp. Sta. (Tokyo) 1:375–402.

——— 1932. Further studies on the specific traits of rice in regard to its nutritive behavior in solution culture [in Japanese with English summary]. J. Imp. Agr. Exp. Sta. (Tokyo) 2:1–32.

Makino, S., and Nakahara, H. 1953. Cytological studies of tumors. X. Further observations on the living tumor cells with a new hanging-drop method. Cytologia (Tokyo) 18:128–132.

Maramorosch, K. 1956. Multiplication of aster yellows virus in *in vitro* preparations of insect tissues. Virology 2:369–376.

Maramorosch, K., and Jensen, D. D. 1963. Harmful and beneficial effects of plant viruses in insects. Ann. Rev. Microbiol. 17:495–530.

Maramorosch, K., Mitsuhashi, J., Streissle, G., and Hirumi, H. 1965. Animal and plant viruses in insect tissues *in vitro*. Bacteriol. Proc. 1965, p. 120. (Abstr. 65th Ann. Meeting Am. Soc. Microbiol.)

Martignoni, M. E. 1960. Problems of insect tissue culture. Experientia 16:125–128.

Martignoni, M. E. 1962. Insect tissue culture: A tool for the physiologist. p. 89–110. *In* V. J. Brookes [ed.] Insect physiology. Oregon State University Press, Corvallis.

Mitsuhashi, J. 1964a. Axenic rearing of insect vectors of plant viruses. Ann. N.Y. Acad. Sci. 118:384–386.

——— 1964b. Insect tissue culture up to present [in Japanese]. Plant Protection 18:359–366, 419–422.

——— 1965a. Aseptic rearing and tissue culture of leafhoppers [in Japanese]. Japan. J. Appl. Entomol. Zool. 9:69.

——— 1965b. *In vitro* cultivation of the embryonic tissues of the green rice leafhopper, *Nephotettix cincticeps* Uhler (Homoptera: Cicadellidae). Japan. J. Appl. Entomol. Zool. 9:107–114.

——— 1965c. Preliminary report on the plant virus multiplication in the leafhopper vector cells grown *in vitro*. Japan. J. Appl. Entomol. Zool. 9:137–141.

——— 1965d. Aseptic rearing of leafhoppers and planthoppers (Homoptera: Cicadellidae and Delphacidae). Kontyû 33:271–274.

——— 1965e. Multiplication of plant virus in the leafhopper vector cells grown *in vitro*. U.S.-Japan Joint Seminar, Tokyo, 60–69.

——— 1965f. Insect tissue culture: A review [in Japanese]. Biol. Sci. (Tokyo) 17:105–117.

——— 1965g. Multiplication of rice dwarf virus in the green rice leafhopper cells grown *in vitro* [in Japanese]. Ann. Phytopathol. Soc. Japan 31:385.

——— 1966a. Tissue culture of the rice stem borer, *Chilo suppressalis* Walker (Lepidoptera: Pyralidae). II. Morphology and *in vitro* cultivation of hemocytes. Appl. Entomol. Zool. 1:5–20.

——— 1966b. Chromosome numbers of the green rice leafhopper, *Nephotettix cincticeps* Uhler (Homoptera: Cicadellidae). Appl. Entomol. Zool. 1:103–104.

——— 1966c. Multiplication of *Chilo* iridescent virus in the *Chilo suppressalis* tissues cultivated *in vitro* (Lepidoptera: Pyralidae). Appl. Entomol. Zool. 1:199–201.

——— 1966d. Tissue culture in insects [in Japanese]. Kagaku-to-Seibutsu 4:608–613.

Mitsuhashi, J., and Maramorosch, K. 1963a. A successful method for rearing leafhopper vectors of plant viruses under aseptic conditions. Proc. 16th Int. Congr. Zool., Washington, 1963, 1:3.

——— 1963b. Aseptic cultivation of four virus transmitting species of leafhoppers (Cicadellidae). Contrib. Boyce Thompson Inst. 22:165–173.

——— 1964. Leafhopper tissue culture: Embryonic, nymphal, and imaginal tissues from aseptic insects. Contrib. Boyce Thompson Inst. 22:435–460.

Nakahara, H. 1953. Cytological studies of tumors. XI. Observations of the multipolar division in tumor cells of ascites tumor of rats by phase microscopy. J. Faculty Sci. Hokkaido Univ. Ser. VI. 11:473–480.

Nasu, S. 1965. Electron microscoptic studies on transovarial passage of rice dwarf virus. Japan. J. Appl. Entomol. Zool. 9:225–237.

Nowell, P. C. 1960. Phytohemagglutinin, an initiation of mitosis in cultures of normal human leukocytes. Cancer Res. 20:462–466.

Orenski, S. W., Mitsuhashi, J., Ringel, S. M., Martin, J. F., and Maramorosch, K. 1965. A presumptive bacterial symbiont from the eggs of the six-spotted leafhopper, *Macrosteles fascifrons* Stål. Contrib. Boyce Thompson Inst. 23:123–126.

Půža, V. 1963. Beobachtung der Bildung zwei zweikerniger Zellen aus einer Zelle der Gewebekultur ohne experimentellen Eingriff. Experientia 19:529–530.

Řeháček, J. 1963. Propagation of tick-borne encephalitis virus in tick tissue cultures. Ann. Epiphyties 14:199–204.

——— 1965. Cultivation of different viruses in tick tissue cultures. Acta Virol. (Prague). 9:332–337.

Řeháček, J., and Pešek, J. 1960. Propagation of eastern encephalomyelitis (EEE) virus in surviving tick tissues. Acta Virol. (Prague) 4:241–245.

Rodriguez, J. G. 1966. Axenic arthropoda: Current status of research and future possibilities. Ann. N.Y. Acad. Sci. 139:53–64.

St. Amand, G. A., Anderson, N. G., and Gaulden, M. E. 1960. Cell division. (IV)

Acceleration of mitotic rate of grasshopper neuroblasts by agmatine. Exp. Cell Res. 20:71–76.

Stillwell, E. F. 1943. The production of multipolar mitoses in normal embryonic chick cells. Science 98:264.

——— 1947. The influence of temperature variation upon the occurrence of multipolar mitoses in embryonic cells grown *in vitro*. Anat. Record 99:227–238.

Tiner, J. D. 1966. Collection and storage of axenic inoculum of plant parasitic nematodes in the laboratory. Ann. N.Y. Acad. Sci. 139:111–123.

Ting, K. Y., and Brooks, M. A. 1965. Sodium: potassium ratios in insect cell culture and the growth of cockroach cells (Blattariae: Blattidae). Ann. Entomol. Soc. Amer. 58:197–202.

Tokumitsu, Y., and Maramorosch, K., 1966. Survival of aphid cells *in vitro*. Exp. Cell Res. 44:652–655.

Vago, C., and Flandre, O. 1963. Culture prolongée de tissus d'insectes et de vecteurs de maladies en coagulum plasmatique. Ann. Epiphyties 14:127–139.

Wyatt, G. R. 1961. The biochemistry of insect hemolymph. Ann. Rev. Entomol. 6:75–102.

Wyatt, G. R., Loughheed, R. C., and Wyatt, S. S. 1956. The chemistry of insect hemolymph. Organic components of the hemolymph of the silkworm, *Bombyx mori* and two other species. J. Gen. Physiol. 39:853–868.

Wyatt, S. S. 1956. Culture *in vitro* of tissue from the silkworm, *Bombyx mori* L. J. Gen. Physiol. 39:841–852.

# Insect Diseases Induced by Plant-Pathogenic Viruses*

D. D. JENSEN

*Division of Entomology,*
*University of California, Berkeley*

## I. INTRODUCTION

For many years the relationship between plant-pathogenic viruses and insects was regarded as a passive one. Insects were assumed to ingest virus with the sap of infected host plants but were thought to be unaffected by the virus—even though some species were known to serve as virus vectors. This early misconception was gradually dispelled by evidence showing that plant viruses may have very complex relationships with their vectors, and sometimes even with nonvectors. Certain viruses multiply in their insect as well as in their plant hosts. Some are transmitted through the eggs of their vectors to the progeny—indeed, a few viruses could probably exist indefinitely in some of their vector species in the absence of plant hosts of the virus. Viruses have also been shown to play a beneficial role in the biology of insects by altering the host plant or the insect in such a way that the reproductive capacity of the insect is increased or that diseased plants can serve as food hosts for leafhoppers that die on healthy plants of the same species.

Of special interest in recent years has been the increasing evidence that at least a few viruses cause tangible and sometimes pathologic effects upon their insect vectors. These include cytological, proliferative, and metabolic changes that possibly contribute to such harmful effects as premature death and sterility or reduced fecundity of the susceptible vectors.

Previous review papers (Jensen, 1963; Maramorosch and Jensen, 1963) treated beneficial and neutral as well as some harmful effects of plant viruses in insects. The present discussion will be confined to a

* This study was aided by U.S. Public Health Service Research Grant No. AI–03490 from the National Institute of Allergy and Infectious Diseases to the University of California, Berkeley.

brief review of diseases or abnormal changes that occur in insects as direct or indirect effects of plant pathogenic viruses. Special emphasis will be given to the developments that have taken place since the earlier reviews.

Because electron microscopy is revealing a number of different plant viruses in the cells and tissues of the respective vectors, this phase of the subject will be considered only in general terms. Viruses apparently may invade most if not all of the insect tissues and organs of the vector. Reported virus sites include the salivary glands, blood cells, fat body, Malpighian tubules, intestinal tract, ovarian tubules, and the nervous system.

Shikata (1966) reported that, like wound tumor virus, rice dwarf virus particles of identical size and form were isolated from both the rice plant and the leafhopper host. By means of electron microscopy, he found that the virus clusters in the insect cells occupied a relatively small amount of space ". . . causing neither death nor appreciable damage, as would be the case if they filled the whole area in a cell." In contrast, large masses of virus particles were found in the infected plant cells. However, Shikata also concluded that the rice dwarf virus particles appeared to have a closer morphological relationship to some of the insect viruses than to the plant viruses.

There is evidence, nevertheless, that in some instances insect vector cells appear to be almost completely filled with virus. Such aggregations and dense concentrations of virus in some of the vector tissues might logically be expected to cause cytological changes and to alter the physiology of the infected insects. Recent and current research indicates that such is the case. Hirumi, Granados, and Maramorosch (1967), for example, observed such an association in the nervous system of the vector, *Agallia constricta*, carrying wound tumor virus. Degenerating ganglion cells were found to carry virions, either isolated or in tubular formation. The cytoplasmic matrix of these infected cells was transformed into homogeneous fine granules.

> "The membrane structure of cytoplasmic organelles, such as the endoplasmic reticulum, the nuclear membrane, and the Golgi apparatus, were indistinct. No normal mitochondria were found in the cytoplasm. The ribosomes of the cell were loosely coagulated. The nuclear substance was also changed to a homogeneous fine granular substance."

Tissues that serve as virus multiplication sites might be expected to be particularly adversely affected.

## II. INCLUSION BODIES

Virus-induced inclusion bodies, in addition to other disease symptoms, have been known in plants for many years. Sporadic efforts have been made during the past 40 years to find similar effects in viruliferous vectors of plant viruses. Dobroscky (1929, 1931) sought but found no cytological effects in the aster leafhopper (*Macrosteles fascifrons*) carrying aster yellows virus. Blattny (1931) and Hartzell (1937) reported effects in viruliferous aphids (*Myzus persicae*) and leafhoppers (*Macropsis trimaculata*), respectively, that have never been confirmed.

### *A. Zakuklivanie Virus*

In Russia, Sukhov and others published several papers on a disease of cereals, known as zakuklivanie, oat mosaic, pupation disease, or pseudo-rosette disease of oats. The causal virus is transmitted by *Delphacodes* (*Delphax*) *striatella* (Fallen). Intracellular polymorphous protein inclusions were reported (Sukhov, 1940a) to occur in diseased plants either as needle-like crystals or as semiliquid, loop-like structures from which the crystals were believed to have been derived. The crystals, which could be concentrated in large masses from diseased plant extracts (Sukhov *et al.*, 1943) were considered to be virus. Inclusions, called X bodies, were also reported from the cells of the intestine, fat tissue, and salivary glands of the vector (Sukhov, 1940b). These bodies were described as being somewhat irregular in shape and of a semiliquid, viscous consistency. Sukhov also reported the lumen of the intestine to contain many aciculate crystals of protein nature. These and the insect tissue inclusions were considered to be different forms of the virus. Crystals could be detected in the intestine after 6 hr of feeding on a diseased plant (Sukhov, Vovk, and Alexeeva, 1943) and disappeared in 7–8 days after the insects were placed on a healthy plant. Insects feeding for 2 days through a wax film into a dense suspension of purified virus preparation filled their intestine with crystals similar to those occurring in insects fed on diseased plants (Sukhov, 1943b). Following an incubation period, 17 of 30 such insects transmitted virus compared with 2 of 92 controls. More recently, Sukhov (1956) reaffirmed his interpretation of his experiments as indicating that zakuklivanie virus exists in two forms in both host plant and vector.

### *B. Winter Wheat Mosaic Virus*

This virus was reported (Sukhov, 1943a) to cause the formation of numerous needle-like crystals in the intestine of the vector, *Delto-*

*cephalus striatus* L., and also in the leaf cells of infected wheat plants. However, the crystals could be detected only after the intestine or the plant cells were placed in an acid solution at *p*H 4.

### *C. Western X-Disease Virus (WXV)*

Crystalline inclusions, similar to those reported by Sukhov from the intestines of *Delphacodes striatella,* have been found in *Colladonus montanus* after feeding on celery plants infected with western X-disease virus (Lee and Jensen, 1963).

The needle-like crystals occur in the lumen of the alimentary tract and range in size from 70 to 295 μ in length and 3–20 μ in width. Crystals rarely appeared until after 16 days of feeding on a diseased plant. However, by 30 days, 50–90% of the leafhoppers had crystalline inclusions. The crystals gradually disappeared from leafhoppers after transfer to healthy celery plants. Thus, after 20 days on healthy plants only 8 of 20 individuals retained crystals compared with 19 of 20 for controls that remained on diseased plants.

These crystals are long, slender, pointed at each end, and in cross section appear to be diamond shaped or hexagonal. They are soluble in 0.01 M carbonate and in 0.01 M phosphate buffer, *p*H 8.0, but are stable in 0.01 N HCl. They dissolve slowly in distilled water.

The nature and origin of the crystals are not yet adequately known, but it is improbable that they are the virus itself. Whitcomb, Jensen, and Richardson (1967) found that crystals did not occur in leafhoppers infected with western X-disease virus by injection. Similar crystals were found in *C. montanus* nymphs after they had fed on *Plantago major* infected with aster yellows virus. They were also found in the intestine of *Fieberiella florii* (Stål) after feeding on WXV-diseased celery. This leafhopper is also a vector of the virus.

## III. CYTOPATHOLOGY AND HISTOPATHOLOGY IN LEAFHOPPERS

Under this heading will be discussed the apparent direct or indirect effects of viruses on the cells and tissues themselves. This is in contrast to the appearance within the insect cells of individual virions or the various arrangements of virions in tubular formation or in massive aggregations.

### *A. Aster Yellows Virus*

The first extensive study made of possible cytological changes in a virus vector was by Littau and Maramorosch (1956, 1960). A comparison was made of tissues and cells from both viruliferous and nonviruliferous aster leafhoppers and it was found that the nuclei in the fat body cells of some of the infective insects, particularly the males, were stellate or sharply irregular in form rather than round as in many nonviruliferous insects. Continuous access to diseased plants appeared to be necessary for more than a few individuals to develop the more extreme stellate nuclei. However, feeding on diseased plants was not the sole cause of the changes because if virus-free insects were injected with infectious leafhopper extracts, some of them developed cytological effects lacking in controls injected with healthy insect extracts. That the cytological effects were in some way related to the virus was further suggested by the fact that a severe, celery-infecting strain of aster yellows virus had no cytological effect on the aster leafhopper.

### *B. Rice Viruses*

Nasu (1963) reported several cytological abnormalities in viruliferous vectors of rice viruses. In *Delphacodes striatella* he found that many embryos died early due to rice stripe virus and others died when they were fully formed and ready to hatch. The number of mycetocytes was also reduced in infected embryos of this species. He attributed these abnormalities to virus disturbances of maternal physiology.

*Rice dwarf virus* was found by Nasu to cause effects in the black-streaked green rice leafhopper (*Nephotettix* sp.). These were primarily in the fat body and the mycetome, but not in the salivary glands, the alimentary canal, or the Malpighian tubules. He found the fat body to assume different forms depending upon the life stage of the leafhoppers. "The fat body of infected adults contains more vacuoles in the cells, the shape of the nuclei being often modified into a star-shape. This is very similar to that observed in infected nymphs." This virus was also reported to cause abnormal hardening of the mycetome in *Nephotettix* sp.

*Rice yellow-dwarf virus* also has been found to cause abnormalities in its vectors. Takahashi and Sekiya (1962) reported that in *Nephotettix cincticeps,* the nuclei of the fat cells first enlarged and then shrank as a result of the virus and that vacuoles increased so greatly that entire cells appeared to be completely reticulated. They also described cytochemical changes, expressed in the Feulgen reaction and the susceptibility to methyl green staining, which they attributed to an increase and subsequent decrease in the DNA and RNA content.

*Rice stripe virus.* Okuyama (1962), in a careful histological study of viruliferous and virus-free *Delphacodes striatella,* found no differences in the morphology of blood cells, reproductive organs, salivary glands, the alimentary tract, epidermal cells or the fat body of virus-free insects and those carrying rice stripe disease virus. However, chemical analyses revealed a reduction in the glycogen content (Bauer-Feulgen reaction) and the carbohydrate content (periodic acid–Schiff reaction) of isolated cells of the fat body and mycetome of viruliferous leafhoppers. Okuyama also observed a decrease in the longevity of viruliferous vectors, as well as a reduction in fecundity.

### *Winter Wheat Mosaic Virus*

Lomakina, Razvyaskina, and Shubnikova (1963) summarized the cytological and histochemical changes they found in the fat body of the vector of winter wheat mosaic virus as follows:

> "Cicadas *Psammotettix striatus* Fall were infected with winter wheat mosaic virus by feeding on infected plants. The virus-carrying cicadas showed considerable changes in the structure and metabolic chemistry of fat body cells. There were changes of shape and structure of nuclei; a significant reduction in the amount of RNA in the cytoplasm and nucleoli; an alteration in the distribution of DNA in nuclei; depletion of polysaccharides in the cytoplasm. The above histochemical changes adversely affected the development and structure of nymphs from virus-carrying cicadas. Peculiar inclusions were found in the cytoplasm of fat body cells of virus-infected cicadas. The inclusions were devoid of any histochemically detectable nucleic acids, but showed the presence of histidine."

### *Western X-Disease Virus (WXV)*

The discovery that this virus is lethal to its insect vector (*Colladonus montanus*) (Jensen, 1958, 1959a) suggested the probability that the virus must be damaging the insect in ways that might be detectable microscopically. Leafhopper vectors, acquiring virus by feeding on diseased plants or by injection with infectious leafhopper extracts, have been found to die earlier than control insects. Death due to virus occurs primarily between 35 and 70 days following injection (Jensen *et al.*, 1967). A search for histopathological symptoms caused by WXV was made by periodically examining, under the compound microscope, samples of sectioned leafhoppers that had been previously injected with infectious or with healthy hemolymph (Figs. 1–5). Leafhoppers sacrificed for this purpose were fixed, embedded, sectioned, mounted, and stained with acid fuchsin and Mallory's triple stain (Whitcomb *et al.*, 1967).

The injected insects were caged singly on healthy celery plants to determine which ones were transmitting virus. Individual mortality records were also kept of the leafhoppers not harvested for microscopic

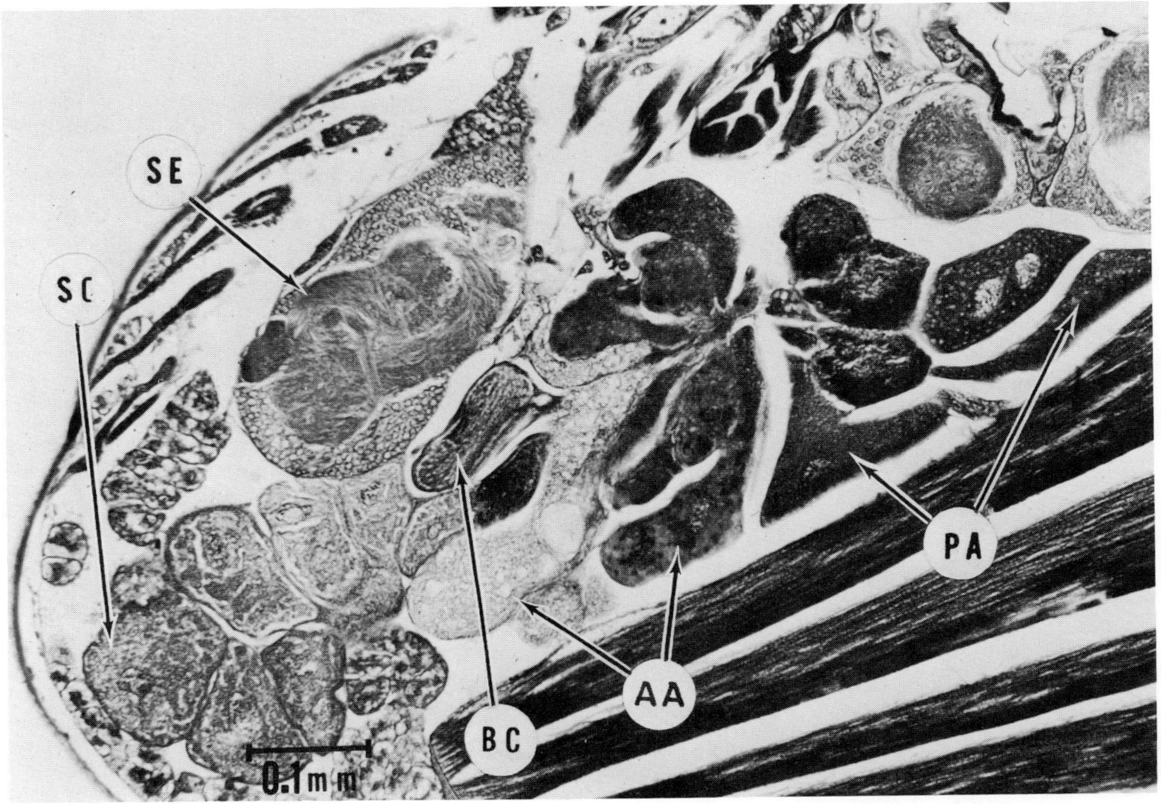


*Fig. 1.* Sagittal section through head of healthy *Collandonus montanus*, approximately 4–5 weeks after molting to adult stage. Salivary glands: SC, serous cell; PA, posterior lobe acini; BC, basal serous cells; AA, posterior acini of anterior lobe. SE, supra-esophageal ganglion. ×88.

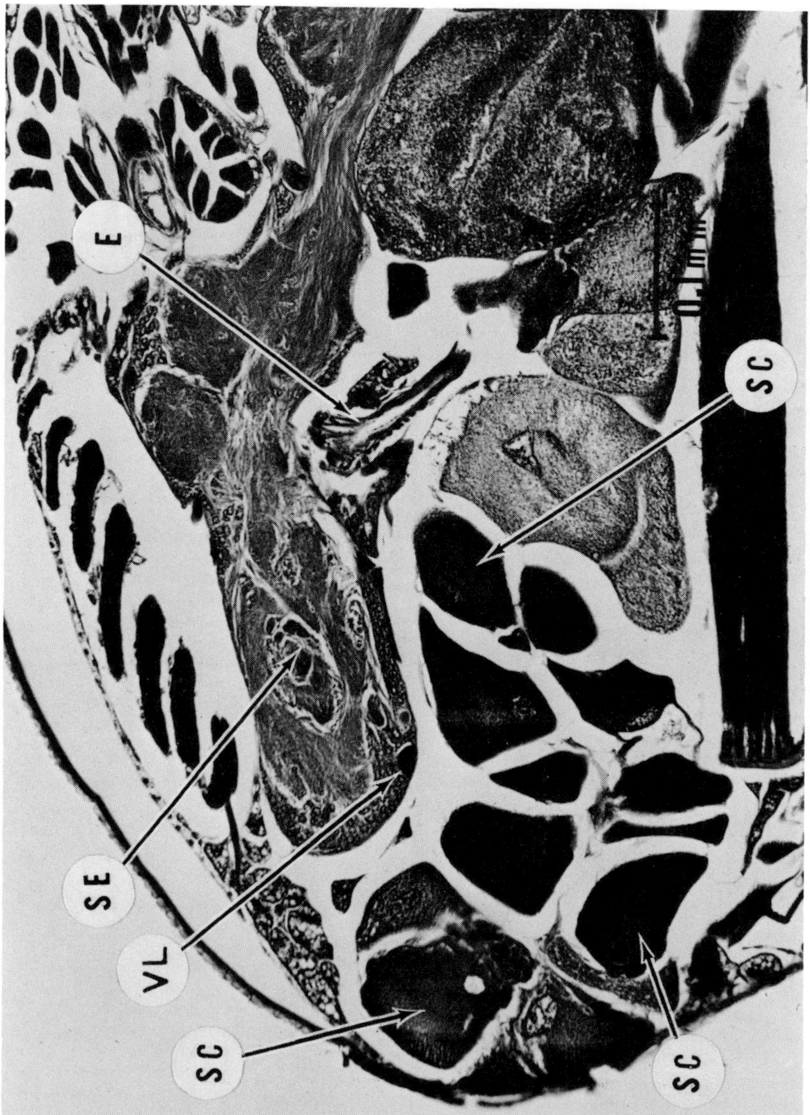


*Fig. 2.* Sagittal section through head of *C. montanus*, 46 days after injection with western X-disease virus. All serous cells (SC) show advanced pathology. SE, supraesophageal ganglion; VL, viral lesion; E, esophagus. ×88

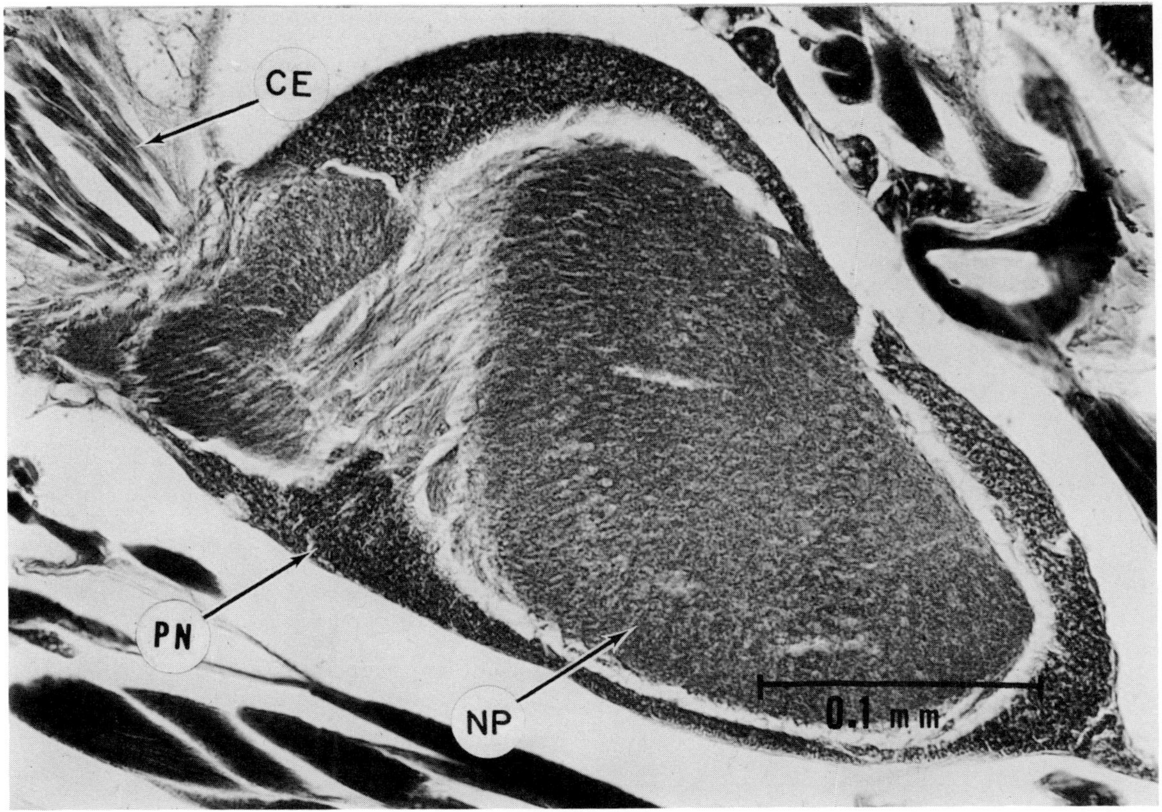


*Fig. 3.* Optic lobe of *C. montanus* 48 days after injection with hemolymph from healthy leafhopper. CE, compound eye; PN, perineurium; NP, neuropil. × 200.

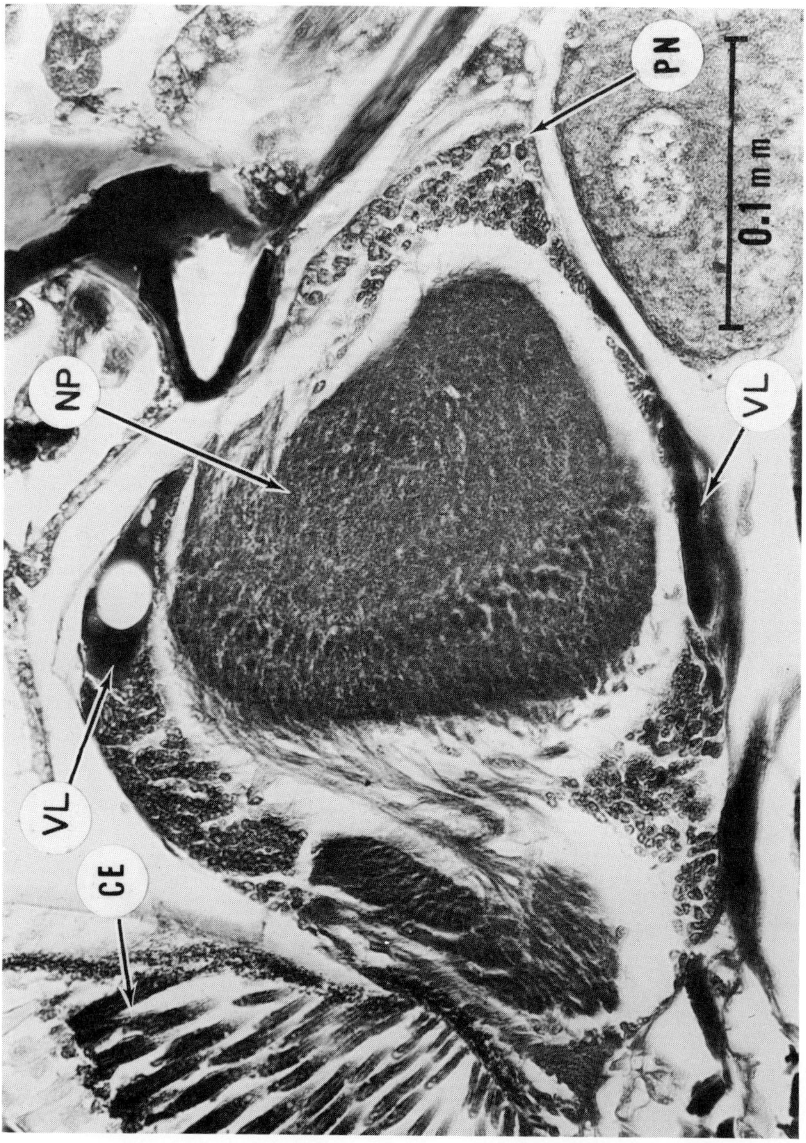


*Fig. 4.* Optic lobe of *C. montanus* 48 days after injection with hemolymph from a viruliferous vector. VL, Viral lesion; CE, compound eye; PN, perineurium; NP, neuropil. $\times 200$

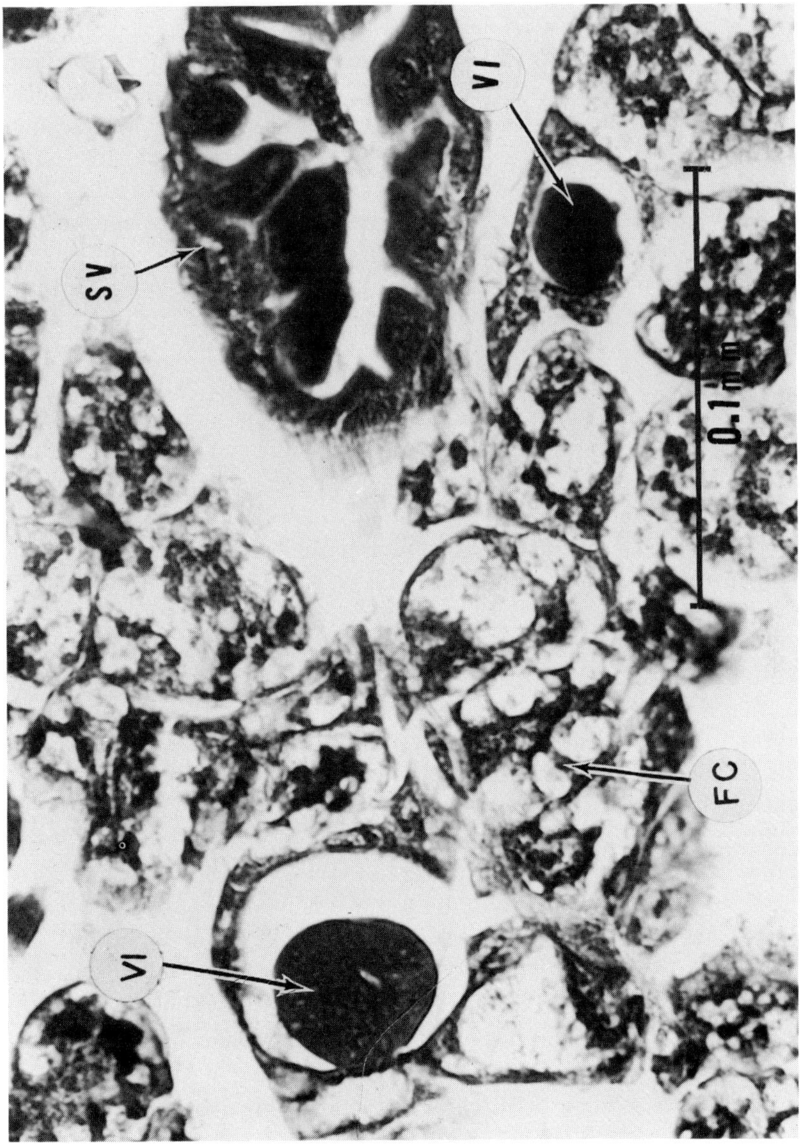


*Fig. 5.* Inclusion bodies in fat tissue of *C. montanus.* VI, Viral inclusion; SV, second ventriculus; FC, fat body cell. $\times$ 30.

examination. This permitted the plotting of survival curves. The mortality curve for viruliferous insects had a positive correlation with the severity of the histopathological symptoms.

As early as 14 days after leafhopper nymphs were caged on WXV-diseased celery plants, small lesions could be detected in the epithelium of the first and second ventriculus. One to two weeks later the lesions were more numerous, enlarged, and stained a deep blue or bluish gray (Whitcomb *et al.*, 1968). At this time cytopathological changes could also be found in the filter complex and in the Malpighian tubules. The pathology of the Malpighian tubules became severe and was apparently irreversible, whereas that of the ventriculus became less acute after 4–6 weeks. Insects receiving virus by injection developed cytopathological symptoms in the Malpighian tubules, but not in the rest of the alimentary tract.

Twenty-six days after leafhoppers were injected with WXV, or 35 days after virus acquisition by feeding, lesions developed in the perineurium of the brain and in the other cephalic and thoracic ganglia. They took stain densely and were particularly conspicuous in the optic lobe (Fig. 4).

Perhaps the most conspicuous cytopathological symptoms were found in the anterior lobe of the salivary gland where the cytoplasm of the affected serous cells was dense and darkly stained (Fig. 2) compared to the loose, lightly staining cytoplasm (Fig. 1) of healthy cells (Whitcomb *et al.*, 1967). These effects were suggested at 19 days after injection with virus and were conspicuous at 26 days (Whitcomb *et al.*, in press). Diseased serous cells eventually became completely filled with homogeneous chromatic material and, occasionally, became detached from the duct.

Pathology in the adipose tissue (Fig. 5) of leafhoppers that had acquired WXV was expressed in terms of cell abnormalities and also in the rate of depletion of adipose tissue (Whitcomb *et al.*, in press). Infected fat cells were often characterized by an accumulation of granules and nuclei at the cell periphery, the center of such cells becoming vacuolate and poorly stained. Virus infections tended to reduce the rate of fat cell depletion if the virus was acquired by injection. The insects that acquired the virus by feeding on diseased plants for 14 days showed an accelerated depletion of fat. However, if they were transferred to healthy plants, the depletion rate diminished and a month after first exposure to diseased celery, the viruliferous leafhoppers appeared to be utilizing fat more slowly than were the nonviruliferous controls. It was concluded ". . . that the physiological stress of feeding on a diseased plant increased the rate of depletion, but that viral replication, taking place in the absence of further stress, slowed depletion as it did in injected insects."

Evidence of WXV pathology was also found in other tissues of *C. montanus* (Whitcomb *et al.*, in press). Sheath cells of the mycetome developed intense basophilia and in some cases there was premature degeneration of the mycetome. Also affected were urate cells, the collaterial gland, the corpus cardiacum, pericardial cells, and connective tissue. Lesions in a particular tissue frequently occurred in apposition with lesions in a different adjacent tissue.

Another type of symptom of particular interest was occasionally found in insects receiving WXV but never in the controls (Whitcomb and Jensen, in press). These were described as proliferative symptoms and included:

> "nuclear divisions without cellular divisions in adipose and salivary tissue, irregular cellular divisions in the acini of the anterior lobe of the salivary glands, and proliferative outgrowths from the esophageal valve, spermatheca, and rectal pad. Several foci of proliferative pathology were found in each of the affected insects, suggesting that a systemic stimulus was involved in the etiology."

## IV. CYTOPATHOLOGY IN APHIDS

Viruses causing diseases in sugar beets, peas, and potatoes have been reported by Schmidt (1959, 1960) to cause cytological changes in the aphid vector, *Myzus persicae.* Potato leaf-roll virus and, in some cases, pea enation mosaic virus caused enlargement of the nuclei of the stomach cells. Instead of a large normal nucleolus there were several small bodies. No abnormalities were seen in the salivary glands. Aphids carrying yellow-net virus or beet yellows virus showed gradations in the cytoplasm of the fat body cells from granular to reticulate structures and the nuclei had forms similar to those described by Littau and Maramorosch (1956, 1960). Schmidt suggested that some of the apparent effects may have been due to the sectioning because similarities were sometimes found in the controls.

### *A. Barley Yellow-Dwarf Virus*

A study was made by Rutschky and Campbell (1964) to determine if barley yellow-dwarf virus caused cytological changes in one of its vectors, the English grain aphid, *Macrosiphum granarium* (Kirby). This is a persistent, circulative virus in the vector, and the authors investigated the possibility that fat body morphology might be used to determine the presence or absence of virus in the vector. However, they found no evidence of virus-induced changes. They did find stellate nuclei, but their occurrence was a function of the developmental stage of the aphid rather than of the presence of virus. The percent of stellate nuclei in the trophocytes ranged from 5% in first instar nymphs to 70% in viviparous adults.

## V. METABOLIC EFFECTS DUE TO VIRUS

The first publications dealing with the effects of plant viruses on the metabolism of their insect vectors were from Japan (Yoshii and Kiso, 1957a, 1957b, 1958; Yoshii, Kiso, and Kikumoto, 1959). They found that the virus causing dwarf disease of orange caused similar metabolic effects in the delphacid planthopper vector, *Geisha distinctissima* Wal., and in the host plant leaves. For example, oxygen consumption and total phosphorus were reduced in both insect and plant host. Yoshii (1959) made similar studies on the effects of rice dwarf virus in *Nephotettix bipunctatus cincticeps*, in which he found oxygen consumption and the respiratory quotient in viruliferous leafhoppers higher than normal.

Ehrhardt (1960) reported that potato leaf-roll virus caused a reduction in oxygen consumption of 30% during the first 30 hr following an 8-hr acquisition feeding period. He inferred that the leveling off of oxygen consumption, following 30 hr of gradual reduction, coincided with the completion of the latent period of the virus in the aphid.

Miller and Coon (1964) reported that barley yellow dwarf virus reduced oxygen consumption in the English grain aphid, *Macrosiphum granarium* (Kirby), by 14%. Viruliferous aphids also tended to: (*1*) develop from birth to adult in shorter than normal time; (*2*) live longer; (*3*) have a longer reproductive period; and (4) therefore produce more progeny than nonviruliferous controls.

## VI. PATHOGENIC VIRUS EFFECTS

Deleterious effects of plant viruses in their insect vectors are, for convenience, discussed separately from the work on histopathology, even though it is probable that certain of the recently discovered cytological and histological abnormalities contribute directly to injurious reactions in these insects. The first pathogens shown to be injurious to insect vectors were the rickettsiae that cause typhus fever. These agents, earlier thought to be viruses, kill the body louse vector but not the flea vector.

### *A. Western X-Disease Virus*

The first virus of plants or animals shown to be deleterious to an insect vector (Jensen, 1958, 1959a) was that causing western X disease of stone fruits and a yellows-type disease in celery, periwinkle, turnip, and radish in the United States. This virus has been found to cause a significant reduction in the life span of one of its vectors, *Colladonus montanus* (Van Duzee), when acquired either by feeding or by needle injection (Jensen *et al.*, 1967). Longevity was first compared between insects that had acquired virus from diseased celery and insects that had fed only on healthy plants. These tests, however, left a variable of nutrition that might explain differences in longevity. To assess the role of the virus itself more exactly, subsequent experiments eliminated the food plant variable and all test insects fed on virus-diseased celery plants. The test leafhoppers were then caged singly on healthy celery plants and transferred at intervals to new celery plants. The longevity and virus transmission record of each insect was kept. The mean longevity of the virus transmitters was then compared with that of the nontransmitters. In a typical experiment, 207 transmitting individuals had a mean longevity of 26 days after transfer from the diseased plant compared to 55 days for those that had failed to transmit virus. The proportion of insects acquiring virus in any given experiment could be predetermined in a general way by changing the access time on the diseased plant.

Many experiments have confirmed the fact that, after the completion of a virus latent period in the vector of approximately 30 days, the leafhoppers acquiring virus by feeding live only approximately half as long as do similar insects that failed to acquire virus.

Since only part of a group of insects acquires virus during any given day on a diseased plant, virus acquisition by feeding has obvious disadvantages in studying the relation of the virus to the vector. Many of these drawbacks can be eliminated, however, if the leafhoppers receive virus by injection instead of by feeding. This can be accomplished by transferring infectious hemolymph directly from diseased to healthy insects or by injecting the test insects with a clarified extract made from viruliferous leafhoppers (Whitcomb *et al.*, 1966). This method has provided further confirmation of the lethality of western X disease virus to its vector without involving any feeding on diseased plants (Jensen *et al.*, 1967).

Leafhoppers injected with virus-containing extracts were found to have shorter mean longevity than the controls that were injected with virus-free extracts. At 35 days after injection in typical experiments there was no difference in the survival of the two groups. However, from this day on, the mortality due to virus increased very rapidly in comparison with the mortality of the control insects. The mean longevity, after injection, of 246 leafhoppers receiving virus was 58 days compared with 77 days for 250 controls injected with healthy extract.

Leafhoppers, injected with hemolymph from viruliferous insects or with only slightly diluted infectious extracts, transmit virus much earlier than do insects that acquire virus by feeding on diseased plants (Whitcomb *et al.*, 1966). Insects acquiring virus by feeding on diseased plants rarely transmit before the 30th day following initial access to the inoculum. In contrast, the virus incubation period in injected insects may be as short as 11 days and most of them are transmitting by the 21st day.

The concentration of the virus injected into healthy leafhoppers determines the duration of the incubation period in the vector and the virus transmission pattern. Injection of four tenfold dilutions of infectious extract into healthy leafhoppers resulted in transmission at progressively later intervals by the insects receiving the more dilute suspensions (Whitcomb *et al.*, 1966). The longevity of the injected insects was also a function of the concentration of the injected virus extracts (Jensen *et al.*, 1967).

### *B. European Wheat Striate Mosaic Virus*

Watson and Sinha (1959) reported that this virus reduced the number of progeny of its vector, *Calligypona* (*Delphacodes*) *pellucida* F. However, Kisimoto and Watson (1965), working with the same virus and vector, subsequently concluded that the cause of the embryonic mortality in this case was due to inbreeding rather than to virus infection.

### *C. Rice Dwarf Virus*

This virus, which has been involved in some of the classical research on virus–vector relations, has also been studied during recent years as a cause of pathogenicity in its leafhopper vectors. Among its vectors to which virus is passed transovarially, are: *Nephotettix cincticeps*, *Nephotettix apicalis*, and *Inazuma* (*Deltocophalus*) *dorsalis* (Motschulsky).

Shinkai (1958, 1960, 1962) reported that the premature death of *I. dorsalis* leafhoppers inheriting rice dwarf virus through the egg may be a cause for the disappearance of the virus from insect colonies relying on transovarial passage. In a representative experiment, 30 nymphs produced by a viruliferous female were reared singly on rice seedlings.

All but three died before reaching the adult stage. Although Shinkai did not report what the longevity was for virus-free control insects, he indicated that the virus must be inherited through the egg, and not acquired through feeding, in order to kill the leafhoppers prematurely. He stated: "Nymphs obtained from crosses between non-viruliferous parents, even when fed with dwarf diseased rice plants, did not die prematurely as in the above case."

The best known vector of rice dwarf virus, *Nephotettix cincticeps*, transmits virus to such a high percentage of its progeny that the virus could survive for long periods of time, if not indefinitely, in the absence of its host plant. This would suggest that the virus is noninjurious to the vector. However, pathogenic effects of the virus in *N. cincticeps* are being found in relation to fecundity, nymphal mortality, and adult longevity.

Satomi *et al.* (personal communication) stated that the mean adult longevity of viruliferous leafhoppers was 19–20 days, compared with 29 days for healthy controls. Nymphal mortality was higher among viruliferous (35%) than among nonviruliferous (21.7%) *N. cincticeps*, but this was much lower than reported by Shinkai (1960) for *Inazuma dorsalis*. The data provided by Satomi *et al.* suggested that continuous access to virus-diseased plants does not increase the mortality significantly among transovarially infected individuals—at least not after the first four days. However, one experiment indicated that virus acquired by feeding does kill some fifth instar nymphs that would survive congenital infection only. Insects that acquire virus only through the egg have higher mortality in the first two days of life than in any single total instar after the first. The per cent mortality tends to decrease with increasing age and is lowest of all in the fifth instar. In contrast, insects that acquired the virus only by feeding on diseased plants had no mortality in the first two days of life and relatively little until the fifth instar in which 29% of the surviving nymphs died. Insects that acquired virus both transovarially and by feeding had high mortality peaks both in the first two days of life and also in the fifth instar. This perplexing result suggests that the virus acquired by feeding may cause the death in the fifth instar of some insects that had already received virus transovarially without having been killed by such virus.

Suenaga (1962) reported similar results with *N. bipunctatus* after the acquisition of rice dwarf virus transovarially or by feeding. Transovarially infected leafhoppers tended to die early in life whereas those that acquired virus by feeding had a high mortality in the fifth instar and very low fecundity.

### *D. Oat Virus*

Jedlinski (personal communication) has been studying a virus of oats that may be causing mortality of eggs and early instar nymphs in leafhopper colonies of *Graminella nigrifrons* (Forbes). The work was done at Urbana, Illinois, for the United States Department of Agriculture.

## VII. FECUNDITY

### *A. Western X-Disease Virus*

Leafhopper colonies of *Colladonus montanus* that had fed on WXV celery plants were observed to produce noticeably fewer progeny than control colonies (Jensen, 1962). The effect of the virus on the fecundity of the females was then tested on an individual basis. Nonviruliferous nymphs were divided into two groups, one group being caged on diseased celery and the other on healthy celery. After approximately 30 days, when all were adults and egg-laying was evident, individual females were caged on healthy celery plants for weekly transfers. The number of eggs laid by each female each week was recorded. The combined results of several experiments, involving 205 females from the viruliferous stock and 130 from the control stocks, were plotted to show the mean number of eggs laid per female per week of life during the experimental period of 6 weeks. Infective females laid 10 eggs per week during the second week and then declined to 2 eggs per week by the end of the sixth week. The mean number per control female rose to 22 by the fourth week and was 15 by the end of the sixth week. Thus, after infective individuals had virtually ceased to lay eggs, the control females were still producing at a rate higher than the maximum for the viruliferous insects. Similar experiments were conducted with leafhoppers that received virus by injection rather than by feeding. The results were confirmatory but the fecundity of the viruliferous leafhoppers was not reduced as sharply as had been the case with those acquiring virus by feeding.

### *B. Rice Dwarf Virus*

Nasu (1963) reported reduced fecundity, due to virus, in two vectors as follows: "The number of eggs laid by infected females is only 32–72% of the number of eggs laid by uninfected females in the green rice leafhopper [*Nephotettix cincticeps*], 55–63% in the black-streaked green rice leafhopper [*Nephotettix* sp. (A)], . . .".

Satomi, Nasu, and Suenaga (personal communication from Dr. H. Suenaga, National Kyushu Agricultural Experiment Station, Chikugo, Japan) have provided information on the number of eggs laid by each of 20 virus-free females and each of 20 females of *Nephotettix cincticeps* carrying rice dwarf virus. Forty-five percent of the viruliferous females were sterile compared with only 15% of the controls. The total numbers of eggs laid by the two groups were 807 and 1426, respectively.

## VIII. TISSUE CULTURE

The developing field of tissue culture holds promise of providing valuable information on the manner in which plant viruses may multiply and cause pathogenic effects in vector and also nonvector cells and tissues.

## BIBLIOGRAPHY

Blattny, C. 1931. Lze zjistiti přitomnost viru působičiho některe chloroby bramborů v jejich přenaseci mišicich? Věstník Kral. Čes. Spol Nauk Praha, Tř. II:1–7.

Dobroscky, I. D. 1929. Is the aster-yellows virus detectable in its insect vector? Phytopathology 19:1009–1015.

———. 1931. Morphological and cytological studies on the salivary glands and alimentary tract of *Cicadula sexnotata* (Fallen), the carrier of aster yellows virus. Contrib. Boyce Thompson Inst. 3(1):39–58.

Ehrhardt, P. 1960. Zum Sauerstoffverbrauch von *Myzus persicae* (Sulz.) vor und nach Aufnahme des Blattrollvirus. Entomol. Exp. Appl. 3:114–117.

Hartzell, A. 1937. Movement of intracellular bodies associated with peach yellows. Contrib. Boyce Thompson Inst. 8(5):375–388.

Hirumi, H., Granados, R. R., and Maramorosch, K. 1967. Electron microscopy of a plant-pathogenic virus in the nervous system of its insect vector. J. Virol. 1(2):430–444.

Jensen, D. D. 1958. Reduction in longevity of leafhoppers carrying peach yellow leaf roll virus. (Abstr.) Phytopathology 48:394.

———. 1959a. A plant virus lethal to its insect vector. Virology 8:164–175.

———. 1959b. Insects, both hosts and vectors of plant viruses. Pan-Pac. Entomol. 35:65–82.

———. 1962. Pathogenicity of Western X-disease virus of stone fruits to its leafhopper vector, *Colladonus montanus* (Van Duzee). Proc. 11th Int. Congress Entomology, Vienna. Aug. 17–25, 1960:790–791.

———. 1963. Effects of plant viruses on insects. Ann. New York Acad. Sci. 105: 685–712.

Jensen, D. D., Whitcomb, R. F., and Richardson, J. 1967. Lethality of injected peach Western X-disease virus to its leafhopper vector. Virology 31:532–538.

Kisimoto, R., and Watson, M. A. 1965. Abnormal development of embryos induced by inbreeding in *Delphacodes pellucida* Fabricius and *Delphacodes dubia* Kirschbaum (Araeopidae, Homoptera), vectors of European wheat striate mosaic virus. J. Invert. Pathol. 7:297–305.

Lee, P. E., and Jensen, D. D. 1963. Crystalline inclusions in *Colladonus montanus* (Van Duzee), a vector of Western X-disease virus. Virology 20:328–332.

Littau, V. C., and Maramorosch, K. 1956. Cytological effects of aster yellows virus on its insect vector. Virology 2:128–130.

Littau, V. C., and Maramorosch, K. 1960. A study of the cytological effects of aster yellows virus on its insect vector. Virology 10:483–500.

Lomakina, L. Ya., Razvyaskina, G. M., and Shubnikova, E. A. 1963. Cytological and histochemical changes in fat body of Cicada *Psammotettix striatus* Fall infected with winter wheat mosaic virus. Voprosy Virusologii 1963(2):168–172.

Maramorosch, K., and Jensen, D. D. 1963. Harmful and beneficial effects of plant viruses in insects. Annual Rev. Microbiol. 17:495–530.

Miller, J. S., and Coon, B. F. 1964. The effect of barley yellow dwarf virus on the biology of its vector, the English grain aphid, *Macrosiphum granarium*. J. Econ. Entomol. 57:970–974.

Nasu, S. 1963. Studies on some leafhoppers and planthoppers which transmit virus diseases of rice plant in Japan [in Japanese, English summary]. Bull. Kyushu Agr. Exp. Sta. 8:153–349.

Okuyama, S. 1962. The propagation of the rice stripe virus in the body of the vector [in Japanese]. Program and Abst. Symp. Vectors Plant Viruses, Sapporo, Japan. Sept. 20, 1962:8–10.

Rutschky, C. W., and Campbell, R. L. 1964. Fat body morphology of English grain aphids infected with barley yellow dwarf virus. Ann. Entomol. Soc. Amer. 57: 53–56.

Schmidt, H. B. 1959. Beitraege zur Kenntnis der Uebertragung pflanzlicher Viren durch Aphiden. Biologischen Zentralblatt. 78:889–936.

———. 1960. Die Ubertragung pflanzlicher Viren durch Insekten. Monatsber. Deut. Akad. Wiss. Berlin 2:214–223.

Shikata, E. 1966. Electron microscopic studies on plant viruses. J. Faculty Agr. Hokkaido Univ. 55:1–110.

Shinkai, A. 1958. Transovarian passage of rice dwarf virus in *Inazuma dorsalis* Motschulsky [in Japanese]. Ann. Phytopathol. Soc. Japan 23:26. (Astr.)

———. 1960. Premature death of *Inazuma dorsalis* Motschulsky which received the rice dwarf virus through eggs [in Japanese]. Ann. Phytopathol. Soc. Japan. 25:42 (Abstr.)

———. 1962. Studies on insect transmission of rice virus diseases in Japan [in Japanese; English summary]. Bull. Nat. Inst. Agr. Sci. Ser. C, No. 14:1–112.

Suenaga, H. 1962. The occurrence of the rice dwarf disease and the ecology of the green rice leafhopper. [in Japanese]. Program and Abstr. Symp. Vectors Plant Viruses, Sapporo, Japan. Sept. 20, 1962:1. (Abstr.).

Sukhov, K. S. 1940a. Intracellular protein inclusions of the new mosaic disease of grain plants (Zakuklivanie). Mikrobiologiia 9(2):188–196.

———. 1940b. X-Bodies in salivary glands of *Delphax striatella* Fallen, the carrier of Zakuklivanie. Compt. Rend. (Dokl.) Acad. Sci. URSS 27(4):377–379.

———. 1943a. A purified protein preparation of winter wheat mosaic virus. Comp. Rend. (Dokl.) Acad. Sci. URSS. 39(2):73–75.

———. 1943b. Proof of infectiousness of a purified protein preparation of oat mosaic virus (Zakuklivanie in oats). Compt. Rend. (Dokl.) Acad. Sci. URSS. 40(4):167–169.

———. 1956. Viruses [in Russian]. Acad. Sci. USSR, Moscow, U.S.S.R. 370 p.

Sukhov, K. S., and Sukhova, M. N. 1940. Interrelations between the virus of a new grain mosaic disease (Zakuklivanie) and its carrier *Delphax striatella* Fallen. Compt. Rend. (Dokl.) Acad. Sci. URSS. 25(5):479–482.

Sukhov, K. S., Vovk, A. M., and Alexeeva, T. S. 1943. Purified protein preparation from the virus of oat mosaic (Zakuklivanie). Compt. Rend. (Dokl.) Acad. Sci. URSS. 41(8):344–346.

Takahashi, Y., and Sekiya, I. 1962. Adipose tissue of the green rice leafhopper, *Nephotettix cincticeps* Uhler, infected with the virus of the yellow dwarf disease of the rice plant. Ann. App. Zool. Entomol. 6:90–94.

Watson, M. A., and Sinha, R. C. 1959. Studies on the transmission of European wheat striate mosaic virus by *Delphacodes pellucida* Fabricius. Virology 8: 139–163.

Whitcomb, R. F., and Jensen, D. D. 1968. Proliferative symptoms in leafhoppers infected with Western X-disease virus. Virology 35:174–177.

Whitcomb, R. F., Jensen, D. D., and Richardson, J. 1966. The infection of leafhoppers by Western X-disease virus. I. Frequency of transmission after injection or acquisition. Virology 28:448–453.

———. 1967. The infection of leafhoppers by Western X-disease virus. III. Salivary, neural, and adipose histopathology. Virology 31:539–549.

———. 1968. The infection of leafhoppers by Western X-disease virus. IV. Pathology in the alimentary tract. Virology 34:69–78.

———. 1968. The infection of leafhoppers by Western X-disease virus. VI. Cytopathological interrelationships. J. Invert. Pathol. (In press).

Yoshii, H. 1959. Studies on the nature of insect-transmission in plant viruses (V). On the abnormal metabolism of the virus-transmitting green rice leafhopper, *Nephotettix bipunctatus cincticeps* Uhler, as affected with rice stunt virus [in Japanese]. Virus 9(4):415–422.

Yoshii, H., and Kiso, A. 1957a. Studies on the nature of insect-transmission in plant viruses (I). Some observations on the unsound metabolism in Satsuma orange affected with the dwarf disease by transmission of green broad-winged plant hopper [in Japanese]. Virus 7(5):306–314.

———. 1957b. Studies on the nature of insect-transmission in plant viruses (II). Some researches on the unhealthy metabolism in the viruliferous plant hopper, *Geisha dinstinctissima* Wal., which is the insect vector of the dwarf disease of Satsuma orange [in Japanese]. Virus 7(5):315–320.

———. 1958. Studies on the nature of insect-transmission in plant viruses (III). Comparative observations on the unhealthy metabolism in Satsuma orange and plant hopper affected with the dwarf disease virus of *Citrus unshu* [in Japanese]. Virus 8(5):385–393.

Yoshii, H., Kiso, A., and Kikumoto, T. 1959. Studies on the nature of insect-transmission in plant viruses (VII). On the metabolic components in the orange virus-transmitting green broad winged planthopper [in Japanese]. Virus 9(5):462–467.

# Isolation and Purification of Vector-Borne Plant Viruses*

Myron K. Brakke

*Plant Pathology Department,*
*University of Nebraska,*
*Lincoln, Nebraska*

Rapid progress has been made in the last few decades in the purification of plant viruses and in the determination of their properties. Besides satisfying curiosity about the nature of the beasts, this information promises to be useful for virus systematics. It is perhaps still too early to use this information to establish the natural relations of viruses, but it is not too early to use this information to place the viruses in provisional groups having several properties in common. Such groups have practical uses, and theoretical interest, because if a virus can be tentatively assigned to a group on the basis of a few known properties, discovery of its unknown properties should be easier.

Vector transmissibility was one of the earliest criteria for virus classification, whereas the properties of the virus (*i.e.*, of the infectious nucleoprotein particles) are often used now. There is now sufficient information to allow examination of the similarities among viruses transmitted by the same type of vector. Before looking at this question, I shall discuss briefly the properties of the viruses that are most useful for systematics, the interrelation of these properties, and their determination. The degree of purification necessary for determination of these properties varies, but often purification is more difficult than the measurement of properties. A short discussion of purification methods follows the discussion of properties.

## I. PROPERTIES OF VIRUSES

### *A. Interrelations*

It is convenient to refer to some properties of viruses as intrinsic. These include molecular weight, shape and spatial organization, and

* Cooperative investigations of the Crops Research Division, Agricultural Research Service, U.S. Department of Agriculture and the Nebraska Agricultural Experiment Station. Published with the approval of the Director as paper No. 2140, Journal Series, Nebraska Agricultural Experimental Station.

chemical composition—percentage of protein, nucleic acid, and other components; amino acid composition and sequence; and nucleic acid base composition and sequence. With some exceptions, the intrinsic properties are not measured directly. However, the properties that can be measured are significant and useful for characterization because they depend upon the intrinsic properties.

It appears now that related viruses have the same shape and spatial organization, molecular weight, percentage of nucleic acid, and nucleic acid base ratio. Accordingly, these properties (and others that depend upon them, such as the sedimentation coefficient, density, diffusion constant, and ultraviolet light absorption) are useful for placing viruses in groups of related strains. Strains of a virus may differ in amino acid composition and sequence and presumably in the sequence of nucleic acid bases. Therefore, these properties and the surface properties that depend on them (serological relations, electrophoretic mobility, and ability to be adsorbed) are potentially useful for distinguishing among virus strains.

It is often simpler to use the measurable properties directly to characterize viruses rather than to calculate the intrinsic properties because the calculation of the intrinsic properties usually depends on assumptions or on a combination of two or more measured properties. Although use of the measured properties for virus characterization is a valid procedure, it must be remembered that some of the measured properties are related because they depend on the same intrinsic property. Two measurable properties that are functions of the same intrinsic property are not independent parameters for virus characterization. If two viruses differ in ultraviolet absorption and buoyant density, it cannot be stated that they differ in two respects, because both of these properties are functions of the nucleic acid content.

Rod-shaped viruses usually have about 5% nucleic acid, whereas small polyhedral viruses usually have 20–40% nucleic acid. Because of this empirical correlation between nucleic acid content and shape, these two properties are not completely independent parameters for virus characterization. Not enough of the large, complex viruses, such as potato yellow-dwarf virus, have been analyzed to permit correlations to be drawn on interrelations of their intrinsic properties.

The properties that can be measured on small amounts of impure virus can be combined to give estimates of the intrinsic properties that determine them. Comparison of values of the intrinsic properties calculated in different ways provides a check on the validity of the assumptions and measurements. The molecular weight of a virus can be calculated from the sedimentation coefficient, diffusion constant, and partial specific volume. If it cannot be measured, the partial specific volume can be calculated from the chemical composition (*i.e.*, percentage of nucleic acid, protein, and lipid). The molecular weight can be estimated from the sedimentation constant and partial specific volume if the shape of the particle is known. The percentage of nucleic acid can be calculated from the ultraviolet absorption spectrum. Nucleic acid can also be isolated and its molecular weight estimated directly from the sedimentation coefficient, or more accurately from the sedimentation coefficient combined with the diffusion coefficient. The percentage of nucleic acid should be given by the ratio of the molecular weight of the nucleic acid to that of the virus.

### *B. Properties Depending on Size and Shape*

1. SIZE AND SHAPE. Lengths of rod-shaped viruses can be measured by electron microscopy in unpurified preparations. In fact, better measurements are often obtained with unpurified than with highly purified preparations (Brandes, 1964). Determination of morphological detail and size of small polyhedral viruses by electron microscopy has not yet proven useful for identification.

2. SEDIMENTATION COEFFICIENT. Sedimentation coefficients depend on the sizes, shapes, and densities of viruses. For spherical particles, they are proportional to the square of the diameters and should be more useful for identification of polyhedral viruses than the diameters. They have not been used as much as they could be. The sedimentation coefficient can be determined on impure virus, provided enough is present to be detected. Clarified plant sap can often be used. As little as 1 $\mu$g can be detected by photometric assay after density gradient centrifugation and 10 $\mu$g is readily detected (Brakke, 1963b). Approximate estimates of the sedimentation coefficient could be made on even less than 1 $\mu$g, if the virus zone could be located by infectivity assay. Traces of infectivity above and below the zone may make the location of the zone uncertain unless a quantitative infectivity assay is used.

The analytical ultracentrifuge provides a more accurate measure of sedimentation coefficients than does density gradient centrifugation, but requires a higher concentration of virus. The values obtained with the analytical ultracentrifuge should be extrapolated to zero virus concentration, especially for rod-shaped viruses. Unfortunately, many values are reported with no concentration specified.

The sedimentation coefficient of brome grass mosaic virus changes with the *p*H, being 87 below *p*H 7 and 79 above *p*H 7 (Incardona and Kaesberg, 1964). Therefore, the buffer and *p*H at which sedimentation coefficients are given should be specified.

3. DIFFUSION COEFFICIENTS. Diffusion coefficients depend on the size and shape of a virus and must be determined if accurate values of the molecular weight are to be calculated from sedimentation coefficients. Highly purified virus is needed for the customary measurements of diffusion coefficients, which are calculated from the change in shape of the boundary between virus solution and buffer. Diffusion coefficients can also be measured by using serological or infectivity assays to determine the rate of diffusion of virus through porous barriers (Ackers and Steere, 1961; Regenmortel, 1966). Smaller amounts of less pure virus are needed for these procedures than for the customary ones.

### *C. Properties Depending on Percentage of Nucleic Acid*

1. ULTRAVIOLET ABSORPTION. The measured ultraviolet light absorption by viruses is due to scattering and to true absorption. The measured absorption above 310 mμ is mainly light scattering and depends on the size and shape of the particle. One can estimate the amount of light scattering at wavelengths below 310 mμ by plotting the logarithm of the absorbance above 310 mμ against the logarithm of the wavelength and extrapolating the resultant straight line to values of less than 310 mμ. Theoretically, this line should have a slope of −4, but in practice it is often less, being −3.6 for TMV and −3.2 for potato virus X (Paul, 1958, 1959a).

The aromatic amino acids of the virus protein absorb ultraviolet light in a band centered at 280 mμ, the nucleic acids at 260 mμ, and the peptide bond below 230 mμ. These bands are not sharp. The aromatic amino acids absorb about one-half to two-thirds as much light of 260 mμ as of 280 mμ and the nucleic acids absorb about half as much at 280 mμ as at 260 mμ. The total absorbance of a virus at a particular wavelength is the sum of the scattering and the absorbance of the component parts of the virus. The absorbance of the component parts is directly proportional to their concentration, just as the total absorbance is directly proportional to the concentration of virus. Therefore, one can calculate the concentration of the component parts if their specific absorbance and their contribution to the total absorbance are known.

Undegraded virus nucleic acid has an absorbance at 260 mμ of about 22 $cm^2\ mg^{-1}$, that is, a 1-mg/ml solution has an absorbance of 22 with a light path of 1 cm. This value is approximate only and depends on the base ratio and secondary structure. Most virus proteins have enough aromatic amino acids to give values at 280 mμ of about 1 $cm^2\ mg^{-1}$. A value of 0.65 $cm^2\ mg^{-1}$ has been reported for protein of carnation latent virus, 1.5 $cm^2\ mg^{-1}$ for protein of potato virus X, and

1.3 $cm^2$ $mg^{-1}$ for TMV protein (Paul, 1958, 1959a; Paul and Wetter, 1964). The nucleic acid content can be calculated from the ratio of absorbance at 280 mμ to that at 260 mμ (commonly called the 260/280 ratio) (Paul, 1959b) or from the absorbance of 260 mμ if the dry weight concentration is known (Englander and Epstein, 1957).

Nucleic acids have a minimum absorbance at about 230 mμ where proteins have a high and rapidly increasingly absorbance. The $A_{230}/A_{260}$ ratio increases rapidly as nucleic acid content decreases and could also be used to calculate the nucleic acid content. However, 230 mμ is on the shoulder of a strong absorbing peak due to the peptide bond, and the measured values depend on instrument parameters such as the band width of the light.

Rod-shaped viruses have about 5% nucleic acid and small polyhedral viruses have 20–40%. Therefore, the specific absorbance at 260 mμ and the 260/280 ratio in practice depend on the shape of the virus (Table 1). The shape can be inferred from the 260/280 ratio or the specific absorbance at 260 mμ. Conversely, if the shape is known, the specific absorbance and 260/280 ratio can be roughly estimated.

Highly purified virus is usually needed for measuring ultraviolet absorbance since many impurities absorb in this region. However, a single gradient column with a virus zone can be photometrically scanned at several wavelengths in a recording spectrophotometer having a flow cell with a reservoir on top (Brakke, unpublished). The gradient column can be pumped repeatedly from the centrifuge tube through the flow cell and into the reservoir and back again, with the absorbance measured at a different wavelength each time. Comparisons of areas or heights of virus peaks can give ratios of absorbance at 260/280 mμ, 260/230 mμ, or other wavelengths. These estimates can be made on small amounts (10–500 μg) of impure virus, though the accuracy is greater with increasing purity.

2. DENSITY. Two measures of density are commonly used. The partial specific volume is the volume increase of solution per gram of added anhydrous virus and is calculated from densities of solutions containing known amounts of purified virus. It is approximately the reciprocal of the anhydrous density of the virus. A considerable amount of highly purified virus is needed for its measurement.

The buoyant density of the virus is the density of the solution in which the virus will not sediment. It can be determined by centrifugation of very small amounts of not necessarily pure virus. The virus can either be centrifuged to equilibrium in a density gradient column, usually of a salt such as CsCl (Meselson *et al.*, 1957), or the buoyant density can be calculated from the decrease in the sedimentation rate

TABLE 1

NUCLEIC ACID CONTENT, ULTRAVIOLET ABSORPTION, AND SEDIMENTATION COEFFICIENTS OF SOME REPRESENTATIVE PLANT VIRUSES.

| | % RNA | $A_{260}$[a] | $A_{280}/A_{260}$[b] | $S_{20,w}$[c] | Reference |
|---|---|---|---|---|---|
| Rod-shaped viruses | | | | | |
| TMV | 5 | 2.7–3.5[d] | 0.83 | 188 | Paul, 1958; Boedtker and Simmons, 1958 |
| Potato virus X | 6.6 | 3.0 | 0.83 | 118 | Paul, 1959a; Reichmann, 1959 |
| Carnation latent virus | 6.0 | | 0.73 | 167 | Paul and Wetter, 1964 |
| Tobacco etch | | 2.4 | 0.8 | 154 | Purcifull, 1966 |
| Polyhedral viruses | | | | | |
| Brome mosaic | 21 | 5.1 | 0.59 | 86 | Bockstahler, and Kaesburg, 1962 |
| Southern bean mosaic | 21 | 5.8 | 0.66 | 115 | Paul, 1959b; Miller and Price, 1946 |
| Turnip yellow mosaic | 35 | 9.2 | 0.58 | 117 | Paul, 1959b; Haselkorn, 1966; DeRosier and Haselkorn, 1966 |
| Tobacco ringspot | 42 | 13 | 0.58 | 128 | Stace-Smith *et al.*, 1965 |

[a] Absorbance for a 1-mg/ml solution of virus with a 1-cm light path.
[b] Ratio of absorbance at 280 mμ to that at 260 mμ, not corrected for scattering.
[c] In Svedberg units.
[d] Numerous values in this range appear in the literature.

with increasing density of the supporting medium. The sedimentation rate as a function of density can be measured either in the analytical ultracentrifuge (Schachman and Lauffer, 1949) or by density gradient centrifugation (Brakke, 1958b).

The buoyant density is always a function of the supporting medium, whether it be calculated from changing sedimentation rates or measured directly from sedimentation to equilibrium in a density gradient, because the buoyant density is the density of the hydrated virus and the substance (sucrose or salt) used to increase the density invariably affects the hydration of the virus. In addition, viruses may bind the salt or other material used to increase the density. Viruses have charged groups and are salts in solution. In CsCl virus will be the cesium salt, which is heavy.

The partial specific volume is an additive function of the partial specific volumes of the component parts of a particle. For most viruses it is a function of the nucleic acid content and can be calculated from the nucleic acid content and vice versa. The buoyant density calculated from measurements of sedimentation rates in heavy water should be the reciprocal of the partial specific volume and a function of the nucleic acid content. Buoyant densities measured in other media are not simple functions of nucleic acid content, but can be used to characterize viruses if values measured under the same conditions are compared. All strains of a virus would be expected to have the same buoyant density.

3. NUCLEIC ACID PROPERTIES. Nucleic acid can be prepared from impure virus and partly characterized if the impurities have no nucleic acid similar to that of the virus. If the virus nucleic acid is infectious, it can be identified, even in the presence of other nucleic acids. We have tested several methods for preparing nucleic acid from small amounts of virus (100 $\mu$g) and have had best results with phenol treatment in 0.1 M ammonium carbonate buffer, *p*H 9.0, 1% sodium dodecyl sulfate, 0.001 M EDTA, 0.1% sodium diethyldithiocarbamate, 0.01% bentonite, and 0.1% 8-hydroxquinoline. The phenol can be removed by dialysis and the nucleic acid characterized by its sedimentation rate in density gradient centrifugation. A few micrograms of nucleic acid is easily detected with ultraviolet optics after density gradient centrifugation or in the analytical ultracentrifuge. The sedimentation rate of single stranded nucleic acid depends markedly on *p*H, salt concentration, and specific ions, but under controlled conditions is empirically related to the molecular weight. Therefore, the molecular weight can be estimated from results of centrifugation. The kind of nucleic acid can be determined by sensitivity to nucleases, with unde-

graded nucleic acid detected by an infectivity assay or density-gradient centrifugation.

Double-stranded nucleic acid is less easily hydrolysed by enzymes than single-stranded nucleic acid, and its sedimentation rate varies less with changes in ionic strength and *p*H than that of single-stranded nucleic acid. Determination of sedimentation rates under different conditions can be used to indicate if the nucleic acid is single or double stranded. If sufficient quantities of pure virus can be obtained to provide a few milligrams, or even less, of pure RNA, the base ratios can be determined. These are very useful to characterize certain large groups of viruses like the turnip yellow-mosaic virus and the squash mosaic virus families.

### *D. Surface Properties*

1. SEROLOGY. Serology has been very useful for identification of viruses. Many useful antisera have been prepared by injecting rabbits with small amounts of impure virus, but several milligrams of pure virus gives better results. Regenmortel (1966) has recently reviewed the serology of plant viruses. However, serology has not been successfully applied to some of the viruses that are difficult to purify. The common serological reactions are variations of the precipitin reaction and require a certain minimum concentration of virus, a few micrograms per milliliter, which some viruses do not attain (Wright and Stace-Smith, 1966). The sensitivity can be greatly increased (sometimes a thousand-fold or more) by adsorption of the virus or the antibody to large particles as in the bentonite flocculation test or the tanned red blood cell test (Saito and Iwata, 1964; Scott *et al.*, 1964; Bercks, 1967). However, all variations of the precipitin reaction require absence of a reaction between plant antigens and corresponding antibodies, an absence which may be obtained by sufficient purification of the immunizing virus, dilution of the antiserum past the endpoint of the plant antigen–antibody system, or adsorption of the antiserum with antigens from healthy plants. In practice it is difficult to use any of these methods with viruses that are present in low concentration and are difficult to purify.

Diffusion tests in agar gels allow one to distinguish the reaction of several antigens in a single system. The reaction due to a virus may be detectable in the presence of a reaction with plant antigens, provided that the virus diffuses, is sufficiently stable, and is present above a minimum concentration. Rod-shaped viruses diffuse with difficulty into agar and not at all if they aggregate. The diffusion of antigens of these rods into agar can be increased if they are disrupted by alkali

(Purcifull and Shepherd, 1964) or by detergent (Hamilton, 1964). Selection of buffers with a low salt concentration and a *p*H higher than 7 and incorporation of citrate or EDTA decreases aggregation of some rod-shaped viruses and improves results of serological tests in agar (Ball, personal communication).

In principle, serological tests that depend on the measurement of a property of the virus should be useful even in the presence of a serological reaction between plant antigens and corresponding antibodies. Unfortunately, the neutralization test has not been highly successful with plant viruses, but recent modifications involving centrifugation to remove virus–antibody aggregates before inoculation promise to be successful, even with impure virus present in very low concentration (Gold and Duffus, 1967). Other serological tests based on a property of the virus are the loss of the virus zone in density gradient centrifugation and appearance of zones due to dimers of virus (Whitcomb and Spendlove, 1967), and aggregation as observed in the electron microscope. Both tests can be more sensitive by an order of magnitude than the usual precipitin test.

2. ELECTROPHORETIC MOBILITY. Electrophoretic mobilities are useful for some situations since they can be used to distinguish between some strains of a virus. The electrophoretic mobility depends on the exposed charged groups and therefore on the amino acid composition. It does not, however, depend on the spatial arrangement of exposed groups as serology does.

Electrophoretic mobilities can be determined on small amounts (a few micrograms) of virus in a gradient column of sugar or glycerine (Brakke, 1955). The effects of the gradient in changing the viscosity and field strength cancel out and the mobility for a given current is the same as in the absence of the gradient, at least up to a sugar concentration of 20% (Matheka and Geiss, 1965). The electrophoretic mobility depends on *p*H, salt concentration, and the presence of specific ions. Values to be compared must be measured under similar conditions.

## II. METHODS OF PURIFICATION

### A. *General Considerations.*

The common methods of purification include salt precipitation and combinations of selective denaturation, precipitation, and adsorption procedures with differential and density gradient centrifugation. These have been reviewed by Steere (1959, 1964); Black (1955);

Markham, (1959); Brakke (1960); and in several chapters in Koprowski and Maramorosch (1967).

1. SOURCE TISSUE. It is important to select a stable strain of the virus that reaches a high concentration, a suitable host, optimum growing conditions, and a proper time of harvest. The age of leaves and plants is important. Many viruses reach their highest concentration when the first, severe systemic symptoms appear. However, hosts with no symptoms may have a high concentration of virus. The highest concentration of virus in some hosts occurs in the inoculated leaves.

2. ASSAY. The importance of the assay in evaluating the source tissue and the various purification procedures can not be overemphasized. If the assay is tedious and inaccurate, it may not be feasible to test those procedures requiring careful control of several conditions as, for example, in the adsorption and elution of viruses.

Infectivity assays should always include dilution curves so that the results can be reliably interpreted as relative virus concentrations. The error in the assay should be determined by statistical analysis since intuitive estimates are often unreliable, especially with small numbers of infections.

Analytical ultracentrifugation and analytical density gradient centrifugation (Brakke, 1963b) are good assay methods of unaggregated virus. The latter method will detect as little as 1 $\mu$g of virus. Both methods will often detect virus in suitably clarified, unconcentrated plant sap, and even without previous clarification if the concentration is high. The electron microscope is useful for following the purification of the easily aggregated rod-shaped viruses. Antisera to virus and to host proteins give good assays.

Each assay depends on a different property of the virus. The selection of the assay may depend on the intended use of purified virus. It may be an advantage to use more than one assay since changes in properties such as specific infectivity of the virus will be detected.

3. SELECTION OF BUFFER. The choice of buffer influences the yield of virus in the sap, the stability of the virus, and its aggregation (for a more extensive discussion, see Brakke, 1967a). The most frequently used buffers are borate, phosphate, Tris, citrate, and glycine. There are specific buffer effects that are presumably due to binding of buffer ions by the virus. Binding of zwitterions would not change the net charge of a macromolecule. With this in mind, Good *et al.* (1966) synthesized several new zwitterionic buffers which proved superior to phosphate and Tris for the Hill reaction and phosphorylation coupled oxidation of succinate by bean. These buffers have not been used for virus purification.

It was common practice in some of the earliest virus purifications to add 3–4 g of disodium phosphate per 100 g of leaves. This increased the *p*H, and probably caused some clarification by formation of calcium phosphate. More recently, more virus has sometimes been obtained in extracts prepared in 0.5 M borate, phosphate or citrate buffers than with lower concentrations (Shepherd and Pound, 1960; Tomlinson, 1964; Scott, 1963). The reason is not known, but it could be that the high buffer concentration disrupts inclusion bodies or prevents binding of virus to cellular components. On the other hand, Taniguchi (1966) found that even relatively low concentrations of salts, particularly $CaCl_2$ and NaCl, increased the binding of TMV to cell debris. In some cases, a second extraction of the pulp yields more virus than the first.

Many viruses are inactivated by oxidation. Reducing agents used to stabilize viruses are 0.01 M $Na_2SO_3$ (Bald and Samuel, 1934), 0.1% thioglycolic acid (Fulton, 1959) and 0.1 M ascorbic acid (Steere, 1964). The acids must be neutralized before use. Sodium diethyldithiocarbamate (DIECA), 0.1%, has been added to bind copper, thus inactivating polyphenoloxidase (Fulton, 1959). Hide powder (1–2%) (Brunt *et al.*, 1964), alkaloids (Thung and Van der Want, 1951), and extraction with organic solvents have been used to prevent inactivation of viruses by tannins (Vaughan, 1956).

The proper hydrogen ion concentration is one of the most important factors in stabilization of viruses. Other cations such as magnesium, calcium, and polyamines will stabilize some viruses (Brakke, 1956, 1963a; Black, 1965). These cations are bound to viruses and decrease their net negative charge, especially at low hydrogen ion concentrations where their stabilization effect can be greatest (Brakke, unpublished results).

Aggregation is a serious problem with almost all rod-shaped viruses and with some others. Tomato spotted wilt and cucumber mosaic viruses, both of which are spherical, sediment rapidly in plant sap and must be either aggregated or bound to, or in, cell organelles. Tomato spotted wilt virus sediments at low speed in unpurified extracts in 0.1 M phosphate buffer with 0.01 M $Na_2SO_3$, but is dispersed if the resulting pellets are suspended in 0.01 M $Na_2SO_3$ (Black, *et al.*, 1963). The aggregation of cucumber mosaic virus can be overcome by treating the extract in 0.5 M citrate, *p*H 6.5, with chloroform and subsequently dialyzing the aqueous phase against 0.005 M borate buffer, *p*H 9.0 (Scott, 1963).

The aggregation of rod-shaped viruses usually increases with purification in contrast to the situation with cucumber mosaic and tomato spotted wilt viruses. Electrostatic repulsion of the rods, which are negative at neutrality, appears to be important in preventing aggregation. Thus, tobacco mosaic virus aggregates more rapidly in 0.1 M salts than at lower concentrations (Boedtker and Simmons, 1958). Generally a low salt concentration, which increases repulsion for a given surface charge, results in less aggregation (*e.g.*, see Delgado-Sanchez and Grogan, 1966). Addition of hydrogen ions, $Mg^{2+}$, $Ca^{2+}$, and polyamines lowers the net negative surface charge and increases the rate of aggregation (unpublished observations). Conversely, EDTA and citrate chelate $Mg^{2+}$ and $Ca^{2+}$ and reduce the rate of aggregation (Boedtker and Simmons, 1958; Reichmann, 1959). The exact bonds responsible for the aggregation are not known, but presumably include hydrogen bonds, hydrophobic bonds, disulfide links, and salt bridges.

To a certain extent, conditions that promote aggregation of the rod-shaped viruses also make them more stable. As the viruses are purified, the range of conditions in which they are stable, but unaggregated, may become narrower. It may be that hydrophilic colloids, or other materials are adsorbed to the viruses in plant sap and both stabilize them and prevent aggregation. Loss of these materials during purification could explain the greater instability and tendency to aggregate. The aggregation of rod-shaped viruses is not instantaneous, but occurs slowly upon incubation and is nearly irreversible, a pattern consistent with the loss of an adsorbed component before aggregation.

4. PURIFICATION FROM VECTORS. Many leafhopper-transmitted viruses multiply in their vectors as do some of the aphid-transmitted viruses. It should be possible to purify these viruses from extracts of their insect hosts as well as from extracts of their plant hosts, but this has seldom been done. Wound tumor and rice dwarf viruses have been purified from both plants and leafhoppers (Brakke *et al.*, 1954; Fukushi *et al.*, 1962). Potato leaf-roll virus has been purified from its aphid vector, but not from plants (Peters, 1967).

Extracts of aphids have been reported to contain virus-like particles of unknown origin which could be mistaken for the virus being sought (Peters, 1965; Moericki, 1963). Peters suggested that these particles could be latent plant viruses, since he recovered similar particles from plants of *Physalis floridana* on which aphids had fed.

### *B. Selective denaturation, Precipitation, and Adsorption Procedures.*

Some viruses have been purified solely by physical means such as zonal centrifugation and electrophoresis (*e.g.*, Brakke *et al.*, 1954; Bancroft and Kaesberg, 1959), but usually some selective denaturation, precipitation, or adsorption procedure is needed. The effectiveness of these procedures is judged better by photometric scanning of centrifuged gradient columns or by analytical ultracentrifugation than by the appearance of pellets or the light scattering of zones in gradient columns. Large impurities and green fragments of chloroplasts are readily detected by the latter methods, but ribosomes and fraction 1 protein are not.

1. ORGANIC SOLVENTS. Schneider (1953) introduced the use of chloroform to denature plant proteins while leaving tobacco mosaic virus intact. Steere (1956) used a 1:1 mixture of chloroform and *n*-butanol in the purification of tobacco ringspot virus. Shepherd and Pound (1960) used 8.5% *n*-butanol in the purification of turnip mosaic virus after they found that higher concentrations destroyed the virus.

The virus is usually sought in the aqueous phase after treatment with organic solvents, but the emulsion at the interface should be tested. Bachrach and Schwerdt (1952) first reported the use of *n*-butanol for virus purification. They found that polio virus was in the interface between *n*-butanol and water at certain salt concentrations and acidities and in the aqueous phase at others. Recently Peters (1967) treated extracts of viruliferous aphids at *p*H 5.0 with chloroform and recovered potato leaf-roll virus from the interface. In the same purification scheme, he made use of the fact that potato leaf-roll virus was at the interface when a solution of the virus was treated with 0.8 volume of a 2:1 mixture of 2-butoxyethanol: 2-ethoxyethanol and 1 volume of 2.5 M phosphate, *p*H 7.5. The virus was eluted from the emulsion with 0.01 M phosphate buffer, *p*H 7.2. This technique was used for purification and concentration of ECHO 7 virus by Kitano *et al.*, (1961).

Many of the organic solvent treatments used to purify virus also disrupt nucleoproteins. Chloroform, for example, was often used to deproteinize nucleic acid preparations before it was found that many viruses were not disrupted by it. These solvents disrupt many plant nucleoproteins and lipoproteins, exposing the virus to different materials (*e.g.*, free nucleic acids and lipids) than previously. Some of these may adsorb to the virus, changing its properties.

*Ether–carbon tetrachloride.* Wetter (1960) purified carnation latent, potato S, potato M, potato X, potato Y, pea streak, white clover mosaic, and bean common mosaic viruses by this method for production of antisera. Undiluted raw sap obtained with a fruit press and containing 0.2% ascorbic acid and 0.2% $Na_2SO_3$ was shaken for 5 min at room temperature with an equal volume of cold diethyl ether which had been kept over ferrous sulfate. The mixture was centrifuged at low speed to break the emulsion, and the virus was further purified from the aqueous phase by differential centrifugation.

*Organic solvent extraction of lyophilized leaves.* Several long flexuous rod-shaped plant viruses have been purified by extraction of lyophilized leaf powder with cold chloroform, ethanol, acetone, and diethyl ether (Rozendaal and Slogteren, 1958; Veken, 1960.) The extracted powder was subsequently suspended in buffer and the virus purified further by centrifugation. Advantages of this method are that the lyophilized powder can be stored either before or after extraction with organic solvents. Also the organic solvents remove tannins as well as lipids.

*Fluorocarbons.* Porter (1956) reported the purification of tobacco mosaic and tobacco ringspot viruses by homogenizing plant extracts with fluorocarbon. Crowley *et al.* (1965) found that lettuce necrotic yellows virus could be purified by treatment of extracts with fluorocarbon, but not by treatment with 8% *n*-butanol, choloroform, a chloroform–butanol mixture, or diethyl ether. Fluorocarbon treatment may be milder than treatment with other organic solvents.

2. ADSORBENTS: *Bentonite.* Bentonite is a negatively charged clay which adsorbs many colloids, particularly positively charged proteins such as ribonuclease. Ribosomes, chloroplast fragments, and fraction 1 (F-1) plant protein are adsorbed by bentonite in the presence of magnesium (Dunn and Hitchborn, 1965). By proper selection of magnesium concentration and amount of bentonite, plant proteins can be adsorbed without adsorbing virus. The optimum amount of bentonite may vary from one extract to another and should be determined by trial and error on an aliquot. With experience, the proper amount can be determined by the appearance of the extract (Lister *et al.*, 1965).

*Calcium phosphate.* The adsorption of viruses by hydrated calcium phosphate depends on the concentration of phosphate buffer (Fulton, 1959). Viruses could be purified by adsorption and elution from calcium phosphate, but usually the calcium phosphate is used to adsorb impurities, not virus. An exception is the recently reported purification of lettuce necrotic yellows virus by adsorption and elution from Hypatite C (McLean and Francki, 1967). The adsorption of virus may be prevented by using a limited amount of calcium phosphate, or by controlling the concentration of phosphate ions. The adsorptive properties of calcium phosphate depend on the way it has been prepared and stored. Many types of preparations of calcium phosphate have been used for protein purification. Fulton (1959) made calcium phosphate by slowly mixing solutions of $CaCl_2$ and $Na_2HPO_4$ and washing the precipitate repeatedly with water. We have had good results with calcium phosphate formed *in situ* by slow, simultaneous addition with stirring of 0.05 volume of 1 M $CaCl_2$ and 0.25 volume of 0.2 M $Na_2HPO_4$ to plant extracts in phosphate buffer.

*DEAE cellulose.* Sehgal (1966) used 5 g of DEAE-cellulose per 100 ml of plant extract to assist in removal of host components in the purification of maize dwarf mosaic virus, a strain of sugar cane mosaic virus. Whitcomb (1965) reported that DEAE–cellulose could be used in the purification of potato yellow-dwarf virus.

*Decolorizing charcoal.* Treatment of plant extracts with decolorizing charcoal (0.1 g/ml) has been reported for purification of unaggregated potato virus X (Corbett, 1961). We have found charcoal to be useful chiefly for removal of brown pigments. Rod-shaped viruses may be lost if the extract is filtered through a thick charcoal pad (Galvez, 1964).

3. MISCELLANEOUS. *Heat, freezing, and acid.* Freezing the leaves or their extract, heating the extract, and acidifying it to *p*H 4–5 are probably the oldest selective denaturation procedures. Incubation of extracts at *p*H 6.0 for 1 hr at 40°C denatures ribosomes, F-1 protein, and chloroplast fragments and is one of the most useful procedures. Precipitation of plant proteins with 30% ethanol or 40% saturated ammonium sulfate has also been much used (Markham, 1959). Selection of buffer can be useful as illustrated by the elimination of ribosomes with citrate or EDTA, which chelate magnesium.

*Polyethylene glycol.* Two aqueous phases may form when two hydrophilic colloids are dissolved in water. Macromolecules will distribute between the phases according to their surface properties, and this distribution may be used for their purification (Albertsson, 1960). In attempting to use this procedure, Hebert (1963) found that viruses were precipitated by low concentrations of polyethylene glycol and NaCl. Rod-shaped viruses were precipitated at lower concentrations of glycol and NaCl than were spheres. For example, tobacco mosaic virus was precipitated by 4% polyethylene glycol and 0.1 M NaCl, and tobacco ringspot virus by 8% polyethylene glycol and 0.3 M NaCl.

Venekamp and Mosch (1963, 1964a, 1964b; Venekamp *et al.*, 1964) have adsorbed viruses to cellulose powder at moderate concentrations of polyethylene glycol and NaCl. Upon stepwise elution with solutions containing successively lower concentrations of polymer and NaCl, plant components were eluted first and virus last. Potato viruses X and Y were reported to be aggregated after purification by this procedure.

*Antihost serum.* Gold (1961) has recommended the precipitation of plant proteins by specific antibodies as an aid in plant virus purification. He used the gamma globulin fraction of antiserum obtained by precipitation with ammonium sulfate. This appears to be a mild method that should find wider use. Coupling the antibodies to a solid support eliminates the possibility of soluble antigen-antibody complexes (Galvez, 1966).

### C. Physical Separation Methods

1. CENTRIFUGATION. Both differential centrifugation (*i.e.*, alternate low- and high-speed cycles) and density-gradient centrifugation are widely used. Differential centrifugation handles larger volumes but is a less efficient procedure than density-gradient centrifugation. Newly developed zonal rotors, however, permit density-gradient centrifigation of larger quantities than in swinging bucket rotors (Anderson, 1966). I have discussed density-gradient centrifugation in detail recently (Brakke, 1967b).

2. ZONE ELECTROPHORESIS. Regenmortel (1964a) reported on the efficacy of various commonly used selective denaturation procedures for removal of F-1 protein, a highly antigenic plant protein of sedimentation coefficient 18S. Since F-1 proteins from most plants cross-react serologically, they should be removed from virus preparations intended for serology. Regenmortel found that emulsification with chloroform, freezing for 6 hr, 8% butanol, and treatment with ether and carbon tetrachloride did not remove F-1 protein, but emulsification with a mixture of chloroform and butanol or precipitation with 30% ethanol was effective. The F-1 protein aggregates readily and is not always removed by centrifugation as well as expected. Regenmortel found that the F-1 protein could be effectively removed from many viruses by zone electrophoresis in a sucrose density-gradient column. Use of a long column with a gradual gradient (Regenmortel, 1964b; Svensson and Valmet, 1959) permits use of larger quantities than the shorter columns originally reported (Brakke, 1955). High concentrations of virus, above a few milligrams per milliliter, may give droplet sedimentation. This disturbance is especially troublesome with rod-shaped viruses.

Solid or gel supports could be used instead of a gradient column for zone electrophoresis, but these may give complications due to molecular sieving and adsorption. Sucrose is easy to remove from viruses, whereas many solid supports give some soluble particles of virus size that may be difficult to remove.

3. AGAR GEL FILTRATION. Beads of 2–8% agar gel can be used in a column for separation of viruses by a molecular sieving effect (Steere, 1962; Cech, 1962: Regenmortel, 1962).

This is a mild method which should be applicable to unstable plant viruses. It is also potentially useful for determining particle sizes and diffusion coefficient (Ackers and Steere, 1961; Steere and Ackers, 1962).

## III. RELATION BETWEEN VECTOR AND VIRUS MORPHOLOGY

Any conclusions on relations between properties of a virus and type of vector must be tentative because of a lack of information about many viruses. Carter (1962) may be consulted for references on vector transmission of these viruses.

### A. *Insect-Transmitted Viruses*

1. BEETLE-TRANSMITTED VIRUSES. The turnip yellow-mosaic virus, turnip crinkle virus, southern bean mosaic virus, squash mosaic virus, and related viruses are typical of this group. They are small polyhedral viruses 25–30 m$\mu$ across that have sedimentation coefficients between 100 and 130 *S*. They occur in high concentrations, and are often accompanied by "incomplete" particles with less nucleic acid than the infectious particle, or no nucleic acid (Bancroft, 1968). These viruses have been purified by a variety of methods. Their properties are summarized by Haselkorn (1966) and by Gibbs *et al.* (1966)

2. THRIPS-TRANSMITTED VIRUSES. Tomato spotted-wilt virus is a large, very unstable virus. It can be partly purified by taking advantage of the fact that it is aggregated in crude extracts in 0.1 M phosphate buffer and sediments rapidly, but is unaggregated at low ionic strength (Black *et al.*, 1963). It is stabilized in plant extracts by reducing agents (Bald and Samuel, 1934), but for unknown reasons seems to become less stable during purification. It has a sedimentation coefficient of about 560*S* (Black *et al.*, 1963), a generally spherical but variable appearance in the electron microscope with a diameter of 68–120 m$\mu$, an envelope (Kammen *et al.*, 1966), and is reported to contain lipid (Best and Katekar, 1964).

MEALY BUG-TRANSMITTED VIRUSES. Cocoa swollen-shoot virus has been difficult to purify, in part because of tannins in extracts of infected trees. The virus is a short rod, about 121 by 28 m$\mu$, and has been partly purified by differential centrifugation from extracts at *p*H 8.0–8.2 containing 0.05 M $Na_2HPO_4$, 0.05 M thioglycolic acid, 0.005 M Na-DIECA, and 1–2% hide powder (Brunt *et al.*, 1964).

3. APHID-TRANSMITTED VIRUSES. Almost all morphological types are represented in the aphid-transmitted viruses. Regardless of morphology, many of these viruses are difficult to purify because of low concentration, instability, and aggregation.

Most of the rod-shaped viruses of the potato virus S, potato virus Y, and beet yellows virus groups of Brandes (1964) are transmitted by aphids. They have been purified for serology (Brandes and Bercks, 1965) but seldom for chemical and physical studies. However, Purcifull (1966) has reported the purification and alkaline degradation of tobacco etch virus, and Paul and Wetter (1964) have studied the physical and chemical properties of purified carnation latent virus. A sedimentation coefficient of 154*S* has been reported for tobacco etch virus (Purcifull, 1966) and for potato virus Y (Delgado-Sanchez and Grogan, 1966). Values of 155*S* and 168*S* have been reported for maize dwarf mosaic virus (Bancroft *et al.*, 1966; Shepherd, 1965) and of 167*S* for carnation latent virus (Paul and Wetter, 1964). The other viruses in this group probably have similar sedimentation coefficients.

Two additional long, flexuous viruses with lengths different from those of the above-mentioned groups have been reported to be aphid transmitted. One causes a necrosis in forage grasses and is 1725 m$\mu$ long (Schmidt *et al.*, 1963) and the other is mallow mosaic virus, 865 m$\mu$ long (Schmidt and Schmelzer, 1964).

The purification of these elongated viruses is still an art, and it seems that procedures that work well in one laboratory are not always successful in others. Consequently, numerous modifications in procedures have been reported for their purification.

Several aphid-transmitted viruses are small to medium-sized polyhedrons. Cucumber mosaic and cauliflower mosaic viruses are manually transmissible to plants but difficult to purify because of the low concentration and, in the case of cucumber mosaic virus, aggregation. Cucumber mosaic virus has a diameter of 28–30 m$\mu$ and a sedimentation coefficient of 98*S* (Scott, 1963; Francki *et al.*, 1966; Regenmortel, 1967), whereas cauliflower mosaic virus and the related dahlia mosaic virus are 50 m$\mu$ in diameter, with sedimentation coefficients of about 200*S* (Pirone *et al.*, 1961; Brunt, 1966).

Alfalfa mosaic virus is a mixture of short rods 18 m$\mu$ in diameter and of four lengths, 20–30, 36, 48, and 58 m$\mu$ (Gibbs, *et al.*, 1963). Three main components are observed on centrifugation with sedimentation coefficients of 73, 89, and 99*S* (Bancroft and Kaesberg, 1960), of which only the 99*S* is infectious (Bancroft, 1961). The structure of the alfalfa mosaic rods is quite different from that of the other short rods such as barley stripe mosaic, tobacco rattle, or tobacco mosaic viruses.

The above viruses are apparently all transmitted by aphids in a "nonpersistent" manner. Several viruses transmitted by aphids in a "persistent" manner have been purified. Barley yellow dwarf virus (Rochow and Brakke, 1964), carrot mottle virus (Watson *et al.*, 1964), and pea enation mosaic virus (Izadpanah and Shepherd, 1966; Gibbs *et al.*, 1966) are polyhedrons of about 30 m$\mu$ diameter. Izadpanah and Shepherd (1966) reported a slightly larger value of 36 m$\mu$ for pea enation mosaic virus. Potato leaf-roll virus has been purified from aphids only and is reported to be a 23-m$\mu$ polyhedron (Peters, 1967). The purification procedure for potato leaf-roll virus is interesting in that the virus was recovered from the interface after chloroform treatment of the extract at *p*H 5.0, and again from the interface after treatment with 2-butoxyethanol and 2-ethoxyethanol. The concentration in the final preparation of the virus was apparently very low since no light-scattering zone was observed in gradient columns.

Barley yellow-dwarf virus also occurs in very low concentration, yielding 25–50 $\mu$g of partly purified virus per liter of plant sap (Rochow and Brakke, 1964). This low concentration is the main difficulty in its purification since it is not particularly unstable (Heagy and Rochow, 1965; Murayama and Kojima, 1965). A sedimentation coefficient of about 117*S* has been reported for barley yellow-dwarf virus.

Pea enation mosaic virus has two sedimenting components of about 106*S* and 122*S* whose relation is unknown since some reports indicate both may be infectious (Izadpanah and Shepherd, 1966; Gibbs *et al.*, 1966). However, Bozarth and Chow (1966) reported that only the bottom component was infectious. They reported values of $S_{20,\,w}$ of 113*S* and 94.5*S* for the two components.

Lettuce necrotic yellows virus is another "persistent" aphid-transmitted virus, but resembles the leafhopper-transmitted viruses in structure. It is a large rod, 66 by 227 m$\mu$ with a sedimentation coefficient of 950*S*, and an envelope (Crowley *et al.*, 1965; Harrison and Crowley, 1965). It occurs in sufficient concentration to be manually transmissible to plants, as does pea enation mosaic virus.

5. LEAFHOPPER-TRANSMITTED VIRUSES. Only a few of these viruses are transmissible by manual inoculation of plants. The only assay for most of these viruses in cell-free extracts has been by injecting leafhopper vectors, which in turn must transmit the virus to plants for its presence to be detected. This slow and imprecise assay has hindered the systematic study of the properties of these viruses and their purification.

Potato yellow-dwarf virus is unusual for this group in that it gives local chlorotic lesions upon inoculation of *Nicotiana rustica* L. Density-gradient centrifugation was developed and tested in the purification of this virus (Brakke, 1951, 1953). Recently, Whitcomb (1965) has shown that some selective denaturation procedures can be used with this virus.

The first electron micrographs of potato yellow-dwarf virus indicated that it was a large particle of variable shape, appearing as a sphere or a short rod (Black *et al.*, 1948; Brakke *et al.*, 1951). Recent pictures confirm the presence of an envelope which was suggested by early pictures (Black *et al.*, 1965). Fixed virus in tissue appears as a 75 by

305-mμ rod, and it is possible that the flattened spheres in purified preparations are distorted structures (MacLeod *et al.,* 1966). Potato yellow-dwarf virus has a sedimentation coefficient of about 900*S* (Brakke, 1958b) and contains lipid (Ahmed *et al.,* 1964). Ultraviolet absorption spectra of purified preparations indicate that it contains a very low concentration of nucleic acid.

Electron microscopy of thin sections of infected tissues indicates that two other leafhopper-transmitted viruses are similar to potato yellow dwarf and lettuce necrotic yellows viruses. These are wheat striate mosaic virus, 65 by 270 mμ (Lee, 1964), and maize mosaic virus, 48 by 242 mμ (Herold *et al.,* 1960, Herold and Munz, 1965). These viruses are similar to certain viruses of warm-blooded animals such as vesicular stomatitis virus (Howatson and Whitmore, 1962).

Wound tumor virus and rice dwarf virus both contain double-stranded RNA (Black and Markham, 1963; Miura *et al.*, 1966), as does Reo virus of warm-blooded animals. They are the only two viruses that have been purified from both plant and arthropod hosts (Brakke *et al.,* 1954; Fukushi *et al.,* 1962). Identical particles are obtained from both hosts.

Wound tumor virus is a polyhedron about 60 mμ across (Bils and Hall, 1962), although it appears larger when impure (Brakke *et al.,* 1954). It has a sedimentation coefficient of about 500*S* and 20% RNA (Black and Markham, 1963). Both potato yellow-dwarf virus and wound tumor virus become unstable during purification, but may be stabilized by a buffer containing magnesium and glycine (Brakke, 1956; Black, 1965). Wound tumor virus loses infectivity during prolonged density-gradient centrifugation, a loss apparently due to difference in buoyancy of the nucleic acid and protein (Black *et al.*, 1967).

Rice dwarf virus is a 70-mμ polyhedron (Fukushi *et al.*, 1962) containing 11% RNA (Toyoda *et al.*, 1964, 1965).It appears to have an envelope which can be removed by ether treatment (Fukushi and Shikata, 1963).

### *B. Viruses Transmitted by Vectors Other Than Insects*

1. MITE-TRANSMITTED VIRUSES. Two of these, rye grass mosaic and wheat streak mosaic viruses, are flexuous rods about 700 mμ long (Mulligan, 1960; Brakke and Staples, 1958). The latter has a sedimentation coefficient of about 165*S*. The concentration of wheat streak mosaic virus is about 10 μg per gram of leaf tissue. An improved selection of buffers for the purification procedure reported earlier (Brakke, 1958a) allows about half of the virus to be recovered after consecutive steps of differential, rate zonal, and "equilibrium" zonal

centrifugation. After heat clarification, the extract is adjusted to *p*H 8.0 and made 0.01 M in sodium citrate for the differential and rate zonal centrifugation. At this stage of purity, the virus is unstable at *p*H 8.0, and a buffer at *p*H 6.5 is used for the "equilibrium" zonal centrifugation. We have also added 1:500 normal rabbit serum to improve the dispersion in the equilibrium zonal centrifugation. Purification beyond this stage becomes increasingly difficult because the range of conditions under which the virus is stable, but unaggregated, becomes narrower as purification proceeds.

2. NEMATODE-TRANSMITTED VIRUSES. These fall into two morphological groups, polyhedrons 25–30 m$\mu$ across (NEPO virus) and rigid rods (NETU viruses) (Cadman, 1963). They appear to be quite stable and have all been purified. Some occur in high concentrations. Barley leaves may contain 5 mg of brome grass mosaic virus per gram of leaves (Proll, 1967).

Tobacco rattle viruses have rigid rods of two lengths; the longer rod, about 190 m$\mu$ (300*S*), is capable of inducing disease by itself (Harrison and Nixon, 1959). The length and sedimentation coefficient of the shorter rods varies with the strain (Harrison and Woods, 1966). Pea early browning virus is very similar to tobacco rattle virus, but its diameter is 6% less than the latter (Harrison and Woods, 1966).

Variants of tobacco rattle virus in which no rods are produced often result from single-lesion isolates (Cadman and Harrison, 1959). Therefore, virus cultures for purification should be maintained by bulk transfer.

3. FUNGUS-TRANSMITTED VIRUSES. Four of the six viruses reported to be transmitted by fungi have been purified, and these four belong to three morphological types. Tobacco necrosis virus and the associated satellite virus are small spherical viruses (Kassanis and Nixon, 1961; Kassanis, 1962). They may be separated by electrophoresis, by density-gradient centrifugation, and by taking advantage of the fact that satellite virus is more stable than some strains of tobacco necrosis virus (Kassanis, 1966).

Potato virus X is a long flexuous rod ($S_{20,\,w}=118S$, 513 m$\mu$ long) which has been reported to be transmitted by *Synchytrium endobioticum* (Nienhaus and Stille, 1965; Reichmann, 1959; Brandes, 1964). Soil-borne wheat mosaic virus is a rigid rod with a length of 160 m$\mu$ (Brandes 1964; Brakke *et al.*, 1965) that appears to be transmitted by *Polymyxa graminis* (Estes and Brakke, 1966). It occurs in moderate concentrations and is stable, but has been difficult to purify because of its extreme tendency to aggregate. The usual selective denaturation procedures aggregate this virus. We have purified it sufficiently for serology by grinding the leaves in 0.5 M borate, *p*H 9.0, followed by differential centrifugation, and finally by centrifugation through a short, steep sucrose gradient in 0.1–0.5% Igepon T-73 (sodium *N*-methyl *N*-oleoyl taurate) in *p*H 6.5 phosphate buffer. The detergent disperses the green impurities so that they remain suspended (Brakke, 1959), while the virus (aggregated or not) sediments to the pellet. We have not yet been able to disperse and purify the virus sufficiently for biophysical studies.

Two other viruses transmitted by *Olpidium brassicae* have not been purified and are probably unstable or present in low concentration. Lettuce big-vein virus is not transmissible by manual inoculation (Campbell and Grogan, 1963). Tobacco stunt virus can be manually transmitted if chelating agents are added (Hiruki, 1964, 1965).

## IV. CONCLUSIONS

It is curious that all of the flexuous rod-shaped viruses longer than 600 mμ are transmitted by aphids in a non-persistent manner, with the exception of two that are transmitted by mites. However, not all viruses transmitted by aphids in a nonpersistent manner are long rods. All of the rod-shaped viruses shorter than 600 mμ are soil borne or have no known natural vectors, with the exception of alfalfa mosaic virus, which is not a typical rod. The viruses that are persistent in insect vectors are polyhedral, or have a complex structure—an inner nucleoprotein core with an outer envelope and an overall shape that is variable, but usually that of a short rod. The nucleoproteins of these complex viruses have not been shown to have helical symmetry, as do the typical rod-shaped viruses.

These generalizations are based on limited data. It will be interesting to see if they continue to hold as more viruses are studied.

## BIBLIOGRAPHY

Ackers, G. K., and Steere, R. L. 1961. A cell for rapidly measuring restriction to diffusion of macromolecules through agar gels of various pore sizes. Nature 192:436–437.

Ahmed, M. E., Black, L. M., Perkins, E. G., Walker, B. L., and Kummerow, F. A. 1964. Lipid in potato yellow dwarf virus. Biochem. Biophys Res. Commun. 17:103–107.

Albertsson, P. A. 1960. Partition of cell particles and macromolecules. Almqvist and Wiksells, Uppsala.

Anderson, N. G. 1966. The development of zonal certifuges and ancillary systems for tissue fractionation and analysis. National Cancer Institute Monograph No. 21. U.S. Government Printing Office, Washington, D.C.

Bachrach, H. L., and Schwerdt, C. E. 1952. Purification studies on Lansing poliomyelitis virus. pH stability, CNS extraction, and butanol purification experiments. J. Immunol. 69:551–561.

Bald, J. G., and Samuel, G. 1934. Some factors affecting the inactivation rate of the virus of tomato spotted wilt. Ann. Appl. Biol. 21:179–190.

Bancroft, J. B. 1961. Association of infectivity with alfalfa mosaic virus, bottom component only. Virology 14:296–297.

——— 1968. *In* L. V. Crawford and M. G. P. Stokers [ed.] Plant viruses: defectiveness and dependence. In the Molecular Biology of Viruses. Cambridge University Press, England. In press.

Bancroft, J. B., and Kaesberg, P. 1959. Partial purification and association of filamentous particles with the yellow mosaic disease of bean. Phytopathology 49:713–715.

——— 1960. Macromolecular particles associated with alfalfa mosaic virus. Biochim. Biophys. Acta 39:519–527.

Bancroft, J. B., Ullstrup, A. J., Messieha, M., Bracker, C. E., and Snazelle, T. E. 1966. Some biological and physical properties of a midwestern isolate of maize dwarf mosaic virus. Phytopathology 56:474–478.

Bercks, R. 1967. Methodische Untersuchungen über den serologischen Nachweis pflanzenpathogener Viren mit dem Bentonit—Flockungstest, dem Latex-Test und dem Bariumsulfat test. Phytopathol. Z. 58:1–17.

Best, R. J., and Katekar, G. F. 1964. Lipid in a purified preparation of tomato spotted wilt virus. Nature 203:671–672.

Bils, R. F., and Hall, C. E. 1962. Electron microscopy of wound tumor virus. Virology 17:123–130.

Black, L. M. 1955. Concepts and problems concerning purification of labile insect-transmitted plant viruses. Phytopathology 45:208–216.

——— 1965. Physiology of virus-induced tumors in plants. *In* Encylopedia of Plant Physiology. Ruhland, W., Ed., Springer-Verlag, Berlin. p. 236–266.

Black, L. M., Brakke, M. K., and Vatter, A. E. 1963. Purification and electron microscopy of tomato spotted-wilt virus. Virology 20:120–130.

Black, L. M., and Markham, R. 1963. Base pairing in the ribonucleic acid of wound tumor virus. Netherlands J. Plant Pathol. 69:215.

Black, L. M., Mosley, V. M., and Wyckoff, R. W. G. 1948. Electron microscopy of potato yellow-dwarf virus. Biochim. Biophys. Acta 2: 121–123.

Black, L. M., Reddy, D. V. R., and Reichmann, M. E. 1967. Virus inactivation by moderate forces during quasi-equilibrium zonal density gradient centrifugation. Virology 31:713–715.

Black, L. M., Smith, K. M., Hills, G. J., and Markham, R. 1965. Ultrastructure of potato yellow-dwarf virus. Virology 27:446–449.

Bockstahler, L. E., and Kaesberg, P. 1962. The molecular weight and other biophysical properties of bromegrass mosaic virus. Biophys. J. 2:1–9.

Boedtker, H., and Simmons, N. S. 1958. The preparation and characterization of essentially uniform tobacco mosaic virus particles. J. Amer. Chem. Soc. 80:2550–2556.

Bozarth, R. F., and Chow, C. C. 1966. Pea enation mosaic virus: purification and properties. Contrib. Boyce Thompson Inst. 23:301–309.

Brakke, M. K. 1951. Density gradient centrifugation: a new separation technique. J. Amer. Chem. Soc. 73:1847–1848.

——— 1953. Zonal separations by density-gradient centrifugation. Arch. Biochem. Biophys. 45:275–290.

——— 1955. Zone electrophoresis of dyes, proteins, and viruses in density gradient columns of sucrose solutions. Arch. Biochem. Biophys. 55:175–190.

——— 1956. Stability of potato yellow-dwarf virus. Virology 2:463–476.

——— 1958a. Properties, assay, and purification of wheat streak mosaic virus. Phytopathology 48:439–445.

——— 1958b. Estimation of sedimentation constants of viruses by density-gradient centrifugation. Virology 6:96–114.

——— 1959. Dispersion of aggregated barley stripe mosaic virus by detergents. Virology 9:506–521.

——— 1960. Density-gradient centrifugation and its application to plant viruses. Advances Virus Res. 7:193–224.

——— 1963a. Stabilization of brome mosaic virus by magnesium and calcium. Virology 19:367–374.

——— 1963b. Photometric scanning of centrifuged density-gradient columns. Anal. Biochem. 5:271–283.

——— 1967a. Miscellaneous problems in virus purification. In Methods in virology. H. Koprowski and K. Maramorosch [ed.]. Vol. II. Academic Press. pp. 119–136.

——— 1967b. Density gradient centrifugation. In Methods in Virology. Vol. II. H. Koprowski and K. Maramorosch, Eds., Academic Press. pp. 93–118.

Brakke, M. K., Black, L. M. and Wyckoff, R. W. G. 1951. The sedimentation rate of potato yellow-dwarf virus. Amer. J. Botany 38:332–342.

Brakke, M. K., Estes, A. P., and Schuster, M. L. 1965. Transmission of soil-borne wheat mosaic virus. Phytopathology 55:79–86.

Brakke, M. K., and Staples, R. 1958. Correlation of rod length with infectivity of wheat streak mosaic virus. Virology 6: 14–26.

Brakke, M. K., Vatter, A. E., and Black, L. M. 1954. Size and shape of wound tumor virus. Abnormal and Pathological Plant Growth. Brookhaven symposia in biology No. 6:137–156.

Brandes, J. 1964. Identifizierung von gestrecken pflanzepathogenen Viren auf morphologischer Grundlage. Biologischen Bundesanstalt für Land-und Forstwirtschaft. Heft 110. Berlin-Dahlem. 130 p.

Brandes, J., and Bercks, R. 1965. Gross morphology and serology as a basis for classification of elongated plant viruses. Advances Virus Res. 11:1–24.

Brunt, A. A. 1966. Partial purification, morphology, and serology of dahlia mosaic virus. Virology 28:778–780.

Brunt, A. A., Kenten, R. H., and Nixon, H. L. 1964. Some properties of cocoa swollen-shoot virus. J. Gen. Microbiol. 36:303–309.

Cadman, C. H. 1963. Biology of soil-borne viruses. Ann. Rev. Phytopathol. 1: 143–172.

Cadman, C. H., and Harrison, B. D. 1959. Studies on the properties of soil-borne viruses of the tobacco-rattle type occurring in Scotland. Ann. Appl. Biol. 47:542–556.

Campbell, R. N., and Grogan, R. G. 1963. Big-vein virus of lettuce and its transmission by *Olpidium brassicae*. Phytopatholgy 53:252–259.

Carter, W. 1962. Insects in relation to plant disease. Interscience, New York. 705 p.

Cech, M. 1962. Some notes on the use of agar gel filtration in the study of plant viruses. Virology 18:487–489.

Corbett, M. K. 1961. Purification of potato virux X without aggregation. Virology 15:8–15.

Crowley, N. C., Harrison, B. D., and Francki, R. I. B. 1965. Partial purification of lettuce necrotic yellows virus. Virology 26:290–296.

Delgado-Sanchez, S., and Grogan, R. G. 1966. Purification and properties of potato virus Y. Phytopathology 56:1397–1404.

DeRosier, D. J., and Haselkorn, R. 1966. Particle weight of turnip yellow mosaic virus. J. Mol. Biol. 19:52–59.

Dunn, D. B., and Hitchborn, J. H. 1965. The use of bentonite in the purification of plant viruses. Virology 25:171–192.

Englander, S. W., and Epstein, H. T. 1957. Optical methods for measuring nucleoprotein and nucleic acid concentrations. Arch. Biochem. Biophys. 68: 144–149.

Estes, A. P., and Brakke, M. K. 1966. Correlation of *Polymyxa graminis* with transmission of soil-borne wheat mosaic virus. Virology 28:772–774.

Francki, R. I. B., Randles, J. W., Chambers, T. C., and Wilson, S. B. 1966. Some properties of purified cucumber mosaic virus (Q strain). Virology 28:729–741.

Fukushi, T., and Shikata, E. 1963. Fine structure of rice dwarf virus. Virology 21:500–503.

Fukushi, T., Shikata, E., and Kimura, I. 1962. Some morphological characters of rice dwarf virus. Virology 18:192–205.

Fulton, R. W. 1959. Purification of sour cherry necrotic ringspot and prune dwarf viruses. Virology 9:522–535.

Galvez, G. E. 1964. Loss of virus by filtration through charcoal. Virology 23: 307–312.

——— 1966. Specific adsorption of plant viruses by antibodies coupled to a solid matrix. Virology 28:171–187.

Gibbs, A. J., Harrison, B. D., and Woods, R. D. 1966a. Purification of pea enation mosaic virus. Virology 29:348–351.

Gibbs, A. J., Hecht-Poinar, E., Woods, R. D., and McKee, R. K. 1966b. Some properties of three related viruses: Andean potato latent, Dulcamara mottle, and Ononis yellow mosaic. J. Gen. Microbiol. 44:177–193.

Gibbs, A. J., Nixon, H. L., and Woods, R. D. 1963. Properties of purified preparations of lucerne mosaic virus. Virology 19:441–449.

Gold, H. A. 1961. Antihost serum improves plant virus purification. Phytopathology 51:561–565.

Gold, A. H., and Duffus, J. E. 1967. Infectivity neutralization—a serological method as applied to persistent viruses of beets. Virology 31:308–313.

Good, N. E., Wingt, G. D., Winter, W., Connolly, T. N., Izawa, S., and Singh, R. M. M. 1966. Hydrogen ion buffers for biological research. Biochemistry 5:467–477.

Hamilton, R. I. 1964. Serodiagnosis of barley stripe mosaic facilitated by detergent. Phytopathology 54:1290–1291.

Harrison, B. D., and Crowley, N. C. 1965. Properties and structure of lettuce necrotic yellows virus. Virology 26:297–310.

Harrison, B. D., and Nixon, H. L. 1959. Separation and properties of particles of tobacco rattle virus with different lengths. J. Gen. Microbiol. 21:569–581.

Harrison, B. D., and R. D. Woods. 1966. Serotypes and particle dimensions of tobacco rattle viruses from Europe and America. Virology 28:610–620.

Haselkorn, R. 1966. Physical and chemical properties of plant viruses. Ann. Rev. Plant Physiol. 17:137–154.

Heagy, J., and Rochow, W. F. 1965. Thermal inactivation of barley yellow dwarf virus. Phytopathology 55:809–810.

Hebert, T. T. 1963. Precipitation of plant viruses by polyethylene glycol. Phytopathology 53:362.

Herold, F., Bergold, G. H., and Weibel, J. 1960. Isolation and elecron microscopic demonstration of a virus infecting corn (*Zea mays* L.). Virology 12:335–347.

Herold, F., and Munz, K. 1965. Electron microscopic demonstration of virus-like particles in *Peregrinus maidis* following acquisition of maize mosaic virus. Virology 25:412–417.

Hiruki, C. 1964. Mechanical transmission of tobacco stunt virus. Virology 23: 288–290.

——— 1965. Transmission of tobacco stunt virus by *Olpidium brassicae*. Virology 25:541–549.

Howatson, A. F., and Whitmore, G. F. 1962. The development and structure of vesicular stomatitis virus. Virology 16:466–78.

Incardona, N. L., and Kaesberg, P. 1964. A *p*H induced structural change in bromegrass mosaic virus. Biophys. J. 4:11–21.

Izadpanah, K., and Shepherd, R. J. 1966. Purification and properties of the pea enation mosaic virus. Virology 28:463–476.

Kammen, A. van, Hentsra, S., and Ie, T. S. 1966. Morphology of tomato spotted wilt virus. Virology 30:574–577.

Kassanis, B. 1962. Properties and behaviour of a virus depending for its multiplication on another. J. Gen. Microbiol. 27:477–488.

——— 1966. Properties and behaviour of satellite virus. *In* A. B. R. Beemster, and J. Dijkstra [ed.] viruses of plants: North Holland Publ. Co., Amsterdam. p. 177–187.

Kassanis, B., and Nixon, H. L. 1961. Activation of one tobacco necrosis virus by another. J. Gen. Microbiol. 25:459–471.

Kitano, T., Haruna, I., and Watanabe, I. 1961. Purification and concentration of viruses by an organic solvent system. Virology 15:503–504.

Koprowski, H., and Maramorosch, K. 1967. Methods in Virus Research, Academic Press, New York. Vols. 1–4.

Lee, P. E. 1964. Electron microscopy of inclusions in plants infected by wheat striate mosaic virus. Virology 23:145–151.

Lister, R. M., Bancroft, J. B., and Nadakavukaren, M. J. 1965. Some sap-transmissible viruses from apple. Phytopathology 55:859–870.

McLean, G. D., and Francki, R. I. B. 1967. Purification of lettuce necrotic yellows virus by column chromatography on calcium phosphate gel. Virology 31: 585–591.

MacLeod, R., Black, L. M., and Moyer, F. H. 1966. The fine structure and intracellular localization of potato yellow-dwarf virus. Virology 29:540–552.

Markham, R. 1959. In F. M. Burnett, and W. M. Stanley [ed.]. The viruses. Vol. II. Academic Press, New York. p. 33–125.

Matheka, H. D., and Geiss, E. 1965. Die Bestimmung des Wanderungsgeschwindigkeit tierischer Virusarten in Dichtegradienten der tragerfreien zonenelektrophorese. Arch. Gesamte Virusforsch. 15:301–326.

Meselson, M., Stahl, F. W., and Vinograd, J. 1957. Equilibrium sedimentation of macromolecules in density gradients. Proc. Natl. Acad. Sci. US 43:581–588.

Miller, G. L., and Price, W. C. 1946. Physical and chemical studies on southern bean mosaic virus. I. Size, shape, hydration and elementary composition. Arch. Biochem. 10:467–477.

Miura, K. I., Kimura, I., and Suzuki, M. 1966. Double-stranded ribonucleic acid from rice dwarf virus. Virology 28:571–579.

Moericke, V. 1963. Über "Virusartige Korper" in organen von *Myzus persicae* (Sulz.). Z. Pflanzenkrankh. pflanzenschutz. 70:464–485.

Mulligan, T. E. 1960. The transmission by mites, host range, and properties of ryegrass mosaic virus. Ann. Appl. Biol. 48:575–579.

Murayama, D., and Kojima, M. 1965. Studies on the properties of potato leaf-roll virus by the aphid-injection method. Ann. Phytopathol. Soc. Japan 30:209–215.

Nienhaus, F., and Stille, B., 1965. Übertragung des Kartoffel-X-Virus durch zoosporen von *Synchytrium endobioticum*. Phytopathol. Z. 54:335–337.

Paul, H. L. 1958. Spektralphotometrische Untersuchungen am tabakmosaik virus. Archiv für Mikrobiologie 30:304–317.

——— 1959a. Spektralphotometrische Untersuchungen am Kartoffel-X-Virus. Arch. Mikrobiol. 32:416–422.

——— 1959b. Die Bestimmung des Nucleinsauregehaltes pflanzlicher Viren mit Hilfe einer spektrophotometrischer Methode. Z. Naturforsch. 14b:427–432.

Paul, H. L., and Wetter, C. 1964. Untersuchungen am Carnation Latent Virus. I. Präparation, physikalische, und chemische Eigenschaften. Phytopathol. Z. 49:401–406.

Peters, D. 1965. The purification of virus-like particles from the aphid, *Myzus persicae*. Virology 26:159–161.

——— 1967. The purification of potato leaf-roll virus from its vector *Myzus persicae*. Virology 31:46–54.

Pirone, T. P., Pound, G. S., and Shepherd, R. J. 1961. Properties and serology of purified cauliflower mosaic virus. Phytopathology 51:541–546.

Porter, C. A. 1956. Evaluation of a fluorocarbon technique for the isolation of plant viruses. Trans. N. Y. Acad. Sci. 18:704–706.

Proll, E. 1967. Untersuchungen über de Vermehrung und Ausbreitung des trespenmosaikvirus (bromegrass mosaic virus) in Gerste. I. Mitteilung. Konzentrations bestimmung des Trespenmosaikvirus und seine Vermehrung in Keimpflanzen. Phytopathol. Z. 58:18–45.

Purcifull, D. E. 1966. Some properties of tobacco etch virus and its alkaline degradation products. Virology 29:8–14.

Purcifull, D. E., and Shepherd, R. J. 1964. Preparation of the protein fragments of several rod-shaped plant viruses and their use in agar-gel diffusion tests. Phytopathology 54:1102–1108.

Regenmortel, M. H. V. van. 1962. Purification of a plant virus by filtration through granulated agar. Virology 17:601–602.

——— 1964a. Purification of plant viruses by zone electrophoresis. Virology 23: 495–502.

——— 1964b. Separation of an antigenic plant protein from preparations of plant viruses. Phytopathology 54:282–289.

——— 1966. Plant virus serology, Advances Virus Res. 12:207–271.

——— 1967. Biochemical and biophysical properties of cucumber mosaic virus. Virology 31:391–396.

Reichmann, M. E. 1959. Potato X virus II. Preparation and properties of purified, non-aggregated virus from tobacco. Can. J. Chem. 37:4–10.

Rochow, W. F., and Brakke, M. K. 1964. Purification of barley yellow dwarf virus. Virology 24:310–322.

Rozendaal, A., and Slogteren, D. H. M. van. 1958. A potato virus identified with virus M and its relationship with potato virus S. Proc. 3rd Conf. Potato Virus Diseases. H. Veenman and Zonen, Wageningen, 1958. p. 20–36.

Saito, Y., and Iwata, Y. 1964. Hemagglutination test for titration of a plant virus. Virology 22:426–428.

Schachman, H. K., and Lauffer, M. A. 1949. The hydration, size, and shape of tobacco mosaic virus. J. Amer. Chem. Soc. 71:536–541.

Schmidt, H. B., Richter, J., Hertzsch, W., and Klinkowski, M. 1963. Untersuchungen über ein virusbedingte Nekrose an Futtergrasern. Phytopathol. Z. 47:66–72.

Schmidt, H. B., and Schmelzer, K. 1964. Electronenmikroskopische Darstellung und Vermessung des Malvenmosaik virus. Phytopathol. Z. 51:516–520.

Schneider, I. R. 1953. Solution of tobacco mosaic virus in the aqueous phase of a chloroform–water emulsion and application of this phenomenon in virus assay. Science 117:30–31.

Scott, H. A. 1963. Purification of cucumber mosaic virus. Virology 20:103–106.

Scott, H., Kahn, R. P., Bozicevich, J., and Vincent, M. M. 1964. Detection of potato virus X in tubers by the bentonite flocculation test. Phytopathology 54:1292–1293.

Sehgal, O. P. 1966. Host range, properties, and partial purification of a Missouri isolate of maize dwarf mosaic virus. Plant Disease Reporter 50:862–866.

Shepherd, R. J. 1965. Properties of a mosaic virus of corn and Johnson grass and its relation to the sugarcane mosaic virus. Phytopathology 55:1250–1256.

Shepherd, R. J., and Pound, G. S. 1960. Purification of turnip mosaic virus. Phytopathology 50:797–803.

Stace-Smith, R., Reichmann, M. E., and Wright, N. S. 1965. Purification and properties of tobacco ringspot virus and two RNA-deficient components. Virology 25:487–494.

Steere, R. L. 1956. Purification and properties of tobacco ringspot virus. Phytopathology 46:60–69.

——— 1959. The purification of plant viruses. Advances Virus Res. 6:1–73.

———1964. In Plant Virology, M. K. Corbett and H. D. Sisler, Eds., University of Florida Press, Gainesville. p. 211–234.

Steere, R. L., and Ackers, G. K. 1962a. Purification and separation of tobacco mosaic virus and southern bean mosaic virus by agar gel filtration. Nature 194:114–116.

——— 1962b. Restricted diffusion chromatography through calibrated columns of granulated agar gel; a simple method for particle size determination. Nature 196:475–476.

Svensson, H., and Valmet, E. 1959. Large-scale density gradient electrophoresis. Part II. A simple experimental technique securing perfectly stable zones and full utilization of the separation capacity of a density gradient column. Science Tools 6:13–17.

Taniguchi, T. 1966. Effect of salts on the adsorption of tobacco mosaic virus to tobacco leaf cell debris. Virology 28:131–134.

Thung, T. H., and Van der Want, J. P. H. 1951. Viruses and tannins. Tijdschr. Plantenziekten 57:173–174.

Tomlinson, J. A. 1964. Purification and properties of lettuce mosaic virus. Ann. Appl. Biol. 53:95–102.

Toyoda, S., Kimura, I., and Suzuki, N. 1964. Rice viruses with special reference to rice dwarf virus. Protein, Nucleic acid, Enzyme (Tokyo) 9:861–867.

——— 1965. Purification of rice dwarf virus. Ann. Phytopathol. Soc. Japan 30:125–230.

Vaughan, E. K. 1956. Attempts to transfer rubus and Fragaria viruses into herbaceous hosts. Tijdschr. Plantenziekten 62:271–273.

Veken, J. A. van der 1960. Some applications of freeze-drying in virological research. Tijdschr. Plantenziekten 66:1–11.

Venekamp, J. H., and Mosch, W. H. M. 1963. Chromatographic studies on plant viruses. I. The isolation of potato virus X by means of various systems of adsorption chromatography. Virology 19:316–321.

——— 1964a. Chromatographic studies on Plant Viruses. II. The use of polyethylene glycol in clarification and purification of tobacco mosaic virus. Virology 22:503–507.

——— 1964b. Chromatographic studies on plant viruses. III. The purification of potato virus X, potato virus Y, tobacco mosaic virus, and potato stem mottle virus by chromatography on cellulose columns with polyethylene glycol-containing solutions as solvents. Virology 23:394–402.

Venekamp, J. H., Mosch, W. H. M., and Erkelens-Nanninga, K. E. 1964. Chromatographic purification of the carnation ringspot, carnation mottle, and tobacco necrosis virus. Phytopathology 54:608–609.

Watson, N., Sergeant, E. P., and Lennon, E. A. 1964. Carrot motley dwarf and parsnip mottle viruses. Ann. Appl. Biol. 54:153–160.

Wetter, C. 1960. Partielle Reinigung einiger gestreckter pflanzenviren und ihre Verwendung als Antigene bie der Immunisierung mittels Freundschem Adjuvans. Archiv. Mikrobiol. 37:278–292.

Whitcomb, R. F. 1965. Clarification of extracts containing potato yellow-dwarf virus. Phytopathology 55:746–748.

Whitcomb, R. F., and Spendlove, R. S. 1967. Density-gradient centrifugation of virus-antibody complexes: A sensitive serological method. Virology 30:752–754.

Wright, N. S., and Stace-Smith, R. 1966. A comparison of the sensitivity of three serological tests for plant viruses and other antigens. Phytopathology 56:944–948

# Purification of Single and Double-stranded Vector-borne RNA Viruses

Naoji Suzuki

*Institute for Plant Virus Research,*
*959 Aobacho, Chiba, Japan*

## I. INTRODUCTION

Since the transovarial passage of rice dwarf virus from generation to generation of *Nephotettix cincticeps* was first reported by Fukushi (1933, 1934), the question of whether viruses really multiply in their insect vectors has been of great concern to plant pathologists. A more detailed study on the transovarial passage of rice dwarf virus covering six generations of leafhoppers has been made by Fukushi (1940). The results led him to hold the view that the rice dwarf virus multiplies in its insect vector. Some virologists, however, have not been convinced that any plant viruses multiply in their vectors (Bawden, 1950). In 1952, the serial passage technique, which supports the evidence for multiplication of plant virus in insect vectors, was applied by Maramorosch. The same technique was also used to demonstrate the multiplication of wound tumor virus in its insect vector, *Agallia constricta* (Black and Brakke, 1952). Since then, two techniques, (*1*) the transovarial passage technique (Fukushi, 1933, 1934, 1940; Black, 1953; Shinkai, 1954, 1958, 1962; Yamada and Yamamoto, 1954, 1955, 1956) and (*2*) the serial passage technique (Maramorosch, 1952; Black and Brakke, 1952; Stegwee and Ponsen, 1958), have usually been applied for demonstrating virus multiplication in insect vectors and many viruses have been proved to be propagative in leafhoppers and aphids.

More recently, some of these viruses have been successfully purified and the infectivity of the purified virus preparations to their specific host plants and insect vectors has been demonstrated. Using highly purified virus preparations, the morphological, physical, and chemical properties of the viruses have been described in detail. This facilitated the use of immunological techniques for the detection of viruses in host plants and insect vectors and also simplified the electron microscopic studies to demonstrate the localization and multiplication of viruses in host plants and insects. These results provided unquestionable evidence for the multiplication of virus' in insect vectors. Wound tumor virus, rice dwarf virus, and pea enation mosaic virus present the best examples.

The present review is mainly concerned with the purification of these three viruses—wound tumor virus, rice dwarf virus, and pea enation mosaic virus—and briefly with recent advances in the studies on chemical and physical properties of the viruses.

The concept of purification of labile, insect-transmitted plant viruses, including wound tumor virus, potato yellow-dwarf virus, and tomato spotted wilt virus, has been discussed by Black (1955). He stated that purification is largely a matter of the separation of evanescent virus particles from other colloidal particles occurring in extracts of host cells and the identification of the presumed virus particles, with specific infectivity, visualized under the electron microscope. He also maintained that the purification of these viruses to such a degree that they are free of extraneous chemical substances must be a subject for future study. This seems to be true even at present, even though there was a great advance in purification techniques during the subsequent 10 years, at least for the labile viruses including rice dwarf virus.

## II. WOUND TUMOR VIRUS (WTV)

Tobacco mosaic virus was first purified by chemical methods (Stanley, 1935). Later, several less stable viruses were purified by differential centrifugation techniques (Stanley and Wyckoff, 1937) and since then purification by differential contrifugation in appropriate solvents at 0–4°C has become a standard procedure. The first of the highly labile insect-transmitted viruses, potato yellow-dwarf virus, was prepared by this technique and was successfully detected with the aid of an electron microscope (Black, Mosley, and Wyckoff, 1948). Better preparations of this virus and other labile viruses have been obtained since Brakke invented a density-gradient centrifugation technique (Brakke, 1951) and demonstrated its value in isolating these viruses. Later he modified zone electrophoresis (Brakke, 1953) so that it may be carried out in a density-gradient column, which permitted still further purification. Brakke, Vatter, and Black (1954) have obtained highly purified preparations of wound tumor virus (WTV) by using differential centrifugation, density-gradient centrifugation, recycling these two techniques, and finally by using zone-electrophoresis in a sucrose density column.

### *A. Sources of Virus*

Both insects (*Agallia constricta*) and diseased sweet clover plants were used as materials for wound tumor virus purification. Leafhoppers, which had begun their acquisition feeding at least a month previously by being caged on diseased crimson clover for a week or more, were weighed, cooled, and macerated with a mortar and pestle in a small amount of solvent containing 0.225 M NaCl, 0.0033 M $KH_2PO_4$, and 0.0067 M $K_2HPO_4$. Extra solvent was added to give the desired dilution and the extract was clarified by centrifuging at 3400 rpm for 20 min in an International 1-SB centrifuge with swinging cups. When plants were used as a virus source, stem tumors caused by the disease on sweet clover plants of clone C10 (Black, 1951) were removed 4–5 months after diseased C10 scions had been grafted onto healthy seedling stocks. Diseased plant tissues were weighed, cooled, and macerated in a small amount of solvent. After the addition of extra solvent to give the desired dilution, the extract was squeezed through cheesecloth and clarified by centrifuging. For density-gradient centrifugation, extracts from stem tumors were clarified by centrifuging at 8000 rpm for 15 min. A solvent containing 0.01% $Na_2SO_3$ and 0.01% Tween 80 (sorbitan monooleate) was used for density-gradient centrifugation.

### *B. Density-Gradient Centrifugation*

Density-gradient centrifugation was performed by the general procedures of Brakke (1953); a special rotor with swinging cups was used to give horizontal centrifugation. Extract from sweet clover stem tumors was used at a dilution of 1:10, and that from viruliferous leafhoppers at a dilution of 1:32. Gradient columns were prepared by layering (*1*) 1, 2.5, 2.5, 2.5 and 1.5 ml of solutions of 0, 80, 150, 207, and 255 g of sucrose per liter; (*2*) 1, 2.5, 2.5, 2.5, and 1.5 ml of solutions of 0, 150, 255, 328, and 380 g of sucrose per liter; (*3*) 1, 1.5, 1.5, 1.5, and 1 ml of solutions of 50, 150, 255, 328, and 380 g of sucrose per liter, respectively. All sucrose solutions contained 0.01 M $Na_2SO_3$. After the sucrose solutions had been layered, the tubes were allowed to stand overnight before extracts were added. One ml of extract (a 0.5-cm layer) was floated on the sucrose density column in each tube, and centrifuged for 3 hr at 9500 *g* at the top and 16,000 *g* at the bottom of a 6-cm column. After examination of the tubes for visible zones, samples for infectivity assay and electron microscopy were removed into a hypodermic needle inserted in the side of the plastic tubes at desired depths.

Recycled preparations were made by removing the zones from several tubes, diluting them with water, spinning down the virus in the high-speed angle head, and resuspending the precipitate in 1 ml of solvent. The preparation usually stood overnight at this stage and on the following day was floated on a density-gradient column and recentrifuged.

Further purification of the virus from tumors was achieved by a combination of zone electrophoresis and density-gradient centrifugation. After the clarification of a tumor extract by centrifuging at 4800 rpm for 30 min, 1-ml aliquots were floated on each of several type-C gradient columns, which were then centrifuged for 2 hr at 10,000 rpm in the swinging cup rotor. The virus zones were removed and floated directly on a second gradient column which contained 4 ml each of solutions of 300, 400, 500, 600, and 700 *g* of sucrose per liter in a 50-ml centrifuge tube. This second gradient column was centrifuged at 45,000 *g* for 2 hr in an angle rotor of the Servall SS-2 centrifuge to give a separation based mainly on densities. The virus zone was removed from this second gradient column and transferred directly to a third density-gradient column in a U-tube for zone electrophoresis. Each leg of the U-tube, which was made from 18-mm diameter Pyrex glass tubing, contained a density gradient column 8 cm deep, made with sucrose solution of concentrations ranging from 300 to 700 *g* per liter. All sucrose solutions used to make this column and also to make the column for the second density-gradient centrifugation contained 0.02 M potassium phosphate buffer at *p*H 6.1. About 400 V was applied to the U-Tube to maintain a current of 1 mA for 9 hr. Samples of the resulting virus zone were removed, and examined in the electron microscope.

### *C. Infectivity Assay of the Purified WTV Preparations*

The correlation of infectivity with the visible zones obtained by the sucrose density-gradient centrifugation was examined by using the leafhopper-injection technique. Six tubes prepared from viruliferous insects and 6 tubes prepared from stem tumors were used for the infectivity assays. Samples were taken from different depths of each tube by puncturing the tube at desired depths with a hypodermic needle. These samples were assayed for infectivity by injection of *Agallia constricta* nymphs. Details of the injection technique of leafhoppers have already been described by Black and Brakke (1952) and details of procedures for testing the injected insects by Brakke,

Maramorosch, and Black (1953). The result clearly indicated that the infectivity was associated with the visible zones and that the concentration of virus in the samples from visible zones was so high that the percentage of infective insects did not decrease with dilution (Brakke *et al.*, 1954).

### *D. Electron Microscopy*

Virus preparations were obtained by single-cycle and recycled density-gradient centrifugations from both viruliferous insects and stem tumors, and also by a combination of zone electrophoresis and density-gradient centrifugation from stem tumors. With increased purification, the observed sizes of the particles in the electron micrographs became smaller and more uniform. Thus, the virus particles in single-cycle insect preparations varied from 80 to 105 mμ in diameter, whereas those in recycled insect preparations measured 75–85 mμ. The virus particles in the preparation obtained by combination of electrophoresis and density centrifugation measured 70–85 mμ across (Brakke *et al.*, 1954).

More recently, Bils and Hall (1962) made a more detailed examination of WTV particles using a highly purified preparation obtained from Black and Markham. The result indicated that the virus particles are about 60 mμ in diameter and have the shape of an icosahedron. The surface consists of subunits about 75 Å in diameter numbering 4 along an edge and 92 in all. A core about 35 mμ in diameter stains heavily with uranyl acetate, indicating that it consists of RNA or nucleoprotein. Disintegration of the virus envelope by the removal of salt releases long strands about 30 Å in diameter that are either RNA in a highly folded or coiled form or an RNA strand perhaps strengthened with combined protein.

### *E. Physical and Chemical Properties*

Black and Markham (1963) obtained a highly purified WTV preparation by a series of steps including rate quasi-equilibrium zonal density-gradient centrifugations. Destruction of virus during purification was reduced by operation at about 4°C and maintenance of the virus in a glycine–$MgCl_2$ solution. The yield of virus from a 500-g sample of root tumors was about 0.75 mg. The RNA content of the virus is about 20%, as indicated by determinations of the nucleic acid and protein content, the ultraviolet absorption curve, and the specific volume of the virus. The virus is about 60 mμ across and has a sedimentation coefficient of about 500*S*. If the RNA content of a virus particle is considered as a single molecule, its molecular weight would be about $1.5 \times 10^7$. Base analyses of the RNA showed that the ratios: adenine/uracil, guanine/cytosine, and (adenine+uracil)/(guanine+cytosine) were 0.96, 0.99, and 1.60, respectively; the corresponding ratios were 0.98, 1.01, and 1.56 in the second experiment. From these data, they suggested the base pairing in a double-stranded WTV–RNA helix,. Similar results, showing the base pairing of WTV–RNA, were obtained by Gomatos and Tamm (1963), who demonstrated that the mole per cent of guanine closely approximates that of cytosine (18.6: 19.1) and the mole per cent of adenine approximates that of uracil (31.1:31.3), while the mole per cent of G+C is about 38% of the total.

Further evidence supporting the base pairing of WTV–RNA was provided by thermal denaturation upon heating in 0.15 M NaCl, 0.015 M sodium citrate solution, at *p*H 7.0. Wound tumor virus–RNA exhibited a hyperchromic effect of 27%. Absorbance began to increase at 87°C, and the $T_m$ was about 90°C, 3°C lower than that of reovirus RNA, possibly because of the lower G+C content as compared with reovirus type-3 RNA which had a G+C content of 44.3%.

X-ray diffraction photographs of WTV–RNA presented by Tomita and Rich (1964) were quite similar to those of reovirus RNA presented by Langridge and Gomatos (1963). The structure of WTV–RNA is similar to that of DNA A form but is slightly different in the number of residues per turn, 10 in WTV–RNA compared with 11 in DNA A form.

These results indicate that WTV greatly resembles reoviruses in shape, fine structure, and also in the double-helical structure of RNA.

### *F. Serology*

Highly purified WTV preparations have made possible the preparation of highly specific antisera of WTV. Gomatos and Tamm (1963) have found no evidence of an immunological relationship between any of the three types of reovirus used and WTV. The precipitin ring test (Whitcomb and Black, 1961) and the precipitin ring time test (Whitcomb, 1964) have been employed for the estimation of relative concentrations of wound tumor-soluble antigen from insect vectors and from infected plants. Nagaraj, Sinha, and Black (1961) reported the use of a fluorescent antibody technique for the detection of virus antigen in smears from insect vectors. Sinha and Reddy (1964) improved the method of conjugating the antiserum with fluorescein isothiocyanate by which they could recover about 50% of the antibody. Sinha, Reddy, and Black (1964) presented a technique by which they could perform fluorescent smear examinations of small droplets of hemolymph taken from individual insects without killing them. Sinha (1965) used fluorescent antibody techniques to detect WTV antigens in the hemolymph, brain, salivary glands, intestine, malpigian tubules, ovaries, fat body, and mycetomes of the viruliferous leafhoppers.

## III. RICE DWARF VIRUS (RDV)

Since the transovarial passage of rice dwarf virus (RDV) was first reported by Fukushi (Fukushi, 1933, 1934, 1940), detailed studies along this line have been made by many workers and the results are discussed by other authors in this book. Partially purified RDV preparations were first obtained by Fukushi, Shikata, and Kimura (1960, 1962) by using differential centrifugation techniques. Rice dwarf virus particles in the preparation were detected by electron microscopy. Infectivity of the RDV preparation was proved by leafhopper injection (Kimura, 1962). Electron micrographs of the virus particles (Fukushi *et al.,* 1962) show that each particle is surrounded by a membranous material. It is not known whether the envelope is an ordinary component of the virus or a substance originating from host responses. Further purification of RDV was achieved by treating the partially purified preparation with phospholipase from snake venom or pancreatin so as to remove the enveloping material and then subjecting the resulting virus suspension to DEAE-cellulose column chromatography (Toyoda, Kimura, and Suzuki, 1965).

### *A. Virus source*

The virus isolate used in our studies (Toyoda *et al.,* 1965; Miura, Kimura, and Suzuki, 1966; Kodama *et al.,* 1966) was furnished by Dr. Shinkai who has preserved this isolate for about 18 years at the National Institute of Agricultural Sciences, Tokyo, and later at the Institute for Plant Virus Research, Chiba. RDV was cultured on rice plants, Variety Norin No. 8, inoculating them by means of viruliferous leafhoppers and incubating the infected seedlings at 28–30°C for 5 weeks in a greenhouse. Leaves and leaf sheaths showing symptoms were collected and about 300 g of fresh weight was used for each purification experiment.

### *B. Differential Centrifugation*

Three hundred g of diseased leaves were chopped into small pieces and then ground in 300–600 ml of M/20 phosphate buffer, at *p*H 6.8, with the aid of a meat chopper. The expressed juice was centrifuged at 1600*g* for 20 min. The supernatant fluid was stirred with the addition of 20% chloroform for 5 min at 5°C and then centrifuged at 1600*g* for 20 min. The supernatant was centrifuged at 26000*g* for 60 min. The resulting pellet was resuspended in M/40 Tris buffer, *p*H 7.2, and centrifuged at 1600*g* for 20 min. The supernatant consisted of a high concentration of virus particles when examined under an electron microscope (fraction A).

### *C. Treatment of Fraction A with Phospholipase*

Preliminary tests to elute RDV particles in fraction A from a DEAE cellulose column with the linear gradient of NaCl solution showed that these were firmly adsorbed by DEAE–cellulose without being eluted with any concentration of NaCl raising up to 1.0 M but were eluted only with 0.5 M NaOH. This suggests that the particles are covered with a strongly acidic substance, possibly the enveloping material shown in the electron micrographs of Fukushi *et al.* (1962). The enveloping material seems to be a lipid from its appearance in the electron micrographs. A trial was made to remove the enveloping material with phospholipase. Phospholipase was prepared from snake venom (*Trimeresurus flavoviridis*) or from pancreatin. Five mg of snake venom was dissolved in 1 ml of Tris buffer, at *p*H 7.2, and treated in boiling water for 7 min. After cooling, it was centrifuged at 3000 rpm for 15 min and the supernatant fluid was used as crude phospholipase preparation. Five mg of snake venom was used for each 5 ml of fraction A obtained from 50 g of diseased leaves. Phospholipase of pancreatin was prepared after Hanahan's method (Hanahan, 1952). Fifteen g of pancreatin was dissolved in 100 ml of M/15 phosphate buffer, at *p*H 6.0, the solution was stirred for 20 min at 4°C, and then centrifuged at 1200 rpm for 45 min. The supernatant was adjusted to *p*H 4.0, treated at 70°C for 10 min, cooled, and then centrifuged at 3000 rpm for 10 min. The supernatant fluid was used as crude enzyme preparation. Three ml of the enzyme solution was used for treating 50 ml of fraction A obtained from 300 g of diseased leaves. Fraction A, to which phospholipase preparation had been added, was incubated at 5° for 48 hr (fraction B).

### *D. Column Chromatography*

DEAE cellulose (Brown Co., 0.93 meq/g) column chromatography was employed for the purification of virus particles from fraction B. Three g of DEAE cellulose were used for each 5 ml of fraction B. The column was conditioned to *p*H 7.2 with M/40 Tris buffer. After adding 5 ml of fraction B, the column was washed with 50–100 ml of 1/40 Tris buffer, *p*H 7.2, and the virus was eluted with a linear gradient of NaCl, rising from 0 to 1.0 M. The optical density at 260 mμ of each 5 ml of the effluent was measured. Electron microscopy was used to detect the presence of virus particles. The virus particles were eluted at a definite range of NaCl concentration, from 0.2 to 0.25 M (Fig. 1) (fraction C).

### *E. Stability of the Purified Virus*

The effluent at this range of NaCl concentration contained a high concentration of virus particles (Fig. 2); however, the electrophoretic (Fig. 3) and the sedimentation (Fig. 4) patterns revealed that the virus preparation thus obtained was very unstable when stored in phosphate or Tris buffer at 4°C, easily releasing nucleic acid and producing empty shells. Virus particles can be stabilized, at least morphologically, by the addition of 0.1 M ammonium acetate (Kodama *et al.*, 1966).

### *F. Infectivity Assay*

The infectivity of the purified virus preparation (fraction C) was tested by injecting the diluted suspensions into nonviruliferous leafhoppers, approximately 1/3000 ml per insect, and feeding them on test rice seedlings (Table 1). The infectivity of the same preparation was compared with that of fraction A, which was prepared only by differential centrifugation techniques (Table 2). The results show that the purified virus preparation retains infectivity and that the infectivity per content of RNA increased conspicuously by the purification.

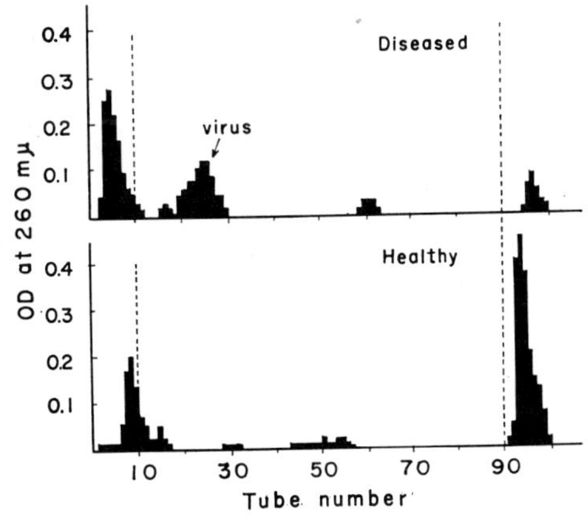


*Fig. 1.* DEAE cellulose column chromatography of RDV preparation partially purified by differential centrifugation (fraction A) followed by phospholipase treatment (fraction B) (Toyoda *et al.*, 1965): (upper) fraction B from infected plants; (lower) control, from healthy plants.

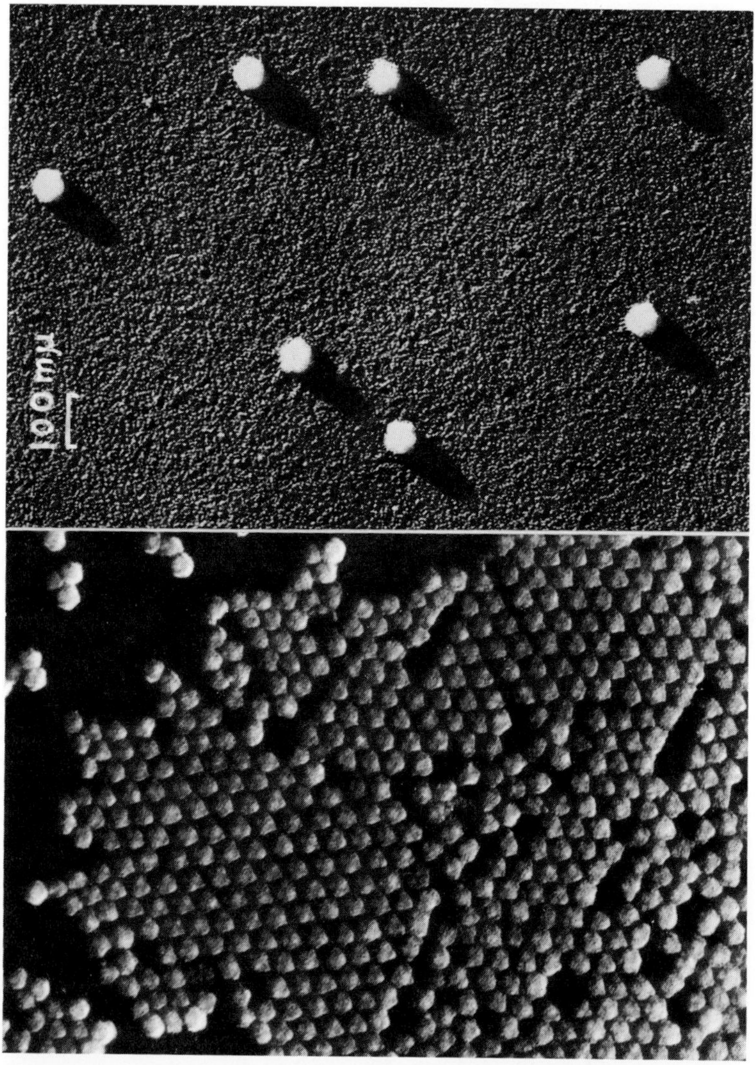


*Fig. 2.* Electron micrographs of purified RDV. (Courtesy of Dr. I. Kimura.)

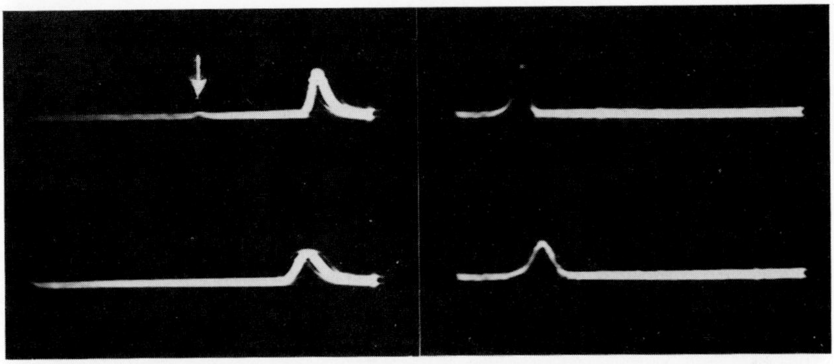

*Fig. 3.* Electrophoretic pattern of purified RDV in M/20 phosphate buffer of ionic strength 0.2 and *p*H 8.2, after 1 (upper) and 2 (lower) hr: r, ascending; d, descending. The small peak (arrow) shows the presence of degradation product (Toyoda *et al.*, 1965).

### *G. Electron Microscopy*

The enveloping material, demonstrated by the electron micrographs of Fukushi *et al.* (1962), seems to be a kind of lipid because it is easily removed by treatment with phospholipase. The removal of this material revealed the surface structure of the virus particles, which were 70 mμ

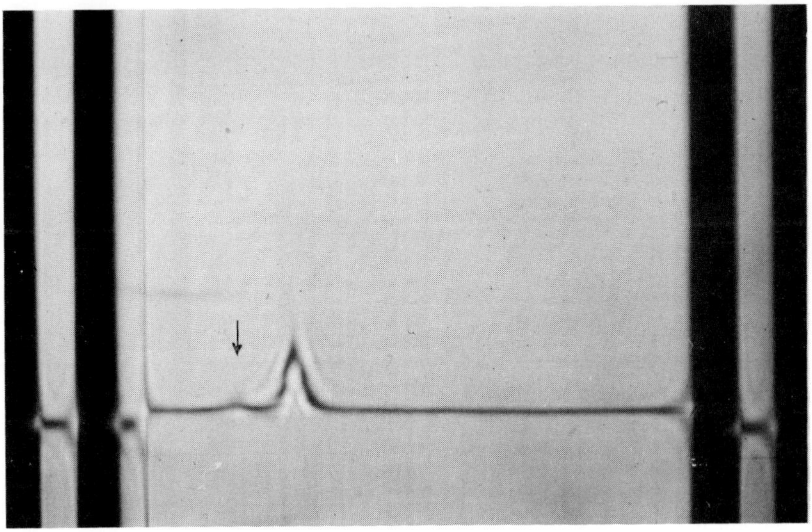

*Fig. 4.* Sedimentation pattern of purified RDV in 0.1 M ammonium acetate, at *p*H 6.5, after 9 min at 16,880 rpm (24,000 *g*) by Hitachi Model UCA-1A, RA-60 rotor. The small peak (arrow) shows the presence of degradation product (Toyoda *et al.*, 1965).

TABLE 1

INFECTIVITY TEST OF PURIFIED RDV [a]

(Toyoda, Kimura, and Suzuki, 1965)

| | Number of leafhoppers | | |
|---|---|---|---|
| | Injected | Survived for 20 days | Transmitted virus to rice seedlings |
| 1 | 60 | 20 | 7 |
| 2 | 60 | 21 | 4 |

[a] Purified virus, obtained from 300 g of diseased leaves, suspended in 50 ml of Tris buffer.

across (Fukushi *et al.*, 1962, 1963a) and icosahedral in shape, with the 92 structural subunits of a hollow cylinder (Kimura and Shikata, 1968). These results indicate that RDV is quite similar to reoviruses and WTV in structure but that the latter is somewhat smaller than the other two.

### *H. Physical and Chemical Properties*

Rice dwarf virus has a sedimentation coefficient of approximately 510*S* (Kawade, personal communication). Its molecular weight has not yet been determined. The RNA content of the virus is about 11% as determined by protein, phosphorus, and ribose contents.

TABLE 2

COMPARISON OF INFECTIVITY BETWEEN FRACTION A AND THE PURIFIED RDV (Toyoda *et al.*, 1965)

| Virus preparation | Number of leafhoppers | | | RNA content, mg/ml |
|---|---|---|---|---|
| | Injected | Survived for 20 days | Transmitted virus to rice seedlings | |
| Fraction A [a] | 60 | 22 | 5 | 0.33 |
| Purified virus | 60 | 22 | 2 | 0.016 |

[a] Prepared from 300 g of diseased leaves by chloroform treatment and differential centrifugation.

### I. *Base Composition of RDV-RNA*

The RNA was analyzed by acid and also by alkaline hydrolyses. The results are shown in Table 3. The ratios G/C, A/U, and A–C/G–U are approximately unity, suggesting that the bases are paired in a double-stranded RNA helix (Miura *et al.*, 1966).

### J. *Thermal Denaturation of RDV-RNA*

On heating to 100°C in a medium of 0.0015 M NaCl, and 0.00015 M sodium citrate, *p*H 7.0, the RNA exhibits a hyperchromic effect. The absorbancy begins to increase at 70°C and the $T_m$ is about 80°C. The $T_m$ may be as high as that of reovirus RNA when measured in a medium of 0.15 M NaCl and 0.015 M sodium citrate because RDV–RNA has a G–C content of about 44% almost equal to that of reovirus RNA.

### K. *X-Ray Diffraction Photographs of RDV-RNA*

X-ray diffraction photographs shown by Sato *et al.* (1966) are similar to those of DNA A form (in which the bases are tilted at about 70° to the helix axis) but somewhat different from DNA, *i.e.*, the number of base residues per turn is 10 as compared with 11 in DNA A form. This structure is similar to that of reovirus RNA (Langridge and Gomatos, 1963) and WTV–RNA (Tomita and Rich, 1964).

### L. *Infrared Dichroism of RDV-RNA*

An infrared absorption spectrum of an oriented film of RDV–RNA at 75% relative humidity was shown by Sato *et al.* (1966). Two strong absorption bands were observed at 1225 $cm^{-1}$ and 1084 $cm^{-1}$; the former was assigned to the $PO_2^-$ antisymmetric stretching vibration and the latter to the $PO_2^-$ symmetric stretching vibration. The former

TABLE 3

BASE COMPOSITION OF RDV–RNA (Miura, Kimura, and Suzuki, 1965)

| Hydrolysis | Base composition, mole %: G | A | C | U | G/C | A/U | Pu/Py | (A+C)/(G+U) | (G+C)/(A+U) |
|---|---|---|---|---|---|---|---|---|---|
| Acid (N HCl, 100°C, 1 hr) | 22.8 | 27.8 | 21.4 | 28.0 | 1.06 | 0.99 | 1.02 | 0.97 | 0.79 |
| Alkaline (0.3 N KOH, 47°C, 18 hr) | 21.8 | 28.4 | 21.6 | 28.2 | 1.01 | 1.00 | 1.01 | 1.00 | 0.77 |

absorption band shows a strong perpendicular dichroism and the latter a strong parallel dichroism. These authors determined the orientation of the $PO^-_2$ group in the DNA helix: the O····O line makes an angle of about 70° and the bisector of ∠OPO makes an angle of about 40° with the helix axis. This is markedly different from the orientation of $PO_2^-$ group in the A and B forms of DNA.

Rice dwarf virus–RNA is more resistant to pancreatic ribonuclease than denatured RDV–RNA, transfer RNA, and ribosomal RNA's from yeast (Miura *et al.*, 1966). Denatured RDV–RNA exhibits hyperchromic effect when treated with 1.8% formaldehyde at 37°C for 0.5 and 22 hr, whereas native RDV–RNA exhibits no change by the same treatment, indicating that $NH_2$ groups in the bases are protected from the attack of formaldehyde by the hydrogen bonding (Miura *et al.*, 1966).

All the available data provide unquestionable evidence that DRV has an RNA of double-helical structure.

### *M. Fluorescent Antibody Technique for the Detection of Virus Antigen*

Kimura and Suzuki (1965) prepared a fluorescein-labeled antibody of RDV which had a titer of 1/512, recovering 0.25 of the titer of the original antiserum. They applied this for the detection of virus antigen in leafhoppers by the smear test. Among 100 leafhoppers tested, 40 were positive in both the smear test and transmission test, 42 were negative in both tests, 14 was positive in the smear test but negative in the transmission test, and 4 were negative in the smear test but positive in the transmission test. In a second experiment, in which the improved smear technique of Sinha and Reddy (1964) was applied, there was a good correlation between the results of the smear test and transmission test except in one instance which was positive in the smear test but negative in the transmission test. The same fluorescent antibody was applied to sections of diseased leaves and leaf sheaths after the sections were fixed with acetone for 30 min. The result indicated that the fluorescent antibody specifically stained inclusion bodies which contain a large number of virus particles, very often packed into a crystal-like arrangement, as indicated by the electron micrographs of Fukushi *et al.* (1962).

## IV. PEA ENATION MOSAIC VIRUS (PEMV)

Although convincing evidence supporting virus multiplication in insect vectors has been provided for several leafhopper-borne viruses, this is not the case with the so-called circulative, aphid-borne viruses.

Detailed reviews on this subject have been published by Maramorosch (1964) and Smith (1965). Pea enation mosaic virus had been known as an aphid-borne circulative virus for many years. The localization of this virus in its insect vector was first demonstrated by Shikata, Maramorosch, and Granados (1966). Partial purification of the virus was performed by Bustrillos (1964) and by Bozarth, Chow, and Gross (1965). Izadpanah and Shepherd (1966) also purified a strain of PEMV by differential centrifugation and density gradient centrifugation, and characterized the virus by electron microscopy. Gibbs, Harrison, and Woods (1966) purified an isolate of PEMV by using almost the same procedures and characterized the virus by electron microscopy. These two authors reported the presence of two components in the purified virus preparations, similar in shape and size but different in appearance and sedimentation coefficient. Both components retain infectivity. The procedure for the purification of PEMV adopted by Izadpanah and Shepherd follows (1966).

### A. *Virus Source*

The Dwarf Telephone variety of pea was used as the host plant to culture the virus for purification. Seedlings 2 weeks after seeding were inoculated on all the leaves and stipules with the sap from infected plants. The best yield of virus was obtained from infected plants 10–12 days after inoculation when the plants were grown at 16–22°C. Only the terminal parts of plants showing definite symptoms were taken as starting material.

### B. *Clarification*

Most of the conventional methods of clarification, such as emulsification with 8% *n*-butanol or addition of bentonite were inadequate because these caused an appreciable loss of infectivity of the extract. Lowering the *p*H of the homogenate to 5.3 with 4 N HCl, followed by incubation for 30 min at 4°C and readjustment to *p*H 6.0 with 4 N NaOH, followed by slow speed centrifugation, gave a partially clarified supernatant fluid with an infectivity endpoint comparable to that of untreated homogenates.

### C. *Differential Centrifugation*

Infected pea tissue was homogenized in 0.2 M acetate buffer, at *p*H 6.0, using 2 ml of buffer per gram of tissue. The homogenate was filtered through cheesecloth and the filtrate was subjected to clarification treatment. The clarified supernatant fluid was centrifuged at 30,000 rpm for 2 hr in a Sorvall SS-34 rotor. The sedimented material was resuspended in 0.1 M acetate buffer, at *p*H 6.0, for 2 hr or longer on a mechanical shaker, and then centrifuged for 20 min at 7000 rpm. The resulting supernatant was centrifuged at 40,000 rpm for 2 hr. The pellet was resuspended in 0.1 M acetate buffer, at *p*H 6.0, for 2 hr or longer on a mechanical shaker and then centrifuged 15–20 min at 6000 rpm. The final step of high- and low-speed centrifugation was repeated and the final volume of resuspended product was 1/20–1/60 that of the original homogenate.

### *D. Density-Gradient Centrifugation*

Further separation of the virus from host components was achieved by rate–zonal density-gradient centrifugation. Gradient columns were prepared by layering 4, 7, 7, 7, and 3 ml of 0.01 M Tris buffer, at *p*H 7.5, containing 100, 200, 300, 400, and 500 mg of sucrose per milliliter, respectively, in $1 \times 3$-in. cellulose nitrate tubes. The gradient columns thus prepared were placed in a cold room until use the following day. Two ml of virus suspension were layered on top of gradient tubes and centrifuged for 3 or more hr at 23,000–25,000 rpm in the SW-25.1 rotor of the Spinco Model L ultracentrifuge. Separation and analysis of the contents of the centrifuged tubes were made by using an ISCO Model D density-gradient fractionator and flow densitometer. The absorbancy at 2536 m$\mu$ was recorded on the chart of a NESCO model JY 120.2 external recorder. After a 4-hr centrifugation at 25,000 rpm, two bands were distinguished in the tubes. The upper band was usually very narrow and had a brownish opalescent appearance and the lower band was prominent and highly opalescent, both being located 2.4–2.5 and 2.7–3.1 cm below the meniscus. These two bands were referred to as the top and bottom components, respectively.

### *E. Infectivity Assay*

Infectivity assays were made by determining the dilution endpoint of virus-containing extracts on a systemic host. Peas of the Dwarf Telephone variety were used for this purpose. Tests on the correlation of infectivity with components in density gradient tubes clearly showed that infectivity was associated with the two visible bands. The top component was about four times as infectious as the bottom component.

### *F. Physical and Chemical Properties*

These top and bottom components have sedimentation coefficients of 106*S* and 122*S*, respectively [these are 95*S* and 115*S* according to Gibbs *et al.* (1966)]. The virus contains approximately 27% RNA. The nucleic acid was found to be of the ribose type with the following base composition: adenine 23.9; guanine, 26.4; cytosine, 24.1; and uracil, 25.6 (Shepherd and Ghabrial, 1966).

### G. Electron Microscopy

Electron micrographs of purified virus preparations and of ultrathin sections of infected tissues always show that the virus particles are spherical. However, the diameter differs according to the authors, namely, 36 mμ by Izadpahah and Shepherd (1966), 30 mμ by Gibbs *et al.* (1966), and 28 mμ by Shikata *et al.* (1966). The estimate of particle diameter by Izadpanah and Shepherd seems to be inaccurate due to the metal shadowing and the partial flattening of the air-dried materials. Particles of the same size and shape were detected in the fat body and gut lumen of the viruliferous pea aphids *Acyrthosiphon pisum*. Pea enation mosaic virus is the first circulative, aphid-borne virus localized by electron microscopy in insect vectors *in situ* (Shikata *et al.*, 1966).

### H. Serology

The virus antiserum prepared by Izadpanah and Shepherd (1966) reacted with healthy pea sap and protein preparations. Aliquots of antiserum were absorbed with 3 volumes of concentrated healthy pea protein prepared by ultracentrifugation and precipitation with ammonium sulfate. After absorption the antiserum reacted specifically with virus to give a somatic-type precipitate with a microprecipitin titer of 1:512. In agar gel diffusion tests, only one precipitin line was formed with partially purified virus, while two lines were formed with extracted sap from infected peas, suggesting the presence of an additional low molecular weight antigen in infected plants (Izadpanah and Shepherd, 1966). Double-diffusion serological tests were made by Gibbs *et al.* (1966) using an antiserum to an American islate of PEMV provided by Shepherd. The top and bottom components of the English isolate formed one and two precipitin lines, respectively; one of the latter was confluent with the former line. The unfractionated PEMV preparations did not react specifically with antisera prepared against cherry necrotic ringspot, cowpea mosaic, prune dwarf, red clover mottle, and true broad bean mosaic viruses. From these results Gibbs *et al.* (1966) concluded that the PEMV preparations contain two kinds of particles of similar shape and size but different sedimentation coefficient and antigenic constitution, and that the top component (90*S*) is actually infective to *Chenopodium amaranticolor*. However, it could not be established whether the 115*S* particle is also infective because this has not been obtained free from 95*S* particles.

## V. DISCUSSION

The development of methods for the separation of discrete protein species has progressed rapidly in recent years and has enabled us to distinguish very small differences in protein properties. Differential and density-gradient centrifugation, absorption and molecular-seive column chromatography, and electrophoresis are some of the powerful tools available to protein chemists. These methods have been adopted by plant virologists for the purification of particular arthropod-borne viruses which are labile when stored in purified state: WTV and RDV, and also for the purification of PEMV, the stability of which has not been reported as yet. The successful purification of these viruses without loss of infectivity, at least for a week during storage at 4°C in appropriate media, enabled us to characterize the virus particles morphologically by electron microscopy, to describe the chemical compositions and physical properties, and to use immunological techniques for virus identification, detection, and estimation.

The virus structures revealed by shadowing and negative staining of purified preparations show that there are striking similarities among reoviruses (Vasquez and Tournier, 1962), WTV (Bils and Hall, 1962), and RDV (Fukushi *et al.*, 1962, 1963; Kimura and Shikata, 1968): they are all icosahedrons with 180 structural subunits, measuring 70, 60, and 70 m$\mu$ in diameter, respectively. The structural similarities between reoviruses and WTV suggested that the nucleic acid of WTV might be double-stranded RNA, and this was indeed the case (Gomatos and Tamm, 1963; Black and Markham, 1963). Recent studies on RDV indicated that this is also the case with RDV (Miura *et al.*, 1966). X-ray diffraction studies of the RNA's of these viruses also showed that these have a double-helical structure resembling the DNA A form but different in the number of base residues per turn: 10 in double-stranded RNA's in contrast to 11 in DNA A form. As already discussed by Gomatos and Tamm (1963), double-stranded RNA of viruses is stable in living host cells against the attack of ribonuclease and other denaturating agents and it can interfere with the distribution of genetic material during replication. This may be one of the factors inducing tumor formation in the case of WTV, but not in RDV, which does not cause tumors on rice plants. In Japan, there is another tumor-causing virus, rice black-streak virus, similar to RDV in shape and size. Its structure and the base composition of its nucleic acid are not known. Before valid generalizations can be made, more information on double-stranded RNA viruses is needed. The RNA of PEMV seems to be a single strand, judging from its base composition.

The morphological characterization of the purified viruses greatly facilitated the electron microscopy of ultrathin sections of viruliferous leafhoppers (Fukushi *et al.*, 1962, 1963b; Nasu, 1965; Shikata and Maramorosch, 1965) and pea aphids (Shikata *et al.*, 1966), and the detection and localization of the virus particles in their vectors.

Fluorescent antibody techniques have been successfully applied for the detection of virus antigens in smears of droplets from viruliferous leafhoppers (Sinha and Black 1962; Sinha and Reddy 1964; Sinha *et al.*, 1964; Kimura and Suzuki, 1965), and for tracing the distribution of WTV antigen in different organs of vectors at intervals after acquisition feeding (Sinha, 1965).

The results of these electron microscopic and immunological studies have provided most convincing evidence for the multiplication of viruses in insect vectors.

Rice stripe virus is known to pass transovarially (Shinkai, 1955, 1962; Yamada and Yamamoto, 1954, 1955, 1956); however, purification of the virus has not been achieved as yet, possibly because of the low concentration of virus in diseased rice plants. Therefore the multiplication in its vector, *Delphacodes striatella,* has not been proved. There are many circulative viruses without evidence of virus multiplication in their vectors (Black, 1953: Maramorosch, 1964; Smith, 1965). Once purification of the viruses has been attained, it will be easy to test whether the viruses are propagative or not.

## BIBLIOGRAPHY


Bawden, F. C. 1950. Plant viruses and virus disease. 3rd ed., Chronica Botanica Co., Waltham, Mass.

Bils, R., and Hall, C. E. 1962. Electron microscopy of wound-tumor virus. Virology 17:123–136.

Black, L. M. 1951. Hereditary variation in the reaction of sweet clover to the wound-tumor virus. Amer. J. Bot. 38:256–267.

——— 1953a. Occasional transmission of some plant viruses through the eggs of their insect vectors. Phytopathology 43:9–10.

——— 1953b. Transmission of plant viruses by cicadellids. Advances Virus Res. 1:69–89.

——— 1955. Concept of problems concerning purification of labile insect-transmitted plant viruses. Phytopathology 45:208–216.

Black, L. M., and Markham, R. 1963. Base-pairing in the ribonucleic acid of wound-tumor virus. Netherlands J. Plant Pathol. 69:215. (Abstr)

Black, L. M., and Brakke, M. K. 1952. Multiplication of wound-tumor virus in an insect vector. Phytopathology 42:269–273.

Black, L. M., Mosley, V. M., and Wyckoff, R. W. G. 1948. Electron microscopy of potato yellow dwarf virus. Biochim. Biophys. Acta 2:121–123.

Bozarth, R. F., Chow, C. C., and Gross, S. 1965. Purification of pea enation mosaic virus. Phytopathology 55:127. (Abstr.)

Brakke, M. K. 1951. Density gradient centrifugation: a new separation technique. J. Amer. Chem. Soc. 73:1847.

——— 1953. Zone electrophoresis of dyes, proteins and viruses in density-gradient columns of sucrose solutions. Arch. Biochem. Biophys. 45:275–190.

Brakke, M. K., Maramorosch, K., and Black, L. M. 1953. Properties of the wound-tumor virus. Phytopathology 43:387–390.

Brakke, M. K., Vatter, A. E., and Black, L. M. 1954. Size and shape of wound-tumor virus. *In* Abnormal and pathological plant growth. Brookhaven Symposia in Biology No. 6 U.S. Brookhaven National Laboratory, Upton, N.Y., p. 137–156.

Bustrillos, A. D. 1964. Purification, serology and electron microscopy of pea enation mosaic virus. Ph.D. Thesis. Michigan State University.

Fukushi, T. 1933. Transmission of virus through the egg of an insect vector. Proc. Imp. Acad. Sci. Tokyo, Japan 9:457:460.

——— 1934. Studies on the dwarf disease of rice plant. J. Faculty Agr. Hokkaido Univ. 37:31–164.

——— 1940. Further studies on the dwarf disease of rice plants. J. Faculty Agr. Hokkaido Univ. 45:83–154.

Fukushi, T., and Shikata, E. 1963a. Fine structure of rice dwarf virus. Virology 21:500–503.

——— 1963b. Localization of rice dwarf virus in its insect vector. Virology 21: 503–505.

Fukushi, T., Shikata, E., and Kimura, I. 1962. Some morphological characters of rice dwarf virus. Virology 18:192–205.

Fukushi, T., Shikata, E., Kimura, I., and Nemoto, M. 1960. Electron microscopic studies on the rice dwarf virus. Proc. Japan Academy 36:352–357.

Gibbs, A. J., Harrison, B. D., and Woods, R. D. 1966. Purification of pea enation mosaic virus. Virology 29:348–351.

Gomatos, P. J., and Tamm, I. 1963. Animal and plant viruses with double-helical RNA. Proc. Nat. Acad. Sci. N.Y. 50:878–885.

Hanahan, D. J. 1952. The enzymatic degradation of phosphatidyl choline in diethyl ether. J. Biol. Chem. 195:199–206.

Izadpanah, K., and Shepherd, R. J. 1966. Purification and properties of the pea enation mosaic virus. Virology 28:463–467.

Kimura, I. 1962. Further studies on the rice dwarf virus. I., II. Ann. Phytopathol. Soc. Japan 27:197–203, 204–213.

Kimura, I., and Shikata, E. 1968. Structural model of rice dwarf virus. Proc. Japan Academy 44:538–543.

Kimura, I., and Suzuki, N. 1965. Purification of rice dwarf virus and application of fluorescent antibody technique for the detection of virus antigen in an insect vector and rice plants [in Japanese]. Shokubutsu-Boeki (Plant Protection) 19:137–140.

Kodama, T., Kimura, I., Tsugita, A., and Suzuki, N. 1966. Homogeneity and stability of purified rice dwarf virus preparation. Ann. Phytopathol. Soc. Japan 32:86. (Abstr.)

Langridge, R., and Gomatos, P. J. 1963. The structure of RNA–reovirus RNA and transfer RNA have similar three-dimensional structure which differ from DNA. Science 141:696–698.

Maramorosch, K. 1952. Direct evidence for the multiplication of aster yellows virus in its insect vector. Phytopathology 45:59–64.

——— 1964. Arthropod transmission of plant viruses. Ann. Rev. Entomol. 8:369–414.

Miura, K., Kimura, I., and Suzuki, N. 1966. Double-stranded ribonucleic acid from rice dwarf virus. Virology 28:571–579.

Nagaraj, A. N., Sinha, R. C., and Black, L. M. 1961. A smear technique for detecting virus antigen in individual vectors by the use of fluorescent antibodies. Virology 15:205–208.

Nasu, S. 1965. Electron microscopic studies on transovarial passage of rice dwarf virus. Japan. J. Appl. Entomol. Zool. 9:225–237.

Sato, T., Kyogoku, Y., Higuchi, S., Mitsui, Y., Iitaka, Y., Tauboi, M., and Mirua, K. 1966. A preliminary investigation on the molecular structure of rice dwarf virus nucleic acid. J. Mol. Biol. 16:180–190.

Shepherd, R. J., and Ghabrial, S. A. 1966. Isolation and some properties of the nucleic acid and protein components of the pea enation mosaic virus Phytopathology 56:900. (Abstr.)

Shikata, E., and Maramorosch, K. 1965. Electron microscopic evidence for the systemic invasion of an insect host by a plant pathogenic virus. Virology 27:461–475.

Shikata, E., Maramorosch, K., and Granados, R. R. 1966. Electron microscopy of pea enation mosaic virus in plants and aphid vectors. Virology 29:426–436.

Shinkai, A. 1954. Transovarial transmission of rice stripe virus in *Delphacodes striatella* Fall. Ann. Phytopathol. Soc. Japan 18:169. (Abstr.)

——— 1958. Transovarial passage of rice dwarf virus in *Inazuma dorsalis* Motsch [in Japanese]. Ann. Phytopath. Soc. Japan 23:26. (Abstr.)

——— 1962. Studies on insect transmission of rice virus diseases in Japan [in Japanese with English summary, p. 108–112]. Bull. Nat. Inst. Agr. Sci. C-14:1–112.

Sinha, R. C. 1965. Sequential infection and distribution of wound-tumor virus in the internal organs of a vector after ingestion of virus. Virology 26:673–686.

Sinha, R. C., and Black, L. M. 1962. Studies on the smear technique for detecting virus antigens in an insect vector by use of fluorescent antibodies. Virology 17:582–587.

Sinha, R. C., and Reddy, D. V. R. 1964. Improved fluorescent smear technique and its application in detecting virus antigens in an insect vector. Virology 24:626–634.

Sinha, R. C., Reddy, D. V. R., and Black, L. M. 1964. Survival of insect vectors after examination of hemolymph to detect virus antigens with fluorescent antibody. Virology 24:666–667.

Smith, K. M. 1965. Plant virus–vector relationship. Advances Virus Res. 11:61–96.

Stanley, W. M. 1935. Isolation of crystalline protein possessing the properties tobacco-mosaic virus. Science 81:644–645.

Stanley, W. M., and Wyckoff, R. W. G. 1937. The isolation of tobacco ring spot and other virus proteins by ultracentrifugation. Science 85:181–183.

Stegwee, D., and Ponsen, M. B. 1958. Multiplication of potato leaf roll virus in the aphid *Myzus persicae* (Sulz.). Entmol. Exp. Appl. 1:291–300.

Tomita, K., and Rich, A. 1964. X-ray diffraction investigations of complementary RNA. Nature 201:1160–1163.

Toyoda, S., Kimura, I., and Suzuki, N. 1965. Purification of rice dwarf virus. Ann. Phytopathol. Soc. Japan 30:225–230.

Vasquez, C., and Tournier, P. 1964. New interpretation of the reovirus structure. Virology 24:128–130.

Whitcomb, R. F. 1964. A comparison of serological scoring with test plant scoring of leafhoppers infected by wound tumor virus. Virology 24:488–492.

Whitcomb, R. F., and Black, L. M. 1961. Synthesis of wound-tumor soluble antigen in an insect vector. Virology 15:136–145.

Yamada, W., and Yamamoto, H. 1954. Transmission of rice stripe virus through the eggs of *Delphacodes striatella* Fall. Ann. Phytopathol. Soc. Japan 18:169. (Abstr.)

——— 1955. Studies on the stripe disease of rice plant. I. [in Japanese]. Okayama Prefecture Agr. Exp. Sta. Special Bull. 52:93–112.

——— 1965. Studies on the stripe disease of rice plant. III. [in Japanese]. Okayama Prefecture Agr. Exp. Sta. Special Bull. 55:35–56.

# Inhibition of Viruses by Vector Saliva

YASUMICHI NISHI

*Kyushu Agricultural Experiment Station,*
*Chikugo, Fuguoka, Japan*

It is known that tobacco mosaic virus (TMV) is stable and easily transmitted mechanically, but not by aphids. The green peach aphid (*Myzus persicae* Sulzer) transmits only cucumber mosaic virus (CMV) from diseased plants doubly infected with CMV and TMV (Hoggan, 1929). When a tobacco plant, infected with TMV, is pricked with a fine needle, and then a healthy tobacco plant is pricked with the same needle, the healthy plant becomes infected with TMV (van Hoof, 1958). Tobacco mosaic virus is thus transmitted easily by needle puncture but not through the stylet of aphids. Day and Irzykiewicz (1954), Nishi (1958), and others suggested that infectivity of TMV may be inhibited by some aphid secretion.

Many nonpersistent viruses are easily transmitted by aphids. They are usually acquired by aphids during short acquisition feeding periods and transmitted in a very short time; however, the viruliferous aphids lose the ability to infect within a short time (Fukushi, 1931; Hoggan, 1933; Watson and Roberts, 1940; Kasai, 1950; Sylvester, 1950, 1954; Bradley, 1952; Hamlyn, 1953; Nishi, 1963).

## I. INHIBITION OF TOBACCO MOSAIC VIRUS BY APHID SALIVA

Miles (1959) reported that there are two types of salivary secretion in aphids: watery saliva and saliva which coagulates to form stylet sheaths. In view of the difficulty of obtaining enough saliva from aphids, a homogenate of plant material previously exposed to aphid feeding was prepared for the test (Nishi, 1958). Three kinds of plants were used: turnip (*Brassica rapa* L.), Japanese radish (*Raphanus sativus* L.), and tobacco (*Nicotiana tabacum* L. var. Xanthi). These plants were exposed to a large number of green peach aphids, homogenated, and centrifuged for 10 min at 10,000 rpm; the supernatant fluid was mixed with an equal quantity of TMV solution. The infectivity of the virus in the mixture was assayed on *Nicotiana glutinosa, Chenopodium album,* or *Phaseolus vulgaris* by means of the half-leaf method. In all trials, sap from the plants which were not infested by aphids was also tested as a control. As shown in Table 1, some inhibitory substance was obviously present in the leaves of the plants which had been infested by a large number of green peach aphids.

The infectivity of TMV was also inhibited by salivary secretion in Chinese cabbage leaves fed upon by *Rhopalosiphum pseudobrassicae* Davis or in cucumber leaves fed upon by *Aphis gossypii* Glover.

Apparently some substance secreted by the aphid and injected into feeding plants strongly inhibits the infectivity of TMV. However, it is not yet known whether the inhibition is due to a direct action of saliva, or due to a by-product produced in the plants as a result of stimulation by aphid saliva. An attempt was made to demonstrate the inhibitory effect of saliva secreted into a sucrose solution by aphids fed through a thin rubber membrane (Nishi, 1963). The upper end of a glass tube, 3 cm in diameter and 10 cm in height, was equipped with a thin rubber sac containing 3% sucrose solution, while the lower end of the tube was covered with a screen. A large number of green peach aphids were placed in the tube for 3 days. The aphids discharged saliva from the stylet into the sucrose solution. Afterwards, the sucrose solution, containing aphid salivary secretions, was mixed with TMV solution. The infectivity of the mixture was assayed on *N. glutinosa* by means of the half-leaf method. As shown in Table 2, TMV was inhibited strongly by the saliva. Even when the sucrose solution containing saliva was dialyzed through a cellophane membrane against running water, the inhibitory substance was not lost.

From Table 2 it is evident that aphid saliva inhibits the infectivity of TMV. The inhibitory effect of the sap of plants on which aphids have fed apparently results from the action of saliva injected by the aphids into the plant tissue.

To determine the effect of aphid saliva on infectivity of TMV, experiments were made using inhibitory sap from turnip and Japanese radish leaves which had been infested by green peach aphid (Nishi, 1963). Tobacco mosaic virus solution was mixed with the same quantity of the inhibitory sap, and kept for 60 min at 10 and 30°C. The infectivity of the mixture was tested every 15 min by means of the half-leaf method, using *N. glutinosa*. The result showed that the infectivity was inhibited strongly and promptly by the inhibitory sap at 10°C, as well as at 30°C, as is shown in Table 3. The inhibition did not depend on the duration of the reaction. The inhibitory substance

TABLE 1

EFFECT OF SAP FROM PLANTS FED FOR 24 HR BY GREEN PEACH APHIDS UPON THE INFECTIVITY OF TMV

| Plants tested | Treatment | | Number of feeding aphids per 1 g of leaf | Number of local lesions per half-leaf of *N. glutinosa* | | | | | | | | Virus inhibition,[a] % |
|---|---|---|---|---|---|---|---|---|---|---|---|---|
| | | | | 1 | 2 | 3 | 4 | 5 | 6 | 7 | Mean | |
| Japanese radish | Aphid feeding | (*A*) | 500 | 11 | 15 | 20 | 22 | 17 | 15 | 6 | 15.1 | 64.8 |
| | Check | (*B*) | 0 | 17 | 30 | 82 | 77 | 23 | 51 | 20 | 42.9 | |
| Turnip | Aphid feeding | (*A*) | 225 | 34 | 23 | 77 | 81 | — | — | — | 53.8 | 52.0 |
| | Check | (*B*) | 0 | 41 | 103 | 92 | 212 | — | — | — | 112.0 | |
| Tobacco | Aphid feeding | (*A*) | 125 | 4 | 22 | 39 | 69 | 87 | — | — | 44.2 | 45.8 |
| | Check | (*B*) | 0 | 48 | 49 | 66 | 112 | 132 | — | — | 81.6 | |

[a] Virus inhibition: $(1 - A/B) \times 100\%$.

TABLE 2

EFFECT OF SUCROSE SOLUTION CONTAINING APHID SALIVA ON TMV INFECTIVITY

| Treatment | Sucrose solution | | Average number of local *lesions on a half leaf* of *N. glutinosa* 1 | 2 | Mean | Virus inhibition,[a] % |
|---|---|---|---|---|---|---|
| No treatment | Aphid feeding | (*A*) | 73.0 | 27.3 | 50.2 | 52.9 |
| | Check | (*B*) | 143.2 | 69.7 | 106.5 | |
| Dialyzed with water | Aphid feeding | (*A*) | 47.8 | 23.7 | 35.8 | 52.7 |
| | Check | (*B*) | 97.0 | 54.3 | 75.7 | |

[a] Virus inhibition: $(1 - A/B) \times 100\%$.

secreted by aphids and injected into feeding plants seems to inactivate TMV.

One half-leaf of *N. glutinosa* was inoculated mechanically by means of carborundum using sap from healthy turnip leaves which had been infested by green peach aphids. The other half of each leaf was inoculated with sap from healthy turnip leaves which had not been infested by aphids, and served as control. The whole surface of the leaves was then promptly washed with water and was inoculated with TMV 5 and 30 min after the sap treatment. There was no difference in the number of local lesions observed on each half-leaf treated with inhibitory sap and with the control sap. In another trial, the surface of each half-leaf of *N. glutinosa* was rubbed, 20 min after TMV inoculation, with absorbent cotton soaked with inhibitory plant sap and the control sap, respectively. Again there was no difference in the number of local lesions produced. Local lesion production on the leaves of *N. glutinosa* was therefore not influenced by the inhibitory substance when the leaves were treated by the inhibitory sap either before or after TMV inoculation.

## II. INHIBITION OF POTATO VIRUS X BY APHID SALIVA

Potato virus X (PVX), which is not aphid borne was also tested for the inhibitory action of aphid saliva (Nishi, 1962). The sap of tobacco plants affected with PVX was mixed with an alcoholic precipitate of sap from Chinese cabbage on which green peach aphids had fed, and the virus infectivity of the mixture was assayed using

TABLE 3

INFLUENCE OF TEMPERATURE AND DURATION OF TREATMENT ON INHIBITION OF TMV INFECTIVITY

| Temp, °C | Duration of treatment, min | Average number of local lesions obtained on 5 half-leaves of *N. glutinosa* | | Virus inhibition,[a] % |
|---|---|---|---|---|
| | | Aphid feeding ($A$) | Check ($B$) | |
| 30 | 0 | 9.4 | 44.6 | 78.9 |
| | 15 | 2.4 | 35.8 | 93.3 |
| | 30 | 9.6 | 56.2 | 82.9 |
| | 45 | 0.8 | 12.2 | 93.4 |
| | 60 | 9.0 | 45.7 | 80.3 |
| 10 | 0 | 5.5 | 28.8 | 80.9 |
| | 15 | 8.2 | 42.0 | 80.5 |
| | 30 | 6.7 | 63.3 | 89.4 |
| | 45 | 5.3 | 38.7 | 86.3 |
| | 60 | 5.6 | 30.2 | 81.5 |

[a] Virus inhibition: $(1 - A/B) \times 100\%$.

*Gomphrena globosa* as a local lesion indicator plant. The results are shown in Table 4. The alcoholic precipitate of the sap from aphid fed plants strongly inhibited the infectivity of PVX as in the case of TMV.

## III. INHIBITION OF NONPERSISTENT VIRUS BY APHID SALIVA

Aphids that acquired a nonpersistent virus rapidly lose the ability to infect when feeding on a healthy plant, or when fasting (Doolittle and Walker, 1928; Hoggan, 1933; Fukushi, 1939; Kasai, 1950; Sylvester, 1950; Hamlyn, 1953). Experiments were made in connection with the effect of aphid saliva on the infectivity of turnip mosaic virus (Nishi, 1963). Green peach aphids were fasted for 24 hr before confinement to mosaic-diseased turnip seedlings. Three acquisition feeding periods, 10 min, 60 min, and 24 hr, were used. After the acquisition feeding, the aphids were placed in a petri dish for post acquisition fasting, which ranged from 0 to 6 hr. The aphids were transferred every hour to fresh turnip seedlings to ascertain their transmission capability. The insects ceased to transmit in 3–5 hr of post acquisition

TABLE 4

EFFECT OF APHID SECRETION ON PVX AND TMV

| Virus | Test plant | Treatment on Chinese cabbage | Average number of local lesions on 3 leaves of *G. globosa* and 5 half-leaves of *N. glutinosa* | | | | | Virus inhibition,[a] % |
|---|---|---|---|---|---|---|---|---|
| | | | 1 | 2 | 3 | 4 | Mean | |
| PVX | *Gomphrena globosa* | Aphid feeding (*A*) | 44.0 | 43.0 | 29.0 | 8.0 | 31.0 | 62.7 |
| | | Check (*B*) | 123.0 | 94.0 | 78.0 | 37.0 | 83.0 | |
| TMV | *Nicotiana glutinosa* | Aphid feeding (*A*) | 8.7 | 6.7 | 9.0 | — | 8.1 | 65.8 |
| | | Check (*B*) | 21.5 | 20.0 | 21.0 | — | 20.8 | |

[a] Virus inhibition: $(1 - A/B) \times 100\%$.

fasting. In another series, radioactive phosphorous-32 ($^{32}P$) was used as a tracer. Green peach aphids were allowed to feed on mosaic-diseased turnip seedlings containing $^{32}P$ as a soluble phosphate, and were treated by the same procedure for post acquisition fasting, as mentioned above. Autoradiographs clearly showed that the aphids transferred $^{32}P$ to the test seedlings, even after 6 hr of post acquisition fasting. As for the acquisition feeding on mosaic diseased turnip seedlings containing $^{32}P$, green peach aphids were allowed to feed on the plant by actually inserting their stylet for 30 sec. Following the acquisition feeding, the aphids were transferred individually to healthy test seedlings for a 30-sec test feeding period, then transferred to another healthy test seedling, and so on. This procedure was repeated until the number of tested seedlings attained 20. The trial was repeated four times. The results indicated that the aphids lost transmission capability in a short time, whereas the autoradiograph showed that the transfer of $^{32}P$ continued to the twentieth seedling, as shown in Table 5.

A solution of turnip mosaic virus was mixed with sap from turnip or Chinese cabbage on which a large number of green peach aphids had fed. The mixture was inoculated to healthy turnip seedlings to test the virus infectivity. The virus infectivity of the mixed solution was lost in about 3 hr. When the virus solution was mixed with sap from noninfested plants, the infectivity was retained for 24 hr or more. Nakazawa (1965) reported that the infectivity of CMV was inhibited by saliva in a sucrose solution on which *Myzus persicae* and *Aphis gossypii* had fed through a thin rubber or Paraffilm membrane. The inhibition by aphid saliva was influenced by the strains of the virus and species of aphid.

Recently, an experiment was made to determine the effect of aphid saliva on the infectivity of the nonpersistent turnip mosaic and alfalfa mosaic viruses by using alcoholic precipitate of sap from Chinese cabbage on which green peach aphids had fed (Nishi, unpublished data). The infectivity of the mixed solution was tested every 30 min by means of the half-leaf method, using *N. tabacum* var. White burley for turnip mosaic virus and *Vigna sequipedalis* var. Kurodane-sanzyaku for alfalfa mosaic virus. An alcoholic precipitate of sap from the plants which had not been infested by aphids was tested as the control. The trial was repeated three times. The infectivities of turnip mosaic virus and alfalfa mosaic virus in the mixture were found to be inhibited gradually with time (Fig. 1).

The results suggest that nonpersistence of turnip mosaic virus and alfalfa mosaic virus in the aphids is due to the action of some sub-

TABLE 5

COMPARISON BETWEEN INFECTIVITY OF TURNIP MOSAIC VIRUS AND TRANSFERABILITY OF $^{32}P$ CARRIED BY THE APHID

| Plant number | Infectivity of the virus | | | | | | | | $^{32}P$ by autoradiography [a] | | | | | | | |
|---|---|---|---|---|---|---|---|---|---|---|---|---|---|---|---|---|
| | 1 | | 2 | | 3 | | 4 | | 1 | | 2 | | 3 | | 4 | |
| | T | S | T | S | T | S | T | S | T | S | T | S | T | S | T | S |
| 1 | 1 | + | 4 | – | 1 | – | 2 | – | 1 | – | | – | 2 | – | 1 | – |
| 2 | 4 | + | 6 | – | 4 | + | 3 | + | 3 | – | | + | 3 | – | 3 | + |
| 3 | 5 | – | 9 | + | 8 | – | 4 | – | 6 | + | 15 | – | 8 | – | 5 | – |
| 4 | 6 | + | 11 | – | 11 | + | 6 | – | 9 | – | 17 | – | 10 | – | 9 | + |
| 5 | 9 | – | 13 | + | 13 | – | 9 | – | 14 | – | 20 | – | 11 | – | 11 | + |
| 6 | 11 | + | 15 | – | 15 | – | 11 | – | 16 | – | 22 | + | 13 | – | 13 | – |
| 7 | 13 | – | 17 | – | 17 | – | 13 | – | 17 | – | 25 | – | 15 | – | 14 | – |
| 8 | 15 | + | 20 | – | 19 | D | 16 | – | 18 | – | 30 | – | 16 | – | 17 | + |
| 9 | 17 | – | 27 | – | 20 | – | 18 | – | 20 | + | 37 | – | 18 | – | 22 | – |
| 10 | 19 | – | 29 | – | 23 | – | 20 | – | 27 | + | 40 | – | 20 | – | 25 | + |
| 11 | 21 | – | 31 | – | 25 | – | 23 | – | 29 | – | 44 | + | 21 | + | 26 | + |
| 12 | 23 | – | 33 | – | 27 | – | 28 | – | 31 | + | 47 | + | 24 | + | 28 | + |
| 13 | 26 | – | 34 | – | 29 | – | 31 | – | 33 | – | 51 | – | 26 | + | 30 | + |
| 14 | 28 | – | 39 | – | 32 | – | 34 | – | 35 | + | 52 | – | 28 | + | 35 | – |
| 15 | 29 | – | 42 | – | 33 | – | 36 | – | 38 | + | 58 | + | 30 | + | 40 | + |
| 16 | 34 | – | 45 | – | 35 | – | 44 | – | 40 | + | 76 | – | 32 | – | 49 | – |
| 17 | 37 | – | 53 | – | 38 | – | 53 | – | 42 | + | 77 | + | 35 | + | 56 | – |
| 18 | 39 | – | 55 | – | 39 | – | 54 | – | 45 | + | 78 | + | 36 | – | 58 | – |
| 19 | 41 | – | 60 | – | 42 | – | 60 | – | 47 | – | 79 | + | 38 | + | 65 | + |
| 20 | 45 | – | 65 | – | 43 | – | 64 | – | 51 | – | 84 | + | 40 | + | 67 | – |

[a] The aphids were fasted for 24 hr and then each given 30-sec acquisition feedings on virus diseased turnip seedlings containing $^{32}P$. (T) is time in minutes after the aphid left the virus source until it inserted the stylet into the test plant. For (S): (+) positive infection and a positive $^{32}P$ autoradiograph, respectively; (—) negative infection and a negative $^{32}P$ autoradiograph, respectively; (D) the plant died during the experiment.

stance contained in aphid saliva. The inhibition is not as prompt as in the case of TMV; hence, there is a period during which a viruliferous aphid with nonpersistent virus can transmit the virus as long as the virus tolerates the inhibiting action of aphid saliva.

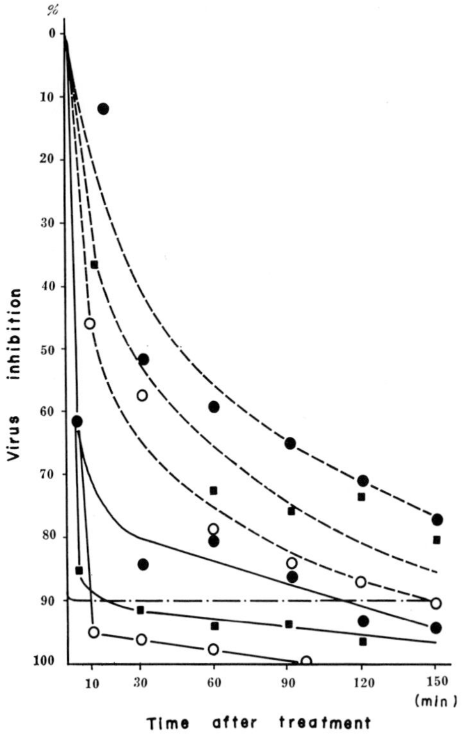


*Fig. 1.* Inhibition of the infectivity of turnip mosaic virus and alfalfa mosaic virus by the alcoholic precipitate of sap from Chinese cabbage on which green peach aphids had fed: (——) turnip mosaic virus; (– – – –) alfalfa mosaic virus; (– · –) tobacco mosaic virus; virus inhibition

$$= \left[ 1 - \frac{\text{number of lesions on treatment}}{\text{number of lesions in check}} \right] \times 100\%.$$

## IV. SOME PROPERTIES OF THE INHIBITORY SUBSTANCE IN APHID SALIVA

Nishi (1963) studied the properties of the inhibitory substance in aphid saliva. Leaves of Chinese cabbage exposed to a large number of green peach aphids were homogenized, and centrifuged for 10 min at 10,000 rpm. The supernatant fluid found to contain the inhibitory substance was heated for 10 min at temperatures ranging from 40 to 90°C (10°C interval) and then was tested against the infectivity of TMV. The results showed that the inhibitory substance in the sap was highly heat stable. A solution containing the inhibitor was treated for 24 hr at 10–15°C and a *p*H range from 2.0 to 13.0. Afterwards the *p*H was adjusted to 6.0, TMV solution was added, and the mixture was assayed for infectivity, using *N. glutinosa* as the indicator. The inhibitory substance in the solution remained active in all cases tested. When a mixture of TMV solution and the inhibitor, which showed an extremely low infectivity, was diluted with water, virus infectivity was regained, proving that the inhibitory action was reversible. The inhibitory substance could be precipitated with alcohol or ether. The inhibitory substance in solution did not pass the cellophane membrane of a Visking tube when dialyzed against running water or a buffer solution.

## V. PURIFICATION OF THE INHIBITORY SUBSTANCE IN APHID SALIVA

Purification of the inhibitory substance was attempted using chromatographic columns of DEAE cellulose and CM cellulose (Nishi, 1963). The inhibitory substance, precipitated with alcohol, was resuspended in distilled water. The inhibitor suspended in water was adsorbed in the DEAE cellulose column, eluted with distilled water, mixed with alcohol, and again precipitated. The precipitate was resuspended in water adjusted to *p*H 4.7 with 0.01 M acetate buffer solution, and then adsorbed to a CM cellulose column. The inhibitory substance was eluted in steps with *p*H 4.7 acetate buffer, *p*H 6.3 phosphate buffer, *p*H 8.0 phosphate buffer, and *p*H 10.0 carbonate buffer solutions. Then the elute was collected in 5-ml aliquots using a fraction collector. The ultraviolet absorption of each fraction at 280 m$\mu$ was determined by a spectrophotometer. Sap from healthy noninfested plants was treated by the same procedure and used as a control. Assays of the inhibitory effects on TMV infectivity on *N. glutinosa* showed that several fractions inhibited the infectivity of TMV (Fig. 2).

These fractions also showed strong ultraviolet absorption, which seemed to be due to protein and nucleic acid, as judged by the ultraviolet spectra at 230–310 m$\mu$ (Fig. 3).

## VI. ELECTRON MICROSCOPIC EXAMINATION

Electron microscopic examinations revealed that TMV particles treated with alcohol-precipitated inhibitory substance were covered with a diffuse granular membrane (Nishi, 1963). When the virus-inhibitor mixture was diluted with distilled water, and virus infectivity was regained, the virus particles showed a distinct contour.

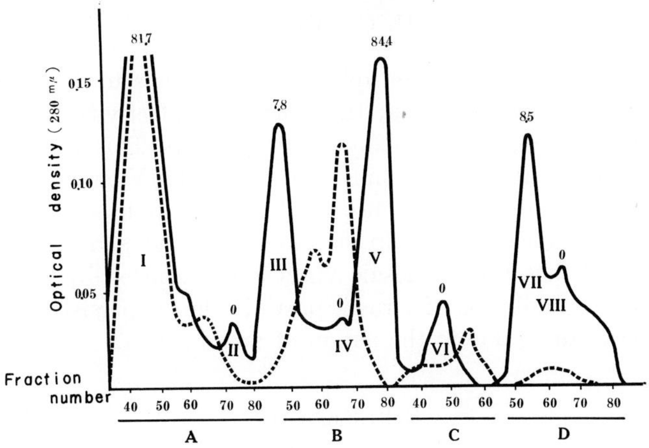


*Fig. 2.* Optical density at 280 mμ of each elute from a CM cellulose column, showing the difference between the sap from an aphid-infested plant (solid line) and one from a noninfested plant (broken line): *A,* eluted with *p*H 4.7 acetate buffer; *B,* eluted with *p*H 6.3 phosphate buffer; *C,* eluted with *p*H 8.0 phosphate buffer; *D,* eluted with *p*H 10.0 carbonate buffer. The numbers in the figure represent the percentage of TMV inhibition.

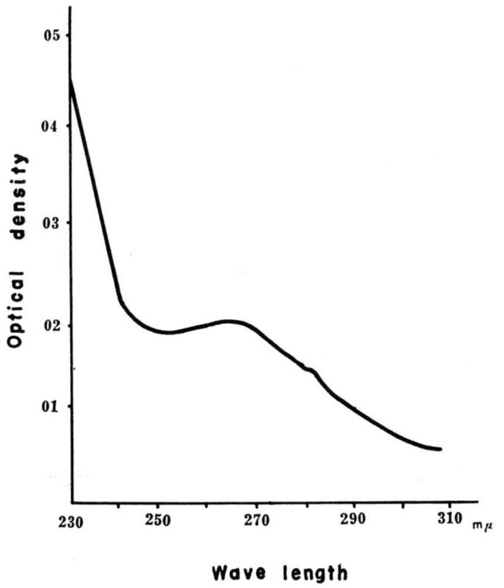


*Fig. 3.* Optical density of the fraction, peak V shown in Fig. 2, in the successive wavelength.

## VII. CONCLUSION

An inhibitory substance in aphid saliva adheres to TMV particles without destroying the virus; therefore, the infectivity of the virus is rapidly reduced, both in plants and in vectors. This apparently accounts for the inability of aphids to transmit TMV, as well as potato virus X. Virus infectivity is also inhibited by aphid saliva, but only gradually in the case of turnip mosaic virus, a nonpersistent virus. It is postulated that turnip mosaic virus is transmitted by aphids only as long as the inhibition of virus is partial. The nonpersistence of virus also appears to be due to the action of aphid saliva.

## BIBLIOGRAPHY

Bradley, R. H. E. 1952. Studies on the aphid transmission of a strain of henbane mosaic virus. Ann. App. Biol. 39:78–97.

Day, M. F., and Irzykiewicz, H. 1954. On the mechanism of transmission of nonpersistent phytopathogenic viruses by aphids. Australian J. Biol. 7:251–273.

Doolittle, S. P., and Walker, M. N. 1928. Aphid transmission of cucumber mosaic. Phytopathology 18:143.

Fukushi, T. 1931. On the modes of transmission of the mosaic disease of tobacco [in Japanese]. J. Sapporo Soc. Agr. Forest 22:305–320.

Fukushi, T. 1939. The relation of aphids to the transmission of legume mosaics [in Japanese]. J. Sapporo Soc. Agr. Forest 30:399–418.

Hamlyn, B. M. G. 1953. Quantitative studies on the transmission of cabbage black ringspot virus by *Myzus persicae* Sulz. Ann. App. Biol. 40:393–402.

Hoggan, I. A. 1929. The peach aphid (*Myzus persicae* Sulz.) an agent in virus transmission. Phytopathology 19:109–123.

——— 1933. Some factors involved in aphid transmission of the cucumber mosaic virus to tobacco. J. Agr. Res. 47:689–704.

Hoof, H. A. van 1958. Investigation of the biological transmission of a nonpersistent virus. Mededel. Inst. Plantenziekten Onderz. Wageningen. 761:96.

Kasai, T. 1950. Transmission of the mosaic disease of Japanese radish by *Myzus persicae* Sulz. [in Japanese]. Ann. Phytopathol. Soc. Japan 15:3–6.

Miles, P. W. 1959. Secretion of two types of saliva by an aphid. Nature 183:756.

Nakazawa, K. 1965. Some factors in relation to aphid transmission of cucumber mosaic virus [in Japanese]. Ann. Phytopathol. Soc. Japan 31:388–390.

Nishi, Y. 1958. Inhibitory action of the pressed juice from some plants on which aphids were fed against the infectivity of tobacco mosaic virus [in Japanese]. Ann. Phytopathol. Soc. Japan 23:185–188.

——— 1962. On the effect of substances secreted by aphid on the infectivity of potato X virus [in Japanese]. Proc. Assoc. Plant Protection Kyushu 8:56–57.

——— 1963. Studies on the transmission of plant viruses by aphids [in Japanese]. Bull. Kyushu Agr. Exp. Sta. 8:351–408.

Sylvester, E. S. 1950. Transmission of *Brassica nigra* virus by the green peach aphid. Phytopathology 40:743–745.

——— 1954. Aphid transmission of nonpersistent plant viruses with special reference to the *Brassica nigra* virus. Hilgaldia 23(3):53–98.

Watson, M. A., and Roberts, F. M. 1940. Evidence against the hypothesis that certain plant viruses are transmitted mechanically by aphides. Ann. App. Biol. 27:227–233.

# Disease Control Through Vector Control

L. Broadbent

*Bath University of Technology, England*

Many attempts have been made to control plant virus diseases before their causal viruses, their sources and vectors have been elucidated. Despite occasional successes from these "shots in the dark," it cannot be stressed too often that effective control depends upon an adequate knowledge of disease epidemiology.

Several different methods of control are discussed below; some concern the elimination of the vectors, but many others the prevention of infection by changing cultural practices or areas of crop production. Often several methods have to be applied successively or at the same time, and even then it may be difficult to completely stop the virus from spreading. Unless the disease is particularly devastating, however, the grower can often increase crop yield and profitability far beyond the cost and effort of the control measures used, even though complete control is not achieved.

## I. CONTROL BY CULTURAL CHANGES

Agrotechnical methods of control, as they are called in eastern Europe, are being increasingly investigated because elimination of the insect vectors of viruses is often difficult and expensive, and insecticides seldom prevent the introduction of virus into a crop by viruliferous insects. The most effective method, where it is possible, is to limit the sources of virus in an area by quarantine restrictions or schemes for the distribution of healthy planting material.

### *A. Quarantine and Certification*

Arthropods are not the only virus vectors, and quarantine regulations are designed to control the spread of viruses by man, one of the most important animal vectors. Viruses are often introduced into new areas in budwood, scions, or rootstocks of trees and shrubs, or in tubers, rhizomes, corms, bulbs, or other parts of herbaceous plants that can be reproduced vegetatively. Spread can be dramatic if an efficient vector exists in the new area. Thus, tristeza virus was not a serious disease of citrus trees in China, South Africa, or the United States because tolerant varieties were planted or efficient vectors were absent, but when many infected sweet orange trees on lemon rootstocks were imported into South America during 1930–1931, the virus was spread to millions of trees on sour orange and other nontolerant rootstocks by an abundant local vector, *Aphis citricides* Kirk.

Few countries quarantine all living plant imports. Most rely on certificates of health from the exporting country, but these are notoriously unreliable where virus diseases are concerned because it is often not possible to rely on visual disease symptoms and much of the imported material is dormant. Thresh (1960) outlined a scheme for intercepting virus-infected cacao, suggesting that the movement of rooted cuttings should be restricted and the use of budwood encouraged because it can be quickly transported by air. Healthy test plants would be top-worked with the buds, and the material would be considered healthy after three or four symptomless flushes of new growth had been produced. Such budwood certification is now the chief means used for controlling tristeza disease in South America (Fernandez Valiela, 1963).

Somewhat naturally, most schemes for the production of healthy tree propagating material are concerned with fruit crop trees. The mother tree scheme for top-fruit trees in England was adopted to decrease the risk of the wrong cultivar being supplied and the risk of bud sports being perpetuated in scions, but also to provide virus-free or tested material (Posnette, 1962). Propagators are provided with the best scion wood of commercial cultivars that is available, and stocks are frequently tested and replaced by better ones. Fridlund (1965) described the U.S.A. Interregional Research Project for obtaining and preserving virus-free or tested fruit tree clones, which has been centered at Prosser, Washington since 1955. Over 200 *Prunus* clones are maintained in isolated desert country, and budwood is distributed throughout the U.S.A. for research or foundation stock for the industry. In Australia, also, the control program for apple virus diseases involves indexing cultivars and rootstocks and the eventual distribution of clean stocks. Short-term measures include the annual examination of budwood sources in the main nurseries to ensure that cuttings are taken from symptomless trees (Shea, 1964). The danger of distributing viruses of economic importance in ornamental trees and shrubs is also now widely realised: Tanaka and Hirose (1966) found little cherry virus in 13 of 31 flowering cherry cultivars that they indexed, and they recommended the use of seed-propagated rootstocks by nursery men. Cammack (1966) also found this virus to be commonly carried symptomlessly in scion mother trees of ornamental *Prunus* on commercial nurseries in Britain.

Strawberry and other soft-fruit nucleus stock schemes are operated in many parts of the world, and recently such a scheme has been started in Britain for ornamental plants, concentrating at first on carnations and chrysanthemums. Stocks are grown from meristem tips, sometimes from heat-treated plants (*e.g.*, Hollings and Stone, 1965), and thoroughly tested for freedom from viruses before being multiplied for distribution to the industry. Stout (1962) described similar schemes for several crops in California, such as potato, strawberry, cherry, citrus fruits, grape, and garlic.

Certification schemes for potato have been developed in all the major seed-potato growing countries of the world during the last 30 years, being initiated because of the considerable yield losses resulting from infection with leaf-roll and Y viruses. The Scottish certification scheme was started in 1932, and no epidemic of aphid-borne virus has developed since 1945, probably because of the decrease in virus sources throughout the country and the concentration of the seed crops in areas where aphids arrive late and the crops are lifted early, measures which are discussed further below (Todd, 1961). Although foundation stocks are often tested serologically or by inoculation to test plants, British certification schemes depend upon visual inspection, the three common grades being Stock Seed (fewer than 4 diseased plants per acre), "A" certificate with fewer than 0.5% diseased, and "H" certificate with fewer than 1% diseased. Such high standards are difficult to maintain in continental areas where aphids are much more numerous than in Britain, but even here they are more numerous in some areas than others; Hollings (1955, 1960) developed a useful method of estimating if much virus spread were likely to occur, based on the earliness and spread of aphid infestation. In the U.S.A. and elsewhere, plants from stocks to be certified are either grown under glass after breaking tuber dormancy, or in areas such as Florida where potatoes will grow during the winter, so that their health can be assessed before the main stocks are planted (Folsom, 1952). The West Germans use field inspection, a callose color tests for leaf roll; serological tests for X, S, and some other viruses; and single-eye cultures under artificial light, but even so they find it difficult to maintain high standards, the stock seed having a limit of 4% severe virus disease, and seed for ware crops, 10% (Kabiersch, 1962). Under these circumstances the Igel-Lange and similar tests for leaf roll are useful and are now widely adopted in Europe (*e.g.*, Arenz and Hunnius, 1963; Keller and Bérces, 1966) but they have not yet proved reliable enough for adoption in

Britain (Broadbent and Heathcote, 1960; Govier, 1963). Recent literature shows that much effort is being expended in trying to find reliable chemical indexing methods for viruses for which antisera are not available, but the results are often disappointing (*e.g.*, Hecht and Arenz, 1965; Wenzl, 1965).

### *B. Isolation*

Certification schemes often rely upon the growing of virus-tested stocks in areas distant from likely sources of virus or of insect vectors. When virus spread is largely from one crop to another, the greater the distance between them, the greater will be the dispersion of vectors leaving the diseased one. Minimum distances for adequate isolation are difficult to determine because much depends on chance. Aphids have been known to be blown over hundreds of miles, and there is circumstantial evidence that persistent viruses such as potato leaf roll are sometimes introduced into Britain from the continent of Europe. However, stylet-borne viruses would be inactivated during prolonged flight, or during occasional probes by aphids on nonsusceptible plants or weeds. When susceptible plants are separated from each other by immune plants, virus spread is considerably lessened, especially if the intervening plants are suitable hosts for the vectors. This is an important consideration in areas of market gardening or truck-crop production, where the overlapping of crops in time is common. Successive crops of lettuce are often planted in England in small areas, new crops being adjacent to old ones, and lettuce or cucumber mosaic viruses are spread to the new crops, reaching epidemic proportions when aphids are numerous. The effectiveness of a short distance of isolation was shown when a grower broke the cycle by omitting winter lettuce one year, then concentrated his early crops in one area, and the next series in another about 0.5 mile away, not returning to the first area until all the old plants had been harvested. Disease incidence in these crops averaged about 3%, whereas that on neighboring farms was about 60% (Broadbent *et al.*, 1951).

In the past seed of biennial crops was often produced in the same area as the crops and this resulted in high incidences of disease. Pound (1946) found 0.2–1% of diseased cabbage seed plants when these were raised 7 or 20 miles from old seed crops, in contrast to 47 and 73% when they were raised 1100 or 110 yards, respectively, from them. Seed plants have been a major source of viruses in the sugar beet crop (Hansen, 1950; Hull, 1952). Watson *et al.* (1951) found that the incidence of sugar beet mosaic was much greater in fields within 100 yards of a seed crop than in fields further away, but yellows virus, which persists much longer in the aphid vectors, spread much further. Since 1950 beet stecklings in Britain have been raised in areas isolated from beet and mangold root crops. Although the stecklings may later be taken into root-growing areas for seed production, insecticidal treatment, and a certification scheme, whereby only beds with fewer than 1% diseased plants are used, have greatly decreased the spread of viruses from the seed crops (Hull, 1965). In Hungary, Szirmai (1957) found as short a distance as 1.5 km between seed and root crops prevented most virus spread.

Martin and Kantack (1960) in Louisiana found that the aphid-transmitted internal cork virus in sweet potato could be controlled by isolating plantings at least 100 yard from infected crops: incidence in isolated crops averaged 1% at harvest and 3–9% after storage, in contrast to 11–23% to harvest in nonisolated crops and 18–58% after these had been stored.

Weeds are frequently reservoirs of viruses, which are carried into crop plants from them. Fortunately, a relatively short distance of isolation may decrease this danger considerably, as Simons (1957) found with pepper veinbanding mosaic which was transmitted from nightshade (*Solanum gracile* Link) to 90% of the pepper plants 1.8 m from the weeds, but to only 10% when these were 15 m away. Removal of nightshade from an area up to 210 m from the pepper plants, however, showed that a few aphids might carry the virus even further, especially during warm spring weather. Simons *et al.* (1959) concluded that most aphid flights are shorter than 30 m but as the insects could often remain infective for an hour, longer flights could occasionally result in infection. Similar work by Wellman (1937) with southern celery mosaic virus showed that it was carried from weeds to 85–95% of celery plants 1–9 m away, but only to 12% at 22 m and 4% at 36 m. During three years no plant was infected in plots 72 m from a weed source.

Although isolation is an effective control measure for stylet-borne viruses and is usually insisted upon in certification schemes, it is less reliable for the control of persistent viruses, as is shown by the transmission of beet curly-top virus from wild hosts in the desert or foothills of the western U.S.A. into beet, tomato, cucumber, beans, and other crops in the cultivated valleys (Wallace and Murphy, 1938).

### *C. Interruption of Crop Cycles*

When a crop, rather than wild plants, is the main source of a virus, and arthropods spread the virus successively from old to new crops, control has sometimes been achieved by omitting a crop from the cycle, as discussed above in relation to lettuce mosaic. This measure succeeds best if growers in an area coordinate the break, or if the farm is isolated from others growing similar plants. Effective control of tobacco leaf curl was obtained by Hopkins (1932) by a three-week fallow after destroying the tobacco crop remains; this prevented the white fly vectors obtaining the virus from infected suckers of the previous crop and carrying it to the new plants. Severin and Freitag (1938) found western celery mosaic virus made summer celery crops unprofitable because numerous aphids from withering weeds invaded celery fields and spread virus; they persuaded growers over a wide area not to grow celery during the summer, thus breaking the cycle of infection and controlling the disease during the remainder of the year. Regrowth of cotton plants in the Sudan when fields were irrigated for the next crop enabled cotton leaf-curl virus to be carried over from one season to the next. Ploughing or grazing did not eliminate the old plants, so Tarr (1951) recommended the grubbing of these by hand and then fallowing the land for a season. Flock and Deal (1959) showed that a beet-free period during the summer decreased beet leafhopper populations and the spread of leaf-curl virus. Unfortunately, small populations of viruliferous insects were maintained on several weeds in irrigated areas of California, so that summer weed control was also advocated.

All stages of *Aceria tulipae* Keifer, the eriophyid mite vector of wheat streak mosaic virus, can live long enough on old wheat leaves, buried or on the soil surface, to attack the next wheat crop if this is sown within six days of ploughing (Slykhuis, 1955; Staples and Allington, 1956). The disease was almost eliminated by delaying the sowing of winter wheat until nearby spring wheat had been harvested and by destroying early volunteer wheat which, germinating before or soon after harvest, favors the mites.

As mentioned earlier, beet seed crops have been a major source of viruses for the sugar beet root crops in Britain; although their importance relative to other sources is now thought by most workers to be small, Ribbands (1964) considered them to be the main virus source and advocated growing beet and mangold seed crops in alternate years only. Hull (1965) considers that infected mangolds stored over winter in clamps (heaps of roots covered with straw and soil) are now the most important source of beet viruses. Aphids overwinter in the clamps on small shoots and carry virus to the young beet and mangold root crops in the spring (Broadbent *et al.*, 1949). Propaganda among growers to clear the clamps before sugar beet seed is sown has only been partially successful, and many clamp remains linger on into the early summer.

One cycle that has proved particularly difficult to break is that of tomato crops under glass, and the spread of tomato (tobacco) mosaic virus (TMV) successively from crop to crop. Man is the main vector of virus from plant to plant and from house to house, but many crops become infected initially when the roots of young plants come into contact with infected root debris from the previous crop that has survived in the soil. Doolittle (1928) found that TMV remained active in glasshouse soils in the United States for at least 70 days, and Jones and Burnett (1935) reported that tomatoes planted within five weeks of clearing infected plants become infected, whereas those planted after other crops had been grown for six months between tomato crops remained healthy. Hoggan and Johnson (1936) recovered infective debris after eight months, and in Britain Broadbent *et al.* (1965) found root debris to a depth of at least 100 cm remained infective for 22 months (the longest period tested). Although steaming the soil inactivates TMV to a depth of about 30 cm, there is no practical method of reaching lower depths. The only reliable method of breaking this cycle, if successive tomato crops are to be grown, or if they are interspersed with short-term crops, is to cover the soil with an impermeable surface such as polyethylene, and grow the plants in an uncontaminated medium above this (Wheeler, 1962). Root debris has been found to be still infective under such a surface after two years (Broadbent *et al.*, 1965).

Cadman (1963) discussed the possibilities of controlling soil-borne viruses by crop rotation. This has proved effective in freeing soils from wheat mosaic virus, and he suggested that some of the nematode-transmitted viruses might also be controlled in this way if efficient weed control were practiced also, as the vectors depend on sources of infection within the crops or their weeds.

Even when it is not practicable to break a cycle by omitting a crop or by delayed planting, it is most important to destroy crop remains as soon as possible. Growers are often unaware of this and allow lettuce, brassica, and other unharvested plants to linger on. If these plants seed, aphids are attracted to their flowers, which are often yellow. Also, aphids may build up bigger populations on virus-infected than on healthy plants, which is another reason for destroying the remains of infected crops. This is not always easy with perennial crop plants; thus, infected potato tubers may remain in the ground and grow under cover of cereal or other crops for several years (Doncaster and Gregory, 1948).

### *D. Protected Cropping*

Crops are seldom grown under glass to escape from virus vectors, although this is becoming more common for quarantine testing or the production of virus-tested nucleus stocks. The exclusion of arthropods from glasshouses is not difficult, but it often necessitates special ventilation and cooling devices. Workers may take vectors in on their clothing or, with contact-transmitted viruses, may act as vectors themselves. Tobacco mosaic virus is the most easily transmitted virus of this type and tomato isolates of the virus readily contaminate the body and clothing, remaining infective for several months on clothes kept in the light, and for several years if they are stored in darkness (Broadbent and Fletcher, 1963). For tomato crops under glass, clean overalls can be kept in each house, to be worn by the worker or visitor before moving among the plants. Workers should never visit a healthy crop after a diseased one without washing well and changing their outer clothing. Nursery managers, commercial representatives, and advisory officers can be efficient vectors of TMV and should take adequate precautions. Workers sometimes wear the same clothing for tending successive crops, and this practice should be discouraged unless the clothing has been cleaned.

Tomato mosaic virus is difficult to remove from sap-engrained hands. Berkeley (1942) found that washing with soap and water in a pail was often ineffective and recommended washing in a solution of trisodium orthophosphate; even this, followed by a scrub with soap and water, did not always inactivate virus under the nails (Broadbent, 1963).

Virus can also be spread under glass on tools used by workers. Heat sterilization is the best way of inactivating such virus, and Freytag (1965) has described a knife whose blade was kept hot by an attached propane torch; this was used to prevent virus spread among orchid plants when removing flower spikes. Franklin (1966) sterilized knife blades used for the same purpose by drawing them across a moist sponge in the neck of a bottle containing 2% formaldehyde and 2% sodium hydroxide, whereas Mulholland (1962) used trisodium orthophosphate to stop the spread of TMV on trimming knives. The knives had hollow handles filled with the solution, which slowly trickled over the blades.

Although TMV is common in smoking tobacco, the strains occurring there seldom infect tomato (Broadbent, 1962; Broadbent and Fletcher, 1966). Nevertheless, transmission can occur, especially when grafting tomato plants; therefore, workers should wash their hands after using tobacco and should not smoke while tending plants that might be susceptible.

Man is the most important vector of TMV in tomato crops, but birds have been shown to transmit it on their plumage and should be excluded from houses by netting the vents (Broadbent, 1965).

Cloth or plastic fiber cages are sometimes used to exclude viruliferous insects from valuable plants. Large cages were used by Jones and Riker (1931) to protect asters from aster yellows virus, and Linn (1940) protected lettuce and endive seedlings from yellows virus in cloth-covered cold frames or screened glasshouses prior to transplanting, obtaining a delay of 14 days in the appearance of disease in the crop and a decrease in its incidence from 36% in field-sown crops to 7%. Screened houses are used in Arkansas for increasing virus-free clones of strawberries (Fulton and Seymour, 1957), and in England for growing mosaic-free lettuce seed plants.

### *E. Barriers and Cover Crops*

Arthropod-transmitted viruses usually spread more easily into, and faster within, pure stands of susceptible host plants than in mixed ones, and nonsusceptible cover crops are sometimes used to screen the undersown susceptible stecklings (first-year plants) in sugar beet seed production. The beet makes little growth before the barley cover is cut, but rapid growth afterwards and Hull (1952) found 12% of undersown plants with beet yellows compared with 100% in open beds. Hansen (1950) obtained similar results when beet rows alternated with rows of barley, and recommended growing cereals on areas of land where beet had been clamped to isolate any volunteers that grew. No protection was obtained when the barrier rows were about 7 yards apart. Viado and Matthysse (1959) found that a cover crop of *Centrosema plumieri* decreased the incidence of abaca mosaic from 33 to 12%, and gave better control than insecticidal sprays applied both to the crop and to nearby weed hosts. Radish mosaic spread rapidly in newly sown radish crops in Japan, but its incidence was greatly decreased by catch-cropping the radish between rows of rice or trefoil (Shirahama, 1957).

Because barrier rows at intervals are more easily managed than cover crops, more use has been made of them. Initially the barriers were of wire mesh or cloth, and Kunkel (1929) found that such screens 1.2–2.4 m high around plots of asters 3×7.5 m cut the incidence of yellows from 80% in unfenced plots to 20%. Similarly, Linn (1940) found that 1.8-m high cloth screens delayed the introduction of leafhopper-borne yellows virus from weeds to lettuce and decreased final incidence from 35 to 12%.

Hull (1952) reported that French sugar beet growers were using 1-m wide barriers of oats, maize, hemp, or sunflower around 20×5-m beds of beet stecklings, and Bonnemaison (1961) found plots screened with 2 rows of maize and 3 of sunflower contained about 21% infected plants when unscreened plots had 56%. In another experiment on plot size, unscreened ones contained 54% yellows-infected plants; those with barriers every 5 m, 28%; every 10 m, 31%; and every 15 m, 36%. Similar methods were used by Broadbent (1957) to protect brassica seedbeds from cauliflower mosaic virus; three rows of barley, which is not colonized by the aphid vectors and is not susceptible to the virus, were more effective than beans or other brassicas, and decreased mosaic incidence by about 80%. The barriers were effective when sown at the same time as the brassica seed between every 12 rows of seedbed, with single cross rows of cereal at intervals in long seedbeds.

Barrier crops are likely to be more effective in protecting plants from stylet-borne than from more persistent viruses because aphids will often lose the nonpersistent virus if they probe the barrier plants. Normally they will fly away over the crop plots when they leave the barrier because they fly upwards towards the brightest light, but if they should subsequently land within the barriers there is an increased chance of them being nonviruliferous. Simons (1957, 1958, 1960) investigated the effectiveness of sunflower and other barriers in Florida to protect crops such as pepper, celery, and tomato from pepper veinbanding mosaic and potato Y viruses, which were prevalent in weeds. Barriers around pepper plots 45 m from a weed source did not prevent infection but decreased and delayed it by about ten days. A low single barrier 38 cm high was nearly as effective as a single 3-m barrier, but more effective were 15-m wide strips of nonsusceptible crops such as beans, corn, or cucumber around the fields, especially when the barriers were also sprayed with parathion. The primary spread of potato Y virus from *Solanum nigrum* to pepper was decreased 50% by one row of sunflowers around three sides of the plots, 70% by a 15-m barrier of beans, and 85% when the beans were sprayed weekly with parathion. In the latter plots incidence of disease after six weeks was only half that in unprotected plots whereas all the plants within the sunflower barriers were infected by then. Wide (9-m) barriers of cereal or timothy grass also effectively decreased the incidence of green petal virus in enclosed strawberry plots in Canada (Mackinnon *et al.,* 1964).

A form of soil barrier was used by Tomlinson and Garrett (1966) to protect lettuce from the soil-borne, *Olpidium* transmitted, big vein virus. Seedlings were grown in uncontaminated compost in small peat pots and planted out into contaminated soil when four weeks old. Big vein symptoms developed in 3/256 plants, in contrast to 185/256 when the four-week old seedlings were pricked out directly into the soil. In a further experiment, two-week old seedlings pricked out into contaminated soil were all diseased within 65 days, in contrast to only 25% of those put in clean soil in small peat pots, which in turn were put in contaminated soil.

### *F. Elimination of Virus Sources*

This may do nothing to decrease the number of vectors but it may considerably decrease the proportion that is viruliferous. If infected plants within a crop are the only source of the virus, the final incidence of disease may be directly proportional to the initial incidence. Seed-borne lettuce mosaic virus in two nearby crops initially infected 0.5 and 3% of the seedlings, and at harvest these crops showed, respectively, 14 and 85% mosaic diseased plants (Broadbent *et al.*, 1951). Multiple infections usually occur, however, and in similar trials Zink *et al.* (1956) found 3.4, 7.6, and 29.5% mosaic in plots which initially showed 0, 0.1, and 1.6%, respectively.

When virus is introduced into a crop from the outside, it can be brought either from other crop plants or from weeds and other wild hosts. Viruses are often spread from one crop to another when arthropod vectors leave old infected crops to seek alternative hosts. The susceptible crop need not be colonized by the vectors, as when pea aphids carry yellow bean mosaic virus from red clover to beans (Crumb and McWhorter, 1948).

Recent work has shown that weeds and wild plants are commonly infected with viruses, often showing few or no symptoms (*e.g.*, MacClement and Richards, 1956; Schwarz, 1959; Lovisolo and Benetti, 1960). Occasionally they are the principal source of a virus, especially if they are perennial; thus both in the United States and Yugoslavia *Sorghum halepense* Pers. is the main reservoir of maize mosaic virus and several aphid species breed on the Johnson grass, carrying virus to nearby maize (Sill, 1966; Sutic and Tosic, 1966). Weeds also are important in the epidemiology of soil-borne, nematode-transmitted viruses, which are carried in the seeds of some weeds, thus insuring the survival of viruses that have relatively immobile vectors (Noordam, 1955; Lister, 1960; Cadman, 1963).

Schmelzer (1963) has pointed out that many variegated cultivars of ornamental woody plants are infected with soil-borne viruses, such as tomato black ring and arabis mosaic, and recommended that they should no longer be propagated and planted because they may serve as infection sources for crop plants.

Many attempts have been made to limit the spread of viruses by eliminating alternative hosts. Beet curly-top virus, which seriously affects beet, tomato, cucumber, beans, and spinach in the western United States, is transmitted by *Circulifer tenellus* (Baker), often during transient feeding when the leafhoppers move from overwintering wild hosts in the desert or mountain foothills, or from weeds in cultivated valleys. Keener (1956) found the elimination of weeds and volunteer crop plants, such as sugar beet, among cantaloupes useful in limiting the spread of the virus, but Piemeisel (1954) advocated a more radical approach. Weed hosts occurred mainly on open range grazing land that had been overstocked and otherwise mismanaged; if these areas could be rehabilitated to productive grass grazing, weeds would be eliminated.

Another leafhopper-transmitted virus of importance in the United States, aster yellows, also overwinters in weeds on the borders of cultivated fields. Wellman (1937) eradicated plantains and other weeds within 30 m of prospective lettuce plots and decreased the subsequent incidence of yellows in the lettuce to 2–4%, in contrast to 11–44% in plots with adjacent weeds. He also controlled southern celery mosaic by eradicating weeds for a distance of 22 m around seed beds and fields before transplanting celery and about five times during the growing season.

Anderson (1959) found 21 species of plants growing near pepper fields in Florida that were infected with viruses to which pepper is susceptible, and Simons, Orsenigo, Stall, and Thayer (1959) showed that their elimination with herbicides before planting the peppers was more effective than the subsequent use of insecticides on the crop. They stressed the necessity to clear whole areas from weeds and hoped that all growers would cooperate in the work.

Roguing, or removing infected plants from a crop, has long been used to maintain healthy potato stocks in areas where aphids are few, as in parts of Scotland and Ireland. Early roguing is essential, as it must be done before virus has spread to many healthy plants within the crop; as most spread is to adjacent plants, these are sometimes removed as well, even though they are symptomless (Bagnall, 1953). In many potato growing areas, however, aphids arrive before symptoms of disease are obvious. In southern England, for example, much of the season's spread of virus occurs early because of the activity of the colonizing winged aphids, and roguing is useless unless it is combined with insecticidal treatment (Broadbent *et al.*, 1950; Broadbent *et al.*, 1961). In the past roguing has usually been done by laboriously digging out infected plants and removing them from the field, but Barnes (1959) found spraying diseased plants with mixtures of diesel oil and growth regulating compounds to kill the foliage was efficient and less costly than hand roguing.

Zink *et al.* (1957) also used diesel oil, mixed with parathion, to rogue lettuce crops; it was slightly more effective than hoeing, probably because aphids were killed, whereas those on hoed plants might move to nearby healthy ones. One roguing of lettuce mosaic soon after thinning was ineffective, presumably because much spread had already occurred, but roguing two or three times decreased disease by about 50%.

Another combination of roguing with insecticidal treatment to prevent aphids moving from the infected plants was described by Adams (1962). This successfully limited the spread of strains of cucumber mosaic virus in Honduras banana crops. Each farm was inspected three times a year and any infected plant was sprayed with nicotine, together with the 25 acres around it. Then the infected plants and the 80 nearest to them were dug out, chopped up, and sprayed with malathion. The 25 acres were sprayed again 20 and 40 days later, the cleared area was checked for regrowth, and not replanted for at least six months. In addition, plantains, an alternative host of the virus, were eradicated from all banana farms and their neighborhood. This ambitious program had decreased the incidence of mosaic on a 14,000-acre plantation from about 21 infected plants per 1000 acres in 1956 to 0.1 in 1960 and prevented its introduction on any new farms.

Roguing also enabled banana bunchy top virus to be controlled in Australia (Cann, 1952). Over 90% of banana production had ceased in New South Wales between 1922 and 1925 because of the disease, so in 1927 the Department of Agriculture destroyed all plantations, and permitted their replanting with virus-free suckers in 1928. The movement of plants in and between all banana-growing areas was strictly regulated and frequent inspections made, any infected plant being sprayed with kerosene to kill the aphid vectors, then dug up and destroyed.

Roguing has been successfully used with several other perennial crops. Large-scale elimination programs were started in the early 1930's to control peach mosaic and phony peach diseases (Persons, 1952; List *et al.*, 1956). Phony peach disease was not widespread and it was soon eliminated from 6 of the 17 states in which it was found; wild plum was found to be a host of the virus and was included in the roguing program, which subsequently prevented serious outbreaks. Trees infected with peach mosaic virus were rogued in Colorado from 1934 onwards, 36% of those removed by 1955 being taken out during the first two years. About 1% of all trees had been removed by 1955, but the virus continued to spread slowly, perhaps from symptomless carriers.

The most extensive roguing program for virus disease control has been that in Ghana and Nigeria for the control of swollen shoot of cacao. During the period 1946–1957 about 63 million trees were destroyed in Ghana. Spread of the virus by mealybugs is slow, so theoretically it can be controlled by eradication, but the high incidence of disease when the work started made it a tremendous job. After experience of the initial clearances, which were considered too drastic, Thresh (1959) in Nigeria recommended removing all trees within 4.5, 9, or 13.5 m of outbreaks of fewer than 6, 6–50, or 50–200 trees, respectively, followed by regular inspection of the peripheral trees. At that time in Ghana all swollen-shoot outbreaks were treated within 4–6 weeks of discovery, the infected trees only being removed, although attempts were made to persuade farmers to remove one ring of trees around the infected one as well. All outbreak areas were inspected every two months until no further infected trees were found during a period of two years.

Since then epidemiological studies have shown that the reduction in cacao yield may be caused by an interaction between swollen-shoot virus and the deterioration of the tree canopy under which cacao is grown, as a result of fungus (*Calonectria*) and capsid attack. Regular spraying with insecticide and cultural measures brought about the recovery of the canopy and minimized the effect of the virus on cacao yield (Longworth, 1963; Tinsley, 1964).

Early roguing has been effective with some contact-transmitted viruses, such as barley stripe mosaic, decreasing disease incidence in two cultivars from 85 to 12%, and 46 to 0% (Inouye, 1962), but it is not worthwhile when viruses are easily spread by handling or machinery, as TMV in tobacco and tomato crops (Wolf, 1933). Even when all diseased young tomato seedlings were removed from the propagating houses of ten tomato growers (a total of 187 out of 374,000), spread had already occurred, and seven of 22 glasshouses planted with symptomless seedlings contained infected plants soon afterwards (Broadbent and Fletcher, 1966).

### *G. Planting and Harvesting Dates*

Altering traditional planting or harvesting times may enable a crop to avoid virus vectors, or increase the plants' resistance to infection. Many plants show increased resistance as they age, and early planting decreased the incidence of groundnut rosette (Smartt, 1961), sugar beet curly top (Wallace and Murphy, 1938) and beet yellows (Bennett, 1960). Susceptibility of Majestic potatoes to leaf-roll virus decreased within two months of emergence from about 80% of plants infected during late June to about 6% in mid-August (Cadman and Chambers, 1960); similarly, 38% of sugar beet plants sown on April 1 contracted yellows, in contrast to 53% of those sown on April 15 and 70% of those sown on May 1 (Hansen, 1950). Steudel (1952) found that the number of aphids per plant, as well as the incidence of yellows, increased with successively later sowings. In Poland, Birecki *et al.* (1964) found that early planting and the use of sprouted tubers decreased virus disease incidence, more with leaf roll than rugose mosaic, but the spread of virus Y was limited by haulm destruction, which was practicable when sprouted tubers were planted early.

In many parts of Europe and America seed production is based partly on early haulm destruction, before virus brought into healthy crops by aphids dispersing in late summer from diseased ones can penetrate to the tubers (Schultz *et al.*, 1944; Münster, 1958; Kabiersch, 1962; Wright and Hughes, 1964). In some countries chemical or mechanical destruction of the haulm must be completed within 3–5 days of a warning issued by the certifying authority and there must be no regrowth if a certificate of health is to be gained. This system of aphid avoidance is best developed in the Netherlands: the aphid forecast on which date of haulm destruction is based is determined after a consideration of aphid overwintering, the time of spring flight from *Prunus* spp. and on trap catches (Hille Ris Lambers, 1955). Catches from two traps per potato field from many localities (220 traps were in use in 1955) are examined daily and a report on the presence of *Myzus persicae* Sulzer is sent to the Regional Inspection Services. The later infection occurs, the longer the virus takes to pass into the tubers (Beemster, 1961). This method allows growth to proceed for as long as possible and is better than lifting on a fixed date, as was done in parts of the United States (Anon., 1955). Hille Ris Lambers pointed out that Friesian seed potato growers realized the advantages of early lifting to minimize leaf roll infection as long ago as 1810.

When insects have well-defined dispersal periods, it may be possible to plan crop growth to avoid them. Stubbs (1948) produced healthy carrot crops in Australia by delaying sowing until the spring dispersal of *Cavariella aegopodii* Scop. was over, and Harpaz (1961) controlled maize rough dwarf virus, spread by *Calligypona marginata,* a hopper that breeds on grasses, moving to young maize seedlings in irrigated fields when the grasses wither. The insects do not breed on maize, and remain few during four summer months. The incidence of disease was less than 3% when sowing was delayed until late May, in contrast to about 45% in crops sown early.

### *H. Field Size, Plant Size, and Plant Spacing*

Incoming insects often land on the outer edges of fields, and if they bring virus with them, or sometimes if they acquire it after they land

(Zink *et al.*, 1956), more plants are infected near the margins than in the centers of fields. Some insects, such as *Aphis fabae* Scop., may stay in the area where they first land (Taylor and Johnson, 1954) but others such as *M. persicae* move from plant to plant for a few days and distribute themselves throughout the crop; the latter are more efficient at spreading viruses. Doncaster and Gregory (1948) found that gradients of introduced leaf roll or rugose mosaic (virus Y) in potato crops declined rapidly between the tenth and twentieth rows from the edge of the field, and Storey and Godwin (1953) found that when a brassica crop was infected with cauliflower mosaic virus by aphids leaving a nearby crop, most disease occurred in the first 50 rows adjacent to the source.

Van der Plank (1948a, 1949) pointed out that the highly infected outer zone formed a greater proportion of a small field than a large one and that the amount of infection entering per acre from nearby sources was approximately inversely proportional to the square root of the field area. Consequently, unless subsequent spread within the field is rapid, disease can often be controlled by making fields larger and fewer. He stated that in South Africa maize streak virus often destroyed all the crops in small fields, whereas many plants in large fields escaped infection. On the other hand, the chance of a pathogen entering a field may well increase with increasing field size (Waggoner, 1962), so van der Plank's thesis may not hold good in all instances.

Van der Plank (1948b) also postulated that more aphids were likely to land on large than on small plants, because of the larger catchment area and the tendency of the insects to land first on the taller plants. He found this to be so with potatoes, and Broadbent (1957) showed that many more large plants than medium-sized or small ones were infected in brassica seedbeds; obviously, large plants should not be transplanted into the cropping area. The tendency for aphids to alight on taller plants and on outer rows has been utilized for the control of disease in seedbeds and other small areas by surrounding them with nonsusceptible barrier plants, which were discussed previously.

Hull (1964) quoted a report by Harper (1928) that close spacing of groundnuts decreased the incidence of rosette, and this is probably the first observation that close planting would reduce virus disease incidence, because the more plants there are per unit area, the smaller will be the proportion infected by the same number of viruliferous insects. Groundnut rosette is introduced by winged *Aphis craccivora* Koch., and there is little subsequent spread of the virus within the crop. Storey and Ryland (1950; see Hull, 1964) found a decrease in the *number* of infected plants per unit area, as well as in the percentage, with an increasing density of planting and Hull confirmed that closely spaced plants were not visited by alatae as frequently as widely spread ones. Storey (1935) early noticed that the delayed weeding, usually practised by the peasant cultivators in East Africa, prevented rosette from seriously affecting their groundnut crops. The weeds not only increased plant density but in addition acted as a cover crop. A'Brook (1964) in several experiments also confirmed the earlier work, and found that the densest plantings had few or no aphid-infested plants after 80% groundcover was achieved. Many more aphids were caught on horizontal traps in the crop in "thin" plantings than in dense ones, and A'Brook considered that Kennedy *et al.* (1961) were probably correct in suggesting that the checker-board effect of earth and widely spaced plants may stimulate an optomotor response in aphids and induce landing. Broadbent (1948) and Steudel (1953) found that a greater proportion of potato and sugar beet plants were visited by aphids when they were widely spaced than when they crowded together, and Way and Heathcote (1966) showed that the number of aphids per plant and per acre was inversely related to plant density in crops of field beans; with equal numbers of plants per acre, fewer aphids developed on plants in rows 28 cm apart than in those 56 cm apart. Pea leaf roll was greatly decreased by increasing the density of planting, but unlike the aphids, infection was related to the number of plants per unit length of row, rather than to the number per unit area.

Van der Plank and Anderssen (1944) also utilized the spacing effect to decrease the incidence of spotted wilt virus in tobacco crops; this is carried into the crops by thrips and, again, there is little subsequent spread. They calculated that if the incidence of disease were 50% with single-plant spacing, it would decrease to 9% by transplanting two plants per hill, and to 1% by three, if the distance between hills were not altered. Later demoval of diseased and surplus plants showed that this method did indeed provide a satisfactory method of control, and it was adopted by Shapovalov *et al.* (1941) to decrease curly-top virus in tomatoes and by Kovachevski (1964) to control stolbur disease in the same crop and pepper. Kovachevski obtained some control with insecticides of the vector, *Hyalesthes obsoletus*, but this method was not economical and better control of the disease in pepper was obtained by closer spacing and double planting.

Higher seed rates within the row and decreasing row widths in sugar beet crops and brassica seedbeds also considerably decreased the incidence of virus diseases when the viruses were introduced from outside (Blencowe and Tinsley, 1951; Broadbent, 1957), and Hull (1965) stated that yellows was more prevalent in sparsely planted sugar beet crops than when the ground was well covered. High seeding rates, again, decreased the percentage incidence of oat blue dwarf and aster yellows in flax (Frederiksen, 1962) and of barley yellow dwarf (Slykhuis *et al.*, 1960), and Ramakrishnan (1963) found that both the rate of spread and percentage of redgram plants in the field infected with sterility mosaic virus were inversely proportional to the number of plants. It thus seems that maximum crop density, consonant with good husbandry, is desirable to minimize the effects of insect-borne viruses.

### *I. Plant Nutrition*

Many workers have shown that the better plants are fed and grow, the more likely they are to become infected with viruses. Increase in the nitrogen supply increased both aphid numbers per plant and the susceptibility of potato plants to infection with leaf roll and Y viruses, and a deficiency in potash had a similar effect (Janssen, 1929). Organic fertilizers, such as dung, increased susceptibility to viruses in potato and cauliflower crops more than inorganic fertilizers (Broadbent *et al.*, 1952; Broadbent, 1957). Sastry and Nariani (1962) showed that the susceptibility of tobacco plants to leaf-curl virus increased with increasing nitrogen but decreased with increasing phosphorus. However, it is very doubtful if changing nutrition will give any worthwhile control of virus diseases in potato and sugar beet crops (Bawden and Kassanis, 1950; Hull and Watson, 1947; Hull, 1965) and this probably applies to insect-borne diseases of other crops.

### *J. Plant Breeding*

Different cultivars of crops plants often differ greatly in the extent to which they are colonized by vectors and also in their susceptibility to infection with viruses. Attempts to breed immunity to economically important virus diseases have been made with most of the world's important crop plants, but they have seldom been successful.

We are concerned here, however, with the ultimate effects of plant breeding on vector populations or activity. Field resistance to a virus may sometimes occur because vectors fail to infect a particular cultivar: for example, Tarr (1951) found the cotton cultivar Lambert was susceptible to leaf-curl virus in graft transmission tests but the white fly vectors usually failed to infect it in the field. Similarly, Evans (1954) found the groundnut cultivar Mwitunde to be field resistant to rosette virus partly because aphids were less able to acquire virus from it than from other cultivars; such differences did not occur in laboratory tests. With virus Y, also, different potato cultivars differed in the ease with which they were infected, the extent to which virus multiplied in them, and the readiness with which aphids became infective when feeding on them (Bawden and Kassanis, 1946). Bagnall and Bradley (1958) stressed that field trials for resistance should always be done with outdoor crops because of such differences and because some cultivars of potato, for example, were hypersensitive to virus Y in the field but not under glass.

Plant resistance to infestation by insect vectors may also affect virus spread. Lettuce and celery cultivars which, experimentally, were equally susceptible to yellows virus, contracted disease to different extents in the field because of differential feeding by leafhopper vectors (Linn, 1940; Yamaguchi and Welch, 1955).

The color of a plant may affect disease incidence because vectors are differentially attracted to different colors. Twice as many aphids, and 3–5 times as many *M. persicae,* alighted on green or yellow-green lettuce plants as on reddish-brown ones, which in consequence were less frequently infected with mosaic virus (Müller, 1956, 1964); more aphids landed on yellow and green cultivars when the color contrasted with that of surrounding plants, but this did not apply to the brown ones.

## II. PROTECTION OF PLANTS WITH CHEMICALS

A considerable amount of work has been done on spraying tomato crops with milk to try to stop the spread of the contact-transmitted TMV. This followed the demonstration by Johnson (1941) and Fulton (1943) that cow's milk, whether whole or skimmed, contained virus inactivators. The first test on tomatoes in England was apparently successful, but unfortunately no control plants were left unsprayed (Newell, 1954). However, Harrison (1957), Crowley (1958) and Fry and Coleman (1960) all reported that sprayed crops, tended by workers whose hands also were wet with a milk solution, were as readily infected as unsprayed plants handled with dry hands. In contrast, Hare and Lucas (1959) obtained only 10% infected young tomatoes when sprayed with milk or a powdered milk solution and dried before being inoculated with TMV-infected tomato sap, 60% when sprayed soon after inoculation, and 90% when unsprayed. Hein (1961) almost completely prevented the spread of TMV among seedlings by spraying with skimmed milk before each handling; later tests were not quite as successful, 4–12% of the sprayed plants being infected in contrast to 24–64% of the controls (Hein, 1964a). Denby and Wilks (1963, 1965) found that spraying of tomato seedlings with skimmed milk prior to transplanting decreased TMV incidence and increased the yield by 13%. Thus, results have been very variable, and it is doubtful if the partial control sometimes achieved is worth the effort expended.

Skimmed milk and whey were used also by Hagborg and Chelach (1960) to reduce the field spread of stripe mosaic virus in barley, one application decreasing incidence from 33% (water sprayed) to 15% (100% whey) and 23% (100% skimmed milk—not a significant decrease). Hein (1964b) also found that skimmed milk did not prevent the transmission of lettuce mosaic virus by *M. persicae* but that whole milk (3% fat content) did, decreasing infection from 36/50 plants unsprayed to 3/49 sprayed; when the source plant was sprayed, only 2/50 of the test plants were infected, whether they were sprayed or not. It was probably the fat in the milk that prevented the aphids from transmitting virus, and one of the most interesting developments in the control of stylet-borne viruses in recent years has been the demonstration by Bradley (1962, 1963) and others that thin films of oil sprayed on plant surfaces inhibit the acquisition of viruses by aphids. Thus, *M. persicae* transmitted potato virus Y from tobacco to 24/40 unsprayed tobacco plants, but to 0–4/40 others sprayed with one of several mineral or vegetable oils. Later, Bradley *et al.* (1966) sprayed plots of potatoes with a light oil or with oil emulsified in water, starting soon after emergence. Both treatments were effective in decreasing the spread of virus Y by 70–90%, and three sprays were nearly as effective as six. Six sprays, but not three, caused slight foliar damage and decreased yield. A decrease of 30–50% in the field spread of potato virus A was obtained by Allen (1965) with six weekly sprays of an emulsified paraffin oil. This technique has been tested by several workers in Israel: Nitzany (1966) sprayed young pepper plants with a light oil emulsion at 5–6 day intervals, delaying by 2–4 weeks the spread of viruses (mainly potato virus Y). However, yield was not affected, probably because peppers are slow growing, unlike cucumbers where control with a 1% emulsion was effective in preventing the transmission of cucumber mosaic virus from sprayed plants by *Aphis gossypii* (Loebenstein *et al.*, 1964). When both the source and healthy plants were sprayed, spread was almost completely stopped for 5–7 days. In the field, also, weekly sprays were highly effective, high-volume application decreasing mosaic incidence by 50–70%, and low-volume application by 80–90%. The cucumbers were irrigated weekly by overhead sprays, and there were no phytotoxic effects from the oil, presumably because the water prevented its accumulation on the leaves (Loebenstein *et al.*, 1966). A 2% emulsion of light paraffin oil in water also prevented the spread of the semipersistent beet yellows virus from beet to beet when the source plants were sprayed, or when the aphids fed on healthy sprayed plants for a day between transfer from the unsprayed source to unsprayed test plants (Vanderveken and Semal, 1966). Clearly this technique holds considerable promise of solving the intractable problem of preventing the spread of nonpersistent viruses that can be acquired and transmitted within minutes, before the vectors can be affected by insecticides.

## III. ELIMINATION OF THE VECTORS

The methods of virus disease control outlined above have depended mainly on the crop plants being protected from the vectors or being grown in areas where, or at times when, the vectors were few. Is it not possible to eliminate arthropod vectors, or considerably decrease their numbers, with insecticides, by encouraging their predators and parasites, by eliminating their alternative host plants, or by heat?

### *A. Elimination of Vector Host Plants*

Vectors often breed on other crops or on wild hosts that are also sources of virus; these were discussed in Section I-C and I-F. *Myzus persicae* is the most important aphid vector of plant viruses in temperate climates, and in areas with cold winters it overwinters in the egg stage on species of *Prunus*. For this reason, the culture of peaches is forbidden in some of the potato and sugar beet growing areas of Europe (Kabiersch, 1962). A potato seed growing area in Idaho was successfully freed from leaf roll virus within five years when State agencies removed or sprayed all peach and apricot trees, sprayed all incoming plants and cut flowers against aphids, and the potato growers distributed healthy seed free to home gardeners (Bishop, 1963). Hille Ris Lambers (1955) drew attention to the great increase in numbers of *M. persicae* in the north Netherlands, where peach is seldom grown, following the planting of millions of *Prunus serotina* Ehrh. as forest shade trees. More could be done to prevent the planting of alternative hosts of vectors, if an appreciation of virus disease epidemiology were widespread.

### *B. Heat*

In climates where the summer day temperature remains around 32°C or above for long periods, aphids will seldom survive, and advantage has been taken of this to produce lettuce seed free from mosaic virus in Australia (Stubbs, 1954; Stubbs and O'Loughlin, 1962). In the hotter parts of South Africa, van der Plank (1944) found that aphids left potatoes and virus spread was negligible when the mean daily maximum temperature reached 32°C, and July planting is one of the principle means of obtaining reasonably healthy seed potatoes in the U.S.S.R. (Chesnokov, 1964). Gabriel (1965) in Poland also found a high summer temperature to be unfavorable for the spread of potato leaf roll virus.

In Israel, Nitzany *et al.* (1964) increased the temperature around young cucumber plants by applying a wheat straw mulch at germination. This resulted in very low populations of *Bemisia tabaci* Genn., the vector of the destructive bottle gourd mosaic virus, until the plants were large enough to shade the straw. In consequence, there were only 1.5% infected plants 30 days after sowing, in contrast to 26% unmulched plants, and 27% mulched plants were infected after 40 days when all the unmulched ones showed mosaic. Marketable yield was nearly trebled by mulching.

### *C. Predators and Parasites*

Predators and parasites often play a large part in limiting insect populations, but their populations naturally fluctuate with those of their hosts and often lag behind them. In Europe, for example, aphid populations on potato tend to follow a biennial rhythm, a year with many aphids leading to the development of many enemies, being followed by one with few aphids because of the numerous surviving enemies. When aphids are few, the predators and parasites decline in number again, enabling the aphids subsequently to increase once more (Blattný, 1925; Hille Ris Lambers, 1955). For this reason the enemies help to control virus vectors, but seldom adequately to stop virus spread, especially if much of this is by a few insects when colonizing the young plants. Even when the insects are relatively immobile, as are the mealybug vectors of cacao swollen-shoot viruses, attempts to control them by introducing hymenopterous, fungal, and other parasites have been unsuccessful (Thresh, 1958).

A few instances are known where predators and parasites apparently play an important part in limiting virus spread. The groundnut cultivar Mwitunde is field resistant to rosette virus partly because the vector, *Aphis craccivora* breeds more slowly on it than on other cultivars, and as a result the aphids can be eliminated more quickly by their enemies (Evans, 1954). Another instance was postulated by Stubbs (1956), who studied the spread of carrot motley dwarf virus. The aphid vector, *Cavariella aegopodii* Scop., is very numerous in Australia and Britain, where the disease is epidemic, but the aphids are checked in California by a braconid wasp which does not occur in the other countries, and it is probably because of this that the disease is not so important in California.

Although control of virus vectors by insect, fungal, viral, or nematode parasites or predators has not been widely tried or very successful to date, the considerable amount of work on biological control of arthropods being done in different parts of the world makes it premature to conclude that success will not be achieved in the future.

### *D. Elimination of Vectors With Chemicals*

The problem of killing insect pests is simple compared with stopping them spreading viruses, it only being necessary to keep their numbers below the level at which their feeding begins to cause economic loss of the crop. It is difficult to kill aphids quickly enough to prevent them from infecting healthy plants, especially with the stylet-borne viruses that can be acquired and transmitted within minutes, and viruliferous vectors bringing the more persistent viruses into treated crops can often infect plants before they die (Stubbs, 1948; Swenson *et al.*, 1954). McClean (1957) found that *M. persicae* infective with sugar beet yellows virus were not killed in less than 80–100 min when they were transferred successively at 20-min intervals to plants treated with phorate, or demeton, methyl parathion, parathion or malathion, and some still transmitted virus in the fifth period. In a recent study, phorate, parathion, disulfoton, and demeton required at least 51–180 min to kill 90% of *M. persicae* 2 hr after application of the insecticide, and much longer 3 days later (Shanks and Chapman, 1965). Much may depend upon how quickly the insects are incapacitated from feeding. Kollmer *et al.* (1963) found that about 48 hr of feeding on potatoes treated with disulfoton was needed to stop the spread of virus Y by *M. persicae,* but that the aphids were unable to transmit leaf roll virus from treated diseased plants or from untreated ones to treated healthy *Physalis. Myzus persicae* and *Brevicoryne brassicae* L. were not killed after 30 min on cauliflower leaves sprayed with DDT emulsion, although many were temporarily incapacitated, but the spread of cauliflower mosaic and cabbage black ring spot viruses was greatly decreased when infected plants were sprayed (Heathcote and Ward, 1963).

Other evidence that insecticides may sometimes prevent the introduction of virus into crops was obtained by Strong and Rawlins (1959) who caged *Macrosteles fascifrons* Stål, infective with lettuce yellows virus, over lettuce plots within 1 or 2 hr, or 3, 6, or 10 days of spraying the plants with DDT, demeton, malathion or parathion, and then removed the cages after 3 days and cleared remaining leafhoppers from the plants. Plots sprayed 1 or 2 hr before infestation had much less disease (32–45%) than the unsprayed ones (89%), but not those sprayed 3 or more days previously.

Soil-applied insecticides have also been tried to limit the introduction of leafhopper-transmitted viruses. Hills *et al.* (1964) found that phorate placed under cantaloupe seed, or mixed with the seed, partially prevented caged infective *C. tenellus* from transmitting beet curly top virus to the young seedlings.

Spraying sugar beet root crops with systemic insecticides, or using them in granular form, at the beginning of aphid colonization delays yellows virus introduction and its subsequent spread within the field, probably because aphids are prevented from moving from plant to plant as frequently as they do in untreated crops (Hull, 1958; Steudel and Thielemann, 1962). A second spray with demeton-methyl brought little extra benefit in England, except when crops were recolonized and there was late spread. Spraying with organophosphorus insecticides gives an economic yield increase in England when 20% or more of the plants become infected; this level of infection is probable when there is more than one *M. persicae* to four plants before early July. A spray warning for growers is operated, based on three counts of aphids made by the British Sugar Corporation's agricultural staff (Dunning, 1962; Hull, 1965). In 1958, for example, there were 2.5 times as many yellows-infected plants in unsprayed than in sprayed crops in the same area, and a single spray increased yield by almost 8% (Hull, 1959). Even better control was obtained in Switzerland, for beet sprayed with morphothion when *M. persicae* first appeared and again 6–12 days later decreased yellows incidence from 67 to 7%, and from 40 to 6% in two years (Münster and Joseph, 1959). Similarly, Smith (1963) in New Zealand effectively controlled barley yellow-dwarf virus in autumn and early-winter sown wheat by one of several organophosphorus sprays at any time between late winter and late spring to kill the vector, *Rhopalosiphum padi* L.

Although growers would usually prefer to spend money on insecticides to apply on their own crops, it is obviously better to kill the vectors before they leave the old crop or wild hosts, when this is possible. The biggest program of this type was that financed by the sugar beet companies and the Department of Agriculture in California to control the vector of beet curly-top virus, *Circulifer tenellus,* from 1931 onwards. Pyrethrum in diesel oil was sprayed on the foothill desert areas where the insects congregated during the autumn and winter, and on Russian thistle, a principal summer and winter wild host (Cook, 1943). Later DDT in diesel oil was sprayed by airplane or ground machines during the first three weeks of October on 141,000 acres in 1950, and on 150,000 in 1951 (Armitage, 1952). Curly top affected almost all early tomatoes in 1950, but only about 10% were diseased after the first season of spraying. Large-scale spraying of the leafhopper breeding grounds was done also in Idaho, applying DDT soon after the susceptible nymphs had hatched. After the spraying in 1950 and 1953, Douglass *et al.* (1955) estimated that curly top in beans was decreased from an expected 11 to 2%.

Spraying of the weed borders of lettuce fields to kill infective leafhoppers was more effective than spraying of the fields themselves (Hoffman, 1952). When parathion, DDT, or both were sprayed from May to August on the crop and the borders, 7% of the lettuce was infected with aster yellows virus; when the border alone was sprayed, 11% was infected, whereas over half the plants were infected in a nearby field without border control. Similarly, Simons (1957) showed that it was possible to prevent much of the spread of potato virus Y into pepper crops when infected nightshade plants growing nearby were sprayed twice per week with parathion.

When the virus sources are within the crop the chances of limiting spread are considerable if the plants can be given an insecticidal treatment before the vectors arrive. It may be possible to stop altogether the spread of circulatory viruses such as potato leaf roll, and even the spread of stylet-borne viruses may be limited if the aphid vectors are incapacitated before they have made as many flights as they might otherwise have done.

Some of the most intensive studies on such control measures have been made in Maine, U.S.A. and England during the last 20 years, following the introduction of persistent insecticides. Simpson, *et al.* (1964) stated that the average yield of potatoes per acre in the U.S.A. increased by 90% during the nine years after the introduction of DDT and nabam for pest and disease control.

Several aphicides proved to be effective in virtually stopping the spread of leaf roll virus, and limiting that of virus Y, within potato crops in England, and because of its relative cheapness and safety, four sprays of DDT emulsion at 2 lb active ingredient per acre, applied fortnightly at either high or low volume were recommended (Broadbent *et al.*, 1956, 1958). Aphid colonization and virus spread were usually early; therefore, it was essential to start spraying as soon as the potato foliage emerged above the ground (Burt *et al.*, 1964). Although insecticidal treatments often halved the spread of virus Y, they were not so effective in years when aphids were numerous and very active. The spraying of potato crops was tried commercially in many parts of England and stocks were kept within the acceptable bounds of health for four or five years in many areas where they had previously been replaced every one or two years (Broadbent *et al.*, 1960). One rogued and sprayed stock was kept for eight years in an area where untreated stocks might be expected to become entirely infected in three years, and was then discarded because disease incidence reached 10% (Broadbent *et al.*, 1961, 1962). Similar work in Maine (Simpson and Shands, 1949; Simpson *et al.*, 1951, 1964) and in Washington (Stitt and Breakey, 1952; King, 1952) showed that once the considerable spread of virus between crops was limited by the use of insecticides, the position was much the same as in Britain, spread being mainly within crops from plants grown from initially infected tubers. Control in the United States has been more effective than in Britain because growers were willing to cooperate in schemes covering large areas, probably because insect populations were much larger in the United States and insecticides were needed to control feeding damage, apart from virus spread. Many growers in Britain have now realized that even the relatively small aphid populations decrease potato yield and have recently begun to treat a considerable proportion of the crop each year, which also limits the chances of spread between crops. Most growers were not willing to spray four times, because of the cost in labor, damage to the plants, and soil compaction by the spraying machinery, but the use of systemic insecticides such as disulfoton, dimethoate, and phorate has led to a change in attitude. Single applications of such insecticides, usually in the form of granules applied in the soil during planting, kept aphids from colonizing the plants for about three months and were as effective as DDT in controlling virus spread (Burt *et al.*, 1960; Broadbent and Burt, 1962). Similar results, good control of leaf roll and other persistent viruses but often poor control of virus Y, were obtained in Europe and America with soil applied systemic insecticides (Hoyman, 1958; Simpson and Shands, 1961, 1963; Küthe, 1961; Bonnemaison, 1962; Pond, 1964). Systemic insecticides are usually more effective than contact ones because they penetrate into new tissues as the plants grow. Initially, however, the plant takes some time to absorb the chemical and if aphid attack is very early, virus spread may occur before the sap is aphicidal. Tuber-applied systemic insecticides, such as menazon, are more effective than soil-applied ones in these circumstances, but they may not last as long. In seed-tuber production the use of soil- or tuber-applied insecticides is particularly valuable because the roguing period can be prolonged (Simpson and Shands, 1963).

Although potatoes have been the principle crop used in research on the control of virus spread within crops, other workers have found insecticides effective in similar circumstances. Malaguti and Angeles (1957) decreased hoja blanca disease in rice from 23 to 4% by spraying demeton-O-methyl from aircraft 3, 7, and 11 weeks after germination, and de Fluiter (1958) stopped the spread of strawberry viruses with demeton-methyl. Davis *et al.* (1961) found that three parathion sprays at weekly intervals, starting soon after emergence, decreased pea enation mosaic from 54 to 9%. Peay and Oliver (1964) used a new technique to obtain good control of beet curly top virus and its vector *C. tenellus* on beans: they mixed several insecticides, such as phorate, phosdrin, and endothion, with juice from sugar beets that were resistant to the virus or with beet sugar, and applied the mixture as a spray. Mixing 0.25% phosdrin with beet juice gave 93% control of the virus, for example, but they were not certain why the mixtures were more effective than the insecticides alone.

Rawlins and Gonzalez (1966) placed phorate or disulfoton granules 2.5 cm below lettuce seed or broadcast the granules and cultivated them into the soil, in an endeavor to obtain as good control of aster yellows as when the chemicals were mixed with the seed in the furrow. The insecticide sometimes adversely affected germination when sown with the seed, but this was the only treatment that considerably decreased infection within the plots.

Several experiments have been done on the elimination of vectors of soil-borne viruses with chemicals. Methyl bromide or dichloropropane-dichloropropene (D-D) applied to the soil in summer killed over 99% of *Xiphinema diversicaudatum* Micol. in different types of soil to a depth of at least 70 cm and almost stopped infection of strawberries with arabis mosaic virus; Harrison *et al.* (1963) suggested that as so few nematodes survived, the treatment would not need repeating for several years. D-D or quintozene also killed most *Longidorus elongatus* de Man and prevented the spread of tomato black ring and raspberry ringspot viruses in strawberries during a four-year experiment (Murant and Taylor, 1965).

Unfortunately, soil fumigants and nematicides often penetrate only the upper layers of the soil, whereas some vectors and viruses in plant roots may inhabit lower layers, as Hewitt *et al.* (1962) found in California, where *Xiphinema index* Thorne and Allen, the vector of the grapevine strain of arabis mosaic virus, can be found to depths of 2.4 m.

Miyamoto (1961) obtained effective field control of soil-borne barley yellow mosaic (vector unknown) with pyroligneous acid applied to the soil several days before sowing, or with a flame gun giving a temperature of 40–60°C to a depth of 15 cm, which suggests that the virus was near the surface. McKinney *et al.* (1957) also freed soil in containers from wheat and oat mosaic viruses with formaldehyde, chloropicrin, carbon disulfide, D-D, or ethyl alcohol. They postulated that the unknown vectors were closely associated with the plant roots, but even if they were near the surface and could be controlled in the field, it would be difficult to obtain effective control because of the large areas of North America that are infected (Cadman, 1963).

## IV. CONCLUSION

The prospects of virus disease control through vector control are much brighter than they were a few years ago. The increasing interest in integrated control, based on a knowledge of the epidemiology of a disease, has shown that chemical control of arthropod vectors can often be supplemented and made more effective by agronomic and biological control measures. The realization that pesticides should be used as sparingly as possible because of their often long-term effects on other organisms and the possible development of resistance in pests will be an added incentive to find or improve alternative methods of control. Basic studies on methods of transmission may lead to the discovery of other means of control, as they did to the new and potentially important use of oil films to prevent the transmission of stylet-borne viruses by aphids.

Much more control of virus diseases could be achieved now, if more growers were educated to realize their importance and put into practice the measures already worked out by the scientists. This often means cooperation by growers over a wide area, which is difficult to achieve in many countries, even if a strong lead is given by government agencies. Fortunately, the examples that have been quoted augur well for the future.

## BIBLIOGRAPHY

A'Brook, J. 1964. The effect of planting date and spacing on the incidence of groundnut rosette disease and of the vector, *Aphis craccivora* Koch, at Mokwa, Northern Nigeria. Ann. Appl. Biol. 54:199–208.

Adam, A. V. 1962. An effective program for the control of banana mosaic. Plant Disease Reporter 46:366–370.

Allen, T. C. 1965. Field spread of potato virus A inhibited by oil. Plant Disease Reporter 49:557.

Anderson, C. W. 1959. A study of field sources and spread of five viruses of peppers in central Florida. Phytopathology 49:97–101.

Anon. 1955. Potato research in Maine by Maine Agricultural Experimental Station and United States Department of Agriculture. Report U.S. Dep Agr. Potato Advisory Committee, Maine Agr. Exp. Sta.

Arenz, B., and Hunnius, W. 1963. Erfahrungen mit dem Igel-Lange-Test in der Serienarbeit. Bayer. Landwirt. Jahrb. 40:122–136.

Armitage, H. M. 1952. Controlling curly top virus in agricultural crops by reducing populations of overwintering beet leafhoppers. J. Econ. Entomol. 45:432–435.

Bagnall, R. H. 1953. The spread of potato virus Y in seed potatoes in relation to the date of harvesting and the prevalence of aphids. Can. J. Agr. Sci. 33:509–519.

Bagnall, R. H., and Bradley, R. H. E. 1958. Resistance to virus Y in the potato. Phytopathology 48:121–125.

Barnes, G. L. 1959. Herbicidal agents as possible aids for roguing diseased seed potato plants. Amer. Potato J. 36:212–218.

Bawden, F. C., and Kassanis, B. 1946. Varietal differences in susceptibility to potato virus Y. Ann. Appl. Biol. 33:46–50.

Bawden, F. C., and Kassanis, B. 1950. Some effects of host nutrition on the susceptibility of plants to infection by certain viruses. Ann. Appl. Biol. 37:46–57.

Beemster, A. B. R. 1961. Translocation of leaf roll and virus Y in the potato. Proc. 4th Conf. Potato Virus Diseases, Braunschweig, 1960, p. 60–67.

Bennett, C. W. 1960. Sugar beet yellows disease in the United States. Tech. Bull. U. S. Dep. Agr. 1218.

Berkeley, G. H. 1942. Tobacco mosaic in Ontario and Quebec. Sci. Agr. 22:465–478.

Birecki, M., Gabriel, W., and Osinska, J. 1964. The effect of agro-technical treatment on the quality of seed potato (in Polish). Rocznski Nauk Rolniczych Ser. A. 88:235–258, 461–483.

Bishop, G. 1963. Idaho Agr. Sci. 44:4–5. (Quoted by Simpson, Landis, and Shands, 1964.)

Blattný, C. 1925. Het voorspellen van het massaal optreden van schadelijke insecten. Tijdschr. Plantenziekten 21:139–144.

Blencowe, J. W., and Tinsley, T. W. 1951. The influence of density of plant population on the incidence of yellows in sugar beet crops. Ann. Appl. Biol. 38:395–401.

Bonnemaison, L. 1961. Protection des betteraves-racines et des porte-graines contre la jaunisse basee sur la lutte contre les vecteurs. Ann. Epiphyties 12:155–217.

Bonnemaison, L. 1962. Traitement des graines de betteraves, du sol et des semences de pomme de terre en vue de la protection contre les maladies a virus. Phytiatrie-Phytopharmacie 11:85–103.

Bradley, R. H. E. 1962. Aphid transmission of potato virus Y inhibited by oils. Virology 18:327–328.

——— 1963. Some ways in which a paraffin oil impedes aphid transmission of potato virus Y. Can. J. Microbiol. 9:369–380.

Bradley, R. H. E., Moore, C. A., and Pond, D. D. 1966. Spread of potato virus Y curtailed by oil. Nature London 209:1370–1371.

Broadbent, L. 1948. Aphis migration and the efficiency of the trapping method. Ann. Appl. Biol. 35:379–394.

——— 1957. Investigations of virus diseases of brassica crops. Agricultural Research Council Report Series No. 14. Cambridge Univ. Press.

——— 1962. The epidemiology of tomato mosaic. II. Smoking tobacco as a source of virus. Ann. Appl. Biol. 50:461–466.

——— 1963. The epidemiology of tomato mosaic. III. Cleaning virus from hands and tools. Ann Appl. Biol. 52:225–232.

——— 1965. The epidemiology of tomato mosaic. IX. Transmission of TMV by birds. Ann. Appl. Biol. 55:67–69.

Broadbent, L., and Burt, P. E. 1962. Systemic insecticides and virus control in potatoes. Proc. Brit. Insecticides Fungicides Conf., 1961, 1, p. 81–85.

Broadbent, L., Burt, P. E., and Heathcote, G. D. 1956. The control of potato virus diseases by insecticides. Ann. Appl. Biol. 44:256–273.

——— 1958. Insecticidal control of potato virus spread. Proc. 3rd Conf. Potato Virus Diseases, Lisse-Wageningen, 1957. p. 91–105.

Broadbent, L., Cornford, C. E., Hull, R., and Tinsley, T. W. 1949. Overwintering of aphids, especially *Myzus persicae* (Sulzer), in root clamps. Ann. Appl. Biol. 36:513–524.

Broadbent, L., and Fletcher, J. T. 1963. The epidemiology of tomato mosaic. IV. Persistence of virus on clothing and glasshouse structures. Ann. Appl. Biol. 52:233–241.

——— 1966. The epidemiology of tomato mosaic. XII. Sources of TMV in commercial tomato crops under glass. Ann. Appl. Biol. 57:113–120.

Broadbent, L., Gregory, P. H., and Tinsley, T. W. 1950. Roguing potato crops for virus diseases. Ann. Appl. Biol. 37:640–650.

——— 1952. The influence of planting date and manuring on the incidence of virus diseases in potato crops. Ann. Appl. Biol. 39:509–524.

Broadbent, L., and Heathcote, G. D. 1960. Detection of leafroll in potato tubers. Plant Pathol. 9:126–128.

Broadbent, L., Heathcote, G. D., Brown, P. H., and Wheeler, G. F. C. 1961. Home production of seed for early potatoes. 1. Prevention of virus infection by spraying for aphid control. Exp. Hort. 4:8–12.

Broadbent, L., Heathcote, G. D., and Burt, P. E. 1960. Field trials on the retention of potato stocks in England. European Potato J. 3:251–262.

Broadbent, L., Heathcote, G. D., and Wright, R. C. M. 1962. Home production of seed for early potatoes. 4. Loss of yield in virus-infected plants. Exp. Hort. 7:4–6.

Broadbent, L., Read, W. H., and Last, F. T. 1965. The epidemiology of tomato mosaic. X. Persistence of TMV-infected debris in soil, and the effects of soil partial sterilization. Ann. Appl. Biol. 55:471–483.

Broadbent, L., Tinsley, T. W., Buddin, W., and Roberts, E. T. 1951. The spread of lettuce mosaic in the field. Ann. Appl. Biol. 38:689–706.

Burt, P. E., Broadbent, L., and Heathcote, G. D. 1960. The use of soil insecticides to control potato aphids and virus diseases. Ann. Appl. Biol. 48:580–590.

Burt, P. E., Heathcote, G. D., and Broadbent, L. 1964. The use of insecticides to find when leaf roll and Y viruses spread within potato crops. Ann. Appl. Biol. 54:13–22..

Cadman, C. H. 1963. Biology of soil-borne viruses. Ann. Rev. Phytopathol. 1:143–172.

Cadman, C. H., and Chambers, J. 1960. Factors affecting the spread of aphid-borne viruses in potato in eastern Scotland. III. Effects of planting date, roguing, and age of crop on the spread of potato leaf-roll and Y viruses. Ann. Appl. Biol. 48:729–738.

Cammack, R. H. 1966. Little cherry virus in ornamental cherry. Plant Pathol. 15:31–33.

Cann, H. J. 1952. Bunchy top disease of bananas. Agr. Gaz. New South Wales 63:73–76.

Chesnokov, P. G. 1964. Sanitation of seed potato in various zones of U.S.S.R. based on the study of the causes of degeneration. (In Russian) Sb. Tr. Vses. Inst. Rasteniev 1964:3–12.

Cook, W. C. 1943. Evaluation of a field-control program directed against beet leafhoppers. J. Econ. Entomol. 36:382–385.

Crowley, N. C. 1958. The use of skim milk in preventing the infection of glass-house tomatoes by tobacco mosaic virus. J. Australian Inst. Agr. Sci. 24:261–263.

Crumb, S. E., and McWhorter, F. P. 1948. Dusting beans against aphid vectors failed to give economic control of yellow bean mosaic. Plant Disease Reporter 32:169–171.

Davis, A. C., McEwen, F. L., and Schroeder, W. T. 1961. Control of pea enation mosaic in peas with insecticides. J. Econ. Entomol. 54:161–166.

Denby, L. G., and Wilks, J. M. 1963. The effect of tobacco mosaic on the yield of field tomatoes as influenced by sprays of milk and DOSS. Can. J. Plant Sci. 43:457–461.

——— 1965. Milk sprays reduce spread of tobacco mosaic in tomatoes. Res. Farmers 10:10–11.

Doncaster, J. P., and Gregory, P. H. 1948. The spread of virus diseases in the potato crop. Agricultural Research Council Report Series No. 7. London, H.M.S.O.

Doolittle, S. P. 1928. Soil transmission of tomato mosaic and streak in the greenhouse. Phytopathology 18:155.

Douglass, J. R., Romney, V. E., and Jones, E. W. 1955. Beet leafhopper control in weed-host areas of Idaho to protect snapbean seed from curly top. Cir. U. S. Dep. Agr. 960.

Dunning, R. C. 1962. Systemic insecticides and sugar beet yellows. Agriculture (London) 69:356–361.

Evans, A. C. 1954. Rosette disease of groundnuts. Nature (London) 173:1242.

Fernandez Valiela, M. V. 1963. Principales enfermedades de virus de los citrus en la Republica Argentina. Bol. Divulg., Delta d. Paraná 3:3–37.

Flock, R. A., and Deal, A. S. 1959. A survey of beet leafhopper populations on sugar beets in the Imperial Valley, California, 1953–1958. J. Econ. Entomol. 52:470–473.

Fluiter, H. J. de 1958. Bladluisbestrijding ter voorkoming van virusverspreiding in aardbeien. Mededel. Landbouwhoogeschool Opzoekingsstations Gent 23: 745–769.

Folsom, D. 1952. Practical control measures for leaf roll. Amer. Potato J. 29:229–233.

Franklin, G. 1966. An anti-virus moistener bottle. Bull. Amer. Orchid Soc. 35:383–384.

Frederiksen, R. A. 1962. Studies on the transmission, effect and control of two viruses in *Linum usitatissimum* L. Dissertation Abstr. 22:3800–3801.

Freytag, A. H. 1965. A "hot knife" to prevent virus spread in orchid. Bull. Amer. Orchid Soc. 34:501–502.

Fridlund, P. R. 1965. Release of virus-indexed prunus budwood from the interregional repository. Plant Disease Reporter 49:187–188.

Fry, P. R., and Coleman, B. P. 1960. Tobacco mosaic virus in glasshouse tomatoes: effect of skimmed-milk sprays on incidence. New Zealand Com. Grower 16:25.

Fulton, J. P., and Seymour, C. 1957. The Arkansas strawberry certification program. Plant Disease Reporter 41:749–754.

Fulton, R. W. 1943. The sensitivity of plant viruses to certain inactivators. Phytopathology 33:674–682.

Gabriel, W. 1965. The influence of temperature on the spread of aphid-borne potato virus diseases. Ann. Appl. Biol. 56:461–475.

Govier, D. A. 1963. The reaction of the Dische diphenylamine reagent with sap from potato plants infected with potato leaf-roll virus. Virology 19:561–564.

Hagborg, W. A. F., and Chelack, W. S. 1960. Whey as an inhibitor of stripe mosaic of barley. Can. J. Bot. 38:111–116.

Hansen, H. P. 1950. Investigations on virus yellows of beets in Denmark. Trans. Danish Acad. Technical Sci. 1:1–68.

Hare, W. W., and Lucas, G. B. 1959. Control of contact transmission of tobacco mosaic virus with milk. Plant Disease Reporter 43:152–156.

Harpaz, I. 1961. *Calligypona marginata,* the vector of maize rough dwarf virus. FAO Plant Protection Bull. 9:144–147.

Harper, R. G. 1928. Uganda Dep. Agr. Ann. Rept. 1927, p. 17. (Quoted by Hull, 1964.)

Harrison, B. D. 1957. Tobacco mosaic in the tomato crop. Rep. Scottish Hort. Res. Inst. 1956–57, p. 31–32.

Harrison, B. D., Peachey, J. E., and Winslow, R. D. 1963. The use of nematicides to control the spread of arabis mosaic virus by *Xiphinema diversicaudatum* (Micol.). Ann. Appl. Biol. 52:243–255.

Heathcote, G. D., and Ward, J. 1963. The effect of DDT on *Myzus persicae* (Sulz.) and *Brevicoryne brassicae* L. (Aphididae) in relation to the spread of cauliflower mosaic and cabbage black ring spot viruses. Bull. Entomol. Res. 53:779–784.

Hecht, H., and Arenz, B. 1965. Der Nachweis des Blattrollvirus in ober- und unterirdischen Organen der Kartoffel mit Hilfe des Phenoltestes. Phytopathol. Z. 54:147–156.

Hein, A. 1961. Verhinderung der Kontaktübertragung des Tabakmosaik-Virus durch Magermilch. Phytopathol. Z. 42:263–271.

——— 1964a. Weitere Untersuchungen zur Verhinderung der Kontaktübertragung des Tabakmosaik-Virus durch Milchanwendung. Z. Pflanzenkrankh. Pflanzenschutz 71:206–210.

——— 1964b. Die Wirkung eines Milchfilms auf die Übertragung eines nicht-persistenten Virus durch Blattläuse. Z. Pflanzenkrankh. Pflanzenchutz 71:267–270.

Hewitt, W. B., Goheen, A. C., Raski, D. J., and Gooding, G. V. 1962. Studies on virus diseases of the grapevine in California. Vitis 3:57–83. (Quoted by Cadman, 1963.)

Hille Ris Lambers, D. 1955. Potato aphids and virus diseases in the Netherlands. Ann. Appl. Biol. 42:355–360.

Hills, O. A., Coudriet, D. L., and Brubaker, R. W. 1964. Phorate treatments against the beet leafhopper on cantaloupes for prevention of curly top. J. Econ. Entomol. 57:85–89.

Hoffman, J. R. 1952. Leafhopper control to prevent the spread of the virus disease aster yellows in commercial lettuce production. Quart. Bull. Michigan Agr. Exp. Sta. 34 p. 262–265.

Hoggan, I. A., and Johnson, J. 1936. Behavior of the ordinary tobacco mosaic virus in the soil. J. Agr. Res. 52:271–294.

Hollings, M. 1955. Aphid movement and virus spread in seed potato areas of England and Wales, 1950–1953. Plant Pathol. 4:73–82.

——— 1960. Aphid movement and virus spread in seed potato areas of England and Wales. Plant Pathol. 9:1–7.

Hollings, M., and Stone, O. M. 1965. Investigation of carnation viruses. II. Carnation ringspot. Ann. Appl. Biol. 56:73–86.

Hopkins, J. C. F. 1932. Further notes on leaf curl of tobacco in Southern Rhodesia. Rhodesia Agr. J. 29:680–686.

Hoyman, W. G. 1958. Effect of thimet on incidence of virus Y and purple-top wilt in potatoes. Amer. Potato J. 35:708–710.

Hull, R. 1952. Control of virus yellows in sugar beet seed crops. J. Roy. Agr. Soc. 113:86–102.

——— 1958. Sugar beet yellows. The search for control. Agriculture (London) 65:62–65.

——— 1959. Sugar beet yellows in Great Britain, 1958. Plant Pathol. 8:145.

——— 1965. Control of sugar beet yellows. Ann. Appl. Biol. 56:345–347.

Hull, R. and Watson, M. A. 1947. Factors affecting the loss of yield of sugar beet caused by beet yellows virus. II. Nutrition and variety. J. Agr. Res. 37:302–310.

Hull, Roger 1964. Spread of groundnut rosette virus by *Aphis craccivora* (Koch). Nature (London) 202:213–214.

Inouye, T. 1962. Studies on barley stripe mosaic in Japan. Ber. Ohara Inst. Landwirt. Biol. Okayama Univ. 11:413–496.

Janssen, J. J. 1929. Invloed der Bemesting op de Gezondheid van de Aardappel. Tijdschr. Plantenziekten 35:119–151.

Johnson, J. 1941. Chemical inactivation and reactivation of a plant virus. Phytopathology 31:679–701.

Jones, L. K., and Burnett, G. 1935. Virus diseases of greenhouse-grown tomatoes. Washington State Univ. Agr. Exp. Sta. Bull. 308 p. 1–36.

Jones, L. R., and Riker, R. S. 1931. Wisconsin studies on aster diseases and their control. Wisconsin Univ. Agr. Exp. Sta. Res. Bull. 111.

Kabiersch, W. 1962. Seed potato production in Western Germany. Outlook Agr. 3:268–273.

Keener, P. D. 1956. Virus diseases of plants in Arizona. II. Field and experimental observations on curly-top affecting vegetable crops. Arizona Univ. Agr. Exp. Sta. Bull. 271.

Keller, E. R., and Bérces, S. 1966. Check-testing for virus Y and leaf-roll in seed potatoes with particular reference to methods of increasing precision with the A6-leaf test for virus Y. European Potato J. 9:1–12.

Kennedy, J. S., Booth, C. O., and Kershaw, W. J. S. 1961. Host finding by aphids in the field. III. Visual attraction. Ann. Appl. Biol. 49:1–21.

King, L. W. 1952. Washington certified seed potatoes. Amer. Potato J. 29:53–54.

Kollmer, G., Hunnius, W., and Arenz, B. 1963. Über die Wirkung systemischer Insektizidgranulate auf das Verhalten der Pfirsichblattlaus (*Myzus persicae* Sulz.) und die Verhinderung von Virusinfektionen bie Kartoffeln. Bayer. Landwirt. Jahrb. 40:824–836.

Kovachevski, I. C. 1964. Stolbur disease and the means for its control. (In Bulgarian). Gradinarstvo 6:20–23.

Küthe, K. 1961. Ein neuer Weg zur Bekämpfung von Vektoren im Kartoffelbau. Ainsatz von systemischen Saatgutbehandlungsmitteln. Z. Pflanzenkrankh. Pflanzenschutz 68:209–218.

Kunkel, L. O. 1929. Wire-screen fences for the control of aster yellows. Phytopathology 19:100.

Linn, M. B. 1940. The yellows disease of lettuce and endive. Cornell Univ. Agr. Exp. Sta. Bull. 742 p. 1–33.

List, G. M., Landblom, N., and Sisson, M. A. 1956. A study of records from the Colorado peach mosaic suppression program. Colorado Agr. Expt. Sta. Tech. Bull. 59.

Lister, R. M. 1960. Transmission of soil-borne viruses through seed. Virology 10:547–549.

Loebenstein, G., Alper, M., and Deutsch, M. 1964. Preventing aphid-spread cucumber mosaic virus with oils. Phytopathology 54:960–962.

Loebenstein, G., Deutsch, M., Frankel, H., and Sabar, Z. 1966. Field tests with oil sprays for the prevention of cucumber mosaic virus in cucumbers. Phytopathology 56:512–516.

Longworth, J. F. 1963. The effect of swollen shoot disease on mature cocoa in Nigeria. Trop. Agr. Trinidad, 40:275–283.

Lovisolo, O., and Benetti, M. P. 1960. Virus e piante spontanee. II. Nuovi ospiti del virus della maculatura anulare nera del cavolo. Boll. Sta. Patol. Veg. 17:61–70.

McClean, D. M. 1957. Effect of insecticide treatments of beets on transmission of yellows virus by *Myzus persicae*. Phytopathology 47:557–559.

MacClement, W. D., and Richards, M. G. 1956. Virus in wild plants. Can. J. Bot. 34:793–799.

McKinney, H. H., Paden, W. R., and Koehler, B. 1957. Studies on chemical control and overseasoning of, and natural inoculation with, the soil-borne viruses of wheat and oats. Plant Disease Reporter 41:256–266.

Mackinnon, J. P., Collins, W. B., and Colpitts, S. R. 1964. A survey of green-petal virus in New Brunswick and some effects of barriers on spread. Can. Plant Disease Surv. 44:91–95.

Malaguti, G., and Angeles, N. 1957. Ensayos preliminares sobre el uso de insecticidas en el control de los vectores de la "hoja blanca" del Arroz. Agron. Trop. (Maracay, Venezuela) 7:161–163.

Martin, W. J., and Kantack, E. J. 1960. Control of internal cork of sweet potato by isolation. Phytopathology 50:150–152.

Miyamoto, Y. 1961. Studies on soil-borne cereal mosaics. VII. Controlling soil-borne cereal mosaics with special reference to the effect of soil treatment with pyroligneous acid (wood vinegar). Ann. Phytopath. Soc. Japan 26:90–97.

Müller, H. J. 1956. Zur Problematic der Blattlausresistenz landwirtschaftlicher Kulturpflanzen. Sitzber. Deut. Acad Landwirt. Wiss. Berlin 5:1–20.

——— 1964. Über die anflugdichte von Aphiden auf Farbige Salatpflanzen. Entomol. Exp. Appl. 7:85–104.

Münster, J. 1958. Methode zur Beobachtung der Entwicklung der virusübertragenden Blattläuse zwecks Ansetzung des Früherntetermins und dessen Rückwirkungen auf den Ertrag an Saatkartoffeln. European Potato J. 1:31–41.

Münster, J., and Joseph, E. 1959. Lutte contre la jaunisse sur betteraves sucrières a l'aide de traitements systemiqus preventifs. Ann. Agr. Suisse, N.S. 8:579–595.

Mulholland, R. I. 1962. Control of the spread of mechanically transmitted plant viruses. Commonwealth Phytopathol. News 8:60–61.

Murant, A. F., and Taylor, C. E. 1965. Treatment of soil with chemicals to prevent transmission of tomato black ring and raspberry ringspot viruses by *Longidorus elongatus* (de Man). Ann. Appl. Biol. 55:227–237.

Newell, J. 1954. Milk spray cured tomato mosaic. Grower (London) 41:1409.

Nitzany, F. E. 1966. Tests for the control of field spread of pepper viruses by oil sprays. Plant Disease Reporter 50: 158–160.

Nitzany, F. E., Geisenberg, H., and Koch, B. 1964. Tests for the protection of cucumbers from a white fly-borne virus. Phytopathology 54:1059–1061.

Noordam, D. 1955. Onkruiden en de verspreiding van viruszieken. Mededelingen Directeur Tuinbouw 18:639–645.

Peay, W. E., and Oliver, W. N. 1964. Curly top prevention by vector control on snap beans grown for seed. J. Econ. Entomol. 57:3–5.

Persons, T. D. 1952. Phony peach disease—review of organized control from 1929–1951 and the effect of recent developments on future control programs. Phytopathology 42:286–287.

Piemeisel, R. L. 1954. Replacement control: changes in vegetation in relation to control of pests and diseases. Bot. Rev. 20:1–32.

Plank, J. E. van der 1944. Production of seed potatoes in a hot, dry climate. Nature (London) 153:589–590.

——— 1948a. The relation between the size of fields and the spread of plant disease into them. 1. Crowd diseases. Empire J. Exp. Agr. 16:134–142.

——— 1948b. The relation between the size of plant and the spread of systemic diseases. 11. The aphis-borne potato virus diseases. Ann. Appl. Biol. 35:45–52.

——— 1949. The relation between the size of fields and the spread of plant disease into them. 2. Diseases caused by fungi with air-borne spores; with a note on horizons of infection. Empire J. Exp. Agr. 17:18–22, 141–147.

Plank, J. E. van der, and Anderssen, E. E. 1944. Kromnek disease of tobacco; a mathematical solution to a problem of disease. S. African Agr. Dep. Sci. Bull. 240 p. 1–6.

Pond, D. D. 1964. Field control of potato leaf roll virus with systemic insecticides. Amer. Potato J. 41:14–17.

Posnette, A. F. 1962. The mother tree scheme. Rep. E. Malling Res. Sta. 1961. p. 125–127.

Pound, G. S. 1946. Control of virus diseases of cabbage seed plants in western Washington by plant bed isolation. Phytopathology 36:1035–1039.

Ramakrishnan, K. 1963. Control of plant virus diseases. Bull. Nat. Inst. Sci. India 24:78–91.

Rawlins, W. A., and Gonzalez, D. 1966. Incidence of aster yellows in lettuce as affected by placement of systemic insecticides. J. Econ. Entomol. 59:226–227.

Ribbands, C. R. 1964. The control of the sources of virus yellows of sugar-beet. Bull. Entomol. Res. 54:661–674.

Sastry, K. S. M., and Nariani, T. K. 1962. Effect of host plant nutrition on growth and susceptibility of tobacco plants to infection with tobacco leaf curl virus. Indian J. Agr. Sci. 32:288–293.

Schmelzer, K. 1963. Untersuchungen an Viren der Zierund Wildgehölze. Phytopathol. Z. 46:17–52, 105–138, 235–268, 315–342.

Schultz, E. S., Bonde, R., and Raleigh, W. P. 1944. Early harvesting of healthy seed potatoes for the control of potato diseases in Maine. Maine Agr. Exp. Sta. Bull. 427.

Schwarz, R. 1959. Epidemiologische Untersuchungen über einige Viren der Unkraut-und Ruderalflora Berlins. Phytopathol. Z. 35:238–270.

Severin, H. H. P., and Freitag, J. H. 1938. Western celery mosaic. Hilgardia 11:495–558.

Shanks, C. H., and Chapman, R. K. 1965. The effects of insecticides on the behavior of the green peach aphid and its transmission of potato virus Y. J. Econ. Entomol. 58:79–83.

Shapovalov, M., Blood, H. L., and Christiansen, R. M. 1941. Tomato plant populations in relation to curly-top control. Phytopathology 31:864.

Shea, K. N. 1964. Apple virus diseases in Queensland. Queensland Agr. J. 90:682–685.

Shirahama, K. 1957. Studies on radish mosaic disease and its controlling (English summary). Agricultural Improvement Extension Work Conference, Tokyo, Japan.

Sill, W. H. 1966. Maize dwarf mosaic virus in Kansas. Plant Disease Reporter 50:11.

Simons, J. N. 1957. Effects of insecticides and physical barriers on field spread of pepper veinbanding mosaic virus. Phytopathology 47:139–145.

——— 1958. Viruses affecting vegetable crops in the Everglades and adjacent areas of South Florida. Rep. Florida Agr. Exp. Sta. 1957. p. 251.

——— 1960. Factors affecting field spread of potato virus Y in South Florida. Phytopathology 50:424–428.

Simons, J. N., Orsenigo, J. R., Stall, R. E., and Thayer, P. L. 1959. Potato virus Y in peppers and tomatoes. Mimeo Rep. Everglades Exp. Sta. Univ. Fla.

Simpson, G. W., Landis, B. J., and Shands, W. A. 1964. Recent advances relating to the control of certain insects attacking Irish potatoes. Potato Handbook 1964.

Simpson, G. W. and Shands, W. A. 1949. Progress on some important insect and disease problems of Irish potato production in Maine. Maine Agr. Exp. Sta. Bull. 470.

——— 1961. Insecticide applications affect leaf roll spread. Maine Farm Res. 9:3–7.

——— 1963. Systemic insecticides and leaf roll spread in 1961. Maine Farm Res. 11:19–21.

Simpson, G. W., Shands, W. A., Cobb, R. M., and Lombard, P. M. 1951. Control of aphids and leaf roll spread. Maine Agr. Exp. Sta. Bull. 491:50–51.

Slykhuis, J. T. 1955. *Aceria tulipae* Keifer (Acarina: Eriophyidae) in relation to the spread of wheat streak mosaic. Phytopathology 45:116–128.

Slykhuis, J. T., Zillinsky, F. J., Young, M., and Richards, W. R. 1960. Notes on the epidemiology of barley yellow dwarf virus in eastern Ontario in 1959. Plant Disease Reporter Suppl. 262:317–322.

Smartt, J. 1961. The diseases of groundnuts in Northern Rhodesia. Empire J. Exp. Agr. 29:79–87.

Smith, H. C. 1963. Control of barley yellow dwarf virus in cereals. New Zealand J. Agr. Res. 6:229–244.

Staples, R., and Allington, W. B. 1956. Streak mosaic of wheat in Nebraska and its control. Nebraska Univ. Agr. Exp. Sta. Res. Bull. 178 p. 1–41.

Steudel, W. 1952. Der Einfluss der Saatzeit auf Auftreten und Ausbreitung der Vergilbungskrankheit der Beta-rüben. Nachrbl. Deut. Pflanzenschutzdienst (Stuttgart) 4:40–44.

——— 1953. Epidemiologische Studien zur Vergilbungskrankheit im Rheinland 1952. Zucker 4:69–73.

Steudel, W., and Thielemann, R. 1962. Vergleichende Untersuchungen zur Wirkung von Disyston als Saatschutzpuder oder-granulat bei Zuckerrüben. Zucker 15:261–267, 286–291.

Stitt, L. L., and Breakey, E. P. 1952. Evidence that aphid control suppressed virus diseases of potatoes and strawberries in northwestern Washington. Mededel. Landbouwhoogeschool Opzoekingsstations Gent 17:94–110.

Storey, H. H. 1935. Virus diseases of East African plants. iii. Rosette disease of groundnuts. E. African Agr. J. 1:206–211.

Storey, H. H., and Ryland, A. K. 1950. Ann. Rep. E. African Agr. For. Res. Org. 1949. p. 15. (Quoted by Hull, 1964.)

Storey, I. F., and Godwin, A. E. 1953. Cauliflower mosaic in Yorkshire, 1950–1951. Plant Pathol. 2:98–101.

Stout, G. L. 1962. Maintenance of "pathogen-free" planting stock. Phytopathology 52:1255–1258.

Strong, R. G., and Rawlings, W. A. 1959. Field evaluation of four insecticides for the prevention of lettuce-yellows disease with known populations of viruliferous six-spotted leafhoppers. J. Econ. Entomol. 52:686–689.

Stubbs, L. L. 1948. A new virus disease of carrots; its transmission, host range, and control. Australian J. Sci. Res. 1:303–332.

——— 1954. Lettuce mosaic virus disease. J. Dep. Agr. Victoria 52:259–264.

——— 1956. Motley dwarf virus disease of carrot in California. Plant Disease Reporter 40: 763–764.

Stubbs, L. L., and O'Loughlin, G. T. 1962. Climatic elimination of mosaic spread in lettuce seed crops in the Swan Hill region of the Murray Valley. Australian J. Exp. Agr. 2:16–19.

Sutic, D. and Tosic, M. 1966. A significant occurrence of maize mosaic virus on Johnson grass (*Sorghum halepense* Pers.) as a natural host plant. Revue Roumaine Biol. Ser. Bot. 11:219–224.

Swenson, K. G., Davis, A. C., and Schroeder, W. T. 1954. Reduction of pea virus spread by insecticide applications. J. Econ. Entomol. 47:490–493.

Szirmai, J. 1957. Hazai tapasztalatok a Cukorrepa virusos sargasaganak elterjedisirol es leküzdesenek mödjairol. Ann. Inst. Prot. Plant. Hungary 7:393–407.

Tanaka, S., and Hirose, K. 1966. Indexing for little cherry virus in Japanese flowering cherry. Ann. Phytopathol. Soc. Japan 32:23–25.

Tarr, S. A. J. 1951. Leaf Curl Disease of Cotton. Commonwealth Mycol. Inst.

Taylor, C. E., and Johnson, C. G. 1954. Wind direction and the infestation of bean fields by *Aphis fabae* Scop. Ann. Appl. Biol. 41:107–116.

Thresh, J. M. 1958. The control of cacao swollen shoot disease in West Africa. A review of the present situation. Tech. Bull. W. African Cacao Res. Inst. 4. p. 1–36.

——— 1959. The control of cacao swollen shoot disease in Nigeria. Trop. Agr. Trinidad 36:35–44.

——— 1960. Quarantine arrangements for intercepting cocoa material infected with West African viruses. FAO Plant Protection Bull. 8:89–92.

Tinsley, T. W. 1964. The ecological approach to pest and disease problems of cacao in West Africa. J. Roy. Soc. Arts 112:353–365.

Todd, J. M. 1961. The incidence and control of aphid-borne potato virus diseases in Scotland. European Potato J. 4:316–329.

Tomlinson, J. A., and Garrett, R. G. 1966. Big vein disease of lettuce. Rep. Nat. Veg. Res. Sta., Warwick, 1965, 16. p. 73–74.

Vanderveken, J., and Semal, J. 1966. Aphid transmission of beet yellows virus inhibited by mineral oil. Phytopathology 56:1210–1211.

Viado, G. B., and Matthysse, J. G. 1959. A field test on the use of insecticides and cover-cropping and their bearing on the control of abaca mosaic. Philippine Agr. 42:364–373.

Waggoner, P. E. 1962. Weather, space, time and chance of infection. Phytopathology 52:1100–1108.

Wallace, J. M., and Murphy, A. M. 1938. Studies on the epidemiology of curly top in Southern Idaho, with special reference to sugar beets and weed hosts of the vector *Eutettix tenellus*. U.S. Dep. Agr. Tech. Bull. 624.

Watson, M. A., Hull, R., Blencowe, J. W., and Hamlyn, B. G. M. 1951. The spread of beet yellows and beet mosaic viruses in the sugar-beet root crop. 1. Field observations on the virus diseases of sugar beet and their vectors *Myzus persicae* Sulz. and *Aphis fabae* Koch. Ann. Appl. Biol. 38:743–764.

Way, M. J., and Heathcote, G. D. 1966. Interactions of crop density of field beans, abundance of *Aphis fabae* Scop., virus incidence and aphid control by chemicals. Ann. Appl. Biol. 57:409–423.

Wellman, F. L. 1937. Control of southern celery mosaic in Florida by removing weeds that serve as sources of mosaic infection. U. S. Dep. Agr. Tech. Bull. 548. p. 1–16.

Wenzl, H. 1965. Der Nachweis von Blattrollinfektionen bei der Kartoffel mittels der Diphenylamin-Reaktion nach Dische. Pflanzenschutzberichte 32:21–31.

Wolf, F. A. 1933. Roguing as a means of control of tobacco mosaic. Phytopathology 23:831–833.

Wright, N. S., and Hughes, E. C. 1964. Effect of defoliation date on yield and leaf roll incidence in potato. Amer. Potato J. 41:83–91.

Yamaguchi, M., and Welch, J. E. 1955. Varietal susceptibility of celery to aster yellows. Plant Disease Reporter 39:36.

Zink, F. W., Grogan, R. G., and Bardin, R. 1957. The comparative effect of mosaic-free seed and roguing as a control for common lettuce mosaic. Proc. Amer. Soc. Hort. Sci. 70:277–280.

Zink, F. W., Grogan, R. G., and Welch, J. E. 1956. The effect of the percentage of seed transmission upon subsequent spread of lettuce mosaic virus. Phytopathology 46:662–664.

# Repelling Aphids by Reflective Surfaces, A New Approach to the Control of Insect-Transmitted Viruses

FLOYD F. SMITH
*Entomologist*
AND
RAYMON E. WEBB
*Plant Pathologist*
*Agricultural Research Service,*
*U.S. Department of Agriculture,*
*Beltsville, Maryland*

To meet the needs for food by the rapidly increasing human population in many parts of the world, drastic changes in agricultural practices are being adopted to increase crop yields on more limited land areas. Older methods are becoming obsolete or ineffective. Thus, isolating crops from insect vectors has become increasingly difficult and high-speed travel has increased the possibility of new invaders from other countries.

Immense obstacles must be overcome as we strive to make more efficient factories of crop plants. We must somehow provide an ideal climate for growth immediately around the plants, even though we lack control of the gross climate in the fields where crops are raised (Irving, 1966).

Great progress has been made in insect control by the development of several potent new insecticides. Although they kill aphids and other vectors of plant virus diseases, their application often fails to decrease the incidence of such diseases in the crops sprayed. Indeed, sometimes, their use has actually increased the incidence of virus infection. Broadbent (1957, 1964) cited results of attempts to control alate aphids. These aphids transmit stylet-borne viruses within the crop and introduce circulative viruses from an outside source. Attempts have resulted in the failure to control infections of both types of viruses.

We have no evidence that applying any insecticide to a plant will prevent a viruliferous insect from infecting the plant. However, infection levels may be reduced in a given crop. Broadbent (1957) stated further that most viruses will not be controlled by insecticides until we discover new, persistent chemicals that will kill the vectors almost instantly or prevent them from feeding.

In the meantime, by utilizing the striking response that flying aphids make to color and light, as disclosed by Moericke (1954) and Cartier (1966), we may be able to devise methods that will largely prevent aphid feeding on plants. Thus, we may avert the usual serious virus infections that follow heavy aphid flights over susceptible crops.

Moericke (1951) described a yellow water pan trap for capturing flying aphids. Investigators in Europe, Australia, and America have adopted this trap to study aphid dispersal in the field. They have discovered the aphids captured in water pan traps, if promptly removed, are in a better condition for examination than those removed from sticky surface traps.

Broadbent (1948) and Moericke (1950) discussed the strong attraction of the green peach aphid *Myzus persicae* (Sulzer) to yellow. When green peach or other aphids fly low (2–3 ft) over vegetation or the bare ground during calm weather,* they are strongly influenced to land on pure yellow and to a lesser extent on orange or green surfaces. But they ignore blue, violet, white, gray, and black surfaces. The color near a yellow or green area also influences landing behavior; it may even cancel aphid flight to a yellow area.

All aphids, however, do not have the same preference for pure yellow. The aphid *Hyalopterus pruni* (Geoffroy) prefers more muted tones —gray green or milky yellow. *Hyalopterus pruni* also prefers a milky yellow composed of yellow plus white lead to a mixture of yellow plus zinc oxide, although both mixtures appear identical to our eyes. The mixtures differ in their remission of ultraviolet * (UV) (Moericke, 1954). Supplemental radiation increases landings of insects on both mixtures. Mixtures of color and UV will sometimes strongly influence the aphids to land.

Moericke (1954) found that when yellow pans were set on bare soil or on cotton cloths of different colors, aphid response varied according to species. Although most aphids were caught in pans set on bare soil, 30% were captured in pans set on black cloths, 3% on blue cloths, and 7% on white cloths. When a yellow pan was set on a yellow cloth, aphids landed on cloth near the edge, which indicated a strong marginal response to yellow. In these tests, only 5% of the green peach aphids

* Wind velocity less than 2–3 miles per hour.

* Ten to 4% at 380–300$\mu$ for zinc oxide (yellow) and 27–16% for lead white (yellow).

landed on the black cloth, whereas the aphid *Rhopalosiphum padi* (L.) landed 2.5 times more often on the black cloth than on bare soil.

Moericke (1954) cited the results of earlier investigators who observed that sprayed foliage, especially that having a whitish residue, attracted the aphids to land. As a consequence, infestations on the plants were heavier than usual. Moericke suggested that the increased attractiveness was due to increased remission of UV. The positive influence of bare soil with low remission of UV caused the aphids to heavily infest the plants scattered in spots that were otherwise bare or the plants at margins of the field (Moericke, 1954).

According to Moericke's hypothesis (1955) winged aphids, in the process of taking off on a dispersal flight, react positively to sky light and fly upward. After a time, however, a change takes place; their reaction to sky light shifts to a negative one. At this point they descend, fly close to the ground, and finally land.

Kring (1954) reported in 1962 that when he placed unpainted aluminum pans around yellow ones, the aphids avoided the yellow pans. Kring thought that the repellency was due to the light being reflected from the aluminum surfaces.

On the basis of Kring's work on repellency of aluminum, we (Smith *et al.*, 1964; anon., 1964) initiated tests in 1963 at Beltsville, Md., with foliage sprays of aluminum powder and soil coverings with sheets of aluminum or other materials to determine whether these materials would repel flying aphids and thus lower the incidence of virus infection.

In 1963 we tested the first foliage sprays of aluminum on the smooth-leaved *Gladiolus* and also on indian iron weed (*Vernonia anthelmintica*). Based on yellow pan trap catches these sprays repelled about as many aphids as aluminum sheets placed on the soil. However, the spray deposits adhered to the plants for only 2 or 3 days and their repellency was lost thereafter. Aluminum sprays that persisted longer than the first ones because they were prepared with varnish or other stickers, stunted growth of foliage, and reduced production of *Gladiolus* corms. On rough, hairy leaved plants, including cucumber, cantaloupe, and tobacco, the same aluminum sprays left a dull gray deposit which did not repel the flying aphids.

When we tested sheets of aluminum placed on the soil between rows, we obtained maximum aphid repellency (96%) when we covered 50% of the area with the sheets. With 30% coverage of the area, repellency was reduced to 70%. Other tests showed that the aluminum sheets repelled equal numbers of aphids whether the sheets were placed at the ground level or suspended 18 in. above the ground near the tops of the plants. Aluminum suspended vertically over the plants was less effective than that placed in a horizontal position on the ground.

During 1965 we conducted an experiment with yellow pans in field plots of bush squash plants in Florida. In these tests we covered 50% of the field with soil coverings (or mulches) of aluminum foil and black plastic. Throughout the growing period of the plants, small numbers of aphids were flying. The nearest sources of watermelon mosaic virus (WMV) known to us were 300 m away. Over the season the numbers of aphids trapped in the yellow pans indicated that the presence of aluminum sheets reduced aphid catch by 93% and the black plastic by 41%, as compared with the catch of aphids in unmulched check plots. The percentages of plants infected with WMV in the plots at the indicated times after planting were as follows:

| | At indicated weeks after planting | | | |
|---|---|---|---|---|
| Mulch treatment | 7 | 9 | 10 | 11 |
| Aluminum | 0.1 | 0.1 | 1.1 | 4.1 |
| Black plastic | 12.0 | 17.0 | 26.0 | 51.0 |
| Unmulched | 19.0 | 25.0 | 41.0 | 69.0 |

In these tests the squash foliage grew rapidly and concealed the soil mulches. Thus, at the end of 14 weeks the repellency was lost and virus infection became evident in all plots. However, because of the delay in transmission of the virus infection, the squashes in the plots mulched with aluminum produced fruits free of disease for several weeks (Moore *et al.*, 1965).

During a second experiment, conducted during 1965 in plots of bush squash at Beltsville, Md., a heavy dispersal flight of aphids coming from other crops and weeds in the area occurred on August 5–7. At this time the young squash plants were in the three-leaf stage. A local source of WMV was provided by mechanically inoculating 3% of the plants within the test and check plots with the virus. * In four of the plots in randomized blocks we covered 50% of the soil with aluminum mulches (Fig. 1); in four other plots we sprayed the plants five times at weekly intervals, beginning at plant emergence, with parathion at 0.6 g of active ingredient per liter. The percentages of infected

* Standard WMV inoculum was applied in sap from infected leaves. The leaves of test plants were lightly dusted with carborundum powder and the infectious sap wiped across the abraded portions.

*Fig. 1.* Squash plants mulched with 1-m strip of aluminum foil to repel flying aphids. Yellow aphid trap is in foreground.

plants and the kilograms per hectare of squashes harvested in treated plots and in the four check plots were as follows:

| Treatment | Percentage of infected plants at indicated weeks after planting | | | | Kg per hectare of squashes harvested |
|---|---|---|---|---|---|
| | 5 | 6 | 7 | 8 | |
| Aluminum mulch | 2 | 11 | 40 | 94 | 9080 |
| Parathion spray | 69 | 88 | 99 | 100 | 1464 |
| Check | 69 | 85 | 98 | 100 | 1637 |

During the 3-day period of heavy aphid flight, 1970 aphids were trapped per yellow pan in the check (equivalent to 61.7 million aphids per hectare) and sprayed plots but only 3–5 were trapped per pan in the plots mulched with aluminum. The increased yields from the plots mulched with aluminum demonstrated that the degree of protection from virus infection afforded by this type of mulch during the critical early growth period permitted about six times more marketable squashes to be produced than in the other plots.

In parallel experiments with three additional crops, increased yields of cucumbers and muskmelons resulted from mulching with aluminum.

However, watermelons were not protected from early virus infection because the young vines grew rapidly and covered the mulch early in the growing period.

In these experiments the melon aphid [*Aphis gossypii* (Glover)] became established on 46 of 240 plants of cucumber and muskmelon mulched with aluminum, whereas none of these aphids became established in comparable unmulched check plots. Although this species is attracted to yellow pan traps, it may also be equally attracted to reflective aluminum. If so, its behavior may be explained by the "aberrant responses" reported by Moericke for certain other aphid species.

In *Gladiolus* plantings on Long Island, N.Y., Johnson, Bing, and Smith (1967) found that ground mulches of embossed and smooth, highly reflective aluminum on paper and also white plastic on black were equally effective in repelling 90–97% of flying aphids and reducing spread of cucumber mosaic virus (CMV) by 67%. Aluminum strips 60 cm wide between the rows were more highly repellent than 30-cm strips. Repellency was virtually lost from margins of aluminum strips at 60 cm. Wolfenbarger and Moore (1968) also reported delayed virus infection and increased yields in mulched plots of bush squash and tomatoes in their 1967 experiments where the plants were set in the ground through holes in 1.37-m wide strips of reflective aluminum or of white plastic on black.

Heinze (1967) reported significant reduction of lettuce mosaic in lettuce plots mulched between rows with aluminum foil.

Three reports of failure to repel aphids and reduce virus infection are cited. Dickson and Laird (1966) obtained no reduction in WMV infection by placing one 18-cm strip of aluminum foil above cantaloupe plants in rows 2.1 m apart. The winged aphids involved were *Mysus persicae* and other species that do not colonize cucurbits.

Rothman (1967) failed to protect oat seedlings growing through holes in the center of inverted aluminum pans 20.3 cm in diameter and spaced 0.61 m apart. In these experiments Dickson's 18-cm reflective strip covered only 8% of the field surface area, and Rothman's 20.3-cm pans covered about 15% of the plot surface. In both cases the reflective areas were too limited to be effective. In the third case, Hakkaart (1967) installed aluminum strips 10 cm wide covering 39% and 29%, respectively, of the area, in beds of chrysanthemum plants consisting of equal numbers of diseased and healthy plants. Under the severe conditions of high proportion of source plants, narrow reflective strips, and inadequate total reflective surface, new infections by tomato aspermy virus were reduced from 12.5% in checks to 4.5% but chrysanthemum virus B infections were not reduced. Aphids of at least 12 species, caught in yellow pan traps in mulched plots, were reduced to about a third of those caught in unmulched plots. We agree with Hakkaart's conclusions that the prospects are not attractive for utilization of reflective material to reduce virus infection in commercial plantings of chrysanthemums.

As pointed out above (Smith *et al.*, 1964; Johnson *et al.*, 1967) maximum repellency is attained when 50% of the area is covered with fully exposed reflective material. In addition, Rothman caged apterous individuals of *Rhopalosiphum fitchii* (Sanderson) on check plants then allowed them to disperse by crawling.

Evidence to the present indicates that only alate aphids in flight are strongly repelled by reflective surfaces.

## DISCUSSION AND CONCLUSIONS

Many workers have observed the response of aphids and other insects to color and light. Moericke's experiments contributed materially to our knowledge of this response and his results have stimulated other workers to investigate possible ways in which the response may be utilized to protect plants from aphid attack or virus infection. As this review indicates, several investigators have conducted experiments in which flying aphids were repelled from various crops and thereby prevented the introduction or delayed spread of virus diseases. However, more research is needed before we can fully exploit the possibilities and reveal the limitations of this method of insect and disease control. Perhaps the most important need is to find a way of prolonging the period of effective repellency so that the normal harvest season of many crops will not be foreshortened by late season infections. The efficiency of aluminum, white plastic, or other mulches will be enhanced by the present trend to grow quickly maturing, less leafy varieties of crops that are being developed for mechanical harvesting. Growth retardants that reduce shoot elongation but not fruitfulness of plants will also help provide smaller plants that cover less of the reflective surfaces that are required for full-season crop protection.

Further studies (Eastop 1955) may reveal additional aphid species that exhibit unusual response to light and color. Under some conditions those aberrant species that Moericke recognized, as well as the deviating melon aphid which we encountered, may include important vectors of several virus diseases and require mulches of special color for maximum repellency. Studies on flight behavior and aphid migration by several investigators (Taylor, 1965) have revealed valuable background information for use in the exploitation of reflective surfaces as repellents.

The wheat straw mulch, which Nitzany, Geisenberg, and Koch (1964) used in Israel to increase the temperature around young cucumber plants and reduce population of *Bemisia tabaci* (Genu), as well as the spread of bottle gourd virus, may also have acted as a repellent due to reflected light (Moericke, 1955).

Few plant breeders have considered the importance of selecting plants for color variation in their search for plant resistance to insect attack. Müller's studies (1964) showed that many more *Myzus persicae* and other flying aphids alighted on green or yellow-green varieties of lettuce as on bronze varieties and that virus infection was correspondingly reduced.

Cartier (1966) in studies with artificial diets for rearing aphids has shown that nymphs as well as apterous and winged adults of *Macrosiphum euphorbiae* (Thomas) have a strong response to colors. Cartier (1966) further suggested the investigation of laser or coherent light to attract or repel aphids in full daylight at great distances.

Increased yields from aluminum or plastic mulches on vegetable crops and ornamentals (Pearson, Oaland, and Noll, 1959; Sheldrake, 1967) has stimulated the widespread use of mulches in American agriculture. By selecting appropriate mulches, additional benefits may be derived by repelling aphid vectors and reducing or delaying virus disease infections.

Future trends in preventing insect damage will lead to increased effectiveness, efficiency, precision, and specificity in insect control methods. Although population control over wide areas may be feasible against some pests, in other instances it will be necessary to protect the crops in localized areas independent of surrounding conditions.

## BIBLIOGRAPHY


Anonymous 1964. Aluminum repels aphids. Agricultural Research. U.S. Dep. Agr. 12(12):5 June (first publication based on report at Phytopathology Meeting).

Broadbent, L. 1948. Aphis migration and the efficiency of the trapping method. Ann. Appl. Biol. 35:379–394.

——— 1957. Insecticidal control of the spread of plant viruses. Ann. Rev. Entomol. 2:239–354.

——— 1964. The importance of alate aphids in virus spread within crops. Proc. 12th Int. Congr. Entomol., London, July 8–16, p. 523–524.

Cartier, J. J. 1966. Aphid responses to colors in artificial rearings. Bull. Entomol. Soc. Amer. 12(4):378–380.

Dickson, R. C., and Laird, E. F., Jr. 1966. Aluminum foil to protect melons from watermelon mosaic virus. Plant Disease Reporter 50(5):305.

Eastop, V. F. 1955. Selection of aphid species by different kinds of insect traps. Nature 176:936.

Hakkaart, F. A. 1967. Effects of aluminum strips on the spread of two aphid-borne chrysanthenum viruses. Neth. Jour. Plant Path. 73(5/6):181–185.

Heinz, K. 1967. Folien versuche mit Salat zur Abschreckung von virusübertragenden Blattläusen. Nachrichtenbl. Deutsch. Pflanzenschtzd. (Braunschweig) 19: 150–153.

Irving, G. W. 1966. Frontiers for agricultural research to meet U.S. needs. Amer. Soc. Agr. Eng., Winter Meeting, Chicago, Ill. Dec. 9 (unpublished).

Johnson, G. V., Bing, A., and Smith, F. F. 1967. Reflective surfaces used to repel dispersing aphids and reduce spread of aphid-borne cucumber mosaic virus in gladiolus plantings. J. Econ. Entomol. 60(1):16–18.

Kring, J. B. 1964. New ways to repel aphids. Frontiers of Science, Connecticut Agr. Exp. Sta. 17(1):607 Nov.

Moericke, V. 1950. Über das Farbsehen der Pfirsichtblattlaus (*Myzodes persicae* Sulz.). Z. Tierpsychol. 7:265–274.

Moericke, Von V. 1951. Eine Farbfalle zur Kontrolle des Fluges von Blattläusen, inbesondre der Pfirsichblattlaus, *Myzodes persicae* (Sulz). Nachrbl. Deut. Pflanzenschutzdienst 3:23–25.

Moericke, V. 1954. New investigations on color in relation to Homoptera. Proc. 2nd Conf. Potato Virus Diseases, Lisse-Wageningen, June 25–29, p. 55–69.

Moericke, V. 1955. Über das Verhalten phytophager Insekten während des Befallsflugs unter dem Einfluss von weissen Flächen. Z. Pflanzenkrank. Pflanzenpathologie Pflanzenschutz. Band 62 Heft 8–9, p. 588–593.

Moore, W. D., Smith, F. F., Johnson, G. V., and Wolfenbarger, D. O. 1965. Reduction of aphid populations and delayed incidence of virus infection on yellow straightneck squash by the use of aluminum foil. Proc. Florida Hort. Soc. 78:187–191.

Müller, H. J. 1964. Über die Anflugdichte von Aphiden auf farbige Salatpflanzen. Entomol. Exp. Appl. 7:85–104.

Nitzany, F. E., Geisenberg, H., and Koch, B. 1964. Tests for the protection of cucumbers from a whitefly-borne virus. Phytopathology 54:1059–1061.

Pearson, R. K., Odland, M. L., and Noll, C. J. 1959. Effect of aluminum mulch on vegetable crop yields. Pennsylvania Agr. Exp. Sta. Progr. Rep. 205:1–7.

Rothman, G. 1967. Aluminum foil fails to protect winter oats from aphid vectors of barley yellow dwarf virus. Plant Disease Reporter 51(5): 354–355.

Sheldrake, R. 1967. Plastic mulches. N.Y. State College Agr. Cornell Bull. 1180. 8 p.

Smith, F. F., Johnson, G. V., Kahn, R. P., and Bing, A. 1964. Repellency of reflective aluminum to transient aphid virus vectors. Phytopathology 54(7):748. (Abstr.)

Taylor, L. R. 1965. Flight behavior and aphid migration. Proc. North Central Branch Entomol. Soc. Amer. 20:9–19.

Wolfenbarger, D. O., and Moore, W. D. 1968. Mulch treatments of squash and tomatoes with respect to virus infections and yields. Proc. Florida Hort. Soc. 80:217–221.

# Author Index



* Numbers in *italics* indicate pages on which full reference listings appear.

## G

## I



## J

## Q



## R

## T

## Y



## Z

# Subject Index